# Sustainable Solutions for Elemental Deficiency and Excess in Crop Plants

Kumkum Mishra • Pramod Kumar Tandon •
Sudhakar Srivastava
Editors

# Sustainable Solutions for Elemental Deficiency and Excess in Crop Plants

Springer

*Editors*
Kumkum Mishra
Department of Botany
University of Lucknow
Lucknow, Uttar Pradesh, India

Pramod Kumar Tandon
Department of Botany
University of Lucknow
Lucknow, Uttar Pradesh, India

Sudhakar Srivastava
Plant Stress Biology Laboratory,
Institute of Environment and Sustainable Development
Banaras Hindu University
Varanasi, Uttar Pradesh, India

ISBN 978-981-15-8635-4 ISBN 978-981-15-8636-1 (eBook)
https://doi.org/10.1007/978-981-15-8636-1



This Springer imprint is published by the registered company Springer Nature Singapore Pte Ltd.
The registered company address is: 152 Beach Road, #21-01/04 Gateway East, Singapore 189721, Singapore

# Foreword

The world population today survives with insufficient nutrition whether due to lack of food or owing to less nutritious food. A number of essential elements constitute an important part of human nutrition. At the same time, there are elements that are toxic to plants as well as humans even if present in minute quantities. The crop plants suffer from two major problems of elemental nutrition: (1) accumulation of toxic metals in crop plant produce beyond the maximum tolerable limits due to increasing environmental contamination and (2) decreases in the essential element nutrition of crop plant produce over the years. The book entitled "Sustainable Solutions for Elemental Deficiency and Excess in Crop Plants" deals with the elemental nutrition in crop plants in a holistic manner and also discusses sustainable solutions for the problem. The book contains chapters from eminent researchers and academicians in the field providing vast information on the topic. The team of editors and all the authors deserve appreciation for this highly useful compilation. I am hopeful that this book will provide opportunity to students, researchers, and academicians to assimilate the essence and importance of elemental nutrition of crop plants.

North-Eastern Hill University
Shillong, Meghalaya, India

Biotech Park
Lucknow, Uttar Pradesh, India

Pramod Kumar Tandon

# Preface

Metals are spread in almost every sphere of the environment. The metals constitute essential and toxic elements. Essential elements are required for the proper growth and development of plants, animals, and humans. However, the problem of deficiency of metals like zinc, copper, selenium, and iron is quite widespread and plants therefore suffer from reduced crop yields. Conversely, there are instances of excessive cadmium, arsenic, lead, and aluminum contamination whose presence reduces crop yields and affects animal and human health. Hence, metal toxicity and deficiency are the two sides of the same problem that is commonly encountered. In addition, the problem of excessive presence of a toxic metal like arsenic may be exacerbated by the deficiency of an essential element like zinc. Therefore, appropriate nutrient management is a burgeoning problem in the present scenario that needs to be managed so as to sustain crop yields required in the near future. In this connection, it is also relevant to note that climate-change-driven changes in elemental concentrations would also be of more concern in the future.

A number of approaches have been investigated in the past several years to manage elemental concentrations in crop plants. These include fertilizer amendments for zinc, iron, selenium, etc. The approaches also include biological interventions through the employment of bacterial and fungal inoculants that enhance bioavailability and consequently the concentration in plants. There are biotechnological approaches also that have been attempted through the expression of specific genes to regulate metal concentrations. Further, crop-breeding strategies have been specifically tried to develop varieties efficient in zinc, copper, and iron accumulation. Hence, there are a variety of possible solutions that have been studied and are in the process of research. The present book entitled "Sustainable Solutions for Managing Elemental Deficiency and Excess in Crop Plants" aims to shed light on the latest developments and research on solutions for managing the concentrations of essential elements or toxic metals in crop plants.

The book contains a total of 18 chapters, which are divided into three sections: general aspects (4 chapters), elemental nutrition of crop plants (6 chapters), and toxic metals in crop plants (8 chapters). The first chapter of this book discusses about elemental concentrations in the environment encompassing air, water, and soil. The second chapter presents views on the deficiency of essential elements in crop plants.

This chapter encompasses a number of major and trace elements, which are important for plant nutrition, and provides information about their roles and deficiency symptoms. Third chapter reviews the problem of toxic metal accumulation in plants including cereals, vegetables, mushrooms, etc. The fourth chapter encompasses human health issues emanating from deficiency of essential while excess of toxic elements. The first four chapters therefore set an excellent beginning of the book.

In the second section, Chaps. 5–10 are dedicated to issues of essential element nutrition of crop plants. Chapter 5 presents in-depth discussion about nitrogen, phosphorus, and potassium in crop plants and provides details of their sources, requirement for agricultural crops, and their deficiency symptoms in crops. Chapter 6 reviews mechanisms of trace metal uptake and transport in crop plants up to the grains. Chapters 7 and 8 shed light on the aspects of biofortification and demonstrate why biofortification is a sustainable and feasible approach to deal with the problem of elemental deficiencies stressing on agronomic solutions. Chapter 9 discusses biological ways through which elemental nutrition of crop plants can be improved, while Chap. 10 focuses on biotechnological approaches pertinent for the improvement of essential elements in crop plants. Therefore, Chaps. 5–10 present not only the problem but also the prospective solutions for the problem of elemental nutrition.

In the third section, Chaps. 11–18 deal with various aspects of toxic metals in crop plants. Chapter 11 discusses the problem of toxic metals in crops in detail spanning its various aspects including ecological risks and human health hazards. Chapter 12 provides holistic information about physiological, biochemical to molecular responses of plants to toxic metals. Chapter 13 gives details about mechanisms of toxic metal uptake and transport in plants, in particular, focusing on the role of transporters. Chapters 14 and 15 deal with cadmium in particular presenting its wide-ranging aspects from soil contamination, phytotoxicity, to plant responses to deal with cadmium toxicity including the roles played by phytohormones. Chapter 16 discusses agronomic management practices that can be utilized to tackle the toxic metal accumulation in crop plants in a sustainable, feasible, and low-cost manner. Chapters 17 and 18 stress on biological (microbial) and genetic engineering approaches, respectively, that can be applied to reduce toxic metals in crop plants and safeguard plants against metal toxicity.

The book therefore comprises a unique combination of chapters on various aspects and will give the reader a comprehensive knowledge of sustainable solutions for managing elemental deficiency and excess in crop plants. This book would act as a guiding textbook for undergraduate and postgraduate students and as a means to understand the latest research trends for doctoral students as well as for academicians and researchers.

Lucknow, Uttar Pradesh, India — Kumkum Mishra
Lucknow, Uttar Pradesh, India — Pramod Kumar Tandon
Varanasi, Uttar Pradesh, India — Sudhakar Srivastava

# Contents

# Editors and Contributors

## About the Editors

**Kumkum Mishra** Ex. Professor and Assistant Coordinator M.Sc. Environmental Science, University of Lucknow, U.P. (India). She has more than 30 years of research and teaching experience in the field of mutagenesis and genotoxic evaluation of environmental pollutants by plant bioassays. She has handled UGC and DST major projects as Principal Investigator. She has also supervised 14 doctoral, 3 postdoctoral, and 2 M.Phil. candidates. She has published 46 research papers, 2 books, and 4 book chapters.

**Pramod Kumar Tandon** Ex. Professor of Botany and Course Coordinator, M.Sc. Environment Science Programme at the University of Lucknow (India). He has more than 40 years of teaching and research experience in the field of Botany/Environment Science. He has guided 4 postdoctoral scientists, 32 Ph.D., and 9 M.Phil. students. In 2003, he visited Brisbane, Australia for academic purpose. He has published more than hundred research papers, one book, and three book chapters. He was awarded distinguished UGC-BSR Faculty Fellowship in 2011 for his outstanding research in the field of metal toxicity and some other important aspects in plants.

**Sudhakar Srivastava** Assistant Professor in the Institute of Environment & Sustainable Development (IESD), Banaras Hindu University (BHU). He obtained his Ph.D. in Botany from CSIR-National Botanical Research Institute (CSIR-NBRI) and University of Lucknow in 2008. He served as Scientific Officer in Bhabha Atomic Research Centre (BARC), Mumbai from 2009 to 2014 before joining BHU. He has more than seventeen years of research experience in the field of plant–metal interactions. He has published more than 100 research and review articles in high impact international and national journals. He has also published 2 books and 25 book chapters in the books of international repute. He has supervised 2 Ph.D. and 3 M.Phil. students. He is NAAS Associate of National Academy of Agricultural Sciences. He has received Young Scientist Award from National Academy of Sciences, India (NASI), Allahabad in 2011 and Young Scientist Award from Uttar Pradesh Council of Science and Technology (UPCST) in 2013–2014.

## Contributors

**Fathima Afsal** Department of Earth Sciences, Indian Institute of Science Education and Research (IISER) Kolkata, Mohanpur, West Bengal, India
Department of Biological Sciences, Indian Institute of Science Education and Research (IISER) Bhopal, Bhauri, Madhya Pradesh, India

**S. B. Agrawal** Center of Advanced Study in Botany, Institute of Science, Banaras Hindu University, Varanasi, Uttar Pradesh, India

**Mohammad Israil Ansari** Department of Botany, Lucknow University, Lucknow, Uttar Pradesh, India

**Bruno Lemos Batista** Centro de Ciências Naturais e Humanas, Universidade Federal do ABC, São Paulo, SP, Brazil

**Sutapa Bose** Department of Earth Sciences, Indian Institute of Science Education and Research (IISER) Kolkata, Mohanpur, West Bengal, India

**Priti Chauhan** Department of Bioscience and Biotechnology, Banasthali Vidyapith, Vanasthali, Rajasthan, India

**Bruna Moreira Freire** Centro de Ciências Naturais e Humanas, Universidade Federal do ABC, São Paulo, SP, Brazil

**Kiran Gupta** Department of Botany, Lucknow University, Lucknow, Uttar Pradesh, India

**Sofi Javed Hussain** Department of Botany, Aligarh Muslim University, Aligarh, Uttar Pradesh, India

**Harmanjit Kaur** Department of Botany, Akal University, Talwandi Sabo, Bathinda, Punjab, India

**Amit Kumar** Department of Botany, Lucknow University, Lucknow, Uttar Pradesh, India

**Jisha Suresh Kumar** Department of Earth Sciences, Indian Institute of Science Education and Research (IISER) Kolkata, Mohanpur, West Bengal, India
The Faculty of Biology, Ludwig-Maximilians-University of Munich, Planegg, Germany

**Camila Neves Lange** Centro de Ciências Naturais e Humanas, Universidade Federal do ABC, São Paulo, SP, Brazil

**Arnab Majumdar** Department of Earth Sciences, Indian Institute of Science Education and Research (IISER) Kolkata, Mohanpur, West Bengal, India

**Jyoti Mathur** Department of Bioscience and Biotechnology, Banasthali Vidyapith, Vanasthali, Rajasthan, India

**Deepak Kumar Mehrotra** Central Geological Service, Lucknow, Uttar Pradesh, India

**Amit K. Mishra** Department of Environmental Sciences, Dr. Rammanohar Lohia Avadh University, Ayodhya, Uttar Pradesh, India

**Kumkum Mishra** Department of Botany, University of Lucknow, Lucknow, Uttar Pradesh, India

**Pratyush Pingita Mishra** Department of Environmental Sciences, Dr. Rammanohar Lohia Avadh University, Ayodhya, Uttar Pradesh, India

**Neeraj Pandey** Ranjan Plant Physiology and Biochemistry Laboratory, Department of Botany, University of Allahabad, Prayagraj, Uttar Pradesh, India

**Anuradha Patel** Ranjan Plant Physiology and Biochemistry Laboratory, Department of Botany, University of Allahabad, Prayagraj, Uttar Pradesh, India

**Tatiana Pedron** Centro de Ciências Naturais e Humanas, Universidade Federal do ABC, São Paulo, SP, Brazil

**Rodrigo Mendes Pereira** Centro de Ciências Naturais e Humanas, Universidade Federal do ABC, São Paulo, SP, Brazil

**Sheo Mohan Prasad** Ranjan Plant Physiology and Biochemistry Laboratory, Department of Botany, University of Allahabad, Prayagraj, Uttar Pradesh, India

**Shalini G. Pratap** Division of Environment Science, School of Basic Sciences, Babu Banarasi Das University, Lucknow, Uttar Pradesh, India

**Vitória Aparecida Procópio** Centro de Ciências Naturais e Humanas, Universidade Federal do ABC, São Paulo, SP, Brazil

**Amandeep Raju** Ranjan Plant Physiology and Biochemistry Laboratory, Department of Botany, University of Allahabad, Prayagraj, Uttar Pradesh, India

**Bruno Alves Rocha** Instituto de Ciências Ambientais, Químicas e Farmacêuticas, Universidade Federal de São Paulo, Diadema, Brazil

**Gauri Saxena** Department of Botany, Lucknow University, Lucknow, Uttar Pradesh, India

**Anita Singh** Center of Advanced Study in Botany, Institute of Science, Banaras Hindu University, Varanasi, Uttar Pradesh, India
Department of Botany, Institute of Science, Banaras Hindu University, Varanasi, Uttar Pradesh, India

**Jaswant Singh** Department of Environmental Sciences, Dr. Rammanohar Lohia Avadh University, Ayodhya, Uttar Pradesh, India

**Madhulika Singh** Ranjan Plant Physiology and Biochemistry Laboratory, Department of Botany, University of Allahabad, Prayagraj, Uttar Pradesh, India

**Pramod Kumar Singh** Division of Environment Science, School of Basic Sciences, Babu Banarasi Das University, Lucknow, Uttar Pradesh, India

**Shraddha Singh** Nuclear Agriculture and Biotechnology Division, Bhabha Atomic Research Centre, Mumbai, Maharashtra, India

**Dwaipayan Sinha** Department of Botany, Government General Degree College, Paschim Medinipur, West Bengal, India

**Sudhakar Srivastava** Plant Stress Biology Laboratory, Institute of Environment and Sustainable Development, Banaras Hindu University, Varanasi, Uttar Pradesh, India

**Vishnu Chandra Srivastava** Geological Survey of India, Aliganj, Lucknow, Uttar Pradesh, India

**Penna Suprasanna** Nuclear Agriculture and Biotechnology Division, Bhabha Atomic Research Centre, Mumbai, Maharashtra, India

**Pramod Kumar Tandon** Department of Botany, University of Lucknow, Lucknow, Uttar Pradesh, India

**Sanjesh Tiwari** Ranjan Plant Physiology and Biochemistry Laboratory, Department of Botany, University of Allahabad, Prayagraj, Uttar Pradesh, India

**Pradeep Kumar Yadav** Center of Advanced Study in Botany, Institute of Science, Banaras Hindu University, Varanasi, Uttar Pradesh, India

# Part I

# General Aspects

# Elemental Concentrations in Soil, Water and Air

1

Vishnu Chandra Srivastava

**Abstract**

Environment encompasses lithosphere, hydrosphere, atmosphere and biosphere, comprising of different types of chemical elements. The existence of living organism depends on interactions within and in between these chemical elements/components. It leads to disaster whenever this dependency is degraded beyond limit of natural replenishment. Emergence of agriculture was probably the first instant of artificial land use capable of modifying the natural environment. Scientific and technical advances particularly after mid-twentieth century coupled with exponential growth of human population have expanded the scale of human activities enormously which has affected the fertility of soil, quality of water and air adversely. Gangetic Plain in India, primarily underlain by alluvial sediments of quaternary period, ranging in thickness from 200 m to 1000 m is at places, associated with volcanic sediments. These sediments indicate presence of naturally occurring excessive arsenic and fluoride in groundwater. Selenium concentration in Punjab besides boron and manganese in higher concentration is recorded in alluvial soils. In order to meet the demands of food-grain for ever increasing population, high priority was accorded to irrigation engulfing more and more areas under water -logging and flooding, alkalinity of soil and lowering of groundwater level due to excessive drawl of groundwater. This has led to progressive decline of soil fertility. In addition, use of chemical fertilizer and pesticides to increase the production of food grain has affected the good agricultural soil adversely leading to chemical elemental concenteration in the soil and water. Industrial waste and mining have become matter of serious concern when pollutants so generated get unchecked access to natural water resources, cultivated fields and in the atmosphere aggravated the problem of elemental

V. C. Srivastava (✉)
Geological Survey of India, Aliganj, Lucknow, Uttar Pradesh, India





K. Mishra et al. (eds.), *Sustainable Solutions for Elemental Deficiency and Excess in Crop Plants*, https://doi.org/10.1007/978-981-15-8636-1_1

concentration depending upon stability, interaction with the medium and other elements which in turn reaches to human beings and animals.

**Keywords**

Element · Concentration · Soil · Water · Air · Industrial waste · Mining

## 1.1 Introduction

Element is a substance where all atoms have the same number of protons. Elements are chemically the simplest substances and hence cannot be broken down using chemical reactions. Elements can only be changed into other element by using nuclear methods. Elements are defined by the number of protons they possess. For example, the hydrogen (H) element is made from atoms containing a single proton.

There are about 118 known elements in the periodic table. The periodic table of the elements is a tabular display of the chemical elements. First 94 elements occur naturally; though some are found only in trace amounts. Elements are of two types, metal and non-metal. Elements that tend to gain electrons to form anions during chemical reactions are called non-metals. They are electro-negative elements with high ionization energies. They are non-lustrous, brittle and poor conductor of heat and electricity. Non-metal can be gas, liquid or solid. There are 22 non-metals. Out of which, 10 non-metals are solid. Eleven non-metals are gases like hydrogen (H), nitrogen (N), oxygen (O), chromium (Cr) and one is liquid (bromine; Br). Four common non-metals are sulphur (S), nitrogen, selenium (Se) and bromium. About 75% of the elements on the periodic table are metals. Common metals are copper, lead, tin, nickel and zinc. Alloys are made of mixture of at least one metal with either other metal or with non-metal like boron (B), silicon (Si), germanium (Ge), arsenic (As), antimony (Sb), tellurium (Te) and polonium (Po), etc. Earth crust is made up of several elements. By weight the major constituent elements are oxygen 46.6%, silicon 27.7%, aluminium 8.1%, iron 5%, calcium 3.6%, sodium 2.8%, potassium 2.8%, magnesium 2.1% and trace elements. Earth crust is composed of a variety of igneous, metamorphic and sedimentary rocks.

Elements are taken into the human body through food chain, water and through the inhalation of atmospheric dust and gases. These are increasing concern about the effect of elements on the health of the man, animals and the crops. A number of geo-chemical factors such as soils and water pH, organic matter content, Eh conditions and chemical form greatly influence the mobility and bio-availability of elements in the environment. Daily requirement of macro-elements like Ca, Na, P, K, Cl, Mg, S is more than 100 mg, whereas the microelements like Fe, Cu, Zn, Mn, I, Mo, Se, F, Br, Cr, Co, Si are required from less than 1 to 100 mg daily. Carbon, hydrogen and oxygen comprise from 94% to 99.5% of fresh plant tissues. The remaining nutrients, which come from soil, make up the balance of the tissue. The requirement of trace elements like Sn, Ni, Ge, V and W is not yet adequately determined, but their presence in microgram quantities is necessary. Plant growth and development may be retarded if any micronutrient is lacking in the soil or is not adequately balanced with other nutrients. As soil pH increases the availability of micronutrients decreases. Most of the trace elements are essential at lower level for mankind but are fatal when the concentration level crosses their permissible limits. Poisonous elements (Cd, Hg, As, Pb) have negative effects on the human body for example, kidney (Pb, Cd), skeleton (Cd, Al), nervous system (Hg, Al, Pb), fertility (Pb, Hg), heart (Pb, As), etc.

Population, rapid urbanization and industrialization and exponential growth in human activities over the past century have resulted in higher water requirement for all types of water uses: human, thermo-industrial and agricultural. Furthermore, domestic and industrial waters have been indiscriminately discharged to water bodies without any or partial treatment. Consequently, water bodies within or near population centre have already been contaminated seriously with domestic and industrial pollutants. This has posed serious health and environment problems. The problem gets aggravated due to loss of natural water bodies and wetland to authorized and unauthorized construction, overrun of flood-plains, loss of mangrove and carrying capacity of rivers. Most of the rivers in India contain Biological Oxygen Demand (BOD) beyond permissible limits (3mg/L). BOD in River Gomati in Lucknow in Uttar Pradesh during the month of summer (April) was found to be 9.5 mg/L, whereas in the month of monsoon (August) was 8 mg/L, indicating waste dumped in the river (Uttar Pradesh Pollution Control Board).

Crops require 16 elements to grow properly. The elements like carbon, hydrogen and oxygen are derived from air and water. Other remaining nutrients used by the plants come from soil in the form of inorganic salts. Legumes can fix nitrogen from air. Plant growth and development may be retarded if any of these elements is lacking in soil or not adequately balanced with other nutrients. Physical and chemical characteristics of soil affect the availability and uptake of micronutrients. Soil temperature and moisture are important factors As soil pH increases the availability of micronutrient decreases with exception of molybdenum. The deficiency or increase of micronutrients/trace elements in soil has a direct bearing on the agriculture produce which in turn adversely affects the health of live stocks and human beings. It is well exemplified by case study on goitre. Sub-Himalayan belt in India being covered with iodine deficient sandy soil and iodine deficient shallow aquifer (dug wells; the only source of drinking water as well as for irrigation) have contributed iodine deficient food chain resulting in wide spread of goitre endemicity. During last 5–6 decades of soil conservation and with adoption of use of deeper aquifer (iodine 6–8 ppm) for drinking as well as for agriculture irrigation, prevalence of goitre incidence has declined drastically (Srivastava et al. 1995; Srivastava 2008).

## 1.2 Soils

Soils are defined as material formed by the influence of physical, chemical and biological processes acting on a parent material which may be bed rock, unconsolidated sediments or older soils. As mineral breaks down during soil formation, microelements are gradually released. There are two types of microelement/trace elements in the soil. One that is absorbed on to soil colloid and other that is in the form of salts dissolved in the soil solution. Physical character of soil plays a vital role in its elemental concentration that depends on the grain size of the constituent particles from coarse to fine (sand, silt and clay) categories. The type of soil is fixed by relative abundance of these three categories. Porosity and permeability, i.e. grain size of the soil determine where soil will retain water /element or allow to pass through easily. This characteristic determines the bonding that may develop between the element concentration and the soil particles. Sandy soil being more porous and permeable allows the water to leach away or percolate down, whereas the clayey soils being finer textured have more retentive capacity to the water along with the element.

The clay grains comprise flat flakes of micron size. The flakes in turn are made up of atomic scale layers, namely tetrahedral layer and octahedral layer. In tetrahedral layer, oxygen atoms are situated at the vertices of tetrahedrons. The centre of each tetrahedral is occupied by a silica (Si) and alumina (Al) atoms. In octahedral layers, oxygen and hydroxyl ions occupy the vertices of an octahedron. The centre of octahedron may be occupied by magnesium (Mg), alumina (Al) or another element.

Clay minerals are characterized by the manner in which the tetrahedral and octahedral layers are assembled. On this basis, Kaolinite, Montmorillonite and Illite are the clay minerals identified. Kaolinite [$Al_2Si_2O_5$ $(OH)_4$] comprises of paired tetrahedral (Si, Al, O) and octahedral Al OH sheets. This has the least variable composition and is the coarsest grained clay minerals. They are formed from other clay minerals by continued leaching. Montmorillonite consists of units comprising two tetrahedral layers on either side of an octahedral layer. Two contiguous units are separated by an interlayer made up of molecular water, sodium (Na), calcium (Ca). This interlayer permits absorption of other liquids and chemical components such as Na, Ca and thus montmorillonite containing soil can easily undergo chemical changes. Illite has similar structure of a soil as montmorillonite but has K in place of Na (Chakravarti 1994; Pandey 2009).

Chemical character of a soil horizon is extremely important because of its direct relationship with productivity and absorption or retention of elements. Clay minerals can capture cations from water also give up cation to the water. This cation exchange capacity (CEC) is extremely important because of the property. Montmorillonite is helpful in absorbing calcium and aluminium ions from the fertilizer and releasing their ions to the plants as they grow. Therefore, they play a vital role in storing elements and even passing them onto the plants and hence into the food chain.

Soil profile is the succession of zones or horizons beginning at the surface that have altered by normal soil forming processes of leaching and oxidation. These zones are designated as O, A, B and C-horizons. O-horizon is the upper part and consists of mineral layer of maximum organic accumulation. A-horizon is steadily interacting with water infiltrating downward from the surface. The water is ordinarily acidic because it contains carbonic acids derived from the atmosphere and from organic matter present in the O-horizon. These weak acids are the most effective near the surface where they react with existing mineral particles to form clay minerals reactive products. This solution also carries chemical constituents and suspended fine particles downwards. This removal and downward transport is called leaching. A-horizon is called the leached-horizon. At depth leaching becomes less effective and decomposition dominates. This horizon is known as B-horizon. It consists of weathered material with an accumulation of clay, iron or aluminium. Below this, C-horizon is the layer of unconsolidated, weathered parent material. Depending on the layer of the soil profile and also the particular nature of the soil profile, the rate and type of interaction with the elements or pollutants or the media containing the pollutant are fixed (Sri Ram 1994).

## 1.3 Elemental Concentrations in the Soil

Rocks are the fundamental building blocks of the planet surface and different rock assemblages contain the 94 naturally occurring chemical elements found on earth. These naturally occurring elements are not distributed evenly on the surface of the earth and problems can arise when element abundance are too low (deficiency) or too high (toxicity). The elemental concentration in the soils is dependent on the nature of the source, the characteristics of the medium being polluted or enriched and nature of the pollutants. The concentration also depends on the stability of the element, interaction of the element with the medium and the interaction of the element with the other elements.

Soils may be enriched or deficient in certain elements depending upon the source of the parent material. If the parent rock/alluvial soil or older soil is rich in certain elements, then it leads to enrichment of that element due to soil erosion. Physical laws of soil matter through erosion by water and wind action results in the removal of fertile element constituent of surface soil. The significant factors for the soil erosion are clay-ratio, clay–silt ratio, ratio of colloid to moisture equivalent of pH and organic matter, slope, forest cover. Soil erosion by water can be classified as sheet erosion, rill erosion, gully erosion, river-bank erosion, strip erosion and shore erosion. Erosion by wind is the most natural dreaded hazard that spreads in most of the coastal, semi-arid tract of the country. The loss of soil is as much as 50 tonnes per acre while it takes 1000 years to build 3 cm thick layer of soil (Sri Ram 1994).

## 1.4 Reasons of Concentration

Concentration of elements/pollution in the soil

1. Direct fallout from atmospheric source, such as radioactive fallout, thermal fly ash generated from thermal power plants, acid rains, gaseous effluent plumes from the factories chimneys.
2. Application of fertilizers and pesticide.
3. Cycling of chemical element through natural environment. Enrichment of arsenic in particular aquifer is mainly due to arsenic bearing soil at that particular depth as seen in the Gangetic Alluvial Plains of Uttar Pradesh.

## 1.5 Sources of Soil Pollutant

1. Point source:
   - Mine waste (Srivastava et al. 1996)
   - Discharge of industrial waste
   - Domestic and industrial waste from landfill
   - Municipal sewage treatment system
   - Septic tanks and domestic wastes (cesspools) source
2. Non-points:
   - Fertilizers and pesticides from agricultural fields
   - Salt from irrigated agricultural land
   - Oil and gasoline from urban areas
   - Air borne pollutants

## 1.6 Tolerance Limits

No such tolerance limits for element or compound exits in case of the soils. Available literature often gives average range of different elements inorganic or organic compound present in different types of soil.

## 1.7 Water

Water is a prime natural resource and is essential for sustaining life. The hydrosphere (total water available on earth) is approximately 1.4 billion cubic km sufficient, if earth was smooth to cover it to a uniform depth of several kilometres. But unfortunately, more than 99% of the total water percent on the earth is either unavailable or unsuitable for human use because of it salinity (sea water ) or location (ice caps or glaciers). Thus the amount of water on which the entire population depends is less than 1% of the total. In India, the picture is equally grim. The country accounts for 2.2% of the global land and 6% of the total water resources, but it has to support 16% of the world population. Hence, India has one of the lowest per capita availability of water. So far the principal consumption of water has been for irrigation, but the need for domestic and industrial consumption is increasing rapidly and almost every summer the specter of acute water scarcity has become a common feature. In this scenario, any further loss of majority of human diseases can be attributed to excess of elements present in water, either through direct consumption or through food chain.

## 1.8 Elemental Concentrations

The terms like concentration, contamination and pollution are often loosely used and are frequently interchanged. Addition of extraneous matter may contaminate water body. Concentration of an element or matter is the amount of a particular substance within a solution or mixture but the water is called polluted only when level of contamination exceeds the tolerance limit stipulated for that contaminant. Prevention and Control of Pollution Act 1986 defines "pollution that means contamination of water or such alteration of physical, chemical or biological properties of water or such discharge of any sewage or trade effluent or any other liquid, solid or gaseous substance into water, whether directly or indirectly, as may or is likely to create a nuisance or render such water harmful or injurious to public health or safety or to domestic, commercial, industrial, agricultural or other legitimate uses or to life and health of animals or plants or aquatic organism".

Water includes rivers (perennial/ephemeral), water courses (perennial/ephemeral), inland water (natural/ artificial) l, underground water and sea or tidal waters. Water body is considered to be polluted when its level of concentration/contamination renders it unsuitable for human consumption, renders it harmful/injurious to animals/plants/aquatic organism and it becomes unfit for its intended use in household, industry, commerce, agriculture, etc. In other words water pollution refers especially to degradation of water quality, as measured by chemical, physical or biological criteria. Degradation of water is generally considered in respect of its intended use, departure from norms, adverse effect on public health and adverse ecological impact.

The extent of water contamination in Karnataka and Uttar Pradesh is a serious cause of concern. Concentration of chemicals like nitrate (552 mg/L in Raichur), fluoride (4.98 mg/L Rural areas in Bangaluru), iron (4.4 mg/L in Shivamogga) and arsenic in high amounts are found in 22 districts out of 30 districts in Karnataka. Alarming levels of chemical concentration in groundwater are mainly attributed to over exploitation (Rohit 2017). Water in municipal area of Udipi in Karnataka is contminated with pathogenic biological organism maximum 2400 MPN and Nitirate pollution is attrobuted to application of excess chemical fertilizer (Lokesh 1996), and same problem in Mysore city (Ramesh 2004). In Uttar Pradesh, major portion of the State area either suffer from arsenic or fluoride concentration in groundwater beyond permissible limits. High concentration of F in drinking water ranging between 5 and 20 ppm in certain parts of Indo-Gangetic Plain in Unnao district of Uttar Pradesh, from Muktsar (Punjab) being of the order of 42.5 mg/L followed by 32.5 mg/L in New Delhi, Karbi Anglong district, Assam (23.4 mg/L), Selenium concenteration in Punjab (Prasad and Kar 2004) and Rajasthan in India leads to severe fluorosis (Faruqui and Nagar 1995; Rao 1996; Srivastava and Rastogi 2009). Even the low concentration of F ranging between 0.4 and 1.2 ppm can cause acute dental and skeleton fluorosis. Groundwater arsenic concentration and suffering of people resulting out of it have been reported from 20 countries. The magnitude is considered highest in Bangladesh, India, Mongolia, China and Taiwan (Bhattacharya et al. 2002; Acharyya 2003, 2004). Excessive withdrawal of groundwater for agricultural irrigation results in gradual lowering of water table which exposes more sediment surfaces for microbial reductions. Thus, in fact results in release of more arsenic in to groundwater and accentuate the problem of arsenic. High-arsenic coal being used in power plants also leads to health hazards. The arsenic pollutants are identified to be the iron rich clastic grained and is incidental and possibly related to microbial degradation of organic matter (Shanker et al. 2002). Since microbes activate more at the air and water interface, the sediment having higher porosity in the top of sedimentary column will be the ideal sites for proliferation. Such activities will eventually release arsenic faster than sedimentary column starting with impervious layer (Mitra et al. 1996, Misra et al. 1996, Pal et al. 2002).

## 1.9 Common Pollutants and Their Source

### 1.9.1 Sediments

Sediments carrying different types of elements are by far the most common among water pollutants. Sediments comprise of rock and mineral fragment ranging in size from sand particle (less than 2 mm diameter) to silt and clay particles (less than 0.02 mm diameter) or even still finer colloidal particles (Chakravarti 1994; Agarwal 2009).

#### 1.9.1.1 Source

1. From land surface through natural processes of soil erosion and weathering of bed rocks, unconsolidated sediments or older soils. Water erosion is one of the most common geological phenomena and leads to loosening of soil particles/elements and also enriches the elements.
2. From construction sites, agricultural fields, deforested lands, urban sewage, discharges from ceramic industry, brewery, distillery, paper and textile mills, discharges from mines & washing plants, thermal power plants, etc.

#### 1.9.1.2 Effects

1. Impedes the rate of generation of dissolved oxygen (DO) in water bodies by reducing sunlight penetration which in turn reduces photosynthetic activities by aquatic flora, a process by which DO is generated.
2. Decimates the population of phyto and zoo planktons which constitutes the starting point of the aquatic food chain and thereby forces higher forms of aquatic fauna and flora to migrate.
3. It causes aesthetic pollution by discolouring streams and lakes.

### 1.9.2 Oxygen Demanding (DO) Wastes

These types of pollutants comprise mostly organic compounds like protein, urea, carbohydrate (sugar and starch), cellulose, fats, soaps, oil and grease, etc. most of these undergo bacterial decomposition. This process of oxidative decomposition requires a large quantity of oxygen which comes from Demand Oxygen being generated in the water body undergoing pollution. This DO supply in water body is depleted. The amount of oxygen required for this process is known as Biological Oxygen Demand (BOD). BOD and DO have antipathy relation and their proportion varies inversely. Pollution zone extends on either side of the sewage discharge point and is characterized by increasing BOD and decreasing DO. Decomposition zone extends downstream from the pollution zone. This is the place where oxidation decomposition of organic matter of the waste takes place. This zone is characterized by gradually decreasing BOD and gradually increasing DO. Recovery zone is the area beyond the decomposition zone where decomposition of the organic matters is complete having low BOD and high DO. Most of the rivers have capacity of degrading organic waste. But the problem arises when the stream becomes overburdened with biochemical oxygen demanding waste that overpowers its natural cleansing power.

#### 1.9.2.1 Source

- Agricultural waste mainly contributes 33% of BOD in the river.
- Discharge from malty, distillery, paper mills, textiles and oil refinery.
- Urban sewage.

#### 1.9.2.2 Effects

Severely affects DO supply of a water body which is the most important aspect of pollution in natural water bodies. The depletion of DO leads to total breakdown of an aquatic ecosystem.

### 1.9.3 Industrial Waste

With growing urbanization and industrialization, one of the most important environmental concerns is the industrial waste. Elemental concentration due to industrial waste in the form of solid, liquids and gases is the pressing problems linked with industrialization. Mineral resource gets depleted besides defacing the environment and causing pollution through mining concentration piles, tailings, millings and smelting operations. The heavy metals from these sources get enriched in the top organic top soil layer and find their way through to the streams to the ecosystem (Dayal 1996). The hazardous chemical wastes generated from textile, paper, plastic, pesticide, paint industry, medicine, metals, oil, gasoline, tanneries, etc. are indiscriminately dumped posing threat to public health and environment. The hazardous radioactive wastes with insidious character and extreme toxicity present most difficult and pressing problems in waste disposal. It is evident that even the low concentrations of toxic elements are quite toxic and are lethal to aquatic life. Concentration of elements like Pb, Cu, Zn, Cd, Cr, Fe, Mn, Al, Na, Se and Hg should not exceed a certain specific limit. If more than one metal occurs in concentration higher than the limit, then the pollution may be even serious (Srivastava 1981a, b, c, 1990; Rao 2008).

#### 1.9.3.1 Source

- Compound volatilized in the soil and the fills, after placement such as methane, ammonia, hydrogen sulphide, nitrogen gases may enter the atmosphere.

- Heavy metals such as Pb, Cr, Fe, ore retained in the soil.
- Soluble material such as Cl, $NO_3$, $SO_3$ readily pass through the fill and the soil to the groundwater system.
- Over land run off may leach from ore stock pile, mine dumps and slag heaps and transport it to the surface water network.
- Plant residue containing toxic elements left in the field return to the environment through soil and run off processes.

#### 1.9.3.2 Effects

- Health hazards including cancer in human beings.
- Decimation of plank tonic population, thereby causing depletion of higher forms aquatic ecosystem.

### 1.9.4 Synthetic Organic and Inorganic Compounds

These comprise organic-chloride compound, organo-phosphatic compound, organic solvent, organic and inorganic dyes, phenolic compound, ammonical compound, caustic soda, fluorides, cyanides, alkaline cleaners and organic and inorganic acids of various types.

#### 1.9.4.1 Source

- Waste discharge from plastic, medicine, paint, metal, leather and textile industries, oil refineries and mineral beneficiation plants.
- Various types of pesticide weedicides applied on agricultural land.

### 1.9.5 Oil and Grease

These are the industrial pollutants and are of special concern to marine pollution.

#### 1.9.5.1 Source

- Liquid effluents discharged from various industries, oil refineries and mineral beneficiation plants (Bora 1996).
- Seepage from passing ships especially large super tankers.
- Oil spill from sinking tankers and off-shore drilling platforms.

#### 1.9.5.2 Effects

- When oil and grease is discharged into inland static water bodies and sluggish streams it forms a film on the water surface. This retards re-aeration and as a result the water cannot replenish their DO, affecting adversely on aquatic fauna especially plank tonic population.
- Health hazards to human being.
- Pollutes beaches that adversely affect the tourist trade.

## 1.10 Air Pollution

Air is the mixture of gases surrounding the earth. It is not possible to survive for living beings without the air. A physical, biological or chemical alteration to the air in the atmosphere is termed as pollution. Air pollution is defined as the presence of toxic chemicals or compound including those of biological origin in the air, at levels that pose a health risk. It occurs when any harmful gases like ammonia, carbon monoxide, sulphur dioxide, nitrogen oxide, methane, chlorofluorocarbon, particulate matters (organic or inorganic) and biological molecules, dust, smoke are introduced in the atmosphere in turn into plants, animals and human beings. Air pollutants can be in the form of gases, liquids and solids:

## 1.11 Air Pollutants

1. Particulate matter (PM 10 and PM2)
2. Ozone ($O_3$)
3. Nitrogen oxide (NO)
4. Carbon monoxide (CO)
5. Sulphur dioxide ($SO_2$)
6. Carbon dioxide ($CO_2$)
7. Methane
8. Ammonia
9. Chlorofluorocarbon
10. Toxic metals, etc.

Air pollutants can be classified as primary and secondary pollutants. The primary air pollutants are emitted directly from the source, whereas the secondary air pollutants are not directly emitted as such, but form when other primary air pollutants react in the atmosphere. Primary air pollutants are the chemicals that when released pose health problem/risks in their unmodified forms. Vehicular exhaust is a major source of CO. Carbon and other hydrocarbons, VOCs are also significant greenhouse gases via their role in creating ozone.

### 1.11.1 Primary Air Pollutants

- Carbon monoxide
- Nitrogen oxides especially $NO_2$ emitted from high temperature combustion
- Sulphur oxides especially $SO_2$
- Hydrocarbons
- Chlorofluorocarbon

### 1.11.2 Secondary Air Pollutants

- Ozone: Formed when hydrocarbon and nitrogen oxide combine in the presence of sun light.
- Nitrogen dioxide: Formed when nitrogen oxide combines with oxygen in the air.
- Acid rains: Formed when $SO_2$, nitrogen oxide react with water.
- Nutrient enrichment compounds: Secondary particulate matters.

#### 1.11.2.1 Source (Srivastava et al. 1996; Absar et al. 1996)

Air pollution is characterized by the presence of particulate matter and element in the air of the atmosphere. It may be due to:

- Poisonous gases/elements: Cd, Hg, As, Pb and composition to certain metals give rise to different negative effects on kidney (Pb, Cd), skeleton (Cd, Al), nervous system (Hg, Al, Pb), fertility (Pb, Hg), heart (Pb, As).
- Emission from vehicle: different types of automobiles increase in $SO_2$ and butadiene, CO.
- Combustion fossil fuel: Power plants, oil refinery, industrial facility and factories releasing $CO_2$.
- Air conditioner, refrigerators, aerosol sprays: Releases chlorofluorocarbon.
- Dust and dirt: Particulate matter.
- Mining activities: increases dust, release of certain elements.
- Natural event: Volcanoes $SO_2$, HCL, HF, $H_2S$ $CO_2$, particulate matters (dust storms).
- Other areas: Agricultural activities, cities, wood burning.

Dust pollution and smog and smoke are the major source of air pollution. Huge piles of garbage collected from single colonies that are dumped in dumping ground are burnt affecting the larger areas. Burning of paper produces dioxins and leaves produce CO and particulate matters, whereas plastic produces dioxins, furan, volatile, organic compounds and halogenated hydrocarbon. Delhi and adjoining area suffer from stubble burning particularly during winter. Mining areas also contribute lot of dust that increases the level of toxic particulate damaging the soil in the adjoining areas. To properly assess the role of dust, it is better to understand the chemical, physical and mineralogical properties of airborne soil particles and also to understand the processes that generate which are highly nonlinear and extremely sensitive.

Smog is generated with a mixture of smoke, sulphur dioxide with various gases, dust and water vapours in the atmosphere. Every year month of October and November is experiencing dense smog in adjoining areas of Delhi, the capital of India. This dense smog is attributed to burning of stubble, locally known as Parali, in the adjoining areas Uttar Pradesh, Haryana, Punjab states. These states are known as agricultural states. These states are not known earlier for rice cultivation but are known for gram, wheat and coarse cereal crops. This change in cropping pattern from coarse cereal crop to rice cultivation, which requires lot of water to grow generates lot of stubble which is left in the field itself. The left-over stubble from agricultural fields, it is allowed to burn generating lot of smoke which in turn when mixed with water and other gases in the atmosphere during early winter days creates dense smog. This leads to immediate spike in air pollution levels during winter every year. Stubble burning has two-fold of disadvantage. (1) Animals do not get their rightful share of fodder. (2) It also kills friendly bacteria and fungi present in abundance in soil permanently damaging the environment and fertility in the soil. Sharp increase in levels of toxic particulate matters in the air to as much as more than 500 microgram per cu m (Air Quality Index, AQI) prevalent in the adjoining area of Punjab, Haryana, Uttar Pradesh and Delhi. It is considered as severe level of (AQI) affecting heart, lung and swelling in the blood vessels (reported by AIMS, New Delhi).

## 1.12 Dust Pollution

The atmosphere is a daily host to a variety of particles. Many of us have seen particles, or aerosol, in thick hazes of pollution. Aerosols absorb and scatter radiation affecting climate. They can affect our health. Many of them are smaller than 2.5 micrometres, making them easy to inhale. It is well known that certain types of fine particles, cigarette smoke, coal dust or asbestos, can affect our respiratory systems, but many other types can also cause problems. Trade winds can lift fire mineral particles from soils, and then carry those particles to great distances around the planet. Satellite gives excellent picture of the transport of dust and other aerosols on a global scale. The presence of high concentrations of mineral dust over such large areas of earth has implications not only human health, but also metrology, climate, geology and biochemical processes. All these processes are all strongly linked indeed the geological records show that dust mobilization and concentration have varied tremendously through time.

Some of the common diseases associated with various types of dust.

## 1.13 Asbestosis

Asbestos is a fibrous variety of six naturally occurring minerals that have been used in commercial products. Asbestos when inhaled can cause asbestosis, a fibrosis of the lungs, as well as lung cancer and malignant mesothelioma. Chrysotile asbestos is less carcinogenic than amphibole asbestoses. The recent collapse of the WTO towers in the USA also generated large amount of asbestos dust, which has created havoc at the accident site. Because of long delay between exposure to asbestos and onset of the mesothelioma, exposure that occurred even decades ago may place a person at risk in some areas.

## 1.14 Silicosis

Silicosis is caused by inhalation of dust containing free silica or silicon dioxide. The incidence of silicosis depends upon the chemical composition of the dust, size of the particles duration of exposure and individual susceptibility. The higher the concentration of free silica in the dust, the greater is the hazard. Particles between 0.5 and 3 micron are the most dangerous because they reach the interior of the lungs with case. The longer the duration of exposure, the greater is the risk of developing silicosis. Incubation period may vary from a few months to 6 years of exposure. The people associated with granite mining and stone crusher plants are more susceptible to this disease.

## 1.15 Anthracnosis

Anthracnosis also known as black lung is a chronic lung disease caused by inhalation of coal dust. The first phase of the disease is labeled simple pneumoconiosis, which is associated with little ventilatory impairment and usually developed after 12 years of exposure. The second phase is characterized by progressive massive fibrosis. This causes severe respiratory disability and frequent results in premature death. Cigarette smoking aggravates the condition of the disease.

## References

Absar A, Srivastava GC, Pandey SN, Jangi BL, Narayan S, Bajpai IP, Khan MA (1996) Hazards of active Barren Island volcano, Andaman and Nicobar island, India. Symposium on earth sciences in environmental assessment and management, Geological Survey India. Lucknow, pp 161–163

Acharyya SK (2003) Late quaternary stratigraphy of the Bengal Basin, influence of tectonism and eustasy on delta sedimentation and their relation to arsenic problems in area. Proceedings of 4th South Asia Geological Congress New Delhi (GEOSAS-IV), pp 385–399

Acharyya SK (2004) Arsenic levels in groundwater from quaternary alluvium in the Ganga plain and the Bengal basin sub-continent. Insights into influence of stratigraphy. Groundwater Res 8 (1):55–66

Agarwal M (2009) Aquatic toxicology: effects on ground water and the remedies for increased water use efficiency of nutrition. Workshop on ground water scenario and enhancing water efficiency. C.G.W.B., Lucknow, pp 61–69

Bhattacharya P, Jacks G, Frisbie SM, Smith E, Naidu R, Sarkar B (2002) In: Sarkar B (ed) Arsenic in the environment: a global prospective in heavy metals in the environment. Marcel Dekker, New York, pp 147–215

Bora, BC (1996) Oil spill and well blow outs-a review and retrospective of ONGC's experience on country's environment management. Invited paper symposium on earth sciences in environmental assessment and management, Geological Survey India Lucknow, 1996. Invited paper pp I-VI

Chakravarti P (1994) Water and soil pollution: a short review. Lecture delivered in 2nd course in environmental geology, Hyderabad

Dayal B (1996) Mining and mineral based industrial pollutants-an environmental perspective. Symposium on earth sciences in environmental assessment and management, Geological Survey of India, Lucknow, pp 84–85

Faruqui NH, Nagar M (1995) Earth resources and geo environmental hazards in parts of frontal foredeep. Symposium on recent advances in geological studies of northwest Himalaya and Foredeep, pp. 377–378

Lokesh KN (1996) Impact of urbanization and human activities on quality of groundwater in Udupi municipal area, Karnataka. Symposium on earth sciences in environmental assessment and management. Geological Survey of India, Lucknow, pp 102–104

Misra SP, Sinha VP, Tripathi AK, Sharma DP, Dwivedi GN, Khan MA, Yadav ML, Mehrotra RD (1996) Arsenic incidence in Son Valley of gold belt. Symposium on earth sciences in environmental assessment and management Geological Survey of India, Lucknow, pp 163–164

Mitra SN, Bandopadhyay AK, Dhar N (1996) Arsenic pollution in ground water in parts of West Bengal: some observations. Symposium on earth sciences in environmental assessment and management. Geological Survey of India, Lucknow, pp 164–165

Pal T, Mukherjee PK, Sengupta S, Sharma S (2002) Arsenic sinks in quaternary aquifer sediments of West Bengal and their role in groundwater. Proceedings of 4th South Asia Geological Congress, New Delhi, India (GEOAS-IV), pp 369–372

Pandey DS (2009) Identification of hexavalent chromium ground water and its lateral and vertical extension in Kanpur city. Workshop on ground water scenario and enhancing water us efficiency, C.G.W.B. Lucknow, pp 35–46

Prasad S, Kar SK (2004) Geological control of selenium concentration in the soil of Punjab, India. Proceedings of workshop on medical geology (IGCP-454). Spl. Publ. No. 83. Geological Survey of India, pp 306–313

Ramesh NR (2004) Environmental impact of geochemical pollution on Mysore city, Karnataka case history. Proceedings of workshop on medical geology IGCP-454, pp 361–368

Rao D (1996) Fluoride enrichment in the groundwater of Unnao district, U.P. Symposium on earth sciences in environmental assessment and management. Geological Survey of India, Lucknow, p 152

Rao CRM (2008) Different source of water contamination and various remedial measures, methodologies for toxic constituents in relation to health. Geol Surv Ind Med Geol News Lett 1(12):11–13

Rohit BR (2017) Contaminated ground water in 22 of state's 30 districts poses health risk. Times of India, Feb 22, 2017.

Shanker R, Pal T, Mukherjee PK, Sengupta S, Sharma S, Sengupta S (2002) Association of microbes with arsenic bearing siderite concentrations from shallow aquifer sediments of Bengal Delta and its implications. J Geol Soc Ind 58:269–271

Sri Ram K (1994) Soils. Lecture delivered in 2nd course in environmental geology, Hyderabad

Srivastava VC (1981a) Pollution in some industrial towns in U.P.—a menace. Seminar on status of environmental studies in India at Trivandrum

Srivastava VC (1981b) Environmental land management problem and pollution in the Arabian sea along the West Coast, India. Seminar on status of environmental studies in India at Trivandrum

Srivastava VC (1981c) Environmental impacts of Zuari fertilizer factory, Goa. Symposium on earth resource for Goa's development, Panaji, Goa

Srivastava VC (1990) Geo Environmental appraisal of Bundelkhand Region Geological Survey of India, unpublished report 1990

Srivastava VC (2008) Goitre and environmental iodine deficiency in the sub-Himalayan belt and adjoining areas. Geol Sur Ind Med Geol News Lett 1(12):9–10

Srivastava SK, Rastogi R (2009) Arsenic distribution and physiological effects in ground water of Uttar Pradesh. Workshop on ground water scenario and enhancing water use efficiency. CGWB 2009, pp 46–49

Srivastava VC, Prasad M, Sharma OP (1995) Geological abstraction of goitre endemicity in parts of lower Himalaya and adjoining plains between Udhampur and Uttar Pradesh. Symposium on recent advances in geological studies of northwest Himalaya and Foredeep, Geological Survey of India, Lucknow, pp 377–378

Srivastava VC, Roy S, Srivastava MC, Sinha AK, Kothiyal DL, Verma R (1996) Geo Environmental degradation due to impact of mining in Bundelkhand region, Uttar Pradesh. Geol Surv Ind Spl Pub 48(1):235–240

# Deficiency of Essential Elements in Crop Plants

2

Sanjesh Tiwari, Anuradha Patel, Neeraj Pandey, Amandeep Raju, Madhulika Singh, and Sheo Mohan Prasad

**Abstract**

Plants are the eminent source of essential elements which are constructing blocks of living organism in the form of protein, carbohydrate, fats and fibres. Living organism, including plants, requires nutrients for their better growth and developmental processes. Based upon the necessity, nutrients are categorized as macro- and micronutrients. Carbon, hydrogen, oxygen, phosphorus, potassium, nitrogen, sulphur, calcium, iron and magnesium are the various plant macronutrients, while micronutrients include manganese, copper, boron, molybdenum and chlorine. Plants are able to draw these mineral nutrients in a balanced way either from soil organic matter or by the use of organic or inorganic fertilizers. Mineral nutrition is acquired by a complex network of root transporters that regulate import of minerals from the soil solution to plants. The requirement of these nutrients varies from plant to plant, and scarcity of these nutrient elements intervenes the metabolic processes. However excessive uptake induces toxicity and causes poor growth associated with reduction in crop production. Hence, balanced nutrient approach is important for proper crop yield. This chapter describes the essential nutrients, the chemical forms in which they are available to plants, their function in plants, their deficiency system, and recommended nutrient levels in plant tissues of selected crops and future prospective of nutrient management approach.

**Keywords**

Macronutrients · Micronutrients · Deficiency symptoms · Nutrient management

S. Tiwari · A. Patel · N. Pandey · A. Raju · M. Singh · S. M. Prasad (✉)
Ranjan Plant Physiology and Biochemistry Laboratory, Department of Botany, University of Allahabad, Allahabad, Uttar Pradesh, India



K. Mishra et al. (eds.), *Sustainable Solutions for Elemental Deficiency and Excess in Crop Plants*, https://doi.org/10.1007/978-981-15-8636-1_2

## 2.1 Introduction

Agriculture is one of the major factors that governs the economy and is unswervingly associated with the country's development. Cereals including vegetable crops constitute an important fraction to maintain the human health as they are heady source of energy and also provide various minerals and vitamins. Plants require several elemental nutrients (collectively termed as plant ionome) for the better growth and development (Salt et al. 2008). From numerous field studies, it was pragmatic that the mineral composition among different plants species depends on the soil type. The difference in the mineral composition of soils is mainly due to diverse nutrient management practises that change the pH, cation exchange capacity and organic carbon status of the soil (Moharana et al. 2017). Besides all nutrients, plants selectively uptake the essential ones that participate in various physiological and biochemical processes such as during seed germination events and early seedling establishments (Williams and Salt 2009). The plant nutrition is the detailed study of chemical elements and compounds necessary for plant growth. Mineral elements are categorized into two groups, i.e. macro- and microelements on the basis of their requirement in plant species. Nutrients that are required in large quantities are termed as macronutrients, such as nitrogen (N), phosphorus (P), potassium (K), calcium (Ca), magnesium (Mg) and sulphur (S), while iron (Fe), manganese (Mn), boron (B), molybdenum (Mo), copper (Cu), zinc (Zn), chloride (Cl) and cobalt (Co) are required in trace amounts and are thus called micronutrients (Mondal and Bose 2019) (Fig. 2.1). Nutrients such as Zn, Fe and Cu act as a co-factor during photosynthesis and other metabolic processes, and their excessive concentrations lead to toxicity, which results in growth reduction (Nagajyoti et al. 2010).

Increased anthropogenic interventions are deteriorating the property of soil and increasing contamination. Elemental contamination reduces the crop productivity owing to the accumulation of these toxic elements in plant and this in turn affects animal and human health (Cakmak 2002). The deficiency of essential nutrients primarily leads to the disturbance in uptake of other nutrients and thus causes ion imbalance (Mei et al. 2009). The excessive use of phosphate-based fertilizers significantly obstructs the growth of mycorrhizal colony due to reduced uptake of Zn and B and reduced translocation of calcium (Ca) that result in stunted growth (Barker and Pilbeam 2015). Similarly, deficiency of S reduced the synthesis of S-containing proteins, lipid quality as well as accumulation of C18 derivatives (D'Hooghe et al. 2014).

Deficiency of microelements in plants is a global concern that limits the productivity of crops and also affects the animal and human health (Welch and Graham 2004). Indiscriminate use of NPK-based fertilizers in agriculture leads to diminution in the micronutrients bioavailability that exhausts the soil reserve of native micronutrients (Sidhu and Sharma 2010; Shukla et al. 2016). Having considered the above backgrounds, there is an urgent need to regulate agro-techniques, which are the most suitable edaphic conditions that deteriorate the micronutrient deficiency in plants (Fig. 2.2, Table 2.1). This chapter deals with the role of essential macro- and micronutrients, their transportation and deficiency symptoms in plants

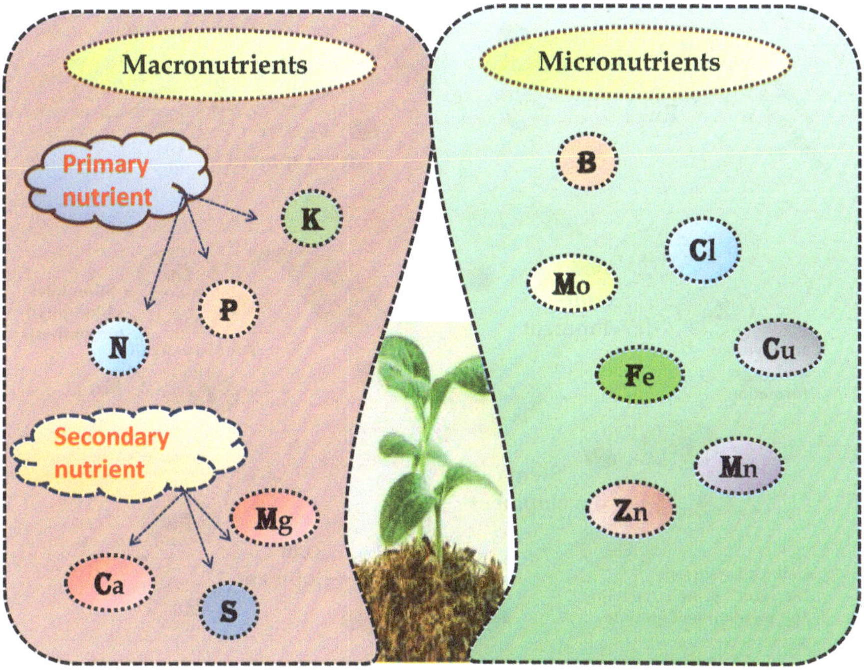


**Fig. 2.1** Representation of micro- and macronutrients needed for plants' growth and development

and also focuses on the nutrient management approach, which is an emerging strategy to relieve deficiency and stressful condition in plants.

## 2.2 Role and Deficiency Symptoms of Essential Nutrients

Micronutrients as well as macronutrients are very indispensable for growth and development of plants and fundamentally take part in various cellular functions such as photosynthesis, respiration, redox reactions in chloroplast and mitochondria, nitrogen fixation, DNA synthesis, transcription and other biosynthetic and derivative reactions (Nouet et al. 2011). An imbalance in either trace amount or in bulk significantly alters the morphological and biochemical parameters and consequently produces visible deficiency symptoms. Hence in the below segment the beneficial role and the deficiency symptoms of various nutrient elements are discussed in detail (Table 2.1).

### 2.2.1 Nitrogen (N)

Being a macronutrient, nitrogen (N) is involved in the biosynthesis of various macromolecules (proteins and nucleic acids). Thus, the balance between uptake

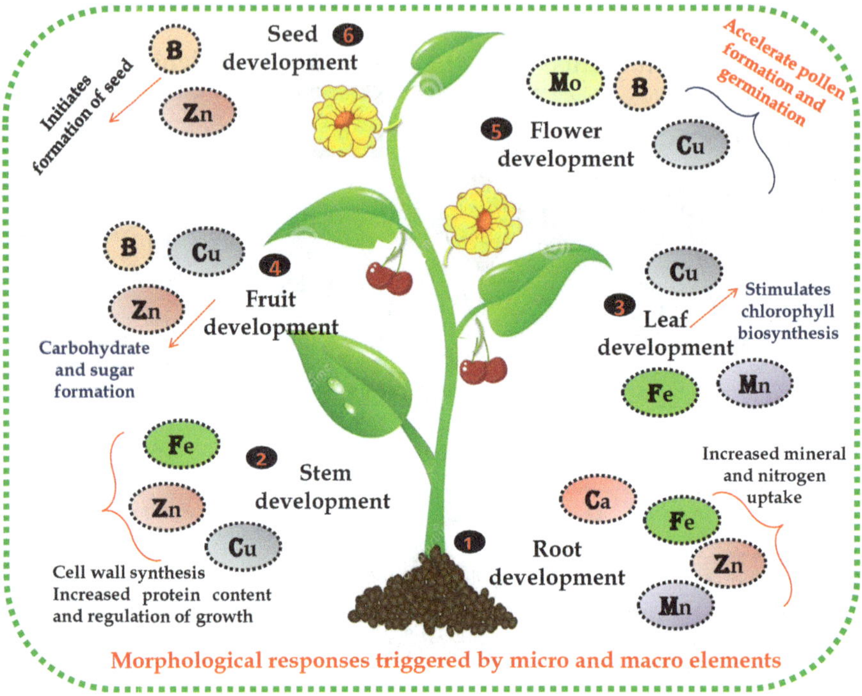


**Fig. 2.2** Schematic representation of morphological responses triggered by micro- and macronutrients in plants

and assimilation of nitrogen is an important physiological process that has a direct impact on the growth and development of plants. Nitrogen is also an integral component of phospholipid and chloroplast in plants (Nawaz et al. 2018). Photoautotroph use nitrate ($NO_3^-$) as a source of nitrogen that is easily available to them and convert it into nitrite ($NO_2^-$) and ammonia ($NH_4^+$) via enzyme nitrate reductase (NR) and nitrite reductase (NiR) (Parween et al. 2011). Furthermore, chlorophyll and amino acid biosynthesis requires N in trace amounts. Exogenous N supplementation increases the quality and quantity of dry matter in leafy vegetables and protein content in cereals and grains (Khan et al. 2019). However under low N availability, plant exhibits overall reduction in biomass associated with unregulated cell division. The yellowing of leaves firstly appears on tips of old leaves and then extends to young leaves and finally leads to cell death. Besides morphological indications, the molecular processes such as translation (protein biosynthesis) are also severely affected by N deficiency, which results in low biomass associated with reduced rate of photosynthesis (Zhao et al. 2005). Study of Zhai and Li (2005) displayed that N deficiency significantly reduced the total crop production. The early seed germination and panicle differentiation also get affected by low N supply to crops (Xiong et al. 2018). Improper N management during crop

**Table 2.1** Showing the potential role, toxicity symptoms of micro- and macronutrients in plants

| S. No. | Macro-/ micronutrients | Beneficial role | Deficiency symptom | Important constituent | References |
|---|---|---|---|---|---|
| 1. | Nitrogen (N) | • Biosynthesis of macromolecules (proteins and nucleic acids)<br>• Chlorophyll biosynthesis and amino acid synthesis<br>• Nitrogen and ammonia assimilation | • Yellowing of leaves<br>• Cell death<br>• Reduced photosynthesis and thus reduced biomass<br>• Inhibition of seed germination<br>• Unregulated cell division<br>• Reduced yield | • Protein<br>• Amino acid<br>• Chloroplast<br>• Phospholipids<br>• Enzymes | Nawaz et al. (2018)<br>Parween et al. (2011)<br>Xiong et al. (2018)<br>Zhai and Li (2005) |
| 2. | Phosphorous (P) | • Regulation of protein synthesis<br>• Stimulates the growth of new tissue<br>• Regulation of cell division<br>• Development of the roots and hastening of maturity<br>• Participates in key structure of plant | • Affects early developmental stages<br>• Stunted or weak growth<br>• Dark to blue-green colouration appears in leaves<br>• Purpling of stems<br>• Reduced $CO_2$ assimilation | • Primary component of the energy molecule ATP (adenosine triphosphate)<br>• Nucleotide<br>• Component of DNA, RNA and phospholipids | Hammond and White (2008)<br>MacDonald et al. (2011) |
| 3. | Potassium (K) | • Regulation of growth and development<br>• Maintains membrane potential<br>• Up-regulates enzymes NR, GDH, GS-GOGAT | • Reduced photosynthesis<br>• Inhibits the process of translation<br>• Inhibits seed germination<br>• Yellowing of leaves<br>• Leaf curling<br>• Poor root development<br>• Poor yield<br>• Plants are more prone to disease<br>• Inhibits activity of aquaporins | • Responsible for many vital processes such as water and nutrient transportation<br>• Protein and starch synthesis | Wang et al. (2013)<br>Çokkizgin and Bölek 2015 |

(continued)

**Table 2.1** (continued)

| S. No. | Macro-/ micronutrients | Beneficial role | Deficiency symptom | Important constituent | References |
|---|---|---|---|---|---|
| 4. | Cobalt (Co) | • Involved in regulation of physiological processes<br>• Involved in catalysis<br>• Regulates developmental processes<br>• Elongation of coleoptile<br>• Opening of hypocotyls<br>• Expansion of leaf<br>• Regulates bud development<br>• Provides resistance<br>• Delayed senescence | • Yellowing of leaves<br>• Dwarf size<br>• Early reproductive stage<br>• Interveinal chlorosis<br>• Reduced dry matter | • Important constituent of vitamin $B_{12}$<br>• Necessary for activation of enzyme involved in nitrogen fixation | Howell and Skoog (1975)<br>Pilon-Smits et al. (2009)<br>Farooq et al. (2012) |
| 5. | Copper (Cu) | • Regulates cellular metabolism<br>• Activates enzymes<br>• Co-factor in several enzymes<br>• Facilitates translocation of iron<br>• Involved in protein trafficking | • Twisted and distorted leaf<br>• Chlorosis<br>• Necrosis<br>• Less formation of plastocyanin<br>• Inhibition of PS I<br>• Reduced photosynthetic pigments<br>• Disintegration of thylakoid membrane | • Important constituent of structural organization<br>• Co-factor in several enzymes<br>• Important constituent of superoxide dismutase, cytochrome oxidase, amino oxidase and ascorbate oxidase<br>• Component of plastocyanin and polyphenol oxidase | Marschner (1995) |
| 6. | Calcium (Ca) | • Involved in signalling<br>• Maintains cell wall structure<br>• Membrane integrity | • Localized cell death<br>• Loss of membrane integrity<br>• Reduction in growth and plant height<br>• Chlorosis<br>• Rapid senescence<br>• Inhibition of antioxidant | • Act as a secondary molecule<br>• Important constituent of cell wall | Stael et al. (2001)<br>White and Broadley (2003) |

| | | | | | |
|---|---|---|---|---|---|
| | | | system<br>• Increased oxidative stress | | |
| 7. | Boron (B) | • Maintains cell integrity<br>• Regulates cell division<br>• Regulates cell elongation<br>• Assists nitrogen fixation<br>• Synthesis of assimilatory powers | • Brittle leaves<br>• Inhibition of root and shoot meristem<br>• Reduced root growth<br>• Discolouration of root<br>• Reduced shoot growth<br>• Reduction of nodulation<br>• Dieback of apical meristem tissue<br>• Abnormal leaf growth<br>• Inter-veinal necrosis<br>• Reduced formation of reducing power<br>• Alteration in metabolic processes<br>• Shortening of internode<br>• Causes stem crack in cereal crops<br>• Inhibition of fertilization<br>• Reduction rate of seed formation | • Important constituent of cell membrane rhamnogalacturonan (RGs), a polysaccharide<br>• Important for synthesis of ATP, NADPH, NADH and RNA | Brown et al. (2002)<br>Marschner (2012)<br>Will et al. (2011) |
| 8. | Zinc (Zn) | • Acts as a co-factor<br>• Binds with oxidative radicle<br>• Acts as co-factor for alcohol dehydrogenase (ADH) and carbonic anhydrase (CA) | • Morphological alteration in plants<br>• Development of chlorosis<br>• Cell death<br>• Poor root development<br>• Short intermodal growth<br>• Down-regulation of | • Essential component of plant cell membrane<br>• Involves in cell repair system<br>• Associated with post-translational changes<br>• Involves in protein trafficking | Rafique et al. (2012)<br>Nagajyoti et al. (2010) |

(continued)

**Table 2.1** (continued)

| S. No. | Macro-/ micronutrients | Beneficial role | Deficiency symptom | Important constituent | References |
|---|---|---|---|---|---|
| | | | enzymatic activity<br>• Decreased biomass<br>• Altered metabolic processes | | |
| 9. | Magnesium (Mg) | • Acts as a co-factor<br>• Needed for ribosomal stability<br>• Biosynthesis of chlorophyll<br>• Regulates photosynthesis<br>• Activation of RuBisCo<br>• Regulates cellular functions<br>• Regulation of carbon assimilation | • Primary symptoms<br>• Yellowing of leaves<br>• Discolouration of green vein in to red tissues<br>• Inter-veinal necrosis<br>• Grain falling<br>• Fruit expansion<br>• Decreased carbon fixation<br>• Reduced biomass | • Essential component of chlorophyll<br>• Stackness of thylakoid membrane<br>• Key component of cell membrane<br>• Component of enzymes | Yang et al. (2012)<br>Jezek et al. (2015) |
| 10. | Manganese (Mn) | • Regulates growth and development<br>• Regulates cellular function such as photosynthesis<br>• Nitrogen fixation<br>• DNA synthesis<br>• Regulates biosynthetic processes<br>• Acts as a catalyst<br>• Responsible for the release of oxygen<br>• Regulates biosynthesis of protein, carbohydrate, lipids and lignin | • Alteration in photosynthesis<br>• Chlorosis of leaves<br>• Stunted growth<br>• Reduced biomass<br>• Affects the PS II proteins D1, PsbP and PsbQ<br>• Down-regulation of antioxidant<br>• Excessive accumulation of reactive oxygen species<br>• Less production of secondary metabolites<br>• Increased transpiration | • Important cellular component<br>• Raises enzyme activity<br>• Important component of glutathione synthase and superoxide dismutase<br>• Key player in oxygen-evolving complex (OEC) of photosystem II (PS II) | Hänsch and Mendel (2009)<br>Nouet et al. (2011)<br>Millaleo et al. (2010)<br>Schmidt et al. (2016)<br>Marschner (2012) |

| | | | | | |
|---|---|---|---|---|---|
| 11. | Iron (Fe) | • Acts as a co-factor<br>• Transports oxygen<br>• Protein stability | • Reduction of growth<br>• Poor yield<br>• Dropping quality of crop plant<br>• Swollen root tip<br>• Lateral root development<br>• Reduced pigment synthesis<br>• Disturbance in thylakoid membrane<br>• Affects PS II photochemistry<br>• Inhibits activity of Cyt $b_6f$<br>• Reduced photosynthesis | • Important component of iron–sulphur (FeS)-rich proteins of PS I and ferredoxins | Couturier et al. (2013)<br>Abadía et al. (2011)<br>Laganowsky et al. (2009) |
| 12. | Molybdenum (Mo) | • Acts as a co-factor<br>• Activates several enzymes such as nitrogenase, nitrate reductase, xanthine dehydrogenase, aldehyde oxidase, and sulphate oxidase | • Stunted growth<br>• Chlorosis<br>• Yellowing of veins<br>• Involuted leaf<br>• Reduced growth<br>• Reduced biomass<br>• Disturbance in equilibrium N uptake and assimilation<br>• Early senescence<br>• Pale colouring of leaves | • Structural component of pterin | Yohe et al. (2016) |
| 13. | Sulphur (S) | • Regulated growth and development under stress and un-stressed condition<br>• Reduces ROS generation and oxidative stress<br>• Regulation of phytohormone<br>• Maintains Na/K ratio | • Lowers photosynthetic pigment content<br>• Alters photosynthesis<br>• Reduces activity of RuBisCo<br>• Inhibits the protein synthesis<br>• Poor yield and growth | • Integral part of different biologically active compounds such as some amino acids, phytohormones and coenzymes<br>• Integrally involved in cysteine, methionine, sulphoxides and glucosinolates | Fatma et al. (2016)<br>Takahashi et al. (2011)<br>Khan et al. (2014) |

(continued)

**Table 2.1** (continued)

| S. No. | Macro-/ micronutrients | Beneficial role | Deficiency symptom | Important constituent | References |
|---|---|---|---|---|---|
| 14. | Nickel (Ni) | • Activates enzyme involved in developmental processes | • Targets metabolic responses in plants<br>• Affects growth directly<br>• Marginal chlorosis<br>• Premature senescence<br>• Poor seed setting | • Integral component of metallo-enzymes urease and hydrogenase | Polacco et al. (2013)<br>Wood et al. (2006) |
| 15. | Sodium (Na) | • Facilitates pyruvate uptake<br>• Enhances metabolic activity<br>• Maintains osmotic balance<br>• Aids chlorophyll synthesis<br>• Regulates stomatal opening<br>• Concentrates carbon dioxide<br>• Modulates osmotic pressure<br>• Involved in signalling | • Reduced growth<br>• Disturbance in membrane transport<br>• Stomatal dysfunction | • Important constituent of biological system<br>• Membrane transport | Gattward et al. (2012)<br>Marschner (2012)<br>Gattward et al. (2012) |
| 16. | Chloride (Cl) | • Regulates enzyme activity asparagine synthetase, and vacuolar proton-pumping ATPase<br>• Balances the electrical charges<br>• Plays major role in signal transduction and perception<br>• Involves in cell elongation<br>• Stimulates growth<br>• Reduces water consumption<br>• Enhances performance of chloroplast<br>• Suppression of plant disease<br>• Opening and closing of stomata | • Depolarization of membrane<br>• Cell expansion, defective pollen growth and shorter root hairs<br>• Reduces chloroplast action<br>• Regulates leaf cell size | • Photosynthesis<br>• Osmotic adjustment<br>• Participates actively in photosynthesis and metabolic processes | Michard et al. (2017)<br>Colcombet et al. (2009)<br>Zhang et al. (2017) |

cultivations significantly declines uptake of N allied with decrease in its assimilation into amino acids, which gets recovered after N alteration in fields (Fu and Yang 2012). Further, low N supplies to plants significantly increase the specific transporters for the uptake of ammonium and nitrate into root cells. In *Arabidopsis*, uptake of ammonia is mediated by ammonium transporters (AMT) (Ludewig et al. 2007) while nitrate is taken up by nitrate transporters (NRTs) (Krapp et al. 2014), which is found up-regulated and plays central part in efficient N uptake under low availability (Gu et al. 2013).

### 2.2.2 Phosphorus ($PO_4^{2-}$)

Phosphorus (P) is an important nutrient and major component of plant dry mass and constitutes 0.2% of plant's dry matter. It contributes in biosynthesis of assimilating powers, i.e. ATP and NADPH along with nucleic acids, membrane phospholipids and glycolipids, all of which play prominent role in photosynthesis (Hammond and White 2008). Plants take up P from soil in the form of phosphates ($PO_4^{2-}$) via phosphate transporters. The study by MacDonald et al. (2011) states that around 30% of the world's arable soils is low in P and needs additional P-containing fertilizers to improve the production and yield of crop plants. Phosphate rocks are natural pool for the P in soil, but excessive use of P in some areas (especially in Asian countries) is making agricultural soil P deficient (Baker et al. 2015). This deficiency is mainly due to immobilization of P, and thus higher concentration of P is required for uptake in plants and only 20% of total P is actively taken up by plants (Cordell and White 2015). Due to requirement of P in early developmental stages, its deficiency leads to stunted or weak growth and dark to blue-green colouration appears on older leaves associated with purpling of stems. Further, phosphate deficiency significantly reduces $CO_2$ assimilation and subsequently enhances the accumulation of starch in cells (Heldt et al. 1977). Besides this, under low P supply, secretion of the organic exudates occurs from roots such as acetic acid, aconitic acid, citric acid, malic acid, fumaric acid, lactic acid, oxalic acid and succinic acid along with phytase and phosphatase (White and Veneklaas 2012). Further, some genes that encode for ribonucleases (RNA), purple acid phosphatases (PAPs) and phosphate transporters (PTs) are also up-regulated in different plant species under P deficient conditions (Miller et al. 2001).

### 2.2.3 Potassium (K)

Potassium (K) is one of the major nutrients present in cationic form $K^+$ in plants, which directly regulates the growth and development and also maintains the membrane potential (Wang et al. 2013). The physiological and biochemical events such as rate of photosynthesis, translation of mRNA into proteins, ion homeostasis, seed germination and maintenance of the osmolality of plants are also affected under the influence of $K^+$ (Çokkizgin and Bölek 2015). In general three types of membrane transport proteins, namely $K^+$ uptake permeases (KT/HAK/KUP), $K^+$ transporter (Trk/HKT) family and the cation proton anti-porters (CPA), facilitate the transport of $K^+$ (Gierth and Maser 2007) in plants. Active form of K is present in soil in combination as potassium chloride (KCl), potassium nitrate ($KNO_3$), potassium sulphate ($K_2SO_4$) and potassium carbonate ($K_2CO_3$) (Kafkafi et al. 2001). Furthermore, the enzymatic activity of ATP synthase is up-regulated under high $K^+$ supply owing to its direct effect on membrane-bound $H^+$-ATPase that leads to acidification (Arend et al. 2004). Similarly, the opening and closure of stomata is necessary for the uninterrupted synthesis of energy in the form of ATP and is highly regulated by the concentration of $K^+$ along with the other ions ($Cl^-$ or $Ca^{++}$) and hormone abscisic acid (ABA) (Taiz and Zeiger 2010). Symptoms of plants under $K^+$ deficiency include growth retardation, yellowing of leaf margins and curling, chlorophyll degradation, poor development of root system and reduction in the crop production. Plants become more prone to diseases and pest infestation under adverse environmental conditions (Wang et al. 2013). Nitrogen metabolic enzymes such as NR, GDH and GS-GOGAT are down-regulated under $K^+$ deficient conditions (Qu et al. 2011) and subsequently decrease the expression of three nitrate transporter genes (AtNRT2;1, AtNRT2;3 and AtNRT2;6) in *Arabidopsis* plant (Armengaud et al. 2004). Low $K^+$ supply also hampers the osmolality of plants and generates water stress by decreasing the water uptake via inhibition of the activity of aquaporins (Kanai et al. 2011). Transcriptome analysis proved that the expression of genes in ion transporter families, such as AtHAK5 transporter (high-affinity $K^+$ transporter5 of *A. thaliana*) (Gierth et al. 2005), OsHAK1 in *Oryza sativa* (Banuelos et al. 2002) and CaHAK1 in *Capsicum annuum* (Martinez-Cordero et al. 2004), was markedly up-regulated under $K^+$ deficiency.

### 2.2.4 Sulphur (S)

After the role of major nutrients such as N, P and K, sulphur (S) is regarded as fourth important nutrient and is directly involved in regulating the plants' growth and development under normal as well as stress conditions (Fatma et al. 2016). Sulphur is present in nature both in inorganic and organic forms and is an integral part of different biologically active compounds such as some amino acids, phytohormones, coenzymes, etc. (Takahashi et al. 2011). Roots absorb S in form of sulphate (form of S present in soil) by involving sulphate transporters (SLUTRs) activated by establishment of proton motive force generated by movement of $H^+$ at the expense of ATP because SLUTR is a type of co-transporter protein ($H^+$-sulphate) (Hawkesford 2003). Around 12–16 genes are involved in the expression of SLUTRs in plants, and in *Arabidopsis thaliana* 12 types of SLUTRs are reported. SULTRs are categorized into four families of SLUTRs (SULTR1, SULTR2, SULTR3 and SULTR4) (Takahashi et al. 2012). The expression site is quite different as epidermis and root cortex express SLUTR1 and SLUTR2 mediating initial uptake of S from soil; SLUTR3 expresses in phloem cells while remobilization and collection of S in vacuoles are performed by SLUTR4 (Yoshimoto et al. 2003; Kataoka et al. 2004).

Cysteine (Cys), methionine (Met), sulphoxides and glucosinolates are active S-containing compounds involved in alleviating the stress conditions (Khan et al. 2014). Sulphur regulates growth of plants mainly by decreasing the ROS production and preventing the oxidative damage (Khan et al. 2015). It is also involved in regulation of ABA, JA, NO and ethylene (Fatma et al. 2016). Further, S also reduces the accumulation of toxic metals and maintains the K/Na ratio and enhances the rate of photosynthesis by enhancing the activity of RuBisCo (Lunde et al. 2008; Matraszek et al. 2016). Under some specific soil conditions, S deficiency occurs that hampers the overall plant metabolic process by lowering the chlorophyll content and rate of photosynthesis (Kastori et al. 2000), reducing the activity of enzyme RuBicCo and protein synthesis (Lunde et al. 2008). Under S starving conditions, plants might express gene sdi1 (sulphate deficiency-induced gene) that participates in maintaining the endogenous sulphate level as reported in *A. thaliana* (Howarth et al. 2003). Similarly, plants up-regulate the expression of genes responsible for SLUTR protein in root cortical cells under the control of promoter sulphur-responsive element (SURE) that improves the endogenous GSH and cysteine or other S-containing active compounds (Maruyama-Nakashita et al. 2005). Decreased S inputs have increased the incidence of S-deficiency in crops, causing decline in yield and yield-associated attributes.

### 2.2.5 Magnesium (Mg)

All research is focused on to increase the crop production by using NPK fertilizers, and magnesium (Mg) might be neglected due to its less efficient role in crop production. It is termed as "a forgotten element in crop production" (Cakmak and Yazici 2010). Magnesium is a macronutrient and persists in soil in divalent form ($Mg^{++}$) that is present on an average concentration of 20,000 ppm in the earth's crust, making it the eighth most abundant element. In many metabolic processes, several enzymes require $Mg^{++}$ as a co-factor, and stability of ribosomes and chlorophyll biosynthesis also require $Mg^{++}$ as central metal ion (Masuda 2008). Magnesium has a potential role in the formation of pigments (Chl a) and is directly linked with photosynthesis (Trankner et al. 2018). Being a divalent cation, Mg interacts with negative charge present in thylakoid membrane and allows thylakoid membrane to stack strongly and thus increases the stability of chloroplast (Puthiyaveetil et al. 2017). Further, Mg is also required for the activation of RuBisCo (Ribulose-1,5-bisphosphat-carboxylase/-oxygenase) by making complex with carbamate group of RuBisCo (Portis 2003). Deficiency of $Mg^{++}$ in plants is due to the competition for transport with other cations (hydrogen: $H^+$; potassium: $K^+$; calcium: $Ca^{++}$) (Koyabashi et al. 2013). Due to high mobility, the concentration of $Mg^{++}$ is high in either phloem or in shoots than roots. $Mg^{++}/H^+$ antiporter named MHX mediates the transport, which is present in vacuolar membrane of xylem parenchyma cells (Shaul et al. 1999). When the Mg content comes below the range of 0.15–0.35% of total dry weight in vegetative parts, plants become Mg deficient and primary symptoms appear on leaves (Masuda 2008). Leaves of plants become yellowish in color due to loss of chlorophyll and veins of leaves change from green to reddish in color. Under Mg deficiency, firstly older leaves showed inter-veinal necrosis associated with reduced grain falling and fruit expansion (Hariadi and Shabala 2004), which results in reduced grain falling and fruit expansion. Decrease in the total chlorophyll contents is correlated with decrease in the carbon fixation and utmost biomass formation under Mg limiting conditions as reported in maize (Jezek et al. 2015) and citrus (Yang et al. 2012). Alteration in the basal photosynthesis leads the generation of ROS that subsequently damages the cell components and up-regulates antioxidant system (Tang et al. 2012). Thus Mg is needed for maintaining important physiological processes.

### 2.2.6 Zinc (Zn)

Due to involvement as co-factor in various enzymes and binding with free radicals, zinc is a vital plant micronutrient and is actively taken up by roots of the plants as $Zn^{2+}$ cation (Barker and Pilbeam 2015). The amount of zinc (Zn) is generally 100 ppm in plant dry matter. Enzymes such as alcohol dehydrogenase (ADH) and carbonic anhydrase (CA) need Zn as co-factor. Total concentration of Zn in the soil depends upon numerous factors like soil pH, the occurrence of other macro- and micronutrients, soil microorganisms etc. For example, increase in the pH of soil significantly reduces the Zn solubility and its uptake via plant roots and leads to the Zn deficiency. Zinc deficiency is frequent in acidic or calcareous soils, which is prevalent in South and Southeast Asia, Southern Africa and Central America (Hotz and Brown 2004). Deficiency of Zn in plants causes drastic changes in morphology, including chlorosis in green leaf, leaf deaths associated with reduced root development, short internodes (Rafique et al. 2012) and increased susceptibility towards pathogen attacks (Helfenstein et al. 2015). There is down-regulation of Cu/Zn-SOD activity resulting in over-accumulation of ROS. Further, rate of carbon fixation is also inhibited under Zn deficiency due to its involvement in the activity of carbonic anhydrase (CA) (Cakmak and Yazici 2010). Transporters that transport Zn are up-regulated under Zn deficiency as reported in *Arabidopsis* (Wintz et al. 2003). Decreased growth and metabolic function, and induction of oxidative stress are all indicators of Zn toxicity (Nagajyoti et al. 2010).

### 2.2.7 Iron (Fe)

In the earth's crust, iron (Fe) is the fourth most abundant element present in the form of ferrous ($Fe^{2+}$) and ferric ($Fe^{3+}$) oxides and hydroxides. Plant-based diet is the only source of iron for world's population (Nikolic et al. 2016). Iron is basically present in the form of $Fe^{3+}$ in the rhizosphere and is not directly absorbed by plants. Its uptake varies among different plant species. Non-graminaceous plants (tomato or *Arabidopsis*) secrete protons by activating membrane $H^+$-ATPase resulting in acidification of rhizosphere, and $Fe^{3+}$ is reduced into $Fe^{2+}$ mediated by ferric reduction oxidases. $Fe^{2+}$ transport into roots occurs by iron-regulated transporter (IRT1) (Vert et al. 2002). Other graminaceae plants (rice, barley and maize) secrete phytosiderophores (PS) that chelate the $Fe^{3+}$ and transport via yellow stripe-like (YSL) transporters (Kobayashi and Nishizawa 2012). Iron regulates different physiological and biochemical activities such as photosynthesis, respiration and nitrogen fixation and also acts as a co-factor for enzyme activation. Iron co-factors also participate in the transport of oxygen or regulation of protein stability. The iron–sulphur (FeS)-rich proteins of PSI and ferredoxins are abundant in chloroplast and FeS clusters (complex I and II). A mix of FeS and haem (complex III) or haem and copper (complex IV) are mitochondrial proteins (Couturier et al. 2013). Deficiency of iron is a very concerning threat to both plants and human beings as it decreases the overall growth, production and quality of crop plants (Abadía et al. 2011). Under Fe deficiency, root tips become swollen and increase in the surface area of root occurs by formation of root hairs or lateral roots. Inhibition of the pigment biosynthesis, hampered structure of thylakoid membrane and ETC, decrease in the number of active reaction centres or LHC and activity of Cyt $b_6f$ complex are observed that result in reduction in the efficiency of photosynthesis (van Leeuwe and Stefels 2007; Laganowsky et al. 2009). Characteristics feature of Fe deficient plants are down-regulation of PSI core units (PsaC and PsaD) as noticed in rice (Yadavalli et al. 2012). Enzymes of glycolysis and TCA cycle and synthesis of different secondary metabolites (carboxylates, phenolics) in roots are found to be increased under Fe starvations (Zocchi 2006; Rodríguez-Celma et al. 2013).

### 2.2.8 Copper (Cu)

Copper is one of essential micronutrients in plants. It commonly occurs in two oxidation states $Cu^{+2}$ and $Cu^{+}$. In histidine side chains, $Cu^{+}$ cation is frequently linked with nitrogen, while $Cu^{+}$ cation is linked with the sulphur in cysteine or methionine. Many metalloproteins involved in electron transport in chloroplast and mitochondria have copper as structural element. Cu also functions as co-factor in several enzymes like Cu/Zn-superoxide dismutase (Cu/Zn SOD), cytochrome c oxidase, amino oxidase, ascorbate oxidase, polyphenol oxidase etc. Cu is responsible for metabolic processes inside cells such as oxidative phosphorylation and iron transportation and also signals for transcription protein trafficking machinery (Yruela 2009). In *A. thaliana*, six high affinity transporters COPT1–6 transport the copper ion into the cytosol, and P-type ATPases like HMA1, HMA5, PAA1, PAA2 and RAN1 are involved in efflux mechanism. The intercellular transportation comes with the help of metallochaperones like ATX1, CCS1 and CCH (Huffman and O'Halloran 2001; Burkhead et al. 2009; Pilon et al. 2009; Puig and Peñarrubia 2009). In *A. thaliana*, methionine residue previous to TDM1 and an Mx3M motif in TMD2 of COPT transporters family are essential in copper transport (Peñarrubia et al. 2010). Transcriptional activation of copper deficiency-responsive genes, including COPT1 and 2, is regulated by SPL7 transcription factor and also modulates a group of Cu-microRNAs such as miR397, 398, 408 and 857, which causes the degradation of messenger RNAs (mRNAs) encoding the non-essential Cu proproteins (Abdel-Ghany and Pilon 2008; Yamasaki et al. 2009).

In plant tissue the average content of copper is 10 μg/g dry weight (Yruela 2005). Under deficient condition, it leads to the activation of several morphological changes. The first visible symptom of copper deficiency appears at the tip of young leaves, which get twisted or distorted and show chlorosis or even necrosis. This symptom of young leaves increases downward along the leaf margins (Marschner 1995). As copper is an important component of plastocyanin, the deficiency causes less formation of plastocyanin resulting in affected photosystem I (PSI) electron transport. The photosystem II is also affected due to deficiency of copper in chloroplast. The thylakoid membrane is changed, and the modification in ambience of photosystem II acceptor site occurs due to the absence of 29 kDa polypeptide, which may be the component of a minor chlorophyll protein (CP29) of PSII. So the symptoms include disintegration of thylakoid membrane, reduced photosynthetic pigment (chlorophyll and carotenoid), lesser synthesis of plastoquinone and C18 unsaturated fatty acid (Yruela 2005).

### 2.2.9 Calcium (Ca)

Among various nutrients, calcium (Ca) has a dual role in plants as it is directly used as nutrient and is also involved in signalling mechanisms of various biochemical pathways by acting as secondary signalling molecule. The cellular concentration of Ca is relatively low as compared to other nutrients, and it is present in the range of 0.1–80 mM in cell walls and organelles while cytoplasmic levels of Ca are maintained at ~100 nM (Stael et al. 2001). Calcium is mainly involved in maintaining the cell wall structure and membrane integrity and is a well-known second messenger (Hepler 2005). Under adverse environmental conditions, such as drought, salinity, pathogen attack, metal stress or nutrient deficiency, there is rapid increase in the concentration of $Ca^{2+}$ ion via plasma membrane ion channels that restores the basal physiological processes (Manishankar et al. 2018; Toyota et al. 2018). Signal cascades mediated by $Ca^{2+}$ depend upon its mobility and sequestration from both internal and external $Ca^{2+}$ stores. Localized cell death and disturbance in membrane integrity is the characteristic feature of $Ca^{2+}$ deficiency and is tightly regulated by transport of $Ca^{2+}$ (Ng and Mcainsh 2003). Plants growing in $Ca^{2+}$ deficient conditions are more susceptible towards pathogen attacks, reduction in growth of apical meristems, chlorosis of leaves, rapid senescence induced by ethylene, change in the fruits characteristics and softening of tissues (White and Broadley 2003). Further, plants in Ca deficient conditions also showed a significant decrease in the antioxidant defence systems resulting in the oxidative stress associated with reduction in crop yields. In study of Lahaye and Epstien (1969), it was demonstrated that supplementation of 1 mM $Ca^{2+}$ minimizes the NaCl toxicity in beans by decreasing the $Na^{+}$ accumulation. Key event of $Ca^{2+}$ signalling cascades is the phosphorylation of calcium-dependent protein kinases (CPK), CPK-related protein kinases (CRK) and calcineurin B-like proteins (CBL). Hence, deficiency in

$Ca^{2+}$ significantly alters the phosphorylation process of these kinases and hampers the growth and development of plants (Hashimoto and Kudla 2011).

### 2.2.10 Cobalt (Co)

Cobalt is a non-essential element for the plant but is reported to be beneficial as it is involved in the catalysis of several physiological functions. The availability of Co and its compound is beneficial for the herbivorous animals, as it is a component of vitamin B12 that is an important constituent of various molecules' building blocks (Komeda et al. 1997). The critical Co concentrations in plants are as low as 0.1–10 μg/g dw (Bakkaus et al. 2005). Cobalt is a transition element with seven possible oxidation states, but $Co^{+2}$ and $Co^{+3}$ are commonly occurring; they can catalyse fenton reactions. Many of the developmental processes are influenced by $Co^{2+}$ ions like stem and coleoptiles elongation, opening of hypocotyls, leaf expansion and bud development (Howell and Skoog 1975). In leguminous plants, the rhizobium requires Co for synthesis of cobalamin (vitamin B12), and this cobalamin activates several enzymes involved in nitrogen fixation. Co also delays leaf senescence by inhibiting the ethylene biosynthesis and makes the seeds more drought resistant (Pilon-Smits et al. 2009). The divalent Co cation is transported to plant cell via broad range transporters such as P-type ATPase transporter as reported in *Lycopersicon esculentum* and HMA3 in *A. thaliana* (Fahimirad and Hatami 2017). Due to less mobility of Co ions, it is also distributed by various organic complexes (Vatansever et al. 2017).

The deficiency symptom of cobalt is rarely visible in plants, but they are similar to the symptoms seen due to nitrogen deficiency that includes yellowing of leaves and dwarf plants size. Deficiency also leads to early flowering and fruiting, i.e. plant goes early into the reproductive phase. The decrease in dry matter growth, female flowers and fruit yields are reported when *Cucurbita pepo* seeds were deficient in cobalt. While in high concentrations, inter-veinal chlorosis in new leaves followed by white leaf margins and tips are reported (Farooq et al. 2012).

### 2.2.11 Boron (B)

Boron is a micronutrient that maintains the cell integrity by enhancing the level of rhamnogalacturonan (RGs) II (a cell wall component) and is important in cell division and cell elongation (Brown et al. 2002) and also assists the nitrogen fixation (Marschner 2012). The average concentration of B is 0.2–800 ppm, and requirement is 20 ppm. Natural reservoir of B in soil is boric acid that leaches during heavy rainfall (Barker and Pilbeam 2015). The deficiency of B is very common in plants in alkali soils majorly found in the parts of China, India and Argentina (Ahmad et al. 2012) and limits the growth and development. It is considered as the second most common deficiency worldwide (Graham 2008). Xylem participates in the B transportation. Aerial parts and firstly young leaves develop symptoms of B deficiency with brittle leaves and cause inhibition in the growth of root and shoot meristems (Poss et al. 1999). Major symptoms of B deficiency are reduced root growth, brown root discolouration, reduced nodulation, reduced shoot growth, dieback of apical meristem tissues, abnormal leaf growth and inter-veinal necrosis (Will et al. 2011). Due to involvement of B in the formation of various assimilating powers such as ATP, NADH, NADPH and RNA, deficiency of B inhibits many metabolic processes (Haneklaus et al. 2007). The internode regions of stem get shortened under low B concentration and give rosette presence, and bloated stems, leading to the disorder known as "stem crack" in celery. In roots that behave as storage organ such as sugar beets, the disorder known as "heart rot" is due to the death of dividing cells in the proliferating region because of B deficiency (Hopkins and Huner 2008). In some crop plants, B deficiency also caused reduction in seed production by inhibiting the process of fertilization.

### 2.2.12 Manganese (Mn)

Among micronutrients, manganese (Mn) is also divalent cation ($Mn^{++}$) that regulates the growth and development of plants by mediating many cellular functions (Hänsch and Mendel 2009). Enzymes such as glutamine synthetase (GS) and Mn-SOD require Mn as co-factor for the activation and also regulate the biochemical process such as photosynthesis, nitrogen fixation, DNA synthesis, transcription and other biosynthetic reactions (Nouet et al. 2011). Manganese (Mn) is a vital plant micronutrient with an essential function as a catalyst in the oxygen-evolving complex (OEC) of photosystem II (PSII) that mediates the photolysis of water and releases the oxygen (Millaleo et al. 2010). Further, increase in the biosynthesis of macromolecules such as protein, carbohydrate, lipids and lignin is under the influence of Mn and results in better plant biomass production (Marschner 2012). Uptake of Mn by plants depends on different kind of membrane transporters such as NRAMP (natural resistance-associated macrophage protein), ZIP (ZRT/IRT-like protein), P-type ATPases and VIT (vacuolar iron transporter) (Mills et al. 2008; Cailliatte et al. 2010; Milner et al. 2013). High pH and high soil porosity, which are common in calcareous soils, lead the oxidization of Mn into Mn oxides, which are now unavailable to plants and generate $Mn^{++}$ deficiency. Alteration in photosynthesis, chlorosis of leaves, stunted growth and reduced biomass are some visual symptoms of plants that face $Mn^{++}$ deficiency (Millaleo et al. 2010). $Mn^{++}$ deficiency majorly decreases the PSII proteins such as D1, PsbP and PsbQ as stability of these proteins needs Mn as co-factor (Bang et al. 2015; Schmidt et al. 2016). Prolonged Mn deficiency leads to the excessive accumulation of ROS due to down-regulation of Mn-SOD within the chloroplasts (Allen et al. 2007). Plants that experience $Mn^{++}$ starvation become more susceptible to fungal pathogens as they are unable to secrete secondary metabolites including lignin (Marschner 2012). The thick waxy cuticle layer of plants is degraded, and increase in the transpiration is noticed in plants under $Mn^{++}$ deficiency (Hebbern et al. 2009). It is concluded that for the maximal efficiency of PSII, biogenesis and repair of PSII proteins requires Mn as co-factor.

### 2.2.13 Molybdenum (Mo)

Molybdenum (Mo) is a heavy metal as well as an important trace element present in soil, and plants require Mo for specific biochemical processes. Being a transition metal, the oxidation states of Mo vary from 0 to VI; among them the most stable and naturally occurring is VI. Plants generally absorb Mo from soil in the form of $MoO_4^{2-}$, and the presence of sulphate ions along with Mo significantly decreases the Mo uptake. Average concentration of Mo in plants varies from 0.10 to 10 ppm, and maximum permissible limit is 0.10 ppm (Kaiser et al. 2005). Uptake of Mo in prokaryotes and eukaryotes is mainly mediated by ATP-binding cassette (ABC) protein family. Plants express a number of putative ABC type proteins, e.g. *Arabidopsis* express around 129 types of these transporters (Sanchez-Fernandez et al. 2001). Although Mo is an active metal, it is majorly present as an integral part of pterin known as molybdenum co-factor (Moco) and binds with enzymes, which require Mo as co-factor present in plants and animals (Williams and Frausto da Silva 2002). Nitrogen metabolism of plants is highly influenced by Mo availability, and Mo acts as co-factor for activation of several enzymes such as nitrogenase, nitrate reductase, xanthine dehydrogenase, aldehyde oxidase and sulphite oxidase (Yohe et al. 2016). In acidic soils the Mo availability decreases, and in plants some visual symptoms appear. Stunting growth, chlorosis of leaves, yellowing the veins of leaf and whiptail (the leaf lamellae became involuted) are some peculiar features (Chang 1999). Mo deficiency results in reduced growth, hampered nodulation, creates imbalance between N uptake and its assimilation, reduces dry matter accumulation and reduces crop productivity (Kumar Rao et al. 2004). Mo deficiency in plants leads to accumulation of nitrates in leaves, not being assimilated into proteins, and leaves get senescence at early stages, and leaves become pale, scorched, cupped and rosettes appearance.

### 2.2.14 Nickel (Ni)

Besides heavy metal properties, nickel (Ni) is added in the list of essential nutrients to plants, humans and environment (Duda-Chodak and Baszczyk 2008). The Ni is a natural component of soil and water that constitutes approximately 0.008% of the content of the earth's crust (Hedfi et al. 2007). First evidence of Ni as an essential nutrient was described by Eskew et al. (1983) in *Glycine max* where leaf tips accumulated excessive concentration of urea under Ni depletion and similarly, barley plants failed to produce viable grains under Ni deficiency (Brown et al. 1987). In agriculture soil, Ni is present in the range of 0.2–450 mg/kg, and average concentration is approximately 20 mg/kg (Cempel and Nikel 2006). In natural environment, Ni is present in combination with oxygen or sulphur as oxides or sulphides and most frequent form is hydrated Ni complex ($NiH_2O^{+6}$) (Seregin and Kozhevnikova 2006) and is mobile in nature. Besides its major role in growth and development, Ni is an integral component of two metallo-enzymes such as urease and hydrogenase that play prominent role in N metabolism. Hydrolysis of urea into two molecules of ammonia and one carbon dioxide is mediated by urease (Polacco et al. 2013), while formation of proton and electron via oxidation of $H_2$ is governed by hydrogenase (Brazzolotto et al. 2016) and increases the growth of nitrogen-fixing plant species. Metabolism of Ni is critical for various biochemical, physiological and growth responses by activating important enzymes such as glycosylase, deformylase, superoxide dismutase and methyl-CoM reductase (Kutman et al. 2013). In some cases Ni formed coordination compounds with chelators such as Ni-DMG that have various medicinal uses. Deficiency of Ni negatively affects the growth and metabolism of plants. Nickel deficiency symptoms are characterized by marginal chlorosis of youngest leaves, premature senescence, diminished seed set and also induced meristematic necrosis (Epstein and Bloom 2005; Wood et al. 2006). Many plant processes such as plant growth, plant senescence, N metabolism and Fe uptake are negatively affected under Ni starvation. Various factors such as high content of Zn, Cu, Mn, Fe or Ca induce the Ni deficiency in soil.

### 2.2.15 Sodium

Sodium is an abundant element of the earth's crust comprising nearly 3%. It is present in the land surface as well as water bodies but is plentiful in the seas and oceans reaching over 5% (w/w). Among the terrestrial plants, only halophytes are able to tolerate and also thrive on high sodium levels in the soil and water media. But for other plants, $Na^+$ ion is utilized as a micronutrient in key metabolic activities and they benefit from its presence in low concentration. But beneficial effects of $Na^+$ are seen in much higher concentrations in comparison to other nutrients such as $NH_4^+$, $K^+$, or $NO_3^-$, and for halophytes, much higher levels are seen (Flowers and Colmer 2008). In Chenopodiaceae family, sodium ions are beneficial for plants even when significant amount of potassium ions is present in growth medium and thus effects are seen without the depletion of $K^+$ (Subbarao et al. 2003).

It is also widely accepted that maximum of the C4 species require $Na^+$ as an essential micronutrient (Pessarakli 2001; Subbarao et al. 2003). C4 metabolism is influenced by $Na^+$ by facilitation of pyruvate uptake into chloroplasts. Sodium ions facilitate the uptake of pyruvate by chloroplast and thus regulate the C4 metabolism because C4 plant species have methyl-erythritol-phosphate (MEP) pathway that needs import of pyruvate into chloroplasts (Weber and von Caemmerer 2010).

Sodium ions at low levels are very beneficial especially in low $K^+$ conditions; this is due to the fact that structurally and chemically $Na^+$ and $K^+$ are very similar (Amtmann and Sanders 1999). Therefore many of the roles of $K^+$ including some of the metabolic activities can be interpreted by $Na^+$ at low availability of $K^+$. But those functions where enzymes rely on $K^+$ as co-factor cannot be replaced by $Na^+$ as different ionic sizes lead to different binding energies and co-ordination and thus specific selectivity is created. Subbarao et al. (1999) saw that replacement of 95% of leaf tissues' $K^+$ by Na in beet plant resulted in no significant impact on osmotic potential. Therefore, it is found that $Na^+$ replacing $K^+$ is possible for osmotic function (Gattward et al. 2012). But in case of starch synthetase enzyme, which requires about 50 mM $K^+$ for its optimum function, equimolar concentrations of $Na^+$ were found to be about 20% effective, but not inhibitory. But cations like $NH_4^+$, $Rb^+$ and $Cs^+$ were effective about 80% (Lindhauer and De Fekete 1990). Therefore it is plausible that while not as effective as $K^+$, it is eminent that $Na^+$, at moderate levels in the cytosol, such as those found in various studies including saline conditions, are able to help functioning of the enzymes as long as $K^+$ concentrations in the cytosol remain the same (Carden et al. 2003; Kronzucker et al. 2008). Furthermore, it is seen that various transport proteins previously thought to be ion specific like KT/HAK/KUP, NHX and AKT proteins are found in transporting either $Na^+$ or $K^+$ under different instances (Kronzucker and Britto 2011). One major function of $K^+$ spared by $Na^+$ as in some special cases is the replacement of $K^+$ in stomatal guard cells (Roelfsema and Hedrich 2005). It is also worth noting that authors have also reported of superior stomatal performance in plants cultivated on high $Na^+/K^+$ ratios than on $K^+$ alone and improved water-use efficiency These studies indicate that $Na^+$ may not only fill in for $K^+$ in the particular case of stomatal guard cell function but can also outperform it (Marschner 2012; Gattward et al. 2012).

### 2.2.16 Chloride

The chloride ($Cl^-$) anion found in soils is the dominant form of chlorine element. In the context of agronomy, earlier it was considered as toxic element to plants mainly because of excessive accumulation of $Cl^-$ in salinity stress, thus causing toxicity in sensitive organs and also due to the belief of $Cl^-$ and $NO_3^-$ as antagonistic molecules. Due to this, Cl uptake and accumulation in roots of the plants occur because of loss in nitrate ($NO_3^-$), an important source of nitrogen (N) for higher plants. However, unlike $NO_3^-$, $Cl^-$ is one of the 16 essential elements that are needed for plant growth. As it is needed in small quantities by the plants, generally about 50–100 μM, it is known as micronutrient (Broyer et al. 1954). There are many functions in the cell where $Cl^-$ ion is needed by the plants. It acts as an essential co-factor for oxygen evolution in the chloroplast of the PSII. For maintaining the coordination structure of Mn(4)Ca cluster, two molecules of $Cl^-$ are required (Kawakami et al. 2009). $Cl^-$ also helps in keeping the OEC (oxygen evolving complex) active by facilitating the proton flux from the water oxidation complex to the thylakoid lumen (Raven 2017). Chloride is also involved in regulating enzymes' activity like asparagine synthetase and vacuolar proton-pumping ATPase (Rognes 1980; Churchill and Sze 1984). By new study, amylase activity regulation by $Cl^-$ is also proposed (Metzler 1979). In glycophyte plants, micromolar amounts are required to ensure many of these cellular functions. About 0.2–0.4 mg $g^{-1}$ dry weight is the minimum requirement of $Cl^-$ for ample growth of plants, which rightfully belongs to the category of micronutrient (Marchener 2012; Johnson et al. 1957).

Chloride is abundantly present in almost all the environment and thus is able to fulfil plants' requirements (White and Broadley 2001). $Cl^-$ is involved in balancing the electrical charges of cations like $Ca^+$, $K^+$ and $H^+$ (protons) as a non-assimilating highly mobile anion and having a major say in the regulation of pH gradients by stabilizing the electric potential of the cell membranes (White and Broadley 2001; Hänsch and Mendel 2009). High concentrations of $Na^+$ and $Ca^+$ ions get accumulated in the vacuoles where $Cl^-$ acts as a major counter anion. By stimulating anion efflux, chloride causes membrane depolarization and has a major role in signal perception and transduction (Michard et al. 2017; Colcombet et al. 2009). Genetic studies have also revealed $Cl^-$ regulation in cell elongation. When DTX33 and DTX35 genes were altered, which are encoded for vacuolar channels involving compartmentalization of $Cl^-$ during cell expansion, defective pollen growth and shorter root hairs were found (Zhang et al. 2017). $Cl^-$ also causes a unique effect on tobacco plants physiology in reducing leaf transpiration by declining stomatal conductance ($g_S$) (Franco-Navarro et al. 2016). This was however done by the declining stomatal density instead of reduced stomatal opening, linked to leaf cells enlargement in $Cl^-$ treated plants (Franco-Navarro et al. 2019). Therefore, $Cl^-$ simultaneously stimulates growth and reduces water consumption, thus improving the efficiency of water use by the plants (Franco-Navarro et al. 2016, 2019). It is further seen that lower $g_S$ is not resulted into lower photosynthetic capacity because mesophyll diffusion conductance to $CO_2$ is increased by $Cl^-$ (Franco-Navarro et al. 2019). Therefore, a major role of $Cl^-$ nutrition on performance of chloroplast is seen as exposure of chloroplasts to intercellular space of mesophyll cells through its higher surface area.

## 2.3 Mitigating Approaches (Nutrient Management)

Micronutrient deficient soil not only limits the crop production for overgrowing population but also affects the human (around half of the world's population) as well as animal health associated directly with consumptions (Welch and Graham 2004). In developing countries like India, excessive use of chemical fertilizers results in depletion of natural micronutrients of the soil (Shukla et al. 2016). Use of chemical fertilizers surprisingly enhances the crop production, but on the ground level it is an inflated nutrient management approach and shows better performance on specific soil conditions (White and Broadley 2009). Further, the proportion of total agriculture land is fixed, and goal regarding increased crop production must be attained by accepting current crop agricultural practices, predominantly sufficient and balanced nutrition and use of bio-fertilizers. Modern agricultural practices include long-term integrated management (INM), which is a combination of moderate use of chemical fertilizers along with various bio-supplements such as farmyard manure (FYM), green manure (GM) and paddy straw (PS), which provides array of various nutrients to nutrient deficient soils and also retains the native nutrient status of soil by transfer

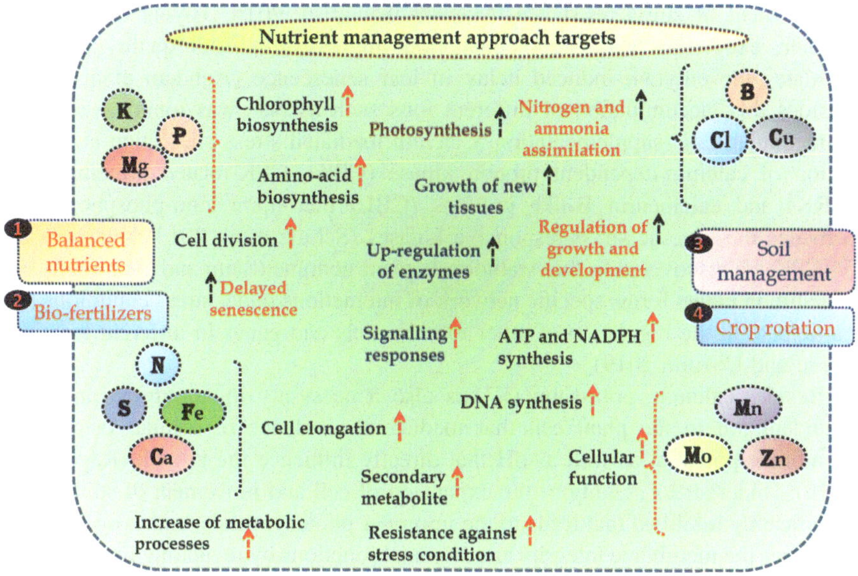


**Fig. 2.3** Triggered action in plants by nutrient supplementation

of micronutrients (Khaliq et al. 2017; Saha et al. 2018). A number of studies demonstrated that the application of INM in Indo-Gangetic plain (IGP) improves the crop yield as well as physical, chemical and biological properties of soil. Minimum nutrient uptake that is necessary for the growth and development is either provided by soil or by foliar supplementation, and there are around 17 essential nutrients out of total 92 natural elements in the plant system. During the entire life, a plant continuously faces various abiotic and biotic stresses that lead to reduction of crop productivity.

Plants that experience low nutrient condition are expected to produce less biomass per unit area as compared to plants under well-nourished conditions. Well-maintained conditions significantly increase the plant productivity as well as maintain the soil fertility for a very long time (Waraich et al. 2011). Majority of studies were conducted to evaluate the potential role of exogenously supplemented nutrients in improving the plant productivity as N and S have beneficial effects on improved salinity tolerance (Singh et al. 2019), and metal tolerance (Singh et al. 2018) in *Solanum melongena*. In the upcoming years, to fulfil the food demand of over-growing population, it is pragmatic that the use of fertilizers will increase by 2-folds that contribute in both crop production and sustaining soil fertility. Hence, nutrient management in the agricultural fields played an essential role in improving the biotic and abiotic stress tolerance in crop plants (Fig. 2.3).

Among various nutrient management strategies, exogenous calcium ($Ca_2^+$) supplementation significantly increases the stress tolerance of crop plants due to involvement as signalling molecule in the activation of calcium-regulated protein kinases, initiation of downstream phosphorylation signalling and finally, achievement of stress resistance (Manishankar et al. 2018; Toyota et al. 2018). Further, exogenous $Ca_2^+$ supplementation improves the fruit quality and also regulates the ethylene-induced delay in leaf senescence (Aghdam et al. 2012). Besides this, accumulation of different ions such as $Na^+$ was found to decrease extracellular $Ca_2^+$ supplementation. Calcium mediated stress alleviation is involve action of calcium-dependent protein kinase (CPK), CPK-related protein kinase (CRK) and calcineurin B-like proteins (CBL) that upon auto-phosphorylation activates the specific stress-responsive kinases (Schulz et al. 2013) and there are 34 CPK are discovered in the Arabidopsis plant genome (Saito and Uozumi 2019). Calcium in plants forms specific network of interactions under stress conditions that participate in the regulation of other ion channels and genes in different locations (Saito and Uozumi 2019).

Besides calcium, potassium ($K^+$) is also a necessary micronutrient and most abundant ion in the plant cell that mediate the balance of cellular osmolarity, membrane potential as well as pH that directly influence the growth (Ragel et al. 2019). Under stress conditions the expansion of cell and movement of stomata are significantly inhibited that leads to the improper photosynthesis, and exogenous $K^+$ improves the membrane integrity and stomatal functions by regulating the guard cell turgor pressure. Along with the other ions such as $NO_3$and Cl, movement of $K^+$ across the cell membrane is necessary for the better growth of pollen tube (Liu et al. 2016). Entry site of $K^+$ inside the cell is mediated by xylem as well as a variety of transporters such as Shaker-type $K^+$ channel AKT1 and the KT/KUP/HAK-type transporter HAK5 (Alemán et al. 2011) and also depends upon CBL1 and CIPK 23 (Ragel et al. 2015). Under low $K^+$ conditions, the CIPK proteins undergo phosphorylation and maintain $K^+$ homeostasis (Liu et al. 2012). Once imported into the xylem, $K^+$ travels long distance from root to shoot to be exported into appropriate aerial tissues. In addition, $K^+$ is transported from green cells into the phloem, performs specific functions, enhances the crop productivity and maintains the soil fertility.

To fulfil the food demand of growing population, second green revolution must be achieved, which could be only done by the use of fertilizers. In some cases, excessive use of nutrients may also behave as damaging the aquatic ecosystem by forming nutrient-rich condition termed as eutrophication. Further, to minimize this effect the crop rotation is the best method as well as testing of soil N quality is also necessary at regular intervals. Nitrogen (vital micronutrient) plays an essential role in utilization of absorbed light energy and photosynthetic carbon metabolism and also major components of various cell components like DNA, RNA and proteins (Razaq et al. 2017). Supplemented N is directly not utilized by plants, but it must be converted to fixed forms such as nitrate ($NO_3$), nitrite ($NO_2$) and ammonium salts ($NH_3$) mediated by the variety of N-fixing microbes that participates in the protection against pathogens as well as in reproduction (Hacquard et al. 2015). The N is the major constituent of organic waste produced by poultry manure, and the end-products of livestock production are better source of N in the agricultural fields. Under drought stress, exogenous N supplementation enhances the physiological responses via elevating N and chlorophyll concentrations associated with increased photochemistry of PS II (Xu et al. 2015) and also improved water loss by increasing the water use efficiency (WUE) and increases the biomass production and accumulates proline (Gesslere et al. 2017). The degrees of adaptive changes of xylem upon drought were enhanced by adequate-N application, attributing to an increased vessel frequency and promoted hydraulic safety. Currently, nitrogen in the form of nitric oxide (NO) is a highly reactive, gaseous molecule involved in secondary signalling having regulatory as well as antioxidant properties (Da Silva et al. 2017). NO positively regulates variety of stress factors such as temperature, chemicals, salinity, radiations, water, mechanical and oxidative stresses (Li et al. 2013). In brief, when plants are scarce of various stresses, NO regulates growth and developmental processes in plants.

Moreover, boron (B) is a micronutrient involved in physiological and biochemical processes such as cell elongation, cell division, cell wall biosynthesis, membrane function, nitrogen (N) metabolism, leaf photosynthesis and uracil synthesis (Rehman et al. 2012). To overcome the negative effects induced by reactive oxygen species, B increases the activity of antioxidants, thereby alleviating oxidative damage. Further, exogenous B supplementation strengthens sugar transport between cells that actively participate in the seed germination and grain formation and also maintains the CHO metabolism and enhances the accumulation of secondary metabolites particularly phenolic in the leaves (Waraich et al. 2011). Foliar application of B in stressed plants improves the overall reproductive fitness (leads to rapid pollen tube growth in *Lilium formosanum*) and results in better yield and quality (Holdaway-Clarke et al. 2003; Tahir et al. 2009). Including above micronutrients, several other nutrients such as Mn, Mg, Cu, Fe, C, H, O, Cl and S also play a significant role in minimizing the stress-induced toxic effects that results in high yield and also the desired quality. On the one hand, clear relations could be demonstrated between the nutrient supply and physiological processes, which are important for crop quality.

## 2.4 Future Prospective

The plant species in its entire life experience various stress factors (temperature, heat, drought, mineral stress, heavy metals, diseases and insect pests) due to their ecological habitation and always struggle for their survival and existence. Agriculture soil is the source of essential micro- and macronutrients that are required by plants in varying amounts for the growth and development from seed germination to reproductive stages. Upsurge in the inorganic or organic pollutants associated with climate change significantly limits the nutrient availability for plants either inhibiting their transport or replacing the essential nutrient from metabolic processes and marks in nutrient deficiency in plants. Plants that experience deficiency of mineral nutrients such as Mg, Mn, Cu, B, Cl, S or Co show reduction in the rate of photosynthesis associated with leaf chlorosis, and overall growth is troubled. Deficiency of Mo and Ni hampers the assimilation of nitrogen into amino acids or protein synthesis. Osmotic adjustment of plants is maintained by the concentrations of Na, K or Cl, and in the signalling responses Ca acts as a secondary signalling molecule. The shortages of nutrient are being improved through soil and/or foliar application during either the active plant growth or seed germination. Exogenous supplementation (foliar spray) of essential nutrients is becoming common in most of the crops, where environmental conditions do not allow the farmers to apply fertilizers through soil medium and this is a cost-effective and immediate remedial strategy to measure the deficiency during growth stage. Thereby, the farmers could fetch greater harvest through foliar application of micronutrients, i.e. increased productivity from 10% to 20% across different horticultural and arable crops.

## References

Abadía J, Vázquez S, Rellán-Álvarez R, El-Jendoubi H, Abadía A, Álvarez Fernández A (2011) Towards a knowledge based correction of iron chlorosis. Plant Physiol Biochem 49:471–482

Abdel-Ghany SE, Pilon M (2008) MicroRNA-mediated systemic down-regulation of copper protein expression in response to low copper availability in Arabidopsis. J Biol Chem 283:15932–15945

Aghdam MS, Hassanpouraghdam MB, Paliyath G, Farmani B (2012) The language of calcium in postharvest life of fruits, vegetables and flowers. Scien Horti 144:102–115

Ahmad W, Zia MH, Malhi SS, Niaz A, Saifullah U (2012) Boron deficiency in soils and crops: a review. In: Crop plant. Akash Goyal, InTechOpen, London, pp 77–114

Alemán F, Nieves-Cordones M, Martínez V, Rubio F (2011) Root $K^+$ acquisition in plants: the Arabidopsis thaliana model. Plant Cell Physiol 52:1603–1612

Allen MD, Kropat J, Tottey S, Campo JAD, Merchant SS (2007) Manganese deficiency in *Chlamydomonas* results in loss of photosystem II and MnSOD function, sensitivity to peroxides, and secondary phosphorus and iron deficiency. Plant Physiol 143:263–277

Amtmann A, Sanders D (1999) Mechanisms of $Na^{(+)}$ uptake by plant cells. Adv Bot Res 29:75–112

Arend M, Monshausen G, Wind C, Weisenseel MH, Fromm J (2004) Effect of potassium deficiency on the plasma membrane $H^+$-ATPase of the wood ray parenchyma in poplar. Plant Cell Environ 27:1288–1296

Armengaud P, Breitling R, Amtmann A (2004) The potassium dependent transcriptome of Arabidopsis reveals a prominent role of jasmonic acid in nutrient signaling. Plant Physiol 136:2556–2576

Baker A, Ceasar SA, Palmer AJ, Paterson JB, Qi W, Muench SP, Baldwin SA (2015) Replace, reuse, recycle: improving the sustainable use of phosphorus by plants. J Exp Bot 66:3523–3540

Bakkaus E, Gouget B, Gallien JP, Khodja H, Carrot F, Morel JL et al (2005) Concentration and distribution of cobalt in higher plants: the use of micro-PIXE spectroscopy. Nuclear Instruments Methods Phys Res 231:350

Banuelos MA, Garciadeblas B, Cubero B, Rodriguez-Navarro A (2002) Inventory and functional characterization of the HAK potassium transporters of rice. Plant Physiol 130:784–795

Barker AV, Pilbeam DJ (2015) Handbook of plant nutrition. CRC, Boca Raton

Brazzolotto D, Gennari M, Queyriaux N, Simmons TR, Pécaut J, Demeshko S (2016) Nickel-centred proton reduction catalysis in a model of [NiFe] hydrogenase. Nat Chem 8:1054–1060

Brown PH, Welch RM, Cary EE, Checkai RT (1987) Beneficial effects of nickel on plant growth. J Plant Nutr 10:2125–2135

Brown PH, Bellaloui N, Wimmer MA, Bassil ES, Ruiz J, Hu H, Pfeffer H, Dannel F, R€omheld V (2002) Boron in plant biology. Plant Biol 4:205–223

Broyer TC, Carlton AB, Johnson CM, Stout PR (1954) Chlorine—A micronutrient element for higher plants. Plant Physiol 29:526–532

Burkhead JL, Gogolin RKA, Abdel-Ghany SE, Cohu CM, Pilon M (2009) Copper homeostasis. New Phytol 182:799–816

Cailliatte R, Schikora A, Briat JF, Mari S, Curie C (2010) High affinity manganese uptake by the metal transporter NRAMP1 is essential for *Arabidopsis* growth in low manganese conditions. Plant Cell 22:904–917

Cakmak I (2002) Plant nutrition research: priorities to meet human needs for food in sustainable ways. Plant Soil 247:3–24

Cakmak I, Yazici AM (2010) Magnesium: a forgotten element in crop production. Better Crops 94:23–25

Carden DE, Walker DJ, Flowers TJ, Miller AJ (2003) Single cell measurements of the contributions of cytosolic $Na^+$ and $K^+$ to salt tolerance. Plant Physiol 131:676–683

Cempel M, Nikel G (2006) Nickel: a review of its sources and environmental toxicology. Polish J Environ Stud 15:375–382

Chang SS (1999) Micronutrients in crop production of Taiwan. Proceedings of international workshop on micronutrient in crop production, November 8–13, National Taiwan University, Taipei, Taiwan ROC

Churchill KA, Sze H (1984) Anion-sensitive, h+−pumping atpase of oat roots—Direct effects of cl-, no3-, and a disulfonic stilbene. Plant Physiol 76:490–497

Çokkizgin H, Bölek Y (2015) Priming treatments for improvement of germination and emergence of cotton seeds at low temperature. Plant Breed Seed Sci 71:121–134

Cordell D, White S (2015) Tracking phosphorus security: indicators of phosphorus vulnerability in the global food system. Food Secur 7:337–350

Couturier J, Touraine B, Briat JF, Gaymard F, Rouhier N (2013) The iron-sulfur cluster assembly machineries in plants: current knowledge and open questions. Front Plant Sci 4:259

D'Hooghe P, Dubousset L, Gallardo K, Kopriva S, Avice JC, Trouverie J (2014) Evidence for proteomic and metabolic adaptations associated to alterations of seed yield and quality in sulphur-limited *Brassica napus* L. Mol Cell Proteomics 13:1165–1183

Da Silva CJ, Fontes EPB, Modolo LV (2017) Salinity-induced accumulation of endogenous H2S and NO is associated with modulation of the antioxidant and redox defense systems in *Nicotiana tabacum* L. cv. Havana. Plant Sci 256:148–159

de Bang TC, Petersen J, Pedas PR, Rogowska-Wrzesinska A, Jensen ON, Schjoerring JK, Jensen PE, Thelen JJ, Husted S (2015) A laser ablation ICP-MS based method for multiplexed immunoblot analysis: applications to manganese-dependent protein dynamics of photosystem II in barley (*Hordeum vulgare* L.). Plant J 83:555–565

Duda-Chodak A, Baszczyk U (2008) The impact of nickel on human health. J Elem 13:685–696

Epstein E, Bloom AJ (2005) Mineral nutrition of plants: principles and perspectives, 2nd edn. Sinauer Associates, Sunderland

Eskew DL, Welch RM, Cary EE (1983) Nickel: an essential micronutrient for legumes and possibly all higher plants. Science 222:621–623

Fahimirad S, Hatami M (2017) Heavy metal-mediated changes in growth and phytochemicals of edible and medicinal plants. In: Ghorbanpour M, Varma A (eds) Medicinal plants and environmental challenges. Springer, Cham

Farooq M, Wahid A, Siddique KHM (2012) Micronutrient application through seed treatments: a review. J Soil Sci Plant Nutr 12(1):125–142

Fatma M, Masood A, Per TS, Khan NA (2016) Nitric oxide alleviates salt stress inhibited photosynthetic performance by interacting with sulfur assimilation in mustard. Front Plant Sci 7:521

Flowers TJ, Colmer TD (2008) Salinity tolerance in halophytes. New Phytol 179:945–963

Franco-Navarro JD, Brumos J, Rosales MA, Cubero-Font P, Talon M, Colmenero-Flores JM (2016) Chloride regulates leaf cell size and water relations in tobacco plants. J Exp Bot 67:873–891

Franco-Navarro JD, Rosales MA, Álvarez R, Cubero-Font P, Calvo P, Díaz-Espejo A, Colmenero-Flores JM (2019) Chloride as macronutrient increases water use efficiency by anatomically-driven reduced stomatal conductance and increased mesophyll diffusion to CO2. Plant J 99:815–831

Fu J, Yang JC (2012) Research advances in high-yielding cultivation and physiology of super rice. Rice Sci 19:177–184

Gattward JN, Almeide AAF, Souza JO, Gomes FP, Kronzucker HJ (2012) Sodium-potassium synergism in *Theobroma cacao*: stimulation of photosynthesis, water-use efficiency and mineral nutrition. Physiol Plant 146:350–362

Gessler A, Schaub M, McDowell NG (2017) The role of nutrients in drought-induced tree mortality and recovery. New Phytol 214:513–520

Gierth M, Ma¨ser P (2007) Potassium transporters in plants involvement in $K^+$ acquisition, redistribution and homeostasis. FEBS Lett 581:2348–2356

Gierth M, Maser P, Schroeder JI (2005) The potassium transporter AtHAK5 functions in $K^+$ deprivation-induced high-affinity $K^+$ uptake and AKT1 $K^+$ channel contribution to $K^+$ uptake kinetics in *Arabidopsis* roots. Plant Physiol 137:1105–1114

Graham RD (2008) Micronutrient deficiencies in crops and their global significance. In: Alloway BJ (ed) Micronutrient deficiencies in global crop production. Springer, Dordrecht, pp 41–61

Gu R, Duan F, An X, Zhang F, von Wiren N, Yuan L (2013) Characterization of AMT-mediated high-affinity ammonium uptake in roots of maize (*Zea mays* L.). Plant Cell Physiol 54:1515–1524

Hacquard S, Garrido-Oter R, González A, Spaepen S, Ackermann G, Lebeis S, McHardy AC, Dangl JL, Knight R, Ley R, Schulze-Lefert P (2015) Microbiota and host nutrition across plant and animal kingdoms. Cell Host Microbe 17(5):603–616

Hammond JP, White PJ (2008) Sucrose transport in the phloem: integrating root responses to phosphorus starvation. J Exp Bot 59:93–109

Haneklaus S, Bloem E, Schnug E, Kok LJ, Stulen I (2007) Handbook of plant nutrition. CRC Press, Boca Raton

Hänsch R, Mendel RR (2009) Physiological functions of mineral micronutrients (Cu, Zn, Mn, Fe, Ni, Mo, B, Cl). Curr Opin Plant Biol 12:259–266

Hariadi Y, Shabala S (2004) Screening broad beans (*Vicia faba*) for magnesium deficiency. I. Growth characteristics, visual deficiency symptoms and plant nutritional status. Funct Plant Biol 31:529–537

Hashimoto K, Kudla J (2011) Calcium decoding mechanisms in plants. Biochimie 93:2054–2059

Hawkesford MJ (2003) Transporter gene families in plants: the sulphate transporter gene family redundancy or specialization? Physiol Plant 117:155–163

Hebbern CA, Laursen KH, Ladegaard AH, Schmidt SB, Pedas P, Bruhn D, Schjoerring JK, Wulfsohn D, Husted S (2009) Latent manganese deficiency increases transpiration in barley (*Hordeum vulgare*). Physiol Plant 135:307–316

Hedfi A, Mahmoudi E, Boufahja F, Beyrem H, Aissa P (2007) Effects of increasing levels of nickel contamination on structure of offshore nematode communities in experimental microcosms. Bull Environ Contam Toxicol 79:345–349

Heldt HW, Chon CJ, Maronde D, Herold A, Stankovic ZS, Walker DA, Kraminer A, Kirk MR, Heber U (1977) Role of orthophosphate and other factors in the regulation of starch formation in leaves and isolated chloroplasts. Plant Physiol 59:1146–1155

Helfenstein J, Pawlowski ML, Hill CB, Stewart J, Lagos-Kutz D, Bowen CR, Frossard E, Hartman GL (2015) Zinc deficiency alters soybean susceptibility to pathogens and pests. J Plant Nutrit Soil Sci 178:896–903

Hepler PK (2005) Calcium: a central regulator of plant growth and development. Plant Cell 17:2142–2155

Holdaway-Clarke TL, Weddle NM, Kim S, Robi A, Parris C, Kunkel JG, Hepler PK (2003) Effect of extracellular calcium, pH and borate on growth oscillations in *Lilium formosanum* pollen tubes. J Exp Bot 54:65–72

Hopkins WG, Huner NPA (2008) Introduction to plant physiology, 4th edn. Wiley, New York, pp 73–74

Hotz C, Brown KH (2004) Assessment of the risk of zinc deficiency in populations and options for its control. Food Nutrit Bull 25:194–195

Howarth JR, Fourcroy P, Davidian JC, Smith FW, Hawkesford MJ (2003) Cloning of two contrasting high-affinity sulfate transporters from tomato induced by low sulfate and infection by the vascular pathogen Verticillium dahliae. Planta 218:58–64

Howell RW, Skoog F (1975) Effect of adenine and other subsurfaces on growth of excised pisum epicotyls cultured in vitro. Am J Bot 49:645–649

Huffman DL, O'Halloran TV (2001) Function, structure, and mechanism of intracellular copper trafficking proteins. Annu Rev Biochem 70:677–701

Jezek M, Geilfus CM, Bayer A, Mühling KH, Struik PC (2015) Photosynthetic capacity, nutrient status, and growth of maize (*Zea mays* L.) upon MgSO4 leaf-application. Front Plant Sci 5:1–10

Johnson CM, Stout PR, Broyer TC, Carlton AB (1957) Comparative chlorine requirements of different plant species. Plant Soil 8:337–353

Kafkafi U, Xu G, Imas P, Magen H, Tarchitzky J (2001) Potassium and chloride in crops and soils: the role of potassium chloride fertilizer in crop nutrition; IPI research topics No. 22; International Potash Institute, Horgen, p 220

Kaiser BN, Gridley KL, Brady JN, Phillips T, Tyerman SD (2005) The role of molybdenum in agricultural plant production. Ann Bot 96:745–754

Kanai S, Moghaieb RE, El-Shemy HA, Panigrahi R, Mohapatra PK, Ito J, Nguyen NT, Saneoka H, Fujita K (2011) Potassium deficiency affects water status and photosynthetic rate of the vegetative sink in green house tomato prior to its effects on source activity. Plant Sci 180:368–374

Kastori R, Plesnicar M, Arsenijevic-Maksimovic I, Petrovic N, Pankovic D, Saka Z (2000) Photosynthesis, chlorophyll fluorescence, and water relations in young sugar beet plants as affected by sulfur supply. J Plant Nutr 23:1037–1049

Kataoka T, Watanabe-Takahashi A, Hayashi N, Ohnishi M, Mimura T, Buchner P (2004) Vacuolar sulfate transporters are essential determinants controlling internal distribution of sulfate in *Arabidopsis*. Plant Cell 16:2693–2704

Kawakami K, Umena Y, Kamiya N, Shen JR (2009) Location of chloride and its possible functions in oxygen-evolving photosystem II revealed by X-ray crystallography. Proc Natl Acad Sci USA 106:8567–8572

Khaliq A, Zafar M, Abbasi MK, Hussain I (2017) Soil-plant micronutrients dynamics in response to integrated fertilization under wheat–soybean cropping system at Rawalakot, Pakistan. Arch Agron Soil Sci 64:640–653

Khan NA, Khan MIR, Asgher M, Fatma M, Masood A, Syeed S (2014) Salinity tolerance in plants: revisiting the role of sulfur metabolites. J Plant Biochem Physiol 2:2

Khan MIR, Nazir F, Asgher M, Per TS, Khan NA (2015) Selenium and sulfur influence ethylene formation and alleviate cadmium-induced oxidative stress by improving proline and glutathione production in wheat. J Plant Physiol 173:9–18

Khan S, Yu H, Li Q, Gao Y, Sallam BN, Wang H, Liu P, Jiang W (2019) Exogenous application of amino acids improves the growth and yield of lettuce by enhancing photosynthetic assimilation and nutrient availability. Agronomy 9:266

Kobayashi T, Nishizawa NK (2012) Iron uptake, translocation, and regulation in higher plants. Annu Rev Plant Biol 63:131–152

Kobayashi NI, Iwata N, Saito T, Suzuki H, Iwata R, Tanoi K, Nakanishi TM (2013) Application of 28Mg for characterization of Mg uptake in rice seedling under different pH conditions. J Radioanal Nucl Chem 296:531–534

Komeda H, Kobayashi M, Shimizu S (1997) A novel transporter involved in cobalt uptake. Proc Natl Acad Sci 94:36–41

Krapp A, David LC, Chardin C, Girin T, Marmagne A, Leprince AS (2014) Nitrate transport and signalling in Arabidopsis. J Exp Bot 65:789–798

Kronzucker HJ, Britto DT (2011) Sodium transport in plants: a critical review. New Phytol 189:54–81

Kronzucker HJ, Szczerba MW, Schulze LM, Britto DT (2008) Non-reciprocal interactions between $K^+$ and $Na^+$ ions in barley. J Exp Bot 59:2793–2801

Kumar R, Harris JVDK, Johansen DC, Musa AM (2004) Low cost provision of molybdenum (Mo) to chickpeas grown in acid soils. Proceedings of the Zvarevashe, CARE stakeholder workshop, report IDG/00/18, Silsoe Research Institute, Bedford

Kutman BY, Kutman UB, Cakmak I (2013) Nickel-enriched seed and externally supplied nickel improve growth and alleviate foliar urea damage in soybean. Plant Soil 363:61–75

Laganowsky A, Gómez SM, Whitelegge JP, Nishio JN (2009) Hydroponics on a chip: analysis of the Fe deficient *Arabidopsis* thylakoid membrane proteome. J Proteome 72:397–415

Lahaye PA, Epstein E (1969) Salt toleration by plants: enhancement with calcium. Science 166:395–396

Li ZG, Yang SZ, Long WB, Wang GX, Shen ZZ (2013) Hydrogen sulfide may be a novel downstream signal molecule in nitric oxide induced heat tolerance of maize (*Zea mays* L.) seedlings. Plant Cell Environ 36:1564–1572

Lindhauer MG, De Fekete MAR (1990) Starch synthesis in potato (*Solanum tuberosum*) tubers: activity of selected enzymes in dependence of potassium content in storage tissue. Plant Soil 124:291–295

Liu LL, Ren HM, Chen LQ, Wang Y, Wu WH (2012) A protein kinase CIPK9 interacts with calcium sensor CBL3 and regulates K+ homeostasis under low-$K^+$ stress in *Arabidopsis*. Plant Physiol 161:266–277

Liu L, Zheng C, Kuang B, Wei L, Yan L, Wang T (2016) Receptor-like kinase RUPO interacts with potassium transporters to regulate pollen tube growth and integrity in rice. PLoS Genet 12:1–23

Ludewig U, Neuhduser B, Dynowski M (2007) Molecular mechanisms of ammonium transport and accumulation in plants. FEBS Lett 581:2301–2308

Lunde C, Zygadlo A, Simonsen HT, Nielsen PL, Blennow A, Haldrup A (2008) Sulfur starvation in rice: the effect on photosynthesis, carbohydrate metabolism, and oxidative stress protective pathways. Physiol Plant 134:508–521

MacDonald GK, Bennett EM, Potter PA, Ramankutty N (2011) Agronomic phosphorus imbalances across the world's croplands. Proc Natl Acad Sci USA 108:3086–3091

Manishankar P, Wang N, Köster P, Alatar AA, Kudla J (2018) Calcium signaling during salt stress and in the regulation of ion homeostasis. J Exp Bot 69:4215–4226

Marschner H (1995) Mineral nutrition of higher plants. Academic Press, London

Marschner H (2012) Marschner's mineral nutrition of higher plants, 3rd edn. Academic Press, London

Martinez-Cordero MA, Martinez V, Rubio F (2004) Cloning and functional characterization of the high-affinity $K^+$ transporter HAK1 of pepper. Plant Mol Biol 56:413–421

Maruyama-Nakashita A, Nakamura Y, Watanabe-Takahashi A, Inoue E, Yamaya T, Takahashi H (2005) Identification of a novel cis-acting element conferring sulfur deficiency response in *Arabidopsis* roots. Plant J 42:305–314

Masuda T (2008) Recent overview of the Mg branch of the tetrapyrrole biosynthesis leading to chlorophylls. Photosyn Res 96:121–143

Colcombet J, Mathieu Y, Peyronnet R, Agier N, Lelievre F, Barbier-Brygoo H, Frachisse J-M (2009) R-type anion channel activation is an essential step for ROS-dependent innate immune response in Arabidopsis suspension cells. Funct Plant Biol 36:832–843

Matraszek R, Hawrylak-Nowak B, Chwil S, Chwil M (2016) Interaction between cadmium stress and sulphur nutrition level on macronutrient status of *Sinapis alba* L. Water Air Soil Pollut 227:355

Mei Y, Lei S, FangSen X, JianWei L, YunHua W (2009) Effects of B, Mo, Zn, and their interactions on seed yield of rapeseed (*Brassica napus* L.). Pedosphere 19:53–59

Metzler DE (1979) Biochemistry: the chemical reactions of living cells. Academic Press, New York

Michard E, Simon AA, Tavares B, Wudick MM, Feijo JA (2017) Signaling with ions: the keystone for apical cell growth and morphogenesis in pollen tubes. Plant Physiol 173:91–111

Millaleo R, Reyes-Díaz M, Ivanov AG, Mora ML, Alberdi M (2010) Manganese as essential and toxic element for plants: transport, accumulation and resistance mechanisms. J Soil Sci Plant Nutr 10:476–494

Miller SS, Liu J, Allan DL, Menzhuber CJ, Fedorova M, Vance CP (2001) Molecular control of acid phosphatase secretion into the rhizosphere of proteoid roots from phosphorus-stressed white lupin. Plant Physiol 127:594–606

Mills RF, Doherty ML, López-Marqués RL, Weimar T, Dupree P, Palmgren MG, Pittman JK, Williams LE (2008) ECA3, a Golgilocalized P2A-type ATPase, plays a crucial role in manganese nutrition in *Arabidopsis*. Plant Physiol 146:116–128

Milner MJ, Seamon J, Craft E, Kochian LV (2013) Transport properties of members of the ZIP family in plants and their role in Zn and Mn homeostasis. J Exp Bot 64:369–381

Moharana PC, Sharma BM, Biswas DR (2017) Changes in the soil properties and availability of micronutrients after six-year application of organic and chemical fertilizers using STCR-based targeted yield equations under pearl millet-wheat cropping system. J Plant Nutr 40:65–176

Mondal S, Bose B (2019) Impact of micronutrient seed priming on germination, growth, development, nutritional status and yield aspects of plants. J Plant Nutri 42(1):1–23. https://doi.org/10.1080/01904167.2019.1655032

Nagajyoti PC, Lee KD, Sreekanth TVM (2010) Heavy metals, occurrence and toxicity for plants: a review. Environ Chem Lett 8:199–216

Nawaz MA, Chen C, Shireen F, Zheng Z, Sohail H, Afzal M, Ali MA, Bie Z, Huang Y (2018) Genome-wide expression profiling of leaves and roots of watermelon in response to low nitrogen. BMC Genomics 19:456

Ng CKY, Mcainsh MR (2003) Encoding specificity in plant calcium signalling: hot-spotting the ups and downs and waves. Annal Bot 92:477–485

Nikolic M, Nikolic N, Kostic L, Pavlovic J, Bosnic P, Stevic N, Savic J, Hristov N (2016) The assessment of soil availability and wheat grain status of zinc and iron in Serbia: implications for human nutrition. Sci Total Environ 553:141–148

Nouet C, Motte P, Hanikenne M (2011) Chloroplastic and mitochondrial metal homeostasis. Trends Plant Sci 16:395–404

Parween T, Jan S, Mahmooduzzafar M, Fatma T (2011) Alteration in nitrogen metabolism and plant growth during different developmental stages of green gram *Vigna radiata* L.in response to chlorpyrifos. Acta Physiol Plant 33:2321–2328

Peñarrubia L, Andrés-Colás N, Moreno J, Puig S (2010) Regulation of copper transport in Arabidopsis thaliana: a biochemical oscillator? J Biol Inorg Chem 15:29–36

Pessarakli M (2001) Handbook of plant and crop stress. Biol Plant 44:304–304

Pilon M, Cohu CM, Ravet K, Abdel-Ghany SE, Gaymard F (2009) Essential transition metal homeostasis in plants. Curr Opin Plant Biol 12:347–357

Pilon-Smits EA, Quinn CF, Tapken W, Malagoli M, Schiavon M (2009) Physiological functions of beneficial elements. Curr Opin Plant Biol 12:267–274

Polacco JC, Mazzafera P, Tezotto T (2013) Opinion-nickel and urease in plants: still many knowledge gaps. Plant Sci 200:79–90

Portis ARJ (2003) Rubisco activase—Rubisco's catalytic chaperone. Photosyn Res 75:11–27

Poss JA, Grattan SR, Grieve CM, Shannon MC (1999) Characterization of leaf boron injury in salt stressed eucalyptus by image analysis. Plant Soil 206:237–245

Puig S, Peñarrubia L (2009) Placing metal micronutrients in context: transport and distribution in plants. Curr Opin Plant Biol 12:299–306

Puthiyaveetil S, van OOrt B, Kirchhoff H (2017) Surface charge dynamics in photosynthetic membranes and the structural consequences. Nat Plants 3:17020

Qu C, Liu C, Ze Y, Gong X, Hong M, Wang L, Hong F (2011) Inhibition of nitrogen and photosynthetic carbon assimilation of maize seedlings by exposure to a combination of salt stress and potassium-deficient stress. Biol Trace Elem Res 144:1159–1174

Rafique E, Rashid A, Mahmood-Ul-Hassan M (2012) Value of soil zinc balances in predicting fertilizer zinc requirement for cotton-wheat cropping system in irrigated Aridisols. Plant Soil 361:43–55

Ragel P, Ródenas R, García-Martín E, Andrés Z, Villalta I, Nieves-Cordones M (2015) CIPK23 regulates HAK5-mediated high-affinity $K^+$ uptake in *Arabidopsis* roots. Plant Physiol 169:01401

Ragel P, Raddatz N, Leidi EO, Quintero FJ, Pardo JM (2019) Regulation of $K^+$ nutrition in plants. Front Plant Sci 10:281

Raven JA (2017) Chloride: essential micronutrient and multifunctional beneficial ion. J Exp Bot 68:359–367

Razaq M, Zhang P, Shen H, Salahuddin (2017) Influence of nitrogen and phosphorous on the growth and root morphology of Acer mono. PLoS One 12(2):e0171321. https://doi.org/10.1371/journal.pone.0171321

Rehman S, Hussain N, Tariq M, Hussain M, Nasir M, Ayaz M (2012) Response of wheat to exogenous boron supply at various growth stages. Sarhad J Agric 28:411–414

Rodríguez-Celma J, Lin WD, Fu GM, Abadía J, López-Míllán AF, Schmidt W (2013) Mutually exclusive alterations in secondary metabolism are critical for the uptake of insoluble iron compounds by *Arabidopsis* and *Medicago truncatula*. Plant Physiol 162:1473–1485

Roelfsema MRG, Hedrich R (2005) In the light of stomatal opening: new insights into 'the Watergate'. New Phytol 167:665–691

Rognes SE (1980) Anion regulation of lupin asparagine synthetase—Chloride activation of the glutamine-utilizing reactions. Phytochemistry 19:2287–2293

Saha S, Saha BN, Ray M, Mukhopadhyay SK, Halder P, Das A, Chatterjee S, Pramanick M (2018) Integrated nutrient management (INM) on yield trends and sustainability, nutrient balance and soil fertility in a long-term (30 years) rice-wheat system in the Indo-Gangetic plains of India. J Plant Nutr 41:2365–2375

Saito S, Uozumi N (2019) Calcium-regulated phosphorylation systems controlling uptake and balance of plant nutrients. Front Plant Sci 11:44. https://doi.org/10.3389/fpls.2020.00044

Salt DE, Baxter I, Lahner B (2008) Ionomics and the study of the plant ionome. Ann Rev Plant Biol 59:709–733

Sanchez-Fernandez R, Emyr Davies TG, Coleman JOD, Rea PA (2001) The *Arabidopsis thaliana* ABC protein superfamily, a complete inventory. J Biol Chem 276:30231–30244

Schmidt SB, Jensen PE, Husted S (2016) Manganese deficiency in plants: the impact on photosystem II. Trends Plant Sci 21:622–632

Schulz P, Herde M, Romeis T (2013) Calcium-dependent protein kinases: hubs in plant stress signaling and development. Plant Physiol 163:523–530

Seregin L, Kozhevnikova A (2006) Physiological role of Nickel and its toxic effects on higher plants. Russian J Plant Physiol 53:257–277

Shaul O, Hilgemann DW, de-Almeida-Engler J, Montagu M, Inz D, Galili G (1999) Cloning and characterization of a novel $Mg^{2+}/H^+$ exchanger. EMBO J 18:3973–3980

Shukla AK, Tiwari PK, Pakhare A, Prakash C (2016) Zinc and iron in soil, plant, animal and human health. Ind J Fertil 12:133–149

Sidhu GS, Sharma BD (2010) Diethylene triamine penta acetic acid–extractable micronutrients status in soil under a rice-wheat system and their relationship with soil properties in different agroclimatic zones of Indo-Gangetic plains of India. Commun Soil Sci Plant Anal 41:29–51

Singh M, Singh VP, Prasad SM (2019) Nitrogen alleviates salinity toxicity in Solanum lycopersicum seedlings by regulating ROS homeostasis. Plant Physiol Biochem 141:466–476

Singh R, Parihar P, Prasad SM (2018) Simultaneous exposure of sulphur and calcium hinder as toxicity: up-regulation of growth, mineral nutrients uptake and antioxidants system. Ecotoxicol Environ Saf 161:318–331

Stael S, Wurzinger B, Mair A, Mehlmer N, Vothknecht UC, Zeige M (2001) Plant organellar calcium signalling: an emerging field. J Exp Bot 63:1525–1542

Subbarao GV, Wheeler RM, Stutte GW, Levine LH (1999) How far can sodium substitute for potassium in red beet? J Plant Nutr 22:1745–1761

Subbarao GV, Ito O, Berry WL, Wheeler RM (2003) Sodium—a functional plant nutrient. Crit Rev Plant Sci 22:391–416

Tahir M, Tanveer A, Shah TH, Fiaz N, Wasaya A (2009) Yield response of wheat (*Triticum aestivum* L.) to boron application at different growth stages. Pak J Life Soc Sci 7:39–42
Taiz L, Zeiger E (2010) Plant physiology, 5th edn. Sunderland, Sinauer
Takahashi H, Kopriva S, Giordano M, Saito K, Hell R (2011) Sulfur assimilation in photosynthetic organisms: molecular functions and regulations of transporters and assimilatory enzymes. Annu Rev Plant Biol 62:157–184
Takahashi H, Buchner P, Yoshimoto N, Hawkesford MJ, Shiu SH (2012) Evolutionary relationships and functional diversity of plant sulfate transporters. Front Plant Sci 2:119
Tang N, Li Y, Chen LS (2012) Magnesium deficiency-induced impairment of photosynthesis in leaves of fruiting *Citrus reticulata* trees accompanied by up-regulation of antioxidant metabolism to avoid photo-oxidative damage. J Plant Nutr Soil Sci 175:784–793
Toyota M, Spencer D, Sawai-toyota S, Jiaqi W, Zhang T, Abraham KJ (2018) Glutamate triggers long-distance, calcium-based plant defense signaling. Science 80:1112–1115
Tränkner M, Tavakol E, Jákli B (2018) Functioning of potassium and magnesium in photosynthesis, photosynthate translocation and photoprotection. Physiol Plant 163:414–431
Van Leeuwe MA, Stefels J (2007) Photosynthetic responses in *Phaeocystis antarctica*towards varying light and iron conditions. Biogeochemistry 83:61–70
Vatansever R, Ozyigit II, Filiz E (2017) Essential and beneficial trace elements in plants, and their transport in roots: a review. Appl Biochem Biotechnol 181(1):464–482
Vert G, Grotz N, Dédaldéchamp F, Gaymard F, Guerinot ML, Briat JF, Curie C (2002) IRT1, an *Arabidopsis* transporter essential for iron uptake from the soil and for plant growth. Plant Cell 14:1223–1233
Wang M, Zheng Q, Shen Q, Guo S (2013) The critical role of potassium in plant stress response. Int J Mol Sci 14:7370–7390
Waraich EA, Ahmad R, Ashraf MY, Saifullah U, Ahmad M (2011) Improving agricultural water use efficiency by nutrient management in crop plants. Acta Agri Scand B Plant Soil Sci 61:291–304
Weber APM, von Caemmerer S (2010) Plastid transport and metabolism of C3 and C4 plants—comparative analysis and possible biotechnological exploitation. Curr Opin Plant Biol 13:257–265
Welch RM, Graham RD (2004) Breeding for micronutrients in staple food crops from a human nutrition perspective. J Exp Bot 55:353–364
White PJ, Broadley MR (2001) Chloride in soils and its uptake and movement within the plant: a review. Ann Bot 88:967–988
White PJ, Broadley MR (2003) Calcium in plants. Annal. Bot. 92:487–511
White PJ, Broadley MR (2009) Biofortification of crops with seven mineral elements often lacking in human diets–iron, zinc, copper, calcium, magnesium, selenium and iodine. New Phytol 182:49–48
White PJ, Veneklaas EJ (2012) Nature and nurture: the importance of seed phosphorus content. Plant Soil 357:1–8
Will S, Eichert T, Fernandez V, Mohring J, Muller T, Romheld V (2011) Absorption and mobility of foliar-applied boron in soybean as affected by plant boron status and application as a polyol complex. Plant Soil 344:283–293
Williams RJP, Frausto da Silva JJR (2002) The involvement of molybdenum in life. Biochem Biophys Res Commun 292:293–299
Williams L, Salt DE (2009) The plant ionome coming into focus. Curr Opin Plant Biol 12:247–249
Wintz H, Fox T, Wu YY, Feng V, Chen W, Chang HS, Zhu T, Vulpe C (2003) Expression profiles of Arabidopsis thaliana in mineral deficiencies reveal novel transporters involved in metal homeostasis. J Biol Chem 278:47644–47653
Wood BW, Reilly CC, Nyczepir AP (2006) Field deficiency of nickel in trees: symptoms and causes. Acta Hortic 721:83–98

Xiong Q, Tang G, Zhong L, He H, Chen X (2018) Response to nitrogen deficiency and compensation on physiological characteristics, yield formation, and nitrogen utilization of rice. Front Plant Sci 9:1075

Xu N, Guo W, Liu J, Du N, Wang R (2015) Increased nitrogen deposition alleviated the adverse effects of drought stress on *Quercus variabilis* and *Quercus mongolica* seedlings. Acta Physiol Plant 37:107

Yadavalli V, Neelam S, Rao ASVC, Reddy AR, Subramanyam R (2012) Differential degradation of photosystem I subunits under iron deficiency in rice. J Plant Physiol 169:753–759

Yamasaki H, Hayashi M, Fukazawa M, Kobayashi Y, Shikanai T (2009) SQUAMOSA promoter binding protein–like7 is a central regulator for copper homeostasis in Arabidopsis. Plant Cell 21:347–361

Yang GH, Yang LT, Jiang HX, Li Y, Wang P, Chen LS (2012) Physiological impacts of magnesium-deficiency in Citrus seedlings: photosynthesis, antioxidant system and carbohydrates. Trees Struct Funct 26:1237–1250

Yohe SL, Choudhari HJ, Mehta DD, Yohe SL, Choudhari HJ, Mehta DD, Dietrich PJ, Detwiler MD, Akatay CM, Stach EA, Miller JT, Delgass WN, Agrawal R, Ribeiro FH (2016) High-pressure vapor-phase hydrodeoxygenation of lignin-derived oxygenates to hydrocarbons by a PtMo bimetallic catalyst: product selectivity, reaction pathway, and structural characterization. J Catal 344:535–552

Yoshimoto N, Inoue E, Saito K, Yamaya T, Takahashi H (2003) Phloem-localizing sulfate transporter, Sultr1;3, mediates redistribution of sulfur from source to sink organs in Arabidopsis. Plant Physiol 131:1511–1517

Yruela I (2005) Copper in plants. Braz J Plant Physiol 17:145–156

Yruela I (2009) Copper in plants: acquisition, transport and interactions. Funct Plant Biol 36:409–430

Zhai BN, Li S (2005) Response to nitrogen deficiency and compensation on growth and yield of winter wheat. J Plant Nutr Fertil Sci 11:308–313

Zhang H, Zhao FG, Tang RJ, Yu Y, Song J, Wang Y, Li L, Luan S (2017) Two tonoplast MATE proteins function as turgor-regulating chloride channels in *Arabidopsis*. Proc Natl Acad Sci USA 114:E2036–E2045

Zhao D, Reddy KR, Kakani VG, Reddy VR (2005) Nitrogen deficiency effects on plant growth, leaf photosynthesis, and hyperspectral reflectance properties of sorghum. Europ J Agron 22:391–403

Zocchi G (2006) Metabolic changes in iron-stressed dicotyledoneus plants. In: Bartonand LL, Abadía J (eds) Iron nutrition in plants and rhizospheric microorganisms. Springer, Dordrecht, pp 359–370

# The Toxicity and Accumulation of Metals in Crop Plants

3

Sudhakar Srivastava, Pramod Kumar Tandon, and Kumkum Mishra

**Abstract**

Metals are ubiquitously present in the environment and are crucial for life on earth. This is because several metals perform life saving biological functions. However, there are other metals, which are non-essential. The entry of toxic metals into plants causes physiological and biochemical disturbances and also affects molecular responses. The growth and development of plants are affected that ultimately reduces the yield and quality of plant produce. Another aspect of this issue is that through plant produce, metals gain entry into humans and cause several ailments including cancer. Hence, the concentration of toxic metals in plants and plant based food products needs to be regulated effectively. The present chapter gives an overview of metal toxicity to plants and the status of metal accumulation in various crop plants, fruits, vegetables, etc.

**Keywords**

Arsenic · Cadmium · Food · Grains · Metals

## 3.1 Introduction

The elements having metallic properties and atomic number of higher than 20 are referred to as heavy metals. However, metals have been categorized according to their preferential binding to different binding ligands available and their reactivity as

S. Srivastava (✉)
Plant Stress Biology Laboratory, Institute of Environment and Sustainable Development, Banaras Hindu University, Varanasi, Uttar Pradesh, India
e-mail: sudhakar.iesd@bhu.ac.in

P. K. Tandon · K. Mishra
Department of Botany, University of Lucknow, Lucknow, Uttar Pradesh, India



K. Mishra et al. (eds.), *Sustainable Solutions for Elemental Deficiency and Excess in Crop Plants*, https://doi.org/10.1007/978-981-15-8636-1_3

class A, B and intermediate (Nieboer and Richardson 1980). Metals include both essential metals [zinc (Zn), iron (Fe), nickel (Ni), cobalt (Co), copper (Cu), etc.] and toxic metals [cadmium (Cd), lead (Pb), arsenic (As), chromium (Cr), mercury (Hg), etc.] depending on their requirement for normal growth and metabolism by plants (Pilon et al. 2009; Afonne and Ifediba 2020). Toxic metals are actually a great threat to the natural ecosystem and ultimately to the environment. Thus proper steps should be taken for saving and correcting the soil ecosystem by removing the excess concentration of heavy metals such as Cd, Pb, Cr, Hg, As, etc. by the number of physico-chemical methods and also through proper legislation (Wuana and Okieimen 2011; Li et al. 2019). Some plant species of *Cynodon, Phragmites, Typha Pteris, Thalspi,* etc. are reported to accumulate high amounts of toxic metals. Such species are often referred to as hyperaccumulators. Due to their hyperaccumulation characteristic, they may be used for removing toxic metals from the contaminated environment (Rascio and Navari-Izzo 2011).

One of the major sources of various heavy metals in the environment is agricultural chemicals, such as fertilizers and pesticides, fungicides, weedicides, etc., which are used to increase and protect grain yield in almost all the major crops. With the increasing population, the need for crop production has also increased manifold and this has necessitated even extensive use of various chemicals in agriculture (Ray et al. 2013). Chemical fertilizers are also used in areas where essential nutrients are present in deficient amounts so as to overcome the deficiency of essential elements. In order to overcome a number of plant diseases, frequent use of pesticides and fungicides is practiced while weedicides are needed to clear the field of weeds for optimum crop growth. However, various agricultural chemicals contain some toxic metals such as Cd, As, Pb, Cr, etc. as contaminant and therefore cause metal contamination in the field (Antoniadis et al. 2019; Afonne and Ifediba 2020; Shukla et al. 2020).

Agricultural soils may also get contaminated through erosion and sediment deposition from contaminated land, landfill site leachate, livestock manure, sewage sludge based amendments, wastewater reuse, and fly ash deposition (Dwivedi et al. 2007; Rai et al. 2019). Other sources of metals are discharges of municipal and domestic sewages, and industrial wastes dumped into waterways, i.e., river, ponds, lakes, and sea. There are several industries like batteries, chip manufacturing, steel, electroplating, textiles, leather, e-waste, etc. which are the sources of a number of metals in the environment (Wuana and Okieimen 2011; Li et al. 2019; Rai et al. 2019). Mining processes and automobile emissions as well as coal use in electricity generation are also the sources of several metals. Particular matter emission from vehicular and industrial pollution results in soil and plant deposition ultimately (Rai et al. 2019). Apart from this, natural geological processes like volcanic eruptions and biogeochemical weathering of rocks are an important source of metals in the environment. Arsenic contamination in Southeast Asian countries is mostly attributable to the natural weathering of rocks due to biogeochemical processes (Shukla et al. 2020).

Owing to the widespread contamination of metals, most of the agricultural crops including cereals and vegetables are prone to metal toxicities. These plants absorb metals present in the soil and irrigation water and even through the air (Rai et al. 2019; Upadhyay et al. 2019; Shukla et al. 2020). The uptake and accumulation of metals vary from plant to plant. A number of plant dependent factors determine their potential for metal accumulation; for example root architecture (Srivastava et al. 2019), expression of a number of transporters involved in the uptake and transport of metals from root to shoot (Clemens and Ma 2016; Awasthi et al. 2017; Das et al. 2020), the ability of plant to transform metal species from one to other form like for As and Se (Chauhan et al. 2019; Srivastava and Shukla 2019; Guarino et al. 2020), the potential of plants to detoxify and tolerate metal stress, vacuolar sequestration of metals and metal homeostasis (Park et al. 2012; Peng and Gong 2014; Song et al. 2014). Apart from plant dependent features, the bioavailability of metal to plant determines its eventual accumulation in plants. There are a number of factors like pH, redox potential, soil porosity, soil type, water level, etc. which affect metal bioavailability to plants (Majumdar and Bose 2017, 2018; Upadhyay et al. 2019). In the case of As, it is known to be accumulated in very high levels in rice. One important factor responsible for such high As accumulation in rice is flooded cultivation that leads to the development of anaerobic conditions. This in turn results in the prevalence of reduced form of As, i.e., arsenite [As(III)] (Awasthi et al. 2017), which is efficienty taken up by highly expressed silicic acid transporters of rice plants. In aerobic conditions, prevailing during cultivation of wheat and other crops, As exists mainly in the oxidized form, arsenate [As(V)], whose uptake in plants is faces strong competition from phosphate and hence, crops other than rice show less As accumumulation. However, in aerobic conditions, divalent metals like Cd become more bioavailable (Yuan et al. 2019). Further, soil microbial activity, mycorrhizal inoculations in plants, and release of organic acids, nicotianamine, etc. by plants also alter metal bioavailability to plants and thus the uptake and accumulation of metals (Takahashi et al. 2003; Schwab et al. 2008; Poonam et al. 2017; Upadhyay et al. 2019). About 80% of land plants have mycorrhizal inoculation in their roots. Mycorrhiza make elements like phosphate available to plants and in return receive carbohydrates from plants (Poonam et al. 2017). Nicotianamine is well known to be excreted by a number of plants for enhancing the availability of Fe and for facilitating its uptake in chelated form (Ishimaru et al. 2010). Plant roots release a number of acids to the soil like citric acid, malic acid, oxalic acid to the soil that generates acidic pH in the rhizospheric zone around the roots (Schwab et al. 2008). This keeps most of the elements in free bioavailable form around roots and hence, metals are easily taken up by plants.

## 3.2 Accumulation of Toxic Metals in Crop Plants: Present Status

A number of studies from throughout the world have reported the accumulation of various toxic metals in higher than recommended ranges. These include market or field based survey experiments analyzing the level of various essential as well as toxic metals in crop plants, fruits, vegetables, etc. (Arora et al. 2008; Tosic et al.

2015; Rubio et al. 2018; Upadhyay et al. 2020). The permissible maximum level (mg/kg) is 0.2 mg/kg As and 0.4 mg/kg Cd in rice grains while 0.2 mg/kg Pb and 0.1 mg/kg Cd for other cereal grains. For leafy vegetables, the limit of Cd is 0.2 mg/kg while that of Pb is 0.3 mg/kg (FAO/WHO 2019; Afonne and Ifediba 2020). However, such limits have not been set for all types of foods for all metals. Hence, more research is needed to set guidelines for maximum permissible levels to ensure human safety in the future. A brief list of various toxic elements present in different food items is given in Table 3.1. Rice, wheat, and maize constitute some of the most important cereal grains, while potato and tomato are used widely among vegetables. The metal contamination of such widely used staple food items is alarming (Table 3.1). It is important to note that food products based on plant produce, fruits, grains, etc. are used for feeding babies and young children throughout the world (Upadhyay et al. 2020). Further, the status of metal exposure to humans is such that even breast milk is not safe and there are plenty of reports about infant exposure to toxic metals via breast milk (Rebelo and Caldas 2016). Mushrooms are very widely consumed owing to their nutritional properties and elemental levels. However, mushrooms are also known to accumulate various elements in the toxic range, e.g., As, Cd, Pb, etc. (Rashid et al. 2018). Hence, the level of toxic metals in crop plants is a prevalent problem throughout the world. The situation has been aggravated in the past few decades owing to ever-increasing metal contamination. The need of the hour is to develop easy low cost methods for routine analysis of metals in food items. This would also help in the determination of maximum allowable limits of metals in food items.

## 3.3 Metal Phytotoxicity and Stress Responses of Plants

The accumulation of non-essential metals as well as essential metals in excess causes a number of harmful effects in plants ranging from morphological, anatomical, physiological, biochemical, molecular, to metabolic changes. Some metals exist in different chemical species and the speciation of such metals affects their toxicity to plants. Chromium can exist as CrIII and CrVI and CrVI has been reported to be more toxic than CrIII (Chatterjee et al. 2011). Arsenic exists as inorganic [arsenate (AsV), arsenite (AsIII)] and organic [monomethylarsonic acid (MMA), dimethylarsinic acid (DMA), etc.] (Upadhyay et al. 2020). Similarly, selenium (Se) and mercury (Hg) exist in a variety of different inorganic and organic species (Chauhan et al. 2019). Several metals exist as monovalent and divalent states like Cu and Fe and are redox active. The different chemical species of metal have variable toxicity in plants and humans and hence, a basic mechanism of metal tolerance is the conversion of more toxic metal species to less toxic ones.

Transporters maintaining physiological concentration of heavy metals include zinc-iron permease (ZIP) heavy metal transporter ATPase, natural resistant associated macrophage protein (NRAMP) cation diffusion facilitator (CDF) and ATP binding cassette (ABC) transporter which are found at the plasma membrane and on the tonoplast membrane of cell (Park et al. 2012; Singh et al. 2015; Clemens

**Table 3.1** Reported concentrations/content of toxic metals in cereals, vegetables, fruits, mushrooms, and baby food products from various locations

| Name of plant/food | Country/ location | Concentration of metal (mg/kg dry weight, dw) | References |
|---|---|---|---|
| *Cereals* | | | |
| Wheat grains (field collection) | India | 1.04 (Cd), 13.97 (Co), 4.53 (Cr), 6.08 (Cu), 16.98 (Pb), 50.87 (Zn) | Sharma et al. (2018) |
| Rice grains (field collection) | India | 0.99 (Cd), 15.21 (Co), 19.98 (Cr), 69.89 (Cu), 17.13 (Pb), 35.71 (Zn) | Sharma et al. (2018) |
| Maize grains (field collection) | India | 1.09 (Cd), 15.13 (Co), 2.48 (Cr), 43.87 (Cu), 18.28 (Pb), 39.17 (Zn) | Sharma et al. (2018) |
| Mustard seeds (field collection) | India | 1.05 (Cd), 13.46 (Co), 2.45 (Cr), 7.61 (Cu), 16.34 (Pb), 59.33 (Zn) | Sharma et al. (2018) |
| Wheat grains (normal Se and high Se fields collection) | India | High Se mean–Normal Se mean 106.50.18 (Se), 6.775–9.34 (Al), 0.010–0.020 (As), 0.030–0.040 (Cd), 0.002–0.004 (Hg), 0.175–1.230 (Ni), 0.110–2.120 (Pb) | Skalnaya et al. (2017) |
| Rice grains (normal Se and high Se fields collection) | India | High Se mean–Normal Se mean 21.41–0.25 (Se), 4.135–5.150 (Al), 0.170–0.110 (As), 0.008–0.040 (Cd), 0.002–0.002 (Hg), 0.120–1.330 (Ni), 0.080–1.820 (Pb) | Skalnaya et al. (2017) |
| Maize grains (normal Se and high Se fields collection) | India | High Se mean–Normal Se mean 24.43–0.380 (Se), 1.8353.320 (Al), 0.007–0.010 (As), 0.005–0.030 (Cd), 0.008–0.002 (Hg), 0.155–2.370 (Ni), 0.070–1.960 (Pb) | Skalnaya et al. (2017) |
| Mustard seeds (normal Se and high Se fields collection) | India | High Se mean–Normal Se mean 121.0–1.090 (Se), 109.5–47.87 (Al), 0.070–0.020 (As), 0.080–0.070 (Cd), 0.007–0.002 (Hg), 0.7402.070 (Ni), 0.255–2.190 (Pb) | Skalnaya et al. (2017) |
| Rice grains (market collection of 10 types of rice) | Bangladesh | Maximum 1.616 (Cd), 0.01 (Cr), 0.08 (Pb), 0.70 (As), 1.63 (Mn) | Real et al. (2017) |
| Rice (*Oryza sativa*) | Pakistan | 3.17 (Cr), 0.44 (As), 0.16 (Cd), 0.28 (Pb) | Nawab et al. (2017) |
| Kidney beans (*Phaseolus vulgaris*) | Pakistan | 2.04 (Cr), 0.06 (As), 0.07 (Cd), 0.33 (Pb) | Nawab et al. (2017) |
| Chick peas (*Cicer arietinum*) | Pakistan | 2.27 (Cr), 0.10 (As), 0.06 (Cd), 0.29 (Pb) | Nawab et al. (2017) |
| Rice grains | Turkey | 0.232 (Cd), 0.15 (Co), 1.71 (Cu), 6.9 (Mn), 0.034 (Mn), 12.0 (Zn) | Sofuoglu and Sofuoglu (2017) |
| Bulgur (prepared from wheat grains) | Turkey | 0.008 (Cd), 0.016 (Co), 3.72 (Cu), 14.1 (Mn), 0.023 (Mn), 14.7 (Zn) | Sofuoglu and Sofuoglu (2017) |
| Rice grains | India | 0.29–0.95 (As) | Upadhyay et al. (2019) |

(continued)

**Table 3.1** (continued)

| Name of plant/food | Country/ location | Concentration of metal (mg/kg dry weight, dw) | References |
|---|---|---|---|
| Maize grains | Egypt | 0.03–0.55 (Pb), 0.009–0.112 (Cd) | El-Hassanin et al. (2020) |
| *Vegetables* | | | |
| Spinach (*Spinacia oleracea*) | India | 10 (Zn), 2.9 (Cr), 0.09 (Cu), 3.2 (Ni), 0.13 (Co), 3.1 (Pb) | Chary et al. (2008) |
| Amaranth (*Amaranthus graecizans*) | India | 8 (Zn), 2.4 (Cr), 1.4 (Cu), 3.1 (Ni), 0.09 (Co), 2.9 (Pb) | Chary et al. (2008) |
| Brinjal (*Solanum melongena*) | India | 4.5 (Zn), 1.1 (Cr), 0.7 (Cu), 3.1 (Ni), 3.0 (Pb) | Chary et al. (2008) |
| Ladies finger (*Abelmoschus esculentus*) | India | 3.7 (Zn), 1.4 (Cr), 0.6 (Cu), 2.4 (Ni), 3.6 (Pb) | Chary et al. (2008) |
| Coriander leaves (*Coriandrum sativum*) | India | 5.4 (Zn), 2.1 (Cr), 1.2 (Cu), 2.7 (Ni), 0.03 (Co), 2.7 (Pb) | Chary et al. (2008) |
| Mint leaves (*Mentha spicata*) | India | 6.5 (Zn), 1.4 (Cr), 1.1 (Cu), 2.4 (Ni), 2.2 (Pb) | Chary et al. (2008) |
| Radish (*Raphanus sativus*) | India | 117 (Fe), 22.5 (Zn), 12.8 (Mn), 5.96 (Cu) | Arora et al. (2008) |
| Spinach (*Spinacia oleracea*) | India | 309 (Fe), 33.1 (Zn), 69.4 (Mn), 16.5 (Cu) | Arora et al. (2008) |
| Turnip (*Brassica rapa*) | India | 197 (Fe), 29.3 (Zn), 18.2 (Mn), 16.1 (Cu) | Arora et al. (2008) |
| Carrot (*Daucus carota*) | India | 216 (Fe), 46.4 (Zn), 17.4 (Mn), 16.8 (Cu) | Arora et al. (2008) |
| Cauliflower (*Brassica oleracea*) | India | 215 (Fe), 40.2 (Zn), 41.3 (Mn), 5.23 (Cu) | Arora et al. (2008) |
| Tomato (*Solanum lycopersicum*) | Bangladesh | Maximum 0.001 (Cd), 0.02 (Cr), 0.025 (Pb) | Real et al. (2017) |
| Red Amaranth (*Amaranthus gangeticus*) | Bangladesh | Maximum 0.001 (Cd), 0.044 (Pb) | Real et al. (2017) |
| Cauliflower (*Brassica oleracea*) | Bangladesh | Maximum 0.016 (Cr) | Real et al. (2017) |
| Kochu (*Colocasia antiquorum*) | Bangladesh | Maximum 0.072 (As) | Real et al. (2017) |
| Potato (*Solanum tuberosum*), agricultural field | Bangladesh | 0.68 (Cr), 1.3 (Ni), 2.4 (Cu), 0.07 (As), 0.10 (Cd), 0.43 (Pb) | Islam et al. (2015) |
| Green Amaranth (*Amaranthus hybridus*), agricultural field | Bangladesh | 1.3 (Cr), 3.2 (Ni), 2.9 (Cu), 0.15 (As), 0.32 (Cd), 1.2 (Pb) | Islam et al. (2015)[a] |
| Red Amaranth (*Amaranthus gangeticus*), agricultural field | Bangladesh | 1.5 (Cr), 3.6 (Ni), 2.6 (Cu), 0.12 (As), 0.25 (Cd), 0.97 (Pb) | Islam et al. (2015)[a] |

(continued)

**Table 3.1** (continued)

| Name of plant/food | Country/ location | Concentration of metal (mg/kg dry weight, dw) | References |
|---|---|---|---|
| Bottle gourd (*Lagenaria siceraria*), agricultural field | Bangladesh | 0.67 (Cr), 3.2 (Ni), 3.2 (Cu), 0.83 (As), 0.09 (Cd), 0.41 (Pb) | Islam et al. (2015)[a] |
| Tomato (*Solanum lycopersicum*), agricultural field | Bangladesh | 0.63(Cr), 0.81 (Ni), 1.6 (Cu), 0.21 (As), 0.07 (Cd), 0.21 (Pb) | Islam et al. (2015)[a] |
| Pumpkin (*Cucurbita maxima*), agricultural field | Bangladesh | 0.67 (Cr), 2.1 (Ni), 2.7 (Cu), 0.22 (As), 0.06 (Cd), 0.20 (Pb) | Islam et al. (2015)[a] |
| Bean (*Phaseolus vulgaris*), agricultural field | Bangladesh | 0.82(Cr), 0.89 (Ni), 2.1 (Cu), 0.11 (As), 0.08 (Cd), 0.95 (Pb) | Islam et al. (2015)[a] |
| Lentil (*Lens culinaris*), agricultural field | Bangladesh | 0.68 (Cr), 1.7 (Ni), 1.9 (Cu), 0.75 (As), 0.03 (Cd), 0.31 (Pb) | Islam et al. (2015)[a] |
| Tomato (*Solanum lycopersicum*) | Pakistan | 1.98 (Cr), 0.14 (As), 0.18 (Cd), 0.11 (Pb) | Nawab et al. (2017) |
| Potato (*Solanum tuberosum*) | Pakistan | 2.19 (Cr), 0.15 (As), 0.15 (Cd), 0.41 (Pb) | Nawab et al. (2017) |
| Pea (*Pisum sativum*) | Pakistan | 2.23 (Cr), 0.12 (As), 0.06 (Cd), 0.40 (Pb) | Nawab et al. (2017) |
| Ladyfinger (*Abelmoschus esculentus*) | Pakistan | 2.72 (Cr), 0.21 (As), 0.27 (Cd), 0.57 (Pb) | Nawab et al. (2017) |
| Onion (*Allium cepa*) | Pakistan | 2.73 (Cr), 0.25 (As), 0.05 (Cd), 0.39 (Pb) | Nawab et al. (2017) |
| *Mushrooms* | | | |
| *Lactarius deliciosus* | Spain | 0.16 (Cr), 1.64 (Cu), 18.2 (Al), 0.006 (Cd), 0.08 (Pb) | Rubio et al. (2018)[a] |
| *Pholiota nameko* | Spain | 0.10 (Cr), 1.73 (Cu), 19.3 (Al), 0.002 (Cd), 0.08 (Pb) | Rubio et al. (2018)[a] |
| *Lentinula edodes* | Spain | 0.15 (Cr), 1.53 (Cu), 16.8 (Al), 0.009 (Cd), 0.09 (Pb) | Rubio ct al. (2018)[a] |
| *Pleurotus ostreatus* | Spain | 0.17 (Cr), 1.99 (Cu), 18.1 (Al), 0.004 (Cd), 0.1 (Pb) | Rubio et al. (2018)[a] |
| *Agaricus bisporus* | Spain | 0.07 (Cr), 1.54 (Cu), 18.2 (Al), 0.002 (Cd), 0.07 (Pb) | Rubio et al. (2018)[a] |
| *Pleurotus highking* | Bangladesh | 0.56 (As), 0.35 (Cd), 0.27 (Cr), 14.2 (Cu), 0.40 (Pb), 0.126 (Hg) | Rashid et al. (2018) |
| *Pleurotus ostreatus* | Bangladesh | 0.45 (As), 0.41 (Cd), 0.30 (Cr), 13.2 (Cu), 0.22 (Pb), 0.124 (Hg) | Rashid et al. (2018) |
| *Agaricus bisporus* | India | 0.05 (Cr), 0.05 (Ni), 0.05 (As), 13.61 (Fe), 0.05 (Pb), 3.85 (Zn) | Sinha et al. 2019 |
| *Fruits* | | | |
| Walnut (production areas, field) | China | 0.056 (Pb), 0.015 (As), 0.007 (Cd), 0.184 (Cr), 0.0005 (Hg) | Han et al. (2018) |

(continued)

**Table 3.1** (continued)

| Name of plant/food | Country/ location | Concentration of metal (mg/kg dry weight, dw) | References |
|---|---|---|---|
| Walnut (supermarket) | Serbia | 0.150 (Pb), 0.157 (Cr) | Tosic et al. (2015) |
| Walnut (supermarket) | Chile | 0.009 (Pb), 0.064 (As), 0.0014 (Cd), 0.083 (Cr) | Kafaoglu et al. (2014) |
| Walnut (supermarket) | Spain | 0.705 (Cd), 1.454 (Cr) | Moreda-Pineiro et al. (2016) |
| Banana (*Musa acuminata*) | Pakistan | 2.55 (Cr), 0.15 (As), 0.07 (Cd), 0.35 (Pb) | Nawab et al. (2017) |
| Tangerine (*Citrus tangerina*) | Pakistan | 2.58 (Cr), 0.12 (As), 0.06 (Cd), 0.33 (Pb) | Nawab et al. (2017) |
| Apple (*Malus domestica*) | Pakistan | 2.18 (Cr), 0.10 (As), 0.06 (Cd), 0.32 (Pb) | Nawab et al. (2017) |
| Guava (*Psidium guajava*) | Pakistan | 2.19 (Cr), 0.06 (As), 0.04 (Cd), 0.05 (Pb) | Nawab et al. (2017) |
| Pineapple | Nigeria | 0.057 (As), 0.00005 (Hg), 0.00011 (Cu) | Ezeonyejiaku and Obiakor (2017)[a] |
| Orange | Nigeria | 0.044 (As), 0.00002 (Hg), 0.00011 (Cu) | Ezeonyejiaku and Obiakor (2017)[a] |
| Guava | Nigeria | 0.020 (As), 0.0009 (Hg), 0.0002 (Cu) | Ezeonyejiaku and Obiakor (2017)[a] |
| *Baby foods/food items* | | | |
| Porridge (fish and vegetables) | Spain | 0.10 (As), 0.006 (Pb), 0.16 (Cu), 2.66 (Fe), 0.30 (Mn) | Skrbic et al. (2016) |
| Infant and follow-on formula | Spain | 2.76 (Cu), 187.77 (Fe), 0.24 (Mn) | Skrbic et al. (2016) |
| Porridge (corn and rice) | Serbia | 0.56 (Cu), 164.85 (Fe), 2.01 (Mn) | Skrbic et al. (2016) |
| Porridge (grains and honey) | Serbia | 1.07 (Cu), 207.63 (Fe), 4.58 (Mn) | Skrbic et al. (2016) |
| Infant and follow-on formula | Serbia | 1.47 (Cu), 79.13 (Fe), 0.14 (Mn) | Skrbic et al. (2016) |
| Crisped rice | UK | 0.21 (As) | Sun et al. (2009) |
| Puffed rice | UK | 0.24 (As) | Sun et al. (2009) |
| Rice noodles | UK | 0.12 (As) | Sun et al. (2009) |
| Rice crackers | UK | 0.28 (As) | Sun et al. (2009) |
| Rice malt | UK | 0.21 (As) | Sun et al. (2009) |

[a]Unit is mg/kg on fresh/wet weight basis

and Ma 2016). One of the major mechanisms of metal toxicity is the induced overproduction of reactive oxygen species (ROS) like superoxide radicals and hydrogen peroxide ($H_2O_2$) and consequently cause oxidative stress to plants (Chatterjee et al. 2011; Srivastava et al. 2011; Awasthi et al. 2017). This is due to the interference of redox active metals with redox reactions of cell and also owing to the conversion of non-redox active metals from one chemical form to other (Mylona et al. 1998). The excessive consumption of GSH as a reductant and other redox molecules like NADPH and NADH in the process disturbs redox balance and causes excessive ROS production (Mylona et al. 1998). ROS are highly reactive and attack various biomolecules in cells including lipids in membranes, proteins, and DNA and RNA (Srivastava et al. 2011). The damage to lipid structure in membranes and proteins has been observed through increased malondialdehyde (MDA) and carbonyl content, respectively, in a number of studies (Hartley-Whitaker et al. 2001; Srivastava et al. 2007; Chauhan et al. 2017; Awasthi et al. 2018). ROS also attack DNA and RNA and cause changes in nucleotides and consequently may induce mutations and affect normal functions. Other indirect mechanism of ROS production includes the role of ROS producing enzymes such as NADPH oxidases, glycolate oxidase, and ascorbate oxidase (Cuypers et al. 2009).

To deal with uncontrolled ROS production and to avoid oxidative stress, plants are equipped with a number of enzymatic and non-enzymatic antioxidants. These include superoxide dismutase (SOD), ascorbate peroxidase (APX), glutathione peroxidase, and catalase (CAT) among enzymes and ascorbate, glutathione (GSH), phenolics, and flavonoids among non-enzymatic molecules (Alscher et al. 2002; Shigeoka et al. 2002; Kováčik et al. 2011; Srivastava et al. 2016; Chauhan et al. 2017). SOD is considered to be the primary enzyme in the antioxidant defense of plants as it dismutates superoxide radicals to hydrogen peroxide, which is then broken down to water and oxygen by a number of peroxidises in cell (Alscher et al. 2002). One of the important pathways of ROS detoxification is ascorbate–glutathione cycle (Foyer and Noctor 2011) which involves both enzymes (APX, dehydroascorbate reductase, DHAR, monodehydroascorbate reductase, MDHAR, GR) and metabolites (ascorbate, glutathione) for ROS quenching.

The detoxification mechanisms include the production of metal binding peptides like GSH, phytochelatins, and metallothioneins (Zagorchev et al. 2013; Kumar and Trivedi 2016). The meal transporters and chelators overcome metal toxicity by chelating and sequestering them in plant vacuoles where essential and toxic metabolites are stored (Mendoza-Cózatl et al. 2011). Kumar et al. (2015) have reported the role of metal transporters and cys-rich metal binding peptides in As metal uptake, transport, and detoxification. The excessive consumption of GSH for metal detoxification and metal-induced effects on ascorbate and GSH metabolism in the presence of high metal concentration leads to uncontrolled ROS generation and toxicity to plants.

A number of biochemical parameters have been monitored and found to be modulated upon metal exposure. These include responses of a number of antioxidant enzymes, biosynthetic enzymes of metabolic pathways including carbon, nitrogen, sulfur, nucleotide metabolisms (Pathare et al. 2013; Srivastava et al. 2009, 2013a, b,

2019; Awasthi et al. 2018). Metallothioneins are found in certain eukaryotic organisms including fungi, invertebrates, mammals, and even some prokaryotes. These contain small cysteine rich, low molecular weight cytoplasmic binding proteins or polypeptides in family metallothioneins (Cobbett and Goldbrough 2002; Hassinen et al. 2011; Benatti et al. 2014). A five-carbon α-amino acid named proline, which behaves as a compatible and metabolic osmolyte, is a constituent of cell wall, an antioxidant, free radical scavenger, and macromolecules stabilizer (Liang et al. 2013; Emamverdian et al. 2015). An increase in antioxidant molecules such as ascorbate, peroxidase, dismutase, catalase, glutathione reductase, superoxide dismutase, vitamin C (ascorbic acid), vitamin E (α-Tocopherol), glutathione, carotenoids (β-carotein) provides a defense mechanism (Tiwari and Lata 2018). A study on effect of Cr on antioxidant potential of *Catharanthus roseus* varieties and production of their anticancer alkaloids such as vincristine and vinblastine revealed that Cr adversely affected foliar contents of total chlorophyll, Chl. a, and Chl. b in C. roseus. Growth performance was also found to be retarded due to Cr in excess amounts. However, this study also indicated the enhancement of two important anticancer alkaloids, i.e., Vincristine and Vinblastine under Cr stress (Rai et al. 2014). Cadmium at increasing concentration and exposure durations caused a reduction in protein, chlorophyll content, and biomass in *Bacopa monnieri*. However, it was observed that bacosite A and bacopside I contents were increased by Cd stress up to 10 μM Cd (Gupta et al. 2014). Similarly, exposure to As was found to cause growth reduction in *Ocimum* sp. and *Withania somnifera* (Siddiqui et al. 2013, 2015). It is also reported that the activity of antioxidant enzymes is directly related to the steady level of ROS in the cell and the augmentation of antioxidant defense plays an important part in the regulation of oxidative stress (Mishra and Tandon 2013).

Displacement of essential metal ions or blocking of functional groups takes place due to the toxic metals. The disintegration of cytoplasmic membrane also takes place which in turn shows a negative effect on important functions such as photosynthesis, respiration, and also enzymatic activities (Emamverdian et al. 2015). Metal stresses affect various physiological processes of plants like leaf water potential, relative water content (RWC), water status, photosynthetic efficiency, respiration, and transpiration (Mobin and Khan 2007; Foyer and Noctor 2009). These processes affect plant growth and metabolism at a holistic level and eventually reduce the growth and halt optimum development of plants. Photosynthetic efficiency has been found to decrease upon metal stress in a number of plants. Photosynthesis is the crucial process to gain biomass and generate energy in the form of ATP. The reduced levels of ATP as well as negative effects on adenine and pyridine metabolisms have been found to correlate to the stress exerted by As in *Brassica juncea* (Srivastava et al. 2013a). Metal stresses also affect the water balance of plants that in turn has a profound impact on all biological processes (Barcelo and Poschenrieder 1990).

Various metal-induced effects on plants have also been monitored through changes in a number of genes, proteins, and metabolites in a number of omics studies with respect to various metals (Chakrabarty et al. 2009; Yu et al. 2012; Srivastava et al. 2013a, b, 2015; Chauhan et al. 2020). Such omics approaches provide extensive database to understand holistic responses of plants towards various metals in different concentration, time, tissue, and cell specific manner. These studies also shed insights into the role of a number of regulator factors like transcription factors, hormones, microRNAs, kinases, calcium signaling, etc. in the process of metal stress tolerance in plants (Rao et al. 2011; Zhou et al. 2012; Srivastava et al. 2013b; Chen et al. 2014; Raghuram et al. 2014; Steinhorst and Kudla 2014; Tang et al. 2014; Zhao et al. 2014; Bukhari et al. 2015; He et al. 2016; Jalmi et al. 2018).

Owing to such diverse effects on plant metabolism, crop production and produce quality are negatively affected under metal stress (Dwivedi et al. 2012). In rice, it has been found that As accumulation in rice grains affects grain quality, reduces the level of essential amino acids while increases that of non-essential amino acids, causes a decline in level of essential nutrient elements like Se Cu, etc. and also affects grain size and weight (1000 grain weight) (Dwivedi et al. 2012; Upadhyay et al. 2019). In the case of Cd and other metals also, a decline in the quality of crop produce and yield has been noticed in a number of studies (Huang et al. 2008).

Biotechnology can be a good tool in mitigating the problem of heavy metal toxicity in plants. Biotechnology in the form of transgenic plants and evolution of use of newer plants breeding techniques are able of getting a solution to overcome heavy metal toxicity problems by generating metal resistant varieties. Also, through proper use of microbes, heavy metal toxicity can be effectively controlled (Tiwari and Lata 2018). These aspects shall be discussed in depth in other chapters of the book.

## 3.4 Conclusions

In conclusion, the entry of toxic metals in plants is a serious issue. The metals entering into plants produce a variety of toxic responses and affect the normal growth and development of plants. The accumulation of toxic metals in crop plants affects the quality of their produce as well as yield. The presence of various toxic metals in various food crops and food products is a major concern throughout the world that threatens the health and safety of millions of people. The issue needs societal, governmental, and scientific measures to tackle it effectively and in a sustainable manner.

**Acknowledgements** SS would like to thank funding supports from Board of Research in Nuclear Sciences (BRNS) (35/14/15/2018-BRNS/10395) and ASEAN-India STI Cooperation, Science and Engineering Research Board (AISTC, SERB) (CRD/2018/000072) for supporting the ongoing research activities of the lab.

## References


Afonne OJ, Ifediba EC (2020) Heavy metals risks in plant foods – need to step up precautionary measures. Curr Opin Toxicol 22:1–6

Alscher RG, Erturk N, Heath LS (2002) Role of superoxide dismutases (SODs) in controlling oxidative stress in plants. J Exp Bot 53:1331–1341

Antoniadis V, Golia EE, Liu Y, Wang S, Shaheen SM, Rinklebe J (2019) Soil and maize contamination by trace elements and associated health risk assessment in the industrial area of Volos, Greece. Environ Int 124:79–88

Arora M, Kiran B, Rani S, Rani A, Kaur B, Mittal N (2008) Heavy metal accumulation in vegetables irrigated with water from different sources. Food Chem 111:811–815

Awasthi S, Chauhan R, Srivastava S, Tripathi RD (2017) The journey of arsenic from soil to grain in rice. Front Plant Sci 8:1007

Awasthi S, Chauhan R, Dwivedi S, Srivastava S, Srivastava S, Tripathi RD (2018) A consortium of alga (Chlorella vulgaris) and bacterium (Pseudomonas putida) for amelioration of arsenic toxicity in rice (*Oryza sativa*): a promising and feasible approach. Environ Exp Bot 150:115–126

Barcelo J, Poschenrieder C (1990) Plant water relations as affected by heavy metal stress: a review. J Plant Nut 13:1–37

Benatti RM, Yookongkaew N, Meetam M, Guo WJ, Punyasuk N, AbuQamar S et al (2014) Metallothionein deficiency impacts copper accumulation and redistribution in leaves and seeds of Arabidopsis. New Phytol 202:940–951

Bukhari SA, Shang S, Zhang M, Zheng W, Zhang G, Wang TZ et al (2015) Genome-wide identification of chromium stress-responsive micro RNAs and their target genes in tobacco (*Nicotiana tabacum*) roots. Environ Toxicol Chem 34:2573–2582

Chakrabarty D, Trivedi PK, Misra P, Tiwari M, Shri M, Shukla D et al (2009) Comparative transcriptomic analysis of arsenate and arsenite stresses in rice seedlings. Chemosphere 74:688–702

Chary NS, Kamala CT, Raj DSS (2008) Assessing risk of heavy metals from consuming food grown on sewage irrigated soils and food chain transfer. Ecotoxicol Environ Saf 69:513–524

Chatterjee J, Tandon PK, Singh RR (2011) Oxidative damages by chromium to growth and metabolism of radish. Indian J Agric Biochem 24(2):100–104

Chauhan R, Awasthi S, Indoliya Y, Chauhan AS, Mishra S, Agrawal L, Srivastava S, Dwivedi S, Singh PC, Mallick S, Chauhan PS, Pande V, Chakrabarty D, Tripathi RD (2020) Transcriptome and proteome analyses reveal selenium mediated amelioration of arsenic toxicity in rice (*Oryza sativa* L.). J Hazard Mater 390:122122

Chauhan R, Awasthi S, Srivastava S, Dwivedi S, Pilon-Smits EA, Dhankher OP, Tripathi RD (2019) Understanding selenium metabolism in plants and its role as a beneficial element. Crit Rev Environ Sci Technol 49:1937–1958

Chauhan R, Awasthi S, Tripathi P, Mishra S, Dwivedi S, Niranjan A et al (2017) Selenite modulates the level of phenolics and nutrient element to alleviate the toxicity of arsenite in rice (Oryza sativa L.). Ecotox Environ Safe 138:47–55

Chen YA, Chi WC, Trinh NN, Huang LY, Chen YC, Cheng KT et al (2014) Transcriptome profiling and physiological studies reveal a major role for aromatic amino acids in mercury stress tolerance in rice seedlings. PLoS One 9:e95163

Clemens S, Ma JF (2016) Toxic heavy metal and metalloid accumulation in crop plants and foods. Annu Rev Plant Biol 67:489–512

Cobbett C, Goldsbrough P (2002) Phytochelatins and metallothioneins: roles in heavy metal detoxification and homeostasis. Ann Rev Plant Biol 53:159–182

Cuypers A, Smeets K, Vangronsveld J (2009) Heavy metal stress in plants. In: Hirt H (ed) Plant stress biology: from genomics to systems biology. Wiley-VCH Verlag GmbH & Co. KGaA, Weinheim, pp 161–178

Das N, Bhattacharya S, Maiti MK (2020) Biotechnological strategies to reduce arsenic content in rice. In: Srivastava S (ed) Arsenic in drinking water and food. Springer, Singapore, pp 445–460
Dwivedi S, Mishra A, Tripathi P, Dave R, Kumar A, Srivastava S, Chakrabarty D, Trivedi PK, Adhikari B, Norton GJ, Nautiyal CS, Tripathi RD (2012) Arsenic affects essential and non-essential amino acids differentially in rice grains: inadequacy of amino acids in rice based diet. Environ Int 46:16–22
Dwivedi S, Tripathi RD, Srivastava S, Mishra S, Shukla MK, Singh R, Rai UN (2007) Growth performance and biochemical responses of three rice (*Oryza sativa* L.) cultivars grown in fly-ash amended soil. Chemosphere 67:140–151
Emamverdian A, Ding Y, Mokhberdoran F, Xie Y (2015) Heavy metal stress and some mechanisms of plant defense response. Sci World J 2015:1–18. Article ID 756120
Ezeonyejiaku CD, Obiakor MO (2017) A market based survey of horticultural fruits for arsenic and trace metal contamination in Southeast Nigeria and potential health risk implications. Journal of Health & Pollution 7:40–50
FAO/WHO Codex Alimentarius International Food Standards: General standard for contaminants and toxins in food and feed (CXS 193-1995). 2019. http://www.fao.org/fao-who-codexalimentarius/en/
Foyer CH, Noctor G (2011) Ascorbate and glutathione: the heart of the redox hub. Plant Physiol 155:2–18
Foyer CH, Noctor G (2009) Redox regulation in photosynthetic organisms: signaling, acclimation, and practical implications. Antiox Redox Signal 11:861–905
Guaorino F, Miranda A, Castiglione S, Cicatelli A (2020) Arsenic phytovolatilization and epigenetic modifications in Arundo donax L. assisted by a PGPR consortium. Chemosphere 251:126310
Gupta P, Khatoon S, Tandon PK, Rai V (2014) Effect of cadmium on growth, Bacoside a, Bacoside I of *Bacopa monnieri* (L.), a memory enhancing herb. Sci World J 2014:1–6. Article ID 824586
Han Y, Ni Z, Li S, Qu M, Tang F, Mo R, Ye C, Liu Y (2018) Distribution, relationship, and risk assessment of toxic heavy metals in walnuts and growth soil. Environ Sci Pollut Res 25:17434–17443
Hartley-Whitaker J, Ainsworth G, Meharg AA (2001) Copper- and arsenate-induced oxidative stress in *Holcus lanatus* L. clones with differential sensitivity. Plant Cell Environ 24:713–722
Hassinen VH, Tervahauta AI, Schat H, Kärenlampi SO (2011) Plant metallothioneins–metal chelators with ROS scavenging activity? Plant Biol 13:225–232
He X, Zheng W, Cao F, Wu F (2016) Identification and comparative analysis of the microRNA transcriptome in roots of two contrasting tobacco genotypes in response to cadmium stress. Sci Rep 6:32805
Huang DF, Xi LL, Wang ZQ, Liu LJ, Yang JC (2008) Effects of irrigation patterns during grain filling on grain quality and concentration and distribution of cadmium in different organs of rice. Acta Agron Sin 34:456–464
Ishimaru Y, Masuda H, Bashir K, Inoue H, Tsukamoto T, Takahashi M, Nikanishi H, Aoki N, Hirose T, Ohsugi R, Nishizawa NK (2010) Rice metal-nicotianamine transporter, OsYSL2, is required for the long-distance transport of iron and manganese. Plant J 62:379–390
Islam MS, Ahmed MK, Habibullah-Al-Mamum M (2015) Metal speciation in soil and health risk due to vegetables consumption in Bangladesh. Environ Monit Assess 187:288
Jalmi SK, Bhagat PK, Verma D, Noryang S, Tayyeba S, Singh K, Sharma D, Sinha AK (2018) Traversing the links between heavy metal stress and plant signalling. Front Plant Sci 9(12):1–21
Kafaoglu B, Fisher A, Hill S, Kara D (2014) Chemometric evaluation of trace metal concentrations in some nuts and seeds. Food Addit Contam Part A Chem Anal Control Expo Risk Assess 31:1529–1538
Kováčik J, Klejdus B, Hedbavny J, Zon J (2011) Significance of phenols in cadmium and nickel uptake. J Plant Physiol 168:576–584
Kumar S, Trivedi PK (2016) Heavy metal stress signaling in plants. In: Ahmad P (ed) Plant metal interaction- emerging remediation techniques. Elsevier, Amsterdam, pp 585–603

Kumar S, Dubey RS, Tripathi RD, Chakrabarty D, Trivedi PK (2015) Omics and biotechnology of arsenic stress and detoxification in plants: current updates and prospective. Environ Int 74:221–230

Liang X, Zhang L, Natarajan SK, Becker DE (2013) Proline mechanisms of stress survival. Antioxid Redox Signal 19:998–1011

Li C, Zhou K, Qin W, Tian C, Qi M, Yan X, Han W (2019) A review on heavy metals contamination in soil: effects, sources, and remediation techniques. Soil Sediment Contam Int J 28:380–394

Majumdar A, Bose S (2017) Toxicogenesis and metabolism of arsenic in rice and wheat plants with probable mitigation strategies. In: Knezevic R (ed) Arsenic: risks of exposure, behavior in the environment and toxicology. Nova Science Publishers, New York, pp 149–166

Majumdar A, Bose S (2018) A glimpse on uptake kinetics and molecular responses of arsenic tolerance in Rice plants. In: Mechanisms of arsenic toxicity and tolerance in plants, edited by Mirza Hasanuzzaman, Kamrun Nahar, Masayuki Fujita. Springer, Singapore, pp 299–315

Mendoza-Cózatl DG, Jobe TO, Hauser F, Schroeder JI (2011) Long-distance transport, vacuolar sequestration, tolerance, and transcriptional responses induced by cadmium and arsenic. Curr Opin Plant Biol 14:554–562

Misra S, Tandon PK (2013) Nickel induced morphological and biochemical changes in cauliflower (*Brassica oleracea* L.) plants. Indian J Agric Biochem 26(2):199–201

Mobin M, Khan NA (2007) Photosynthetic activity, pigment composition and antioxidative response of two mustard (Brassica juncea) cultivars differing in photosynthetic capacity subjected to cadmium stress. J Plant Physiol 164:601–610

Moreda-Pineiro J, Herbello-Hermelo P, Dominguez-Gonzalez R, Bermejo-Barrera P, Moreda-Pineiro A (2016) Bioavailability assessment of essential and toxic metals in edible nuts and seeds. Food Chem 205:146–154

Mylona PV, Polidoros AN, Scandalios JG (1998) Modulation of antioxidant responses by arsenic in maize. Free Rad Biol Med 25:576–585

Nawab J, Farooqi S, Xiaoping W, Khan S, Khan A (2017) Levels, dietary intake, and health risk of potentially toxic metals in vegetables, fruits, and cereal crops in Pakistan. Environ Sci Pollut Res 25:5558–5571

Nieboer E, Richardson DHS (1980) The replacement of the nondescript term 'heavy metals' by a biologically and chemically significant classification of metal ions. Environ Pollut B 1:3–26

Park J, Song WY, Ko D, Eom Y, Hansen TH, Stokholm M, Lee TG, Martinoia E, Lee Y (2012) The phytochelatin transporters AtABCC1 and AtABCC2 mediate tolerance to cadmium and mercury. Plant J 69:278–288

Pathare V, Srivastava S, Suprasanna P (2013) Evaluation of effects of arsenic on carbon, nitrogen, and sulfur metabolism in two contrasting varieties of *Brassica juncea*. Acta Physiol Plant 35:3377–3389

Peng JS, Gong JM (2014) Vacuolar sequestration capacity and long-distance metal transport in plants. Front Plant Sci 5:19

Pilon M, Cohu CM, Ravet K, Abdel-Ghany SE, Gaymard F (2009) Essential transition metal homeostasis in plants. Curr Opin Plant Biol 12:347–357

Poonam, Srivastava S, Pathare V, Suprasanna P (2017) Physiological and molecular insights into rice-arbuscular mycorrhizal interactions under arsenic stress. Plant Gene 11:232–237

Raghuram B, Sheikh AH, Sinha AK (2014) Regulation of MAP kinase signaling cascade by microRNAs in Oryza sativa. Plant Signal Behav 9:e972130

Rai PK, Lee SS, Zhang M, Tsang YF, Kim KH (2019) Heavy metals in food crops: health risks, fate, mechanisms and management. Environ Int 125:365–385

Rai V, Tandon PK, Khatoon S (2014) Effect of chromium on antioxidant potential of *Catharanthus roseus* varieties and production of their anticancer alkaloids: vincristine and vinblastine. Biomed Res Int 1-10:934183

Rao KP, Vani G, Kumar K, Wankhede DP, Misra M, Gupta M et al (2011) Arsenic stress activates MAP kinase in rice roots and leaves. Arch Biochem Biophys 506:73–82

Rascio N, Navari-Izzo F (2011) Heavy metal hyperaccumulating plants: how and why do they do it? And what makes them so interesting? Plant Sci 180:169–181

Rashid MH, Rahman MM, Correll R, Naidu R (2018) Arsenic and other elemental concentrations in mushrooms from Bangladesh: health risks. Int J Environ Res Public Health 15:919

Ray DK, Mueller ND, West PC, Foley JA (2013) Yield trends are insufficient to double global crop production by 2050. PLoS One 8:e66428

Real MIH, Azam HM, Majed N (2017) Consumption of heavy metal contaminated foods and associated risks in Bangladesh. Environ Monit Assess 189:651

Rebelo FM, Caldas ED (2016) Arsenic, lead, mercury and cadmium: toxicity, levels in in breast milk and the risks for breastfed infants. Environ Res 151:671–688

Rubio C, Martinez C, Paz S, Gutierrez AJ, Gonzalez-Weller D, Revert C, Burgos A, Hardisson A (2018) Trace element and toxic metal intake from the consumption of canned mushrooms marketed in Spain. Environ Monit Assess 190:237

Schwab AP, Zhu DS, Banks MK (2008) Influence of organic acids on the transport of heavy metals in soil. Chemosphere 72:986–994

Sharma S, Nagpal AK, Kaur I (2018) Heavy metal contamination in soil, food crops and associated health risks for residents of Ropar wetland, Punjab, India and its environs. Food Chem 255:15–22

Shigeoka S, Ishikawa T, Tamoi M, Miyagawa Y, Takeda T, Yabuta Y, Yoshimura K (2002) Regulation and function of ascorbate peroxidase isoenzymes. J Exp Bot 53:1305–1319

Shukla A, Awasthi S, Chauhan R, Srivastava S (2020) The status of arsenic contamination in India. In: Srivastava S (ed) Arsenic in drinking water and food. Springer Nature, Singapore, pp 1–12

Siddiqui F, Krishna SK, Tandon PK, Srivastava S (2013) Arsenic accumulation in *Ocimum* spp. and its effect on growth and oil constituents. Acta Physiol Plant 35:1071–1077

Siddiqui F, Tandon PK, Srivastava S (2015) Analysis of arsenic induced physiological and biochemical responses in a medicinal plant. *Withania somnifera* Physiol Mol Biol Plants https://doi.org/10.1007/s12298-014-0278-7

Singh S, Parihar P, Singh R, Singh VP, Prasad SM (2015) Heavy metal tolerance in plants: role of transcriptomics, proteomics, metabolomics, and ionomics. Front Plant Sci 6:1143

Sinha SK, Upadhyay TK, Sharma SK (2019) Heavy metals detection in white button Mushroom (Agaricus Bisporus) cultivatedin state of Maharashtra, India. Biochem Cell Arch 19:3501–3506

Skalnaya MG, Jaiswal SK, Prakash R, Prakash NT, Grabeklis AR, Zhegalova IV, Zhang F, Guo X, Tinkov AA, Skalny AV (2017) The level of toxic elements in edible crops from seleniferous area (Punjab, India). Biol Trace Elem Res 184(2):523–528

Skrbic B, Zivancev J, Jovanovic G, Farre M (2016) Essential and toxic elements in commercial baby food on the Spanish and Serbian market. Food Addit Contam Part B 10(1):1–12

Sofuoglu SC, Sofuoglu A (2017) An exposure-risk assessment for potentially toxic elements in rice and bulgur. Environ Geochem Health 40:987–998

Song WY, Yamaki T, Yamaji N, Ko D, Jung KH, Fujii-Kashino M, An G, Martinoia E, Lee Y, Ma F (2014) A rice ABC transporter, OsABCC1, reduces arsenic accumulation in the grain. Proc Natl Acad Sci U S A 111:15699–15704

Srivastava S, Shukla K (2019) Microbes are essential components of arsenic cycling in the environment: implications for the use of microbes in arsenic remediation. In: Arora PK et al (eds) Microbial metabolism of xenobiotic compounds. Springer Nature, Singapore, pp 217–227

Srivastava S, Pathare VS, Sounderajan S, Suprasanna P (2019) Nitrogen supply influences arsenic accumulation and stress responses of rice (*Oryza sativa* L.) seedlings. J Hazard Mater 367:599–606

Srivastava S, Akkarakaran JJ, Sounderajan S, Shrivastava M, Suprasanna P (2016) Arsenic toxicity in rice (Oryza sativa L.) is influenced by sulfur supply: impact on the expression of transporters and thiol metabolism. Geoderma 270:33–42

Srivastava S, Akkarakaran JJ, Suprasanna P, D'Souza SF (2013a) Response of adenine and pyridine metabolism during germination and early seedling growth under arsenic stress in Brassica juncea. Acta Physiol Plant 35:1081–1091

Srivastava S, Mishra S, Tripathi RD, Dwivedi S, Trivedi PK, Tandon PK (2007) Phytochelatins and antioxidant systems respond differentially during arsenite and arsenate stress in Hydrilla verticillata (L.f.) Royle. Environ Sci Technol 41:2930–2936

Srivastava S, Srivastava AK, Suprasanna P, D'Souza SF (2009) Comparative biochemical and transcriptional profiling of two contrasting varieties of Brassica juncea L. in response to arsenic exposure reveals mechanisms of stress perception and tolerance. J Exp Bot 60:3419–3431

Srivastava S, Suprasanna P, D'Souza SF (2011) Redox state and energetic equilibrium determine the magnitude of stress in Hydrilla verticillata upon exposure to arsenate. Protoplasma 248:805–815

Srivastava S, Srivastava AK, Suprasanna P, D'Souza SF (2013b) Identification and profiling of arsenic stress-induced microRNAs in Brassica juncea. J Exp Bot 64:303–315

Srivastava S, Srivastava AK, Sablok G, Deshpande T, Suprasanna P (2015) Transcriptomics profiling of Indian mustard (*Brassica juncea*) under arsenate stress identifies key candidate genes and regulatory pathways. Front Plant Sci 6:646

Steinhorst L, Kudla J (2014) Signaling in cells and organisms–calcium holds the line. Curr Opin Plant Biol 22:14–21

Sun GX, Williams PN, Zhu YG, Deacon C, Carey AM, Raab A, Feldmann J, Meharg AA (2009) Survey of arsenic and its speciation in rice products such as breakfastcereals, rice crackers and Japanese rice condiments. Environ Int 35:473–475

Takahashi M, Terada Y, Nakai I, Nakanishi H, Yoshimura E, Mori S, Nishizawa NK (2003) Role of nicotianamine in the intracellular delivery of metals and plant reproductive development. Plant Cell 15:1263–1280

Tang M, Mao D, Xu L, Li D, Song S, Chen C (2014) Integrated analysis of miRNA and mRNA expression profiles in response to cd exposure in rice seedlings. BMC Genomics 15:835

Tiwari S, Lata C (2018) Heavy metal stress, signaling, and tolerance due to plant-associated microbes: An overview. Front Plant Sci 9:452

Tosic SB, Mitic SS, Velimirovic DS, Stojanovic GS, Pavlovic AN, Pecev-Marinkovic ET (2015) Elemental composition of edible nuts: fast optimization and validation procedure of an ICP-OES method. J Sci Food Agric 95:2271–2278

Upadhyay MK, Shukla A, Yadav P, Srivastava S (2020) A review of arsenic in crops, vegetables, animals and food products. Food Chem 276:608–618

Upadhyay MK, Majumdar A, Barla A, Bose S, Srivastava S (2019) An assessment of arsenic hazard in groundwater-soil-rice system in two villages of Nadia district, West Bengal, India. Environ Geochem Health 41:2381–2395

Wuana RA, Okieimen FE (2011) Heavy metals in contaminated soils: a review of sources, chemistry, risks and best available strategies for remediation. Inter Schol Res Notices 2011:402647

Yu LJ, Luo YF, Liao B, Xie LJ, Chen L, Xiao S, Li JT, Hu SN, Shu WS (2012) Comparative transcriptome analysis of transporters, phytohormone and lipid metabolism pathways in response to arsenic stress in rice (Oryza sativa). New Phytol 195:97–112

Yuan C, Li F, Cao W, Yang Z, Hu M, Sun W (2019) Cadmium solubility in paddy soil amended with organic matter, sulfate, and iron oxide in alternative watering conditions. J Hazard Mater 378:120672

Zagorchev L, Seal CE, Kranner I, Odjakova M (2013) A central role for thiols in plant tolerance to abiotic stress. Int J Mol Sci 14(4):7405–7432

Zhao FY, Wang K, Zhang SY, Ren J, Liu T, Wang X (2014) Crosstalk between ABA, auxin, MAPK signaling, and the cell cycle in cadmium-stressed rice seedlings. Acta Physiol Plant 36:1879–1892

Zhou ZS, Zeng HQ, Liu ZP, Yang ZM (2012) Genome-wide identification of Medicago truncatula microRNAs and their targets reveals their differential regulation by heavy metal. Plant Cell Environ 35:86–99

# Effect of Deficiency of Essential Elements and Toxicity of Metals on Human Health

4

Deepak Kumar Mehrotra

**Abstract**

In recent years, a lot of studies are in advanced stage in assessing the risk posed by the deficiency and toxicity of different elements/metals on human health. It has also been established that many areas of globe can be identified as endemic areas due to excesses/deficiency of different metals/elements and affecting the health of inhabitants. In fact the occurrence of various elements or metals in the environment is the result of various geological cycles of earth, which are responsible for mobilization of earth material along with different chemical compounds in the soil system, thereby making those available to human population depending upon their source and mode of intake. The deficiency/toxicity of certain elements in the soil system may lead to severe health problems. Major functions of the cells of human body are dependent on the rational availability of Hydrogen, Oxygen, Nitrogen, Carbon, K, Na, Ca, Cl, Fe, etc. while the toxicity/deficiency of elements like I, F, As, Se, Hg, etc. may lead to serious health hazards. Besides radiation hazards, urban and industrial waste toxicity should be effectively controlled to minimize the toxic level of metals jeopardizing the human health.

**Keywords**

Arsenic · Fluoride · Mercury · Radiation hazard · Selenium

D. K. Mehrotra (✉)
Geological Survey of India, Lucknow, Uttar Pradesh, India





K. Mishra et al. (eds.), *Sustainable Solutions for Elemental Deficiency and Excess in Crop Plants*, https://doi.org/10.1007/978-981-15-8636-1_4

## 4.1 Introduction

A complex nexus of minerals and trace elements can be observed in human body in everyday life. But majority of humans do not understand the process of interaction with these naturally occurring elements through many processes regarding their benefits or toxicity. In recent years, the studies carried out in different parts of the world show that there are many naturally endemic areas, which can be linked to trace element excesses or chronic poisoning.

The occurrence of essential or trace elements on earth is the product of geological cycles of the earth. These geologic cycles have/had mobilized large volumes of earth material (along with chemical compounds and elements) into environment. These elements are absorbed by organic matter content of soil and water, thereby absorbing different elements and making them bioavailable to human and animal population. The health of human population in a particular area has a direct bearing on the availability of essential life forming elements in bedrock, soil, water and atmosphere.

The elements that play a pivotal role in human health can broadly be classified into:

1. Bulk elements
2. Trace elements/metals
3. Toxic elements/metals

About 99% of the body consists of bulk elements of which $O_2$, C, $H_2$, and $N_2$ constitute 96% while Na, Mg, K, Cl, S, P, Ca etc. (macroelements) constitute the rest. The deficiency of bulk elements results in abnormal biological function or development that is prevented by dietary supplementation of that element. Trace elements/metals such as Fe, Se, Zn etc. are present in the body in very small quantity and are essential for regulating the dynamic processes of life being the constituents of biomolecules mainly enzymes. Toxic elements/metals such as As, Pb, Hg, Al, F etc. interfere with the functioning of main elements and undermine human health.

Paracelsus (1493–1541) a German-Swiss physician defined a basic law of toxicology, 'Any increase in the amount or concentration of element causes increasing negative biological effects, which may lead to inhibition of biological functions and eventually to death'. However, many other elements are essential for life. Therefore deleterious biological effects can result from either increasing or decreasing concentration of various trace elements. All of the elements that affect our health are found in nature and form the basis of our existence as living creature.

The potential environmental and health effects of elements and metals are of critical importance in order to:

1. Ensure that metals are produced, used and disposed of with environmentally sound ways.
2. To minimize exposure to toxic levels of poisonous metals and metal species as the hazards are strongly dependent on their mode of occurrence with environment.

**Table 4.1** Abundance by mass of major and minor elements in the human body

| Element | Mass percent | Elements | Mass percent |
|---|---|---|---|
| Oxygen | 65.0 | Magnesium | 0.50 |
| Carbon | 18.0 | Potassium | 0.34 |
| Hydrogen | 10.0 | Sulphur | 0.26 |
| Nitrogen | 3.0 | Sodium | 0.14 |
| Calcium | 1.4 | Chloride | 0.14 |
| Phosphorus | 1.0 | | |

## 4.2 Functions of Essential Elements

Human body comprises roughly 70% of water by weight (Table 4.1). All the functions of cells and organs depend on water. That is why qualities of $H_2$ and $O_2$ present in the body constitute an important issue.

1. *Hydrogen* occurs in 2 states—$H^+$ (cation), $H^-$ (anion) and H- (a co-valently bound state and makes possible many redox reactions).
2. *Oxygen* is transported in the blood by haemoglobin and is actively involved in the production of energy-rich molecules, such as ATP (adenosine triphosphate), besides required for many other functions as life without oxygen cannot be sustained.
3. *Carbon* forms numerous organic compounds for the body like carbohydrates, lipids, proteins and nucleic acids. These compounds are essential for the sustenance of body functions.
4. *Nitrogen* though abundantly available in the atmosphere, only with its fixed form nitrogen becomes available for the biological functions.
5. Potassium, sodium and chlorine generally control the following properties in all cells of all organisms:
   (a) Osmotic pressure
   (b) Membrane potential
   (c) Condensation of polyelectrolyte
   (d) Required ionic strength for activity
6. Calcium forms an integral part of bone and teeth and is required for several other metabolic functions in the body.
7. Magnesium plays an important role in the photosynthesis of plants, thereby affecting the agricultural produce. Besides, the role of magnesium in the cellular function is manifold.

As discussed above the minerals required for the normal metabolism and physiological function of animals and humans can be categorized as below:

| | Element | Function |
|---|---|---|
| 1 | Ca, P, Mg, F | Bone and membrane structure |
| 2 | Na, K, Cl | Water and electrolyte balance |

(continued)

| | | |
|---|---|---|
| 3 | Zn, Cu, Se, Mn, Mo | Essential constituents of enzymes |
| 4 | Fe | Oxygen carrier for metabolism |
| 5 | I, Cr | Hormone activities |

Unlike other essential nutrients, the mineral elements cannot be derived from the biosynthesis of food, plants or animals—they must be obtained from soils and pass thorough food systems to humans in food forms. Therefore, local deficiencies of minerals in soils can produce deficiency in local food systems, which clinically impact the people dependent on those systems.

### 4.2.1 Calcium

The human body contains more than 1 kg of calcium, 99% of which is in the skeleton. Bone calcium is in constant turnover, with mineralization and mobilization of bone minerals occurring continually in the healthy bones of both children and adults. Ca level in human body is regulated by vitamin D metabolites across the various systems of body. Deficiency of Ca in young children results in deformation of growing bones called *Rickets*. In adult it results in *Osteomalacia*, characterized by increased fracture risk.

Excessive Ca intake can lead to renal stone formation as well as impanels utilization of Fe, Zn, Mg and P.

### 4.2.2 Phosphorus

The human body contains approximately 700 g of phosphorus, about 85% of which is in bones. As an element, phosphorus is also important for metabolic processes in the body. Deficiency of phosphorus may result in nervous signs, renal dysfunction and smooth muscle problems.

### 4.2.3 Magnesium

The human body contains 20–28 g of magnesium. About 60–65% of that amount is in bone, 25–30% in muscle and the balance in other tissues. Deficiency of Mg leads to neuromuscular and gastrointestinal malfunctioning.

### 4.2.4 Sodium

The human body contains about 1.3 g of sodium and 90% of which is in extracellular space. It is freely absorbed along with $K^+$ and $Cl^-$ as an energy source for variety of solutes. Its deficiency leads to muscle cramps, headache, poor appetite and dehydration showing the signs of fatigue.

### 4.2.5 Potassium

It is present in the human body in the range of 110–150 g. In contrast to sodium, $K^+$ is found primarily (98%) in the intracellular compartments. Its deficiency causes skeleton muscular weakness, nausea, vomiting and constipation besides impanelling renal function.

### 4.2.6 Chloride

Like sodium, it is also freely absorbed in the body. It maintains balance between the amount of fluid inside and outside of body cells. It helps to maintain proper blood volume and blood pressure and pH of the body fluids.

### 4.2.7 Iron

It is present in the human body approx. 5 g and is responsible for transporting oxygen and electrons. It acts as a reducing agent in a large number of enzymatic reactions. The loss of iron in the body is very low about 0.6 mg/day. Its deficiency leads to anaemia, fatigue and reduced physical performance. In pregnancy, iron deficiency leads to increased risk of premature delivery, low birth weight and infant and maternal mortality.

### 4.2.8 Zinc

The human body contains 2–2.5 g of zinc, 55% of which is in muscle, 30% in bone and balance distributed in other tissues. Zinc deficiency is characterized by poor growth, dwarfism, diarrhoea, osteoporosis and impanelled immune function.

### 4.2.9 Copper

Human body contains approx. 120 mg of copper widely distributed in tissues and fluids. Its deficiency may cause increased susceptibility to infection and poor growth. The toxicity of copper causes liver cirrhosis.

### 4.2.10 Iodine

The human body contains approx. 5 mg of iodine, which functions only in thyroid hormone. Iodine deficiency in adults results in thyroid hypertrophy or goitre and in children as cretinism. These deficiency diseases comprise a global health problem.

### 4.2.11 Selenium

The human body contains about 20 mg of selenium widely distributed in all the tissues. As an essential compound of selenocysteine proteins, selenium functions in antioxidant protection and thyroid hormone regulation. The effects of deficiency of selenium (alone) in the human body are not well defined.

### 4.2.12 Manganese

Mn is widely distributed in the human body in non-calcified tissues with the greatest concentration in liver. There is little evidence of Mn deficiency in humans.

### 4.2.13 Molybdenum

Mo is widely distributed in the body with the greatest concentration in liver, kidney, adrenal gland and bone. Like Mn, the Mo deficiency evidences on humans are also not well known.

### 4.2.14 Chromium

Cr enhances the action of insulin and has been shown to restore glucose tolerance. The absorption of Cr by the body is very low, and it accumulates in liver, kidney, spleen and bone.

### 4.2.15 Fluoride

The ionic form of fluoride accumulates in the calcified tissues of human body because of its high affinity with calcium. Its deficiency may cause dental cavities while its excess can result in fluorosis of enamel and bone.

From the above, it is well illustrated that deficiency and toxicity of certain elements/minerals may occur due to environmental influences. The deficiency/toxicity of certain minerals in soils has resulted in disease in population of that area. The essential and toxic level of element/metals/minerals needs to be in balance to prevent adverse health effects to the organism.

### 4.2.16 Iodine Deficiency

Iodine has long been known as an essential element for humans, and mammals in general, where it is concentrated in thyroid gland. Deprivation of iodine results in a series of Iodine Deficiency Disorder (IDD), the most common of which is endemic goitre, a condition in which the thyroid gland becomes enlarged in an attempt to be more efficient. Iodine deficiency during foetal development and in the first year of life can result in cretinism, a disease that causes stunted growth and general development along with brain damage.

The lithosphere is generally depleted in iodine and by the process of weathering of rocks gets into the soil. The oceans represent the largest reservoir of iodine on earth, and virtually all iodine in the terrestrial environment is derived from oceans by way of the atmosphere. The volatalized iodine from the oceans is deposited on land by wet and dry precipitation and consequently to the roots.

It has been observed that content of iodine is comparatively higher in the rocks of marine origin, low lands and areas lying close to the sea. The iodine content is less in metamorphic, calcareous and granitic rocks. Concentration of iodine and its enrichment in soils at lower level is observed because of addition of material brought from higher slopes. Iodine is also contributed by rains, water logging and biological activities within the soil. There is a relationship between iodine content and texture of the soil. The clayey soils are richer in the element (1–9 ppm) than the sandy soils (trace to 1 ppm). Sandy soils being porous and permeable allow the water to leach away or percolate down, whereas the clayey soils being finer textured have more retentive capacity to hold the water bodies. In the humus soils, it is found to be 25 ppm. Human being gets iodine from air, terrestrial plants, land animals, fresh water, marine animals and marine plants. Drinking water and eating food usually supply 0.05–1.0 mg of iodine, estimated to make the normal daily requirement. Goitre—a condition arising due to deficiency of iodine whenever the deficiency affects more than 9% to 10% of the total population in the area, it is known as endemic goitre. One Indian endemic goitre belt stretches across the entire sub-Himalaya (300–1000 mts) and lesser Himalaya (1000–3000 mts) region owing to various geological, geomorphological and geochemical factors.

The goitre endemicity is more pronounced in the lesser Himalaya (1000–3000 mt altitude) where the rocks like dolomite, limestone, granites and crystallines having less iodine or are completely devoid of it are exposed. The rocks of Siwalik Group, mainly of sand and alluvial plain in Terai region, are covered with sandy soils and attributes to the Goitre endemicity in the inhabitants of flood plain areas. Soil from active flood plain areas and newer alluvium contains iodine traces because of sandy nature while soil from older alluvium areas contains iodine ranging from 6 to 8 ppm because of its clayey nature, which has a retention property. Higher concentration of iodine is found in the water obtained from the deeper aquifers like tube wells whereas water obtained from the dug wells contains only traces of iodine. Changing the soil texture from sandy to clayey and adding seaweeds and peat would help to retain more iodine in the soil, thereby ameliorating iodine deficiency (Prasad and Srivastava 2004).

### 4.2.17 Fluoride Deficiency

Fluoride is an essential element in the human diet. Deficiency in fluoride has been long linked to the incidence of dental caries while excessive ingestion of fluoride has been linked to the development of dental fluorosis and skeletal fluorosis (Edmunds and Smedley 2005).

Drinking water is particularly sensitive in this respect because large variations in fluoride concentration exist in water supplies in different areas ranging from 0.1 to 10 $mgl^{-1}$. The value around 1 mg $l^{-1}$ in drinking water is required as safe. Prolonged use of drinking water with concentration above 1.5 mg $l^{-1}$ is considered to be detrimental to health leading to dental and skeletal fluorosis. The range of fluoride intake as per WHO standard is 0.5–1.5 mg $l^{-1}$. The deficiency/excess intake of fluoride results in Fluoride Endemicity.

About 96% of the fluoride in the body is found in bones and teeth. Fluoride is beneficial between 0.8 and 1.0 mg $l^{-1}$ for calcification of dental enamel of children within 8 years of age. High fluoride concentrations are most often associated with ground water resources, which accumulate fluoride for rock dissolution. The geological linkage of fluoride being released in aqueous system is as vague as its distribution or its point of concentrations. The various studies carried out so far identified major geological linkage grouped into three broad classes, viz.

1. Fluoride enrichment in the drainage basin for source rocks like fluoride-bearing granite, bantonites and some alkaline rocks and phosphatic fertilizers.
2. The quaternary sediments with volcanic ash containing high fluoride are geochemically reactive with Na replacing Ca in aqueous system resulting in enriched pockets.
3. Use of fertilizers in agriculture field.

Thus hydrogeochemical aspects of fluoride in water play an important role in the human population inhabiting that particular area (Sinha Roy 2004).

Chemistry of hot springs provides reliable clues to the reactions responsible for chemical evaluation of thermal water and affinities between their acquired major cations and anions, i.e. $Na^+$ and $HCO_3$—respectively. The affinity between Na and F, as indicated by the chemistry of hot springs in selected areas, has been found to exist under normal conditions and helps in identifying areas prone to fluorosis endemicity.

It is a common character in fluorosis-affected areas, particularly in Indo-gangetic plains where deeper aquifers tapped by tube wells have more or less normal fluoride content whereas dug wells and normal hand pumps discharge high fluoride water. It is obvious that fluoride enrichment in shallow ground water is due to addition from the surface rather than transport of fluoride ions from deeper levels. It is quite likely that a closed system of ground water circulation gets established in these areas. Alkali soil formed during dry season is washed and taken into solutions during the monsoon period. These solutions charged with Na and its associated ions (F and $HCO_3$) infiltrate downwards affecting the chemical quality of ground water at least in top 30–40 m. The following dry season again promotes capillary action and dispersal of sodium salts on the surface.

Typically, only a small amount of the element present naturally in soils (e.g. $<10$ $mgkg^{-1}$) is easily soluble while the remaining reside within a variety of minerals. Of the anthropogenic inputs to soils, high fluoride concentration is found in phosphate fertilizers. The amount of fluoride absorbed by the soils varies with soil type, soil pH, salinity and fluoride concentration condition, and uptake by acid soils can be up to 10 times that of alkaline soils. Fine-grained soils also generally retain fluoride better than sandy types.

### 4.2.18 Fluoride Endemicity in Unnao District, Uttar Pradesh, India

Fluoride endemicity has been recorded mainly in the waterlogged and 'Reh' infested areas in rural parts of Unnao District, Uttar Pradesh, India. Concentration of fluoride in drinking water responsible for the large incidence of dental and skeletal fluorosis in the area is evidently controlled by a number of factors like geology, geomorphology, geohydrology and chemical composition of water and dietary exposures. The region is occupied by the older alluvium covering waterlogged area 'Reh' infested areas, which is a part of low gradient zone. Reh consisting predominantly of $Na_2CO_3$ with $Na_2SO_4$.NaCl gets accumulated at the ground surface during the summer season due to capillary action and thereby due to physico-chemical changes brought about by seasonal and hydrological factors result in concentrating $F^-$ in ground water. Fluoride concentration bears an inverse relationship to total hardness and calcium hardness (Faruqui 2004).

### 4.2.19 Fluoride Endemicity in Rajasthan, India

Rajasthan, India, is the most seriously affected state by high fluoride where the climate is mostly arid with low and variable annual rainfall. Fluoride concentration in the ground water of many parts of Rajasthan has increased to hazardous proportions ($>2$ to $>12$ ppm) in the confined to unconfined aquifers of unconsolidated sediments (Maithani et al. 1998).

### 4.2.20 Arsenic Toxicity

In environment As may be mobilized through contaminating natural processes such as weathering and erosion, biological activity and volcanic emissions. Human exposure to As may be through a number of pathways, including air, food, water and soil. The relative impact of these factors depends on the level of circumstances. But of the potential sources of As available, drinking water poses one of the greatest threats to human health. High value of As concentration in ground water is detrimental to human health. The problem has been reported in Bangladesh, India,

Taiwan, Thailand, China, Hungry, Vietnam, Nepal, Myanmar, Mexico, Argentina and Chile.

As has a complex chemistry and forms both organic and inorganic compounds. The toxicity of As confirms to the following order: Arsines > arsenites > arsenic oxides (organic trivalent compound) > arsenates (inorganic pentavalent compound) > arsenic compounds > metallic arsenic. There are two views on release/mobilization of As.

1. Due to oxidation of sulphides present in the aquifer by the atmospheric oxygen that enters the ground water due to lowering of water table.
2. Due to reductive dissolution of ferric oxy-hydroxide that contains absorbed As. There is why presence of high concentration of Fe in the ground water is linked with high concentration of As.

Arsenic is often present in varying concentrations in minerals like arsenian pyrite $[Fe(S,As)_2]$, arsenopyrite (FeAsS), realgar (AsS) and orpiment ($As_2S_3$). As the chemistry of As follows sulphur closely, its greatest concentration tends to occur in sulphide minerals such as pyrites ($FeS_2$) (Smedley and Kinniburgh 2005).

Concentration of As in soils can vary in different terrains. Acid sulphate soils and mangrove swamps can be relatively rich in arsenic.

Additional As inputs to soils may be derived locally from industrial sources and agricultural sources such as pesticides and phosphate fertilizers.

Some anthropogenic sources of As to the soil are fossil fuel combustion and mining activity in the area.

## 4.3 Case Study of Bangladesh and West Bengal

Bangladesh and West Bengal, India, comprising Bengal Delta show characteristic presence of numerous active meander channels and also detached ox-bow channels. Other than the active channels, the rest of soil-covered part of the delta shows curvilinear low-lying areas due to palaeo-channels. These are the results of lateral migration of numerous meander channels distributed throughout the area. Therefore the areas where active channels or palaeo-channels migrate laterally, the high As value in the ground water of those areas has been observed. It can be inferred that high As value in ground water of Bengal delta is the effect of meander channel migration or active and/or palaeo-channel migration. The relationship between the As distribution pattern iñ ground water and As toxicity in human health does not have simple correlation. This is due to the reason that the measured As value in ground water is a total of $As^{3+}$, $As^{5+}$ and particulate As, whereas toxicity depends on the value of $As^{3+}$ (toxic form of As) (Acharyya et al. 1999).

Long-term exposure to As by ingestion of foodstuff and drinking water can cause cancer in the skin, lungs, bladder and kidney. It can also cause other skin changes such as thickening and pigmentation. Increased health issues are reported from the areas where people ingest As at concentration of $>50\ \mu g l^{-1}$ (WHO guidelines for permissible limit of As in ground water is 10–50 μg $l^{-1}$). Soluble As can have immediate toxic effects. Ingestion of large amounts through foodstuff obtained from As-rich soils can lead to gastrointestinal problems and damage to circulating and nervous system.

## 4.4 Selenium Deficiency and Toxicity

Selenium (Se) is essential to human health in trace amounts, which has one of the narrowest range between dietary deficiency (<40 μg $l^{-1}$) and toxic levels (>400 μg $l^{-1}$).

Geology exerts a fundamental control on the concentration of Se in the soils. However, health outcomes are dependent not only on the total Se contents of rocks and soils but also on the amount of Se taken up into the plants, i.e. the bioavailable selenium. Even soils containing adequate amounts of selenium can still produce selenium-deficient crops if the selenium is not in the form ready for plant uptake. The bioavailability of Se to the plant is determined by a number of biophysio-chemical parameters.

These include prevailing pH and redox condition, the chemical form, soil texture, mineralogy and organic matter content. Selenium is much more bioavailable under oxidizing alkaline condition and much less in reducing acid condition. Further, Se is less available to the plants grown on clayey soils because it tends to stick on fine particles of clay. The capacity of clays and iron oxides to adsorb Se is strongly influenced by pH conditions. Soil organic matter also has a large capacity to remove Se from the soil solution as a result of organometallic complexes. The reason why some plants are better at accumulating Se than others depends upon selenium metabolism (Fordyce 2005).

Food crops tend to have low tolerance to Se toxicity, and most crops have the potential to accumulate the Se to the toxic level for humans. In general root crops contain high Se concentrations than other crops, and plant leaves often contain higher concentration than the tuber.

Selenium deficiency has been known to be the cause of heart disorder (Keshan disease) and bone and joint condition (Kashin-Beek disease) in humans in various parts of China. The deficiency also affects thyroid metabolism. The role of Se in many other diseases like cancer (Se acting as antioxidant) and muscular dystrophy etc. are still debatable and researches are still on.

Selenium toxicity to humans depends on the chemical form, concentration and on a number of compounding factors. Therefore, Se toxicity is not that well known in human as Se deficiency as the human population can adapt or tolerate high Se intakes without major clinical symptoms. There are certain cases of Se toxicity where excess intake causes loss of hair and fingernails, numbness in fingers or toes and circulatory problems.

## 4.5 Mercury Toxicity

Hg is known to be responsible for a reduction of micro-biological activity vital to the treatment of food chain in the soils. To prevent ecological effect due to Hg in organic soils, the range has been set as 0.7–0.3 $mgkg^{-1}$ for the total Hg content in the soil. Hg is naturally emitted from volcanic eruption, weathering of Hg-bearing rocks and as vapours from oceans and forest fires. It is then dispersed through air and water and settles on ground directly or after absorption in rain and snow. The deposited Hg is again re-emitted naturally from land and water bodies. In general river waters, coastal waters and ground water comprise the sink of Hg of which the ground water sinks are more prevalent and are of greater threat to local population. The coastal seawater especially of Arabian Sea is a major sink of Hg (Chakrabarti 2004).

Hg enters human body through inhalation of Hg vapours, ingestion of Hg-contaminated drinking water and food, especially certain types of fish caught from Hg-contaminated aquatic environments. Such fish from both marine and fresh water contain extremely high concentration of methyl mercury, which is 50 to 100 times more toxic than any other form of Hg. Hg is a potent neuro-toxic, and even low levels of exposure can permanently damage human nervous system. Higher levels of exposure consequently damage lung and kidney. Methyl mercury ingested as food through fish is a poison for nervous system, may have unborn baby in pregnancy and can cause cancer (Kaladharan and Pillai 1999).

## 4.6 Radiation Hazards

Radiations caused by natural process are of very low level. However, its biological effects are observed when ionizing radiations interact with living tissues by transferring energy to molecules of cellular matter. Radioactivity is contributed in natural material due to Uranium (U), Thorium (Th) and Potassium ($K^{40}$) present naturally in the geological material. The anomalous concentration of radioactive material at certain places may give rise to radiation enhancement. The decaying radio nucleotides attach to available surfaces and form radioactive aerosols, which are major pollutants of the soil. The decaying $^{222}Rn$ gets deposited on the plant surfaces and ingested by human being by consumption of foodstuff and drinking water.

In natural surroundings certain radioactive substances occur in abundance far greater than the natural level of concentration for e.g. vast stretches of coastal Brazil and India where the beach sands are enriched in placer deposits of 'black sand', which contain radio-nucleotides that have long-lasting ill effects on human health. Damage to the human body depends on the dose of radiation to which he or she is exposed through different pathways, which range from direct exposure to soil complex. Exposure to small areas of the body leads to localized tissue damage (necrosis) and gangrene. Symptoms of absorption of radionucloides by the human body manifest rather delayed and may cause severe vascular damage, neurological disturbances, cerebral oedema, lung cancer and haemorrhage (Saxena 2004).

## 4.7 Urban Waste Toxicity

Treatment of agricultural soil with urban waste has long been a practice of disposal of solid and liquid refuse in many cities. Occurrence of trace elements like lead, cadmium, chromium etc., which are highly toxic in native have been reported in the waste as well as in the treated soil. Intake of these elements through food chain even in small dose for a long period has deleterious effect on health as many of the waste-fed areas cater to the need of supply of vegetables for the adjoining urban belts.

Due to long dumping of waste materials, soil organic matter forms an important part of the total soil system. It consists of very complex mixture of elemental compounds, from easily degradable materials such as carbohydrates and proteins to humic substances, which are secondary products very resistant to further degradation and thus form the most stable components of the soil organic matter.

## 4.8 Industrial Waste Toxicity

In many areas soil cover is subjected to application of agro-industrial waste. These solid/liquid wastes interact with soil and bring out series of changes in textural properties and micronutrient levels. Increase in phosphorus level in finer fraction has been observed in effluent irrigated plots. Organic carbon, copper and zinc and some other elements show a change and fixation of nutrients in effluents treated land. These elemental changes percolate directly by process of bio-magnification in the agriculture procedure and inflict health hazards to the consumers of the agriculture produce. The areas where the permeability of soil is more, easy inflow of effluents in the soil profile is more likely to change the soil chemistry, thereby ultimately affecting the soil productivity of the area.

## 4.9 Conclusion

Air, water and soil play a vital role for life on this planet of the naturally occurring element found on earth. Many elements are useful to plants, animals and human health in small dose as ppm/ppb level, and these elements enter the human body via water and soil system (our dietary source). The inability of the environment to provide the correct mineral balance can lead to serious health problems. The presence of certain elements in excess of prescribed limit may constitute severe health problems. Whereas the deficiency of certain elements in the soil system, which is the dietary source of human population, can lead to many deformities.

In general the trace elements are important in environmental health because they perform several vital functions in organisms. Exposure of toxic level of certain elements is one of the major problems in human health. The toxicity of metal in an environment in a system is dependent on certain parameters such as its chemical form, molecular, atomic and ionic species and ion exchange capacity in the soil system. Thus the detailed research in collaboration with medical practitioners, soil scientists, geologists, agronomists and scientists of related field may play an important role in understanding and ameliorating human health hazards.

## References

Acharyya SK, Chakrabarti P, Lahiri S, Raymahashay BC, Guha S, Bhowmik A (1999) Arsenic poisoning in the Ganga delta. Nature 401:545

Chakrabarti P (2004) Arsenic and mercury pollution - a geomedical perspective in the Indian context. Proceeding of the Workshop on Medical Geology, IGCP-454, GSI, Nagpur, India, Special Publication No 83, pp 20–28

Edmunds M, Smedley P (2005) Fluoride in natural waters. In: Essentials of medical geology, vol 12. Springer, Dordrecht, pp 301–329

Faruqui NH (2004) Endemic fluorosis in Unnao, Kanpur Dehat and Kanpur Nagar Districts, U.P.: Some geological constraints. Proceeding of the Workshop on Medical Geology, IGCP-454, GSI, Special Publication No 83, Nagpur, India, pp 102–103

Fordyce F (2005) Selenium deficiency and toxicity in the environment. In: Essentials of medical geology, vol 15. Springer, Dordrecht, pp 353–415

Kaladharan P, Pillai VK (1999) Mercury in seawater along the western coast of India. Indian J Mar Sci 28:338–340

Maithani PB, Gurjar R, Rahul B, Balaji BK, Ramachandran S, Singh R (1998) Anomalous fluoride in groundwater from western part of Sirohi district, Rajasthan and its crippling effects on human health. Curr Sci 74(9):773–777

Prasad M and Srivastava VC (2004) Goitre endemicity in Deoria and Gonda districts of Uttar Pradesh and its geological linkage. Proceedings of the workshop on medical geology, I. G.C.P.-454, G.S.I. Special Publication No 83, Nagpur, India, pp 139–143

Saxena VP (2004) Radioactive minerals, radiations and health hazards. Proceedings of the workshop on medical geology, I. G.C.P.-454, G.S.I. Special Publication No 83, Nagpur, India, pp 46–58

Sinha Roy S (2004) Geological and hydrogeochemical controls of fluoride distribution in the ground waters of parts of Rajasthan. Proceedings of the Workshop on Medical Geology, I. G. C.P.-454, G.S.I. Special Publication No 83, Nagpur, India, pp 79–95

Smedley P, Kinniburgh DG (2005) Arsenic in ground water and the environment. Essent Med Geol 11:263–299

# Part II

# Elemental Nutrition of Crop Plants

# An Overview of Nitrogen, Phosphorus and Potassium: Key Players of Nutrition Process in Plants

5

Dwaipayan Sinha and Pramod Kumar Tandon

**Abstract**

Elements play an important role in the physiology and overall growth of the plant. Depending upon the amount required by the plants for their growth, they can be broadly divided into macro- and microelement. Plants generally absorb these elements through the root system from the soil. Humans have been dependent on plants since the advent of civilization for food and medicine. With the increase in population there has been an increase in demand for the food and which resulted in the intensification of agriculture. Needless to mention, this often results in scarcity of available nutrients from the soil, thereby stressing the necessity adding excess nutrients from outside. Thus the relevance of fertilization comes into the picture and its importance has been gradually felt by the scientists since the last 200 years. Various types of fertilizers containing essential elements are now being added to agricultural lands for betterment of yield. This chapter is an attempt to highlight the various aspects of three essential macroelements required by the plants, namely nitrogen (N), phosphorus (P) and potassium (K). The chapter deals with the requirement of these three elements from the agronomic point of view and the present status of the fertilization process involving the mentioned elements.

**Keywords**

Nitrogen · Phosphorus · Potassium · Fertilizer · Agriculture

D. Sinha (✉)
Department of Botany, Government General Degree College, Paschim Medinipur, West Bengal, India

P. K. Tandon
Department of Botany, University of Lucknow, Lucknow, India



K. Mishra et al. (eds.), *Sustainable Solutions for Elemental Deficiency and Excess in Crop Plants*, https://doi.org/10.1007/978-981-15-8636-1_5

## 5.1 Introduction

Elements play a pivotal role in the overall physiology of the plant. The plant acquires elements from the soil largely through their root system which are then transported and translocated in their desired destination inside the plant body (Paez-Garcia et al. 2015). This uptake of elements from the soil is facilitated by various transporters present in the cells of root and root hairs of the plants (Nussaume et al. 2011; Kimura et al. 2019). Once inside the plant body the elements play their individual role all of which are related to overall growth and productivity of the plant. Plants require 17 elements for their growth and depending upon the amount of requirement they can be classified into macro- and micronutrients. The macronutrients of the plants are those which are present at greater than 1000 mg per kg of the dry weight of the plant. These elements include carbon, hydrogen, oxygen, calcium, potassium, magnesium, nitrogen, sulphur and phosphorus out of which carbon, hydrogen and oxygen constitute roughly 95% of plant dry matter. Other elements which are present at less than 100 mg per kg of dry weight are called micronutrient. They include chlorine, boron, copper, iron, manganese, molybdenum, nickel and zinc (Pilon-Smits et al. 2009). The mineral nutrient elements play some vital role in plant and may be broadly classified into (1) constituent of plant cell wall, (2) aids in osmotic relation and maintenance of turgor pressure of the cell, (3) process of energy transfer, (4) participation in enzyme catalysed reactions, (5) participation in reproduction (Pandey 2018). This chapter is an attempt to overview the different aspects of three essential marcoelements, namely nitrogen (N), phosphorus (P) and potassium (K) and their relationship with the plants.

## 5.2 Elemental Nutrition in Plants-Historical Aspects

There has been a long history of the nutritional aspect of plants and constant investigation by humans on the nutritional processes. The practices of plant nutrition are mentioned in the Odyssey (eighth century BC) by Homer where the use of manure, compost, straw, animal residues, river and pond silt, green manure, ash, bones, marl, lime and gypsum has been mentioned as agents of soil fertilization (Antonokiewicz and Labetowicz 2016). Marcus Porcius Cato (234-149 BCE) also mentioned strategies of soil fertilization through composting in his book "De AgriCultura", while CaisPliniusSecundus (CE 23-79) described the process of manuring and recommended the use of green manures (Blakemore 2018). However efforts to decipher the secret of plant nutrition gained pace from the late medieval age onwards. It was Flemish physician and chemist Van Helmont (1577–1644) who had concluded that water provided plants the elements required for growth. Further extension of the concepts was provided by Claude Perrault (1613–1688) towards the end of seventeenth century. He proposed that the roots were the functional organs responsible for sequestering "juices" from the earth for their nutritive purpose. The eighteenth century proved to be a breakthrough in the field of plant nutrition. At that time a couple of theories were proposed by two German scientists namely "The humus theory for plant nutrition" by Albrecht Daniel von Thaer (1752–1828) and the "Theory of mineral plant nutrition" by Justus von Liebig (1803–1873) (Métioui et al. 2016). The humus theory was presented by Albrecht Daniel von Thaer in his work "The Principles of Agriculture". He proposed that the plants mainly draw their nutrients from the humus or the soil organic matter. It was further postulated that the driest matter of the plants is derived from soil nutritive juices containing fractions of soil organic matter (Feller et al. 2012). "Theory of mineral plant nutrition" was proposed by Justus von Liebig. He proposed the "law of minimum" which directly indicated the relevance of mineral in growth and yield of plants (Van der Pleog et al. 1999). Throughout the nineteenth century and the early twentieth century experienced the import of guano and nitrate fertilizers from South America and this was considered to be one of the driving force for the growth of fertilizer industry (Clark and Foster 2009; Espie and Ridgway 2020). One of the greatest achievement in the early twentieth century is the discovery of industrial production of ammonia from atmospheric nitrogen which was the first basic product of the fertilizer industry (Galloway et al. 2013). These discoveries eventually paved the path of chemicalization and stressed the need for chemical fertilizers which can bring about fertility to the soil and increase crop productivity.

## 5.3 Nitrogen as an Essential Macronutrient Source

Nitrogen is the main component of earth's atmosphere (Luo et al. 2018). Existing as a diatomic molecule having one of the strongest known triple bonds which results in its unreactivity in normal atmospheric condition (Howie et al. 2016). Thus only a few organisms can pick up molecular nitrogen due to its stability (Galembeck and Dos Santos 2019). The formation of nitrogen compounds in the atmosphere occurs spontaneously during lightning when nitrogen reacts with oxygen to form various oxides. These nitrogen oxides (NOx) react with the moisture in the atmosphere to form various acids of nitrogen and come down on the earth crust with rains. The acids rapidly react with minerals present in the earth crust to form various nitrates and nitrite salts and thus get integrated into the soil (Wong et al. 2017). These contribute to overall nitrogen input in the earth crust. In addition to it, nitrogen is also fixed from the atmosphere by microbes in the form of ammonium ion which is then sequentially oxidized into nitrate. This oxidized nitrogen forms an important source of nitrogen for eukaryotic primary producers (Zerkle and Mikhail 2017). The microbiota of guts of animals are also responsible for adding up nitrogen and for this reason animal faeces have been regarding as an important source of nitrogen and fertilizing agent (Aiysha and Latif 2019). An important source of nitrate is the saltpeter (Clements et al. 2014) which played crucial role during world war in Europe and triggered the production of soluble nitrate compounds that can act as fertilizers (DiNicolantonio and O'Keefe 2017). The most important source of nitrogenous compounds for agriculture is the Haber–Bosch process of ammonia production (Vicente and Dean 2017). Ammonia is used as an ingredient of fertilizer to promote growth of agricultural crops (Wendeborn 2020).

Nitrogen is absorbed by plants in a combination of two forms, namely the nitrate form ($NO_3^-$) and ammoniacal form ($NH_4^+$) (Abbasi et al. 2017). Most of the crops prefer nitrogen in the form of nitrate (Liu et al. 2014; Hu et al. 2014). However paddy and few other higher plants prefer nitrogen in the form of ammonia (Duan et al. 2007; Yang et al. 2017a). Most of the fertilizers contain nitrogen either in ammoniacal form or in nitrate form. The urea fertilizers contain an amide which is swiftly converted by soil microorganisms into ammoniacal form and finally into nitrate (Staley et al. 2018). Based on the forms of nitrogen, nitrogenous fertilizers may be classified into: (a) nitrate fertilizers, (b) ammoniacal fertilizers, (c) ammoniacal-nitrate fertilizers and (d) amide fertilizers. The various forms of nitrogen fertilizers are tabulated in Table 5.1.

So far as global production is concerned, the USA tops the list of the countries producing ammonium sulphate in 2017 producing 2.96 million tonnes which accounted for 25.71% of the world's ammonium sulphate production. Other countries including Russia, Japan, Canada, Indonesia account for 60.93% of the total production. World's total ammonium sulphate production in 2017 was estimated to be 11.5 million tonnes (Knoema website® 2017a). The production of ammonium sulphate by the top eight countries is depicted in Fig. 5.1. Russia is the leading country in terms of ammonium nitrate production with a production of 9.86 million tonnes in the year 2017 accounting for 45.46% of the world's ammonium nitrate production. The total ammonium nitrate production in the world was estimated to be 21.6 million tonnes in 2017 (Knoema website® 2017b). Figure 5.2 describes the quantity of ammonium nitrate produced by the top eight countries in the world. The USA is the leading producer of monoammonium phosphate. In the year 2017, the total production of monoammonium phosphate by the USA was 5.18 million tonnes which accounted for 69.38% of the world's total production, while the total production of the world was 7.47 million tonnes (Knoema website® 2017c). Production of monoammonium phosphate by the top eight countries is depicted in Fig. 5.3.

**Table 5.1** Various forms of nitrogen fertilizers

| Fertilizer form | Compound | Formula | Percentage of nitrogen |
|---|---|---|---|
| Ammoniacal form | Ammonium sulphate | $(HH_4)_2SO_4$ | 20.5 |
| | Ammonium chloride | $NH_4Cl$ | 26 |
| | Monoammonium phosphate | $(HH_4)H_2PO_4$ | 11–20 |
| | Diammonium phosphate | $(NH_4)_2HPO_4$ | 18 |
| | Anhydrous ammonia | $NH_3$ | 82 |
| | Aqua ammonia | $NH_3+H_2O$ | 20–24.6 |
| Nitrate form | Calcium nitrate | $Ca(NO_3)_2$ | 15.5 |
| | Sodium nitrate | $NaNO_3$ | 16 |
| | Potassium nitrate | $KNO_3$ | 13 |
| Ammoniacal nitrate forms | Ammonium nitrate | $NH_4NO_3$ | 33.5 |
| | Cal Nitro (ammonium nitrate + limestone) | $NH_4NO_3$ + Ca $(CO_3)_2$ | 26 |
| | Ammonium sulphate nitrate (ASN) | $((NH_4)_3(NO_3)(SO_4))$ | 26 |
| Amide forms | Urea | $CO(NH_2)_2$ | 45 |
| | Calcium cyanamide | $CaCN_2$ | 19.8 |

Adapted from: Kumar et al. (2013), Agropedia; Mengel (2020), Nitrogenous Fertilizers and Agronomy guide, Purdue University Cooperative Extension Service

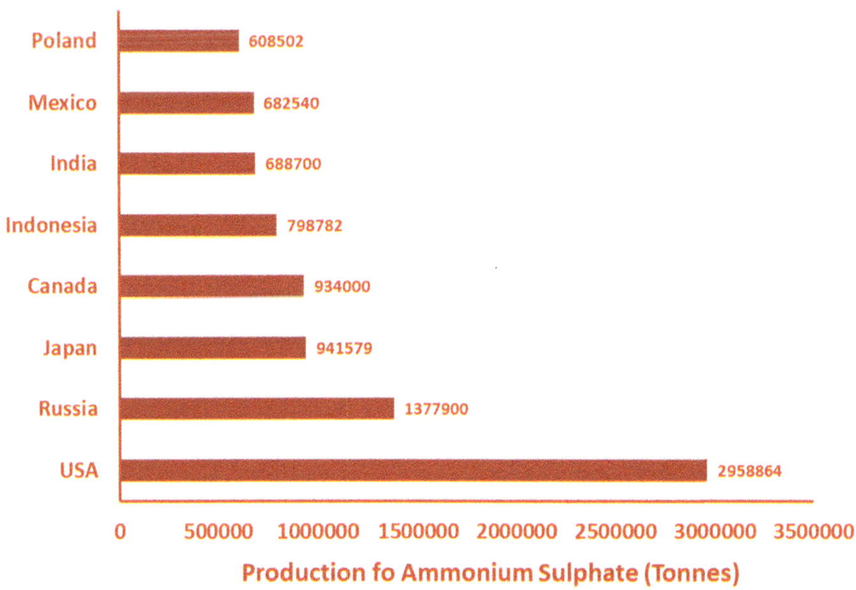


**Fig. 5.1** Production of ammonium sulphate by top 8 countries in the year 2018 (Adapted from: Knoema® website 2017a)

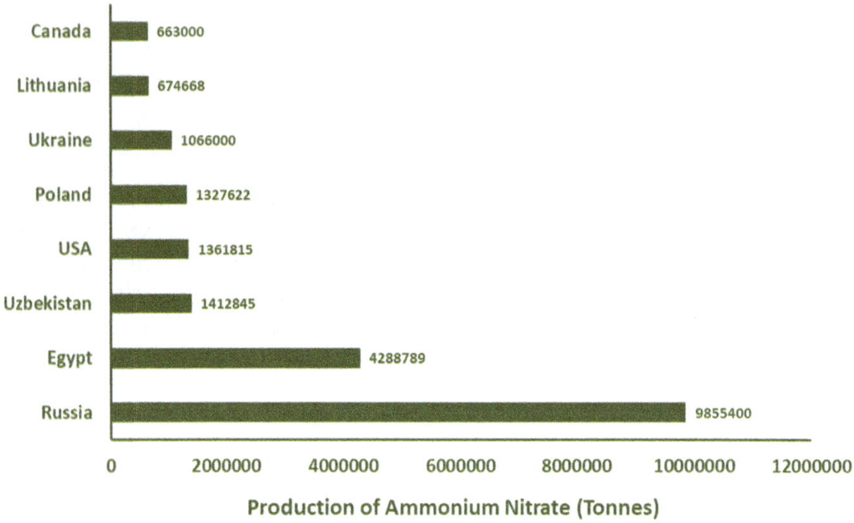


**Fig. 5.2** Production of ammonium nitrate by top 8 countries in the year 2018 (Adapted from: Knoema® website 2017b)

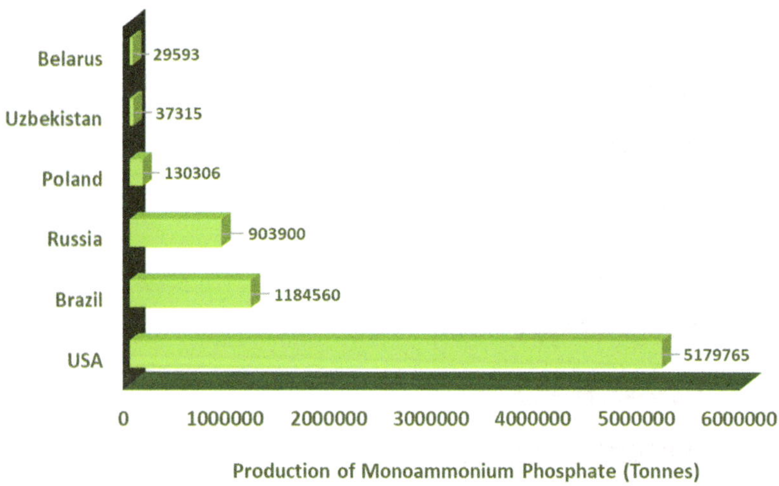


**Fig. 5.3** Production of monoammonium phosphate by top 6 countries in the year 2018 (Adapted from: Knoema® website 2017c)

### 5.3.1 Nitrogen Requirement by Agricultural Crops

Nitrogen is the most abundant mineral required by the plant and acts as an important determinant of plant growth (Prinsi and Espen 2015). The element is the key component of cellular biomolecules such as nucleic acids, proteins, chlorophyll and plant growth regulators (Nguyen et al. 2015). Nitrate is absorbed from the soil by the plants mainly in the form of nitrate ($NO_3^-$) or ammonium ($NH_4^+$) by the root system of the plant (Hu et al. 2014). However the nitrogen supply in the soil often becomes limited prompting the farmers to use nitrogenous fertilizers to adjust the deficiency (Muñoz-Huerta et al. 2013). All plants require a balanced amount of nitrogen for their optimum growth and development (Gastal and Lemaire 2002). This requirement often varies in different plants and is often supplemented by the use of nitrogenous fertilizers if required. In addition to it, symbiotic microbes are also capable of fixing nitrogen in the nodules of leguminous plants, thereby incorporating nitrogen into the biological system (Sulieman and Tran 2014; Mus et al. 2016). In this section, the requirement of nitrogen by selected crops will be discussed.

Reports from China indicate that the rice plant requires 21.10 kg of nitrogen to yield one ton of rice. Moreover, it has been reported in Huai River basin, the consumption of nitrogen by dryland rice varied from 60 to 100 kg per hectare for a yield of 3.2 to 4.1 tonnes per hectare. In Southern China, a yield of 7.5 tonnes per hectare of irrigated rice was achieved by the addition of nitrogen at rates of 60–120 kg per hectare (Che et al. 2016). It is also reported that for irrigated rice in Sahelian West Africa, the internal efficiency ranges from 48 and 112 kg grain per kg of nitrogen (Haefele et al. 2003). In another study it is reported that basmati rice of India requires 40 kg of nitrogen per hectare of cultivable (Aulakh et al. 2016). Another report states that Pakistan achieved maximum yield of super basmati of 4.2 ton per hectare of land through the application of 157 kg per hectare of nitrogen (Manzoor et al. 2006). Report from Myanmar indicates that input of nitrogenous nutrient at rate of 5 and 36 kg per hectare resulted in a production of 1.2–2.3 tonnes of rice in the rainfed lowland area, while application of 76–110 kg nitrogenous fertilizer per hectare resulted in an average production of 2.8 to 3.5 tons (Matsuda 2011). Another study from Bangladesh indicated that application of nitrogen at 60 kg per hectare resulted in the highest yield of grain (5.36 tonnes per hectare) of rice variety BUdhan1 (Haque and Haque 2016). Another study revealed that application of nitrogen at 60 kg per hectare resulted in highest panicle length, filled grain per panicle and gain yield in Morichsail variety of rice (Jahan et al. 2014).

Wheat is another important food crop extensively cultivated in India. It has been reported in a study from Haryana, India that the recommended dose of nitrogen for optimum yield and chapatti quality of wheat is 130 kg per hectare with an equal application of 50 kg per hectare at seedling and early tillering stage and the lesser rate at the first node stage (Coventery et al. 2011). In another study done in Nadia district of West Bengal, It was reported that to obtain 1 ton of grain, the nitrogen requirement varies from 8.3 to 29.6 kg (Maiti et al. 2006). Another report from Punjab province in Pakistan states that application of nitrogen at 120 kg per hectare increased the yield of wheat up to 5.12 tonnes per hectare (Majeed et al. 2015). A study from China reported the effect of irrigation and nitrogen application on Yumai 49–198, a winter wheat cultivar from Huanghai area of China. It was observed that the highest grain yield was observed with nitrogen application at 300 kg per hectare. This was accompanied by a significant increase in grain protein and the total essential and non-essential amino acid content (Zhang et al. 2017). Another study reported the use of nitrogen fertilizer in the management of dryland wheat in the USA. In the study, using hard red spring wheat (cv. Choteau), it was observed that application of 90 kg of nitrogen per hectare of land significantly increased grain yield, protein content and nitrogen uptake (Walsh et al. 2018).

Next to wheat, maize is another cereal that is popularized in a large number of regions across the globe. Several studies have been done to evaluate the nitrogen requirement of maize. A report from a field experiment from Peshawar valley of Pakistan states that application of nitrogen at rate of 200 kg per hectare to hybrid maize was found to be most beneficial in terms of productivity and cost effectivity of the farmer. Application of nitrogen at 150 kg per hectare was most beneficial for the local varieties of maize (Amanullah et al. 2016a). In another study from China, using spring maize (Zhengdan 958), it was found that application of 304 kg per hectare of nitrogen leads to maximum yield of 10506 kg per hectare (Yin et al. 2014). A study from Ethiopia using hybrid long maturing variety (BH661) of maize, it was found that application of 92 kg of nitrogen per hectare of land resulted in highest yield (Abebe and Feyisa 2017). Another report from Bangladesh states that application of 180 kg of nitrogen per hectare resulted in a higher amount of grain and maximum yield (Tajul et al. 2013).

Soybean is one of the important legume contributing to 25% of global edible oil and two-thirds of world's protein concentrate for feeding livestock (Agarwal et al. 2013). A study from China estimated the nitrogen requirement of Soybean using quantitative evaluation of the fertility of tropical soils (QUEFTS) model. It was found that to produce 1000 g of soybean seeds, 55.4 kg of nitrogen is required for the aboveground parts which corresponded to internal efficiency of 18.1 per kg of nitrogen (Yang et al. 2017b). Another report from China indicates that application of nitrogen at 50 kg per hectare resulted in optimum grain yield, pod number per plant and grain number per plant (Gai et al. 2017). In a recent report from Brazil, it was stated that soybean plants can be inoculated with bacterial strains SEMIA 587 and 5019 (*Bradyrhizobium elkanii*), 5079 (*Bradyrhizobium japonicum*) and 5080 (*Bradyrhizobium diazoefficiens*) for a more efficient nitrogen fixation and improvement in yield (De Souza et al. 2019). There are also reports of microbial inoculation to enhance nitrogen uptake in Soybean plant in Ghana. A study reported that treatment of soybean plants with rhizobial inoculants, namely Legumefix and Biofix as well as 100 kg nitrogen per hectare resulted in increase in shoot dry weight, grain yield and nodule dry weight of soybean plant (Ulzen et al. 2016). Another report from Siaya County of western Kenya also states that inoculation rhizobial inoculant legume fix resulted in an increase in grain yield of up to 4000 kg per hectare (Thuita et al. 2018).

### 5.3.2 Nitrogen Deficiency

Nitrogen input through nitrogenous fertilizer forms the single largest input of the mineral in croplands (Liu et al. 2010). Deficiency of nitrogen results in reduced photosynthesis, reduction in contents of chlorophyll and accessory pigments (Cetner et al. 2017), stunted growth and chlorosis (Sett and Soni 2013). In addition to it, nitrogen deficiency results in reduced cell size (Yanagida et al. 2011), volume and protein content (Ding et al. 2005) and reduces the number and size of chloroplasts (Makino and Ueno 2018; Omondi et al. 2019). In rice, nitrogen deficiency results in paling of old leaves and sometimes all leaves along with chlorosis in the tip. The leaf colouration becomes light green (Chen et al. 2014). In addition to it, rice plants suffering from nitrogen deficiency also results in small leaves and leaf etiolation from the tip (Sun et al. 2018). This is accompanied by dwarfing of plants and lowering of grain yields (Zhang et al. 2015). In wheat, the specific symptoms of nitrogen deficiency symptoms first appear on the oldest leaves with the new leaves appearing comparatively green. Consequently older leaves become paler than the new ones due to chlorosis with the symptoms initially beginning at the tip and then extending down the leaves to the base (Snowball and Robson 1991). A study reported that deficiency of nitrogen resulted in decreased activity of superoxide dismutase, guaiacol peroxidase and catalase as well as a higher concentration of reactive oxygen species in the peduncles (Kong et al. 2013). Another study indicates that nitrogen deficiency results in higher phenolic content and high degree of cross-linking in cell walls of wheat roots which results in thickening of cell walls (Meychik et al. 2017). In maize, deficiency of nitrogen results in stunted growth and reduction in leaf photosynthesis. In addition to it, in the harvest time, the plant height, leaf area and shoot biomass were less than that of the control plants which received nitrogen enriched nutrient solution throughout the study (Zhao et al. 2003). Another study reports that nitrogen deficiency resulted in reduction of grain yields and plant weight in maize (Ding et al. 2005). A study also reports that nitrogen deficiency results in strong metabolic shifts and metabolite profiles. Deprivation of nitrogen resulted in selective downregulation of processes involved in nitrate reduction and amino acid assimilation. Decrease in nitrogen availability also resulted in accumulation of phosphorus along with downregulation of genes usually involved in phosphate starvation response (Schlüter et al. 2012). Another report states that nitrogen deficiency in maize resulted in greater starch concentrations in leaves due to more and larger starch granules in bundle sheath cells (Ning et al. 2018). In soybean, a a study reports that nitrogen deficiency resulted in a drastic decrease in the content of galactolipids, monogalactosyldiacylglycerol and digalactosyldiacylglycerol in leaves (Narasimhan et al. 2013). In another study, it was indicated that nitrogen deficiency resulted in n 8- and 15-fold increases in the secretion of daidzein and genistein, respectively, in soybean (Sugiyama et al. 2016).

## 5.4 Phosphorus as an Essential Macronutrient Source

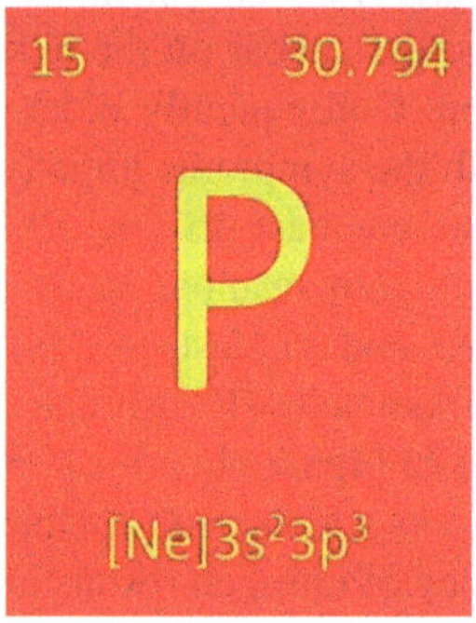


In most of the living systems, phosphorus is represented by phosphates (Razzaque 2011). The main source used by fertilizer company for industrial production of phosphates is hydroxyapatite, a member of the apatite group (Xiong et al. 2018). In nature, phosphorus are generally present in various forms of calcium phosphate minerals and are produced through a wide range of environmental procedures namely geological (igneous apatite), geochemical and/or geomicrobiological (phosphorite) and biological (biological apatite). The igneous apatites nucleate and crystallize from molten, phosphate-rich rock resulting in the formation of luorapatite ($Ca_5F_2[PO_4]_3$), chlorapatite ($Ca_5Cl_2[PO_4]_3$) or hydroxyapatite ($Ca_5[OH]_2[PO_4]_3$) (Omelon et al. 2013). The most commonly used phosphatic fertilizers include diammonium phosphate (DAP), monoammonium phosphate (MAP), NPKs and single super phosphate (SSP) (Indorama Corporation website 2017). As per reports of Knoema®, India is the largest producer diammonium phosphate in the world. As of 2017, the total DAP production by India was 4.65 million tonnes that accounts for 44.90% of the world's DAP production. The total world's DAP production was estimated to be at 10.4 million tonnes in 2017 (Knoema® website 2017d). The amount of DAP production by top 8 countries of the world is depicted in Fig. 5.4.

As per a report from Global Industry Analysts, Inc, amid the COVID-19 crisis and looming economic recession, the worldwide phosphate market will grow by a projected 12.5 million metric tonnes propelled by a CAGR OF 2.4% (Global Industry Analysts, Inc report 2020). China tops the list of mine production of phosphate in the year 2018 with a whopping 140 million metric tonnes (Williams 2019). The top phosphate producers of the world are depicted in Fig. 5.5.

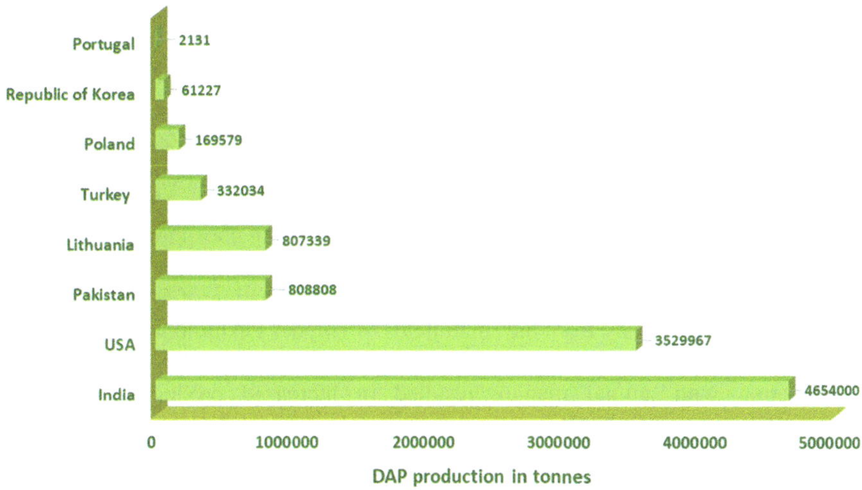


**Fig. 5.4** Quantity of DAP produced by top 8 countries (Adapted from Knoema® website 2017d)

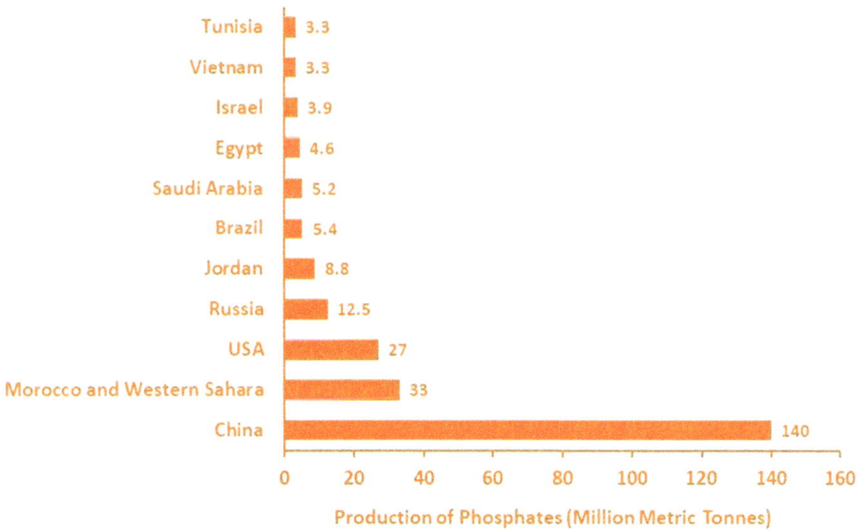


**Fig. 5.5** Production of phosphates by top 10 countries. Tunisia and Vietnam and tied at 3.3 million metric tonnes. (Adapted from: Investing news website 2013)

### 5.4.1 Phosphorus Requirement by Agricultural Crops

Phosphorus is another essential element required for the development and growth of plants and constitutes up to 0.2% of dry weight in plants (Alori et al. 2017). It is one of the essential macronutrient required for the synthesis of nucleic acid, membrane build up and stability, energy metabolism and many other critical physiological and biological processes during plant growth and development (Hasan et al. 2016).

Phosphorus is poorly available to the soil due to its extremely low diffusion rate ($10^{-12}$ to $10^{-15}$ $m^2/s$) (Shen et al. 2011). It has been reported by IRRI that for optimum nutrition, rice plants take up 6.4 kg $P_2O_5$ (2.8 kg P) per ton of grain yield (4.4 kg $P_2O_5$ in grain and 2.0 kg $P_2O_5$ in straw) (Rice Knowledge Bank website 2020a). A study from Japan recommended that for rice, phosphorus fertilization requires to be restricted to 20 kg per hectare per year which corresponds to 46 kg per hectare per year as $P_2O_5$ (Nagumo et al. 2013). In a study from Malaysia, it was found that the optimum yield of red rice was obtained when phosphorus in the form of triple phosphate was applied at a rate of 35 kg per hectare. The yield was not significantly affected by any further increase in the application of fertilizer (Masni and Wasli 2019). A study from the small fields of Khyber Pakhtunkhwa of Pakistan revealed that application of phosphorus at the rate of 90 kg per hectare of land in combination with animal manure resulted in increased productivity of rice (Amanullah et al. 2016b). In another study in Nizamabad district of Telangana, India, it was shown that the highest grain yield of 6.41 tonnes per hectare was achieved in kharif rice when $P_2O_5$ is applied at a rate 85 kg per hectare (Archana et al. 2017). In another study from Bangladesh it is reported that application of phosphorus in the form of triple phosphate at the rate of 10 kg per hectare resulted in the yield of 4850 kg per hectare and straw yield of 5125 kg per hectare in demonstration plots. It was also found that the yield was higher in demonstration plots than in non-demonstration plots (Razzaque and Rafiquzzaman 2007). In another experiment from Japan, it was found that application of superphosphate at the rate of 60 kg per hectare of land along with poultry manure resulted in a yield of 6.90 and 7.42 tons per hectare of land in consecutive years of 2017 and 2018 (Moe et al. 2019).

Wheat is considered to be one of the most productive and important crops of the twenty-first century (Curtis and Halford 2014). A very interesting study was performed in northwestern Pakistan to assess the influence of residual phosphorus on wheat productivity under rice-wheat cropping system. It was found that heavier grains (40.54 g/1000 grains) were recorded when wheat plants were grown in plots that had received phosphorus at the rate of 120 kg per hectare during its previous rice cultivation. A study from China reported that wheat plants require phosphorus at 100 kg per hectare of land for optimal growth (Deng et al. 2018). It was also observed that the yields were greater when wheat were grown in plots which had received higher levels of phosphorus (120 and 80 kg per hectare, respectively) (Amanullah and Inamullah 2016). In a study from Toba Tek Singh district of Pakistan, it was observed that application of 81 kg of $P_2O_5$ per hectare of land resulted in a yield of 3.94 mega grams (Mg) per hectare of wheat crop cultivar Inquilab-91 (Rahim et al. 2010). In another study it was reported that 90 kg per hectare of $P_2O_5$ is optimum for maximal yield of wheat in the agroclimatic condition of Sindh, Pakistan (Khan et al. 2008). Similar results were also obtained from a pot experiment study of wheat. It was shown that application of 80 kg per hectare of single super phosphate resulted in the highest number of tillers per plant, plant height, spike length, number of grains per spike, grain yield and straw yield (Khan et al. 2010). In another experiment conducted in Bahauddin Zakariya University, Multan, Pakistan reported that application of 60–120 kg of phosphorus in the form of $P_2O_5$ resulted in maximal improvement of growth parameters as well as the yield of wheat crop (Hussain et al. 2008).

The phosphorus requirement of maize has also been elaborately investigated. For maize diammonium phosphate and triple phosphates are used for fertilization. Depending upon the soil phosphorus index, maize plants require 0–185 kg per hectare of diammonium phosphate or 0–85 kg per hectare of triple phosphate ($P_2O_5$). Soil with high phosphorus index does not require any additional phosphorus fertilization (Potash Development Association website 2008). In an experiment in the agronomy research farm of The University of Agriculture Peshawar-Pakistan, it was observed that application of phosphorus at higher rates resulted in a significant increase in yield and yield components of maize under semi-arid conditions (Amanullah and Khan 2015). In another experiment conducted in China Agricultural University, it was found that optimum grain yield of 7.12 Mg per hectare was observed on the application of phosphorus at a rate of 75 kg per hectare (Deng et al. 2014). In another study from Maharashtra, India it was found that the application of phosphorus at 100 kg per hectare resulted in higher growth attributes including yields (Kwadzo et al. 2016). In an experiment conducted by Department of Soil Science and Land Management, Federal University of Agriculture Abeokuta (FUNAAB), Ogun State, Nigeria, it was reported that application rate of 30–45 kg of phosphorus per hectare of land resulted in a maximum dry yield of maize (Ogunsola and Adetunji 2016).

The effect of phosphorus on the yields of soybean has also been studied by various groups of researchers. In one study conducted by University for Development Studies, Nyankpala in the Guinea Savannah agro-ecological zone, it was found that application of phosphorus at 45 kg per hectare along with rhizobium inoculant resulted in maximal plant height, canopy spread, number and weight of nodules, number of pods and total grain yield (Ahiabor et al. 2014). Another study from Indonesia reported that maximal grain yield of soybean was obtained upon application of 125 kg per hectare of $P_2O_5$ (Kuntyastuti and Suryantini 2015). A study conducted at Main Agricultural Research Station, Dharwad, India reported that optimal yield of soybean was obtained upon treatment of 80 kg per hectare of phosphorus in conjugation with 60 kg per hectare of nitrogen (Raghuveer et al. 2017).

### 5.4.2 Phosphorus Deficiency

Phosphorus deficiency is a very common nutritional factor that limits agricultural production worldwide (Wissuwa 2003). It is reported that suboptimal levels of phosphorus in soils may result in a reduction in crop yields by 5–15%. Phosphorus deficiency is more critical in highly withered soil and in calcareous-alkaline soil (Shenoy and Kalagudi 2005). A typical phenotypic response of phosphorus deficiency in plants are stunted shoot growth and branching, dark to blue green colouration of leaves, weaker and thin stems, reduced tillering, imperfect pollination, fewer flowers, delayed maturity, poor grain quality and low yield (Ajmera et al. 2019). The general morphological symptoms related to phosphorus deficiency in

**Fig. 5.6** Deficiency symptoms of phosphorus in rice plants (**a**) in rice fields, (**b**) close up of individual plant (source: Rice Knowledge Bank Website 2020b)

rice are stunted growth, reduced tillering, narrowing and shortening of older leaves, erect having dirty green colouration. The stems are thin and spindly with retarded development (Rice knowledge Bank Website 2020b) (Fig. 5.6).

A study with rice reported that deficiency of phosphorus resulted in a reduction in lateral root density, length of small and large lateral roots. In addition to it, there is an increase in root hair length and density along with a reduction in nodal cross-sectional area (Vejchasarn et al. 2016). In another study it was reported that phosphorus deficiency resulted in reduced accumulation of all nutrients except sulphur and copper in the aboveground biomass of the plants and also results in a reduction in biomass accumulation by as much as 30% (Rose et al. 2016). It was also reported that phosphorus deficiency results in the initiation of root aerenchyma in rice (Pujol and Wissuwa 2018). In case of wheat, the phosphorus deficiency results in pale and olive green wilted seedlings. In matured leaves, chlorosis initiates at the tip and migrates down the leaf on a front, while the base of the leaf and the rest of the plant remains dark green. The necrosis of this chlorotic area is rapid with tips turning orange and dark brown and shrivelling while the rest part turns yellow (Department of Primary Industries and Regional Development, Government of Australia 2017a). Maize is another crop which is widely cultivated throughout the globe but is frequently subjected to phosphorus deprivation (Lin et al. 2013). In maize, phosphorus deficiency has a negative impact on leaf area index along with the reduction in absorbance of photosynthetically active radiation (PAR) by the canopy. In addition to it, the plants deprived of phosphorus exhibits reduction in growth accompanied by a delay in the emergence of adventitious roots (Pellerin et al. 2000). Phosphorus deprivation also resulted in reduction of elongation rates of leaf (Plénet et al. 2000). A recent study reports the impairment of root and shoot growth in the case of phosphorus deficiency. This is accompanied by a decline in phosphorus concentration (Klamer et al. 2019). Another recent study states the upregulation of gibberellic acid synthesis genes such as AN1, GA20ox1 and GA20ox2 and downregulation of gibberellic acid inactive genes such as GA2ox1 and GA2ox2 in maize plants subjected to low phosphate concentration (Zhang et al. 2019a). In soybean, phosphorus deficiency results in yellowing of leaves, while some veins still remained green. The yellow leaves progressively turned red and then violet finally resulting in collapse (Rosolem and Tavares 2006). Deficiency of phosphorus also results in reduced biomass and phosphorus content in seedlings (Zhou et al. 2016; Singh et al. 2018) and alteration in the balance of diurnal starch accumulation and utilization (Qiu and Israel 1992). A study reported that phosphorus deficiency in soybean resulted in impairment of symbiotic nitrogen fixation by delaying the onset of nodule function and decreasing nodule development (Qiao et al. 2007).

## 5.5 Potassium as an Essential Macronutrient Source

The necessity of potassium for the growth of plants was first indicated by Justus Leibig in 1840 (Galembeck and Dos Santos 2019). Though the reserves of potassium in soil are large but most of them are in unavailable form resulting in the requirement of potassium fertilizers (Zörb et al. 2014). This results in a high demand of potassium by agricultural crops. This led to search of potassium deposits in the soil which was eventually discovered during the 1850s at Stassfurt in Germany. Two most important minerals unearthed were hydrated double salts carnallite, $KCl{\cdot}MgCl_2{\cdot}6H_2O$, and kainite, $MgSO_4{\cdot}KCl{\cdot}3H_2O$. Consequently in 1861, a factory was set up in Stassfurt to produce potash salts from these deposits which produced 20,000 tonnes of the salts the following year. By 1909, annual production of potassium fertilizer in Germany soared to more than seven million tonnes (Freemantle 2016). Another potash deposit was accidentally discovered in

**Table 5.2** Different types of potassium fertilizers

| S.No. | Name of fertilizer | Chemical formula | Percentage of potassium |
|---|---|---|---|
| 1. | Muriate of potash | KCl | 50 |
| | | KCl | 41 |
| | | KCl | 33 |
| 2. | Potassium sulphate | $K_2SO_4$ | 43 |
| 3. | Potassium nitrate | $KNO_3$ | 18 |
| 4. | Sulphate potash magnesia | $K_2SO_4.MgSO_4$ | 18 |
| 5. | Kainite | $KCl+NaCl+MgSO_4$ | 10 |

Adapted from: Scherer 2005

Saskatchewan, Canada in the process of drilling oil in the 1940s. In 1958, the Potash Company of America became the first potash producer in Canada with the establishment of a conventional potash mine at Patience Lake which is still operational. At present Canada exports 95% of its potash to over 50 countries around the world (Western Potash Corporation Website 2020). The reserves of potash in Ural mountains amounts to 210 billion tonnes and are owned by Uralkali (Investing News website 2013). The company produced 11.1 million tonnes of potassium chloride with a sales volume of 9.8 million tonnes in the year 2019 (Uralkali website 2020). The main constituents of potash include sylvite (KCl), carnallite [$KCl{\cdot}MgCl_2$.6 ($H_2O$)], kainite ($MgSO_4{\cdot}KCl{\cdot}3H_2O$) and langbeinite ($2MgSO_4{\cdot}K_2SO_4$) (Garrett 1996). The deposit formation of potash is of two types, namely the underground deposits naturally protected from underground waters and the salt beds or lakes in the arid regions of the world (Sun and Ma 2018; Zhang et al. 2019b). The potassium fertilizers are of two main categories in which the $K^+$ ions are combined either with chloride (muriate of potash) or sulphate (sulphate of potash). Potassium chloride is available in three different grades, namely 50% K, 41% K and 33% K. The last two variants contain substantial quantities of sodium chloride (NaCl) and are recommended as $K^+$ fertilizers for natrophilic crops (Scherer 2005). The different types of potassium fertilizers are tabulated in Table 5.2.

Canada tops the list among the country in the production of muriate of potash. As per the report of Knoema®, Canada produced 20.3 million tonnes of potassium chloride in the year 2017 and accounted for 44.98% of the world's potassium chloride production (Knoema website 2017e). The production of muriate of potash (potassium chloride) by the top eight countries of the world is depicted in Fig. 5.7.

### 5.5.1 Potassium Requirement by Agricultural Crops

Potassium is an essential element that affects most of the biochemical and physiological processes related to plant growth and metabolism (Wang et al. 2013). Potassium is of great importance in physiology of plants. It performs critical functions related to activation of enzyme, osmotic adjustment, turgor generation, cell expansion, regulation of membrane electric potential and pH homeostasis (Ragel

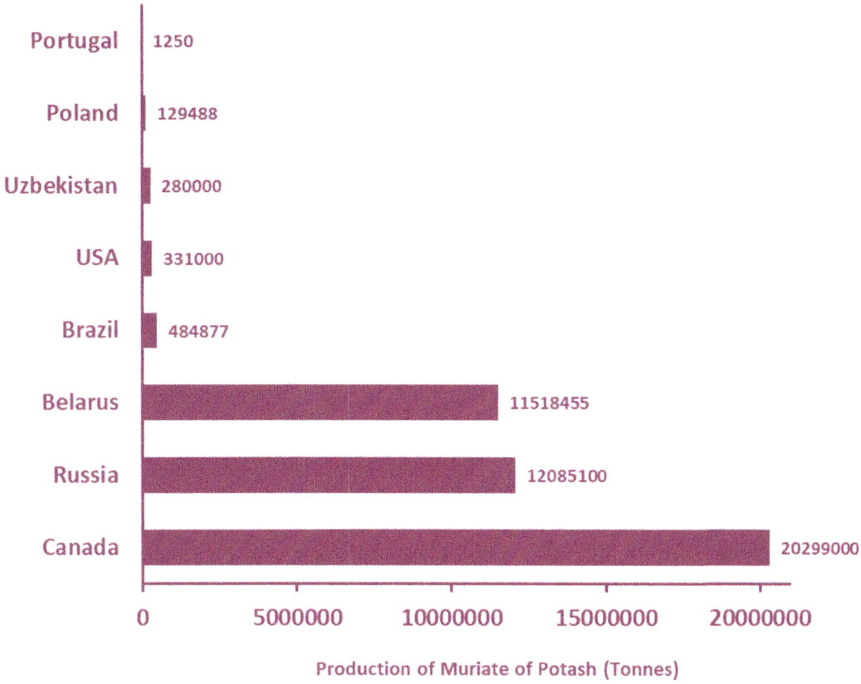


**Fig. 5.7** Production of muriate of potash by top 8 countries in the yield 2017 (Adapted from Knoema website 2017e)

et al. 2019). A study on the hybrid rice growing in coastal saline soils of West Bengal, India recommended a dose of 101.5 kg of potassium oxide ($K_2O$) per hectare for achieving higher productivity during the wet season (Banerjee et al. 2018). A report from a pot experiment study indicated the combined application of nitrogen and potassium at the rate of 23 and 30 kg per hectare, respectively, resulted in maximum grain yield (Kumar et al. 2013). In another study, it was reported that potassium fertilizer applied at the rate of 37.5 muriate of potash per hectare of land during the heading time resulted in best head rice yield (HRY) when harvested between 25 and 30 days after 50% flowering (DAFF) (Atapattu et al. 2018). A report from Pakistan states that application of potassium fertilizer at the rate of 375 kg per hectare resulted in improvement of growth and yield parameters in wheat (Arif et al. 2017). A study from Bangladesh also reported that application of potassium at 60 kg per hectare resulted in a maximal yield of wheat subjected to irrigation by municipal wastewater (Mojid et al. 2012). A report from Bangladesh recommends 72 kg of potassium oxide ($K_2O$) for the production of wheat in light soil (Hossain et al. 2015). In a study carried out in Tigray Region, northern Ethiopia, it was found that application of 30 kg of $K_2O$ per hectare of land resulted in highest apparent potassium recovery and agronomic efficiency in wheat (Brhane et al. 2017). One recent study reported that application rate of potassium ($K_2O$) at 150 mg and 250 mg per kg of soil helped in achieving higher yield of winter wheat under shading at early filling stage (SE) by alleviating the damage done on the photosynthetic apparatus by SE (Wang et al. 2020). Studies on the requirement of potassium by maize plants have also been undertaken by a number of researchers. In one recent study it was reported that potassium application at the rate of 75 kg per hectare resulted in maximum performance of maize plants from yield point of view under water stress condition (Ul-Allah et al. 2020). In another study, it was found that application of 100 kg per hectare of potassium resulted in maximum plant height, leaf area index, thousand grains weight, grains per year and grain yield in maize crops (Liaqat et al. 2018). In case of soybean, a study reported that combined treatment of phosphorus at 175 kg per hectare and potassium (in form of muriate of potash) at 120 kg per hectare resulted in highest number of filled pods per plant, length of the pods, number of seeds per pods and highest number of seeds and this constitutes the recommended dose for optimum yield of soybean (Khanam et al. 2016). In an experiment with intercropping system using maize and soybean in China it was found that joint application of phosphorus at 17 kg per hectare and potassium at 112.5 kg per hectare resulted in higher seed yields of soybean (Xiang et al. 2012). In mixed cropping system where maize is intercropped with soybean, it was found that application of potassium at 80 kg per hectare for maize and 60 kg per hectare for soybeans resulted in an accelerated biomass accumulation and distribution of other essential nutrients in the plant parts (Ahmed et al. 2020).

### 5.5.2 Potassium Deficiency

The visual symptoms of potassium deficiency are stunted growth of plants accompanied by yellowing of leaf margins (Hasanuzzaman et al. 2018). In addition to it, the older leaves of plants suffering from potassium deficiency undergo necrosis due to evacuation of potassium ions to the younger leaves (Cochrane and Cochrane 2009). In rice potassium deficiency results in decrease in concentration of potassium in shoots and roots. In addition to it, there is an increase in activities of antioxidant enzymes, namely superoxide dismutase, ascorbate peroxidase, glutathione reductase and catalase (Liu et al. 2013). Another study reveals that potassium deficiency results in decrease in growth of the roots. Both roots volume, roots surface area and numbers of laterals were decreased in plants growing under potassium deprivation. Moreover, potassium deficiency also resulted in damage of cellular organelles and membranes along with precipitation of dark particles in root cell walls (Jia et al. 2008). In wheat, potassium deficiency results in paler and weaker plants. In this case, the older leaves are first affected starting with necrosis and death of leaf tip followed by progressive yellowing and death downwards. There is a contrast in colouration of leaf margins (yellow) and green centre. The yellowing leaf tip and leaf margins often generate a green "arrow" like design towards leaf tip (Department of Primary Industries and Regional Development, Government of Australia 2017b).

In maize, potassium deficiency first appears on the lower leaves. The leaf symptoms include yellowing to necrosis ultimately leading to tissue death in the

**Fig. 5.8** Picture of potassium deficient maize plant (Source: Sawyer, Iowa State University, Integrated crop management 2018)

outer margins but located from the leaf tip to midrib in a v shape. The yellowing consequently covers the entire leaf. Upon persistence of potassium deficiency, the lower leaves die back and the leaf symptoms proceed towards the apex of the plant and accompanied by reduction in growth (Sawyer, Iowa State University, Integrated Crop Management 2018) (Fig. 5.8).

A study reported that potassium deficiency results in reduction of total length, root surface area, the root diameter and root volume of root system in maize (Du et al. 2017). Another experiment reported that content of chlorophyll a, b and a + b decreased due to deficiency of potassium in maize (Zhao et al. 2016). Another recent report states that potassium deficiency resulted in damage of chloroplast and photosynthetic reaction centres (PSII) along with increased superoxide and hydrogen peroxide levels. The anatomical structures of the leaves were also affected due to deficiency of potassium and the manifestations included smaller thickness of leaf, lower epidermis cells and vascular bundle area. Other physiological parameters including chlorophyll content, net photosynthetic rate, stomatal conductance, photochemical quenching and electron transport rate of PSII were also reduced due to potassium deficiency (Du et al. 2019).

In soybean, the potassium deficiency symptom can be observed in the leaves involving yellowing of leaflet margins with mild deficiency which may turn brown and necrotic with extreme deficiency (Mallarino, Iowa State University, Integrated Crop Management 2018) (Fig. 5.9). A study reported that potassium deficiency resulted in reduction in net photosynthetic rate, transpiration rate and stomatal conductance. This was accompanied by reduction in RUBISCO activity and dry weight of soybean plants (Wang et al. 2015).

**Fig. 5.9** Picture of potassium deficient soybean plant (**a**) young plants, (**b**) matured plants (Source: Mallarino, Iowa State University, Integrated crop management 2018)

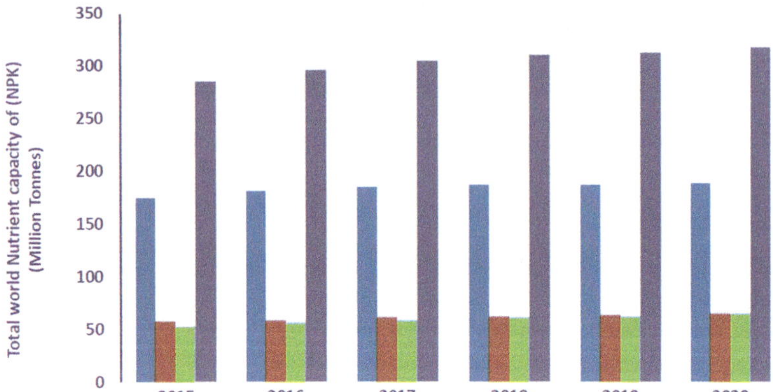


| | 2015 | 2016 | 2017 | 2018 | 2019 | 2020 |
|---|---|---|---|---|---|---|
| Ammonia (N) | 174.781 | 181.228 | 185.222 | 186.804 | 186.92 | 188.31 |
| Phosphoric Acid (P2O5) | 57.422 | 58.385 | 60.955 | 61.955 | 63.036 | 64.677 |
| Potash (K2O) | 52.942 | 55.974 | 58.111 | 61.576 | 62.136 | 64.486 |
| Total | 285.145 | 295.587 | 304.288 | 310.335 | 312.092 | 317.473 |

**Fig. 5.10** Total world nutrient capacity of nitrogen, phosphorus and potassium, 2015–2020 (thousand tonnes) (Source: FAO, United Nations, Rome 2017)

## 5.6 Nitrogen(p), Phosphorus (p) and Potassium-Demand and Supply

As per the FAO bulletin, the consumption of three main fertilizer across the world, namely nitrogen (N), phosphorus ($P_2O_5$) and potassium ($K_2O$) was estimated to reach 186.67 million metric tons in 2016. In addition to it, the demand of N, $P_2O_5$ and $K_2O$ is likely to grow annually on average by 1.5, 2.2 and 2.4%, respectively, from 2015 to 2020. It is also expected that over the next 5 years, the capacity of production of fertilizers, intermediates and raw materials is expected to increase (FAO 2017).

The FAO bulletin states that the global total nutrient capacity including N, $P_2O_5$ and $K_2O$ was 285.15 million tonnes in 2015 with a total supply of 245.77 million tons. The detailed estimated nutrient capacity of ammonia ($NH_3$), phosphoric acid ($H_3PO_4$) and potash ($K_2O$) from 2015 to 2020 is illustrated in Fig. 5.10, while the world supply is illustrated in Fig. 5.11.

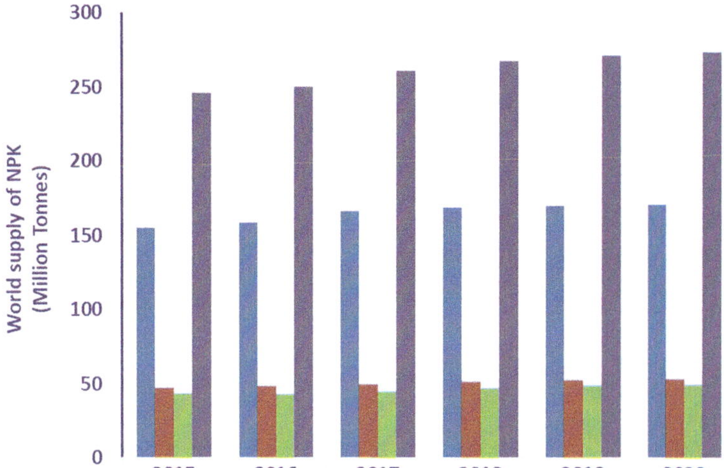


| | 2015 | 2016 | 2017 | 2018 | 2019 | 2020 |
|---|---|---|---|---|---|---|
| Ammonia (N) | 154.773 | 158.85 | 166.402 | 168.987 | 169.693 | 170.761 |
| Phosphoric Acid (P2O5) | 47.424 | 48.394 | 49.558 | 51.19 | 52.361 | 53.078 |
| Potash (K2O) | 43.571 | 42.772 | 44.868 | 47.249 | 48.898 | 49.545 |
| Total | 245.768 | 250.016 | 260.828 | 267.426 | 270.952 | 273.384 |

**Fig. 5.11** World supply of nitrogen, phosphorus and potassium, 2015–2020 (thousand tonnes) (Source: FAO, United Nations, Rome 2017)

## 5.7 NPK Fertilizer: A Brief Overview

NPK fertilizer is the combination of three macronutrients required by plants, namely nitrogen (N), phosphorus (P) and potassium (K). It is used as a fertilizer in agriculture industry to make plants healthy from nutritional point of view and meet the demand of healthy crops (Databridge Market Research website 2019). The NPK fertilizers are composed of nitrogen, phosphorus (in form of $P_2O_5$) and potassium (in form of $K_2O$) and these constituents may vary in their ratios. As, for example, a 12-32-16 grade of NPK complex fertilizer indicates the presence of 12 % nitrogen (N), 32 % phosphorous ($P_2O_5$) and 16 % potash ($K_2O$) (FAO website 2020). The individual constituent of selected NPK formulations is tabulated in Table 5.3.

So far as production is concerned, Russia tops the list of among NPK producers followed by India, Indonesia, Vietnam and Poland. In the year 2017, the production of NPK fertilizer by Russia was 8.41 million tons which accounts for 32% of total NPK production of the world. The other countries include India, Indonesia, Vietnam and Poland account for 76.72% of the production (Knoema website 2017f). The year wise production of NPK fertilizers for selected countries from 2010 to 2017 is tabulated in Table 5.4.

**Table 5.3** Constituents of selected NPK formulations (Source: FAO website 2020)

| | NPK complex 15-15-15 | NPK complex 17-17-17 | NPK complex 19-19-19 |
|---|---|---|---|
| Moisture content by weight | 1.5% (Max) | 1.5% (Max) | 1.5% (Max) |
| Total nitrogen content by weight (on dry basis) | Minimum 15% | Minimum 17% | Minimum 19% |
| Neutral ammonium citrate soluble phosphate (as $P_2O_5$) content by weight | Minimum 12% | Minimum 17% | Minimum 19% |
| Water soluble potash (as $K_2O$) content by weight | Minimum 15% | Minimum 13.6% | Minimum 19% |
| Particle size | Not less than 90% of the material shall pass through 4 mm IS sieve and be retained on 1 mm IS sieve. Not more than 5% shall be below 1 mm size. | Not less than 90% of the material shall pass through 4 mm IS sieve and be retained on 1 mm IS sieve. Not more than 5% shall be below 1 mm size. | Not less than 90% of the material shall pass through 4 mm IS sieve and be retained on 1 mm IS sieve. Not more than 5% shall be below 1 mm size. |

According to a leading market research report, the global NPK fertilizer (food-grade) market was valued at 2.31 billion US dollars in the year 2017 and is projected to reach 2.90 billion US dollars by the year 2023 growing at compound annual growth rate (CAGR) of 4.0% from the year 2018 (Research and Markets website 2019). The same organization also reported that this market is likely to reach a value of 3.92 billion US dollars with a 6.1% CAGR (CISION PR Newswire 2019). Another report states that the NPK fertilizer (feed-grade) was estimated to be valued around 5.4 billion US dollars in the year 2018 and is projected to reach 6.6 billion US dollars by the year 2023 growing at CAGR of 4.1% from 2018 (Markets and Markets website 2019a). The key players in the production of NPK fertilizers include Borealis AG (Austria), AkzoNobel (Netherlands), Yara International ASA (Norway), PetróleoBrasileiro S.A. (Brazil) and Agrium Inc. (Canada). In addition to it, The Mosaic Company (US), Israel Chemicals Ltd. (Israel), EuroChem (Switzerland), PotashCorp (Canada), K+S AKTIENGESELLSCHAFT (Germany), Alltech (US), PhosAgro (Russia), Haifa Chemicals (Israel), Aditya Birla Chemicals (India) and SKW Stickstoffwerke Piesteritz (Germany) are the other players that hold a significant share in the NPK fertilizers (feed-grade and food-grade) market (Markets and Markets website 2019b). A report from Ken Research Private limited states that the complex fertilizer production in Asia is likely to grow at a CAGR of 3.1% during the span of 2018–2022 whereas the consumption of complex fertilizers is likely to incline at a CAGR of 1.9% during the same period (Ken Research Website 2018). Study done by the same organization also suggests that Kingenta,

**Table 5.4** Detailed of the production of NPK fertilizers in selected countries from 2010 to 2017 (tonnes) (Source: Knoema website, f)

| | 2017 | 2015 | 2015 | 2014 | 2013 | 2012 | 2011 | 2010 |
|---|---|---|---|---|---|---|---|---|
| Russia | 8407500 | 6310000 | 6080600 | 5641700 | 44700 | 3115000 | 3253800 | 3205600 |
| India | 3846100 | 3655000 | 3744700 | 3754100 | 3104500 | 480300 | 506000 | 494000 |
| Indonesia | 3282957 | 2764687 | 3001087 | 2716098 | | 2893868 | 2213491 | 1853172 |
| Vietnam | 3276100 | 3081000 | 3304000 | 3387100 | 3372300 | | | |
| Poland | 1343362 | 1407953 | 1376489 | 1217933 | | | | |
| Republic of Korea | 1339597 | 1273238 | 1410913 | 1560501 | 1409707 | 1639316 | 1248324 | 1288861 |
| Japan | 1182693 | 1163715 | 1097430 | 1141338 | 1207014 | 1207634 | 1212210 | 1220684 |
| Belarus | 877806 | 819079 | 779816 | 630561 | 426726 | 385122 | 342639 | 314654 |

**Table 5.5** Year wise consumption of NPK fertilizer in India (Source: CEIC website 2019)

| S.No. | Year | Consumption of NPK fertilizer ( kg per hectare) |
|---|---|---|
| 1. | 2008 | 115.27 |
| 2. | 2009 | 127.21 |
| 3. | 2010 | 135.27 |
| 4. | 2011 | 146.32 |
| 5. | 2012 | 130.79 |
| 6. | 2013 | 131.36 |
| 7. | 2014 | 118.49 |
| 8. | 2015 | 127.45 |
| 9. | 2016 | 130.66 |
| 10. | 2017 | 123.41 |
| 11. | 2018 | 128.02 |

Coromandel International, Binh Dien Fertilizer, Petrokimia Gresik and Thai Central Chemicals will continue to be major producers of NPK fertilizers in China, India, Vietnam, Indonesia and Thailand, respectively (Open PR website 2020). In India, the consumption of NPK fertilizer was reported to be 128.020 kg per hectare of land in March 2018 which is an increase with respect to the consumption in 2017. The year wise consumption of NPK fertilizer from 2008 to 2018 is tabulated in Table 5.5.

## 5.8 Conclusive Remarks

According to the reports of UN population prospects, the population of world is projected to be 34% from 6.8 billion to 9.1 billion by 2050 (FAO). This increase in population will automatically increase the demand for food. It is projected that the global demand for cereals for both food and animal feed is likely to reach 3 billion tonnes by 2050 (FAO). In order to satisfy the demand of food by 2050, the world's food production should increase by 70% which implies an extension of farmed land or an intensification of the production on the currently farmed land (Aznar-Sánchez et al. 2020). This intensification can be obtained by applying high levels of fertilizers for increasing the yield of crops (Withers et al. 2018). Thus the fertilizer production is likely to have an uptrend in the coming years. Though the nitrogen present in the nitrogenous fertilizers is ultimately recycled back to the environment in gaseous form through the action of biogeochemical cycle, however, phosphate and potash are exhaustible. Thus judicious use of phosphate and potash is very important in the coming years. Efforts should be taken to recycle the phosphate and potash fertilizer post-application in the agricultural fields which can be a topic of future research. In addition to it, the focus should also be given to use of biofertilizers which can very well fix, solubilize and transport essential nutrients from the atmosphere and deliver to the crops. Thus an overall balance is required in the application of fertilizers and recycling the same. A multidisciplinary approach towards the fertilizer management coupled with the promotion of green manures seems to be the most optimal approach towards sustainable development for the overall benefit of mankind.

## References

Abbasi HN, Vasileva V, Lu X (2017) The influence of the ratio of nitrate to ammonium nitrogen on nitrogen removal in the economical growth of vegetation in hybrid constructed wetlands. Environments 4:24

Abebe Z, Feyisa H (2017) Effects of nitrogen rates and time of application on yield of maize: rainfall variability influenced time of N application. Int J Agron:1545280

Agarwal DK, Billore SD, Sharma AN, Dupare BU, Srivastava SK (2013) Soybean: Introduction, improvement, and utilization in India—problems and prospects. Agric Res 2:293–300

Ahiabor BDK, Lamptey S, Yeboah S, Bahari V (2014) Application of phosphorus fertilizer on Soybean (Glycine max L. (Merril)) inoculated with rhizobium and its economic implication to farmers. Am J Exp Agric 4(11):1420–1434

Ahmed A, Aftab S, Hussain S, Cheema HN, Liu W, Yang F, Yang W (2020) Nutrient accumulation and distribution assessment in response to potassium application under maize–soybean intercropping system. Agronomy 10:725

Aiysha D, Latif Z (2019) Insights of organic fertilizer micro flora of bovine manure and their useful potentials in sustainable agriculture. PLoS One 14(12):e0226155

Ajmera I, Hodgman TC, Lu C (2019) An integrative systems perspective on plant phosphate research. Genes 10(2):139

Alori ET, Glick BR, Babalola OO (2017) Microbial phosphorus solubilization and its potential for use in sustainable agriculture. Front Microbiol 8:971

Amanullah, Inamullah (2016) Residual phosphorus and zinc influence wheat productivity under rice-wheat cropping system. SpringerPlus 5:255

Amanullah, Khan A (2015) Phosphorus and compost management influence Maize (Zea mays) productivity under semiarid condition with and without phosphate solubilizing bacteria. Front Plant Sci 6:1083

Amanullah, Iqbal A, Ali A, Fahad S, Parmar B (2016a) Nitrogen source and rate management improve maize productivity of smallholders under semiarid climates. Front Plant Sci 7:1773

Amanullah, Khan SU, Iqbal A, Fahad S (2016b) Growth and productivity response of hybrid rice to application of animal manures, plant residues and phosphorus. Front Plant Sci 7:1440

Antonokiewicz J, Labetowicz J (2016) Chemical innovation in plant nutrition in a historical continuum from Ancient Greece and Rome until modern times. Chem Didact Ecol Metrol 21 (1-2):29–43

Archana K, Reddy TP, Anjaiah T, Padmaja B (2017) Effect of levels of phosphorus and its time of application on soil nutrient status and yield of rice grown on P accumulated soil. Int J Curr Microbiol Appl Sci S4:92–99

Arif M, Tasneem MT, Bashir F, Yaseen G, Anwar A (2017) Evaluation of different levels of potassium and zinc fertilizer on the growth and yield of wheat. Int J Biosen Bioelectron 3 (2):242–246

Atapattu AJ, Prasantha BDR, Amaratunga KSP, Marambe B (2018) Increased rate of potassium fertilizer at the time of heading enhances the quality of direct seeded rice. Chem Biol Technol Agric 5:22

Aulakh CS, Kaur P, Walia SS, Gill RS, Sharma S, Buttar GS (2016) Productivity and quality of basmati rice (*Oryza sativa*) in relation to nitrogen management. Indian J Agron 61(4):467–473

Aznar-Sánchez JA, Velasco-Muñoz JF, López-Felices B, Román-Sánchez IM (2020) An analysis of global research trends on greenhouse technology: towards a sustainable agriculture. Int J Environ Res Public Health 17(2):664

Banerjee H, Ray K, Dutta SK, Majumdar K, Satyanarayana T, Timsina J (2018) Optimizing potassium application for hybrid rice (Oryza sativa L.) in coastal saline soils of West Bengal, India. Agronomy 8:292

Blakemore RJ (2018) Critical decline of earthworms from organic origins under intensive, Humic SOM-Depleting Agriculture. Soil Syst 2(2):33

Brhane H, Mamo T, Teka K (2017) Potassium fertilization and its level on wheat (Triticum aestivum) yield in shallow depth soils of Northern Ethiopia. J Fertil Pestic 8(2):1000182

CEIC (2019) India chemical fertilizers: NPK consumption: per hectare: all India. https://www.ceicdata.com/en/india/chemical-fertilizers-nitrogen-phosphate-and-potash-npk-consumption-per-hectare-by-regions/chemical-fertilizers-npk-consumption-per-hectare-all-india. Accessed 22 June 2020

Cetner MD, Kalaji HM, Goltsev V et al (2017) Effects of nitrogen-deficiency on efficiency of light-harvesting apparatus in radish. Plant Physiol Biochem 119:81–92

Che SG, Zhao BQ, Li YT, Yuan L, Lin ZA, Hu SW, Shen B (2016) Nutrient uptake requirements with increasing grain yield for rice in China. J Integr Agric 15(4):907–917

Chen L, Lin L, Cai G, Sun Y, Huang T, Wang K, Deng J (2014) Identification of nitrogen, phosphorus, and potassium deficiencies in rice based on static scanning technology and hierarchical identification method. PLoS One 9(11):e113200

CISION PR newswire (2019) Global NPK fertilizers market report 2019-2023. https://www.prnewswire.com/news-releases/global-npk-fertilizers-market-report-2019-2023-300899966.html. Accessed 22 June 2020

Clark B, Foster JB (2009) Ecological imperialism and the global metabolic rift: unequal exchange and the guano/nitrates trade. Int J Comp Sociol 53(3-4):311–334

Clements WT, Lee SR, Bloomer RJ (2014) Nitrate ingestion: a review of the health and physical performance effects. Nutrients 6(11):5224–5264

Cochrane TT, Cochrane TA (2009) The vital role of potassium in the osmotic mechanism of stomata aperture modulation and its link with potassium deficiency. Plant Signal Behav 4 (3):240–243

Coventery DR, Yadav A, Poswal RS, Sharma RK, Gupta RK, Chookar RS, Gill SC, Kumar V, Kumar A, Mehta A, SGL K, Cummins JA (2011) Irrigation and nitrogen scheduling as a requirement for optimizing wheat yield and quality in Haryana. Field Crops Res 123:80–88

Curtis T, Halford NG (2014) Food security: the challenge of increasing wheat yield and the importance of not compromising food safety. Ann Appl Biol 164(3):354–372

Databridge market research website (2019) Global NPK fertilizers market – industry trends and forecast to 2026. https://www.databridgemarketresearch.com/reports/global-npk-fertilizers-market#. Accessed 22 June 2020

De Souza GK, Sampaio J, Longoni L, Ferreira S, Alvarenga S, Beneduzi A (2019) Soybean inoculants in Brazil: an overview of quality control. Braz J Microbiol 50(1):205–211

Deng Y, Chen K, Teng W, Zhan A, Tong Y, Feng G, Cui Z, Zhang F, Chen X (2014) Is the inherent potential of maize roots efficient for soil phosphorus acquisition? PLoS One 9(3):e90287

Deng Y, Teng W, Tong YP, Chen XP, Zou CQ (2018) Phosphorus efficiency mechanisms of two wheat cultivars as affected by a range of phosphorus levels in the field. Front Plant Sci 9:1614

Department of Primary industries and regional development, Government of Australia (2017a) Diagnosing phosphorus deficiency in wheat. https://www.agric.wa.gov.au/mycrop/diagnosing-phosphorus-deficiency-wheat. Accessed 22 June 2020

Department of Primary industries and regional development, Government of Australia (2017b) MyCrop. https://www.agric.wa.gov.au/mycrop/diagnosing-potassium-deficiency-wheat. Accessed 22 June 2020

Ding L, Wang KJ, Jiang GM et al (2005) Effects of nitrogen deficiency on photosynthetic traits of maize hybrids released in different years. Ann Bot 96(5):925–930

DiNicolantonio JJ, O'Keefe JH (2017) The history of the salt wars. Am J Med 130(9):1011–1014

Du Q, Zhao X, Jiang C, Wang X, Han Y, Wang J, Yu H (2017) Effect of potassium deficiency on root growth and nutrient uptake in maize (*Zea mays* L.). Agric Sci 8:1263–1277

Du Q, Zhao XH, Xia L, Jiang CJ, Wang XG, Han Y, Wang J, Yu HQ (2019) Effects of potassium deficiency on photosynthesis, chloroplast ultrastructure, ROS, and antioxidant activities in maize (*Zea mays* L.). J Integr Agric 18(2):395–406

Duan YH, Zhang YL, Ye LT, Fan XR, Xu GH, Shen QR (2007) Responses of rice cultivars with different nitrogen use efficiency to partial nitrate nutrition. Ann Bot 99(6):1153–1160

Espie P, Ridgway H (2020) Bioactive carbon improves nitrogen fertiliser efficiency and ecological sustainability. Sci Rep. 10(1):3227

FAO (2017) World fertilizer trends and outlook to 2020: Summary report, FAO: Rome. http://www.fao.org/3/a-i6895e.pdf. Accessed 22 June 2020

Feller C, Blanchart E, Bernoux M, Lal R, Manlay R (2012) Soil fertility concepts over the past two centuries: the importance attributed to soil organic matter in developed and developing countries. Arch Agron Soil Sci 58(S1):S3–S21

Food and Agricultural Organisation of the United Nations (2020) NSP-fertilizer specification. http://www.fao.org/agriculture/crops/thematic-sitemap/theme/spi/plantnutrition/fertspecs/en/#c29994. Accessed 22 June 2020

Freemantle M (2016) Potash on a megascale. Chemistry world. https://www.chemistryworld.com/features/potash-on-a-megascale/9558.article. Accessed 22 June 2020

Gai Z, Zhang J, Li C (2017) Effects of starter nitrogen fertilizer on soybean root activity, leaf photosynthesis and grain yield. PLoS One 12(4):e0174841

Galembeck A, Dos Santos LP (2019) NPK: essentials for sustainability. Quím Nova 42(10):1199–1207

Galloway JN, Leach AM, Bleeker A, Erisman JW (2013) A chronology of human understanding of the nitrogen cycle. Philos Trans R Soc B: Biol Sci 368(1621):20130120

Garrett DE (1996) Origin of potash deposits. In: Potash. Springer, Dordrecht, pp 1–80

Gastal F, Lemaire G (2002) N uptake and distribution in crops: an agronomical and ecophysiological perspective. J Exp Bot 53(370):789–799

Global Industry Analysts, Inc Report 2020 Report ID: 338437, June 2020. Research and Markets. https://www.researchandmarkets.com/research/r79m28/global_phosphates?w=12. Accessed 22 June 2020

Haefele SM, Wopereis MCS, NdiayeMK BSE, Isselmou MO (2003) Internal nutrient efficiencies, fertilizer recovery rates and indigenous nutrient supply of irrigated lowland rice in Sahelian West Africa. Field Crops Res 80:19–32

Haque MA, Haque MM (2016) Growth, yield and nitrogen use efficiency of new rice variety under variable nitrogen rates. Am J Plant Sci 7:612–622

Hasan MM, Hasan MM, Teixeira da Silva JA, Li X (2016) Regulation of phosphorus uptake and utilization: transitioning from current knowledge to practical strategies. Cell Mol Biol Lett 21:7

Hasanuzzaman M, BorhannuddinBhuyan MHM, Nahar K, Hossain MS, Mahmud JA, Hossen MS, Masud AAC, Moumita, Fujita M (2018) Potassium: a vital regulator of plant responses and tolerance to abiotic stresses. Agronomy 8:31

Hossain A, Jaime A, Da Silva T, Bodruzzamman M (2015) Rate and application methods of potassium in light soil for irrigated spring wheat. Songklanakarin J Sci Technol 37(6):635–642

Howie RT, Turnbull R, Binns J, Frost M, Dalladay-Simpson P, Gregoryanz E (2016) Formation of xenon-nitrogen compounds at high pressure. Sci Rep 6:34896. https://doi.org/10.1038/srep34896

Hu Y, Fernández V, Ma L (2014) Nitrate transporters in leaves and their potential roles in foliar uptake of nitrogen dioxide. Front Plant Sci 5:360

Hussain N, Khan MB, Ahmad R (2008) Influence of phosphorus application and sowing time on performance of wheat in calcareous soils. Int J Agric Environ Biotechnol 10(4):399–404

Indorama Corporation Website (2017) Phosphate fertilizer. http://www.indorama.com/products/phosphate-fertilizers. Accessed 22 June 2020

Investing News Website (2013) World class potash deposit. https://investingnews.com/daily/resource-investing/agriculture-investing/potash-investing/world-class-potash-deposits/. Accessed 22 June 2020

Jahan MS, Sultana S, Ali MY (2014) Effect of different nitrogen levels on the yield performance of aromatic rice varieties. Bull Inst Trop Agr Kyushu Univ 37:47–56

Jia YB, Yang XE, Feng Y, Jilani G (2008) Differential response of root morphology to potassium deficient stress among rice genotypes varying in potassium efficiency. J Zhejiang Univ Sci B 9 (5):427–434

Kawdzo KA, Narkhede WN, Khazi GS (2016) Effect of fertility levels on growth, yield and soil fertility status of maize (*Zea mays* L.) in vertisol of Maharashtra. J Appl Nat Sci 8(4):1779–1785

Ken Research Private Limited (2018) Spending water soluble fertilizer, country growth complex Asia report. https://www.kenresearch.com/blog/2018/03/spending-water-soluble-fertilizer-country-growth-complex-asia-report-ken-research/. Accessed 22 June 2020

Khan P, Imtiaz M, Aslam M, Shah SKH, Nizamuddin, Memon MY, SUH S (2008) Effect of different nitrogen and phosphorus ratios on the performance of wheat cultivar khirman. Sarhad J Agric 24(2):233–240

Khan MB, Lone MI, Ullah R, Kaleem S, Ahmed M (2010) Effect of different phosphatic fertilizers on growth attributes of wheat (*Triticum aestivum* L.). J Am Sci 6(12):1256–1262

Khanam M, Islam MS, Ali MH, Chowdhury IF, Masum SM (2016) Performance of soyabean under different levels of phosphorus and potassium. Bangladesh Agron J 19(1):99–108

Kimura Y, Ushiwatari T, Suyama A, Tominaga-Wada R, Wada T, Maruyama-Nakashita A (2019) Contribution of root hair development to sulfate uptake in arabidopsis. Plants 8(4):106

Klamer F, Vogel F, Li X et al (2019) Estimating the importance of maize root hairs in low phosphorus conditions and under drought. Ann Bot 124(6):961–968

Knoema® (2017a) Ammonium Sulphate production. https://knoema.com/atlas/topics/Agriculture/Fertilizers-Production-Quantity-in-Nutrients/Ammonium-sulphate-production. Accessed 22 June 2020

Knoema® (2017b) Ammonium Nitrate production. Available from: https://knoema.com/atlas/topics/Agriculture/Fertilizers-Production-Quantity-in-Nutrients/Ammonium-nitrate-production. Accessed 22 June 2020

Knoema® (2017c) Monoammonium phosphate production. https://knoema.com/atlas/topics/Agriculture/Fertilizers-Production-Quantity-in-Nutrients/Monoammonium-phosphate-production. Accessed 22 June 2020

Knoema® 2017d, Diammonium phosphate (DAP) – production. https://knoema.com/atlas/topics/Agriculture/Fertilizers-Production-Quantity-in-Nutrients/Diammonium-phosphate-production. Accessed 22 June 2020

Knoema® (2017e). Potassium chloride (muriate of potash) (MOP) – production. https://knoema.com/atlas/topics/Agriculture/Fertilizers-Production-Quantity-in-Nutrients/Potassium-chloride-production. Accessed 22 June 2020

Knoema® (2017f) NPK fertilizers – production. https://knoema.com/atlas/topics/Agriculture/Fertilizers-Production-Quantity-in-Nutrients/NPK-fertilizers-production. Accessed 22 June 2020

Kong L, Wang F, Si J, Feng B, Zhang B, Li S, Wang Z (2013) Increasing in ROS levels and callose deposition in peduncle vascular bundles of wheat (*Triticum aestivum* L.) grown under nitrogen deficiency. J Plant Interact 8(2):109–116

Kumar V (2013) Nitrogenous fertilizers. Agropedia. http://agropedia.iitk.ac.in/content/nitrogenous-fertilizers. Accessed 22 June 2020

Kumar R, Kumar S, Kumar A, Singh H, Kumar S (2013) Effect of top dressing nitrogen and potassium on yield and yield components of rice (*Oryza sativa* L). Agrways 1(2):90–94

Kuntyastuti H, Suryantini (2015) Effect of phosphorus fertilization on soil phosphorus level and yield of soybean (*Glycine max* L.) in paddy soil. J Exp Biol Agric Sci 3(1):1–9

Liaqat W, Jan MF, Ahmad H, Rehan W, Rafiullah KEH, Khan N (2018) Impact of mineral potassium on phenology, growth and yield of maize hybrid. Int J Environ Sci Nat Res 8 (5):168–171

Lin HJ, Gao J, Zhang ZM et al (2013) Transcriptional responses of maize seedling root to phosphorus starvation. Mol Biol Rep 40(9):5359–5379

Liu J, You L, Amini M, Obersteiner M, Herrero M, Zehnder AJ, Yang H (2010) A high-resolution assessment on global nitrogen flows in cropland. Proc Natl Acad Sci U S A 107(17):8035–8040. https://doi.org/10.1073/pnas.0913658107

Liu CH, Chao YY, Kao CH (2013) Effect of potassium deficiency on antioxidant status and cadmium toxicity in rice seedlings. Bot Stud 54(1):2

Liu XY, Koba K, Makabe A, Liu CQ (2014) Nitrate dynamics in natural plants: insights based on the concentration and natural isotope abundances of tissue nitrate. Front Plant Sci 5:355

Luo G, Junium CK, Izon G, Ono S, Beukes NJ, Algeo TJ, Cui Y, Xie S, Summons RE (2018) Nitrogen fixation sustained productivity in the wake of the Palaeoproterozoic Great Oxygenation Event. Nat Commun 9(1):978

Maiti D, Das DK, Pathak H (2006) Fertilizer requirement for irrigated wheat in eastern India using the QUEFTS simulation model. Sci World J 6:231–245

Majeed A, Muhmood A, Niaz A, Javid S, Ahmad ZA, Shah SSH, Shah AH (2015) Bed planting of wheat (Triticum aestivum L.) improves nitrogen use efficiency and grain yield compared to flat planting. Crop J 3(2):118–124

Makino Y, Ueno O (2018) Structural and physiological responses of the C4 grass Sorghum bicolor to nitrogen limitation. Plant Prod Sci 21(1):39–50

Mallarino A, Iowa State University, Integrated Crop Management (2018) Extension and outreach. https://crops.extension.iastate.edu/cropnews/2018/08/soybean-potassium-deficiency-symptoms-during-early-and-late-growing-stages. Accessed 22 June 2020

Manzoor Z, Awan TH, Zahid MA, Faiz FA (2006) Response of rice crop (Super Basmati) to different nitrogen level. J Anim Pl Sci. 16(1–2):52–55

Markets and Markets (2019a) NPK fertilizers (feed-grade and food-grade) market by application (bakery products, meat & meat products, dairy products, and beverages), form (powder, liquid, and others), type (nitrogen, phosphorus, potassium, and others), and region-global forecast to 2023. https://www.marketsandmarkets.com/Market-Reports/npk-fertilizers-market-197135673.html. Accessed 22 June 2020

Markets and Markets (2019b) Market leader - NPK fertilizers (feed-grade and food-grade) market. Markets and Markets, b. Available from: https://www.marketsandmarkets.com/ResearchInsight/npk-fertilizers-market.asp. Accessed 22 June 2020

Masni Z, Wasli ME (2019) Yield performance and nutrient uptake of red rice variety (MRM 16) at different NPK fertilizer rates. Int J Agron 2019:5134358

Matsuda M (2011) Intensification level of rice farming in Myanmar: implication for its sustainable development. Environ Dev Sustain 13:51–64

Mengel DB Types and uses of nitrogen fertilizers for crop production. agronomy guide. Purdue University Cooperative Extension Service. https://www.extension.purdue.edu/extmedia/AY/AY-204.html. Accessed 22 June 2020

Métioui A, Matoussi F, Trudel L (2016) The teaching of photosynthesis in secondary school: a history of the science approach. J Biol Educ 50(3):275–289

Meychik NR, Nikolaeva YI, Kushunina MA (2017) Effect of nitrogen deficiency on the ion-exchange properties of cell wall polymers from wheat roots. Moscow Univ Biol Sci Bull 72:74–78

Moe K, Htwe AZ, Thu TTP, Kajihara Y, Yamakawa T (2019) Effects on NPK status, growth, dry matter and yield of rice (*Oryza sativa*) by organic fertilizers applied in field condition. Agriculture 9:109

Mojid MA, Wyseure GCL, Biswas SK (2012) Requirement of nitrogen, phosphorus and potassium fertilizers for wheat cultivation under irrigation by municipal wastewater. J Soil Sci Plant Nutr 12(4):655–665

Muñoz-Huerta RF, Guevara-Gonzalez RG, Contreras-Medina LM, Torres-Pacheco I, Prado-Olivarez J, Ocampo-Velazquez RV (2013) A review of methods for sensing the nitrogen status in plants: advantages, disadvantages and recent advances. Sensors 13(8):10823–10843

Mus F, Crook MB, Garcia K et al (2016) Symbiotic nitrogen fixation and the challenges to its extension to nonlegumes. Appl Environ Microbiol 82(13):3698–3710

Nagumo T, Tajima S, Chikushi S, Yamashita A (2013) Phosphorus balance and soil phosphorus status in paddy rice fields with various fertilizer practices. Plant Prod Sci 16(1):69–76
Narasimhan R, Wang G, Li M, Roth M, Welti R, Wang X (2013) Differential changes in galactolipid and phospholipid species in soybean leaves and roots under nitrogen deficiency and after nodulation. Phytochemistry 96:81–91
Nguyen G, Rothstein S, Spangenberg G, Kant S (2015) Role of microRNAs involved in plant response to nitrogen and phosphorous limiting conditions. Front Plant Sci 6:629
Ning P, Yang L, Li C, Fritschi FB (2018) Post-silking carbon partitioning under nitrogen deficiency revealed sink limitation of grain yield in maize. J Exp Bot 69(7):1707–1719
Nussaume L, Kanno S, Javot H, Marin E, Pochon N, Ayadi A, Nakanishi TM, Thibaud MC (2011) Phosphate import in plants: focus on the PHT1 transporters. Front Plant Sci 2:83
Ogunsola KE, Adetunji MT (2016) Effects of phosphorus and sulphur on dry matter yield of Maize (Zea mays) in some soils at Abeokuta, Ogun state, Nigeria. J Trop Agric Food Environ Ext 15 (2):1–8
Omelon S, Ariganello M, Bonucci E, Grynpas M, Nanci A (2013) A review of phosphate mineral nucleation in biology and geobiology. Calcif Tissue Int 93(4):382–396
Omondi JO, Lazarovitch N, Rachmilevitch S, Yermiyahu U, Sperling O (2019) High nitrogen availability limits photosynthesis and compromises carbohydrate allocation to storage in roots of Manihot esculenta Crantz. Front Plant Sci 10:1041
Open PR website (2020) Asia complex fertilizer market is expected to reach over USD 18 billion by 2022: Ken Research. https://www.openpr.com/news/1905713/asia-complex-fertilizer-market-is-expected-to-reach-over-usd-18. Accessed 22 June 2020
Paez-Garcia A, Motes CM, Scheible WR, Chen R, Blancaflor EB, Monteros MJ (2015) Root traits and phenotyping strategies for plant improvement. Plants 4(2):334–355
Pandey N (2018) Role of plant nutrients in plant growth and physiology. In: Hasanuzzaman M, Fujita M, Oku H, Nahar K, Hawrylak-Nowak B (eds) Plant nutrients and abiotic stress tolerance. Springer, Singapore
Pellerin S, Mollier A, Plénet D (2000) Phosphorus deficiency affects the rate of emergence and number of Maize adventitious nodal roots. Agron J 92(4):690–697
Pilon-Smits EA, Quinn CF, Tapken W, Malagoli M, Schiavon M (2009) Physiological functions of beneficial elements. Curr Opin Plant Biol 12(3):267–274
Plénet D, Etchebest S, Mollier A, Pellerin S (2000) Growth analysis of maize field crops under phosphorus deficiency. Plant Soil 223:119–132
Potash Development Association (2008) 17. Forage maize – fertiliser requirements. https://www.pda.org.uk/pda_leaflets/17-forage-maize-fertiliser-requirements/#phosphate-requirement. Accessed 22 June 2020
Prinsi B, Espen L (2015) Mineral nitrogen sources differently affect root glutamine synthetase isoforms and amino acid balance among organs in maize. BMC Plant Biol 15:96
Pujol V, Wissuwa M (2018) Contrasting development of lysigenous aerenchyma in two rice genotypes under phosphorus deficiency. BMC Res Notes 11:60
Qiao Y, Tang C, Han X, Miao S (2007) Phosphorus deficiency delays the onset of nodule function in Soybean. J Plant Nutr 30(9):1341–1353
Qiu J, Israel DW (1992) Diurnal starch accumulation and utilization in phosphorus-deficient soybean plants. Plant Physiol 98(1):316–323
Ragel P, Raddatz N, Leidi EO, Quintero FJ, Pardo JM (2019) Regulation of $K^+$ nutrition in plants. Front Plant Sci 10:281
Raghuveer, Hosmath JA, Keerti, Chandranath HT (2017) Effect of different levels of nitrogen and phosphorus on growth and yield parameter of soybean (Glycine max L. Merrill). Int J Pure App Biosci 5(4):1686–1690
Rahim A, Ranjha AM, Waraich EA (2010) Effect of phosphorus application and irrigation scheduling on wheat yield and phosphorus use efficiency. Soil Environ 29(1):15–22
Razzaque MS (2011) Phosphate toxicity: new insights into an old problem. Clin Sci 120(3):91–97

Razzaque MA, Rafiquzzaman S (2007) Comparative analysis of T. Aman rice cultivation under different management practice in coastal area. J Agric Rural Dev 5(1&2):64–69

Research and Markets (2019) NPK (nitrogen, phosphorous, and potassium) fertilizers - global market outlook (2018-2027), statistics market research consulting pvt Ltd, 2019, ID: 4844717. https://www.researchandmarkets.com/reports/4844717/npk-nitrogen-phosphorous-and-potassium?utm_source=GNDIY&utm_medium=PressRelease&utm_code=8p89mj&utm_campaign=1302084+-+NPK+Fertilizers+Market+Set+to+Reach+%243.92+Billion+by+2027+-+Comprehensive+Industry+Insights+%26+Opportunities&utm_exec=joca220prd. Accessed 22 June 2020

Rice Knowledge Bank Website (2020a) Phosphorus (P). http://www.knowledgebank.irri.org/training/fact-sheets/nutrient-management/item/phosphorus. Accessed 22 June 2020

Rice Knowledge Bank Website (2020b) Phosphorus (P) deficiency. http://www.knowledgebank.irri.org/training/fact-sheets/nutrient-management/deficiencies-and-toxicities-fact-sheet/item/phosphorous-deficiency. Accessed 22 June 2020

Rose TJ, Kretzschmar T, Liu L, Lancaster G, Wissuwa M (2016) Phosphorus deficiency alters nutrient accumulation patterns and grain nutritional quality in rice. Agronomy 6:52

Rosolem CA, Tavares CA (2006) Phosphorus deficiency symptoms in soybean. Rev Bras Ciênc Solo 30(2):385–389

Sawyer JE, Iowa State University, Integrated Crop Management (2018) Extension and outreach. https://crops.extension.iastate.edu/encyclopedia/potassium-deficiency-symptoms-corn. Accessed 22 June 2020

Scherer HW (2005) Fertilizers and fertilization. In: Hillel D (ed) Encyclopedia of soils in the environment. Elsevier, Amsterdam, pp 20–26

Schlüter U, Mascher M, Colmsee C, Scholz U, Bräutigam A, Fahnenstich H, Sonnewald U (2012) Maize source leaf adaptation to nitrogen deficiency affects not only nitrogen and carbon metabolism but also control of phosphate homeostasis. Plant Physiol 160(3):1384–1406

Sett R, Soni B (2013) Foliar nitrogen, phosphorus and potassium content in trees in environmentally toxic plastic industry area. J Environ Sci Eng 55(2):167–174

Shen J, Yuan L, Zhang J et al (2011) Phosphorus dynamics: from soil to plant. Plant Physiol 156 (3):997–1005. https://doi.org/10.1104/pp.111.175232

Shenoy VV, Kalagudi GM (2005) Enhancing plant phosphorus use efficiency for sustainable cropping. Biotechnol Adv 23(7-8):501–513

Singh SK, Reddy VR, Fleisher DH, Timlin DJ (2018) Phosphorus nutrition affects temperature response of soybean growth and canopy photosynthesis. Front Plant Sci 9:1116

Snowball K, Robson AD (1991) Nutrient deficiencies and toxicities in wheat. A guide for field identification. CIMMYT, Mexico

Staley C, Breuillin-Sessoms F, Wang P, Kaiser T, Venterea RT, Sadowsky MJ (2018) Urea amendment decreases microbial diversity and selects for specific nitrifying strains in eight contrasting agricultural soils. Front Microbiol 9:634

Sugiyama A, Yamazaki Y, Yamashita K, Takahashi S, Nakayama T, Yazaki K (2016) Developmental and nutritional regulation of isoflavone secretion from soybean roots [published correction appears in Biosci Biotechnol Biochem. 2017 Nov;81(11):2225]. Biosci Biotechnol Biochem 80(1):89–94

Sulieman S, Tran LS (2014) Symbiotic nitrogen fixation in legume nodules: metabolism and regulatory mechanisms. Int J Mol Sci 15(11):19389–19393

Sun MG, Ma LC (2018) Potassium-rich brine deposit in Lop Nor basin, Xinjiang, China. Sci Rep 8 (1):7676. https://doi.org/10.1038/s41598-018-25993-6

Sun Y, Zhu S, Yang X, Weston MV, Wang K, Shen Z, Xu H, Chen L (2018) Nitrogen diagnosis based on dynamic characteristics of rice leaf image. PLoS One 13(4):e0196298

Tajul MI, Alam MM, Hossain SM, Naher K, Rafii MY, Latif MA (2013) Influence of plant population and nitrogen-fertilizer at various levels on growth and growth efficiency of maize. Sci World J. 2013(1):193018

Thuita M, Vanlauwe B, Mutegi E, Masso C (2018) Reducing spatial variability of soybean response to rhizobia inoculants in farms of variable soil fertility in Siaya County of western Kenya. Agric Ecosyst Environ 261:153–160

Ul-Allah S, Ijaz M, Nawaz A, Sattar A, Sher A, Naeem M, Shahzad U, Farooq U, Nawaz F, Mahmood K (2020) Potassium application improves grain yield and alleviates drought susceptibility in diverse maize hybrids. Plant 9(1):75

Ulzen J, Abaidoo RC, Mensah NE, Masso C, AbdelGadir AH (2016) Bradyrhizobium inoculants enhance grain yields of soybean and Cowpea in Northern Ghana. Front Plant Sci 7:1770

Uralkali website (2020) Press release. https://www.uralkali.com/press_center/press_releases/item43020/. Accessed 22 June 2020

Van der Pleog RR, Böhm W, Kirkham MB (1999) On the origin of the theory of mineral nutrition of plants and the law of the minimum. Soil Sci Soc Am J 63:1055–1062

Vejchasarn P, Lynch JP, Brown KM (2016) Genetic variability in phosphorus responses of rice root phenotypes. Rice 9(1):29

Vicente EJ, Dean DR (2017) Keeping the nitrogen-fixation dream alive. Proc Natl Acad Sci U S A 114(12):3009–3011. https://doi.org/10.1073/pnas.1701560114

Walsh OS, Shafian S, Christiaens RJ (2018) Nitrogen fertilizer management in dryland wheat cropping systems. Plants 7(1):9

Wang M, Zheng Q, Shen Q, Guo S (2013) The critical role of potassium in plant stress response. Int J Mol Sci 14(4):7370–7390. https://doi.org/10.3390/ijms14047370

Wang XG, Zhao XH, Jiang CJ, Li CH, Cong S, Wu D, Chen YQ, Wang CY (2015) Effects of potassium deficiency on photosynthesis and photoprotection mechanisms in soybean (*Glycine max* (L.) Merr.). J Integr Agric 14(5):856–863

Wang Y, Zhang Z, Liang Y, Han Y, Han Y, Tan J (2020) High potassium application rate increased grain yield of shading-stressed winter wheat by improving photosynthesis and photosynthate translocation. Front Plant Sci 11:134

Wendeborn S (2020) The chemistry, biology, and modulation of ammonium nitrification in soil. Angew Chem Int Ed Engl. 59(6):2182–2202

Western Potash Corporation Website (2020) Saskatchewan Potash. https://www.westernpotash.com/about-potash/prairie-potash. Accessed 22 June 2020

William G (2019) Top phosphate mining production by country. Phosphate, investing news. https://investingnews.com/daily/resource-investing/agriculture-investing/phosphate-investing/top-phosphate-producing-countries/. Accessed 22 June 2020

Wissuwa M (2003) How do plants achieve tolerance to phosphorus deficiency? Small causes with big effects. Plant Physiol 133(4):1947–1958

Wong ML, Charnay BD, Gao P, Yung YL, Russell MJ (2017) Nitrogen oxides in early earth's atmosphere as electron acceptors for life's emergence. Astrobiology 17(10):975–983

Withers P, Rodrigues M, Soltangheisi A, de Carvalho TS, Guilherme L, Benites VM, Gatiboni LC, de Sousa D, Nunes RS, Rosolem CA, Andreote FD, Oliveira A Jr, Coutinho E, Pavinato PS (2018) Transitions to sustainable management of phosphorus in Brazilian agriculture. Sci Rep 8(1):2537

Xiang DB, Yong TW, Yang WY, Wan Y, Gong WZ, Cui L, Lei T (2012) Effect of phosphorus and potassium nutrition on growth and yield of soybean in relay strip intercropping system. Sci Res Essays 7(3):342–351

Xiong L, Wang P, Hunter MN, Kopittke PM (2018) Bioavailability and movement of hydroxyapatite nanoparticles (HA-NPs) applied as a phosphorus fertiliser in soils. Environ Sci: Nano 5:2888–2898

Yanagida M, Ikai N, Shimanuki M, Sajiki K (2011) Nutrient limitations alter cell division control and chromosome segregation through growth-related kinases and phosphatases. Philos Trans R Soc Lond B Biol Sci 366(1584):3508–3520

Yang HC, Kan CC, Hung TH, Hsieh PH, Wang SY, Hsieh WY, Hsieh MH (2017a) Identification of early ammonium nitrate-responsive genes in rice roots. Sci Rep 7(1):16885

Yang F, Xu X, Wang W, Ma J, Wei D, He P, Pampolino MF, Johnston AM (2017b) Estimating nutrient uptake requirements for soybean using QUEFTS model in China. PLoS One 12(5): e0177509

Yin G, Gu J, Zhang F, Hao L, Cong P, Liu Z (2014) Maize yield response to water supply and fertilizer input in a semi-arid environment of Northeast China. PLoS One 9(1):e86099

Zerkle AL, Mikhail S (2017) The geobiological nitrogen cycle: from microbes to the mantle. Geobiology 15(3):343–352

Zhang Y, Tan L, Zhu Z, Yuan L, Xie D, Sun C (2015) TOND1 confers tolerance to nitrogen deficiency in rice. Plant J 81(3):367–376

Zhang P, Ma G, Wang C, Lu H, Li S, Xie Y, Ma D, Zhu Y, Guo T (2017) Effect of irrigation and nitrogen application on grain amino acid composition and protein quality in winter wheat. PLoS One 12(6):e0178494

Zhang X, Wang B, Zhao Y, Zhang J, Li Z (2019a) Auxin and GA signaling play important roles in the maize response to phosphate deficiency. Plant Sci 283:177–188

Zhang X, Fan Q, Li Q, Du Y, Qin Z, Wei H, Shan F (2019b) The source, distribution, and sedimentary pattern of K-rich Brines in the Qaidam Basin, Western China. Minerals 9(11):655

Zhao D, Raja RK, Kakani V, Read JJ, Carter GA (2003) Corn (Zea mays L.) growth, leaf pigment concentration, photosynthesis and leaf hyperspectral reflectance properties as affected by nitrogen supply. Plant Soil 257:205–218

Zhao X, Du Q, Zhao Y, Wang H, Li Y, Wang X, Yu H (2016) Effects of Different Potassium Stress on Leaf Photosynthesis and Chlorophyll Fluorescence in Maize (*Zea Mays* L.) at Seedling Stage. Agric Sci 7:44–53

Zhou T, Du Y, Ahmed S, Liu T, Ren M, Liu W, Yang W (2016) Genotypic differences in phosphorus efficiency and the performance of physiological characteristics in response to low phosphorus stress of soybean in Southwest of China. Front Plant Sci 7:1776

Zörb C, Senbayram M, Peiter E (2014) Potassium in agriculture--status and perspectives. J Plant Physiol 171(9):656–669

# The Mechanisms of Trace Element Uptake and Transport Up To Grains of Crop Plants

6

Pramod Kumar Singh, Shalini G. Pratap, and Pramod Kumar Tandon

**Abstract**

Trace elements are vital for the operation of metabolic pathways that promote growth and structural integrity. Trace elements are essential for plant growth, i.e., iron, zinc, copper, manganese, cobalt, nickel, boron, molybdenum, and chlorine. Unfortunately, the first two i.e. iron and zinc deficiency in humans has increased on a global scale. The main reasons for these two micronutrients deficiency are dietary intake of food with low levels of Fe and Zn. Biofortification approaches would result in the enrichment of these two elements on target tissue to a considerable extent. Absorption of trace elements is a complex physiological trait which is mainly governed by element transporters and metal chelators of the plant system. Different factors like crop growth stage, edaphic factors, season, etc. also influence the element efficiency of a particular crop. The absorption and transport of trace elements in crop plants are based on the thermodynamics of adsorption on charged solid surfaces embedded in a solution phase of charged ions and metal-binding ligands together with redox systems in the case of iron, zinc, and some other elements. During the normal condition, constitutive absorption systems function in nutrient uptake while in deficiency inducible turbo systems function which increases the supply of a particular nutrient. Crops are divided into two groups, i.e., dicotyledonous crops and monocotyledonous crops. Fe and Zn uptakes in dicotyledonous crops have a turbo system that is an upregulated version of the constitutive system which consists of a membrane-bound reductase and an ATP-driven hydrogen ion extrusion pump while monocotyledonous crops have a constitutive system with an inducible system

P. K. Singh (✉) · S. G. Pratap
Division of Environment Science, School of Basic Sciences, Babu Banarasi Das University, Lucknow, Uttar Pradesh, India

P. K. Tandon
Department of Botany, Lucknow University, Lucknow, Uttar Pradesh, India



K. Mishra et al. (eds.), *Sustainable Solutions for Elemental Deficiency and Excess in Crop Plants*, https://doi.org/10.1007/978-981-15-8636-1_6

remarkably different from dicot that uses the mugenic acid class of phytosiderophores and chelators. Biofortification of these metals involves through enhancing uptake and translocation of Fe and Zn by introducing genes responsible for the biosynthesis of mugineic acid family phytosiderophores (MAs). MAs played an important role in iron transport from an iron-rich soil into the iron-starved crops. Iron is stored in crops after absorption as phytoferritin and transported to active sites by transport-specific ligands. In the grains, Fe and Zn are transported by the phloem sap system through chelation of heavy metals which is governed by three or four genes each that control chelation, membrane transport, and deposition as phytate.

**Keywords**

Plant nutrition · Micronutrients · Adsorption · Transport · Phytosiderophores · Chelation

## 6.1 Introduction

The essential micronutrients which are required by plants i.e., iron (Fe), zinc (Zn), copper (Cu), manganese (Mn), cobalt (Co), nickel (Ni), boron (B), molybdenum (Mo), and chlorine (Cl) (Marschner 1993). Mo and Cl are present in soils as anions and undoubtedly require active transport across the plasmalemma of plant root cells for uptake while B is either an anion or neutral molecule in most soils which is fairly permeable across biological membranes (Stangoulis et al. 2001). The remaining six micronutrients are generally absorbed as divalent ions via divalent ion channels which are specific for each element (Welch 1995).

Fe is essential for plant growth. At the same time, Fe is highly reactive and toxic via the Fenton reaction. Consequently, plants tightly control Fe homeostasis and react to Fe deficiency as well as Fe overload. The ability of plants to respond to Fe availability ultimately affects human nutrition, both in terms of crop yield and the Fe concentration of edible tissues. Deficiency of Fe is a major worldwide agricultural problem. Fe is not easily available in neutral to alkaline soils which cover nearly 30% of global cultivated soils with low iron availability and present in insoluble oxidized forms (Guerinot and Yi 1994; Mori 1999). Dicots and monocots have developed the Strategy I response to iron deficiency stress (Romheld and Marschner 1986; Marschner and Römheld 1994). This response includes acidification of the rhizosphere by releasing protons, subsequent induction of $Fe^{+3}$ chelate reductase activity that reduces $Fe^{+3}$ to $Fe^{+2}$, and acquisition of $Fe^{+2}$ across the plasma membrane of root epidermal cells (Romheld 1987). In the nutrition of chlorophyll-containing plants, Fe has multiple roles. It is a tightly bound component in a wide group of hem-proteins, which include the cytochromes system, catalase, peroxidase, and hemoglobin. Their distribution in plant tissues and cell fractions is quite varied. Deficiency of Fe suppresses cell division and leaf primordial production. Iron may be involved in some reaction of photosynthesis but its role in chlorophyll synthesis is still unsolved although correlation has been found to exist between the chlorophyll content of leaves and their Fe content (Jacobson and Oertli 1956; Hewitt 1961). Fe can be considered to be at least moderately mobile in plants (Branton and Jacobson 1962; Brown Yamaguchi and Leal-Diez 1965). While Zn is also a crucial element for the survival of most organisms including humans (Hambidge 2000). Zn acts as a cofactor of more than 300 proteins (Coleman 1998; Lopez-Millan et al. 2005) and present in all six enzyme classes i.e. oxidoreductase, transferase, hydrolases, lyases, isomerases, and ligases which regulates the activities, conformational stabilization, and folding of various proteins (enzymes). Zn also acts in membrane integrity and stabilization (Cakmak 2000; Disante et al. 2010), alleviation of oxidative stress (Cakmak 2000), and as an intracellular second messenger (Yamasaki et al. 2007). In spite of these Zn is involved in several plant physiological processes such as hormone regulation (e.g., tryptophan synthesis, a precursor of IAA), signal transduction via mitogen-activated protein kinases (Lin et al. 2005; Hansch and Mendel 2009), repair processes of PS II complex during photoinhibition (Bailey et al. 2002; Hansch and Mendel 2009), and maintenance of $CO_2$ concentration in the mesophyll. Peck and McDonald (2010) confirm the participation of Zn in the regulation of Rubisco activity along with alleviation from adverse effects of heat stress in wheat.

Because iron and zinc deficiencies are extremely widespread in humans and are also common in some farm animals, this article focuses on sustainable management of Fe and Zn through fortification and their uptake, transport, and loading into grains that constitute the staple foods of most of the human race. What is known about the uptake, transport, and loading of the other transition elements is generally analogous to iron and zinc.

## 6.2 Soil as a Medium of Plant Growth and Custodian of Plant Nutrient

Soil consists of primary minerals (i.e., sand, silt, and clay), organic matter, water, and air. Thus soil has three-phase systems: there is a solid phase consisting of innumerable minerals and organic substances; the liquid phase consisting of the soil moisture or water in which relatively small amounts of the solid phase dissolve; and the gaseous phase, the soil air which fills the pore space not occupied by the soil moisture. But a soil serving as a medium for plant growth is more than a three-phase system which is a living phase that must always be taken into account in any consideration of soil as a medium for crop production. The pH of the systems which involve many chemical reactions and, particularly biological activity and life is tremendously important. Crop plants as well as plants in general vary greatly as regards the reaction of the nutrient medium in which they grow best. Some species of plants grow best in acidic medium, some in an alkaline medium, and others in a neutral medium or nearly neutral medium. The reason for these reaction preferences is due to the direct and indirect influences of soil reaction. The nutrient elements are held or stored in soils in three degrees or categories of availability; (a) readily available, (b) moderately available, and (c) slowly available. To produce good crops, the supply of readily available forms of nutrients must be high enough so that most of which required by crops can be obtained directly from these forms.

Iron plays an important role in soil chemistry being involved in numerous reactions affecting the chemical and physical properties of soils. Due to its interaction with other elements, it may considerably affect the availability of micro- as well as macro-nutrients. The availability of Fe to plants is greatly affected by various soil properties and also by interaction with other elements. The amount of plant-available Fe in soils depends largely on soil pH and redox potential. In acid soils, Fe is usually in ferrous forms which are available to plants, but in neutral or alkaline soils it is oxidized to ferric iron which may be so insoluble that plants have difficulty in absorbing enough, e.g., at pH 7, the solubility of $Fe^{+3}$ is only $10^{-17}$ moles per liter while that of $Fe^{+2}$ is $10^{-4}$ (Granick 1958). Increasing oxidation potential leads to the oxidation of Fe from ferrous to ferric forms and, therefore, decreases its availability. The oxidation potential increases with increasing pH and with increasing aeration so that the reaction $Fe^{+2}$ to $Fe^{+3}$ due to both of these factors is likely to proceed towards $Fe^{+3}$ and to decreasing Fe availability. Improvements in soil structure may have a similar influence (De Kock 1955; Mokady 1961; Wiklander 1958). Increasing acidity, as well as depletion in soil structure, is likely to cause a reaction towards the reduction of iron. Therefore, the portion of ferrous iron of the total Fe usually increases towards the lower soil horizons due to the decreasing aeration. In poorly drained soils the reduction of ferric to ferrous iron may be effected by decomposing organic matter, and some wet soils may contain so much ferrous iron at low pH values that they are toxic to plant roots (Bear 1965). Although the deficiency of Fe may occur on both acid and alkaline soils, it is perhaps most common on calcareous soils as a lime induced Fe chlorosis. In addition to high pH decreasing the Fe availability, high concentrations of calcium in soil solutions may not only decrease the uptake of Fe on these soils, but there is also evidence of inactivation of Fe within the plant due to abundance of Ca or Ca and P (Mc George 1949; Thorne 1957; Taper and Leach 1957; Olsen 1958; Brown et al. 1959a, b; Miller et al. 1960). Fe deficiency has been induced through both high Ca and P applications, but also through K deficiency. In all cases, the deficiency could be corrected with potassium application (Bolle-Jones 1955).

Zn is also more available in acid than in alkaline soils, the range of least availability being about pH 6.0–7.0. Increased soil pH stimulates Zn adsorption to cation exchange sites of soil constituents (e.g., metal oxides, clay minerals) which reduce Zn availability in soil solution. Moreover, at a pH range of 5.5–7.0, Zn concentration in soil solution (soluble Zn) decreases significantly by 30-fold to 45-fold for each unit increase in soil pH. It is one of the reasons for enhancing the risk of Zn deficiency in crop plants (Marschner 1993). At high pH, the formation of insoluble calcium zincates is favored and Zn may be less available. Therefore liming of acid soils decreases the availability of Zn and may produce Zn deficiency (Thorne et al. 1951). In soils of sodium alkalinity, the availability of Zn may increase with increasing pH due to the formation of relatively soluble sodium zincates (Bear 1955). In several studies, however, no apparent connection between the soil pH and Zn uptake by plants could be established (Tisdale and Nelson 1966). Soil moisture is another physical factor that affects Zn uptake by plant roots via diffusion. The role of soil moisture is very critical in soils with low Zn availability (Rattan and Deb 1981; Marschner 1993). The solubility of Zn and ratio of $Zn^{+2}$ to organic-Zn ligand complexes increases at lower pH especially in soils of low soluble organic matter. Higher levels of soil organic matter enhance the pool of phytoavailable Zn in soil (Catlett et al. 2002; Obrador et al. 2003). In addition to soil pH, the redox potential of soil strongly influences the speed and intensity of the humification process, the redox status of the rhizosphere, and hence the mobility of Zn. In general, the physiological growth and functioning of plants take place at a redox potential range of +300 to +700 mV (Volk 1993). However, reduced conditions (+350 mV) limit the growth of many plants by depriving the oxygen pool of roots. Plant roots modify redox potential of the rhizosphere in two ways: first is through secretion of various reductants and chelating compounds in root exudates and second is through association with the microbial community (Hartmann et al. 2009). Both of the mechanisms influence availability, solubility, and oxidation–reduction state of Zn in soil solution. In conclusion, soil type, mineral and clay types, soil biota and plant uptake, etc., collectively determine Fe and Zn distribution in soil–root–plant fluxes.

## 6.3 Mechanism of Uptake and Transport of Iron From Roots to Stem

The contact theory distinguishes between contact intake and contact depletion of plant nutrients. Soil clays played important role in exchange reaction with root surface, and replenishment of the soil solution by replaceable metallic cations, i.e., Fe and Zn subsequent on their depletion by plants, is a prima facie evidence of the significance of the soil particle-size distribution upon nutrient availability. Plants absorb most of the minerals from the soil via their specialized organs, i.e., roots. The availability of Fe and Zn depends upon physicochemical properties of soil, the activity of plant roots, and micro flora in the rhizosphere and other non-edaphic factors. Some fraction of soil Fe and Zn exists either as insoluble complexes or in adsorbed and exchangeable form. However, another fraction exists in a water-soluble form which is freely available to plants. Root activity also makes the exchangeable form available for uptake to some extent via ion exchange and release of organic acids, etc. The major determining factor affecting soil Fe and Zn distribution is soil pH, which affects the solubility of these elements in soil solution.

At moderate and high soil pH values, Fe deficiency in crops was observed due to the low availability of Fe in the presence of oxygen. These insoluble forms of iron in the soil are solubilized either by complexation of ferric iron [Fe (III)] with chelators to ferrous iron [Fe (II)] due to lowering of the pH value. The mechanisms to adopt these two strategies vary between species by plant roots. The strategy I that is constitutive system adopted by all dicotyledonous plants together with the nongraminaceous monocotyledonous plants while Strategy II in Gramineae plants (Marschner 1995). Both groups have a constitutive system that is adequate to supply plants that are grown in fertile soils having plenty of available forms of Fe. The constitutive system consists of a membrane-bound ferric reductase that is linked to a divalent ion transporter or channel and an ATP-driven proton-extrusion pump. Single amino acid substitutions in the sequence of this channel protein create specificity for the various divalent cations (Rogers et al. 2000). These two membrane functions can supply adequate iron to most plants in healthy soil. However, leaf chlorosis symptoms occur in iron-deficient soils, in which additional mechanisms of iron acquisition are induced to restore iron status in plants. Induced responses are restricted in both strategies to the apical zones of the roots and are fully shut down again within 1 d of restoration of normal iron status. Strategy I plants respond to signals of low iron status by up-regulating the ferric reductase and the proton-extrusion pump. Also, many Strategy I plants have a mechanism for excreting iron-binding ligands and soluble reluctant, which are commonly phenols. All these changes are designed which expressed in the apical zones of the roots where the adaptations are associated with changes in root morphology and the appearance of transfer cells with invaginated membranes. The reductase is stimulated by a low pH level and thereby by the proton-extrusion pump such that its function is effectively inhibited by bicarbonate in high-pH soils. This is the basis for the severe iron chlorosis that is seen in dicotyledonous plants from high-pH soils. Insensitivity to bicarbonate is a feature of Strategy II plants, which induce an entirely new mechanism of mobilizing iron under iron stress. An induced turbo system that is an upregulated version of the constitutive system, Strategy II plants synthesize and release to the soil nonprotein amino acids are known as phytosiderophores (PS) or phytometallophores which recognizes that these amino acids can chelate most of the transition metals and not just iron. These form strong soluble chelates with ferric ions in the soil, and because they are soluble and less positively charged, they are free to diffuse towards the root in soil–water films. Additionally, Strategy II plants have constitutively a highly specific transporter protein [the genes encoding for this transporter most likely belong to the natural resistance associated macrophage protein (NRAMP) family (Curie et al. 2000; Thomine et al. 2000) or the interferon-g-responsive transcript (IRT-1) family (Eide et al. 1996)]. This highly specific transporter protein, which is not present in Strategy I plants, recognizes and transports its specific ferric chelate across the membrane. In the cytoplasm, the ligand is separated from the metal by reduction of the latter, which is then stored in phytoferritin or transported in the plant with ferrous-specific ligands such as nicotianamine. Graminaceous species contain the various members of the PS family in unique ratios: generally, the small-grain cereals such as barley, wheat, oat, and rye have the greatest expression, which explains their remarkable adaptation to the high-pH soils that are usually found in the semi-arid winter cereal-cropping belts of the world. The PS pathway appears to be a major vehicle for the entry of iron into the biosphere from the lithosphere. Curiously, the release of PS from the roots is diurnal and peaks a few hours after sunrise. As in Strategy I plants, the synthesis of PS is quickly suppressed when the plants are restored to adequate iron status, which suggests that these inducible systems are energetically demanding. PS also binds zinc, copper, and manganese and can enhance their absorption along with that of iron. However, with the possible exception of zinc, the mechanism is not induced by deficiency of these other transition metals in the plant. The constitutively expressed extrusion of protons, reductants, and metal-binding ligands will enhance the absorption of all the divalent cations. Inducible systems for upregulated absorption of micronutrients are best understood for iron, and indeed, although the existence of an inducible system in the gut of humans is generally accepted, its nature is not as clearly understood as that in plants and bacteria. The latter has an inducible system that involves the synthesis of members of the hydroxamate group of ferric-binding ligands.

## 6.4 Mechanism of Uptake and Transport of Zinc from Roots to Stem

Variations in Zn requirement and Zn concentration of closely related species may be due to differences in their Zn uptake, transport of root absorbed Zn to shoots, and in Zn sequestration. But downstream of such Zn differences is the tight regulation of Zn absorption for the survival of root cells and Zn homeostasis for normal functioning of physiological processes (transpiration, nutrient absorption, etc.) in which root participates. Zn is taken up mainly as a divalent cation ($Zn^{+2}$ion) by plant roots. However, in some cases, organic ligands-Zn complexes are also absorbed by plant roots. Depending upon the ligands secreted by plant roots, two physiological strategies are involved in the uptake of divalent cations like $Zn^{+2}$. The strategy I involves efflux of reductants, organic acids, and $H^+$ ions, which enhance the solubility of Zn-complexes (Zn phosphates, hydroxides, etc.) and release $Zn^{+2}$ ions for absorption by root epidermal cells. The organic acids released either in root exudates/mucilage or directly by epidermal cells include citric acid, malic acid, oxalic acid or tartaric acid, etc. Strategy II involves the efflux of phytosiderophores (phytometallophores) which form stable complexes with Zn and their subsequent influx into root epidermal cells. However, this absorption mechanism (i.e., strategy II) is restricted to cereal roots. Phytosiderophores are low molecular weight organic compounds (particularly nonprotein amino acids such as nicotianamine, deoxymuigenic acid, avenic acid, etc.) that possess a high binding affinity for their respective metals resulting in their chelation and acquisition. Biosynthesis of these compounds involves condensation of S-adenosylmethionine molecules, causing the formation of nicotianamine, which is a precursor of phytosiderophores.

For an understanding of mineral absorption, one should keep in mind that roots are not just static organs. Plant roots release various organic acids, amino acids, sugars, protons, even some mineral ions, etc. in the rhizosphere that facilitate their adequate functioning and growth. Zn is absorbed as divalent metal ion $Zn^{2+}$ through mass flow and diffusion mechanisms by roots. Passive Zn uptakes by these mechanisms involve the participation of water (solvent) molecules and differences in Zn concentrations across the root cell-plasma membrane (RCPM). The main driving force in $Zn^{2+}$ uptake (cation uptake) is the hyperpolarization of RCPM which is mediated through the activity of the RCPM $H^+$-ATPase system. The RCPM $H^+$-ATPase system actively pumps $H^+$ ion extracellularly at the expense of

ATP hydrolysis. Release of H+ ion in the rhizosphere causes hyperpolarization of RCPM on one hand while reduces the soil pH on the other hand which results in increased cation uptake rate. But unlike water, charged Zn ions are not able to cross cell membranes freely (Alberts et al. 2007), so these divalent cations are transported by specific transporter proteins (Guerinot 2000; Clemens et al. 2002). These proteins are not in close association with ATP breakdown which confirms passive uptake of Zn rather than active. Furthermore, $Zn^{2+}$ uptake also occurs by non-selective cation channels associated with the passive flux of diverse groups of cations (Demidchik et al. 2002). This additional driving force in the uptake of many metal cations is likely due to their very low cytoplasmic activity, which is a result of metal sequestration and their binding to intracellular sites (i.e., Zn finger proteins, organic acids, enzymes, etc.).

## 6.5 Interactive Effect of Zinc with Iron During Uptake

Transport of metal ions in plants consists of events such as uptake from soil solution to root, movement from root to shoot, and ultimately their (metal ion) supply from shoot to various sink organs of the plant. These functions are performed by specific transporter proteins that regulate both intercellular and intracellular transport. Low abundance of the metal ions/trace elements in soil explains the evolutionary development of high affinity metal transporter in plants. A number of genes have been identified till date that either encodes Zn transporter proteins or regulates their expression at transcriptional, post-transcriptional, and translational levels. Most important families of transporter proteins involved in Zn transport are: the ZIP (Zinc-, Iron-Permease family/ZRT-, IRT-like proteins) family, the HMA (Heavy Metal ATPases) family, and the MTP (Metal Tolerance Protein) family. Among these, members of ZIP family perform the function of Zn influx into the cytosol while HMA family participates in efflux of Zn to the apoplast. MTP family is involved in sequestration of Zn into intracellular compartments such as vacuole, endoplasmic reticulum (Palmer and Guerinot 2009).

Root Zn uptake from rhizosphere is greatly influenced by Fe divalent cations which compete with Zn for the same ligands binding site such as in phytosiderophores, transporter proteins like IRT1. Some metal transporters like IRT1 and non-selective cation channels transport $Fe^{+2}$ cations in addition to Zn. Although plants have a basal metal tolerance for these cations but at higher levels, they may have detrimental effects on plant growth. Regarding Fe–Zn interactions, both metal ions have common transporter proteins required for their absorption. Expression studies of Zn deficiency response genes in Fe-deficient mutant (fit mutant) demonstrated low expression of MTP3 and HMA3 suggesting Fe deficiency induces Zn accumulation (Colangelo and Guerinot 2006). The involvement of Fe in transcriptional regulation of the ZIF1 gene has also been demonstrated (Haydon et al. 2012). In contrast, high levels of Zn cause physiological Fe deficiency which is visible as stunted plant growth and chlorotic leaves.

## 6.6 Path of Transport of Trace Elements in Roots

### 6.6.1 Radial Transport

$Fe^{+2}$ and $Zn^{+2}$ are transported radial within the root and pass through several tissues e.g., epidermis, cortex, endodermis, and pericycle before reaching the xylem for transport to shooting. The diameter of these tissue layers decreased when moving towards the stellar region in the center. The radial pattern of transport accelerates water flow in its path to xylem and concentration increases faster at smaller radii for the same flux (lesser solvent volume), both of which help in a large accumulation of ions (metal cations) in the pericycle, which ultimately loads the xylem (Claus et al. 2013). Although specialized transporter proteins are present on cell membranes of respective tissues, yet tight regulation of their expression and activity (rate of Zn transport in per unit time) ultimately decides the intracellular zinc content of root tissues. Several factors such as zinc compartmentation in roots, translocation and use in shoots, sink growth stage, etc. also affect xylem loading of Zn in roots and ultimately lead to differences in symplastic Zn levels.

During the transport of mineral ions to xylem, two main routes are followed: (1) Symplastic route; (2) Apoplastic route. The cytoplasms of adjacent cells in root tissues are connected by cytoplasmic bridges (plasmodesmata) in the cell wall, forming a symplastic continuum without membrane barriers. In addition to this, water and mineral ions can also move via cell wall and intercellular spaces, the apoplast, which also has significant contributions to root mineral transport (Steudle 1994). But the apoplastic route is interrupted in regions of endodermis due to deposition of suberin lamellae in the Casparian strip. Suberin is highly impermeable to water and charged ions. So mineral ions and water molecules need to pass the membrane before the Casparian strip and are transported further through a symplastic route (Yang and Jie 2005). Some of the $Zn^{2+}$ and $Fe^{2+}$ ions are also delivered extracellularly to stellar apoplast in regions where the Casparian band is not fully formed (i.e., near root apex and region of lateral root initiation) as depicted by White et al. (2002) and Ranathunge et al. (2005). Both symplastic and apoplastic fluxes contribute to net Zn and Fe fluxes in the shoot. The apoplastic Zn transport involves entry of Zn from the cell wall plasma membrane interface into the cytosol, making this process less selective than symplastic transport. Contrary to this, symplastic transport controls the selectivity and magnitude of nutrient delivery (Clemens et al. 2002).

### 6.6.2 Transport of Trace Elements in Xylem and Phloem

After crossing the barrier of Casparian strip in root endodermis, metal ions enter simplistically in living cells of pericycle and xylem parenchyma bordering the xylem. Another barrier in metal ion transport occurs at this step of nutrient transfer which is known as xylem loading. It is the key determining step in the root export of metal cations. Continuous activity of $H^+$-ATPase in xylem parenchyma causes membrane hyperpolarization which restricts the movement of positive ions out of the cytosol. Thus, the loading of metal cations from xylem parenchyma/pericycle cells to apoplastic xylem is an active process (Sondergaard et al. 2004). The active efflux of symplasmic $Zn^{2+}$ is mediated by specific transporters (HMA family or P1B-type ATPase) present on pericycle cells/xylem parenchyma. Enhanced activity and higher expression of plasma membrane $H^+$-ATPase in xylem parenchyma is also responsible for the acidic nature of xylem sap (pH 5–6.5). In xylem sap, metal ions are transported mainly as metal complexes with asparagines, histidine, organic acids, and nicotianamine which was studies on chloronerva mutant of tomato that revealed these amino acid regulates metal distribution between apoplastic and symplastic compartments (Pich and Scholz 1991; Takahashi et al. 2003). Energization of nutrients uptake at the site of xylem unloading has also been reported because of dilution of xylem sap by a continuous supply of water through transpirational pull (Sondergaard et al. 2004). Specialized cells are present at such transport interfaces to enhance the symplastic transfer of solutes. These mainly involve transfer cells and vessel associated cells. Both cell types are associated with the sieve element companion cell complex of the phloem. Transfer cells are considered to be the most specialized cell type for membrane transport. Greater number of mitochondria, abundant $H^+$-ATPases, and numerous plasma membrane invaginations of these cells towards the tracheary elements help in nutrient transfer in and out of xylem. On the other hand, vessel associated cells are a special class of xylem parenchyma cells which are associated with numerous large pits to neighboring xylem vessels (Sondergaard et al. 2004). Once Fe and Zn enter the phloem, further translocation to various plant organs and developing sinks is mediated by short and long-distance pathways. Moreover, the mobility of Fe and Zn is higher in phloem than xylem due to increased concentration of chelating solutes (peptides, organic acids, etc.) in phloem sap. Metals are transported either in ionic form or as complex with chelators like Zn nicotianamine, Zn-malate, Zn-histidine complexes in phloem tissues wise versa Fe is also transported nicotianamine (NA). NA is a non-proteogenic amino acid ubiquitous in higher plants, synthesized by the condensation of 3 molecules of S-adenosylmethionine in a reaction catalyzed by nicotianamine synthase (NAS). NA complexes with $Fe^{2+}$ and $Fe^{3+}$; it has a higher affinity for $Fe^{3+}$, but forms a more stable complex with $Fe^{2\ +\ 0}$.73 NA also readily binds $Cu^{2+}$, $Ni^{2+}$, $Co^{2+}$, $Zn^{2+}$, and $Mn^{2+}$, in decreasing order of affinity (Curie et al. 2009). Although xylem contains a lower concentration of solutes, nevertheless is important in nutrient transfer to various organs. Young sink tissues such as developing grains, tubers, etc. are mainly fed by phloem.

## 6.7 Mineral Nutrition During Ontogeny of Plants

Crops showed three fairly distinct stages of metabolism from the period of germination to flowering. The conspicuous stages in the nutrition of the vegetative plant comprise an initial anabolic phase (I) in which intake of inorganic nutrients and synthesis of proteins is rapid. In the second phase (II), the accumulation of carbohydrates accelerates while the rate of protein synthesis gradually diminishes. As flowering is approached, a third or catabolic phase (III) becomes evident in which hydrolysis of reserves begins to overbalance synthesis and a general internal redistribution of nutrients is initiated. Though, conditions of the environment and nutrient supply determine to a considerable degree the exact time of the shift from predominantly anabolic to catabolic activity, the latter characteristic is associated with flowering and commonly initiated before anthesis (Dennison 1945). Some studies (Briggs et al. 1920; Murneek and Wittwer 1942; Wittwer 1943; Biddulph and Brown 1945; Whyte 1946;) indicated that the phenomenon of synopsis or sporogenesis represents a turning point in nutritional metabolism (Hornberger 1882; Knowles and Watkins 1931).

(a) In monoecious species such as corn, it has been observed that the origin of the staminate and pistillate organ is associated with a transitory but systemic acceleration of anabolic over catabolic processes including more rapid nutrient absorption and dry weight gain.
(b) Under normal conditions, the metabolic stimulus associated with synapsis is brief and soon gives way to a reduction in anabolic processes during the ensuing phase of blossoming or anthesis. The flowering phase is usually characterized by subsidence of anabolic activity as well as the inauguration of fundamental modifications (Lugg and Weller 1948) and redistribution of the organic and inorganic nutrient component.
(c) The decline of anabolic activity associated with blossoming gradually gives way to what is usually the final resurgence of absorption of mineral nutrients and acceleration of organic synthesis in vegetative tissues. The anabolic stimulus is associated with the fusion of male and female nuclei in syngamy and the very early enlargement of young fruits (Wittwer 1943; Whyte and Murneek 1948).

The fruiting stage has its origin in syngamy. The early stage of fruit enlargement is commonly associated with marked increments in absorption by roots and accelerated anabolism of the younger parts of the shoot. Gains in nitrogen and potassium become appreciable higher. Absorption of phosphorus and Fe, though fairly steady at first, tend to rise, some time to surprisingly high level at maturation supervenes. There is an increase in the accumulation of protein carbohydrates, the latter usually being greater in terms of dry weight gains. As would be expected, an increasing amount of organic and inorganic reserves are diverted from vegetative to reproductive organs as more fruits are set and their enlargement accelerates. Seeds and fruits are highly selective in the elements which they accumulate from leaves and stems. Fe moves to the seeds, most likely via the phloem, as the flow of the xylem is driven by transpiration, and seeds do not transpire (Grusak 1994). Fe is received in developing seeds from the roots and from senescent leaves. The level of remobilization from shoot to seed varies by species: rice transports only 4% of shoot Fe to the seeds (Marr et al. 1995) versus wheat, that transports 77% of shoot Fe to the seeds (Garnett and Graham 2005). The timing and regulation of senescence have been shown to have a significant effect on Fe accumulation in the seeds. NAC family of transcription factors with RNAi was found to delay senescence in wheat by over 3 weeks, and to decrease seed Fe by over 30% (Uauy et al. 2006). Generally, crop breeding has been often selected for improved grain maturation time, but ignored nutrient accumulation in the grain as a desirable trait. Many staple crops are agronomically productive, but they have low levels of nutrients like Fe and Zn in the seed. Thus, fortification is necessary for sustainable crop production as well as nutrient supplement for human health. Cereal seeds provide more than 50% of the world's energy intake (Bewley 1997) and are a large part of the diet in many developing countries. As the plant based diet offers relatively low amounts of bioavailable Fe and Zn, large portions of the developing world suffer from these metal deficiency, including over 60% of all children in Africa and Southeast Asia (de Benoist et al. 1993). This chapter has focused on understanding how nutrients (particularly Fe and Zn) are transported to seeds, and how this can be increased. This shows the importance of determining how Fe and Zn levels are sensed at the tissue and intracellular level, and how this ultimately affects metal allocation to the seed.

## References

Alberts B, Johnson A, Walter P, Lewis J, Raff M, Roberts K (2007) Molecular biology of the cell, 5th edn. Taylor & Francis, Garland Science, New York

Bailey S, Thompson E, Nixon PJ, Horton P, Mullineaux CW, Robinson C, Mann NH (2002) A critical role for the var-2 FtsH homologue of *Arabidopsis thaliana* in the photo system II repair cycle in vivo. J Biol Chem 277:2006–2011

Bear FE (1955) Chemistry of the soil. Reinhold, New York, p 373

Bear FF (1965) Soils with crop growth. Reinhold, London, p 297

Bewley DB (1997) Seed germination and dormancy. Plant Cell 9:1055–1066

Biddulph O, Brown DH (1945) Growth and phosphorus accumulation in cotton flowers as affected by meiosis and fertilization. Am J Bot 32:182–188

Bolle-Jones EW (1955) The interrelationships of iron and potassium in the potato plant. Plant Soil 6:129–171

Branton D, Jacobson L (1962) Iron transport in pea plants. Plant Physiol 37:539–545

Briggs GE, Kidd F, West C (1920) A quantitative analysis of plant growth. Ann Appl Biol 7:103–123

Brown JC, Holmes RS, Tiffin LO (1959a) Hypotheses concerning iron chlorosis. Soil Sci Soc Am Proc 23:321–324

Brown JC, Tiffin LO, Holmes RS (1959b) Internal activation of iron in soybeans as affected by root growth medium. Soil Sci 87:89–94

Brown Yamaguchi S, Leal-Diez J (1965) Evidence for translocation of iron in plants. Plant Physiol 40:35–38

Cakmak I (2000) Possible roles of zinc in protecting plant cells from damage by reactive oxygen species. New Phytol 146:185–205

Catlett KM, Heil DM, Lindsay WL, Ebinger MH (2002) Soil chemical properties controlling zinc $^{2+}$ activity in 18 Colorado soils. Soil Sci Soc Am J 66:1182–1189

Claus J, Bohmann A, Chavarria Krauser A (2013) Zinc uptake and radial transport in roots of *Arabidopsis thaliana*: a modeling approach to understand accumulation. Ann Bot 112:369–380

Clemens S, Palmgren MG, Kramer U (2002) A long way ahead: understanding and engineering plant metal accumulation. Trends Plant Sci. 7:309–315

Colangelo EP, Guerinot ML (2006) Put the metal to the petal: metal uptake and transport throughout plants. Curr Opin Plant Biol 9:322–330

Coleman JE (1998) Zinc enzymes. Curr Opin Plant Biol 2:222–234
Curie C, Alonso JM, Le Jean M, Ecker JR, Briat JF (2000) Involvement of Nramp1 from *Arabidopsis thaliana* in iron transport. Biochem J 347:749–755
Curie C, Cassin G, Couch D, Divol F, Higuchi K, Le Jean M, Misson J, Schikora A, Czernic P, Takahashi M (2009) Metal Movement Within the Plant: Contribution of Nicotianamine and Yellow Stripe 1-like Transporters. Ann Bot 103:1–11
de Benoist B, Mc Lean E, Egli I, Cogswell M (1993) Worldwide prevalence of anaemia WHO Global Database on Anaemia. World Health Organization, Geneva
De Kock PC (1955) Iron nutrition of plants at high pH. Soil Sci 79:167–175
Demidchik V, Davenport RJ, Tester M (2002) Nonselective cation channels in plants. Annu Rev Plant Biol 53:67–107
Dennison RA (1945) Growth and nutrient responses of Little Turkish Tobacco to long and short photoperiods. Plant Physiol 20:183–199
Disante KB, Fuentes D, Cortina J (2010) Response to the drought of Zn-stressed *Quercus suber* L. seedlings. Environ Exp Bot 70:96–103
Eide D, Broderius M, Fett J, Guerinot ML (1996) A novel iron-regulated metal transporter from plants identified by functional expression in yeast. Proc Natl Acad Sci USA 93:5624–5628
Garnett TP, Graham RD (2005) Distribution and remobilization of iron and copper in wheat. Ann Bot 95:817–826
Granick S (1958) Iron metabolism in animals and plants. In: Lamb CA et al (eds) Trace Elements. Academic Press, New York
Grusak MA (1994) Iron transport to developing ovules of *Pisum sativum*. Plant Physiol 104:649–665
Guerinot ML, Yi Y (1994) Iron: Nutritious, noxious, and not readily available. Plant Physiol 104:815–820
Guerinot ML (2000) The ZIP family of metal transporters. Biochim Biophys Acta 1465:190–198
Hambidge M (2000) Human zinc deficiency. J Nutr 130:1344–1349
Hansch R, Mendel RR (2009) Physiological functions of mineral micronutrients (Cu, Zn, Mn, Fe, Ni, Mo, B, Cl). Curr Opin Plant Biol 12:259–266
Hartmann A, Schmid M, van Tuinen D, Berg G (2009) Plant driven selection of microbes. Plant Soil 321:235–257
Haydon MJ, Kawachi M, Wirtz M, Hillmer S, Hell R, Kramer U (2012) Vacuolar nicotianamine has critical and distinct roles under iron deficiency and for zinc sequestration in Arabidopsis. Plant Cell 24:724–737
Hewitt EJ (1961) The present status of research on the importance of iron, manganese, zinc and boron in crop nutrition. In: Summary and discussion of the papers; Co-op. research project on trace elements of the Sub-Comm. Agriculture Research of ECA meeting in Dublin, May 1959. FAO, Rome
Hornberger R (1882) Chemische Untersuchungen iiber das Wachsthum der Maispflanze. Landw Jahrb 11:359–523
Jacobson L, Oertli JJ (1956) The relation between iron and chlorophyll contents in chlorotic sunflower leaves. Plant Physiol 31:199–204
Knowles F, Watkins JE (1931) The assimilation and translocation of plant nutrients in wheat during growth. J Agric Sci 21:612–637
Lin CW, Chang HB, Huang HJ (2005) Zinc induces mitogen-activated protein kinase activation mediated by reactive oxygen species in rice roots. Plant Physiol Biochem 43:963–968
Lopez-Millan AF, Ellis DR, Grusak MA (2005) Effect of zinc and manganese supply on the activities of superoxide dismutase and carbonic anhydrase in *Medicago truncatula* wild type and raz mutant plants. Plant Sci 168:1015–1022
Lugg JW, Weller RA (1948) Protein in senescent leaves of *Trifolium subterraneum*: partial amino-acid composition. Biochem J 42:412–414
Marr KM, Batten GD, Blakeney AB (1995) Relationships between minerals in Australian brown rice. J Sci Food Agricul 68:285–291

Marschner H, Römheld V (1994) Strategies of plants for the acquisition of iron. Plant Soil 165:261–274
Marschner H (1993) Zinc Uptake from Soils. In: Robson AD (ed) Zinc in soils and plants. Kluwer, Dordrecht, pp 59–77
Marschner H (1995) Mineral Nutrition of Higher Plants, 2nd edn. Academic Press, London
Mc George JT (1949) A study of lime-induced chlorosis in Arizona orchards. University of Arizona Technical Bulletin, Tucson, p 117
Miller GW, Brown JC, Holmes RS (1960) Chlorosis in soybean as related to iron, phosphorus, bicarbonate, and cytochrome oxidase activity. Plant Physiol 35:619–625
Mokady R (1961) The effect of the partial pressure of oxygen in field soils on lime-induced chlorosis. Plant Soil 15:377–386
Mori S (1999) Iron acquisition by plants. Curr Opin Plant Biol 2:250–253
Murneek AE, Wittwer SH (1942) Relation of sexual reproduction to development of horticultural plants. I. General effects of flower and fruit production. Proc Am Soc Horticul Sci 40:201–204
Obrador A, Novillo J, Alvarez JM (2003) Mobility and availability to plants of two zinc sources applied to a calcareous soil. Soil Sci Soc Am J 67:564–572
Olsen C (1958) Iron absorption in different plant species as a function of the pH value of the solution. CR rrav Lab, r:arlsberg 31:41–53
Palmer CM, Guerinot ML (2009) Facing the challenges of Cu, Fe and Zn homeostasis in plants. Nat Chem Biol 5:333–340
Peck AW, McDonald GK (2010) Adequate zinc nutrition alleviates the adverse effects of heat stress in bread wheat. Plant Soil 337:355–374
Pich A, Scholz G (1991) Nicotianamine and the distribution of iron into apoplast and symplast of tomato (*Lycopersicon esculentum* Mill.). II. Uptake of iron by protoplasts from the variety Bonner Beste and its nicotianamine-less mutant chloronerva and the compartmentation of iron in leaves. J Exp Bot 42:1517–1523
Ranathunge K, Steudle E, Lafitte R (2005) A new precipitation technique provides evidence for the permeability of Casparian bands to ions in young roots of corn (*Zea mays* L.) and rice (*Oryza sativa* L.). Plant Cell Environ 28:1450–1462
Rattan RK, Deb DL (1981) Self-diffusion of zinc and iron in soils as affected by pH, $CaCO_3$, moisture, carrier and phosphorus levels. Plant Soil 63:377–393
Rogers EE, Eide D, Guerinot ML (2000) Altered selectivity in an Arabidopsis metal transporter. Proc Natl Acad Sci USA 97:12356–12360
Romheld V, Marschner H (1986) Evidence for a specific uptake system for iron phytosiderophore in roots of grasses. Plant Physiol 80:175–180
Romheld V (1987) Different strategies for iron acquisition in higher plants. Physiol Plant 7:231–234
Sondergaard TE, Schulz A, Palmgren MG (2004) Energization of transport processes in plants. Roles of the plasma membrane $H^+$-ATPase. Plant Physiol 136:2475–2482
Stangoulis JCR, Reid RJ, Brown PH, Graham RD (2001) Kinetic analysis of boron transport in Chara. Planta 213:142–146
Steudle E (1994) Water transport across roots. Plant Soil 167:79–90
Takahashi M, Terada Y, Nakai I, Nakanishi H, Yoshimura E, Mori S, Nishizawa-Naoko K (2003) Role of nicotianamine in the intracellular delivery of metals and plant reproductive development. Plant Cell 15:1263–1280
Taper CD, Leach W (1957) Studies in mineral nutrition III. The effects of calcium concentration in culture solution upon the absorption of iron and manganese by dwarf kidney beans. Canal J Bot 357:773–777
Thomine S, Wang R, Ward JM, Crawford NM, Schroeder JI (2000) Cadmium and iron transport by members of a plant metal transporter family in Arabidopsis with homology to Nramp genes. Proc Natl Acad Sci USA 97:4991–4996
Thorne DW (1957) Zinc deficiency and its control. Adv Agron 9:31–65

Thorne DW, Wann FB, Robinson W (1951) Hypotheses concerning lime-induced chlorosis. Soil Sci Soc Am Proc 15:254–258

Tisdale SL, Nelson WI (1966) Soil Fertility and Fertilizers. Macmillan, New York, p 694

Uauy C, Distelfeld A, Fahima T, Blechl A, Dubcovsky J (2006) A NAC Gene Regulating Senescence Improves Grain Protein, Zinc, and Iron Content in Wheat. Science 5803:1298–1301

Volk NJ (1993) The effect of oxidation-reduction potential on plant growth. J Am Soc Agron 31:665–670

Welch RM (1995) Micronutrient nutrition of plants. Critical Review of Plant Science 14:49–82

White PJ, Whiting SN, Baker AJM, Broadley MR (2002) Does zinc move apoplastically to the xylem in roots of *Thlaspi caerulescens*. New Phytol 153:199–211

Whyte RO, Murneek AE (1948) Vernalization and Photoperiodism. Chronica Botanica Co., Waltham

Whyte RO (1946) Crop Production and Environment. Faber and Faber Ltd, London

Wiklander, L. (1958). The soil. *Handbuch der Pflan'enphysiologie* (ed. W. Ruhland)

Wittwer SH (1943) Growth hormone production during sexual reproduction of higher plants: with special reference to synapsis and syngamy. Missouri Agriculture Experiment Station Research Bulletin, Columbia, p 371

Yamasaki S, Sakata-Sogawa K, Hasegawa A, Suzuki T, Kabu K, Sato E, Kurosaki T, Yamashita S, Tokunga M, Nishida K, Hirano T (2007) Zinc is a novel intracellular second messenger. J Cell Biol 177:637–645

Yang HQ, Jie YL (2005) Uptake and transport of calcium in plants. J Plant Physiol Mol Biol 31:227–234

# 7 Biofortification of Crop Plants: A Practical Solution to Tackle Elemental Deficiency

Bruna Moreira Freire, Rodrigo Mendes Pereira, Camila Neves Lange, and Bruno Lemos Batista


**Abstract**

Malnutrition englobes overnutrition and undernutrition. One in four children suffer from chronic undernutrition and approximately 820 million people have a caloric deficit. The effects of malnutrition are transgenerational and they have an impact from the individual to the national level. Although globally there is sufficient food for all, several countries have inadequately domestic food production. Moreover, the deficit in micronutrient achieves about 3 billion people worldwide due to the lower levels in food or availability of these micronutrients for absorption by the intestines. Therefore, agronomic sciences have an important role in providing nutritious food (quality) rather than adequate calories (quantity). In this scenario, biofortification is a notable tool to improve individual nutritional status. Biofortification is the use of the most appropriate biotechnological or traditional breeding practices for micronutrient enrichment (such as vitamins and chemical elements) of staple crops. From the chemical elements considered essential, the deficiencies of calcium, copper, iodine, iron, magnesium, selenium, and zinc are the most common. Several studies for biofortification were conducted focusing the use of agronomic approaches (use of fertilizers in soils, irrigation water, and hydroponic cultivation systems, or by the foliar application during plant growth), conventional breeding, and genetic approaches (the ancient breeding to the modern genetic engineering employing synthetic genes), and the plant growth-promoting microorganisms (PGPM) approaches (use of microorganisms in soil/plant rhizosphere during plant growth). These biofortification approaches have disadvantages and advantages and are dependent on important variables such as farming practices and soil properties. Moreover, biofortification must be associated with the plant-resistance to stress during



B. M. Freire · R. M. Pereira · C. N. Lange · B. L. Batista (✉)
Centro de Ciências Naturais e Humanas, Universidade Federal do ABC, São Paulo, SP, Brazil
e-mail: bruno.lemos@ufabc.edu.br





K. Mishra et al. (eds.), *Sustainable Solutions for Elemental Deficiency and Excess in Crop Plants*, https://doi.org/10.1007/978-981-15-8636-1_7

cultivation, yield improvement, food color/palatability, and the bioavailability of the nutrients after human ingestion. The highest number of publications on biofortification are from countries that are among the main food producers in the world (the USA, India, China, Australia, and Brazil), evidencing the importance of this technique in contributing to more nutritious food, especially for the poverty population.

Keywords

Zinc · Iodine · Breeding · Genetic · Fertilizers · Bioavailability · Hidden hungry · Farming practices

## 7.1 Why Crop Biofortification Is Necessary?

According to the Food and Agriculture Organization of the United Nations (FAO 2019) after continuously declining for over a decade, global hunger is on the rise again, especially in regions where economic slowdowns occurred. It is also worth noting that this tendency would be more pronounced in the next following years as a consequence of the coronavirus (COVID-19) pandemic and its economic impacts (Hafiz et al. 2020).

Malnutrition englobes both undernutrition (wasting, stunting, underweight, and mineral- and vitamin-related malnutrition) and overnutrition (overweight, obesity, and diet-related noncommunicable diseases) (Dukhi 2020). Around 820 million are in a state of caloric deficit (FAO 2019); nearly one in four children suffer from chronic malnutrition; 52 million children suffer from acute malnutrition; and two billion adults are overweight (Meybeck et al. 2018). Malnutrition is a root cause of many health disorders and it is an imbalance of needs and intake of nutrients and calories. The main reasons for this problem are unavailability or lack of access of food; poor diet (due to a person's inability to select, take-in, and absorb the nutrients in the food); vulnerability of an individual (i.e., increased micronutrient needs during certain life stages, including pregnancy; and health problems such as diseases, infections, or parasites that can spread in unhealthy environments with poor water, sanitation, and hygiene conditions) and finally, decrease in the micronutrient content of common crops due to productivity demands and climate change (Stein 2010; Von Grebmer et al. 2014; Nelson et al. 2018). According to Nelson et al. (2018), the challenge in 2050 will be providing nutritious diets rather than adequate calories and future studies and policies should emphasize nutritional quality by increasing the availability and affordability of nutrient-dense foods and improving dietary diversity (Nelson et al. 2018). However, in the past, the main focus of the professionals from the agriculture field was on yield increasing without balancing the nutritional qualities of staple crops (Stein 2010). Production systems must, therefore, align with nutritional and health goals (Geyik et al. 2020).

Chronic malnutrition effects are transgenerational and they have an impact at the individual, community, and national levels in the short- and long term (Reinhardt and Fanzo 2014). Thus, dietary solutions that could manage to balance nutritional, economic, environmental, and health pillars are a great challenge for a sustainable future and will require the efforts of agriculturists, public health professionals, educators, nutritionists, policy-makers, and food industries (Tilman and Clark 2014).

Minerals and vitamins malnutrition are defined as hidden hunger and its effects hold significant and immediate negative consequences for the cognitive and physical development of children, however in long term may cause profound consequences in health, on productivity and economic potential in later adulthood (Ruel-Bergeron et al. 2015; Biesalski 2013). In diets from nutritionally vulnerable groups, the co-occurrence of deficiencies from more than one micronutrient is common (Ruel-Bergeron et al. 2015). The most affected continents are Africa and South Asia, nevertheless, it occurs globally, especially to underprivileged people (FAO 2019). In a recent study, Geyik et al. (2020) have explored the spatiotemporal trends in dietary nutrient content and adequacy of primary production based on the production of 174 commodities covering a period of 1995–2015 for 177 countries. The authors highlighted that while total production can adequately provide the global human population with all nutrients except vitamin A, more than 120 countries have inadequate domestic production (Geyik et al. 2020).

Mineral nutrients are fundamentally metals and other inorganic compounds (Gupta and Gupta 2014). Adequate mineral intake is needed for the maintenance of normal organism functions. However, about 3 billion people worldwide have a micronutrient deficient diet (Khush et al. 2012). Factors contributing to this scenario are low concentrations or low bioavailability of these nutrients in food (El-Ramady et al. 2014).

Considering the hidden hungry is a serious public health concern, many strategies have been developed to overcome this problem (Khush et al. 2012). No single intervention will offer a "silver bullet" to micronutrient deficiencies, but there are some strategies commonly employed, such as supplementation, dietary diversification, food fortification, and biofortification (Bouis and Saltzman 2017; Khush et al. 2012; White and Broadley 2009). Although been defended by nutritionists, dietary diversification is a contradictory strategy since people tend to return to their old habits (Khush et al. 2012). The World Health Organization (WHO) highlights food fortification and nutrient supplementation as strategies to combat malnutrition (WHO 2019). The Consultative Group on International Agricultural Research (CGIAR) emphasizes the importance of biofortification through breeding and biotechnological approaches (Khush et al. 2012). By making staple foods more nutritious, people can overcome malnutrition without changing their habits.

According to Nestel et al. (2006), the definition of biofortification is the process of development of micronutrient-rich staple crops using the most suitable traditional breeding practices and recent biotechnology to develop staple crops (Nestel et al. 2006). It is important to biofortified staple foods even if they accumulate micronutrients in a relatively low rate since they are consumed regularly in larger quantities for many vulnerable populations in the way that they can enhance the micronutrient status of these populations (Junqueira-Franco et al. 2018). However, the feasibility of biofortification depends on: (1) nutrients bioavailability for plants and humans; (2) nutrients stability after harvesting of the crop (not degrade during processing, storage, and preparation); (3) the acceptance of the crop sensory qualities by producers and consumers in the target regions; (4) provide high yielding and profitability to the producers (Sharma et al. 2017). This process should be comparatively cost-effective, sustainable, and long-terms of delivering more micronutrients (Saltzman et al. 2013).

There is an estimative that by the end of 2018, 7.6 million farming households were growing biofortified planting material, benefiting around 38 million people (HarvestPlus 2019). According to Herrington et al. (2019), the selection of the regions, crops, and micronutrients to prioritize biofortification should be based on production, consumption, and micronutrient deficiency using country-level data (Herrington et al. 2019). Also, the continuous search for new techniques, or the improvement of the existing biofortification techniques, is essential to continue this positive scenario and expand the food biofortification around the world. In this chapter, an overview of the biofortification of crop plants will be described, and several studies showing a wide variety of biofortification approaches will be discussed to demonstrate the main challenges and trends.

## 7.2 Mineral Requirements in Human Nutrition

Considering dietary minerals, there are more than twenty elements considered essential for human body maintenance (Williams 2005). Adequate mineral intake is needed for the maintenance of normal organism functions. However, about 3 billion people worldwide have a micronutrient deficient diet (Khush et al. 2012). Factors contributing to this scenario are low concentrations or low bioavailability of these nutrients in food (El-Ramady et al. 2014). The hidden hunger or micronutrient deficiencies resulting from unbalanced diets is a high priority issue that impedes human and economic development (Khush et al. 2012; Valença et al. 2017). Nowadays, the big challenge is increasing the productivity and the concentration of micronutrients in food crops (El-Ramady et al. 2014).

Around the world, the most common and devastating mineral deficiencies involve calcium, copper, iodine, iron, magnesium, selenium, and zinc. The main functions, as well as the problems related to deficiency and/or excess, are presented in Table 7.1 (White and Broadley 2009; Khush et al. 2012). It is worth mentioning that worldwide, starchy food crops such as rice, maize, wheat, cassava, and legumes are the main focus of biofortification programs. It occurs because these foods are prevalent in the diet of the majority world population, especially for the most vulnerable populations who do not have access to supplements, diverse diets, and commercially fortified foods (Saltzman et al. 2013). Studies have shown that some crops such as

**Table 7.1** Main functions of minerals in the human organism and some problems related to inadequate intake of calcium, copper, iodine, iron, magnesium, selenium, and zinc in the human diet

| Element | Description |
|---|---|
| Calcium (Ca) | It is the most abundant mineral in the human body, and it is present mainly in the skeleton. It plays many essential functions, such as supporting the structure and hardness of bones and teeth, being also vital for muscle movement, enzymes, hormones release, and blood movement through blood vessels (Weaver 2012; NIH 2020a; Gharibzahedi and Jafari 2017). Besides, nerves need Ca to transmit messages between different parts of the body. The average recommended intake for this element in adults is 1000 mg $day^{-1}$ (NIH 2020a). In general, Ca ingestion around the world is below the recommended intake, which increases the risk of many diseases (Weaver 2012). Insufficient Ca intake leads the body to take it from bone to keep healthy levels in the blood. Calcium deficiency can cause osteoporosis and fractures due to the decrease in bone mass. Other possible consequences are convulsions, numbness and tingling in the fingers, and abnormal heart rhythm (NIH 2020a). |
| Copper (Cu) | The essentiality of Cu is linked to brain development, maintenance of immune and nervous systems, and gene activation (NIH 2020b; Gharibzahedi and Jafari 2017). Copper is a constituent of various enzymes, which take part in many metabolic reactions. These cuproenzymes are involved in energy production and utilization, synthesis of proteins of blood vessels, and connective tissues (NIH 2020b). The recommended intake of Cu for adults is 900 μg $day^{-1}$ (IOM 2000; NIH 2020b). Some effects of Cu deficiency are extreme tiredness, high cholesterol levels, weak and brittle bones. Connective tissue disorders, loss of balance, and coordination can also occur. People with a diet deficient in Cu are at increased risk of infection (NIH 2020b). |
| Iodine (I) | It is an essential mineral for thyroid function, being constituent of the hormones T3 and T4 (Gonzali et al. 2017). These hormones are relevant for the body's metabolism, growth, development, reproduction, nerve and muscle function, production of blood cells, among others (Gharibzahedi and Jafari 2017). During infancy and pregnancy, thyroid hormones are essential for proper brain and bone development (IOM 2000; NIH 2019a). The recommended intake of I is 150 μg $day^{-1}$ for adults (NIH 2019a). Iodine deficiency is a widespread problem, affecting both developing and developed countries (Gonzali et al. 2017). In children, cognitive development and mental health can be compromised. Among pregnant women, some possible consequences are spontaneous abortion, stillbirth, and congenital abnormalities (NIH 2019a). According to the World Health Organization (WHO), iodine deficiency is the most prevalent cause of brain damage in the world (WHO 2013). |
| Iron (Fe) | It is responsible for oxygen transport, antioxidant activity, hormone synthesis, neurodevelopment, connective tissues synthesis, and energy metabolism (Aggett 2012; Gharibzahedi and Jafari 2017). Iron deficiency is the most common and widespread nutritional disorder in the world, leading to severe anemia. It is estimated that iron deficiency anemia (IDA) affects 2 billion people around the world, mainly in developing countries. Meantime, it is the only nutrient deficiency that is also prevalent in industrialized countries (WHO 2019). The recommended intake of Fe varies between 8 and 18 mg $day^{-1}$ for adults, depending on gender (NIH 2020c). Among the consequences of IDA in adults are weakness, irritability, and reduced work productivity. In children, IDA can lead to susceptibility to disease, impaired physical and mental development, and increased mortality risk. In developing countries, IDA affects around 40% of preschool children (Khush et al. 2012; WHO 2019). |

(continued)

**Table 7.1** (continued)

| Element | Description |
|---|---|
| Magnesium (Mg) | It is the fourth most abundant cation in the organism and is needed for over 300 metabolic reactions (Volpe 2012). It takes part in the synthesis of DNA, protein, and bone (Gharibzahedi and Jafari 2017). Beyond that, it is important for the regulation of blood sugar levels, blood pressure, muscle, and nerve function. The recommended intake of Mg varies between 310 and 420 mg day$^{-1}$ for adults, depending on gender (NIH 2019b). Magnesium deficiency leads to nausea, vomiting, loss of appetite, weakness, and fatigue. In extreme cases, the symptoms can include personality changes, seizures, abnormal heart rhythm, and muscle cramps (NIH 2019b). Hypertension, cardiovascular disease, and type 2 diabetes mellitus can also be related to Mg deficiency (Volpe 2012). |
| Selenium (Se) | It is an essential trace element in the human diet, being necessary for a narrow concentration range (Skalickova et al. 2017; Jones et al. 2017). The Recommended Dietary Allowance (RDA) for Se in adults is 55 µg day$^{-1}$, while the tolerable upper intake level (UL) is set at 400 µg day$^{-1}$(IOM 2000). It plays an essential role in the formation of selenoproteins and selenoenzymes, which are very important due to its antioxidant properties aiding on the body protection from the damaging effects of free radicals, potentially toxic elements (e.g., mercury), and other dangerous substances (Skalickova et al. 2017; Yu et al. 2005; Gharibzahedi and Jafari 2017). It is also involved in several metabolic processes such as the production and regulation of thyroid hormones, regulation of redox status, increasing the resistance of the immune system, and reducing risks of some chronic diseases (Skalickova et al. 2017; IOM 2000). Approximately one in seven people around the world have a low Se intake (Jones et al. 2017). Selenium availability in the diet is controlled by some factors, namely geographical location, soil concentration, interactions in the soil-plant system, seasonal changes, and food processing (Jones et al. 2017; Navarro-Alarcon and Cabrera-Vique 2008). Selenium deficiency leads to several metabolic disorders. Some possible effects of inadequate Se intake are Keshan disease (human cardiomyopathy in Se deficient children) and Kashin–Beck disease (human cartilage disease). Selenium deficiency may also increase predisposition to other illnesses (IOM 2000). |
| Zinc (Zn) | It has functions related to growth, physical and cognitive development, and immune function. It is essential for genetic expression, cell division, and programmed cell death. Also, it plays a vital role in the function of many enzymes, such as copper-zinc superoxide dismutase, alcohol dehydrogenase, and other enzymes in the nervous system (Mafra and Cozzolino 2004; Gharibzahedi and Jafari 2017). The recommended intake of Zn varies between 8 and 12 mg day$^{-1}$ for adults, depending on gender (NIH 2020d). Approximately one-third of the world's population has a zinc-deficient diet. Zinc deficiency can harm the immune system, beyond causing oxidative damage (Mafra and Cozzolino 2004; Khush et al. 2012). Other consequences include neuropsychological impairment, hypogeusia, hypogonadism and dermatitis (Mafra and Cozzolino 2004). |

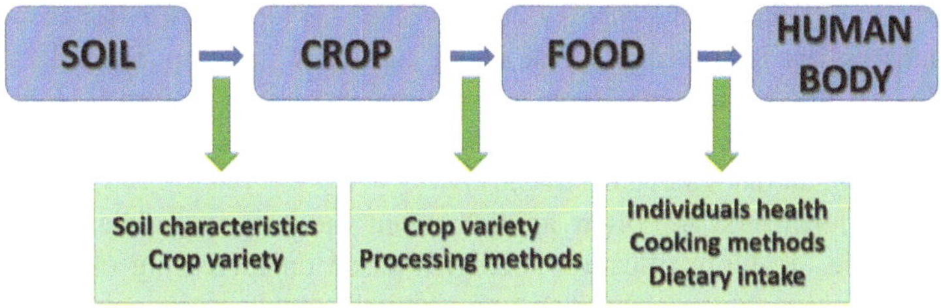


**Fig. 7.1** Schematic representation of nutrients pathway from the soil to the human body, highlighting the main factors that have an impact on elemental bioavailability in each step. Adapted from Valença et al. (2017) and Mayer et al. (2011)

rice, cassava, and maize, in general, present more effective results when submitted to biofortification (Díaz-Gómez et al. 2017).

## 7.3 Biofortification Approaches

Biofortification approaches are usually used to increase the bioavailable mineral content of food crops, and some techniques have been developed and applied for this purpose (White and Broadley 2009; Khush et al. 2012). Despite this, biofortification techniques can be employed with different goals. Crops production in mineral-deficient soils may compromise the growth and yield, for example, and biofortification is also useful to solve these issues (Chugh and Dhaliwal 2013). This fact is because these elements are also essential for the proper development of plants. Thus, biofortification strategies have been also studied aiming yield improvement, resistance to stress, and food palatability (Valença et al. 2017; Gonzali et al. 2017; White and Broadley 2009; Navarro-Alarcon and Cabrera-Vique 2008).

It is important to note that there is a path followed by the minerals from the soil to the human body, passing through the crop and the food, and biofortification strategies should be carefully selected considering each application (Valença et al. 2017). In this path, many factors can influence elemental bioavailability, as shown in Fig. 7.1. In this way, some challenges must be overcome to be successful in biofortification (Valença et al. 2017; White and Broadley 2009). The first challenge is related to the presence and bioavailability of elements in the soil (Valença et al. 2017). It is necessary to be aware of the chemical forms of elements that plant roots can acquire, for example. The biological and physicochemical properties of the soil influence the chemical forms of the elements that will be present in the rhizosphere solution. In this way, the phytoavailability may be affected, limiting the accumulation of these species by crops (White and Broadley 2009). Another critical issue is that different plant varieties can accumulate mineral elements in a wide concentration range, therefore the crop variety needs to be carefully selected for effective biofortification (White and Broadley 2009; Valença et al. 2017). After absorbed by roots, nutrients are translocated to the edible tissues of the crop. Some factors that can influence this process are crop variety and processing methods. Finally, the human ability to absorb nutrients is influenced by individuals' health, dietary intake, and cooking methods (Valença et al. 2017).

Although fertilizers are often applied when the soil is deficient in mineral elements, there are biofortification strategies based on increasing element uptake from soils. These techniques focus on improving the uptake of nutrients by the roots and their redistribution to edible tissues (White and Broadley 2009; Durán et al. 2013). On the other hand, there are agronomic approaches based on fertilizer application in leaves, seeds, as well as in irrigation water and hydroponic cultivation systems. These strategies emerged to circumvent the limitations related to the complex reactions of minerals in the soil and enhance the plant biofortification process. Some minerals have low mobility in the soil depending on the chemical conditions of the soil (pH, composition, etc.) and end up becoming unavailable to plants. In brief, several processes have been used to promote the biofortification of crop plants such as conventional and mutational breeding, genetic engineering, agronomic approaches, among others (Garg et al. 2018; Bouis and Welch 2010; Hirschi 2009; Saltzman et al. 2013). In this chapter, biofortification strategies will be classified into three categories: agronomic, conventional breeding and genetic, and plant growth-promoting microorganisms (PGPM) approaches.

It is worth mentioning that the development of studies to evaluate the strategies for the biofortification of foods has grown significantly in the last 20 years, and approximately 1918 documents were published from January 2000 to March 2020 as can be seen in Fig. 7.2. By the end of March 2020, more than 100 documents had already been published, which shows that this upward trend is expected to continue given the relevance of crop plant biofortification today.

Regarding the percentage of publications by country or territory, the USA (15%), India (10%), China (6%), Australia (5%), and Brazil (4%), stand out as they account for about 40% of the publications presented in Fig. 7.2. Indeed, these countries with a higher number of publications on crop plant biofortification, have a large fertile territory and are the main food producers in the world. In Fig. 7.3, a choropleth map showing the percentage of publications for the biofortification of foods by country or territory is presented.

### 7.3.1 Agronomic Approaches

The agronomic approaches are based on the application of chemical substances containing minerals (fertilizers) during plant growth aiming to increase micronutrient concentrations in edible tissues (Valença et al. 2017; White and Broadley 2009). The most common agronomic approach used for crop plant biofortification is the

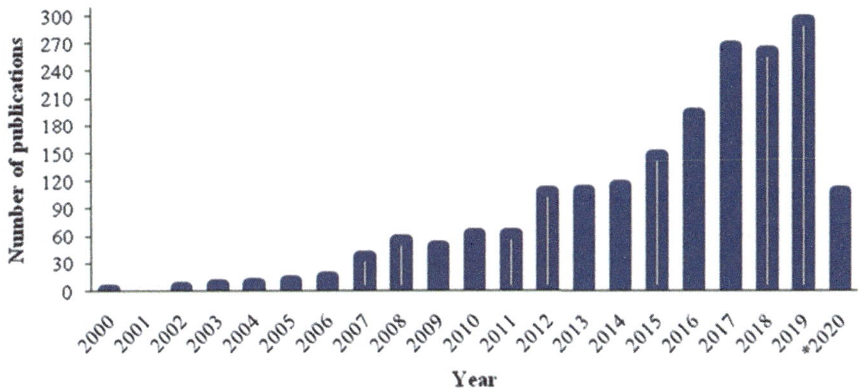


**Fig. 7.2** Number of publications for the biofortification in foods per year, in the last 20 years (from January 2000 to March 2020). The search was conducted on the Scopus database using the following search equation: [TITLE-ABS-KEY ("biofortification") AND TITLE-ABS-KEY ("food" OR "cereal" OR "legume" OR "crop" OR "fruit" OR "staple food"]

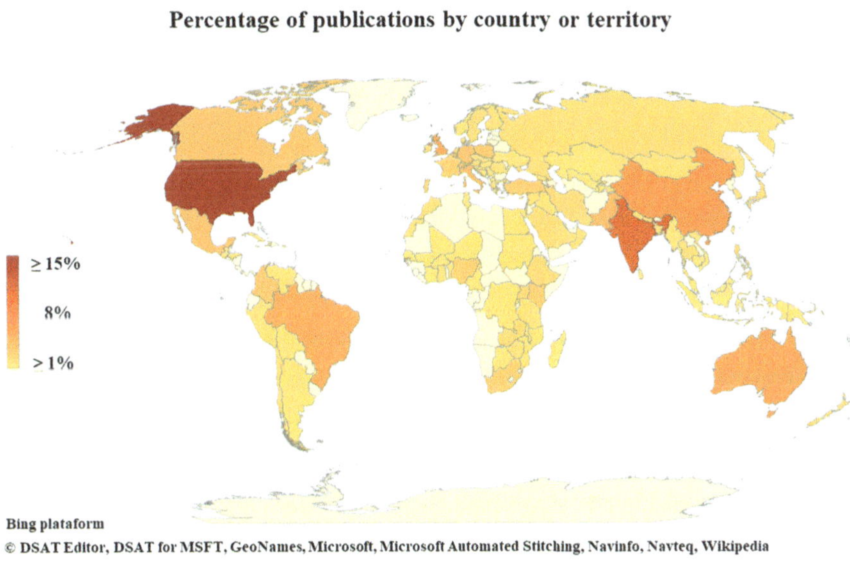


**Fig. 7.3** Choropleth map showing the percentage of publications for the biofortification of foods by country or territory (from January 2000 to March 2020). Data were obtained on the Scopus database using the following search equation: [TITLE-ABS-KEY ("biofortification") AND TITLE-ABS-KEY ("food" OR "cereal" OR "legume" OR "crop" OR "fruit" OR "staple food")]

application of fertilizers in the soil. Solutions of inorganic salts are the most widespread for this purpose. However, micronutrients delivered by using these solutions usually have relatively low availability in the soil since they may be fixed as insoluble forms, or still, be easily released and leached down the soil profile

(El-Ramady et al. 2014). On the other hand, the soil application of algal-based iodized organic fertilizer has proved to be an interesting choice for I biofortification in crop plants (Weng et al. 2013, 2008b; Hong et al. 2008, 2009). Also, some studies have shown that the application of organic amendments such as biosolids biochar and hyperaccumulator plants can also be efficient and advantageous for Fe, Se, and Zn biofortification in crop plants (Gartler et al. 2013; Bañuelos et al. 2015; Ramzani et al. 2016). Nevertheless, the efficiency of soil fertilizer application is dependent on several factors, especially those related to management practices and soil factors, which affect the mobility of elements in soil and their bioavailability for plants.

In this way, the natural process in which plants absorb nutrients through the leaves has been extensively explored in agriculture for crop plant biofortification by foliar application of fertilizers. The foliar application consists of the foliar spray or application of nutrients on aboveground plant parts to supply traditional soil applications of fertilizers. It may be considered one of the most important approaches used for delivering nutrients in suitable concentrations to plants, improving their nutritional status, the crop yield as well as their quality (Alshaal and El-Ramady 2017). This type of application is much less influenced by external factors than soil fertilizer application and, therefore, it has been the target of several studies aiming at the biofortification of crop plants. The use of nanoparticles containing elements that are intended to increase the concentration in the crop plants should be emphasized among the fertilizers used for foliar application. Recent studies have shown that, in addition to biofortification with essential elements, nanoparticles application has promoted the mitigation of toxic elements, such as Cd and Pb, present in the cultivation soil (Hussain et al. 2018, 2020).

Other strategies that have also been successfully used for crop plant biofortification are the application of fertilizers in irrigation water, hydroponic systems, and seeds before cultivation (De Figueiredo et al. 2017; Smoleń et al. 2014, 2015, 2018; Smoleń and Sady 2012; Trolove et al. 2018; Rizwan et al. 2019). In general, agronomic biofortification is simpler and less expensive in the short term when compared with genetic approaches. On the other hand, fertilizer application must be done regularly and may cause damages to the environment, beyond increasing labor and cost in the long term. Some studies showing the application of common agronomic strategies are shown in Table 7.2. In short, these studies demonstrate the main challenges and trends of agronomic approaches for crop plant biofortification.

There is scientific evidence that the nutritional quality of staple crops can be improved by using agronomic biofortification. Valença et al. (2017) stated that these techniques are useful tools for enhancing micronutrient content in edible parts of food crops. Some factors that can influence the success of these approaches are the soil composition, application method, plant species, which can affect mineral mobility and accumulation, and the nutrient accumulation on plant tissues. Thus, some strategies may be limited by geographical locations and crop types, so they may not be applied universally. The efficiency of micronutrient fertilization can be optimized by using integrated soil fertility management, such as combination with organic and NPK fertilizers and selection of improved crop varieties, which can more effectively capture nutrients and accumulate them in consumed parts (Valença et al. 2017).

**Table 7.2** Studies demonstrating the most common agronomic approaches used for crop plant biofortification

| Application | Crop plants | Elements | Results | Reference |
|---|---|---|---|---|
| Foliar | Lettuce | Se | Effects of Se biofortification of lettuce was studied on the toxicity and transfer of Hg from soil contaminated with $HgCl_2$ to the terrestrial food chain, with slug as a primary consumer. Selenium foliar application increased Se concentration in the lettuce and did not affect Hg bioaccumulation. Besides, Se application increased Hg bioavailability for slugs. | Kavčič et al. (2020) |
| | Potato | Se | The foliar application of $Na_2SeO_4$ and $Na_2SeO_3$ for Se biofortification of potato was assessed. Selenate was more efficient than $Na_2SeO_3$ for Se accumulation, and the highest tuber Se concentration was obtained at the tuber bulking stage. Moreover, the major species in tubers treated with inorganic Se was selenomethionine (up to 80% of total Se) suggesting the foliar application is appropriate for the production of Se-rich potatoes. | Zhang et al. (2019) |
| | Rice | Se | Nanoparticles of Se and Si were evaluated for plant growth, metals accumulation, and Se biofortification of rice. The concentrations of Cd and Pb in brown rice were significantly decreased, and the combined application of Se and Si nanoparticles | Hussain et al. (2020) |

(continued)

**Table 7.2** (continued)

| Application | Crop plants | Elements | Results | Reference |
|---|---|---|---|---|
| | | | was the treatment more effective for this purpose (up to 62% and 52% for Cd and Pb, respectively). Moreover, Se and Si nanoparticle foliar applications increase the grain yield, rice biomass, and Se concentrations in brown rice, improving rice quality. | |
| | Rice | Se and Zn | The authors found significant differences in Se and Zn biofortification capability of two Mozambican rice cultivars evaluated in this study. The genotype labeled IR-87684-23-2-3-2 responded better to biofortification. A preferential Se and Zn accumulation in the outer part of the grain was also observed. Selenium and zinc concentrations in whole grain (IR genotype) ranged from 12.6 (control) to 17.9 mg $kg^{-1}$ and from 0.0 (control) to 6.1 mg $kg^{-1}$, respectively. | Mangueze et al. (2018) |
| | Wheat | B, Cu, Fe, Mn and Zn | The foliar application was useful for growth and yield parameters. Besides, foliar application of micronutrients at tillering, jointing, and booting stages allowed the enriching of wheat grains with B, Cu, Fe, Mn, and Zn, with an uptake up to 48% (about 12 g $ha^{-1}$), 74% (38 g $ha^{-1}$), 44% | Aziz et al. (2019) |

(continued)

**Table 7.2** (continued)

| Application | Crop plants | Elements | Results | Reference |
|---|---|---|---|---|
| | | | (196 g ha$^{-1}$), 44% (193 g ha$^{-1}$) and 43% (272 g ha$^{-1}$), respectively. | |
| | Wheat | Se | The foliar spray was effective for Se biofortification of wheat. The dosage of 21 g ha$^{-1}$ of Se was the most suitable (Se absorption efficiency by wheat grains was of about 3%). | Lara et al. (2019) |
| Foliar and hydroponic system | Lettuce | I and Se | Iodine and Se foliar application resulted in better results (higher I and Se concentration in lettuce) in comparison with the hydroponic system application. Iodine and Se concentrations were up to about 75 and 780 mg kg$^{-1}$ in leaves, and up to 85 and 95 mg kg$^{-1}$ in roots, respectively. | Smoleń et al. (2014) |
| Foliar, hydroponic system and soil | Nectarine, plum, potato, and tomato | I | For soil and foliar I application, the highest I concentration was 0.95 and 1.43 μg kg$^{-1}$ for plum and nectarine, respectively. For potato and tomato, the highest I concentration was of 8.94 and 14.4 μg kg$^{-1}$, respectively. Nectarine and plum trees accumulated lower I concentration in their edible tissues compared with potato and tomato. Besides, hydroponic culture proved to be the most efficient system for I biofortification of tomato, since accumulated up to 242 μg kg$^{-1}$ fresh fruit. Iodine was preferably stored in the leaves. | Caffagni et al. (2012) |

(continued)

**Table 7.2** (continued)

| Application | Crop plants | Elements | Results | Reference |
|---|---|---|---|---|
| | | | Only a small I fraction was transported to plum tree branches and fruit, as well as to potato stems and tubers. | |
| | Tomato | I | Tomato plants were treated with radioactive iodine ($Na^{125}I$), and the results indicate that tomato is an excellent crop for I biofortification. Iodine was taken up better when supplied to the roots using hydroponically grown plants. Nevertheless, a considerable I concentration was also stored after foliar application, which suggests that I is also transported through the phloem. Besides, according to the authors, tomato plants can tolerate higher I concentrations. Iodine is stored both in the vegetative tissues and fruits in concentrations that are more than sufficient for the human diet. | Landini et al. (2011) |
| Foliar and soil | Kohlrabi, lettuce, and radish | I | $KIO_3$ and KI solutions (concentrations of up to 15 kg $ha^{-1}$) were evaluated for I biofortification of kohlrabi, lettuce, and radish. Iodine concentration in the edible plant parts increased with the addition of the I fertilizer application. Better results (higher I accumulation and lower growth impairment) | Lawson et al. (2015) |

(continued)

**Table 7.2** (continued)

| Application | Crop plants | Elements | Results | Reference |
|---|---|---|---|---|
| | | | were observed for the $KIO_3$ foliar application. | |
| | Maize and wheat | Zn | Foliar and soil application was performed using $ZnSO_4.7H_2O$. Zinc concentration in grain increased up to 37% for maize and 89% for wheat through foliar Zn application. Besides, foliar Zn application increased up to 22% Fe concentration in maize grain. Soil Zn application does not affect the Zn concentration in both grains. | Wang et al. (2012) |
| | Pea | Zn | Foliar Zn application in combination with soil Zn application promoted increases in grain Zn concentration up to threefold. The effect of processing (freezing and cooking) was also studied in fortified grains and a decrease of about 30% in grain Zn concentration was observed. The combination of soil and foliar application could be a good option for biofortifying field peas. | Poblaciones and Rengel (2016) |
| | Rice | Zn | Zinc concentration increased by 25% and 32% by foliar (about 32 mg $kg^{-1}$) and foliar + soil (about 35 mg $kg^{-1}$) Zn applications, respectively, and only 2.4% by soil (about 19 mg $kg^{-1}$) Zn application. | Phattarakul et al. (2012) |
| | Wheat | Zn | Nanoparticles of ZnO were studied as an alternative for Zn biofortification and | Hussain et al. (2018) |

(continued)

**Table 7.2** (continued)

| Application | Crop plants | Elements | Results | Reference |
|---|---|---|---|---|
| | | | mitigation of Cd. In short, ZnO NPs decreased Cd concentration in wheat by 30–77%, and 16–78% with foliar and soil application, respectively, when compared to the control. Also, its application increased Zn concentration as well as enhance the growth, photosynthesis, and grain yield. | |
| Hydroponic system | Beans | Se and Zn | Biofortification with Se and Zn did not affect the Fe bioavailability, and it proved to be an interesting alternative to increase the food quality of beans. While Se and Zn concentration had an increase of up to threefold twofold, respectively, when compared to those observed for control. | De Figueiredo et al. (2017) |
| | Chinese cabbage | I | NaI and $NaIO_3$ solutions containing I concentrations of up to 5.0 mg $L^{-1}$ and an organic iodine fertilizer (seaweed composite) were evaluated for I biofortification of lettuce. Results show that I uptake by cabbage was more effective using $NaIO_3$ when low I concentration (<0.5 mg $L^{-1}$) was applied. On the other hand, I uptake was also useful using NaI when I concentration of 0.5 mg $L^{-1}$ or higher was applied. NaI and $NaIO_3$ provided a quicker supply for | Weng et al. (2008a) |

(continued)

**Table 7.2** (continued)

| Application | Crop plants | Elements | Results | Reference |
|---|---|---|---|---|
| | | | cabbage uptake, but higher I concentrations were toxic to plant growth. In short, the seaweed composite provided a more sustainable I biofortification for cabbage. | |
| | Lettuce | I | KI and $KIO_3$ solutions containing I concentrations of up to 240 μM were evaluated for I biofortification of lettuce. Based on the results, I concentrations of up to 40 μM using KI were the most appropriate because these concentrations did not reduce biomass when compared to control plants. Also, in these conditions, it was observed the highest foliar I accumulation, and the treated plants show a significant increase in antioxidant compounds. | Blasco et al. (2008) |
| | Lettuce | I | I concentrations of up to 129 μg $L^{-1}$, applied as iodate ($IO_3^-$) or iodide ($I^-$), was evaluated for I biofortification of lettuce in a winter and summer trial. I application did not affect plant biomass, produce quality, or water uptake. Nevertheless, increases in I concentration significantly enhanced I biofortification of the plant, and I concentrations in plant tissue were up to fivefold higher with $I^-$ application. The outer | Voogt et al. (2010) |

(continued)

**Table 7.2** (continued)

| Application | Crop plants | Elements | Results | Reference |
|---|---|---|---|---|
| | | | leaves presented the higher I concentrations. The highest I concentration rates in both trials resulted in a total I concentration of 653 and 764 μg $kg^{-1}$ leaf fresh weight. | |
| | Lettuce | I | The effect of 5-iodosalicylic acid (5I-SA) on the growth, chemical composition, and efficiency of iodine biofortification of lettuce was evaluated. A strong toxic effect on lettuce was observed only when the highest I concentrations as 5I-SA (40 μmol $L^{-1}$) were applied. Iodine concentration of up to 8 mmol $L^{-1}$ as 5I-SA resulted in higher I transfer factor values than those obtained after the application of $KIO_3$ or $KIO_3$ plus 5I-AS. | Smoleń et al. (2017) |
| | Pepper | I | The I hydroponic system application using 0.25–5.0 mg $L^{-1}$ KI solutions made it possible to obtain concentrations of up to 1330 μg $kg^{-1}$ fresh weight, matching the World Health Organization recommendations about dietary iodine allowance. Besides, low-moderate levels (0.25–1.0 mg $L^{-1}$) improved the fruit quality, which demonstrates that pepper can be used as a candidate crop for iodine biofortification. | Li et al. (2017) |

(continued)

**Table 7.2** (continued)

| Application | Crop plants | Elements | Results | Reference |
|---|---|---|---|---|
| | Potato | I and Se | The influence of salicylic acid on I and Se biofortification of potato plants was evaluated in this study. The evaluated treatments had no significant effect on the yield of tubers. The simultaneous application of I, Se, and salicylic acid caused a significant increase in the content of Se in the roots but had no significant effect on the content of Se in the tubers. Besides, this application promoted a decrease of Se concentration in the leaves and petioles. | Smoleń et al. (2018) |
| | Rice | I | KI and $KIO_3$ solutions (concentrations ranging from 1 to 100 μmol $L^{-1}$) were evaluated for I biofortification of rice. However, the authors highlight that none of the treatments provided I enough in the edible parts of rice plants to meet the human dietary requirement. Results suggesting differences in uptake or translocation between I forms since $KIO_3$treatments had more I partitioning to the roots (56%) on average than did the KI treatments (36%). | Mackowiak and Grossl (1999) |
| | Spinach | I | KI and $KIO_3$ solutions (concentrations of up to 100 μmol $L^{-1}$) were evaluated for I biofortification of spinach. The solution-to-spinach leaf transfer factors for plants treated | Zhu et al. (2003) |

(continued)

**Table 7.2** (continued)

| Application | Crop plants | Elements | Results | Reference |
|---|---|---|---|---|
| | | | with KI and $KIO_3$ were up to 20.7 and 23.7, respectively. However, the solution-to-spinach leaf transfer factors had the opposite effects according to the substance used: while it increased with the KI concentration, it decreased with the increase in the $KIO_3$ concentration. Besides, the distribution coefficients of I between leaves and roots were higher for plants treated with $KIO_3$ iodate than those treated with KI. | |
| | Tomato | I | The influence of salicylic acid on I biofortification of tomato using KI or $KIO_3$ was investigated. Salicylic acid combined with KI or $KIO_3$ increased by about 37% and 157%, respectively, the I accumulation in fruits. Although fruits of plants treated with KI contained significantly more I, the treatment with $KIO_3$ was the best for nutritional value. | Smoleń et al. (2015) |
| Irrigation water and seed | Spinach | I | $KIO_3$ and KI solutions were applied for I biofortification using two treatments: pre-sowing fertilization and fertigation. The latter proved to be much more effective for I biofortification of spinach (up to 339% and 498% for KI and $KIO_3$, respectively). Also, I biofortification decreased Na and Zn concentration in | Smoleń and Sady (2012) |

(continued)

**Table 7.2** (continued)

| Application | Crop plants | Elements | Results | Reference |
|---|---|---|---|---|
| | | | spinach, irrespective of the chemical form (KI and $KIO_3$) and method (pre-sowing fertilization or fertigation) used. Besides, pre-sowing fertilization reduced Pb accumulation in spinach. | |
| Seed | Broccoli, green radish and purple radish sprouts | Se | The effect of Se addition on the nutrient composition of broccoli, purple radish and green radish sprouts was investigated for Zn biofortification. In short, Se concentration in sprouts increased exponentially with time, when high Se supply was applied (up to 9 μmol Se $g^{-1}$ seed). | Trolove et al. (2018) |
| | Cucumber, lettuce, and tomato | Se | The Se biofortification of cucumber, lettuce, and tomato, based on the use of Se-enriched peat during the pre-transplanting stage was evaluated. The proposed method using Se concentrations of up to 20 mg $kg^{-1}$ of dry peat induced a Se-enrichment in transplants without adverse effects on plant growth. Edible parts of Se-enriched plants at the end of cropping cycle showed Se concentrations ranging from 29–48 μg $kg^{-1}$ for cucumber, 23–53 μg $kg^{-1}$ for lettuce, and 15–20 μg $kg^{-1}$ for tomato, which was up to 3.4-fold, 8.5-fold, and 1.6-fold, respectively, higher than the control. | Businelli et al. (2015) |

(continued)

**Table 7.2** (continued)

| Application | Crop plants | Elements | Results | Reference |
|---|---|---|---|---|
| | Wheat | Fe and Zn | Nanoparticles of Fe and ZnO were evaluated for Fe and Se biofortification as well as on the plant growth and Cd mitigation of wheat. When the higher nanoparticle treatments were applied, Cd concentration in the grains was below the threshold level of Cd for cereals (0.2 mg $kg^{-1}$). The application of ZnO and Fe nanoparticles increased the Zn and Fe concentrations in roots, shoots, and grains. | Rizwan et al. (2019) |
| | Wheat | Zn | Nanoparticles of ZnO were studied as an alternative for Zn biofortification of wheat, and its effects were compared to the $ZnSO_4$ application. Nanoparticles of ZnO were more effective than the $ZnSO_4$ application for Zn biofortification of wheat grains but less effective at increasing leaf Zn. At moderate concentrations of ZnO NPs and $ZnSO_4$, the grain yield and biomass significantly increased. At high concentrations, $ZnSO_4$ was more toxic than ZnO NPs. | Du et al. (2019) |
| Soil | Aubergine, cucumber, and radish | I | Iodine soil application was performed using mixtures of granular kelp and diatomite with different iodine concentrations (prepared by varying the kelp/diatomite ratio). Iodine concentrations in both leaf and fruit/ rhizome tissues | Weng et al. (2008b) |

(continued)

**Table 7.2** (continued)

| Application | Crop plants | Elements | Results | Reference |
|---|---|---|---|---|
| | | | increased with the increase of I concentrations in iodized fertilizer applied. The use of diatomite helped to increase the durability of the iodized fertilizer. The organic mixture is a safe and interesting alternative for I biofortification of aubergine, cucumber, and radish. | |
| | Barley, maize, potato, tomato, and wheat | I | $KIO_3$ and KI solutions (concentrations of up to 23 mmol $L^{-1}$ and 6 mmol $L^{-1}$, respectively) were evaluated for I biofortification of barley, maize, potato, tomato, and wheat. The plants tolerated higher I concentrations as $IO_3^-$ than $I^-$ in the root environment, and barley showed the lowest biomass reductions. In turn, maize showed the most significant biomass decrease due to I toxicity. In all cases, the KI application provides much higher accumulation efficiency than the $KIO_3$ application. | Caffagni et al. (2011) |
| | Beetroot, broccoli, carrot, corn, courgette, leek, lettuce, onion, radish, spinach, and tomato | Zn | The efficiency of a biosolids/biochar soil amendment in Zn biofortification of vegetables was evaluated. The biomass and Zn concentration of most species significantly increased using the biosolids and biosolids + biochar treatments. The highest increase in Zn | Gartler et al. (2013) |

(continued)

**Table 7.2** (continued)

| Application | Crop plants | Elements | Results | Reference |
|---|---|---|---|---|
| | | | concentrations was observed for beetroot, which was of up to 178 and 1200 mg $kg^{-1}$ in the bulbs and leaves, respectively. Based on results, the mixture of biosolids and biochar is an efficient approach for Zn biofortification of crops with edible leaves as well as beetroot. | |
| | Bitter tomato, African eggplant, and Turkey berry | K | The potassium fertilizer application was assessed for the biofortification of vegetables. While the yield of turkey berry was significantly affected by type, rate, and interactive effect of type and rate of fertilizer application, bitter tomato, and African eggplant were affected only by the rate of fertilizer application. Potassium concentrations in leaves were higher than in fruits of all the vegetables, and the highest K concentrations for bitter tomato (2130 mg $kg^{-1}$ dry weight) and turkey berry (1883 mg $kg^{-1}$ dry weight) was observed when a KCl solution was used. However, the highest K concentration for African eggplant (1801 mg $kg^{-1}$ dry weight) was obtained using a sulfate of Potash solution. | Adu et al. (2018) |
| | Broccoli, and carrots | Se | Soils amended with ground shoots of the Se-hyperaccumulator | Bañuelos et al. (2015) |

(continued)

**Table 7.2** (continued)

| Application | Crop plants | Elements | Results | Reference |
|---|---|---|---|---|
| | | | *Stanleya pinnata* were evaluated for Se biofortification of broccoli, and carrots. Application of higher rates of *S. pinnata* increased total plant Se concentrations inside edible parts. Se-enriched *S. pinnata* proved to be valuable as a soil amendment for Se biofortification of broccoli and carrots. | |
| | Cabbage, coriander, hot pepper, long cowpea, eggplant, potherb mustard, Chinese cabbage, tomato, cucumber, and spinach | I | A novel approach for I biofortification of vegetables with algal-based iodized organic fertilizer was proposed. Ten species of vegetables were tested, and, in general, the I absorption in these vegetables increased with the increasing amount of the algal-based iodized organic fertilizer used. Besides, I uptake by leaf vegetables was significantly higher than that of fruit vegetables. Iodine concentration decreased from root, leaf, stalk, to fruit. | Weng et al. (2013) |
| | Carrot | I and Se | I and Se soil application did not affect yield, but the plants of all genotypes evaluated in this study accumulated both elements in leaves and roots. The concentration of I and Se in roots increased about eightfold and fivefold, respectively. | Smoleń et al. (2019) |
| | Carrot, celery, onion, pak choi, spinach, and water spinach | I | $KIO_3$ was used as fertilizer to evaluate I biofortification of six vegetables. Iodine soil | Dai et al. (2004) |

(continued)

**Table 7.2** (continued)

| Application | Crop plants | Elements | Results | Reference |
|---|---|---|---|---|
| | | | application had a significant effect on the biomass of edible parts of pak choi and spinach. In contrast, it had no significant effect on that of carrots, water spinach, celery, and onion. Iodine concentrations in edible parts of vegetables and the transfer factors of soil-to-edible parts of vegetables significantly increased with I soil application, with emphasis on spinach. Thus, spinach was considered as an efficient vegetable for iodine biofortification. | |
| | Celery, pak choi, pepper, and radish | I | Iodine soil application was evaluated for I biofortification using inorganic (KI) and organic (seaweed fertilizer) forms. The I concentrations in soil decreased with time and with plant growth as well. Iodine from KI and seaweed decreased by 50% and 60% of the applied dose, respectively. Iodine accumulation in the edible portion was ranked as follows: pak choi > celery > radish > pepper. Seaweed fertilizer proved to be a better choice for I biofortification of celery, pak choi, pepper, and radish | Hong et al. (2009) |
| | Chinese cabbage, lettuce, tomato, and carrot | I | Both inorganic iodine (KI) and organic seaweed iodine, were evaluated for I biofortification of | Hong et al. (2008) |

(continued)

**Table 7.2** (continued)

| Application | Crop plants | Elements | Results | Reference |
|---|---|---|---|---|
| | | | cabbage, lettuce, tomato, and carrot. Iodine concentrations in vegetables increased with both I application. Iodine accumulation in the edible portion was ranked as follows: cabbage > lettuce > carrot > tomato. Iodine concentration in cabbage was twofold and fourfold higher than that of lettuce and carrot, respectively, and 20-fold higher than that of tomato. In turn, I distribution in vegetable tissues was: root > leaf > stem > fruit, except for carrot. Organic seaweed iodine application demonstrates more potential for durability than KI. | |
| | Cowpea | Zn | Zn concentrations reached 42 mg $kg^{-1}$ (red clay) and 45 mg $kg^{-1}$ (sandy) in cowpea with Zn soil application against grain Zn concentrations of 36 mg $kg^{-1}$ and 31 mg $kg^{-1}$ measured in cowpea grown with no Zn soil application on red clay and sandy soils, respectively. In general, Zn soil application under integrated soil fertility management increased grain yield and grain Zn content. | Manzeke et al. (2017) |
| | Green beans, and lettuce | I | The I soil application using the two lower concentrations (0.10 and 0.25 mg $L^{-1}$) stimulated the growth of both plants. I concentration in | Dobosy et al. (2020) |

(continued)

**Table 7.2** (continued)

| Application | Crop plants | Elements | Results | Reference |
|---|---|---|---|---|
| | | | edible parts of green bean and lettuce was up to 0.6 and 5.2 mg $kg^{-1}$ dry weight, respectively. | |
| | Lettuce | Se | Sodium selenate and selenite were evaluated for Se biofortification of lettuce. Results indicated that $Na_2SeO_4$ was less toxic form, and it induced greater biomass, higher Se accumulation, and more antioxidant compounds than did $Na_2SeO_3$ application. Sodium selenate concentration of 40 μmol $L^{-1}$ proved the most suitable for lettuce plants. | Ríos et al. (2008) |
| | Onion | Zn | Zn soil application using Zn chelated by EDTA and/or DTPA promoted the better results (up to 7.80 mg $kg^{-1}$ of total Zn concentration; up to 5.16 mg $kg^{-1}$ of soluble Zn concentration, and the highest plant biomass and chlorophyll and carotenoid contents). | Almendros et al. (2015) |
| | Spinach | I | KI and $KIO_3$ solutions containing I concentrations of up to 2 mg $kg^{-1}$ were evaluated for I biofortification of spinach. Biomass productions were not significantly affected, while I concentrations increased with the increasing addition of KI and $KIO_3$. Potassium iodate application provides much higher I concentrations in tissue plants than KI application. Moreover, | Dai et al. (2006) |

(continued)

**Table 7.2** (continued)

| Application | Crop plants | Elements | Results | Reference |
|---|---|---|---|---|
| | | | the soil-to-leaf transfer factors were about tenfold high when plants were grown with $KIO_3$. In short, $KIO_3$ may be considered as potential fertilizer for the I biofortification of spinach. | |
| | Tomato | I | Several substances containing I (KI and $KIO_3$, as well as of organic iodine compounds—5-ISA (5-iodosalicylic acid), 3,5-diISA (3,5-diiodosalicylic acid), 2-IBeA (2-iodobenzoic acid), 4-IBeA (4-iodobenzoic acid) and 2,3,5-triIBeA (2,3,5-triiodobenzoic acid)) were applied to evaluate its uptake by tomato plants. Only 2,3,5-triIBeA harmed plant development. Also, 2-IBeA and 4-IBeA were the most active compounds for transferring iodine to fruits and leaves, respectively. | Halka et al. (2019) |
| | Tomato | I | $KIO_3$ and KI solutions (concentrations of up to 10 mmol $L^{-1}$) were evaluated for I biofortification of tomato. Both treatments promoted a significant increase in the I concentration in the fruits that did not affect plant growth and development. Besides, I soil application did not affect fruit appearance and quality, even with the highest concentrations applied. | Kiferle et al. (2013) |

(continued)

**Table 7.2** (continued)

| Application | Crop plants | Elements | Results | Reference |
|---|---|---|---|---|
| | | | Results suggest that tomato is a suitable crop for I biofortification. | |
| | Wheat | Fe | Fe (as sulfate) soil application combined with biochar and S promoted the highest Fe concentration in grain (up to 1.4-fold). This approach was efficient in improving growth and grain Fe biofortification of wheat in pH affected calcareous soil. | Ramzani et al. (2016) |
| | Wheat | Se | Grain Se concentration increased up to 26 ng $g^{-1}$ fresh weight, for each gram of Se $ha^{-1}$ applied as $Na_2SeO_4$, while yield and harvest index were not affected by Se fertilization. | Broadley et al. (2010) |
| | Wheat | Se and Zn | Twenty Brazil wheat accessions (including 15 varieties and 5 cultivars) were used in this study, and Zn and Se concentrations in grains exhibited about twofold and 1.5-fold difference, respectively, between these wheat accessions. The soil Zn application enhanced grain Zn concentration in all accessions up to threefold. The soil Se and Zn application improved Se and Zn concentration in grain and promoted the additional accumulation of Fe. | Souza et al. (2014) |

The elements most targeted for crop plant biofortification are micronutrients, such as I, Se, and Zn. This may be associated with the fact that these elements are probably more efficiently absorbed by plants. Moreover, they are extremely important for the human organism and are usually found in very low concentrations in crop plants. According to Gonzali et al. (2017), I biofortification of food crops can be a cost-effective approach to control I deficiency with a bioavailable source. In many plant species, such as potato and lettuce, the agronomic approach is sufficient to increase I content. The most common administration ways are in the soil, as a foliar spray or in hydroponic solutions. The chemical form varies since there are studies with the application of organic and inorganic species. Doses and timing of application must be evaluated for each specie (Gonzali et al. 2017).

In turn, Se biofortification by agronomic strategies such as fertilizer application is an efficient way to produce Se-enriched food products (Wan et al. 2018). However, attention is needed since the levels that characterize deficiency, essentiality, and toxicity of this element are very close (Navarro-Alarcon and Cabrera-Vique 2008). The chemical form of Se influences its bioaccessibility. Fortunately, agricultural methods to improve Se bioaccessibility in food products can be used (Wan et al. 2018). Moreover, the major forms of Se in the diet are highly bioavailable (IOM 2000). According to Wan et al. (2018), agronomic strategies may help supply the daily needs of this element, mainly in Se deficiency regions. Other studies have shown that agronomic Se biofortification of cereals is effective to increase Se intake in animals and humans (Valença et al. 2017). On the other hand, processing methods such as heating and milling may decrease Se content in food due to volatilization and solubilization (Wan et al. 2018; Navarro-Alarcon and Cabrera-Vique 2008).

Zinc biofortification of edible crops has been identified as a strategy to improve the intake of this element. For this purpose, agronomic strategies namely Zn-fertilizers application have been employed and showed to increase Zn content in roots, stems, and leaves without compromising yield. Zinc fertilizers showed promising results when applied either in the soil or in leaves and also in combination with nitrogen fertilizers (White and Broadley 2011). There is evidence that nitrogen availability is a key component of Zn biofortification (Hefferon 2015).

Nevertheless, some studies have shown that other nutrients, such as B, Ca, Cu, Fe, K, Mg, Mn, among others, have also been evaluated for plant biofortification (Aziz et al. 2019; Rizwan et al. 2019; Adu et al. 2018). It is important to mention that in addition to being essential elements for the human organism, they are also very important for the proper development of plants. For this reason, in most cases, the agronomic approaches for crop plant biofortification also improve yield and/or food quality, as described in some studies in Table 7.2.

### 7.3.2 Conventional Breeding and Genetic Approaches

Breeding and genetic engineering are the main tools employed in this type of biofortification (Gonzali et al. 2017). Genetic engineering can employ synthetic genes (Khush et al. 2012). In general, these approaches are more complex and laborious than agronomic ones (Gonzali et al. 2017), but are sometimes needed when conventional methods are insufficient to obtain substantial enhancement of the target element (De Steur et al. 2017). The two methods aim to achieve plant lines carrying genes that result in the most efficient accumulation of bioavailable minerals. However, plant breeding achieves this by crossing the best performing plants and selecting those with favorable traits over many generations, whereas genetic engineering accesses genes from any source and introduces them directly into the crop (Gómez-Galera et al. 2010).

Plant breeding started more than 10,000 years ago with the selection of seeds to domestication, as occurred to the crops of maize (*Zea mays* L.), wheat (*Triticum aestivum* L.), barley (*Hordeum vulgare* L.), among others (Dudley 1997). With Mendel's laws, genetic principles began to be applied to plant breeding, ranging from the introduction, phenotypic selection on natural variants, selection with controlled mating, to marker-assisted selection for desirable genes (Allard 1999). In the beginning, plant breeding was performed unconsciously and deliberately by farmers when they kept some plant of the harvest for planting or sowing their next crop. Besides that, natural selection occurred during the genetic diversity of the crop in new environments, during domestication and subsequent dispersion. Then, the hybridization and genetic-based process were added (Bradshaw 2016). Conventional breeding is possible only between closely related (sexually compatible) individuals, thus relies on natural variation of the target compound within parental lines (De Steur et al. 2015). To increase mineral content by breeding is challenging since numerous genes may be involved in elemental uptake by the roots, translocation throughout the plant, and deposition in edible tissues. Moreover, other factors such as environmental conditions and cultural practices can modify gene expression and alter mineral accumulation by plants (Bouis and Welch 2010).

According to Saltzman et al. (2017) more than 30 countries have officially released biofortified varieties developed using the conventional plant breeding approach, and at least an additional 20 countries have commenced the testing of these varieties and they provide considerable amounts of bioavailable micronutrients, and consumption of these varieties may help to reverse the micronutrient deficiency status among target populations (Saltzman et al. 2017).

Genetic approaches refer to developing crops with improved abilities to acquire and accumulate minerals in edible parts. Modified varieties can also present increased concentrations of "promoter" substances, which stimulate mineral absorption and reduced concentrations of "antinutrients", substances that negatively interfere with nutrient absorption. However, food's taste and color may be affected by changes in the concentration of promoters and antinutrients, so these strategies must be cautiously evaluated (White and Broadley 2009). Research and development phases and the regulatory approval process for genetically modified (GM) crops are often time-consuming and expensive. However, after establishment, enhanced crops become sustainable (Khush et al. 2012). Then, in the long term, these strategies can be cost-effective (Gonzali et al. 2017), besides that, they can increase micronutrient concentrations in the desired tissue, such as cereal endosperm, to reduce post milling losses through the outer layers (De Steur et al. 2017). Various genomic approaches, such as quantitative trait loci (QTL) mapping, marker-assisted selection (MAS), marked-assisted recurrent selection (MARS), genome-wide selection (GS), and next-generation sequencing (NGS) have been widely employed for the biofortification.

Multiples genetic approaches are commonly employed to achieve the best results on mineral biofortification. For instance, Masuda et al. (2012) combined three transgenic approaches to produce Fe-biofortified rice: (1) enhancement of Fe storage in grains via expression of the Fe storage protein ferritin using endosperm-specific promoters; (2) enhancement of Fe translocation through overproduction of the natural metal chelator nicotianamine and (3) enhancement of Fe flux into the endosperm through the expression of the Fe(II)–nicotianamine transporter OsYSL2 expression under the control of an endosperm-specific promoter and sucrose transporter promoter. The authors reported that the Fe concentration of polished seeds increased up to sixfold in greenhouse cultivation and 4.4-fold in paddy field cultivation (Masuda et al. 2012).

Johnson et al. (2011) reported that Fe concentrations were increased, reaching 14 mg $kg^{-1}$, in rice grains by GM. Besides that, Fe was unlikely to be bound by phytic acid and therefore likely to be more bioavailable in human diets (Johnson et al. 2011). Conventional breeding is also an option since there is a natural genetic variation in Zn concentrations of edible crops. Other approaches use genetic engineering to develop modified plants with increased abilities to acquire and accumulate Zn. Still, higher Zn concentrations in edible plant parts can be reached with the development of crops with more tolerance to high Zn levels in tissues. There are already genetically modified plants that have higher concentrations of Zn in the edible parts compared to traditional varieties (White and Broadley 2011). It was noted that plants modified to increase Fe accumulation have also presented increased Zn concentrations. It may indicate a cross-talk between Fe and Zn transport pathways (Hirschi 2009). Connorton and Balk (2019) reviewed several GM crops for iron biofortification, including cassava, maize, wheat, rice, soybean, and sweet potato. The authors also mentioned that several quantitative trait loci and transgenes increase both iron and zinc, due to overlap in transporters and chelators for these two mineral micronutrients (Connorton and Balk 2019).

Considering I biofortification, in some cases, there is a need for genetic engineering strategies to guarantee an effective result. It occurs mainly in cereals because the amount that reaches grains is insufficient to supply human needs. Genetic approaches focusing on reducing I volatilization from leaves or aiming to control the uptake and mobilization of this element through the phloem are promising, but still very scarce. There is a need for reliable protocols for I biofortification of staple crops to enable the dissemination of these practices (Gonzali et al. 2017).

Therefore, the development of genetic biofortification methods must consider the impact that these modifications may have on the accumulation of other elements that are not necessarily the object of the study. Other questions that must be considered are the impact of biofortification on plant metabolism, growth, productivity, environment, and conservation of genetic resources (Garcia-Casal et al. 2017). For instance, enzyme activities may be modified by metal content. Finally, possible alterations of plant stress, interactions with other nutrients, and allergic reactions in humans must be evaluated (Hirschi 2009). The main limitations of genetically modifying crops included consumers acceptance and to fulfill the regulatory requirements for labeling and approving commercialization of these crops.

### 7.3.3 Plant Growth-Promoting Microorganisms Approaches

Some recent strategies do not fit the previous definitions since they do not include the application of fertilizers during plant growth or even conventional breeding and genetic strategies. The use of plant growth-promoting microorganisms (PGPM), especially the plant growth-promoting rhizobacteria (PGPR), is one of the strategies that has grown significantly in the last years aiming at the crop plant biofortification. The PGPM approaches consist of the application of beneficial microorganisms (bacteria, fungi, among others) in cultivation soil. The soil application of these microorganisms increases mineral bioavailability contributing to crop plant biofortification and improve the soil fertility and crop yield (Khan et al. 2019; Rana et al. 2012). In turn, PGPR consists of a varied group of beneficial bacteria that colonize the rhizosphere and plant roots (Glick 1995). In short, the PGPR is the soil bacteria that stimulate the growth of the host through increasing mobility, uptake, and enrichment of nutrients in the plant (Prasanna et al. 2016). Moreover, they contribute to plant growth development by fixing biological nitrogen, enhancing root function, suppressing disease, among other benefits (Glick 1995; Vessey 2003; Hafeez et al. 2006).

The application of PGPR in agriculture is an attractive way to minimize the use of fertilizers and related agrochemicals (Rana et al. 2012). According to De Santiago et al. (2011), agronomic and genetic approaches have a higher cost than PGPR application, present ethical problems, and are non-environmental friendly. In this way, the use of PGPR agents could be an interesting alternative to agronomic and genetic approaches aiming to promote the crop plant growth as well as enhance the uptake of micronutrients by plants (De Santiago et al. 2011; Mora et al. 2015). Vessey (2003) defined PGPR as biofertilizers, i.e., substances that contain living microorganisms and, once applied to plant or soil, colonizes the rhizosphere or the interior of the plants promoting the increase of supply or availability of primary nutrients to the host plant (Vessey 2003). However, some authors consider that the use of PGPR for crop plant biofortification should be carried out as a possible supplementary measure, along with other approaches (Bouis et al. 2003; Blanchfield 2004). In addition to the use of bacteria, other organisms such as fungi have also been used for this purpose (Durán et al. 2013). In Table 7.3 are presented some studies demonstrating the application of microorganism strains to the soil for crop plant biofortification.

According to the studies described in Table 7.3, it is possible to verify that the application of microorganism strains to the cultivation soils, especially for cereals and legumes, is a promising approach for mineral biofortification. Although the combination of agronomic and PGPR approaches can be an advantageous alternative for crop plant biofortification, in some cases only the application of microorganism strains to the soil may promote the same benefits. The application of strains of

**Table 7.3** Studies demonstrating the use of plant growth-promoting microorganisms for mineral biofortification of crop plants

| Crop plants | Elements | Results | Reference |
|---|---|---|---|
| Chickpea | Ca, Cu, Fe, Mg, Mn, and Zn | The potential of plant growth-promoting actinobacteria in increasing seed mineral density of chickpea under field conditions was evaluated. Nineteen isolates of actinobacteria were tested, and for all them, mineral concentration was higher than those observed for uninoculated control treatments. Concentration of Ca, Cu, Fe, Mg, Mn, and Zn were up to 26%, 54%, 38%, 21%, 35%, and 30%, respectively. | Sathya et al. (2016) |
| Chickpea | Fe | Five bacterial isolates were evaluated for improving plant growth and bioavailable Fe concentration in chickpea. Application of the PGPR significantly enhanced the plant height, root length, root fresh and dry weights, shoot fresh and dry weights. Besides, the inoculated plants presented Fe concentration higher than those obtained for uninoculated control plants. Application of PGPR along with $FeSO_4$ (as fertilizer) showed 81% and 75% increase in grain and shoot Fe concentration, respectively, when compared to control (uninoculated plants) | Khalid et al. (2015) |
| Chickpea and pigeonpea | Ca, Cu, Fe, Mg, Mn, and Zn | Seven strains of bacteria were evaluated for improving plant growth and biofortification in chickpea and pigeonpea under field conditions. Evaluated bacteria significantly enhanced the shoot height and root length of both chickpea and pigeonpea over the uninoculated control. Besides, mineral concentration in the harvested grains from the inoculated plants were higher than those observed for uninoculated control treatments—was up to 22% and 11% for Ca, 19% and 8% for Cu, 18% and 12% for Fe, 2% and 39% for Mn, and 23% and 5% for Zn, in chickpea and pigeonpea, respectively. | Gopalakrishnan et al. (2016) |
| Mung bean | Fe | Two strains of bacteria were evaluated, and results showed that both have a high chelating potential for iron. Also, pot study results revealed a significantly increased in vegetative parameters, Fe concentration (up to 3.4-fold), protein (up to 2.5-fold) and carbohydrates (up to 1.5-fold) in inoculated plants, demonstrating the potential of this approach for plant growth and Fe biofortification in mung bean. | Patel et al. (2018) |

(continued)

**Table 7.3** (continued)

| Crop plants | Elements | Results | Reference |
|---|---|---|---|
| Rice | Fe | Three plant growth-promoting rhizobacterial strains isolated from rhizospheric soils were applied to field-grown rice plants for improving Fe concentration of grains. The results showed that grain Fe concentration almost doubled, and the Fe translocation efficiency from roots to shoots to grains significantly enhanced. The authors emphasize that the application of PGPR strains is a promising strategy to combat the problem related to Fe deficiency in rice and consecutively in human masses. | Sharma et al. (2013) |
| Soybean and wheat | Zn | Three strains of bacteria were assessed for improving crop growth, and the mobilization and biofortification of Zn. Results demonstrated that Zn concentration in shoots and roots had an increase of up to 23% and 29% for soybean and 68% and 49% for wheat, respectively. Also, the evaluated strategy increased the yield of both crops and, therefore, can be used for biofertilization and biofortification. | Ramesh et al. (2014) |
| Wheat | Cu, Fe, Mn, and Zn | One bacterial and three cyanobacterial strains were evaluated in a field experiment. Results demonstrated that bacterial strain improves the nutritional quality of wheat grains. The concentration of Cu, Fe, and Mn increased by up to 150%, 105%, and 37%, respectively. The concentration of Zn was similar to those obtained in control (only NPK fertilizer application) for all evaluated strains. | Rana et al. (2012) |
| Wheat | Fe and Se | The effect of bacterial inoculation and selenate fertilization on Se uptake and plant growth was assessed. Inoculation with YAM2 (a bacterium with 99% similarity to *Bacillus pichinotyi*), both in the presence and absence of selenate, showed significantly higher dry weight, shoot length, and spike length compared to uninoculated plants. Se concentration in inoculated plants was considerably higher in wheat kernels (167%) and stems (252%) when compared to uninoculated plants. Similar behavior was observed in Fe concentration for inoculated plants that have an increase in kernels (70%) and stems (147%). | Yasin et al. (2015) |
| Wheat | Fe and Zn | Four bacterial isolates were in vitro studied and, further, in field experiments on two varieties of *Triticum aestivum*. The strains | Shaikh and Saraf (2017) |

(continued)

**Table 7.3** (continued)

| Crop plants | Elements | Results | Reference |
|---|---|---|---|
| | | were selected based on their plant growth-promoting abilities and their superior properties of solubilizing insoluble zinc compounds. Results demonstrated the PGPR improves the yield, micronutrient concentration, and uptake of Zn and Fe about sixfold (18 mg $kg^{-1}$ and 25 mg $kg^{-1}$) in wheat grains. | |
| Wheat | Se | The co-inoculation of native rhizobacteria strains and arbuscular mycorrhizal fungi were assessed for Se biofortification of wheat plants. The inoculated plants presented higher Se concentration in comparison to uninoculated controls. In plants co-inoculated with a mixture of rhizobacteria and arbuscular mycorrhizal fungi, Se concentration in wheat grain was about 24% higher (725 mg $kg^{-1}$) than non-mycorrhizal plants. A synergistic effect between the rhizobacteria and arbuscular mycorrhizal fungi was observed demonstrating a great potential of these rhizosphere microorganisms for cereal biofortification. | Durán et al. (2013) |

bacteria in soil, for example, was effective in increasing the Ca, Cu, Fe, Mg, Mn, and Zn concentration of chickpeas and wheat without the need to add fertilizers avoiding problems related to environmental pollution (Rana et al. 2012). Moreover, the use of fungi strains has also been promising. Durán et al. (2013), for example, observed an increase of 24% in Se concentration in wheat co-inoculate with a mixture of rhizobacteria and arbuscular mycorrhizal fungi. It is important to note that, in both cases, the application of microorganism strains to soil would not have the same success if were performed in mineral-deficient soils. Even so, in these cases, the use of PGPM is an environmentally friendly and low-cost alternative that, associated with agronomic approaches, may provide savings regarding the use of fertilizers.

## 7.4 Is Biofortification a Solution to Tackle Elemental Deficiency?

It is known that hidden hunger or micronutrient deficiency is a worldwide concern, leading to about two billion people who do not have access to supplements or a diversified diet to consequences such as anemia and even death (HarvestPlus 2020). It is also known that biofortification strategies are sustainable and effective tools to improve the nutritional status of staple crops (Díaz-Gómez et al. 2017). In the last years, it was possible to observe significant progress in research and development of biofortified foods, with a variety of new strategies emerging for several nutrients/ crops (Hefferon 2015). However, biofortification efficiency to tackle elemental deficiency in humans is not yet a fully clarified subject, generating controversies among researches. There is a lack of nutritional assessment regarding biofortified foods and their impact on global human health.

Valença et al. (2017) highlighted that, despite the potential of biofortification to increase nutritional content and yield of food crops, more evidence is necessary to prove its influence in human health and its efficacy to alleviate micronutrient deficiencies. Another point is that biofortification strategies must be adapted for different staple crops that are commonly harvested in each region. Moreover, the success of biofortification is related to the correct choice of food preparation and cooking methods that can impact on nutrient bioavailability (Díaz-Gómez et al. 2017). Another challenge that must be overcome is the public perception of biofortification, which may influence the regulation and implementation of genetically modified crops (Hefferon 2015). Thus, before these techniques are widely applied, its influence on nutrient bioavailability must be confirmed (Díaz-Gómez et al. 2017). Beyond that, systematic research and comprehensive feeding trials are needed to clarify the benefits that they can have on human health in the long term. Finally, the impacts of these foods must be assessed in the fields of nutrition, health, environment, and agriculture (Hirschi 2009).

On the other hand, a review conducted by White and Broadley (2009) concluded that biofortification of crop plants has a great potential to improve the nutritional status of humans, without compromising crop yield. Khush et al. (2012) and Díaz-Gómez et al. (2017) agreed that biofortification is a promising tool to alleviate malnutrition in vulnerable populations.

Biofortification is one of the tools to combat hidden hunger by increasing the micronutrient content of staple foods (HarvestPlus 2020). Both Food and Agriculture Organization of the United Nations (FAO) and HarvestPlus, which is part of the CGIAR Research Program on Agriculture for Nutrition and Health (A4NH) and is led by the International Food Policy Research Institute (IFPRI), have been working together in the development, production, and implementation of biofortified staple crops aiming to improve nutrition and health of vulnerable populations. Iron, zinc, and vitamin A are the main focus of these programs where biofortification is carried out through conventional crop breeding. In general, the target foods are stapling crops such as rice, maize, wheat, cassava, beans, and sweet potato. The adoption and expansion of biofortification programs are highly encouraged and supported by the aforementioned agencies (HarvestPlus 2020).

For the success of a biofortified crop, tests must be carried out to scientifically prove that it will indeed contribute to the increase in micronutrient intake. Only after this stage, the biofortified crop can be disseminated and consumed as a safe and effective nutrient source. One of the advantages of these crops is that they can be continuously improved after implementation, since varieties with superior qualities, such as the higher concentration of micronutrients, can be always selected (HarvestPlus 2020).

According to HarvestPlus (2020), biofortified crops of 200 varieties are already officially present in 30 countries (HarvestPlus 2020). For example, in 2019 there were 39 varieties of iron-biofortified beans released in Africa and 21 in Latin America and the Caribbean. These beans, when consumed as a staple, would supply 80% of the estimated average requirement (EAR) for Fe. Also, a total of ten varieties of pearl millet and eight of cowpea biofortified with Fe were released, supplying 80% and 25% of the iron EAR, respectively. In 2019, Zn biofortified crops (11 varieties of wheat, 10 of rice, and 7 of maize) were legalized, providing respectively 50%, 40%, and 70% of the EAR of Zn. The overall climate-adaptiveness and higher yields of biofortified crops contributed to its acceptance by farmers (HarvestPlus 2018). By 2018, biofortified crops such as iron beans and zinc rice were grown by about 7.6 million farmers (HarvestPlus 2018, 2019). Consumers usually have a good acceptance of biofortified crops, enjoying its taste, appearance, odor, and texture (HarvestPlus 2020). A total of 38 million people were growing and consuming biofortified crops in 2018 (HarvestPlus 2018).

Many studies have shown the nutritional and health benefits of biofortified crops, mainly to people who consume then as staple foods. These studies found that nutrients in biofortified crops are as bioavailable as those of traditional varieties. The consumption of these crops can improve micronutrient status, cognitive function, and reduce morbidity, as well as supply 80% of the daily average requirement of Fe and 70% of Zn (HarvestPlus 2020).

A study conducted with Rwandan women suggested that the consumption of Fe-biofortified beans contributed to the improvement of iron status and to prevent and reverse iron deficiency among those women (Haas et al. 2016). Scott et al. (2018) performed an intervention study in 140 Indian boys and girls, aged 12–16 years old, concluding that the consumption of iron-biofortified pearl millet improved Fe status as well as some measures of cognitive performance (memory and attention) (Scott et al. 2018).

Brnić et al. (2015) have compared the zinc absorption from a rice variety fortified with Zn and the same rice variety biofortified with zinc. The results showed that rice biofortification was as good as the postharvest fortification to combat zinc deficiency and biofortified rice presented more bioavailable zinc than conventional rice (Brnić et al. 2015). A study conducted with 6005 participants suggested that the consumption of zinc-biofortified wheat reduces maternal and child morbidity (Sazawal et al. 2018).

In conclusion, scientific research, development, and application studies have suggested that biofortification can contribute to more people having access to a healthy and diverse diet by making staple crops more nutritious. It contributes to the improvement of the nutritional status of vulnerable populations and helps fight hidden hunger. Moreover, there is evidence that farmers and consumers have accepted biofortified foods well (HarvestPlus 2020). Then, biofortification together with other approaches namely supplementation, dietary diversity, and food fortification are complementary strategies to tackle elemental deficiency.

## References


Adu MO, Asare PA, Yawson DO, Nyarko MA, Osei-Agyeman K (2018) Agronomic biofortification of selected underutilised solanaceae vegetables for improved dietary intake of potassium (K) in Ghana. Heliyon 4(8):e00750. https://doi.org/10.1016/j.heliyon.2018.e00750

Aggett PJ (2012) Iron. In: Macdonald IA (ed) Present knowledge in nutrition. Wiley-Blackwell, Hoboken, pp 506–520

Allard RW (1999) Principles of plant breeding. Wiley, New York

Almendros P, Obrador A, Gonzalez D, Alvarez JM (2015) Biofortification of zinc in onions (*Allium cepa L.*) and soil Zn status by the application of different organic Zn complexes. Sci Hortic 186:254–265. https://doi.org/10.1016/j.scienta.2015.02.023

Alshaal T, El-Ramady H (2017) Foliar application: from plant nutrition to biofortification. Environ Biodiversity Soil Secur 1:71–83. https://doi.org/10.21608/jenvbs.2017.1089.1006

Aziz MZ, Yaseen M, Abbas T, Naveed M, Mustafa A, Hamid Y, Saeed Q, Xu M-G (2019) Foliar application of micronutrients enhances crop stand, yield and the biofortification essential for human health of different wheat cultivars. J Integr Agric 18(6):1369–1378. https://doi.org/10.1016/S2095-3119(18)62095-7

Bañuelos GS, Arroyo I, Pickering IJ, Yang SI, Freeman JL (2015) Selenium biofortification of broccoli and carrots grown in soil amended with Se-enriched hyperaccumulator *Stanleya pinnata*. Food Chem 166:603–608. https://doi.org/10.1016/j.foodchem.2014.06.071

Biesalski HK (2013) Hidden hunger in the developed world. In: Eggersdorfer M, Kraemer K, Ruel M et al (eds) The road to good nutrition. Karger Publishers, Basel, pp 39–50. https://doi.org/10.1159/000355992

Blanchfield JR (2004) Genetically modified food crops and their contribution to human nutrition and food quality. J Food Sci 69(1):28–30. https://doi.org/10.1111/j.1365-2621.2004.tb17846.x

Blasco B, Rios JJ, Cervilla LM, Sánchez-Rodrigez E, Ruiz JM, Romero L (2008) Iodine biofortification and antioxidant capacity of lettuce: potential benefits for cultivation and human health. Ann Appl Biol 152(3):289–299. https://doi.org/10.1111/j.1744-7348.2008.00217.x

Bouis HE, Saltzman A (2017) Improving nutrition through biofortification: a review of evidence from HarvestPlus, 2003 through 2016. Glob Food Sec 12:49–58. https://doi.org/10.1016/j.gfs.2017.01.009

Bouis HE, Welch RM (2010) Biofortification—a sustainable agricultural strategy for reducing micronutrient malnutrition in the global south. Crop Sci 50(S1):20–32. https://doi.org/10.2135/cropsci2009.09.0531

Bouis HE, Chassy BM, Ochanda JO (2003) Genetically modified food crops and their contribution to human nutrition and food quality. Trends Food Sci Technol 14(5):191–209. https://doi.org/10.1016/S0924-2244(03)00073-6

Bradshaw JE (2016) Plant breeding: past, present and future. Springer, Cham

Brnić M, Wegmüller R, Melse-Boonstra A, Stomph T, Zeder C, Tay FM, Hurrell RF (2015) Zinc absorption by adults is similar from intrinsically labeled zinc-biofortified rice and from rice fortified with labeled zinc sulfate. J Nutr 146(1):76–80. https://doi.org/10.3945/jn.115.213421

Broadley MR, Alcock J, Alford J, Cartwright P, Foot I, Fairweather-Tait SJ, Hart DJ, Hurst R, Knott P, McGrath SP, Meacham MC, Norman K, Mowat H, Scott P, Stroud JL, Tovey M, Tucker M, White PJ, Young SD, Zhao F-J (2010) Selenium biofortification of high-yielding winter wheat (*Triticum aestivum L.*) by liquid or granular Se fertilisation. Plant Soil 332(1):5–18. https://doi.org/10.1007/s11104-009-0234-4

Businelli D, D'Amato R, Onofri A, Tedeschini E, Tei F (2015) Se-enrichment of cucumber (*Cucumis sativus L.*), lettuce (*Lactuca sativa L.*) and tomato (*Solanum lycopersicum L. Karst*) through fortification in pre-transplanting. Sci Hortic 197:697–704. https://doi.org/10.1016/j.scienta.2015.10.039

Caffagni A, Arru L, Meriggi P, Milc J, Perata P, Pecchioni N (2011) Iodine fortification plant screening process and accumulation in tomato fruits and potato tubers. Commun Soil Sci Plant Anal 42(6):706–718. https://doi.org/10.1080/00103624.2011.550372

Caffagni A, Pecchioni N, Meriggi P, Bucci V, Sabatini E, Acciarri N, Ciriaci T, Pulcini L, Felicioni N, Beretta M (2012) Iodine uptake and distribution in horticultural and fruit tree species. Ital J Agron 7(3):e32. https://doi.org/10.4081/ija.2012.e32

Chugh V, Dhaliwal HS (2013) Chapter 9 - biofortification of staple crops. In: Bhullar GS, Bhullar NK (eds) Agricultural sustainability. Academic, San Diego, pp 177–196. https://doi.org/10.1016/B978-0-12-404560-6.00009-5

Connorton JM, Balk J (2019) Iron biofortification of staple crops: lessons and challenges in plant genetics. Plant Cell Physiol 60(7):1447–1456. https://doi.org/10.1093/pcp/pcz079

Dai J-L, Zhu Y-G, Zhang M, Huang Y-Z (2004) Selecting iodine-enriched vegetables and the residual effect of iodate application to soil. Biol Trace Elem Res 101(3):265–276. https://doi.org/10.1385/BTER:101:3:265

Dai JL, Zhu YG, Huang YZ, Zhang M, Song JL (2006) Availability of iodide and iodate to spinach (*Spinacia oleracea L.*) in relation to total iodine in soil solution. Plant Soil 289(1):301–308. https://doi.org/10.1007/s11104-006-9139-7

De Figueiredo MA, Boldrin PF, Hart JJ, de Andrade MJ, Guilherme LR, Glahn RP, Li L (2017) Zinc and selenium accumulation and their effect on iron bioavailability in common bean seeds. Plant Physiol Biochem 111:193–202. https://doi.org/10.1016/j.plaphy.2016.11.019

De Santiago A, Quintero JM, Avilés M, Delgado A (2011) Effect of *Trichoderma asperellum* strain T34 on iron, copper, manganese, and zinc uptake by wheat grown on a calcareous medium. Plant Soil 342(1):97–104. https://doi.org/10.1007/s11104-010-0670-1

De Steur H, Blancquaert D, Strobbe S, Lambert W, Gellynck X, Van Der Straeten D (2015) Status and market potential of transgenic biofortified crops. Nat Biotechnol 33(1):25–29. https://doi.org/10.1038/nbt.3110

De Steur H, Wesana J, Blancquaert D, Van Der Straeten D, Gellynck X (2017) The socioeconomics of genetically modified biofortified crops: a systematic review and meta-analysis. Ann N Y Acad Sci 1390(1):14–33. https://doi.org/10.1111/nyas.13199

Díaz-Gómez J, Twyman RM, Zhu C, Farré G, Serrano JCE, Portero-Otin M, Muñoz P, Sandmann G, Capell T, Christou P (2017) Biofortification of crops with nutrients: factors affecting utilization and storage. Curr Opin Biotechnol 44:115–123. https://doi.org/10.1016/j.copbio.2016.12.002

Dobosy P, Kröpfl K, Óvári M, Sandil S, Németh K, Engloner A, Takács T, Záray G (2020) Biofortification of green bean (*Phaseolus vulgaris L.*) and lettuce (*Lactuca sativa L.*) with iodine in a plant-calcareous sandy soil system irrigated with water containing KI. J Food Compos Anal 88:103434. https://doi.org/10.1016/j.jfca.2020.103434

Du W, Yang J, Peng Q, Liang X, Mao H (2019) Comparison study of zinc nanoparticles and zinc sulphate on wheat growth: from toxicity and zinc biofortification. Chemosphere 227:109–116. https://doi.org/10.1016/j.chemosphere.2019.03.168

Dudley JW (1997) Quantitative genetics and plant breeding. In: Sparks DL (ed) Advances in agronomy, vol 59. Academic, Cambridge, pp 1–23. https://doi.org/10.1016/S0065-2113(08)60051-6

Dukhi N (2020) Global prevalence of malnutrition: evidence from literature. In: Malnutrition. IntechOpen, London, pp 1–16

Durán P, Acuña JJ, Jorquera MA, Azcón R, Borie F, Cornejo P, Mora ML (2013) Enhanced selenium content in wheat grain by co-inoculation of selenobacteria and arbuscular mycorrhizal fungi: a preliminary study as a potential Se biofortification strategy. J Cereal Sci 57(3):275–280. https://doi.org/10.1016/j.jcs.2012.11.012

El-Ramady HR, Abdalla N, Fári M, Domokos-Szabolcsy É, El-Ramady HR, Abdalla N, Fári M (2014) Selenium enriched vegetables as biofortification alternative for alleviating micronutrient malnutrition. Int J Hortic Sci Technol 20(1-2):75–81. https://doi.org/10.31421/IJHS/20/1-2/1121

Food and Agriculture Organization of the United Nations (FAO) (2019) The State of Food Security and Nutrition in the World: Safeguarding against economic slowdowns and downturns. http://www.fao.org/3/ca5162en/ca5162en.pdf. Accessed March 2020

Garcia-Casal MN, Peña-Rosas JP, Giyose B (2017) Staple crops biofortified with increased vitamins and minerals: considerations for a public health strategy. Ann N Y Acad Sci 1390 (1):3–13. https://doi.org/10.1111/nyas.13293

Garg M, Sharma N, Sharma S, Kapoor P, Kumar A, Chunduri V, Arora P (2018) Biofortified crops generated by breeding, agronomy, and transgenic approaches are improving lives of millions of people around the world. Front Nutr 5(12):1–33. https://doi.org/10.3389/fnut.2018.00012

Gartler J, Robinson B, Burton K, Clucas L (2013) Carbonaceous soil amendments to biofortify crop plants with zinc. Sci Total Environ 465:308–313. https://doi.org/10.1016/j.scitotenv.2012.10.027

Geyik O, Hadjikakou M, Bryan BA (2020) Spatiotemporal trends in adequacy of dietary nutrient production and food sources. Global Food Secur 24:100355. https://doi.org/10.1016/j.gfs.2020.100355

Gharibzahedi SMT, Jafari SM (2017) The importance of minerals in human nutrition: Bioavailability, food fortification, processing effects and nanoencapsulation. Trends Food Sci Technol 62:119–132. https://doi.org/10.1016/j.tifs.2017.02.017

Glick BR (1995) The enhancement of plant growth by free-living bacteria. Can J Microbiol 41 (2):109–117. https://doi.org/10.1139/m95-015

Gómez-Galera S, Rojas E, Sudhakar D, Zhu C, Pelacho AM, Capell T, Christou P (2010) Critical evaluation of strategies for mineral fortification of staple food crops. Transgenic Res 19 (2):165–180. https://doi.org/10.1007/s11248-009-9311-y

Gonzali S, Kiferle C, Perata P (2017) Iodine biofortification of crops: agronomic biofortification, metabolic engineering and iodine bioavailability. Curr Opin Biotechnol 44:16–26. https://doi.org/10.1016/j.copbio.2016.10.004

Gopalakrishnan S, Vadlamudi S, Samineni S, Sameer Kumar CV (2016) Plant growth-promotion and biofortification of chickpea and pigeonpea through inoculation of biocontrol potential bacteria, isolated from organic soils. Springerplus 5(1):1882. https://doi.org/10.1186/s40064-016-3590-6

Gupta UC, Gupta SC (2014) Sources and deficiency diseases of mineral nutrients in human health and nutrition: a review. Pedosphere 24(1):13–38. https://doi.org/10.1016/S1002-0160(13)60077-6

Haas JD, Luna SV, Lung'aho MG, Wenger MJ, Murray-Kolb LE, Beebe S, Gahutu J-B, Egli IM (2016) Consuming iron biofortified beans increases iron status in Rwandan women after 128 days in a randomized controlled feeding trial. J Nutr 146(8):1586–1592. https://doi.org/10.3945/jn.115.224741

Hafeez FY, Yasmin S, Ariani D, Mehboob-ur-Rahman ZY, Malik KA (2006) Plant growth-promoting bacteria as biofertilizer. Agron Sustain Dev 26(2):143–150. https://doi.org/10.1051/agro:2006007

Hafiz H, Oei SY, Ring DM, Shnitser N (2020) Regulating in pandemic: evaluating economic and financial policy responses to the coronavirus crisis. Boston College Law School Legal Studies Research Paper 527. https://doi.org/10.2139/ssrn.3555980

Halka M, Smoleń S, Czernicka M, Klimek-Chodacka M, Pitala J, Tutaj K (2019) Iodine biofortification through expression of HMT, SAMT and S3H genes in *Solanum lycopersicum L.* Plant Physiol Biochem 144:35–48. https://doi.org/10.1016/j.plaphy.2019.09.028

HarvestPlus (2018) Catalyzing Biofortified Food Systems: 2018 Annual Report. HarvestPlus. https://www.harvestplus.org/sites/default/files/publications/153613_HarvestPlus_AR_2018_Proof.pdf. Accessed April 2020

HarvestPlus (2019) Biofortification: the evidence. HarvestPlus. https://www.harvestplus.org/evidence-document. Accessed March 2020

HarvestPlus (2020) Biofortification: a food-systems solution to help end hidden hunger. HarvestPlus. https://www.harvestplus.org/sites/default/files/HarvestPlus-FAO%20brief.pdf. Accessed April 2020

Hefferon KL (2015) Nutritionally enhanced food crops; progress and perspectives. Int J Mol Sci 16 (2):3895–3914. https://doi.org/10.3390/ijms16023895

Herrington C, Lividini K, Angel MD, Birol E (2019) Prioritizing countries for biofortification interventions: biofortification priority index, Second Edition (BPI 2.0). HarvestPlus https://www.harvestplus.org/content/prioritizing-countries-biofortification-interventions-biofortification-priority-index-second. Accessed March 2020

Hirschi KD (2009) Nutrient biofortification of food crops. Annu Rev Nutr 29(1):401–421. https://doi.org/10.1146/annurev-nutr-080508-141143

Hong C-L, Weng H-X, Qin Y-C, Yan A-L, Xie L-L (2008) Transfer of iodine from soil to vegetables by applying exogenous iodine. Agron Sustain Dev 28(4):575–583. https://doi.org/10.1051/agro:2008033

Hong C-lL, Weng H-xX, Yan A-lL, Islam E-uU (2009) The fate of exogenous iodine in pot soil cultivated with vegetables. Environ Geochem Health 31(1):99–108. https://doi.org/10.1007/s10653-008-9169-6

Hussain A, Ali S, Rizwan M, Zia Rehman M, Javed MR, Imran M, Chatha SAS, Nazir R (2018) Zinc oxide nanoparticles alter the wheat physiological response and reduce the cadmium uptake by plants. Environ Pollut 242:1518–1526. https://doi.org/10.1016/j.envpol.2018.08.036

Hussain B, Lin Q, Hamid Y, Sanaullah M, Di L, Hashmi MLR, Khan MB, He Z, Yang X (2020) Foliage application of selenium and silicon nanoparticles alleviates Cd and Pb toxicity in rice (*Oryza sativa L.*). Sci Total Environ 712:136497. https://doi.org/10.1016/j.scitotenv.2020.136497

Institute of Medicine (IOM) (2000) Dietary reference intakes for vitamin C, vitamin E, selenium, and carotenoids. The National Academies Press, Washington, DC

Johnson AA, Kyriacou B, Callahan DL, Carruthers L, Stangoulis J, Lombi E, Tester M (2011) Constitutive overexpression of the OsNAS gene family reveals single-gene strategies for effective iron-and zinc-biofortification of rice endosperm. PLoS ONE 6(9):e24476. https://doi.org/10.1371/journal.pone.0024476

Jones GD, Droz B, Greve P, Gottschalk P, Poffet D, McGrath SP, Seneviratne SI, Smith P, Winkel LHE (2017) Selenium deficiency risk predicted to increase under future climate change. Proc Natl Acad Sci U S A 114(11):2848–2853. https://doi.org/10.1073/pnas.1611576114

Junqueira-Franco MVM, Dutra de Oliveira JE, Nutti MR, Pereira HS, Carvalho JL, Abrams SA, Brandão CF, Marchini JS (2018) Iron absorption from beans with different contents of iron, evaluated by stable isotopes. Clin Nutr ESPEN 25:121–125. https://doi.org/10.1016/j.clnesp.2018.03.120

Kavčič A, Budič B, Vogel-Mikuš K (2020) The effects of selenium biofortification on mercury bioavailability and toxicity in the lettuce-slug food chain. Food Chem Toxicol 135:110939. https://doi.org/10.1016/j.fct.2019.110939

Khalid S, Asghar HN, Akhtar MJ, Aslam A, Zahir ZA (2015) Biofortification of iron in chickpea by plant growth promoting rhizobacteria. Pak J Bot 47(3):1191–1194

Khan A, Singh J, Upadhayay VK, Singh AV, Shah S (2019) Microbial biofortification: a green technology through plant growth promoting microorganisms. In: Shah S, Venkatramanan V, Prasad R (eds) Sustainable green technologies for environmental management. Springer, Singapore, pp 255–269. https://doi.org/10.1007/978-981-13-2772-8_13

Khush GS, Lee S, Cho J-I, Jeon J-S (2012) Biofortification of crops for reducing malnutrition. Plant Biotechnol Rep 6(3):195–202. https://doi.org/10.1007/s11816-012-0216-5

Kiferle C, Gonzali S, Holwerda H, Real Ibaceta R, Perata P (2013) Tomato fruits: a good target for iodine biofortification. Front Plant Sci 4:205. https://doi.org/10.3389/fpls.2013.00205

Landini M, Gonzali S, Perata P (2011) Iodine biofortification in tomato. J Plant Nutr Soil Sci 174 (3):480–486. https://doi.org/10.1002/jpln.201000395

Lara TS, Lessa JH, de Souza KR, Corguinha AP, Martins FA, Lopes G, Guilherme LR (2019) Selenium biofortification of wheat grain via foliar application and its effect on plant metabolism. J Food Compos Anal 81:10–18. https://doi.org/10.1016/j.jfca.2019.05.002

Lawson PG, Daum D, Czauderna R, Meuser H, Härtling JW (2015) Soil versus foliar iodine fertilization as a biofortification strategy for field-grown vegetables. Front Plant Sci 6(450). https://doi.org/10.3389/fpls.2015.00450

Li R, Li D-W, Liu H-P, Hong C-L, Song M-Y, Dai Z-X, Liu J-W, Zhou J, Weng H-X (2017) Enhancing iodine content and fruit quality of pepper (*Capsicum annuum L.*) through biofortification. Sci Hortic 214:165–173. https://doi.org/10.1016/j.scienta.2016.11.030

Mackowiak CL, Grossl PR (1999) Iodate and iodide effects on iodine uptake and partitioning in rice (*Oryza sativa L.*) grown in solution culture. Plant Soil 212(2):133–141. https://doi.org/10.1023/A:1004666607330

Mafra D, Cozzolino SMF (2004) Importância do zinco na nutrição humana. Rev Nutr 17:79–87. https://doi.org/10.1590/S1415-52732004000100009

Mangueze AV, Pessoa MF, Silva MJ, Ndayiragije A, Magaia HE, Cossa VS, Reboredo FH, Carvalho ML, Santos JP, Guerra M, Ribeiro-Barros AI, Lidon FC, Ramalho JC (2018) Simultaneous Zinc and selenium biofortification in rice. Accumulation, localization and implications on the overall mineral content of the flour. J Cereal Sci 82:34–41. https://doi.org/10.1016/j.jcs.2018.05.005

Manzeke MG, Mtambanengwe F, Nezomba H, Watts MJ, Broadley MR, Mapfumo P (2017) Zinc fertilization increases productivity and grain nutritional quality of cowpea (*Vigna unguiculata [L.] Walp.*) under integrated soil fertility management. Field Crop Res 213:231–244. https://doi.org/10.1016/j.fcr.2017.08.010

Masuda H, Ishimaru Y, Aung MS, Kobayashi T, Kakei Y, Takahashi M, Higuchi K, Nakanishi H, Nishizawa NK (2012) Iron biofortification in rice by the introduction of multiple genes involved in iron nutrition. Sci Rep 2(1):543. https://doi.org/10.1038/srep00543

Mayer AB, Latham MC, Duxbury JM, Hassan N, Frongillo EA (2011) A food systems approach to increase dietary zinc intake in Bangladesh based on an analysis of diet, rice production and processing. In: Thompson B, Amoroso L (eds) Combating micronutrient deficiencies: food-based approaches. CAB international and Food and Agriculture Organization of the United Nations, Rome, pp 254–267. https://doi.org/10.1079/9781845937140.0254

Meybeck A, Laval E, Lévesque R, Parent G (2018) Food security and nutrition in the age of climate change. Food and Agriculture Organization of the United Nations (FAO). http://www.fao.org/3/CA1334EN/ca1334en.pdf. Accessed March 2020

Mora ML, Durán P, Acuña J, Cartes P, Demanet R, Gianfreda L (2015) Improving selenium status in plant nutrition and quality. J Soil Sci Plant Nutr 15:486–503. https://doi.org/10.4067/S0718-95162015005000041

National Institutes of Health (NIH) (2019a) Iodine: fact sheet for health professionals. Office of Dietary Supplements.https://ods.od.nih.gov/factsheets/Iodine-Consumer/. Accessed March 2020

National Institutes of Health (NIH) (2019b) Magnesium: fact sheet for health professionals. Office of dietary supplements. https://ods.od.nih.gov/factsheets/Magnesium-HealthProfessional/. Accessed March 2020

National Institutes of Health (NIH) (2020a) Calcium: fact sheet for health professionals. Office of dietary supplements. https://ods.od.nih.gov/factsheets/Calcium-HealthProfessional/. Accessed March 2020

National Institutes of Health (NIH) (2020b) Copper: fact sheet for health professionals. Office of dietary supplements. https://ods.od.nih.gov/factsheets/Copper-HealthProfessional/. Accessed March 2020

National Institutes of Health (NIH) (2020c) Iron: fact sheet for health professionals. Office of dietary supplements. https://ods.od.nih.gov/factsheets/Iron-HealthProfessional/. Accessed March 2020

National Institutes of Health (NIH) (2020d) Zinc: fact sheet for health professionals. Office of dietary supplements. https://ods.od.nih.gov/factsheets/Zinc-HealthProfessional/. Accessed March 2020

Navarro-Alarcon M, Cabrera-Vique C (2008) Selenium in food and the human body: a review. Sci Total Environ 400(1):115–141. https://doi.org/10.1016/j.scitotenv.2008.06.024

Nelson G, Bogard J, Lividini K, Arsenault J, Riley M, Sulser TB, Mason-D'Croz D, Power B, Gustafson D, Herrero M, Wiebe K, Cooper K, Remans R, Rosegrant M (2018) Income growth and climate change effects on global nutrition security to mid-century. Nat Sustain 1 (12):773–781. https://doi.org/10.1038/s41893-018-0192-z

Nestel P, Bouis HE, Meenakshi JV, Pfeiffer W (2006) Biofortification of staple food crops. J Nutr 136(4):1064–1067. https://doi.org/10.1093/jn/136.4.1064

Patel P, Trivedi G, Saraf M (2018) Iron biofortification in mung bean using siderophore producing plant growth promoting bacteria. Environ Sustain 1(4):357–365. https://doi.org/10.1007/s42398-018-00031-3

Phattarakul N, Rerkasem B, Li LJ, Wu LH, Zou CQ, Ram H, Sohu VS, Kang BS, Surek H, Kalayci M, Yazici A, Zhang FS, Cakmak I (2012) Biofortification of rice grain with zinc through zinc fertilization in different countries. Plant Soil 361(1):131–141. https://doi.org/10.1007/s11104-012-1211-x

Poblaciones MJ, Rengel Z (2016) Soil and foliar zinc biofortification in field pea (*Pisum sativum L.*): Grain accumulation and bioavailability in raw and cooked grains. Food Chem 212:427–433. https://doi.org/10.1016/j.foodchem.2016.05.189

Prasanna R, Nain L, Rana A, Shivay YS (2016) Biofortification with microorganisms: present status and future challenges. In: Singh U, Praharaj CS, Singh SS, Singh NP (eds) Biofortification of food crops. Springer, New Delhi, pp 249–262. https://doi.org/10.1007/978-81-322-2716-8_19

Ramesh A, Sharma SK, Sharma MP, Yadav N, Joshi OP (2014) Inoculation of zinc solubilizing *Bacillus aryabhattai* strains for improved growth, mobilization and biofortification of zinc in soybean and wheat cultivated in Vertisols of central India. Appl Soil Ecol 73:87–96. https://doi.org/10.1016/j.apsoil.2013.08.009

Ramzani PMA, Khalid M, Naveed M, Ahmad R, Shahid M (2016) Iron biofortification of wheat grains through integrated use of organic and chemical fertilizers in pH affected calcareous soil. Plant Physiol Biochem 104:284–293. https://doi.org/10.1016/j.plaphy.2016.04.053

Rana A, Joshi M, Prasanna R, Shivay YS, Nain L (2012) Biofortification of wheat through inoculation of plant growth promoting rhizobacteria and cyanobacteria. Eur J Soil Biol 50:118–126. https://doi.org/10.1016/j.ejsobi.2012.01.005

Reinhardt K, Fanzo J (2014) Addressing chronic malnutrition through multi-sectoral, sustainable approaches: a review of the causes and consequences. Front Nutr 1:13. https://doi.org/10.3389/fnut.2014.00013

Ríos JJ, Rosales MA, Blasco B, Cervilla LM, Romero L, Ruiz JM (2008) Biofortification of Se and induction of the antioxidant capacity in lettuce plants. Sci Hortic 116(3):248–255. https://doi.org/10.1016/j.scienta.2008.01.008

Rizwan M, Ali S, Ali B, Adrees M, Arshad M, Hussain A, Zia Rehman M, Waris AA (2019) Zinc and iron oxide nanoparticles improved the plant growth and reduced the oxidative stress and cadmium concentration in wheat. Chemosphere 214:269–277. https://doi.org/10.1016/j.chemosphere.2018.09.120

Ruel-Bergeron JC, Stevens GA, Sugimoto JD, Roos FF, Ezzati M, Black RE, Kraemer K (2015) Global update and trends of hidden hunger, 1995-2011: the hidden hunger index. PLoS ONE 10 (12):e0143497. https://doi.org/10.1371/journal.pone.0143497

Saltzman A, Birol E, Bouis HE, Boy E, De Moura FF, Islam Y, Pfeiffer WH (2013) Biofortification: progress toward a more nourishing future. Global Food Secur 2(1):9–17. https://doi.org/10.1016/j.gfs.2012.12.003

Saltzman A, Birol E, Oparinde A, Andersson MS, Asare-Marfo D, Diressie MT, Gonzalez C, Lividini K, Moursi M, Zeller M (2017) Availability, production, and consumption of crops biofortified by plant breeding: current evidence and future potential. Ann N Y Acad Sci 1390 (1):104–114. https://doi.org/10.1111/nyas.13314

Sathya A, Vijayabharathi R, Srinivas V, Gopalakrishnan S (2016) Plant growth-promoting actinobacteria on chickpea seed mineral density: an upcoming complementary tool for sustainable biofortification strategy. Biotech 6(2):138. https://doi.org/10.1007/s13205-016-0458-y

Sazawal S, Dhingra U, Dhingra P, Dutta A, Deb S, Kumar J, Devi P, Prakash A (2018) Efficacy of high zinc biofortified wheat in improvement of micronutrient status, and prevention of morbidity among preschool children and women - a double masked, randomized, controlled trial. Nutrition 17(1):86. https://doi.org/10.1186/s12937-018-0391-5

Scott SP, Murray-Kolb LE, Wenger MJ, Udipi SA, Ghugre PS, Boy E, Haas JD (2018) Cognitive performance in Indian school-going adolescents is positively affected by consumption of iron-biofortified pearl millet: a 6-month randomized controlled efficacy trial. J Nutr 148 (9):1462–1471. https://doi.org/10.1093/jn/nxy113

Shaikh S, Saraf M (2017) Biofortification of *Triticum aestivum* through the inoculation of zinc solubilizing plant growth promoting rhizobacteria in field experiment.Biocatal Agric. Biotechnology 9:120–126. https://doi.org/10.1016/j.bcab.2016.12.008

Sharma A, Shankhdhar D, Shankhdhar S (2013) Enhancing grain iron content of rice by the application of plant growth promoting rhizobacteria. Plant Soil Environ 59(2):89–94. https://doi.org/10.17221/683/2012-PSE

Sharma P, Aggarwal P, Kaur A (2017) Biofortification: a new approach to eradicate hidden hunger. Food Rev Int 33(1):1–21. https://doi.org/10.1080/87559129.2015.1137309

Skalickova S, Milosavljevic V, Cihalova K, Horky P, Richtera L, Adam V (2017) Selenium nanoparticles as a nutritional supplement. Nutrition 33:83–90. https://doi.org/10.1016/j.nut.2016.05.001

Smoleń S, Sady W (2012) Influence of iodine form and application method on the effectiveness of iodine biofortification, nitrogen metabolism as well as the content of mineral nutrients and heavy metals in spinach plants (*Spinacia oleracea L.*). Sci Hortic 143:176–183. https://doi.org/10.1016/j.scienta.2012.06.006

Smoleń S, Kowalska I, Sady W (2014) Assessment of biofortification with iodine and selenium of lettuce cultivated in the NFT hydroponic system. Sci Hortic 166:9–16. https://doi.org/10.1016/j.scienta.2013.11.011

Smoleń S, Wierzbińska J, Sady W, Kołton A, Wiszniewska A, Liszka-Skoczylas M (2015) Iodine biofortification with additional application of salicylic acid affects yield and selected parameters of chemical composition of tomato fruits (*Solanum lycopersicum L.*). Sci Hortic 188:89–96. https://doi.org/10.1016/j.scienta.2015.03.023

Smoleń S, Ledwożyw-Smoleń I, Halka M, Sady W, Kováčik P (2017) The absorption of iodine from 5-iodosalicylic acid by hydroponically grown lettuce. Sci Hortic 225:716–725. https://doi.org/10.1016/j.scienta.2017.08.009

Smoleń S, Kowalska I, Skoczylas Ł, Liszka-Skoczylas M, Grzanka M, Halka M, Sady W (2018) The effect of salicylic acid on biofortification with iodine and selenium and the quality of potato cultivated in the NFT system. Sci Hortic 240:530–543. https://doi.org/10.1016/j.scienta.2018.06.060

Smoleń S, Baranski R, Ledwożyw-Smoleń I, Skoczylas Ł, Sady W (2019) Combined biofortification of carrot with iodine and selenium. Food Chem 300:125202. https://doi.org/10.1016/j.foodchem.2019.125202

Souza GA, Hart JJ, Carvalho JG, Rutzke MA, Albrecht JC, Guilherme LRG, Kochian LV, Li L (2014) Genotypic variation of zinc and selenium concentration in grains of Brazilian wheat lines. Plant Sci 224:27–35. https://doi.org/10.1016/j.plantsci.2014.03.022

Stein AJ (2010) Global impacts of human mineral malnutrition. Plant Soil 335(1):133–154. https://doi.org/10.1007/s11104-009-0228-2

Tilman D, Clark M (2014) Global diets link environmental sustainability and human health. Nature 515(7528):518–522. https://doi.org/10.1038/nature13959

Trolove SN, Tan Y, Morrison SC, Feng L, Eason J (2018) Development of a method for producing selenium-enriched radish sprouts. LWT 95:187–192. https://doi.org/10.1016/j.lwt.2018.04.048
Valença AW, Bake A, Brouwer ID, Giller KE (2017) Agronomic biofortification of crops to fight hidden hunger in sub-Saharan Africa. Global Food Secur 12:8–14. https://doi.org/10.1016/j.gfs.2016.12.001
Vessey JK (2003) Plant growth promoting rhizobacteria as biofertilizers. Plant Soil 255 (2):571–586. https://doi.org/10.1023/A:1026037216893
Volpe SL (2012) Magnesium. In: Macdonald IA (ed) Present knowledge in nutrition. Wiley online books. Wiley-Blackwell, Hoboken, pp 459–474. https://doi.org/10.1002/9781119946045.ch30
Von Grebmer K, Saltzman A, Birol E, Wiesmann D, Prasai N, Yin S, Yohannes Y, Menon P, Thompson J, Sonntag A (2014) Synopsis of 2014 Global hunger index: The challenge of hidden hunger. International Food Policy Research Institute (IFPRI). https://www.ifpri.org/cdmref/p15738coll2/id/128364/filename/128575.pdf. Accessed March 2020
Voogt W, Holwerda HT, Khodabaks R (2010) Biofortification of lettuce (*Lactuca sativa L.*) with iodine: the effect of iodine form and concentration in the nutrient solution on growth, development and iodine uptake of lettuce grown in water culture. J Sci Food Agric 90(5):906–913. https://doi.org/10.1002/jsfa.3902
Wan J, Zhang M, Adhikari B (2018) Advances in selenium-enriched foods: from the farm to the fork. Trends Food Sci Technol 76:1–5. https://doi.org/10.1016/j.tifs.2018.03.021
Wang J, Mao H, Zhao H, Huang D, Wang Z (2012) Different increases in maize and wheat grain zinc concentrations caused by soil and foliar applications of zinc in Loess Plateau, China. Field Crop Res 135:89–96. https://doi.org/10.1016/j.fcr.2012.07.010
Weaver CM (2012) Calcium. In: Macdonald IA (ed) Present knowledge in nutrition. Wiley online books. Wiley-Blackwell, Hoboken, pp 434–446. https://doi.org/10.1002/9781119946045.ch28
Weng H-X, Hong C-L, Yan A-L, Pan L-H, Qin Y-C, Bao L-T, Xie L-L (2008a) Mechanism of iodine uptake by cabbage: effects of iodine species and where it is stored. Biol Trace Elem Res 125(1):59. https://doi.org/10.1007/s12011-008-8155-2
Weng H-X, Weng J-K, Yan A-L, Hong C-L, Yong W-B, Qin Y-C (2008b) Increment of iodine content in vegetable plants by applying iodized fertilizer and the residual characteristics of iodine in soil. Biol Trace Elem Res 123(1):218–228. https://doi.org/10.1007/s12011-008-8094-y
Weng H, Hong C, Xia T, Bao L, Liu H, Li D (2013) Iodine biofortification of vegetable plants—An innovative method for iodine supplementation. Chin Sci Bull 58(17):2066–2072. https://doi.org/10.1007/s11434-013-5709-2
White PJ, Broadley MR (2009) Biofortification of crops with seven mineral elements often lacking in human diets – iron, zinc, copper, calcium, magnesium, selenium and iodine. New Phytol 182 (1):49 84. https://doi.org/10.1111/j.1469-8137.2008.02738.x
White PJ, Broadley MR (2011) Physiological limits to zinc biofortification of edible crops. Front Plant Sci 2:80. https://doi.org/10.3389/fpls.2011.00080
Williams MH (2005) Dietary supplements and sports performance: minerals. J Int Soc Sports Nutr 2 (1):43. https://doi.org/10.1186/1550-2783-2-1-43
World Health Organization (WHO) (2013) Micronutrient deficiencies: iodine deficiency disorders. https://www.who.int/nutrition/topics/idd/en/. Accessed March 2020
World Health Organization (WHO) (2019) Micronutrient deficiencies: iron deficiency anaemia. https://www.who.int/nutrition/topics/ida/en/. Accessed March 2020
Yasin M, El-Mehdawi AF, Anwar A, Pilon-Smits EAH, Faisal M (2015) Microbial-enhanced selenium and iron biofortification of wheat (*Triticum aestivum L.*) - applications in phytoremediation and biofortification. Int J Phytoremediation 17(4):341–347. https://doi.org/10.1080/15226514.2014.922920
Yu H-J, Liu J-Q, Böck A, Li J, Luo G-M, Shen J-C (2005) Engineering glutathione transferase to a novel glutathione peroxidase mimic with high catalytic efficiency Incorporation of selenocysteine into a glutathione-binding scaffold using an auxotrophic expression system. J Biol Chem 280(12):11930–11935. https://doi.org/10.1074/jbc.m408574200

Zhang H, Zhao Z, Zhang X, Zhang W, Huang L, Zhang Z, Yuan L, Liu X (2019) Effects of foliar application of selenate and selenite at different growth stages on selenium accumulation and speciation in potato (*Solanum tuberosum L.*). Food Chem 286:550–556. https://doi.org/10.1016/j.foodchem.2019.01.185

Zhu YG, Huang YZ, Hu Y, Liu YX (2003) Iodine uptake by spinach (*Spinacia oleracea L.*) plants grown in solution culture: effects of iodine species and solution concentrations. Environ Int 29 (1):33–37. https://doi.org/10.1016/S0160-4120(02)00129-0

# An Overview on Management of Micronutrients Deficiency in Plants Through Biofortification: A Solution of Hidden Hunger

8

Pradeep Kumar Yadav, Anita Singh, and S. B. Agrawal

**Abstract**

Nowadays, malnutrition is one of the major problems, especially for the poor population of developing countries. The major staple crops are found to be deficient in some mineral elements, especially the micronutrients that result in the problem of hidden hunger. There are several promising strategies that are applied in agricultural fields to solve this problem. They enhance the bio-available concentrations of micronutrients in edible crops. One of the recent strategy is biofortification, which can be used to increase the content and/or bioavailability of vital nutrients in food crops through genetic (genetic transformation/plant breeding) and agronomic pathways (application of nutrient fertilizers). These strategies provide more nutritious diets to more people. Along with the traditional agricultural practices, the "omics" technologies can modify the crops by genetic transformation that improves the uptake, transport, and mineral accumulation in hybrid plants. This chapter has detail information about the nutrient constituents and its uptake in the plants along with a critical comparison of the several strategies that have been developed to enhance mineral levels and bioavailability of micronutrients in most of the important food crops.

P. K. Yadav
Center of Advanced Study in Botany, Institute of Science, Banaras Hindu University, Varanasi, Uttar Pradesh, India

A. Singh (✉)
Center of Advanced Study in Botany, Institute of Science, Banaras Hindu University, Varanasi, Uttar Pradesh, India

Department of Botany, Institute of Science, Banaras Hindu University, Varanasi, Uttar Pradesh, India

S. B. Agrawal
Laboratory of Air Pollution and Global Climate Change, Center of Advanced Study in Botany, Institute of Science, Banaras Hindu University, Varanasi, Uttar Pradesh, India



K. Mishra et al. (eds.), *Sustainable Solutions for Elemental Deficiency and Excess in Crop Plants*, https://doi.org/10.1007/978-981-15-8636-1_8

The use of biofortified crops should be promoted by educating the farmers by government agencies, so that they can be included in their diet to solve the problem of malnutrition up to certain extent.

**Keywords**

Food crops · Nutrients · Malnutrition · Agronomic · Biofortification

## 8.1 Introduction

Plants require several mineral elements or nutrients available in nature for their proper growth and development (Watanabe et al. 2007). These mineral elements are classified in to macro and micronutrients. Macronutrients are used in large concentration and further divide into structural (carbon: C, hydrogen: H, nitrogen: N), primary (nitrogen: N, phosphorus: P, potassium: K), and secondary nutrients (calcium: Ca, magnesium: Mg, sulfur: S). The micronutrients like boron, chlorine, manganese, iron, zinc, copper, and molybdenum are required in very less amount. These elements are naturally present in the soil, and taken up by the roots in ionic forms only. However, with the frequent uses of fertilizers, over-cropping, and application of waste water and sewage sludge, the availability of these elements get disturbed. The practices of waste water irrigation and sewage sludge application also lead to the accumulation of several toxic metals. They ultimately affect soil characteristics and availability of different nutrients due to the competition among heavy metals and mineral elements. The deficiencies of micronutrients not only affect the production of crops but also contribute malnutrition due to poor nutritional quality of food crops. It often results into invisible health problems, hence termed as hidden hunger (de Valenca et al. 2017). Mineral nutrients are mainly absorbed by the roots from soil system, however many factors have their effect on nutrient attainment. Sometimes, mineral elements are not in their available forms and also soil properties such as pH, conductivity, bulk density, etc. have their effect on nutrient uptake (Morgan and Connolly 2013).

To certain extent, to manage the deficiency of the micronutrients and to maintain the balance among essential elements, the plants try to cope up by themselves. However, at the time of the severe deficiency of micronutrients, the plants cannot cope with the condition by themselves only; in that case several other strategies are applied to deal with the situation. In this direction, biofortification is one of the processes that can be applied in different ways (agronomic or genetic) to increase the bioavailability and solubility of essential nutrients (Bouis and Saltzman 2011). Genetic biofortification can be achieved by genetic engineering or classical plant breeding (Saltzman et al. 2013), whereas agronomic biofortification can be achieved through fertilizer application either as a soil solution/foliar spray or through fertigation. With this process the essential minerals can be added in the common diets of population by increasing the solubility and availability of nutrients. From an economic point of view, biofortification is also one time investment because it offers a cost-effective, long term, and sustainable approach to manage the hidden hunger. The basic theme of biofortification should be that all the malnourished and poor families get all the essential micronutrients through their diet. There are some organizations like World Health Organization and the counseling Group on International Agricultural Research (CGIAR) have developed several nutritionally enhanced high-yielding biofortified crops (Jha and Warkentin 2020).

With the above context, the present chapter has detailed information on micronutrient uptake and includes different ways to manage deficient level of micronutrients in food crops. It also consists of discussion on biofortification processes in detail to show their positive impact on elemental composition of staple crops.

## 8.2 Uptake and Distribution of Micronutrients in the Plants

Plants naturally absorb different mineral elements from the soil through their root system. They can only take the element in their available forms, i.e., in ionic forms (Table 8.1). The availability of nutrients is dependent upon several soil characteristics like moisture content, bulk density, texture, organic matter content, pH, cation exchange capacity (CEC), and soil biological properties. The nutrient uptake is the natural process through which they enter in to the plants either by roots or by the leaves. For the uptake and distribution of nutrients, several physiological and molecular processes are involved. Among all the micronutrients, Cl and Mo, absorbed as anions, "B" is in neutral or anionic form, Mn, Cu, Zn, and Ni are in divalent cations and Fe can be absorbed as both divalent and trivalent cations (Lambert et al. 2008). As the cytoplasm of plant cell is negatively charged so the type and number of charges play important role in the transportation of micronutrients. The two common pathways include passive and active absorption. In passive absorption, minerals are absorbed without the direct expenditure of metabolic energy. If the transported elements carry a net charge, its movement is influenced by both its concentration gradient and membrane potential. Passive transport can be occurred either through simple diffusion and facilitated diffusion.

### 8.2.1 Simple Diffusion

During simple diffusion, a mineral element simply dissolves in phospholipid bilayer, diffuses across it and no membrane proteins are involved and the direction of movement determined simply by the relative concentration of molecule inside and outside of the cell.

### 8.2.2 Facilitated Diffusion

It involves the movement of mineral solute along the concentration gradient. Facilitated diffusion either occurs through carrier protein or channel protein. It

**Table 8.1** The available forms and function of micronutrients

| Micronutrients | Available form | Transporters involved for uptake | Function |
|---|---|---|---|
| Boron | Absorbed as $BO_3^{3-}$ or $B_4O_7^{2-}$ | NIPs and BOR1 | Boron is required for uptake and utilization of $Ca^{2+}$ membrane functioning, pollen germination, cell elongation, cell differentiation, and carbohydrate translocation. |
| Molybdenum | Obtain in the form of molybdate ion ($MoO_2^{2+}$) | MOT1, MOT2 | It is a component of several enzymes, including nitrogenase and nitrate reductase both of which participate in nitrogen metabolism. |
| Chlorine | Absorbed in the form of chloride anion ($Cl^-$) | CLC, CCC | Along with $Na^+$ and $K^{+\cdot}$ it helps in determining the solute concentration and the anion-cation balance in cells. It is essential for the water-splitting reaction in photosynthesis, a reaction that leads to oxygen evolution. |
| Manganese | Absorbed in the form of manganous ion ($Mn^{2+}$) | NRAMP, ZRT/IRT, YSL | It activates many enzymes involved in photosynthesis, respiration, and nitrogen metabolism. The best defined function of manganese is in the splitting of water to liberate oxygen during photosynthesis. |
| Zinc | Obtain zinc as $Zn^{+2}$ | ZIPs, HMAs, YSLs MTPs, FRD3, ZIF1, NASs | It activates various enzymes, especially carboxylases. It is also needed in the synthesis of auxin. |
| Copper | Absorbed as cupric ion ($Cu^{2+}$) | COPT1, COPT2, COPT3 and COPT4, ZIP2 and ZIP4 | It is essential for the overall metabolism in plants. Like iron, it is associated with certain enzymes involved in redox reactions. |
| Iron | Plant obtains iron in the form of ferric ion ($F^{2+}$, $Fe^{3+}$) | ZIPs, NRAMPs, YS1 and YSLs | It is an important constituent of proteins involved in the transfer of electrons like ferredoxin and cytochromes. It is reversibly oxidized from $Fe^{2+}$ to $Fe^{3+}$ during electron transfer. It activates catalase enzyme, and is essential for the formation of chlorophyll. |

allows polar and charged mineral elements such as carbohydrate, amino acid, nucleoside, and ion to cross the plasma membrane.

Active absorption of minerals directly utilized metabolic energy and during this process, minerals are absorbed in ionic forms against the concentration gradient. The energy required for this process is obtained from the cell's metabolism either directly or indirectly.

The micronutrients are present in very low concentration and available in the charged form so they cannot move across cell membranes with passive absorption only. They can also enter into the roots actively with the help of ATP as energy source in presence of specific proteins (transport ATPase, ABC transporter, etc.). They are present in the plasma membranes of endodermal cells that control the entry of mineral ions into the xylem cell based upon their type and quantity. The transporter proteins involved in the uptake and distribution pattern of some specific micronutrients are discussed below:

1. Boron (B)

Plant absorbed boron (B) as $BO_3^{3-}$ or $B_4O_7^{2-}$ When boron is present in sufficient amount then it is absorbed by simple passive diffusion but under B deficient condition, it is absorbed by the plant with the help of transporters. The molecular genetic studies revealed that there are two types of B transporters, (nodulin-26-like intrinsic proteins) NIPs and BOR1 in *Arabidopsis thaliana* (Takano et al. 2006). NIP5;1 and a boric acid channel facilitate influx of B in the root cells. It is observed that NIPs help in the transfer of B from xylem–phloem to young growing tissues (Tanaka et al. 2008). In *Arabidopsis thaliana* L. the first B transporter, AtBOR1 was reported and studied most intensively. In rice (*Oryza sativa* L), OsBOR1 also helps in the uptake of B just like AtBOR1 (Nakagawa et al. 2007). Kato et al. (2009) reported that under B deficient condition the NIP5;1 transporter gets over expressed and improved elongation of root. With the over-expression of both the transporters such as BOR1 and NIPs, the *Arabidopsis thaliana* plants can be able to grow in B deficient soil.

2. Molybdenum (Mo)

In plants, molybdenum is absorbed as molybdate ($MoO_4^{2-}$). Due to high degree of similarity with $SO4^{-2}$, the uptake and distribution of molybdate ($MoO_4^{2-}$) are supported by the transporters involved in sulfate transporters (Dudev and Lim 2004). The first molybdate-specific transporters (MOT1) were identified in *Arabidopsis thaliana* (Tomatsu et al. 2007). MOT1 is a relative of the sulfate transporter superfamily (Buchner et al. 2004), but does not appear to transport sulfate. The role of MOT1 in molybdate uptake is still unclear as results suggest MOT1 is localized to mitochondria (Baxter et al. 2008). Along with this MOT1, another molybdate transporter MOT2 has also been identified from *Arabidopsis* that also belongs to the sulfate family (Gasber et al. 2011). It localizes in the vacuolar membrane and helps in exporting stored molybdate from the vacuole into the cytosol and finally into maturing seeds. Another molybdate transporter also denoted as MOT2 has been reported from *Chlamydomonas* that does not belong to the sulfate transporter family (Tejada-Jiménez et al. 2011). Researches to find out the homolog of this transporter in higher plants are still under process to understand the uptake of molybdate at the root: soil interface in detail.

3. Chlorine (Cl)

The chlorine is absorbed as chloride ($Cl^-$) and it is transported via $H^+$/anion symporters. They help in the $Cl^-$ uptake and release it into the xylem cell (Roberts 2006). Putative $H^+$/halide transporters include ATP binding cassette (ABC) protein super family and chloride channel (CLC) transporter family (Marmagne et al. 2007; Verrier et al. 2008). At the same time, Na:K/Cl symporters also help in the uptake of $Cl^-$ by the cation chloride co-transporter (CCC) gene family (Colmenero-Flores et al. 2007). Some organic acid transporters also help in the halide fluxes in the plants (White 2001). The $Cl^-$ is mainly accumulated in roots and leaves and little is redistributed via the phloem to fruits or seeds (Muramatsu et al. 1995).

4. Manganese (Mn)

Manganese (Mn) is only available in its reduced form ($Mn^{2+}$) and can be able to transport from soil to root and then to the shoot. In alkaline soil, availability of Mn is decreased by converting $Mn^{2+}$ into insoluble Mn oxides ($MnO_x$) (Stumm and Morgan 1996). There are several transporters involved in the homeostatic network of Mn in plants. The Natural Resistance Associated Macrophage Protein (NRAMP) family, the Zinc-Regulated Transporter/Iron-Regulated Transporter (ZRT/IRT)-related Protein (ZIP) family, and the Yellow Stripe-Like (YSL) are involved in the transportation of $Mn^{2+}$ into the cytosol (Alejandro et al. 2020).

5. Zinc (Zn)

In soil solution, Zn is present in very low amount but has critical importance for plants. The substantial amount of Zn reaches to the xylem cells of root apoplastically (Broadley et al. 2007). It can be transported to the plasma membrane of root cell in the form of $Zn^{2+}$ or as a complex of Zn with phytosiderophore (Ismail et al. 2007). The influx of most of the $Zn^{2+}$into the cytoplasm mediated by ZIPs (ZIP1, ZIP3, and ZIP4; Palmgren et al. 2008), and the Yellow Stripe-Like (YSL) family proteins help in uptake of Zn by the formation of Zn-phytosiderophore complexes (Suzuki et al. 2008). In the xylem cell, the transportation of Zn occurs in the form of $Zn^{2+}$, by binding with organic acids like histidine or nicotianamine (Broadley et al. 2007; Palmgren et al. 2008). In the leaf and phloem cell, influx of $Zn^{2+}$ is mediated by the members of the ZIP family (Ishimaru et al. 2005). In addition, YSL proteins may load Zn into the phloem, where Zn is transported as a Zn-NA complex, or as a complex with small proteins, to sink tissues (Waters and Grusak 2008). Although Zn mobility in the phloem is generally considered to be low, this may not always be the case (Welch 2002; Haslett 2001). During Zn deficiency, uptake, sequestration, and redistribution of Zn get increased by the over-expression of genes responsible for Zn uptake in the plant. These genes encode different proteins such as ZIPs, HMAs, YSLs MTPs,

FRD3, ZIF1, NASs and it also increases the biosynthesis of phytosiderophores to enhance the Zn uptake (Milner and Kochian 2008).

6. Copper (Cu)

Copper can be absorbed as $Cu^{+}$ and $Cu^{2+}$ with the help of copper transporters (CTR) such as COPT1, COPT2, COPT3, and COPT4 and by ZIPs (ZIP2 and ZIP4), respectively (Grotz and Guerinot 2006). Expressions of these transporters get up-regulated under Cu deficient condition (Wintz et al. 2003). Cu is loaded into the xylem cell and transported in a $Cu^{2+}$NA complexed form (von Wiren et al. 1999). In phloem it is by YSL protein and transported as Cu–NA complex (DiDonato Jr et al. 2004). This protein helps in the transportation of Cu–NA complexes and $Cu^{2+}$ and $Fe^{2+}$ cations in their free form (Wintz et al. 2003).

7. Iron (Fe)

The uptake of iron (Fe) is dependent upon types of plants species. In non-graminaceous species, the plant's roots release some organic and phenolic compounds to acidify the rhizospheric zone that increase the $Fe^{+3}$ concentrations in soil solution. Then, with the help of ferric reductases (encoded by members of the ferric reductase oxidase; FRO, gene family), $Fe^{+3}$ get reduced to $Fe^{+2}$ in the epidermal cell of root (Mukherjee et al. 2006). Next, the members of different transporter proteins help in the influx of $Fe^{2+}$ to root cells such as zinc-regulated transporter (ZRT)-, iron-regulated transporter (IRT)- protein (ZIP) family (AtIRT1 in *Arabidopsis*) (Ishimaru et al. 2005). In contrast to this, graminaceous spp. release structural derivatives of mugineic acid, i.e., phytosiderophores that bind with $Fe^{3+}$ and whole complex is absorbed by root cells (Ishimaru et al. 2005). Again, within the xylem, Fe is transported as a $Fe^{3+}$ citrate complex (Abadía et al. 1984; Mukherjee et al. 2006). In *Arabidopsis*, it was reported that one of the member of the multidrug and toxin efflux (MATE) family; FRD3 is present in the root pericycle and help in the transportation of Fe from root to shoot in the form of citrate complex (Puig et al. 2007) and members of the ZIP family help in the uptake of $Fe^{2+}$ by shoot cells. During Fe deficiency, expression of genes ferric reductase oxidase (FROs) get upregulated that encode proteins responsible for the uptake and redistribution of Fe. These include genes encoding ZIPs, NRAMPs, YS1, and YSLs (Grotz and Guerinot 2006; Kramer et al. 1996; Stacey et al. 2008) and enzymes such as nicotianamine synthase (NAS), and phytosiderophores to synthesize nicotinamide (NA) and help in more uptake of iron by the plant's root cell

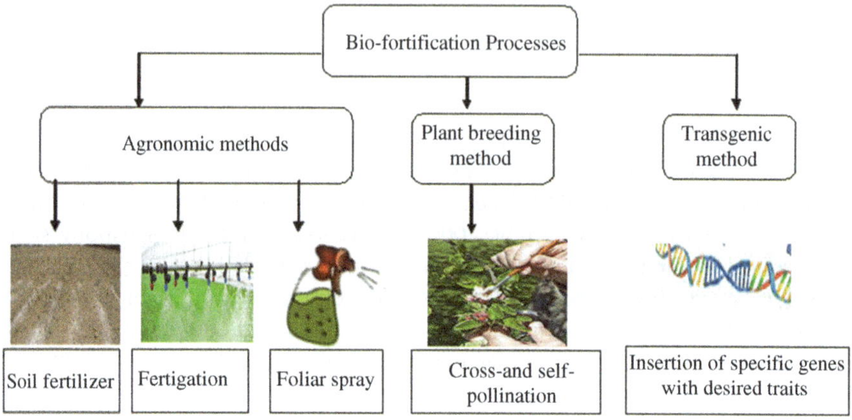


**Fig. 8.1** Different ways of biofortification

## 8.3 Different Ways of Biofortification to Manage Micronutrient Deficiency in Plants

For biofortification, there are three main methods used to manage nutrient deficiency in the plants such as agronomic method, plant breeding method, and transgenic method (Fig. 8.1). Each one is discussed further in detail.

### 8.3.1 Agronomic Approach

Agronomic approach includes application of fertilizers that helps in increasing nutritive values of plants without modifying their genetic setup (Almendros et al. 2015). This technique is able to provide efficient micronutrients concentrations in edible crops and it is one of the immediate and effective approaches (de Valenca et al. 2017). Soil deficiency is reflected by the poor nutrients composition of crops. This problem is aggravated by growing cereal crops on soils potentially deficient in nutrients. Nutrient deficiency in humans is also seen mainly in those regions where crops are grown in nutrient deficient soil (Bilski et al. 2012). Intervention of new agricultural approaches to improve production of micronutrient-rich foods is one of the main areas of research and competent strategy to supplement the nutrients in food products (Pandey et al. 2016). The agronomic biofortification of cereal crops appears to be a rapid and simple solution to manage the deficiency of important elements in soils and plants. With this approach, one should take care that the over fertilization can be toxic to the plants. The potential of agronomic fortification is strongly related with micronutrient bioavailability at three stages: from soil to plants, from plants root to the edible parts, and from edible parts to humans. Agronomic biofortification has positive impact on plant characteristics and nutritional status of plants. In combination with NPK, organic fertilizers, the micronutrients fertilization can improve crop varieties and that particularly highlights the importance of integrated soil fertility management.

Biofortification through agricultural methods includes application of nutrients directly in the soil and water that affect the health status of crops and provide quality food to the human being of plants. Agronomic biofortification is easy and cost-effective technique but more awareness and detail study is required to understand about the forms of fertilizer, mode of applications, and impact on other environmental components.

#### 8.3.1.1 Application of Fertilizer in Soil and Irrigation Water

The simplest way to increase the density of nutrients in the edible crops is by enhancing their availability through different forms of fertilizers, so that plants can uptake the nutrient from soil in a more efficient way (Almendros et al. 2015). The types of nutrients source and soil characteristics have a great influence on agronomic biofortification and that consequently affect the qualitative and quantitative characteristics of food crops. Soils show variation in their mineral composition and phytoavailability of nutrients basically based upon several factors such as pH, water holding capacity, cation exchange capacity of soil, specific surface area, surface charge density, as well as cation exchange capacity (Pinto and Ferreira 2015). Based on adsorption–desorption characteristics of soils, the application of fertilizers leads to enhancement in the concentration of nutrients in the plant's parts (Dai et al. 2009). The composition of fertilizers play important role in providing nutrients as well interactions among them can have positive neutral or even negative effects on yields and nutrient use efficiencies (Saha et al. 2015; Rietra et al. 2015).

##### Inorganic and Organic Fertilizers

With rising expectations toward agricultural production, the importance of micronutrient fertilization has increased tremendously. Soil nutrients especially the microelements are insufficient to meet increased crop requirements that affect both yields and quality of the crops. The standard NPK-based fertilization must often be supplemented by the deficient micronutrients. There are several inorganic forms of micronutrients that are applied with NPK fertilizers to support the growth of plants (Table 8.2a). Within the agronomic biofortification practice, the most common method to enhance the micronutrient levels in the field soil is by adding fertilizers in the form of inorganic salts. It brings good results based upon kind of supplemented micronutrient and the chemical properties of fertilized soil. Zinc sulfate ($ZnSO_4{\cdot}7H_2O$) and copper sulfate ($CuSO_4{\cdot}5H_2O$) are the most tested fertilizers, it has been observed that inorganic fertilizers applied after seed sowing lead to better yield as compared to pre-sowing soil fertilization (Smoleń and Sady 2012). Another interesting approach for the biofortification is the application of inorganic salt to obtain a new formulation, as in the case of Se-enriched peat. Although Se is not required by the plants but its certain level is important for the metabolic activities of human and animals. The peat was enriched by thoroughly

**Table 8.2a** Inorganic form of nutrients applied in the soil along with the NPK fertilizers

| Nutrient | Salt | Form |
|---|---|---|
| Zinc (Zn) | Zinc sulfate | $ZnSO_4.7H_2O$ |
| | Chelate | Zn EDTA |
| | Zinc oxide | ZnO |
| Copper(Cu) | Copper sulfate | $CuSO_4$ |
| Iron (Fe) | Ferrous sulfate | $FeSO_4.7H_2O$ |
| Manganese (Mn) | Manganese sulfate | $MnSO_4.H_2O$ |
| Chlorine (Cl) | Potassium chloride | KCl |

Source: Modified from Jones and Jacobsen (2009)

mixing it with a solution of sodium selenite and then applied during the pre-transplanting stage (Businelli et al. 2015).

Some plants and peat, in the presence of high level of inorganic Se, can metabolize and accumulate it in the form of organic derivatives. This process is important for the plant because it reduces the toxicity of the chalcogen and, at the same time, when the bioaccumulation occurs in edible tissues, it allows enrichment of food with Se that is good for the humans and animals. Moreover, Se biofortification also increases secondary metabolites production in human beings when consumed with the diet. Therefore, biofortification strategies applied to produce Se-enriched foods could help to overcome Se deficiency and its implications on human health and it also improve the nutraceutical (substance, which has physiological benefit or provides protection against chronic disease) value of food.

Similarly, organic fertilizers are also the source of micronutrient for plants. Earlier during the agricultural practices, it was observed that crop yields could be enhanced with the addition of animal manure or plant debris to the soil. A new study reveals that Neolithic farmers when the first developments of farming appeared used livestock manure to enhance crop yields (Bogaard et al. 2013). This practice is in continuation with regular additions of organic matter (from different sources), which is used mainly in organic and integrated farming systems (Kizos et al. 2010). Organic fertilizers are materials whose basic ingredient is organic matter (Adegoke et al. 2016). They traditionally derived from animal excreta (livestock manure, slurry, poultry feces) and vegetable matter (straw, green manures). Naturally occurring organic fertilizers include peat, seaweed, and guano (accumulated excrement of seabirds and bats). Guano is also an effective fertilizer due to its exceptionally high content of nutrients (Hazra 2016). Recently, municipal and industrial wastes are taken into account as organic renewable resources to improve the nutritional status of plants. The most important organic materials that accumulate in industrial countries are sewage sludge, bio-compost and by-products from the food and foodstuff industry provide several nutrients (Table 8.2b). Waste from the food and luxury item industries can also be applied in agricultural fields as organic waste. These organic wastes act as an important secondary source of micronutrients and their availability depend upon soil organisms (Jones and Jacobsen 2009).

**Table 8.2b** Micronutrient content of selected organic fertilizers in mg/kg

| Organic fertilizers | Fe | Cu | Mn | Zn | Source |
|---|---|---|---|---|---|
| Sewage sludge | 2275–3322 | 7–11 | 100–287 | 68–177 | Tennakoon and Bandara (2003) |
| Green manure (Acacia) | 870–994 | 7–9 | 78–92 | 54–610 | Tennakoon and Bandara (2003) |
| Cattle manure | 1075 | 880 | 247 | 44 | Uyanoz (2007) |
| Pig manure | 1416 | 502 | 367 | 563 | Li et al. (2009) |
| Rice straw | 225 | 3.73 | 467 | 49.6 | Li et al. (2009) |
| Sheep manure compost | 1248.9 | 4.0 | 45.6 | 68.9 | Wang et al. (2016) |

### Biofertilizers

Application of biofertilizers is also one of the ways of agronomic biofortification to raise nutrients content in the plants by increasing the solubilization and mobilization rate of elements (Almendros et al. 2015). Biofertilizers can be described as diverse groups of soil-borne microbes, such as root endophytic fungi, mycorrhizal fungi, plant growth-promoting rhizobacteria, and rhizobia that exert positive effects on plant yields and survival through direct and plant-mediated mechanisms. They help in the nitrogen fixation, solubilize the insoluble minerals, produce phytohormones, and also protect the plants from pathogens (Olivares et al. 2015). Microorganisms can be used as substitutes for various chemical fertilizers and improve plant nutrition and health. Usually biofertilizers do not contain a single culture of beneficial microorganism but a mixture of different microorganisms. Soil particularly the rhizospheric zone contains some bacterial species that promote growth of the plants and collectively termed as Plant Growth-Promoting Rhizobacteria (PGPR; *Rhizobia* spp. And *Frankia* spp.). PGPR can facilitate acquisition of resources and modulate the levels of plant hormones. They are able to provide resistance to the plants against various pathogenic agents. The PGPR consortium, named "BioPower" consist of two *Azospirillum lipoferum* strains, two *Pseudomonas* sp. strains and one *Agrobacterium* sp. strain. It has been found to increase the availability of Zn in rice crop (Tariq et al. 2007). Rana et al. (2012) have applied three rhizobacterial strains: *Bacillus* sp., *Providencia* sp., and *Brevundimonas* sp., applied along with NPK fertilizers. The study showed significant enhancement in Fe, Cu, Zn, and Mn content in wheat plant. The application of *Pantoea dispersa* MPJ9 and *Pseudomonas putida* MPJ6 increased the Fe content in mung beans by 3.4 times under Fe deficient soil by producing Fe chelating agent, i.e., siderophores (Ghosh et al. 2019). Ramesh et al. (2014) have inoculated two strains of *Bacillus* aryabhattai (MDSR7 and MDSR14) in Zn deficient soil that improved the Zn uptake in soybeans and wheat crops. Gopalakrishnan et al. (2016) have studied effect of seven strains of PGPR and reported that among all *Enterobacter ludwigii* and *Acinetobacter tandoii* SRI-229 strains showed significant enhancement in Fe, Zn, Cu, and Mn concentrations in chickpeas and pigeon peas. Recently, Singh and Prasanna (2020) have coated chickpea seeds with Zn solution along with Zn solubilizer PGPR, *Enterobacter* sp. MN17 that improved the bioavailability of Zn and consequently the grain yield.

Different PGPR strains have differential abilities to fix or solubilize nutrients within the rhizosphere for promoting growth and yield of the plants (Amaya-Gómez et al. 2020).

The effectiveness of micronutrient fertilizer application on crop biofortification is influenced not only by the fertilizer type but also by the application method (Mao et al. 2014; Melash et al. 2016; de Valenca et al. 2017). In crop plants, micronutrients may be applied to the soil as foliar spray seed treatments, or through fertigation (Farooq et al. 2012; Singh and Prasad 2014; Smoleń et al. 2016). The decision for the selection of method depends upon the requirement of specific nutrients and form of fertilizers (Pankaj and Dewangan 2016). Some of the micronutrients like Fe and Mn were applied through foliar application and that improved the growth of plants (Narwal et al. 2012). Organic fertilizers are spread uniformly in the field and incorporated several days before planting (Piechota et al. 2014). Micronutrient salts can be applied as a granular material or dissolved in liquid fertilizers (Pagani et al. 2013).

Another way of fertilizer application is with irrigation water, i.e., Fertigation, it is made up of two words, i.e., fertilization and irrigation. In this, fertilizer is applied with the irrigation water trough drip irrigation method (Bell and Dell 2008). Through this process fertilizer solution is distributed evenly with irrigation water. With this mode, the availability of nutrients is increased mainly in the rhizospheric zone. During this process, only liquid fertilizer and the fertilizers soluble in water are used. Fertigation is practiced extensively in commercial agriculture and horticulture. It is used to add additional nutrients or to manage nutrient deficiencies detected in plant tissue. It is usually practiced on the high-value crops such as vegetables, fruit trees, and cereals for the purpose of biofortification. The nutrient used in fertigation must be highly soluble in water like monoammonium phosphate (Nitrogen and Phosphorus), poly feed (Nitrogen, Phosphorus, and Potassium), Multi K (Nitrogen and Potassium), Potassium sulfate (Potassium and Sulfur). Some of the nutrients used in fertigation are as follows:

- Ammonium nitrate
- Ammonium sulfate
- Urea
- Monopotassium phosphate
- Potassium sulfate
- Potassium nitrate
- Potassium chloride
- Diammonium phosphate

Through fertigation, the water and fertilizer are evenly supplied to the crops, so there is more possibility of getting 25–50% higher yield. It also minimizes the amount of fertilizers applied and the time, labor, and energy utilized during this process. It leads to the reduction in soil erosion as here nutrients are applied through the drip irrigation (Khalid et al. 2015).

**Agronomic Fortification Through Foliar Application**

In addition to nutrients being added to the soil as fertilizers, some mineral nutrients can be sprayed to the leaves and this process is known as foliar application (Mortvedt 1985). Over to soil fertilization, the foliar spray is found to be more beneficial by enhancing the nutrient uptake and their allocation in the edible plant parts (Lawson et al. 2015; de Valenca et al. 2017). Other advantages of foliar sprays are: uniform distribution is easily obtained, response to the applied nutrient is almost immediate; therefore, deficiencies can be managed easily. Foliar feeding is associated with higher yields and better quality of fruits. The efficiency of nutrient uptake is increased by 8–9 times when nutrients are applied as a foliar spray, as compared to the soil application (de Valenca et al. 2017).

Foliar fertilization has the ability to improve the efficiency and utilization of nutrients, required by the plant for their maximum growth and yield. The main advantage of foliar fertilization is the immediate uptake of applied nutrients. The most important use of foliar application is that only limited amount of micro and macronutrients are applied, which do not cause any kind of phytotoxicity (Oosterhuis and Weir 2010).

It also makes available those nutrients like Zn and Fe which are not available to the plants through root uptake. A foliar application is recommended when environmental conditions limit the uptake of nutrients by roots such as variation in pH, moisture and nutrient imbalances in soil, etc. The availability of micronutrient is decreased at high soil pH and under such circumstances the more efficient way to supply micronutrients to the plant is foliar spray rather than soil application (Adams 1984).

Zhang et al. (2012) have also reported that foliar application of Zn is more effective than soil application. The study showed that foliar application of 0.4% $ZnSO_4$.7H2O resulted 58 and 76% increase in Zn concentration, respectively, in grain and flour of wheat. The foliar Zn application provides an effective way to enhance dietary Zn in the edible products derived from wheat. Rugeles-Reyes et al. (2019) have also reported that the application of zinc (1.5 kg $ha^{-1}$) as foliar spray leads to 279% enhancement in the supply of Zn from plant to the humans as compared to the control. So, the foliar applications of the nutrient are found to be effective in increasing Zn contents in the plant leaf.

### 8.3.2 Plant Breeding Technology

Biofortified crops can be developed by breeding methods, but it is possible only when sufficient genetic variation is present in crop populations for the desired traits. Conventional plant breeding through cross- and self-pollination strategies plays major role in improving agricultural productivity. During conventional plant breeding, in order to produce desirable agronomic traits, the parent plants having high nutrients are crossed over several generations with recipient one (Garg et al. 2018). The most significant, systematic and symbolic program of biofortification through conventional breeding is "Harvest Plus." The main goal of *Harvest Plus program is to develop and promote the production of biofortified food crops that improve the public health.* It basically increased levels of three nutrients (iron, zinc, and pro-vitamin A) in seven staple crops (beans, cassava, maize, rice, wheat, sweet potato, and pearl millet) (Nestel et al. 2006). The Harvest Plus programme is funded principally by grants from foundations, governments, and international agencies, started in 2003, by the Consultative Group on International Agricultural Research (CGIAR). The main vision of this program is to provide more nutritious food to 1 billion people by 2030.

The success rate of plant breeding technique to produce fortified crop is dependent upon their acceptance by the farmers as well as absorption rate of micronutrients in the consumers (Bouis 2003). Only the staple food varieties whose seeds are micronutrient rich are feasible for plant breeding. The micronutrient efficient varieties developed by plant breeding can grow deeper in the mineral deficient soil. The roots of new varieties are more efficient in mobilizing the external minerals and are able to utilize the moisture and nutrients present in the subsoils. This will reduce the application rate of fertilizer as well as irrigation. The efficient uptake of minerals from soil and their loading into the grains lead to higher yield. So, the farmers can easily accept new varieties with mineral dense seeds and higher yield (Bouis 2003). There are several studies where Zn and Fe concentrations are estimated in different varieties of plants to find out the best variety with their highest concentration of nutrients for breeding program (Velu et al. 2015; Garg et al. 2018; Jha and Warkentin 2020). Researchers at IRRI have studied six sets of genotypes ($n = 939$) and evaluated Fe and Zn concentrations. The Fe concentration ranged from 7.5–24.4 μg $g^{-1}$ for Fe, and Zn concentration ranged from 13.5–58.4 μg $g^{-1}$. Among all varieties, Jalmagna, Zuchem, and Xua Bue Nuo had highest concentration of Fe and Zn. The F2-derived populations of these varieties showed that this trait is not found to be pleiotropic for grain-Fe or-Zn concentrations so can be used for further breeding program. Over the common rice variety, Jalmagna is the traditional variety that had nearly 40% more iron concentration. Under Harvest Plus program, Bangladesh Rice Research Institute has developed world's first rice varieties (BRRIdhan 62, BRRIdhan 72, and BRRIdhan 64) with high Zn concentrations (20–22 ppm). In India and Philippines, by crossing a variety having high yield (IR72) with Zawa Bonday; a tall variety, an improved line (IR68144-3B-2-2-3) was developed in order to enhance the level of Fe (21 ppm) in the grain (Palanog et al. 2019).

Similarly, wide range of wheat germplasm is being studied at International Maize and Wheat Improvement Center (CIMMYT; Spanish acronym) with respect to the Fe and Zn concentrations in the whole grain. Among all wheat germplasm, *Triticum dicoccum* with highest concentrations of Fe and Zn can be used for further study (Welch and Graham 2002). Through collaboration with Banaras Hindu University (BHU), Uttar Pradesh (UP), India, in 2014, six varieties of wheat named as BHU 1, BHU 3, BHU 5, BHU 6, BHU 7, and BHU 18 with high concentration of Zn (4–10 ppm) were released under Harvest Plus program (Velu et al. 2015). One variety (WB2) with high concentration of Zn and Fe has been developed by Indian Institute of Wheat and Barley Research, India (Chatrath et al. 2018). Along with this, four varieties (NR 419, 42, 421, and Zincol) and one variety (PBW1Zn) of wheat with high Zn were released by University of Agriculture Faisalabad, Pakistan (Ohly et al. 2019) and Punjab Agricultural University, India (Bhati et al. 2016), respectively. The degrees of genetic variability for Zn and Fe contents in bean seeds have been studied by the researchers of the International Center for Tropical Agriculture (CIAT). More than 1000 varieties of common beans are collected and showed varied levels of Fe from 34–89 μg $g^{-1}$ and of Zn from 21–54 μg $g^{-1}$. Due to presence of sufficient genetic variability, the Fe and Zn concentration can be increased significantly by plant breeding technology (Graham et al. 1999). Several varieties of common bean with high Fe content have been developed under HarvestPlus in Rwanda, Eastern Africa Democratic Republic of Congo also released ten biofortified varieties (COD MLB 001, COD MLB 032, HM 21–7, RWR 2245, PVA 1438, COD MLV 059, VCB 81013, Nain de Kyondo, Cuarentino, Namulenga) with high Fe concentration (Andre et al. 2007). With the help of ICARDA, HarvestPlus biofortification program, some varieties of lentil Barimasur-4, Barimasur-5, Barimasur-6, Barimasur-7, and Barimasur-8 in Bangladesh and ILL 7723, Khajurah-1, Khajurah-2, Shital, Sisir Shekhar, Simal in Nepal and L4704, Pusa Vaibhav in India, Alemaya in Ethiopia and Idlib-2, Idlib-3 in Syria have been released having high Zn and Fe content (Thavarajah et al. 2008).

Along with the cereals and legumes some vegetables varieties are also developed as they are the main source of antioxidants in human diet. Some varieties of potato have been obtained by collecting 1000 genotypes that have more antioxidants and Cu, Fe, Mn, and Zn concentrations (Andre et al. 2007). With the collaboration of International potato center (CIP) and HarvestPlus an advanced variety has been developed by crossing diploid Andean landrace potatoes (high Zn and Fe) with tetraploid clones (disease resistant). One more variety INIA 321 Kawsay with high Zn and Fe content in Peru has been developed under National Institute for Agrarian Innovation's (INIA) Potato Program (Andre et al. 2007). Cassava and cauliflower have a wide range of genotype differences particularly for minerals (iron and zinc), so new varieties can be developed by breeding technique (Chavez et al. 2005). Through breeding technology, the biofortified edible crops can be produced having high essential micronutrients (Table 8.3) that will definitely improve the health and economic conditions of the world' population.

### 8.3.3 Application of Transgenic Method

The application of biotechnology in developing nutrient rich transgenic crops has been started since last 20 years. Genetic engineering is a technique that concerned with the specific genes with desired traits. Once a gene with specific trait has been identified and with the help of marker and promoter genes a new plant with improved nutrient content will be produced by inserting a gene through nonviable virus called *Agrobacterium* as a carrier. GE (Genetic engineering) leads to the development of transgenic crops also known as genetically modified organism (GMOs). When there is less or no genetic variation in nutrients concentration among plant varieties then

**Table 8.3** Biofortification of food crop through breeding

| Plants | Nutrients | Country/Variety | Reference |
|---|---|---|---|
| Rice | Iron | **India, Philippines**: IR68144-3B-2-2-3 (improved line) Jalmagna | IRRI Gregorio et al.(2000) |
| | Zinc | Jalmagna | Gregorio et al. (2000) |
| | Zinc, iron | **Bangladesh:** BRRIdhan 62, BRRIdhan 72, BRRIdhan 64 | CIAT, HarvestPlus; Garg et al. (2018) |
| Cassava | Iron | **Africa**: Cassava clones | Maziya-Dixon et al. (2000) |
| Potato | Zinc, iron, copper, and, manganese | BTD0054-3, BTD0118-5 | Haynes et al. (2012) |
| Wheat | Zinc | **India:** BHU 1, BHU 3, BHU 5, BHU 6, BHU 17, BHU 18 **Pakistan:** NR 419, 42, 421, Zincol | Velu et al. (2015) |
| | Zinc and iron | **India:** WB2 | Chatrath et al. (2018) |
| | Zinc | **India:** PBW1Zn | Bhati et al. (2016) |
| Sorghum | Iron | **India:** ICSR 14001, ICSH 14002 Hybrids: ICSA 661 × ICSR 196, ICSA 318 × ICSR 94, ICSA 336 × IS 3760 | ICAR (2016) |
| | Iron | **Nigeria:** 12KNICSV (Deko)-188 12KNICSV-22 (Zabuwa) | ICRISAT, HarvestPlus (2016) |
| Cow pea | Iron | **India:** Pant Lobia-1, Pant Lobia-2, Pant Lobia-3, Pant Lobia-4 | Singh et al. (2017) |
| Millets | Iron and zinc (Pearl Millet) | **India:** Dhanashakti Hybrid ICMH 1201 (Shakti-1201) | Govindaraj (2019) |
| Lentils | Iron and zinc | **Bangladesh**: Barimasur-4, Barimasur-5, Barimasur-6, Barimasur-7, Barimasur-8 **Nepal**: ILL 7723- Khajurah-1, Khajurah-2, Shital, Sisir, Shekhar and Simal **India**: L4704 and Pusa Vaibhav **Ethiopia**: Alemaya **Syria**: Idlib-2 and Idlib-3 | Darai et al. (2020) |
| Beans | High iron and zinc | **Rwanda:** RWR 2245; RWR 2154; MAC 42; MAC 44; CAB 2; RWV 1129; RWV 3006; RWV 3316; RWV 3317; RWV2887 | Jha and Warkentin (2020) |

Source: Modified from Garg et al. (2018)

this transgenic approach can be an applicable option for developing biofortified crops (Aung et al. 2013). Incorporation of desired traits includes micronutrient enhancement and bioavailability as well as reduction in the anti-nutrients concentrations (that bind with the nutrient and make them unavailable). Genetic modifications particularly affect the redistribution of micronutrients in plant tissues and mainly enhance their concentration in edible portions of cash crop. Several studies were done where crops are genetically modified to improve their micronutrient levels particularly for Fe and Zn as they are found to be more deficient than the other micronutrients. In rice plant, genetic modification is done by over expressing iron (II)-nicotianamine transporter OsYSL2 to enhance translocation of Fe in the endosperm (Masuda et al. 2012). The transgenic rice crop showed 4-times higher levels of iron than the conventional one. Mugineic acid acts as a ferric ion chelator, and its production is increased in the transgenic rice crop by the expression of mugineic acid synthase gene (IDS3). This gene is over expressed by expressing the soybean ferritin gene (SoyferH2). The transgenic rice crop is found to be tolerant in iron deficient soil and showed 2.5 times higher concentration of Fe. In Myanmar, about 70% population is found to be Fe deficient so here Aung et al. (2013) have produced a transgenic line of rice by over expressing the nicotianamine synthase gene HvNAS1 (increases the transportation of Fe), the Fe (II)-nicotianamine transporter gene OsYSL2 (enhances transportation of iron in the endosperm), and the Fe storage protein gene SoyferH2 (increases accumulation of iron in the endosperm). The milling step during rice crop processing removes the nutrient-rich outer layers of the embryo that leads to reduction in the concentration of Fe and Zn. In order to solve this issue, a transgenic cultivar of rice (indica) with high yield has been developed by expressing SoyferH2 gene. These new line showed enhancement (2.6 times) in the ferritin level. By using MxIRT1 (iron transporter gene) from apple trees a transgenic rice crop is produced that exhibited 3 time higher Fe and Zn accumulation (Tan et al. 2015). By over expressing nicotianamine synthase (OsNAS2) and soybean ferritin (SferH-1) genes, a rice plant can be developed having high Fe and Zn concentration in the endosperm (Trijatmiko et al. 2016). It was reported that in wheat Gpc-B1 (GRAIN PROTEIN CONTENT B1) is quantitative trait locus responsible for increasing the translocation of protein to the grain that consequently increased Fe, Zn, and Mn concentrations in grain (Uauy et al. 2006). Ozturk et al. (2016) have supported positive correlations between protein content and concentration of Fe, Zn, and Mn. Through genetic transformation of Gpc-B1 locus from the wild tetraploid wheat *Triticum turgidum* ssp. dicoccoides, the concentration of Zn, Fe, Mn, and protein content can be increased by 10–34% in wheat grain of different recombinant chromosome substitution lines (Distelfeld et al. 2007). For the uptake and translocation of Zn, the most predominant cation transporter families are the members of the ZIP (ZRT, IRT-related protein) and CDF (Cation diffusion facilitator). Some genetically modified varieties of rice with higher level of Zn and Fe, such as IR64 and IR69428, have been produced by over expressing rice ferritin and rice nicotianamine synthase (NAS2) genes at Indian Rice Research Institute (IRRI) from the field trials (Mallikarjuna Swamy et al. 2016). With this technique, several other important staple crops can be transformed to produce biofortified crops (Table 8.4) that have great potential in combating global problem of malnutrition.

**Table 8.4** Biofortification of plants by transgenic approaches

| Micronutrients and crops | Over expressed genes | Plants | References |
|---|---|---|---|
| Enhancement of Fe storage in rice seeds | OsGluB1proSoyferH1[b] OsGlb1 proSoyferH1[b] | *Oryza sativa* Japonica cv. Kitaake | Qu et al. (2005) |
| Enhancement of Zn, Mn, and Fe in wheat | Gpc-B1 locus into different recombinant chromosome substitution lines | *Triticum turgidum* ssp. dicoccoides | Distelfeld et al. (2007) |
| Enhancement of Fe uptake | Barley IDS3 genome fragment | *Oryza sativa* Japonica cv. Tsukinohikari | Masuda et al. (2008) |
| Enhancement of Fe translocation | Ubiquitin pro-OsIRT1 | *Oryza sativa* Japonica cv. Dongjin | Lee and An (2009) |
| Enhancement of Zn and Fe in rice | over expressing ferritin and nicotianamine synthase (NAS2) | IR64 and IR69428 | Zhang et al. (2010) |
| Fe translocation increased | 35S pro-OsIRO2 | *Oryza sativa* Japonica cv. Tsukinohikari | Ogo et al. (2011) |
| Enhancement of Fe translocation | 35S pro- OsNAS1, 2,3 | *Oryza sativa* Japonica cv. Nipponbare | Johnson et al. (2011) |
| Fe content in endosperm | iron (II)-nicotianamine transporter OsYSL2 | *Oryza sativa* | Masuda et al. (2012) |
| Transportation of Fe | Nicotianamine synthase gene HvNAS1 | *Oryza sativa* | Aung et al. (2013) |
| Uptake and root-to-shoot translocation of Zn increased | ZRT/IRT-like protein | *barley* | Tiong et al. (2015) |
| Zn, Fe, and Mn content in wheat grain | Gpc-B1 (GRAIN PROTEIN CONTENT B1) | *Triticum aestivum* | Trijatmiko et al. (2016) |
| Increased the Fe content | AtIRT1 | *Oryza sativa* Japonica cv. Taipei 309 | Boonyaves et al. (2016) |
| Storage of Fe content increased in endosperm | OsNAS1, HvHAATb | *Oryza sativa* L. (cv. EYI 105) | Banakar et al. (2017) |
| Enhancement in Fe concentration | Os DMAS1 | *Oryza sativa* Japonica cv. Dongjin | Bashir et al. (2017) |
| Uptake of Fe increased | OsYSL9 | *Oryza sativa Japonica cv. Tsukinohikari* | Senoura et al. (2017) |
| High accumulation of iron and zinc | OsHMA7transcript levels | *Oryza sativa* | Kappara et al. (2018) |

(continued)

**Table 8.4** (continued)

| Micronutrients and crops | Over expressed genes | Plants | References |
|---|---|---|---|
| Fe content increased | OsYSL1 | *Oryza sativa* Japonica cv. Zhoghua11 | Zhang et al. (2018) |
| Fe content increased in grain | OsNAS2 | *Triticum aestivum* (*cv* Bob White) | Beasley et al. (2019) |
| Zn and Fe concentration increased | IRT1 (iron transporter) and FER1 (ferritin) genes | Cassava | Ghislain et al. (2019) |

## 8.4 Conclusions and Future Prospective

The biofortified crops are not easily accepted by the poor farmers due to lack in the awareness about their benefits on human health. Particularly, in developing countries, malnutrition is one of the inevitable problems due to more population. Because of this, the food provided to them is not having sufficient level of micronutrients that not only harms their health but also increases the susceptibility towards various diseases. Consequently, it leads to considerable loss in Gross Domestic Product and shows devastating effects on socio-economic condition of country. To deal with this situation, biofortification of crop varieties is one of the most sustainable and cost-effective approach. By applying different ways of biofortification whether agronomic or genetic, different biofortified crops can be produced that provide nutrients directly to the common people in their natural form. The biofortified crop varieties act as important sources of nutrients to poor people and also provide nutritional security. In order to enhance the acceptance rate of these biofortified crops, the farmers should be participated in the awareness programs that demonstrate their beneficial roles. The biofortified crops improve the health and nutritional status of the young generation. This process can adequately supply the food to those people, who are underprivileged and low-income households. After developing biofortified crops, there are no further charges, so this strategy can be a sustainable way to manage the hidden hunger particularly in developing and under-developed countries.

**Acknowledgments** We are very grateful to The Head, Department of Botany, Banaras Hindu University, Varanasi for providing necessary facilities. The University Grants Commission, New Delhi is thankfully acknowledged for providing financial assistant to Dr. A. Singh as PI [F.30-431/2018(BSR)].

## References


Abadía J, Monge E, Montañés L, Heras L (1984) Extraction of iron from plant leaves by Fe (II) chelators. J Plant Nutrit 7(1–5):777–784

Adams F (1984) Crop response to lime in the southeastern United States. In: Dinauer RC (ed) Soil Acidity and Liming. ASA-CSSA-SSSA, Madison

Adegoke AA, Awolusi OO, Stenstrom TA (2016) Organic fertilizers: Public health intricacies. In: Larramendy M (ed) Organic Fertilizers-From Basic Concepts to Applied Outcomes. InTech, Rijeka. https://www.intechopen.com/books/organic-fertilizers-from-basic-concepts-toapplied-outcomes/organic-fertilizers-public-health-intricacies

Alejandro S, Höller S, Meier B, Peiter E (2020) Manganese in Plants: From Acquisition to Subcellular Allocation. Front Plant Sci 11:300. https://doi.org/10.3389/fpls.2020.00300

Almendros P, Obrador A, Gonzalez D, Alvarez JM (2015) Biofortification of\ zinc in onions (*Allium cepa* L.) and soil Zn status by the application of different organic Zn complexes. Sci Hortic 186:254–265

Andre CM, Ghislain MP, Bertin O, Mouhssin M, Del Rosario H, Hoffmann L et al (2007) Andean potato cultivars (*Solanum tuberosum* L.) as a source of antioxidant and mineral micronutrients. J Agric Food Chem 55(2):366–378. https://doi.org/10.1021/jf062740i

Aung MS, Masuda H, Kobayashi T, Nakanishi H, Yamakawa T, Nishizawa NK (2013) Iron biofortification of Myanmar rice. Front Plant Sci 4:158

Bashir K, Nozoye T, Nagasaka S, Rasheed S, Miyauchi N, Seki M et al (2017) Paralogs and mutants show that one DMA synthase functions in iron homeostasis in rice. J Exp Bot 68:1785–1795

Banakar R, Alvarez-Fernandez A, Abadia J, Capell T, Christou P (2017) The expression of heterologous Fe (III) phytosiderophore transporter HvYS1 in rice increases Fe uptake, translocation and seed loading and excludes heavy metals by selective Fe transport. Plant Biotechnol J 15:423–432

Baxter I., Muthukumar B., Park H. C., Buchner P., Lahner B., Danku J., et al. (2008). Variation in molybdenum content across broadly distributed populations of *Arabidopsis thaliana* is controlled by a mitochondrial molybdenum transporter (MOT1). PLoS Genet 4:e1000004 https://doi.org/10.1371/journal.pgen.1000004

Beasley JT, Bonneau JP, Sanchez-Palacios JT, Moreno-Moyano LT, Callahan DL, Tako E et al (2019) Metabolic engineering of bread wheat improves grain iron concentration and bioavailability. Plant Biotechnol J doi. https://doi.org/10.1111/pbi.13074

Bell RW, Dell B (2008) Micronutrients for sustainable food, feed, fibre and bioenergy production. International Fertilizer Industry Association, Paris

Bhati KK, Alok A, Kumar A, Kaur J, Tiwari S, Pandey AK (2016) Silencing of ABCC13 transporter in wheat reveals its involvement in grain development, phytic acid accumulation and lateral root formation. J Exp Bot 67:4379–4389

Bilski J, Jacob D, Soumaila F, Kraft C, Farnsworth A (2012) Agronomic biofortification of cereal crop plants with Fe, Z, and Se, by the utilization of coal fly ash as plant growth media. Adv Biores 3:130–136

Bogaard A, Fraser F, Heaton THE et al (2013) Crop manuring and intensive land management by Europe's first farmers. Proc Natl Acad Sci U S A 110(31):12589–12594

Boonyaves K, Gruissem W, Bhullar NK (2016) NOD promoter-controlled AtIRT1 expression functions synergistically with NAS and FERRITIN genes to increase iron in rice grains. Plant Mol Biol 90:207–215

Bouis H, Saltzman A (2011) Improving nutrition through biofortification: A review of evidence from HarvestPlus. Glob Food Sec 12:49–58

Bouis H (2003) Micronutrient fortification of plants through plant breeding: can it improve nutrition in man at low cost? Proc Nutr Soc 62(2):403–411

Broadley MR, White PJ, Hammond JP, Zelko I, Lux A (2007) Zinc in plants. New Phytol 173:677–702. https://doi.org/10.1111/j.1469-8137.2007.01996.x

Buchner P, Takahashi H, Hawkesford MJ (2004) Plant sulphate transporters: co-ordination of uptake, intracellular and long-distance transport. J Exp Bot 55:1765–1773. https://doi.org/10.1093/jxb/erh206

Businelli D, D'Amato R, Onofri A, Tedeschini E, Tei F (2015) Se-enrichment of cucumber (*Cucumis sativus* L.), lettuce (*Lactuca sativa* L.) and tomato (*Solanum lycopersicum* L. Karst) through fortification in pre-transplanting. Sci Horticul 197:697–704

Chatrath R, Tiwari VS, Gupta V, Kumar S, Singh SK, Mishra CN, Venkatesh K, Saharan MS, Singh G, Tyagi BS, Tiwari R, Sharma I, Parkash O, Singh GP (2018) WB 2: A high yielding bread wheat variety for irrigated timely sown conditions of North Western Plains Zone of India. Wheat Barley Res 101:40–44

Chavez AL, Sanchez T, Jaramillo G, Bedoya JM, Echeverry J, Bolanos EA et al (2005) Variation of quality traits in cassava roots evaluated in landraces and improved clones. Euphytica 143 (1–2):125–133. https://doi.org/10.1007/s10681-005-3057-2

Colmenero-Flores JM, Martinez G, Gamba G, Vizquez N, Iglesias DJ, Brumós J, Talón M (2007) Identification and functional characterization of cation-chloride cotransporters in plants. Plant J 50:27

Dai JL, Zhang M, Hu QH, Huang YZ, Wang RQ, Zhu YG (2009) Adsorption and desorption of iodine by various Chinese soils: II Iodide and iodate. Geoderma 153:130–135

Darai R, Sarker A, Pandey MP, Dhakal KH, Kumar S, Sah RP (2020) Genetic Variability and Genotype X Environment Interactions Effect on Grain Iron (Fe) and Zinc (Zn) Concentration in Lentils and Their Characterization under Terai Environments of Nepal. Adv Nutr Food Sci 5 (1):1–12

de Valença A, Bake A, Brouwer I, Giller K (2017) Agronomic biofortification of crops to fight hidden hunger in sub-Saharan Africa. Glob Food Sec 12:8–14

Distelfeld A, Cakmak I, Peleg Z, Ozturk L, Yazici AM, Budak H, Saranga Y, Fahima T (2007) Multiple QTL-effects of wheat Gpc-B1 locus on grain protein and micronutrient concentrations. Physiol Plant 129:635–643

DiDonato RJ Jr, Roberts LA, Sanderson T, Eisley RB, Walker EL (2004) Arabidopsis Yellow Stripe-Like2 (YSL2): a metal-regulated gene encoding a plasma membrane transporter of nicotianamine-metal complexes. Plant J 39:403–414

Dudev T, Lim C (2004) Oxyanion selectivity in sulfate and molybdate transport proteins: An ab initio/CDM study. J Am Chem Soc 126:10296–10305

Farooq M, Wahid A, Siddique KHM (2012) Micronutrient application through seed treatments–A review. J Soil Sci Plant Nutr 2(1):125–142

Garg M, Sharma N, Sharma S, Kapoor P, Kumar A, Chunduri V, Arora P (2018) Biofortified Crops Generated by Breeding, Agronomy, and Transgenic Approaches Are Improving Lives of Millions of People around the World. Front Nutr 5:12

Gasber A, Klaumann S, Trentmann O, Trampczynska A, Clemens S, Schneider S et al (2011) Identification of an Arabidopsis solute carrier critical for intracellular transport and inter-organ allocation of molybdate. Plant Biol 13(5):710–718. https://doi.org/10.1111/j.1438-8677.2011.00448.x

Ghislain M, Muzhingi T, Low JW (2019) Zinc and iron fortifcation in cassava. Nat Biotechnol 37:130–132

Ghosh A, Hasim Reja M, Nalia A, Kanthal S, Maji S, Venugopalan VK, Nath R (2019) Micronutrient Biofortification in Pulses: An Agricultural Approach. Curr J Appl Sci Technol 35:1–12

Gopalakrishnan S, Vadlamudi S, Samineni S, Sameer Kumar CV (2016) Plant growth-promotion and biofortification of chickpea and pigeonpea through inoculation of biocontrol potential bacteria, isolated from organic soils. Springerplus 5(1):1882

Govindaraj M, Rai KN, Cherian B, Pfeiffer WH, Kanatti A, Shivade H (2019) Breeding Biofortified Pearl Millet Varieties and Hybrids to Enhance Millet Markets for Human Nutrition. Agriculture 2019(9):106–117

Graham RD, Senadhira D, Beebe S, Iglesias C, Monasterio I (1999) Breeding for micronutrient density in edible portions of staple food crops. Conventional approaches. Field Crops Res 60:57–80

Gregorio GB, Senadhira D, Htut H, Graham RD (2000) Breeding for Trace Mineral Density in Rice. Food Nutri Bull 21(4):382–385

Grotz N, Guerinot M (2006) Molecular aspects of Cu, Fe and Zn homeostasis in plants. Biochimica et Biophysica Acta (BBA) – Mole. Cell Res 1763(7):595–608

Haslett B (2001) Zinc Mobility in Wheat: Uptake and Distribution of Zinc Applied to Leaves or Roots. Ann Bot 87(3):379–386. https://doi.org/10.1006/anbo.2000.1349

Haynes KG, Yencho GC, Clough ME, Henninger MR, Sterrett SB (2012) Genetic variation for potato tuber micronutrient content and implications for biofortification of potatoes to reduce micronutrient malnutrition. Am J Pot Res 89:192–198

Hazra G (2016) Different types of eco-friendly fertilizers: An overview. Sustainability in Environment 1(1):54–70

ICAR (2016) Best Practices for sorghum cultivation and importance of value addition 123–138

ICRISAT, HarvestPlus (2016) ICRISAT, In-house Happening newsletter no. 1716

Ishimaru Y, Suzuki M, Kobayashi T, Takahashi M, Nakanishi H, Mori S, Nishizawa NK (2005) OsZIP4, a novel zinc-regulated zinc transporter in rice. J Exp Bot 56:3207–3214

Ismail AM, Heuer S, Thomson JT, Wissuwa M (2007) Genetic and genomic approaches to develop rice germplasm for problem soils. Plant Mole Biol 65:547–570

Jha AB, Warkentin TD (2020) Biofortification of Pulse Crops: Status and Future Perspectives. Plan Theory 9:73. https://doi.org/10.3390/plants9010073

Johnson AAT, Kyriacou B, Callahan DL, Carruthers L, Stangoulis J, Lombi E, Tester M (2011) Constitutive overexpression of the OsNAS gene family reveals single gene strategies for effective iron- and zinc-biofortification of rice endosperm. PLoS One 6:e24476

Jones C, Jacobsen J (2009) Micronutrients: Cycling, testing and fertilizer recommendations. Nutrient Management Module 7:1–16

Kappara S, Neelamraju S, Ramanan R (2018) Down regulation of a heavy metal transporter gene influences several domestication traits and grain Fe-Zn content in rice. Plant Sci 276:208–219

Kato Y, Miwa K, Takano J, Wada M, Fujiwara T (2009) Highly boron deficiency-tolerant plants generated by enhanced expression of NIP5;1, a boric acid channel. Plant Cell Physiol 50:58–66

Khalid E., Salah E.H., , Eslam A.S., Abdallah E., Mukhtar Ahmed (2015). Impacts of fertigation via surface and subsurface drip irrigation on growth rate, yield and flower quality of *Zinnia elegans*. Bragantia, 75(1), 96–107

Kizos T, Veikontis G, Marin-Guirao JI (2010) Comparison of organic and integrated farming systems: The case of Sultana table grapes in Korinthos, Greece. J Sus Agri 35(1):7–47

Kramer U, Cotter-Howells JD, Charnock JM, Baker AJM, Smith JAC (1996) Free histidine as a metal chelator in plants that accumulate nickel. Nature 379:635–638

Lambert JJ, Dahlgren RA, Battany M, McElrone A, Wolpert JA (2008) Impact of Soil Properties on Nutrient Availability and Fruit and Wine Characteristics in a Paso Robles Vineyard. In: Proceedings of the 2nd Annual National Viticulture Research Conference • July 9–11

Lawson PG, Daum D, Czauderna R, Meuser H, Hartling JW (2015) Soil versus foliar iodine fertilization as a biofortification strategy for field-grown vegetables. Fron Plant Sci 6:450

Lee S, An G (2009) Over-expression of OsIRT1 leads to increased iron and zinc accumulations in rice. Plant Cell Environ 32:408–416

Li B, Wei M, Shen A, Xu J, Zhang H, Hao F (2009) Changes of yields, soil properties and micronutrients as affected by 17-yr fertilization treatments. J Food, Agri Environ 7 (3–4):408–413

Mallikarjuna Swamy BP, Rahman MA, Inabangan-Asilo MA, Amparado A, Manito C, Mohanty PC, Reinke R, Slamet-Loedin IH (2016) Advances in breeding for high grain Zinc in Rice. Rice 9:49

Mao H, Wang J, Wang Z, Zan Y, Lyons G, Zou C (2014) Using agronomic biofortification to boost zinc, selenium, and iodine concentrations of food crops grown on the loess plateau in China. J Soil Sci Plant Nutr 14(2):459–470

Marmagne A, Vinauger-Douard M, Monachello D, de Longevialle AF, Charon C, Allot M, Rappaport F, Wollman FA, Barbier-Brygoo H, Ephritikhine G (2007) Two members of the Arabidopsis CLC (chloride channel) family, AtCLCe and AtCLCf, are associated with thylakoid and Golgi membranes, respectively. J Exp Bot 58:3385–3393

Masuda H, Suzuki M, Morikawa KC, Kobayashi T, Nakanishi H, Takahashi M, Saigusa M, Mori S, Nishizawa NK (2008) Increase in iron and zinc concentrations in rice grains via the introduction of barley genes involved in phytosiderophore synthesis. Rice 1:100–108

Masuda H, Ishimaru Y, Aung MS, Kobayashi T, Kakei Y, Takahashi M, Higuchi K, Nakanishi H, Nishizawa NK (2012) Iron biofortification in rice by the introduction of multiple genes involved in iron nutrition. Sci Rep 2:543

Maziya-Dixon B, Kling JG, Menkir A, Dixon A (2000) Genetic variation in total carotene, iron, and zinc contents of maize and cassava genotypes. Food Nutrit Bull 21(4):419–422

Melash AA, Mengistu DK, Aberra DA (2016) Linking agriculture with health through genetic and agronomic biofortification. Agri Sci 7:295–307

Milner MJ, Kochian LV (2008) Investigating Heavy-metal Hyperaccumulation using Thlaspi caerulescens as a Model System. Ann Bot 102(1):3–13. https://doi.org/10.1093/aob/mcn063

Morgan JB, Connolly EL (2013) Plant-Soil Interactions: Nutrient Uptake. Nature Education Knowledge 4(8):2

Mortvedt JJ (1985) Micronutrient fertilizers and fertilization practices. In: Vlek PLG (ed) Micronutrients in Tropical Foods. Springer, Dordrecht, pp 221–235

Mukherjee I, Campbell NH, Ash JS, Connolly EL (2006) Expression profiling of the Arabidopsis ferric chelate reductase (FRO) gene family reveals differential regulation by iron and copper. Planta 223:1178–1190. https://doi.org/10.1007/s00425-005-0165-0

Muramatsu Y, Yoshida S, Bannai T (1995) Tracer experiments on the behaviour of radioiodine in the soil-plant-atmosphere system. J Radioanal Nucl Chem 194:303

Nakagawa Y, Hanaoka H, Kobayashi M, Miyoshi K, Miwa K, Fujiwara T (2007) Cell-type specificity of the expression of Os BOR1, a rice efflux boron transporter gene, is regulated in response to boron availability for efficient boron uptake and xylem loading. Plant Cell 19:2624–2635

Narwal R, Dahiya R, Malik R, Kala R (2012) Influence of genetic variability on zinc, iron and manganese responses in wheat. J Geochem Explor 121:45–48

Nestel P, Bouis HE, Meenakshi JV, Pfeiffer W (2006) Biofortification of staple food crops. J Nutr 136:1064–1067

Ogo Y, Itai RN, Kobayashi T, Aung MS, Nakanishi H, Nishizawa NK (2011) OsIRO2 is responsible for iron utilization in rice and improves growth and yield in calcareous soil. Plant Mol Biol 75:593–605

Ohly H, Broadley MR, Joy EJM, Khan MJ, McArdle H, Zaman M, Zia M, Lowe N (2019) The BiZiFED project: Biofortified zinc flour to eliminate deficiency in Pakistan. Nutr Bull 44 (1):60–64

Olivares FL, Aguiar NO, Rosa RCC, Canellas LP (2015) Substrate biofortification in combination with foliar sprays of plant growth promoting bacteria and humic substances boosts production of organic tomatoes. Sci Hortic 183:100–108

Oosterhuis DM, Weir BL (2010) Foliar fertilization of cotton. In: Stewart JMD et al (eds) Physiology of Cotton. Springer Science + Business Media B.V, Dordrecht, pp 272–288. https://doi.org/10.1007/978-90-481-3195-2-25

Ozturk L, Yazici MA, Yucel C, torun, A.A. (2016) Concentration and localization of zinc during seed development and germination in wheat. Physiol Plant 128(1):144–152

Pagani, A., Sawyer, J. E., and Mallarino, A. 2013. Site-specific nutrient management- for nutrient management planning to improve crop production, environmental quality, and economic return. Micronutrient Management, Chap. 7. NRCS, Iowa State University

Palmgren M, Clemens S, Williams L, Krämer U, Borg S, Schjørring J, Sanders D (2008) Zinc biofortification of cereals: problems and solutions. Trends Plant Sci 13(9):464–473

Palanog AD, Calayugan MIC, Descalsota-Empleo GI, Amparado A, Inabangan-Asilo MA, Arocena EC, Sta Cruz PC, Borromeo TH, Lalusin A, Hernandez JE, Acuin C, Reinke R, Mallikarjuna Swamy BP (2019) Zinc and iron nutrition status in the Philippines population and local soils. Front Nutr 6:81

Pandey N, Hossain F, Kumar K, Vishwakarma AK, Muthusamy V, Saha S, Agrawal PK, Guleria SK, Reddy SS, Thirunavukkarasu N, Gupta HS (2016) Molecular characterization of endosperm- and amino acids- modifications among quality protein maize inbreds. Plant Breed 135:47–54

Pankaj SC, Dewangan PK (2016) Biofortification: An alternative for zinc and iron deficiency in cereals—A review. Int J Advan Res 4(5):618–631

Piechota T, Kowalski M, Sawinska Z, Majchrzak L (2014) Assessment of one operation strip tillage and in row liquid organic manure injection in maize. Fragm Agrono 31(1):74–82

Pinto E, Ferreira IMPLVO (2015) Cation transporters/channels in plants: Tools for nutrient biofortification. J Plant Physiol 179:64–82

Puig S, Andrés Colás N, García-Molina A, Peñarrubia L (2007) Copper and iron homeostasis in Arabidopsis: responses to metal deficiencies, interactions and biotechnological applications. Plant, Cell & Environ 30:271–290

Qu LQ, Yoshihara T, Ooyama A, Goto F, Takaiwa F (2005) Iron accumulation does not parallel the high expression level of ferritin in transgenic rice seeds. Planta 222:225–233

Ramesh A, Sharma SK, Sharma MP, Yadav N, Joshi OP (2014) Inoculation of zinc solubilizing Bacillus aryabhattai strains for improved growth, mobilization and biofortification of zinc in soybean and wheat cultivated in Vertisols of central India. Appl Soil Ecol 73:87–96

Rana A, Joshi M, Prasanna R, Shivay YS, Nain L (2012) Biofortification of wheat through inoculation of plant growth promoting rhizobacteria and cyanobacteria. Eur J Soil Biol 50:118–126

Rietra RPJJ, Heinen M, Dimpka C, Bindraban PS (2015) Effects of nutrient antagonism and synergism on fertilizer use efficiency. VFRC Report 2015/5. Virtual Fertilizer Research Centre, Washington, DC

Roberts SK (2006) Plasma membrane anion channels in higher plants and their putative functions in roots. New Phytol 169:647–666

Amaya-Gómez CV, Porcel M, Mesa-Garriga L, Gómez-Álvarez MI (2020) A Framework for the Selection of Plant Growth-Promoting Rhizobacteria Based on Bacterial Competence Mechanisms. Appl Environ Microbiol. AEM.00760-20doi. https://doi.org/10.1128/AEM.00760-20

Rugeles-Reyes S, Cecílio Filho A, López Aguilar M, Silva P (2019) Foliar application of zinc in the agronomic biofortification of arugula. Food Sci Technol 39(4):1011–1017. https://doi.org/10.1590/fst.12318

Saha S, Mandal B, Hazra GC et al (2015) Can agronomic biofortification of zinc be benign for iron in cereals? J Cereal Sci 65:186–191

Saltzman A., Andersson M.S., Asare-Marfo D., Lividini K., De Moura F.F., Moursi M., Oparinde A., Taleon V. Elsevier; (2013). Biofortification Techniques to Improve Food Security. Reference Module in Food Sciences; pp. 1–9. ISBN: 978-0-08-100596-5

Senoura T, Sakashita E, Kobayashi T, Takahashi M, Aung MS, Masuda H et al (2017) The iron-chelate transporter OsYSL9 plays a role in iron distribution in developing rice grains. Plant Mol Biol 95:375–387

Singh MK, Prasad SK (2014) Agronomic aspects of zinc biofortification in rice (Oryza sativa L). Proceedings of the National Academy of Sciences, India Section B: Biological Sciences 84 (3):613–623

Singh, Y.V., Singh, B.B., Masseyand, P., . Singh, P. K (2017). Short duration cowpea varieties for cultivation as a niche crop in various cropping systems for enhanced pulse production. Agric Sci Digest, 37(3): 232–236

Smoleń S, Sady W (2012) Influence of iodine form and application method on the effectiveness of iodine biofortification, nitrogen metabolism as well as the content of mineral nutrients and heavy metals in spinach plants (*Spinacia oleracea* L.). Sci Horticult 143:176–183

Smoleń S, Ledwożyw-Smoleń I, Sady W (2016) The role of exogenous humic and fulvic acids in iodine biofortification in spinach (Spinacia oleracea L.). Plant Soil 402:129–143

Stacey MG, Patel A, McClain WE, Mathieu M, Remley M, Rogers EE, Gassmann W, Blevins DG, Stacey G (2008) The *Arabidopsis* AtOPT3 protein functions in metal homeostasis and movement of iron to developing seeds. Plant Physiol 146:589–601

Stumm W, Morgan JJ (1996) Aquatic chemistry: chemical equilibria and rates. In: Natural waters, 3rd edn. Wiley, New York, p 1022

Suzuki M, Morikawa KC, Nakanishi H, Takahashi M, Saigusa M, Mori S, Nishizawa NK (2008) Transgenic rice lines that include barley genes have increased tolerance to low iron availability in a calcareous paddy soil. Soil Sci Plant Nutr 54:77–85

Takano J, Wada M, Ludewig U, Schaaf G, von Wire'n N, Fujiwara T (2006) The Arabidopsis major intrinsic protein NIP5;1 is essential for efficient boron uptake and plant development under boron limitation. Plant Cell 18:1498–1509

Tan S, Han R, Li P, Yang G, Li S, Zhang P, Wang W-B, Zhao W-Z, Yin L-P (2015) Over-expression of the MxIRT1 gene increases iron and zinc content in rice seeds. Transgenic Res 24:109–122

Tanaka M, Wallace IS, Takano J, Roberts DM, Fujiwara T (2008) NIP6; 1 is a boric acid channel for preferential transport of boron to growing shoot tissues in *Arabidopsis*. Plant Cell 20:2860–2875

Tariq M, Hameed S, Malik KA, Hafeez FY (2007) Plant root associated bacteria for zinc mobilization in rice. Pak J Bot 39:245–253

Tejada-Jiménez M, Galván A, Fernández E (2011) Algae and humans share a molybdate transporter. PNAS 108(16):6420–6425

Tennakoon NA, Bandara SDS (2003) Nutrient content of some locally available organic materials and their potential as alternative sources of nutrients for coconut. Cocos 15:23–30

Thavarajah D, Ruszkowski J, Vandenberg A (2008) High potential for selenium biofortification of lentils (Lens culinaris L.). J Agric Food Chem 56(22):10747–10753. https://doi.org/10.1021/jf802307h

Tiong J, McDonald G, Genc Y, Shirley N, Langridge P, Huang CY (2015) Increased expression of six ZIP family genes by zinc (Zn) deficiency is associated with enhanced uptake and root-to-shoot translocation of Zn in barley (*Hordeum vulgare*). New Phytol 207:1097–1109

Tomatsu H, Takano J, Takahashi H, Watanabe-Takahashi A, Shibagaki N, Fujiwara T (2007) An Arabidopsis thaliana high-affinity molybdate transporter required for efficient uptake of molybdate from soil. Proceedings of the National Academy of Sciences, USA 104:18807–18812

Trijatmiko KR, Dueñas C, Tsakirpaloglou N, Torrizo L, Arines FM, Adeva C, Balindong J, Oliva N, Sapasap MV, Borrero J et al (2016) Biofortified indica rice attains iron and zinc nutrition dietary targets in the field. Sci Rep 6:19792

Uauy C, Distelfeld A, Fahima T, Blechl A, Dubcovsky J (2006) A NAC gene 15 regulating senescence improves grain protein, zinc, and iron content in wheat. 16. Science 314:1298–1301

Singh D, Prasanna R (2020) Potential of microbes in the biofortification of Zn and Fe in dietary food grains. A review. Agron Sustain Dev 40:15

Uyanoz R (2007) The effects of different bio-organic, chemical fertilizers and their combination on yield, macro and micro nutrition content of dry bean (*Phaseolus vulgaris* L.). Int J Agric Res 2 (2):115–125

Velu G, Singh R, Arun B, Mishra V, Tiwari C, Joshi A et al (2015) Reaching out to farmers with high zinc wheat varieties through public-private partnerships-An experience from Eastern-Gangetic Plains of India. AFTNSOJ 1(3):73–75

Verrier PJ, Bird D, Burla B, Dassa E, Forestier C, Geisler M, Klein M, Kolukisaoglu Y, Lee Y, Martinoia E et al (2008) Plant ABC proteins - a unified nomenclature and updated inventory. Trends Plant Sci 13:151–159 von Wiren N, Klair S, Bansal S, Briat JF, Khodr H, Shioiri T et al (1999) Nicotianamine chelates both FeIII and FeII: implications for metal transport in plants. Plant Physiol 119:1107–1114

Wang F, Wang Z, Kou C, Ma Z, Zhao D (2016) Responses of wheat yield, macro- and micro-nutrients, and heavy metals in soil and wheat following the application of manure compost on the North China Plain. PLoS One 11(1):e0146453

Watanabe T, Broadley MR, Jansen S, White PJ, Takada J, Satake K, Takamatsu T, Tuah SJ, Osaki M (2007) Evolutionary control of leaf element composition in plants. New Phytol 174:516–523

Waters BM, Grusak MA (2008) Whole-plant mineral partitioning throughout the life cycle in *Arabidopsis thaliana* ecotypes Columbia, Landsberg erecta, Cape Verde Islands, and the mutant line ysl*1ysl3*. New Phytol 177:389–405. https://doi.org/10.1111/j.1469-8137.2007.02288.x

Welch R, Graham R (2002) Breeding crops for enhanced micronutrient content. Plant Soil 245 (1):205–214. https://doi.org/10.1023/a:1020668100330

Welch RM (2002) Breeding strategies for biofortified staple plant foods to reduce micronutrient malnutrition globally. J Nutrit 132(3):495S–499S

White P (2001) Chloride in Soils and its Uptake and Movement within the Plant: A Review. Ann Bot 88(6):967–988. https://doi.org/10.1006/anbo.2001.1540

Wintz H, Fox T, Wu Y-Y, Feng V, Chen W, Chang H-S et al (2003) Expression profiles of Arabidopsis thaliana in mineral deficiencies reveal novel transporters involved in metal homeostasis. J Biol Chem 278:47644–47653. https://doi.org/10.1074/jbc.M309338200

Zhang L, Cheng Z, Ai C, Jiang X, Bei X, Zhang Y, Clahn RP, Welch RM et al (2010) Nicotinamine, a novel enhancer of rice iron bioavailbility to humans. PLOS ONE 5(4):e101190

Zhang YQ, Deng Y, Chen RY et al (2012) The reduction in zinc concentration of wheat grain upon increased phosphorus-fertilization and its mitigation by foliar Zn application. Plant Soil 361:143–152

Zhang C, Shinwari KI, Luo L, Zheng L (2018) OsYSL13 is involved in iron distribution in rice. Int J Mol Sci 19:3537

# Biological Interventions Towards Management of Essential Elements in Crop Plants

9

Dwaipayan Sinha and Pramod Kumar Tandon

**Abstract**

Nutrition plays an important role in the overall developmental process of plants. Nutrients play a vital role in the nutrition process. Plants absorb their nutrients largely based on two mechanisms, namely the autotrophic mode and the sequestration mode. The former relates to the photosynthetic process, while the latter relates to absorbing nutrients from the atmosphere. In plants the elements play an important role as nutrients and depending upon the requirement and presence within the plant body they are classified into macronutrients and micronutrients. The world is facing an increase in population since the last 300 years which can be correlated to the industrialization and modernization. This has resulted in an increase in global food demand and consequent increase in agricultural intensification. The requirement of fertilizer gradually became more relevant in order to increase the crop productivity from a comparatively limited area of arable land. However the use of chemical fertilizers also has its own limitations. On the one hand most of them are exhaustible and less likely to be recovered post-application and on the other hand excess use of those fertilizers can result in toxicity in the soil leading to elemental stress. Thus to overcome these contrivances, the scientific fraternity is presently shifting their focus from chemicals to greener approach by using the biological agents largely based on microbes. The most unique feature of the microbes is their capability to solubilize, uptake and assimilate nutrients from otherwise hardy substrates which cannot be broken down easily or require rigorous physical and chemical processes. In addition to it, there are a number of microbes which can form a symbiotic relationship with plants and deliver

D. Sinha (✉)
Department of Botany, Government General Degree College, Mohanpur, Paschim Medinipur, West Bengal, India

P. K. Tandon
Department of Botany, University of Lucknow, Lucknow, India



K. Mishra et al. (eds.), *Sustainable Solutions for Elemental Deficiency and Excess in Crop Plants*, https://doi.org/10.1007/978-981-15-8636-1_9

essential nutrients to the host plant in return of space and protection. These properties of microbes have enabled scientist to formulate a number of biofertilizers. This chapter is an attempt to overview the various microbial organisms that are beneficial to the plants with respect to the supply of nutrients. The mechanisms of action of the microbes in uptake and delivery of the nutrients have also been discussed in the process.

**Keywords**

Nutrition · Biofertilizers · Microbes

## 9.1 Introduction

Nutrition plays a vital role in the overall growth and development of a plant. The plant body especially the angiosperms has evolved in a very interesting way for the nutritional process. In order to have balanced nutrition, a plant must rely both on organic and inorganic components of the environment. For this reason, their nutritional strategy can be broadly divided into two major heads, namely the (I) autotrophic mode and (II) the sequestration mode. The autotrophic mode is related to photosynthesis in which the plant prepares its own food in the form of carbohydrates for sustaining its life. This process is highly elaborate and requires the involvement of a number of enzymes which in turn requires a number of metal ions as cofactors for proper functioning. Thus on the one hand the plant performs the photosynthetic process primarily in the leaves which are uniquely adapted for the purpose and on the other hand makes a continuous effort to sequester metal ions and other elements from the soil through the elaborate root system which are designed for that purpose in addition to mechanical function. The specialized connecting tissue, namely xylem and phloem functions as the channel of transportation from leaf to root (organic materials through phloem) and from roots to leaves (minerals and metal ions through xylem) with water being the medium of transportation. Thus metals and elements in the form of minerals are of utmost importance for the normal physiology of the plant and have a direct correlation with overall growth and productivity.

Elements that are required for normal physiological process and completion of the life cycle are termed as essential elements. Plants require a total of 17 elements out of 92 known elements for their normal physiological process and are broadly categorized as macronutrients and micronutrients. Macronutrients include carbon, hydrogen, oxygen, calcium, potassium, magnesium, nitrogen, sulphur and phosphorus out of which carbon, hydrogen, oxygen make up roughly 95% of the dry matters of the plant while the others are generally present at a concentration of $>100$ mg/kg of dry weight. The micronutrients include chlorine, boron copper, iron, manganese, molybdenum, nickel and zinc which are generally present at comparatively less concentrations ($<100$ mg/kg of dry weight) (Pilon-Smits et al. 2009). The mineral nutrients are generally available in low concentration in the soil in which a plant grows and the availability can generally fluctuate in both space and time due to environmental factors such as precipitation, temperature, wind and physicochemical properties such as erosion, soil pH and type of soil, thus to sustain growth the plant adopts and the number of adaptive and flexible mechanisms for successful absorption and distribution of minerals in their body (Maathuis and Diatloff 2013).

## 9.2 Essential Macronutrient

Each nutrient has its own specific role in the overall growth and development of the plant. Of all the necessary elements, nitrogen is required in maximum quantities by the plant and its availability and concentration influences biomass partitioning between roots and shoots (Bown et al. 2010). Plants have adapted a number of means by which they can adapt themselves to absorb nitrogen from the soil (Kraiser et al. 2011). It is a significant contributor to plant biomass and essential component of most of the biomolecules particularly the proteins. Major proportion of nitrogen reserves of plants are accumulated in protein biomass or as inorganic nitrogen in vacuoles (Beatty et al. 2016). It has also been found that the amount and timing of nitrogen application result in alteration of plant nutrient availability and net photosynthesis (Zhao et al. 2008). Increased nitrogen availability results in increased photosynthesis and Rubisco activity which can also be correlated with the absorption of carbon in the photosynthetic biochemical process (Wang et al. 2012). Phosphorus is considered to be the primary nutrient for plant growth (Abdolzadeh et al. 2010) and is required for optimum production (Taliman et al. 2019). It is an essential macroelement that is required for cell division, reproduction, plant metabolism, storage and energy production (Razaq et al. 2017). Sulphur plays an important role in overall metabolism and development of the plant. It is the constituent of cysteine and methionine which are building blocks of proteins in the plant body (Droux 2004). In addition to it, they are also component of glutathione (GSH) (Mendoza-Cózatl et al. 2005) and iron-sulphur (Fe-S) (Couturier et al. 2013) cluster all of which play their specific role in plant's physiology. Calcium forms an important metallic essential macroelement and in its divalent form $Ca^{+2}$, the nutrient is required for mechanical functioning and making the architecture of cell wall and membrane. Calcium ions help in cross-linking of acidic pectin residues during cell wall formation. In cell membrane, calcium ions bind to the phospholipids, thereby stabilizing the lipid bilayer and maintaining the integrity of the membrane. Calcium ions also play an active role in the regulation of membrane permeability in plant cells (Hepler 2005). In addition to it, the plants depend on changes in cytosolic free calcium during transduction of a wide range of biotic and abiotic signals, thereby acting as a messenger molecule (Sanders et al. 2002). It is also involved as a secondary messenger in nitrate signalling pathway (Riveras et al. 2015). Magnesium is another essential macroelement and plays an important role in a number of physiological processes. Magnesium is the central metal atom of the chlorophyll molecule and thus is required for its biosynthesis (Zhang et al. 2018a). It is also a cofactor of a series of enzymes involved in carbon metabolism (Guo et al. 2015).

Finally other elements such as carbon, hydrogen and oxygen are indispensable to the plants as they are integrated into the biomolecules primarily the carbohydrates and also offer the structural framework of the plant. However under certain conditions the availability of these macronutrients becomes limiting resulting in alterations in the physiology of the plant which consequently gets expressed as various morphological anomalies which ultimately affects the production. The various deficiency symptoms of the macronutrients in plants are tabulated in Table 9.1.

## 9.3 Essential Micronutrient

Essential micronutrients are those which are required in smaller quantities by the plant as compared to those of the macronutrients. Important micronutrients include boron (B), chlorine (Cl), copper (Cu), iron (Fe), manganese (Mn), molybdenum (Mo), nickel (Ni) and zinc (Zn). Boron is required for optimum yield, development, growth and quality of crops and thus holds relevance to the plant physiology and agriculture (Shireen et al. 2018). The element plays an important role in the formation of cell wall by forming borate-diol bonding with a pair of two rhamnogalacturonan II molecules (Martínez-Cuenca et al. 2015) and controls tensile strength and porosity of cell walls (Ryden et al. 2003). Boron influences function and localization of plasma membrane proteins, membrane transport and membrane integrity (Matthes et al. 2020). Boron also has a positive influence on pollen germination, pollen tube growths (Wang et al. 2003), carbohydrate metabolism (Ruuhola et al. 2011) and root elongation (Riaz et al. 2018). Chloride is another micronutrient that is required in small quantities for healthy growth of the plant (<50–100 μM in the nutrient media) (Franco-Navarro et al. 2016). Chloride ions play an important role in osmoregulation of growing plant organs (Beauzamy et al. 2014) and play an important role in turgor dynamics (Argiolas et al. 2016). Chloride ion also stimulates plant growth and reduces water consumption, thereby causing an overall improvement in water use efficiency (Nieves-Cordones et al. 2019). Chloride ion also plays an active role in maintaining the osmolarity of guard cells and thereby facilitates the guard cell movement during opening and closing of stomata (Santelia and Lawson 2016; Saito and Uozumi 2019). Copper is an essential micronutrient that is required as a cofactor for numerous plant proteins (Printz et al. 2016). The destination of copper in the plants are mainly mitochondria and chloroplast. In mitochondria, copper is an essential cofactor for cytochrome C oxidase which is directly involved in aerobic respiration. It is a highly conserved enzymatic complex from the family of heme-copper oxidase and consists of two copper sites, namely CuA and CuB (Llases et al. 2019). In chloroplast, copper is associated with Cu-Zn superoxide dismutase (SOD) and plays a role as a cofactor of the antioxidant enzyme (Grene 2002). Copper is also required in the chloroplast as it is the cofactor of plastocyanin, an electron carrier present in thylakoid lumen (Aguirre and Pilon 2016). The element is also required for symbiotic nitrogen fixation where the element is delivered by the host plant to the nodule cell by a copper membrane transporter where cupro proteins are synthesized (Senovilla et al. 2018).

**Table 9.1** Various symptoms related to macronutrient deficiency in plants

| S. No. | Name of the nutrient | Plant species | Morphological and physiological alterations | References |
|---|---|---|---|---|
| 1. | Nitrogen | *Zea mays* | *Morphological*: Reduction in shoot biomass and leaf growth rate. *Physiological*: Reduction in PSII quantum yield. | Schlüter et al. (2012) |
| | | *Hordeum vulgare* | *Morphological*: Smaller leaves in seedlings. Lower shoot and root biomass and shoot/root ratio. *Physiological*: Decrease in chlorophyll, leaf protein and photosynthetic rate. | Comadira et al. (2015) |
| | | *Arabidopsis thaliana* Wassileskija accession | *Morphological*: Delay in flowering time. Reduction in fresh weight, harvested seeds, dry weight of vegetative parts. *Physiological*: Decrease in nitrate reductase activity and increase in activity of glutamine synthase. Increase in nitrogen remobilization to the seeds from vegetative parts. | Lemaître et al. (2008) |
| 2. | Phosphorus | *Trifolium repens* | *Morphological*: Reduction in weight of dry matter and shoot/root ratio. Arrest of root nodule growth and decrease in per unit root weight. *Physiological*: Decline in net photosynthesis per unit area of shoot surface. | Høgh-Jensen et al. (2002) |
| | | *Hordeum vulgare* | *Physiological*: Enhanced electron flow through photosystem I resulting in increased levels of NADPH and decreased ATP production due to reduced ATP synthase activity. | Carstensen et al. (2018) |
| 3. | Sulphur | *Arabidopsis thaliana* | *Physiological*: Inhibition of glucosinolate biosynthesis via involvement of sulphur deficiency induced 1 (SDI1). | Aarabi et al. (2016) |
| | | *Arabidopsis thaliana* | *Cellular*: Heterogeneity in mitochondria with dilation, lower matrix density and fewer cristae. Disintegration of inner and outer mitochondrial membrane. *Physiological*: Reduced capacities of complex I and complex IV of electron transport chain with decreased phosphorylation rates. Decrease in ATP level. | Ostaszewska et al. (2014) |

(continued)

**Table 9.1** (continued)

| S. No. | Name of the nutrient | Plant species | Morphological and physiological alterations | References |
|---|---|---|---|---|
| 4. | Potassium | *Oryza sativa* | *Morphological*: Decrease in root biomass, root length, root volumes, root surface area and number of laterals.<br>*Cellular*: Break of root cell membrane along with deposition of dark coloured substances in root cell cytoplasm. | Jia et al. (2008) |
| | | *Oryza sativa* | *Physiological*: Increase in activities of antioxidant enzymes such as superoxide dismutase, ascorbate peroxidase, glutathione reductase and catalase. | Liu et al. (2013) |
| 5. | Calcium | *Phaseolus vulgaris* | *Morphological*: Hypocotyl collapse. | Helms (1971) |
| | | *Coffea arabica* | *Morphological*: Decrease in shoot and total biomass along with increase in root/shoot ratio. Decrease in total leaf area and leaf area duration. | Ramalho et al. (1995) |
| | | *Spinacia oleracea* | *Morphological*: Decrease in fresh weight, dry weight.<br>*Physiological*: Decrease in chlorophyll content of the leaves. Reduction in photosynthetic rate, oxygen evolution rate and ribulose-1, 5-bisphosphate carboxylase/oxygenase activity. In addition there was a reduction in activities of nitrate reductase, glutamate dehydrogenase, glutamate synthase and glutamic-pyruvic transaminase. | Chao et al. (2008) |
| 6. | Magnesium | *Oryza sativa* | *Morphological*: Senescence of leaves.<br>*Physiological*: Decrease in chlorophyll concentration and starch accumulation. Drastic decrease in nutrient transport as a result of defect in transpirational flow. | Kobayashi et al. (2013) |
| | | *Citrus sinensis* | *Morphological*: Decrease in soluble protein concentration in leaves. Decrease in levels of proteins involved in photosynthesis such as ribulose-1,5-bisphosphate carboxylase (Rubisco), rubisco activase, | Peng et al. (2015) |

(continued)

**Table 9.1** (continued)

| S. No. | Name of the nutrient | Plant species | Morphological and physiological alterations | References |
|---|---|---|---|---|
| | | | oxygen evolving enhancer protein 1, photosynthetic electron transfer-like protein, ferredoxin-NADP reductase (FNR), aldolase thus decreasing leaf photosynthesis. | |

Manganese is another essential metallic micronutrient required by the plant. It forms an essential cofactor of oxygen evolving complex of the photosynthetic machinery and catalyses splitting up of water in photosystem II (Alejandro et al. 2020). The element also acts as a cofactor of SOD enzymes and thus plays a role in scavenging process of superoxide ion (Marques et al. 2014). It is also a cofactor of a number of enzymes in plants such as oxalate oxidase (Moomaw et al. 2013), pyruvate carboxylase (Waldrop et al. 2012), glutamine synthetase (Betti et al. 2012) and arginase (Caldwell et al. 2015) all of which are involved in specific physiological processes in the plant body. It is also a cofactor of RNA polymerase which is involved in the transcription process (Li et al. 2019a). Iron is a cofactor that performs electron transfer and facilitates chemical reactions such as hydroxylations, radical-mediated rearrangements and (de)hydration reactions. They act as cofactors in oxygen transport, oxygen or iron sensing, or regulation of protein stability (Connorton et al. 2017). Iron is also present as iron-sulphur clusters in chloroplasts and plays an important role in electron transfer chain and other metabolic processes including carbon fixation, nitrogen and sulphur assimilation, pigment, amino acid and vitamin biosynthetic pathways (Przybyla-Toscano et al. 2018). In mitochondria, the element is present in the form of iron-sulphur clusters forming reactive centres of respiratory chain complexes (Böttinger et al. 2018). Iron is also present in catalytic centres of cytochrome P450 oxidases (Bak et al. 2011) and ascorbate peroxidases (Smirnoff 2018). Molybdenum is another trace element that is required by the plant and is responsible for plant growth (Qin et al. 2017). It is associated with a number of enzymes such as nitrate reductase (nitrate assimilation), sulphite oxidase (sulphite detoxification), xanthine dehydrogenase (purine dehydration), aldehyde oxidase (synthesis of auxins) and mitochondrial amidoxime reductase (detoxification) (Mendel and Hänsch 2002). Nickel is another essential micronutrient in plants and is associated with a number of enzyme related metabolic process. It is present in the catalytic site of the enzyme urease which hydrolyses urea in plants (Polacco et al. 2013). In addition to it, nickel is also associated with activation of glyoxylase I, an enzyme required for degradation and detoxification of methylglyoxal (Fabiano et al. 2015). Finally zinc is an indispensable micronutrient as it is associated with a number of plant enzymes. Zinc is associated with RNA polymerases and plays a role in polymerization reaction by coordinating with the catalytic aspartates and facilitating the formation of phosphodiester bonds between nucleotide triphosphates (Venkataraman et al. 2018). It is also a cofactor of superoxide dismutase which is involved in reducing oxidative stress (Wang et al. 2016). It is also associated with alcohol dehydrogenase which is responsible for the conversion of ethanol to acetaldehyde (Kumar et al. 2016). Zinc aids in the catalytic activity of carbonic anhydrase for converting carbon dioxide into carbonic acids (Tu et al. 2012). The metals also play a role in protein synthesis and metabolism of carbohydrate, lipid and nucleic acid (Sharma et al. 2013). The various deficiency symptoms of the micronutrients in plants are illustrated in Table 9.2.

## 9.4 Soil and Depletion of Mineral Resources

Soil is the outermost layer of the planet which houses living organisms. Since the advent of life on the earth surface, the soil has been continuously providing a number of facilities to the living beings in terms of food and shelter and this dependence has compounded manyfold with the advent of civilizations. The world is entirely dependent on agriculture for their ever-increasing demand for food along with a burst of population. It is reported that the world population has increased from 1 billion to 7 billion in just a span of 200 years and by the year 2100 the population is expected to reach 10.8 billion (Roser et al. 2020). Concomitant to the increase in world population there has been a 30.76% increase in agricultural land from 37189665.516 $km^2$ in 1961 to 48632687.604 $km^2$ in 2016 (Anonymous n.d.). The increase in agricultural production to compensate the increasing demand for food can be obtained by intensification and expansion of agricultural land. However, expansion of agricultural land in forest areas should be given a thought in order to conserve biodiversity and maintaining a balance in the ecosystem (Zabel et al. 2014). To make the scenario more constrained, the area suitable for agricultural land is decreasing due to increase in urbanization (Seto et al. 2011). This results in a limited scope of expansion of agricultural land though there has been rapid deforestation of rainforests in some regions of the world to make way for agriculture severely threatening ecological balance (Ordway et al. 2019; Peng et al. 2020). Anthropogenic use of soil for infrastructural, industrial and agricultural purposes has severely affected the soil quality and its capacity to provide fundamental ecosystem services (Kopittke et al. 2019). The food and agricultural organization of United Nations reported that 33% of the land is degraded by soil erosion, salinization, compaction, acidification and chemical pollution of soils (FAO and ITPS 2015) and 52% of the agricultural land is affected by soil degradation (ELD 2015). As per reports of Global soil partnership, led by FAO, United Nations, on a global level, 75 million tonnes of soil are lost from arable lands each year which corresponds to loss of agricultural production of an amount of 400 billion USD (GSP 2017). It is reported by United Nations that the global food production can decrease by 12% due to land degradation leading to 30% increase in the price of foods (United Nations, Convention to Combat Desertification 2011). Soil erosion is a major component of soil degradation and possess a threat to crop production in a number of ways. Erosion reduces the agricultural value of land through physicochemical degradation, the major driver of soil fertility decline being nutrient loss through runoff and sediment (Bashagaluke

**Table 9.2** Various symptoms related to micronutrient deficiency in plants

| S. No. | Name of the nutrient | Plant species | Morphological and physiological alterations. | References |
|---|---|---|---|---|
| 1. | Boron | *Arabidopsis* sp. | *Morphological*: Significant reduction of primary root elongation. *Physiological*: Enhancement of ethylene synthesis through upregulation of ACC synthase (ACS11) transcripts. Increased generation of reactive oxygen species (ROS) in the roots. | Camacho-Cristóbal et al. (2015) |
| | | *Gossypium hirsutum* | *Morphological*: Decrease in plant height and shoot biomass along with formation of brown rings in cotton petiole. Slight decrease in leaf area. *Physiological*: Significant decrease in photosynthetic rate, transpiration rate and stomatal conductance. | Li et al. (2017) |
| 2. | Chloride | *All plants* | *Morphological*: Wilting of leaves especially at the leaf margins. Severe deficiency results in shrivelling of young leaves followed by necrosis. *Cellular*: Impairment of cell division and cell extension in roots. | Chen et al. (2013) |
| | | *Durum wheat* | *Morphological*: Spotting of leaf accompanied by reduction in biomass, grain size and grain yield. | Schwenke et al. (2015) |
| 3. | Copper | *Carthamus tinctorius* | *Physiological*: Decrease in chlorophyll content, hill activity, photosynthesis and transpiration. | Pandey and Sharma (1996) |
| | | *Brassica napus* | *Morphological*: Significant reduction in growth and dry weight. | Billard et al. (2014) |
| 4. | Manganese | Plants | *Morphological*: Inhibition of growth and decrease in biomass. *Physiological*: Interveinal chlorosis due to decrease in net photosynthesis and chlorophyll content. | Socha and Guerinot (2014) |
| | | *Hordeum vulgare* | *Morphological*: Reduction in root and shoot biomass. Intravenous chlorosis and necrosis along with slack leaves. | Chen et al. (2019) |

(continued)

**Table 9.2** (continued)

| S. No. | Name of the nutrient | Plant species | Morphological and physiological alterations. | References |
|---|---|---|---|---|
| | | *Hordeum vulgare* | *Physiological*: Reduction in quantum yield and increase in nocturnal leaf conductance of water vapour indicating greater transpiration accompanied by reduction in epicuticular wax. | Hebbern et al. (2009) |
| 5. | Iron | *Glycine max* | *Morphological*: Stunted growth, yellowing and interveinal chlorosis. | Bai et al. (2018) |
| | | *Hordeum vulgare* | *Morphological*: Reduction in shoot and root lengths. *Physiological*: Decrease in chlorophyll content. | Bocchini et al. (2015) |
| 6. | Molybdenum | Brassicaceae | *Morphological*: Mottling, leaf cupping, grey tinting and flaccid leaves in seedling stage. Improper lamina development (whip-tail), leathery leaves and meristem necrosis in matured stage. | Kaiser et al. (2005) |
| | | *Vigna mungo* | *Morphological*: Interveinal chlorosis of young leaves, decolouration of lamina from tip to entire marginal portion. Reduction in total dry matter, number of flowers and seed yield. | Gopal et al. (2016) |
| 7. | Nickel | *Carya illinoinensis* | *Morphological*: Dwarfing of young leaves, smaller in size thickened as compared to normal leaves along with blunted and cupped apical tips. Reddish pigmentation of leaves. *Physiological*: Decreased urease activity. Increased accumulation of free amino acids, lactic acid and oxalic acid. | Bai et al. (2006) |
| | | *Vigna unguiculata* | *Morphological*: Leaf tip necrosis. | Walker et al. (1985) |
| | | *Glycine max* | *Morphological*: Leaf tip necrosis. *Physiological*: Decrease in urease activity. | Krogmeier et al. (1991) |
| 8. | Zinc | Representatives of moraceae and brassicaceae | *Morphological*: Decrease in biomass. *Physiological*: Decrease in chlorophyll content. | Zhao and Wu (2017) |
| | | *Oryza sativa* | *Morphological*: Stunting, reduction in total dry matter, leaf bronzing and high mortality. | Wissuwa et al. (2006) |

et al. 2018). On a global scale, the soil loss of agricultural area due to the erosional process ranges from 23.7 petagrams to 120 petagrams per year (Borrelli et al. 2017).

Soil organic carbon represents the largest carbon store in the terrestrial ecosystem and is an important factor for controlling the productivity of the soil. Thus maintaining of soil organic carbon pool in agricultural land is of utmost importance for ensuring global food security and preventing carbon dioxide emission (Wiesmeier et al. 2016). With the growth in the human population, there has been an exponential growth of the use of soil resources particularly with respect to agricultural use. This domination on the soil by humans resulted in adverse ecological situations like accelerated erosion, desertification, salinization, acidification, compaction, biodiversity loss, nutrient depletion and loss of soil organic matter (SOM). SOM is responsible for a number of processes including cation binding and acid buffering (Gruba and Mulder 2015), maintaining soil moisture and water holding capacity (Williams et al. 2016; Lin et al. 2018) and acting as a reservoir of nutrients (Qaswar et al. 2019). It has been observed that more and more native lands have been converted to agricultural land and this resulted in a decrease in soil organic carbon (Wei et al. 2014; Sanderman et al. 2017). It is estimated that the conversion of native vegetation to agricultural land resulted in a massive change in the content of carbon, nitrogen, phosphorus and sulphur stocks. This conversion resulted in a 43% decrease in carbon, 42% for nitrogen, 27% for phosphorus and 33% for sulphur. This is largely due to increased mineralization of organic matter as a result of elevated soil temperatures, physical disruption of soil aggregates, reduction in soil-inputs of vegetative material and product export. This results in a gradual reduction of fertility and the soil becomes more reliant on the external application of fertilizers (Kopittke et al. 2017). Another role of SOM is retention of water available to plants. This water is frequently termed as the green water and may be defined as the water that is potentially available to the plants and the soil biota after precipitation losses to runoff and occurrence of deep percolation. It is estimated that global flow of green water comprises of 65% of the total global flow of any freshwater. It is further estimated that 90% of water consumption by crop plants comes from green water. However increase in agriculture has resulted in a decrease in available water due to depletion of organic matters (Liu et al. 2015). As mentioned earlier, the soil is a major storehouse of vital elements. It is estimated that 1500 petagrams of total carbon and 136 petagrams of total nitrogen are stored in the uppermost layers of the soil layer and this represents the largest terrestrial and pools of both the elements. It is estimated that 30–35% of the soil carbon store of top 7 cm of the soil is lost within first 30 years of turning forests into agricultural lands (Oertel et al. 2016). Another report states that 186 petagrams of carbon are lost between 1860 and 2010 due to conversion of forest to agricultural lands (Smith et al. 2016). Additionally, in mineral soils conversion of native grasslands and forests to croplands may reduce the organic carbon stocks by 20 to 40 percent (Scharlemann et al. 2014). Thus the major challenge at present is to reduce the loss of organic matters present in the soil as well as adopting new more efficient techniques through which the utilization of essential elements can be optimized by the crop plants.

## 9.5 Bio-Organic Fertilizer: An Introduction

Extensive use of chemical fertilizers have made a serious impact on the environment, human health and food safety. In addition to it, overusage of chemical fertilizers often causes toxicity in the soil which in turn affects the yield of the plant on the one hand and also human health on the other (Ahmed et al. 2017; Kidd et al. 2017; Steffan et al. 2018). Indiscriminate use of chemical fertilizers with a motive of increasing productivity in order to satisfy the growing demand of food supply has resulted in contamination and damage to beneficial microorganisms including insects. In addition to it, the addition of excess chemical fertilizers has made the crops more prone to diseases and reduced soil fertility. It is estimated that by the year 2020, the world requires 321 million tons of food grains for feeding 8 billion populations which corresponds to a nutrient requirement of 28.8 million tons while the availability is only 21.6 million tons making a deficit of 7.2 million tons (Mahanty et al. 2017). Thus there has been a shift in the type of fertilizer usage from the conventional chemical ones to the biofertilizers with a motive of flourishing agricultural productivity in an eco-friendly way. Biofertilizers are generally defined as preparations containing live microbes which help in enhancing the soil fertility either through fixation of atmospheric nitrogen, solubilization of phosphorus or decomposition of organic wastes or by augmentation of plant growth through the production of growth hormones having specific biological activities (Soumare et al. 2020). The biofertilizers have an edge over conventional fertilizers as they keep the soil milieu enriched with all macro- and micronutrients through the process of nitrogen fixation, phosphate and potassium solubilization or mineralization, release of plant growth regulating substances, production of antibiotics and biodegradation of organic matter in the soil (Bhardwaj et al. 2014). The microbial population influences soil fertility in a number of ways. Firstly, symbiotic microorganisms increase the efficiency of nutrient uptake by the plants. Secondly, a wide array of microbial communities participate in decomposition, mineralization and nutrient availability, thereby influencing the effectivity of nutrient cycle. Thirdly the microbial communities mediate synthesis and decomposition of soil organic matter and therefore have a positive effect on the cation exchange capability, nitrogen, phosphorus and sulphur pool and capacity to retain water. Finally the soil structure and water regime are also influenced by the burrowing and particle transport activities of soil microflora (Singh et al. 2011). Biofertilizers are also cost-effective, eco-friendly and prolonged use enhances soil fertility with minimum damage to the ecosystem (Garcia-Gonzalez and Sommerfeld 2016; Goljanian-Tabrizi et al. 2016). A wide array of microorganism have the capability to absorb the nutrients from the soil and transfer the same to the plants for their growth and production. These include fungi, bacteria, blue green algae (BGA) and nematodes each of which has their own strategy in benefiting the plants and maintaining the nature's equilibrium.

## 9.6 Microbiome: A Potential Source of Beneficial Microorganism for Nutritional Management of Plant

### 9.6.1 Plant Growth Promoting Bacteria

Plant growth promoting bacteria (PGPB) are those bacteria that are free living and form symbiotic relationships with plants and colonize some parts of the plant tissue and promote plant growth either by facilitating the uptake of nutrients or through modification of hormone levels. They also act indirectly by minimizing the inhibitory effect of various pathogenic agents on plant growth and development (Mahanty et al. 2017). The example of various growth promoting bacteria is tabulated in Table 9.3, while the generalized action is displayed in Fig. 9.1.

### 9.6.2 *Azolla* Blue Green Alga Symbiosis

Blue green algae (BGA) are a group of prokaryotic photosynthetic autotrophic microorganisms containing photosynthetic pigments and have the capacity of trapping atmospheric nitrogen. Hence they are utilized as biofertilizers in order to enhance nitrogen content in crops and increase in yield. They belong to eight different families and promote plant growth by synthesizing auxin, gibberellic acid and IAA and is also responsible for fixing 20–30 kg of nitrogen per hectare of land in submerged paddy fields (Mishra and Dash 2014). Nitrogen is required in far more large quantities than any other elements by rice plants and it primarily influences the grain yield (Wang et al. 2017). One of the major sources of nitrogen for both leguminous and non-leguminous plants is the biological nitrogen fixation (BNF) (Mahmud et al. 2020). In this aspect, the BGA are undoubtedly the versatile agents of nitrogen fixation. These algae rapidly form a symbiotic association with almost all groups of plants, namely the liverworts, ferns, gymnosperms and angiosperms (Chang et al. 2019). Among all symbiotic relationships, the association of BGA with water fern *Azolla* holds extreme relevance to BNF of cultivated crops. It is considered as a green manure for rice crops due to its high growth rate, nitrogen fixing capacity and ability to scavenge nutrients from soil and water. The fern doubles its biomass in 2–5 days and supplies more than half of the nitrogen requirement. It is also beneficial to the rice fields in maintaining pH, reduction of water temperature, minimizing ammonia volatilization and suppression of weeds and mosquito larvae (Bhuvaneshwari and Singh 2015). *Azolla* traditionally comprises of seven species divided into three sections, namely *A. filiculoides, A. rubra, A. caroliniana, A. microphylla* and *A. Mexicana* from section *Azolla* (distributed throughout the tropical areas of America, New Zealand and Australia), *A. pinnata* in section Rhizosperma (found in Africa, Australia and Asia) and *A. nilotica* in section Tetrasporocarpia (restricted only in Africa). The capacity of *Azolla* as a nitrogen fixer comes from the fact that it houses BGA in a cavity on the dorsal lobe of the leaf. All the strains isolated from different species of *Azolla* belong to *Anabaena azollae* and molecular phylogenetic studies reveal that the bacterial

**Table 9.3** Role of various bacteria as plant growth promoting agent

| S. No. | Type of Biofertilizers used | Beneficiary plant | Role in plant growth promotion | References |
|---|---|---|---|---|
| 1. | *Pseudomonas fluorescens* SS101 | *Nicotiana tabacum* | Initiation of plant growth through production of volatile organic compounds. | Park et al. (2015) |
| 2. | *Pseudomonas fluorescens* | *Pisum sativum* | Production of indole acetic acid (IAA), siderophore and solubilization of phosphate. | Anwar et al. (2019) |
| 3. | *Pseudomonas geniculata* | *Cicer arietinum* | Production of IAA, siderophore, HCN, cellulase, protease and β-1, 3-glucanase. Enhancement of total nitrogen, available phosphorus and percentage of organic carbon. Enhancement of plant growth. | Gopalakrishnan et al. (2015a) |
| 4. | *Rhizobium cellulosilyticum, Rhizobium taibaishanense* | *Glycine max* | Solubilization of phosphate, production of exopolysaccharide (EPS), 1-aminocyclopropane-1-carboxylate (ACC), siderophore and IAA. | Igiehon et al. (2019) |
| 5. | *Rhizobium laguerreae* | *Spinacia oleracea* | Production of IAA. | Jiménez-Gómez et al. (2018) |
| 6. | *Rhizobium panacihumi* | Ginseng-cultivated soil | Production of IAA, solubilization of phosphate and siderophore production. | Kang et al. (2019) |
| 7. | *Bacillus cereus* YL6 | *Glycine max* *Triticum aestivum* *Brassica rapa* | Solubilization of phosphates along with production of IAA and gibberellic acid. | Ku et al. (2018) |
| 8. | *Mesorhizobium* spp | *Cicer arietinum* | Production of IAA, cytokinin, siderophore and solubilization of phosphate. | Brígido et al. (2017) |
| 9. | *Bradyrhizobium japonicum* | *Glycine max* and non-legumes | Synthesis of ethylene, IAA, gibberellin and zeatin. | Boiero et al. (2007) |
| 10. | *Sinorhizobium meliloti* | *Medicago* sp | Production of cytokinin. | Kisiala et al. (2013) |
| 11. | *Microbacterium* sp. P27 | *Phragmites karka* | Production of IAA, ammonia and enhanced 1-aminocyclopropane-1- | Singh and Singh (2019) |

(continued)

**Table 9.3** (continued)

| S. No. | Type of Biofertilizers used | Beneficiary plant | Role in plant growth promotion | References |
|---|---|---|---|---|
| | | | carboxylate deaminase activity. Degradation of indane. | |
| 12. | *Azorhizobium caulinodans* ORS571 | *Triticum aestivum* | Strengthening of leaf and roots along with increase in biomass. | Liu et al. (2017) |
| 13. | *Azoarcus sp* CIB | *Oryza sativa Nicotiana tabacum* | Fixation of nitrogen by nitrogenase, solubilization of phosphate, production of IAA. | Fernández et al. (2014) |
| 14. | *Burkholderia contaminans* KNU17BI1 | *Zea mays* | Fixation of nitrogen. Production of IAA, solubilization of phosphate and zinc. | Tagele et al. (2018) |
| 15. | *Azospirillum brasilense* | *Zea mays* | Improved plant growth and biochemical traits. Increase in nitrogen use efficiency (NUE) under N limiting conditions. | Zeffa et al. (2019) |
| 16. | *Serratia marcescens* CDP-13 | *Triticum aestivum* | ACC deaminase activity, IAA production, phosphate solubilization. | Singh and Jha (2016) |
| 17. | *Klebsiella* sp | *Avena sativa* | Increased shoot length, root length and dry weight. Increased chlorophyll and IAA content. | Sapre et al. (2018) |
| 18. | *Streptomyces* sp | *Cicer arietinum* | Enhanced stover yield, grain yield, total dry matter, pod weight, seed number and seed weight. Increased microbial biomass carbon, activity of dehydrogenase, total nitrogen, phosphorous availability and organic carbon. | Gopalakrishnan et al. (2015b) |
| 19. | *Cellulosimicrobium funkei* | *Phaseolus vulgaris* | Production of IAA, EPS and solubilization of phosphates. Increase in rate of seed germination, lengths of shoots and roots. Increase in chlorophyll content. | Karthik et al. (2016) |

(continued)

**Table 9.3** (continued)

| S. No. | Type of Biofertilizers used | Beneficiary plant | Role in plant growth promotion | References |
|---|---|---|---|---|
| 20. | *Pantoea* sp<br>*Kosakonia* sp | *Lycopersicon esculentum* | Solubilization of phosphates. Enhancement of root elongation. | Chakdar et al. (2018) |

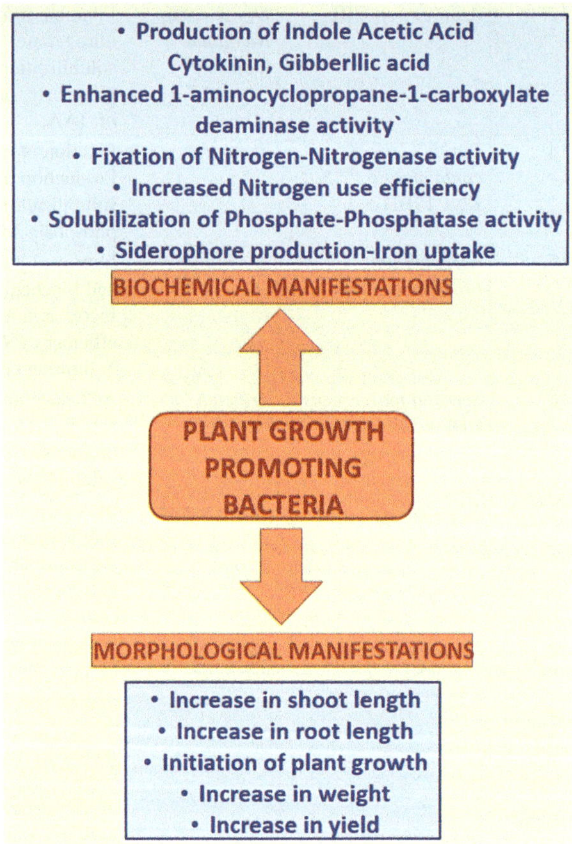


**Fig. 9.1** Schematic representation of the beneficial activities of plant growth promoting bacteria

phylogeny is in parallel to that of the host plant (Qiu and Yu 2003). It is also reported that in addition to *Anabaena*, *Nostoc azollae* is also present as a symbiont in *Azolla*. It is presumed that *Azolla* contributes sugar to *Nostoc azollae*, while the latter releases 40% of the fixed nitrogen into the leaf pocket as ammonia which is then taken up and assimilated by glutamate synthase and glutamine synthetase machinery of the fern (Brouwer et al. 2017). It is estimated that *Nostoc azollae* fixes 0.15–0.17 mg N $h^{-1}$ per gram of dry biomass which is much higher than that of nitrogen fixed by Rhizobia residing in root nodules of soybean. Cumulatively, *Nostoc azollae* supplies 1100–1200 kg N $ha^{-1}$/year for the rapid growth of *Azolla*. This results in high protein content in the biomass of *Azolla* which can range between 200 and 400 g kg $ha^{-1}$ of the dry weight (Brouwer et al. 2018). *Azolla* thus qualifies a unique plant to be used as biofertilizer because it is decomposed in the soil very easily and its nitrogen content is readily available for uptake by the plants. In addition to it *Azolla* also provides a significant amount of minerals such as phosphorus, potassium calcium and magnesium (Kollah et al. 2016). *Azolla* is thus used as a green manure in rice cultivation of China, Vietnam, other Asiatic and African countries. It is calculated that basal application of *Azolla* at 10–12 t/ha increases soil nitrogen by 50–60 kg/ha and this minimizes the necessity of chemical nitrogen fertilizers by 30–35 kg/ha (Chen et al. 2017a). The importance of *Azolla* for rice cultivation was first noted in Northern Vietnam and from there it spread to other countries like USA, Japan, Philippines and China (Razavipour et al. 2018). It is also reported that use of *Azolla* as a biofertilizer improved NUE of paddy plants and this can be attributed to reduction in loss of nitrogen and enhanced uptake by the rice plants (Yao et al. 2018). The use of *Azolla* in rice cultivation also reduced the requirement of urea fertilizer without affecting the grain yield (Malyan et al. 2019). In India, the Central Rice Research Institute, Cuttack popularized *Azolla pinnata* was successfully introduced and popularized as a biofertilizer largely for eastern regions of the country. It was observed that the plant was able to fix 75 mg N/g of dry weight and generated a biomass of 347 ton fresh weight per hectare annually (Yadav et al. 2014). In Jaunpur district of Uttar Pradesh, India, *Azolla pinnata* is used in rice cultivation and resulted in an increase in grain yield (Mishra and Mishra 2007). In Africa, the application of *Azolla* resulted in an increase in plant height, tiller numbers, neck node, and panicle length, number of panicle m-2 and grain yield (Oyange et al. 2019). In another recent study it was reported that application of *Azolla filiculoides* compost also resulted in an increase in numbers of tillers and grain of rice plants (Razavipour et al. 2018). Another recent study reported the effect of *Azolla filiculoides* on the growth of maize crop. It was observed that application of *Azolla filiculoides* extracts improved NUE, nitrogen uptake efficiency and irrigation water use efficiency of the maize plant along with increase in growth and yield attributes (Maswada et al. 2020). In Bangladesh, *Anabaena-Azolla* system is used as a biofertilizer for rice and large-scale production of *Azolla pinnata* var. pinna has been undertaken in Sirajganj and Ghazipur districts of the country (Aziz 2012). Recently, photobioreactors have been developed for mass-scale production of strains of *Anabaena* and *Nostoc* which can be subsequently used as biofertilizers for rice cultivations (Jochum et al. 2018).

### 9.6.3 Arbuscular Mycorrhizal Fungi

Arbuscular mycorrhizal fungi are obligate biotrophs which participates in ancient symbiosis with plants. In this symbiotic association, the fungi provide the plant increased access to soil resources, while the plants inturn provide carbon in the form of sugars and lipid to the fungi which are required for their metabolism and growth (Kokkoris and Hart 2019). Besides nutritional benefits, the arbuscular mycorrhizal fungi also helps the plant to withstand drought stress (Li et al. 2019b), salinity stress (Evelin et al. 2019) and heavy metal stress (Zhang et al. 2018b). In addition to it, the arbuscular mycorrhizal fungi also stimulate the photosynthetic activity (Birhane et al. 2012) and enhance the disease resistance of the plant (Hao et al. 2019). The arbuscular mycorrhizal fungi are the most widely distributed species of endotrophic mycorrhizal fungi and belong to a monophyletic phylum, namely the glomeromycota covering ten out of 11 families, namely Acaulosporaceae, Ambisporaceae, Archaesporacea, Claroidoglomeraceae, Diversisporaceae, Gigasporaceae, Glomeraceae, Pacisporaceae, Paraglomeraceae and Sacculosporaceae (Giovannini et al. 2020). Arbuscular mycorrhizal fungi can form a symbiotic relationship with 90% of flowering plants, ferns and bryophytes (Liu et al. 2014; Rimington et al. 2018). They are mainly composed of hyphae, mycorrhiza and vacuole in the root and hyphae and spore in the soil. They form an extensive hyphal network in the rhizosphere of the plant. These hyphae not only are responsible for the promotion of plant growth, yield and quality of the vegetables but also physicochemical properties of the soil as well as improve the nutrient uptake (Chen et al. 2017b). Thus arbuscular mycorrhiza plays an important role in plant productivity and proper functioning of ecosystem. They act as an important element for sustainable crop improvement programmer (Sosa-Hernández et al. 2019). The various application of arbuscular mycorrhiza for management of mineral nutrition and development of plants is tabulated in Table 9.4.

### 9.6.4 Ectomycorrhizal Fungi

Ectomycorrhiza is a symbiotic association of fungi with feeder roots of higher plants in which both the plants are mutually benefitted and the association is significant for the existence of both the partners. Majority of ectomycorrhizal fungi belongs to class ascomycetes and basidiomycetes (Charya and Garg 2019). Ectomycorrhizal fungi play a major role in the functioning of forest ecosystem and compete to colonize the root of the host plant. They receive carbon from the host plant and in return benefit the host plant by providing minerals, water, resistance towards pathogens and tolerance of heavy metals (Hortal et al. 2017). Gymnosperms and other woody plants are closely associated with ectomycorrhizal association. A study reported that ectomycorrhiza aids in the enrichment of carbon and nitrogen in fine roots of young beech (*Fagus sylvatica*) (Valtanen et al. 2014). Another study reported that *Pinus densiflora* seedlings responded positively to ectomycorrhizal inoculation consisting of *Phialocephala fortinii, Cenococcum geophilum, Mollisia cinerea, Leptodontidium elatius* and *Lachnum pygmaeum*. It was observed that the pine seedlings exhibited greater dry weight and higher height upon inoculation with the fungi (Sim and Eom 2006). In another study, effect of coinoculation of ectomycorrhizal fungi *Suillus variegatus* and *Pisolithus tinctorius* with root entophytic bacteria *Methylobacterium extorquens* DSM13060 on the growth and

**Table 9.4** Beneficial role of arbuscular mycorrhizal fungi on crop plants

| S. No. | Arbuscular mycorrhizal fungi | Beneficiary plant | Developmental effect | References |
|---|---|---|---|---|
| 1. | *Glomus caledonium* 90036, *Acaulospora* spp | *Cucumis sativus* | Elevated biomass and improved yield. | Hu et al. (2010) |
| 2. | *Glomus mossea* | *Sorghum bicolor* | Enhancement of nitrogen uptake by the roots. | Koegel et al. (2013) |
| 3. | *Glomus intraradices* BEG 141 and *G. mosseae* BEG 167 | *Daucus carota* cv. Changfeng, *Allium fistulosum* L. cv. Fengwang | Increase in fresh weights of carrot and average yield of carrot roots. Enhancement of shoot and root fresh weight and yield of green onion. | Wang et al. (2011) |
| 4. | *Bacillus sonorensis* and *Funneliformis mosseae* | *Lycopersicon esculentum* *Capsicum annuum* | Higher germination rate, seedling length and vigour index. Increase in plant height, higher root length stem girth and biovolume index. Higher dry weight. | Desai et al. (2020) |
| 5. | *Rhizophagus intraradices, Funneliformis mosseae* | *Lactuca sativa* | Higher shoot biomass, accumulation of iron proteins, carotenoids and anthocyanins. Enhanced levels of potassium, sugars, ascorbate, magnesium and copper. | Baslam et al. (2013) |
| 6. | *Glomus mosseae* GP11, *Glomus intraradices* GB67 *Glomus viscosum* GC41 *Agrobacterium radiobacter* AR39, *Bacillus subtilis* BA41, *Streptomyces spp* SB14, *Beauveria* spp., *Trichoderma harzianum* TH01, *Pichia pastoris* PP59 | *Lycopersicon esculentum* cv. Faino | Significant increase in growth of the seedling along with increase in biomass. Increase in number and weight of the fruit. | Candido et al. (2015) |

(continued)

**Table 9.4** (continued)

| S. No. | Arbuscular mycorrhizal fungi | Beneficiary plant | Developmental effect | References |
|---|---|---|---|---|
| 7. | *Rhizophagus irregularis* CD1 | *Gossypium hirsutum* | Enhanced expression of phosphate transporter family of genes and concentration of phosphorus in the biomass of cotton. Increase in photosynthetic rate, growth of the plant and number of balls per plant and maturity of the fibre. Significant increase in yield. | Gao et al. (2020) |
| 8. | *Glomus* spp, *Pseudomonas fluorescens* Pf4, *Pseudomonas sp.* 5Vm1K | *Fragaria* x *ananassa* var. *Selva* | Increase in anthocyanin concentration of fruits. | Lingua et al. (2013) |
| 9. | *Claroideoglomus claroideum* | *Fragaria ananassa* cv. *camarosa* | Higher antioxidant capacity. | Parada et al. (2019) |
| 10. | *Funneliformis mosseae* | *Citrullus lanatus* (experiment done under salinity stress) | Increase in plant height, stem diameter, root dry weight and shoot dry weight. Significant reduction of reactive oxygen species and MDA as compared to stressed plants. Increase in praline content. Increase in activity of SOD, APX, CAT, GR, MDHAR and DHAR enzymes. Improvement in chlorophyll and carotenoid contents, net photosynthesis rate, PSII maximum. Reduction in damage of chloroplasts. | Ye et al. (2019) |
| 11. | *Claroideoglomus etunicatum* *Rhizophagus sp.* | *Capsicum annuum* | Higher shoot and root dry weights along with increased | Selvakumar et al. (2018) |

(continued)

**Table 9.4** (continued)

| S. No. | Arbuscular mycorrhizal fungi | Beneficiary plant | Developmental effect | References |
|---|---|---|---|---|
| | RK4 *Funneliformis mosseae* *Gigaspora margarita* S-23 *Claroideoglomus lamellosum* S-11 | | fruit dry weight and number of fruits. High nutrient content with respect to phosphorus, potassium and calcium. | |
| 12. | *Glomus* sp | *Olea europaea* (experiment done under high manganese) | Significant increase in plant shoot length, diameter and total biomass along with increase in dry weight of the plant. Increased uptake of phosphorus. Reduction in manganese concentration. | Briccoli Bati et al. (2015) |
| 13. | *Funneliformis mosseae* and *Diversispora versiformis* | *Chrysanthemum morifolium* (experiment done under salt stress) | Under high salinity levels increase in root length upon inoculation with AMF. Increase in shoot dry weight under non-saline and moderately saline condition. Increase in total dry weight in non-saline and moderate saline condition. Increase in shoot nitrogen content under high saline condition. Increase in shoot phosphorus under non-saline condition. | Wang et al. (2018) |
| 14. | *Funneliformis mosseae*, *Rhizophagus irregularis* | *Malus pumila* | Significant increase in tree height. | Berdeni et al. (2018) |
| 15. | *Acaulospora longula* (BEG ID 8), *Claroideoglomus claroideum* (BEG ID 31), *Glomus intraradices* | *Malus domestica* cv. Golden delicious (experiment performed in presence of nematode | Increase in shoot height, stem diameter, number of leaves and weight of fresh roots in nematode infested | Ceustermans et al. (2018) |

(continued)

**Table 9.4** (continued)

| S. No. | Arbuscular mycorrhizal fungi | Beneficiary plant | Developmental effect | References |
|---|---|---|---|---|
| | | *Pratylenchus penetrans*) | apple seedlings treated with arbuscular mycorrhizal fungi. | |
| 16. | *Rhizophagus irregularis* MUCL 41833 | *Zea mays* | Significant higher uptake of inorganic phosphate. Improved water use efficiency and shoot biomass. | Le Pioufle et al. (2019) |
| 17. | AMF consortium containing mixture of 25 species of order Glom ales belonging to four genus *Glomus, Gigaspora, Acaulospora, Entrophospora* | *Ceratonia siliqua* (experiments performed under drought condition) | Increase in mean shoot and root biomass. Higher accumulation of phosphorus, potassium, sodium and calcium levels. Reduction in oxidative stress as evident from reduction in hydrogen peroxide and MDA levels. | Boutasknit et al. (2020) |
| 18. | *Rhizophagus irregularis* | *Medicago truncatula* (experiment performed in presence of benzo[a] pyrene, a polyaromatic hydrocarbon soil pollutant) | Increased inorganic phosphate uptake along with higher accumulation of phosphorus in roots and shoots. | Calonne-Salmon et al. (2018) |
| 19. | *Glomus iranicum* var. *tenuihypharum* sp. | *Euonymus japonica* | Higher content of phosphorus, calcium and potassium in leaves of mycorrhizal plant which were conventionally watered with a slight increase in height. | Gomez-Bellot et al. (2015) |
| 20. | *Rhizophagus intraradices* HDSF5 | *Oryza sativa.* | Significant increase in nitrogen accumulation in vegetative tissues at tillering stage and translocation of nitrogen into seeds from heading to maturity. Reduced | Zhang et al. (2017) |

(continued)

**Table 9.4** (continued)

| S. No. | Arbuscular mycorrhizal fungi | Beneficiary plant | Developmental effect | References |
|---|---|---|---|---|
| | | | carbon to nitrogen ratio in seeds and increase in protein content. Significant increase in grain yield and increase in grain to straw ratio. Higher carbon accumulation in roots with increased carbon to nitrogen ratio. | |
| 21. | *Gigaspora margarita*, *Funneliformis mosseae*, *Rhizophagus irregularis*, *G. clarum*, *G. deserticola*, *G. monosporum*, *G. brasilianum*, *G. aggregatum* | *Triticum durum* Desf., cv. Anco Marzio *Vicia faba* cv. Dorenza | Positive influence on root biomass, specific root length and root density along with increased uptake of P, Fe and Zn in wheat. Increase in percentage of nitrogen derived from the atmosphere in total nitrogen biomass of faba bean. | Ingraffia et al. (2019) |
| 22. | *Glomus deserticola* | *Osmium basilicum* (experiment performed under salinity stress) | Increase in shoot and root dry matter, shoot height, leaf area and leaf number were significantly higher in mycorrhizal plants. Higher content of potassium, phosphate and calcium in leaves. | Elhindi et al. (2017) |

nutrient uptake of *Pinus* seedling was studied. It was found that the coinoculation of bacteria with ectomycorrhizal fungi resulted in an increase in root and shoot dry weight. This was accompanied by an increase in carbon, nitrogen and potassium in roots (Pohjanen et al. 2014). A study reported the effect of ectomycorrhizal fungi *Astraeus hygrometricus* on uptake of caesium and potassium by *Pinus densiflora* seedlings. The results indicated a betterment of growth of shoots of the seedlings upon treatment of ectomycorrhizal fungi as evident from the lowering of root/shoot ratio. In addition to it, there was an increase in uptake of caesium and potassium by seedlings upon treatment with ectomycorrhizal fungi. Moreover concentration of potassium increased in shoot and whole seedling when caesium was added in the presence of ectomycorrhizal formation (Ogo et al. 2018). Another study reported that ectomycorrhizal fungi *Paxillus involutus* increased phosphorus content and growth of *Picea abies*. In addition to it, the hyphae of ectomycorrhizal fungi translocated a considerable proportion of nitrogen, potassium or magnesium uptake which were also linked to the uptake of phosphorus (Jentschke et al. 2001). It was reported from another study that ectomycorrhizal fungi *Pisolithus tinctorius, Descolea antarctica* stimulated the growth of *Nothofagus dombeyi* and increased foliar nitrogen and phosphorus concentration. It was further observed that ectomycorrhizal fungal inoculation resulted in an increase in activity glutamine oxoglutarate aminotransferase, glutamine synthetase, glutamate dehydrogenase, nitrate reductase and acid phosphomonoesterase all of which are interlinked and involved in carbon, nitrogen and phosphorus transfers (Alvarez et al. 2009).

### 9.6.5 Nematodes

Nematodes are worms that are most ubiquitous organism available on the earth. They may be free living or parasitic to plants, insects, humans or any other animals (Basyoni and Rizk 2016). Their body is unsegmented, long and threadlike. They are generally microbial feeders and play an important role in decomposition (Kiontke and Fitch 2013). They play a major role in linking below and above ground processes such as soil organic matter decomposition and plant production (Wardle et al. 2004). The nematodes feeding on plants release carbon rich exudates in the soil which in turn stimulates microbial growth (Yeates et al. 1999). This leads to an increase in soil organic matter through a mechanism known as the priming effect (Kuzyakov 2002). Nematodes which are bacterivorous, fungivorous and omnivorous feed on the respective microbes and excrete nutrients which mainly comprise of amino acids, ammonium and phosphate ions. Thus some nematodes play a significant role in the augmentation of soil resources to plants and can be very well exploited for their agricultural benefit. It is reported from a study that soil nematodes help in a significant increase of biomass of *Lolium perenne* along with greater availability of net nitrogen and phosphorus (Gebremikael et al. 2016).

## 9.7 Mechanism of Uptake of Nutrients by Plant Through Beneficial Microorganism

The microorganisms which facilitate in the enrichment of nutrients and promote plant growth adopt a number of mechanisms by which they can trap the atmospheric elements and transfer them to the plants. Most of them are highly specific and is dependent on the physiology of the microorganisms. The important strategies adopted by the microorganisms are discussed.

### 9.7.1 Nitrogen Fixation

Nitrogen is the most available element present in the atmosphere and is vital for plant growth. Although 78% of the atmosphere is nitrogen but still it is unavailable to the plants. In order to assimilate into the biological system it requires to be converted to ammonia which is a versatile molecule that possess both organic and inorganic entity. This ammonia is assimilated by plants through the process of BNF. This process is performed by bacteria and are called diazotrophs. These bacteria encode enzyme nitrogenase which catalyses the conversion of gaseous nitrogen into ammonia. This enzyme complex is highly conserved among the diazotrophs (Santi et al. 2013). The diazotrophic bacteria forms various types of associations with the host plants. The various types of association of nitrogen fixing bacteria with the host plants are tabulated in Table 9.5. The enzyme nitrogenase is a two component system and is composed of a MoFe protein (dinitrogenase or component I which reduces nitrogen to ammonia) and the electron transfer Fe protein (dinitrogenase reductase or component II) (Fig. 9.2). The enzyme also requires a reducing source in the form of MgATP which eventually undergoes hydrolysis along with association and dissociation of Fe protein and MoFe protein in a catalytic cycle involving single electron transfer. It was also reported that MoFe protein is equipped with two metal clusters, namely the iron-molybdenum cofactor (Fe-Mo Co) which provides the active site for binding of the substrate and the P-cluster involved in the transfer of electrons from Fe protein to FeMo-Co (Hoffman et al. 2014). The overall reaction catalysed by nitrogenase is as follows:

$N_2 + 8H^+ + 16MgATP + 8e^- \rightarrow 2NH_3 + H_2 + 16MgADP + 16P_i$ (Hu and Ribbe 2015)

The genes involved in nitrogen fixation are called nif genes and are present in both symbiotic and free living microorganisms. They include structural genes which are involved in activation of iron protein, iron-molybdenum cofactor biosynthesis, electron donation and regulatory genes that are necessary for synthesis and function of the enzyme (Ahemad and Kibret 2014). In case of bacterial symbiosis for nitrogen fixation, formation of root nodule is an important step for accommodation of symbiotic bacterium into the host plant. For effective formation of nodules, the bacteria secrete the nodulation factors. They are acylated chitin oligomeric backbone with various substitutions at the (non)reducing-terminal and/or nonterminal residues. They are responsible for formation and deformation of root hairs, intra- and extracellular alkalization, membrane potential depolarization, changes in ion fluxes, early nodulin gene expression and formation of nodule primordial (D'Haeze and Holsters 2002). It is also noteworthy that the hormone ethylene is a negative regulator of nodule formation (Guinel 2015) and it has been reported that some strains of Rhizobium increase the number of nodules of the host by synthesizing rhizobitoxine which is a phytotoxin and inhibits the enzyme 1-aminocyclopropane-1-carboxylate (ACC) synthase which is a key enzyme in ethylene biosynthesis (Sugawara et al. 2006). Some rhizobial strains ACC deaminase which removes ACC, the precursor of ethylene (Tittabutr et al. 2015).

**Table 9.5** Details of nitrogen fixing bacteria and the host plants

| S. No. | Types of association | Bacteria | Host species | Family | References |
|---|---|---|---|---|---|
| 1. | Actinorhizal symbiosis | *Frankia* | *Casuarina glauca*<br>*Casuarina equisetifolia* | Casuarinaceae | Ngom et al. (2016) |
| | | | *Alnus* | Betulaceae | Carro et al. (2015) |
| | | | *Ceanothus* | Rhamnaceae | Oakley et al. (2004) |
| | | | *Elaeagnus umbellata, Elaeagnus angustifolia, Elaeagnus commutata* | Elaeagnaceae | Mirza et al. (2009) |
| | | | *Talguenea quinquenervia,*<br>*Trevoa trinervis*<br>*Discaria trinervis*<br>*Colletia hystrix* | Rhamnaceae | Clawson et al. (1998) |
| | | | *Coriaria arborea* | Coriariaceae | Clawson et al. (1998) |
| | | | *Purshia tridentata*<br>*Dryas drummondii* | Rosaceae | Clawson et al. (1998) |
| 2. | Non-legume endosymbiosis | *Rhizobium* | *Parasponia* | Cannabaceae | van Velzen et al. (2018) |
| 3. | Cyanobacterial-plant endosymbiosis | | | | |
| 4. | Liverwort | *Nostoc punctiforme* | *Blasia pusilla* | Blasiaceae | Adams and Duggan (2008) |
| | | *Noscotc* sp | *Cavicularia densa* | Blasiaceae | |
| 5. | Hornwort | *Nostoc punctiforme* | *Phaeoceros* sp | Notothyladaceae | Adams and Duggan (2008) |
| | | *Nostoc* sp<br>*Calothrix* sp | *Phaeoceros* sp | Notothyladaceae | West and Adams (1997) |
| | | *Nostoc* sp | *Leiosporoceros dussii* | Leiosporocerotaceae | Villarreal and Renzaglia (2006) |
| | | *Nostoc* sp | *Anthoceros fusiformis* | Anthocerotaceae | Costa et al. (2001) |

| | | | | | |
|---|---|---|---|---|---|
| 6. | Pteridophytes | *Anabaena* sp | *Azolla* sp | Salviniaceae | Pereira and Vasconcelos (2014) |
| 7. | Gymnosperms | *Nostoc* sp, *Calothrix* sp, *Scytonema* sp and *Richelia* sp | Cycads | Cycadaceae | Chang et al. (2019) |
| 8. | Angiosperms | *Nostoc* sp | *Gunnera* sp | Gunneraceae | Khamar et al. (2010) |
| 9. | Associative and endophytic nitrogen fixation | *Azoarcus* sp | *Oryza sativa* | Poaceae | Rosenblueth et al. (2018) |
| | | *Burkholderia vietnamensis* | | | Govindarajan et al. (2008) |
| | | *Gluconacetobacter diazotrophicus* | | | Muthukumarasamy et al. (2007) |
| | | *Herbaspirillum huttiense* | | | Andreozzi et al. (2019) |
| | | *Azospirillum brasilense* | *Zea mays* | Poaceae | Zeffa et al. (2019) |
| | | *Azospirillum brasilense* *Azospirillum lipoferum* | *Triticum aestivum* | Poaceae | Kim et al. (2005) |

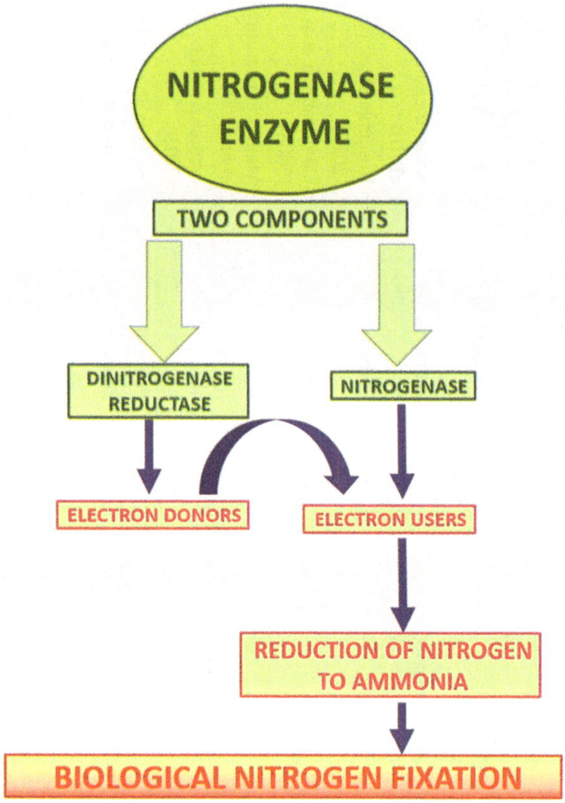


**Fig. 9.2** Schematic representation of nitrogen fixation by plant growth promoting bacteria

In case of cyanobacteria, the symbiotic association is initiated by the hormogonia. They are short filaments that are released from parental filaments of heterocyst and non-heterocyst forming cyanobacteria. They are distinguished from vegetative filaments mainly through their gliding ability and small size of cells. These hormogonia function as infective units during the formation of symbiotic association with plants. A hormogonium inducing factor (HIF) produced by the host plant is low nitrogen condition leading to differentiation of vegetative filaments into hormogonia which then chemotactically migrates towards the host cell infection sites. After infection, the host release hormogonia repressing factor which prevents further proliferation of hormogonia and the cyanobacteria then shifts towards heterocyst formation and subsequent nitrogen fixation (Rai et al. 2000; Meeks and Elhai 2002). The various nitrogen fixing bacteria along with the host plant is illustrated in Table 9.5.

### 9.7.2 Phosphate Solubilization

Most of the phosphorous present in the soil remains in insoluble inorganic forms and hence unavailable to the plants. The plants can absorb phosphates only in monobasic and dibasic forms (Souza et al. 2015). In soil the inorganic phosphorus solubilization occurs by the action of low molecular weight organic acids such as gluconic, formic acid, 2-ketogluconic, citric, oxalic, lactic, isovaleric, succinic, glycolic and acetic acid which are synthesized by the solubilizing bacteria. These organic acids solubilize mineral phosphate as a result of a mineral exchange or chelation of Fe or Al ions associated with phosphate. This results in increased phosphorus availability which is ultimately absorbed by the plants (Pande et al. 2017). A second way of making phosphorus available is the mineralization of phosphates by phosphatases which catalyses the hydrolysis of a wide range of phosphomonoesters and also catalyse trans phosphorylation reactions by transferring phosphoryl group to alcohol in presence of phosphate acceptors. Acid phosphatase enzymes are located in the cell wall of bacteria and in extracellular polymeric substances that surround it (Fig. 9.3)

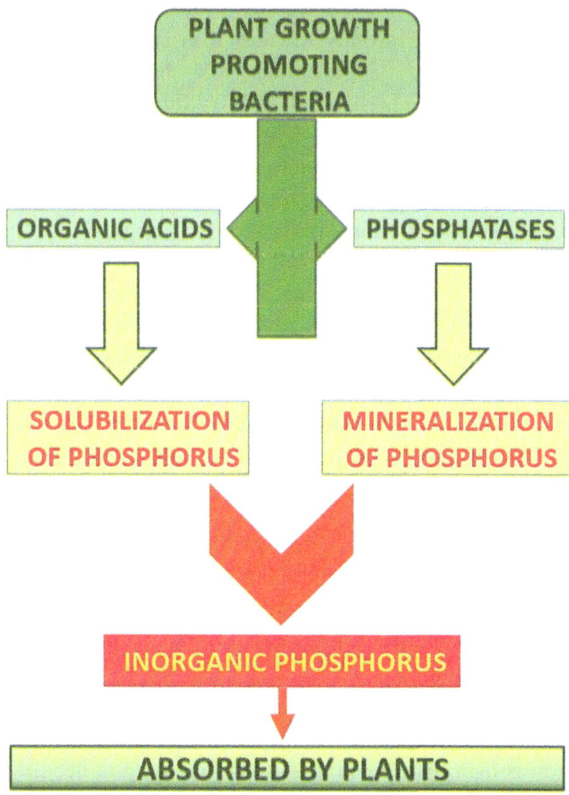


**Fig. 9.3** Schematic representation of phosphate solubilization by plant growth promoting bacteria

(Behera et al. 2017). The acid phosphatases hydrolyse phosphate esters via a two-step process as depicted below.

1. $E + ROPO_3H^- \leftrightarrow E \bullet ROPO_3H^- \leftrightarrow ROH + (E - PO_3^-)$
2. $E - PO_3^- + H_2O \leftrightarrow E + P_i$. (Gandhi and Chandra 2012)

### 9.7.3 Sequestering Iron

Iron is another vital element that is required for plant growth. In aerobic condition, iron is available as $Fe^{+3}$ form and is more prone to form insoluble oxides and hydroxides which makes them unavailable for plants and microbes (Ju et al. 2019). Plants adopt two strategies for the absorption of iron. The first strategy involves the use of microbial ferric reductases which catalyses the conversion of $Fe^{+3}$ into $Fe^{+2}$. This process involves the formation of a weak $Fe^{+2}$-chelate complex which eventually results in dissociation of free $Fe^{+2}$ for transport or incorporation into the cell (Schröder et al. 2003). The second strategy involves secretion of low molecular weight siderophores which are low molecular weight iron chelators having a strong affinity towards iron (Wilson et al. 2016). Siderophores may be extracellular or intracellular but bacteria generally produce extracellular siderophores (Schrettl et al. 2007). Siderophores may be divided into three main families depending on the characteristic functional group, namely the hydroxamates, catecholates and carboxylates (Ahmed and Holmström 2014). In gram negative bacteria, the bacterial siderophores chelate the iron to form a complex which are too large to be transferred through porins. They thus require outer membrane receptors for their uptake into periplasmic space. Once inside the periplasm, they are taken over by periplasmic siderophore binding proteins which are responsible for transporting siderophores to the cytoplasmic membrane transporters for onward transport to the bacterial cell cytoplasm. After binding with the periplasmic binding protein, the siderophores prepares for its further journey towards cytoplasm of the cell. This is brought about by ABC transporter protein complex which is coupled to hydrolysis during the siderophore transportation process. The bacterial ABC transporters comprise of four structural domains: two transmembrane domains which form a channel for passing of ferric siderophore and two nucleotide binding domains that hydrolyse ATP (Krewulak and Vogel 2008). The gram positive bacteria possess a simple siderophore uptaking system. The siderophore binding protein of gram positive bacteria is associated with a permease. Upon binding to a ferric siderophore, there is a conformational change in siderophore binding protein-permease complex which facilitates the transport of ferric siderophores across membranes into the bacterial cytoplasm (Wilson et al. 2016). Inside the microbial cells, iron is reduced to $Fe^{+2}$ and thus its affinity for siderophores is lost. The reduced iron is then distributed into iron containing molecules or stored in ferritins (Aznar et al. 2014). Plants access the iron from bacterial siderophores by direct uptake,

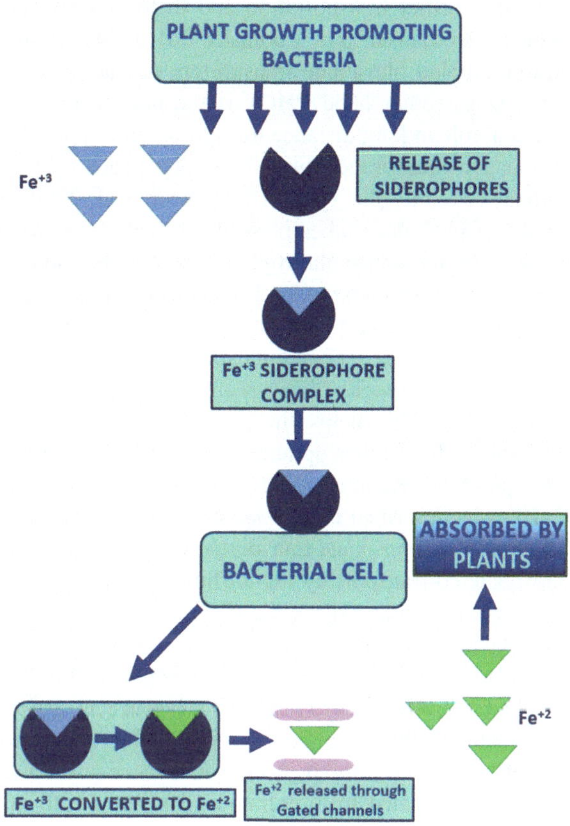


**Fig. 9.4** Schematic representation of iron uptake by plant growth promoting bacteria

chelate degradation or ligand exchange reactions (Kurth et al. 2016). The schematic representation of iron uptake by the bacteria is depicted in Fig. 9.4.

### 9.7.4 Mycorrhiza

In mycorrhizal association, both fungi and host plants share a mutual relationship where both partners are benefitted. The fungi provide minerals that are required by the plants and in return the plants provide them with sugars. The most well-known and extensively explored function of mycorrhiza is uptake of phosphorus (Johri et al. 2015). Plants adapt to two strategies in uptake of phosphorus, namely the direct uptake and the indirect uptake. The direct mechanisms involve the uptake of phosphorus by the root hairs from the surrounding soil (Nestler and Wissuwa 2016). Phosphate transporters are present in the plasma membrane of the cells of epidermis and root hairs which are in direct contact with the soil. Two types of transporters, namely PHT1 and PHO1 are responsible for phosphate uptake from the soils and further transportation and distribution to above ground plant parts and tissues (Młodzińska and Zboińska 2016). PHT1 proteins are plasma membrane proton coupled phosphate transporters that helps in uptake of phosphate from the soil. On the other hand PHO1 is the phosphate transporter family that plays an essential role in long distance transportation of phosphates from roots to shoots. It is also responsible for the regulation of phosphate export from root epidermal cells and cortical cells into the xylem (Srivastava et al. 2018). In the indirect pathway, the uptake of phosphate takes place through the mycorrhiza. In this case phosphate is taken up by the arbuscular mycorrhizal hyphae and then transported to intracellular fungal arbuscules to be catered in the root cortical cells (Smith et al. 2011). Most of the arbuscular mycorrhizal fungi have genes encoding acid phosphatase which enables them to secrete acid phosphatases. It has been reported that the extraradical hypha of *Rhizophagus clarus* secretes acid phosphatase into the rhizosphere (Sato et al. 2019). This helps the fungi to solubilize the phosphates present in the rhizosphere for further uptake. Arbuscular mycorrhizal fungi have transporters on their plasma membrane. The first arbuscular mycorrhizal phosphate transporter, GvPT, was described in *Glomus versiforme* (Harrison and van Buuren 1995). It is a high affinity phosphate transporter and is coupled to proton transport. This proton/phosphate co-transport by high affinity phosphate transporters requires the activity of plasma membrane $H^+$ ATPase which generates a proton motive force and drives the uptake of phosphate across extraradical mycelial membrane. Orthologs of GvPT were also identified in *Rhizophagus irregularis* (GintPT), *Glomus mosseae* (GmPT) and *Gigaspora margarita* (GigmPT) (Ferrol et al. 2019). Once inside the cytosol, the phosphate is then transformed into ATP inside mitochondria and finally to polyphosphate and stored in the vacuoles (Ezawa et al. 2002). The polyphosphate is then translocated from extraradical mycelium to intraradical mycelium through cytoplasmic streaming and/or along a motile tubular vacuole system (Ashford and Allaway 2002; Olsson et al. 2002). Once the polyphosphates reach the arbuscular region, it is likely to get hydrolyzed by fungal polyphosphatases leading to increase in concentration of phosphates in the arbuscules which facilitates efflux (Solaiman et al. 1999; Ohtomo and Saito 2005). After hydrolysis of polyphosphates, the phosphate is transferred to the cytosol of the mycelium via intraradical mycelium-expressed vacuolar Pi transporter PHO91 (Hürlimann et al. 2007). Transport of phosphate to the periarbuscular membrane possibly occurs through orthologous of the yeast plasma membrane Pho89p that mediates bidirectional phosphate transport or through plasma membrane VTC complex which polymerizes cytosolic phosphates into polyphosphates and export to periarbuscular space which will then be subjected to action by plant acid phosphatase (Ezawa and Saito 2018; Ferrol et al. 2019). The symbiotic uptake of phosphate by plants occurs at specialized interface formed in arbuscule colonized cortical cells. The uptake of phosphate by the plant cell is then mediated by arbuscular mycorrhiza induced phosphate transporters. These transporters belong to Pht1 family (Walder et al. 2016).

In ectomycorrhiza, the accusation of free phosphates takes place through plasma membrane phosphate transporter. Most of the ectomycorrhizal fungi have three to five putative phosphate transporters all of which belong to Pht1 family. After

**Table 9.6** Transporters involved in transportation of nutrients

| S. No. | Element | Transporter | Host | References |
|---|---|---|---|---|
| 1. | Nitrogen—Ammonia transporter | GintAMT1, GintAMT2, GintAMT3 | Fungi: *Glomus intraradices* | Calabrese et al. (2016) |
| | | GmAMT4.1 | Plant: *Glycine max* | Kobae et al. (2010) |
| | | LjAMT2;2 | Plant: *Lotus japonicus* | Guether et al. (2009) |
| | | TaAMT2;3a | Plant: *Triticum aestivum* | Jiang et al. (2019) |
| 2. | Phosphorus—Phosphate transporter | GvPT | Fungi: *Glomus versiforme* | Harrison and van Buuren (1995) |
| | | GmosPT | Fungi: *Glomus mosseae* | Benedetto et al. (2005) |
| | | PiPT | Fungi: *Piriformospora indica* | Yadav et al. (2010) |
| | | GigmPT | Fungi: *Gigaspora margarita* | Xie et al. (2016) |
| | | StPT3 | Plant: *Solanum tuberosum* | Rausch et al. (2001) |
| | | MtPT4 | Plant: *Medicago truncatula* | Konečný et al. (2019) |
| | | OsPT8 | Plant: *Oryza sativa* | Dong et al. (2019) |
| | | LjPT4 | Plant: *Lotus japonicas* | Volpe et al. (2016) |
| | | PhPT3, PhPT5 | Plant: *Petunia hybrida* | Liu et al. (2018) |
| 3. | Sulphur—Sulphate transporter | SOA1 | Yeast: *Saccharomyces cerevisiae* | Holt et al. (2017) |
| | | AtSULTR | Plant: *Arabidopsis thaliana* | Giovannetti et al. (2014) |
| | | LjSultr1;2 | Plant: *Lotus japonicus* | |
| 4. | Zinc | ZRT1/ZRT2 | Fungi: *Ustilago maydis* | Martha-Paz et al. (2019) |
| | | From environment to cell: ZIP (ZIP1, ZIP2 and ZIP4) and P-Type ATPase (HMA2). Transport to organelles: CAX (MHX1), CDF (MTP1 and | Most plants | Hefferon (2019) |

(continued)

**Table 9.6** (continued)

| S. No. | Element | Transporter | Host | References |
|---|---|---|---|---|
| | | ZAT1), P-Type ATPase (HMA2 and HMA4) | | |
| 5. | Iron | FET1, FET3 | Yeast: *Saccharomyces cerevisiae* | Tamayo et al. (2014) |
| | | RiFTR1 and RiFTR2 | Fungi: *Rhizophagus irregularis* | Tamayo et al. (2014) |
| | | IRT1 | Plant: Most no graminaceous plants | Kobayashi et al. (2019) |
| | | HaIRT1, HaNramp1 and HaZIP1 | Plant: *Helianthus annuus* | Kabir et al. (2020) |
| 6. | Copper | CTR family (RiCTR1 and RiCTR2) | Fungi: *Rhizophagus irregularis* | Gómez-Gallego et al. (2019) |
| | | COPT1/2 | Plant: Most dicots | Printz et al. (2016) |
| 7. | Magnesium | PiMgT1 | Fungi: *Piriformospora indica* | Prasad et al. (2019) |
| | | MGT6 | Plant: *Arabidopsis thaliana* | Mao et al. (2014) |
| 8. | Potassium | Trk1 | Fungi: *Candida glabrata* | Llopis-Torregrosa et al. (2019) |
| | | SlHAK10 | Plant: *Lycopersicon esculentum* | Liu et al. (2019) |
| | | K+/H+ exchanger—AtCHX20, Medtr7g099800.1 | Plant: *Medicago truncatula* | Garcia et al. (2017) |

absorption, the phosphate is transferred through the extraradical mycelium to ectomycorrhizal roots and the phosphate gets polymerized into polyphosphate and is stored within the vacuoles of the mycelial cells. In the hartig net the fungi and plant have a common apoplastic space lacking a symplastic communication. At the hartig net, hydrolysis of polyphosphates results in an increase of phosphate concentration into the fungal cytosol and consequently it is taken up by the plant cells through their own phosphate transporters (Becquer et al. 2014). Similarly most of the other elements are also transferred through the mycorrhiza to the plant through their own set of transporters. In most cases the transporters are induced in the host plant upon infection by mycorrhiza. A brief outline of the transporters is illustrated in Table 9.6.

## 9.8 Conclusive Remarks

The population of the world is on a constant increase day by day. This has resulted in an increase in agricultural activities which has severely affected soil fertility. In addition to it, the expansion of agricultural land to forest area resulted in degradation and imbalance in the ecosystem. To make things worse, increase in industrialization and urbanization have resulted in stress over available land for agriculture. All these anthropogenic activities have resulted in a gradual decline in fertility of soil. To overcome this problem, chemical fertilizers were implemented. However chemical fertilizers have their own problem and often lead to soil toxicity which in turn leads to several health hazards. Thus the focus has been shifting from the use of chemical fertilizers to the use of biological agents which can provide nutrients to the plants as well as initiate growth and yields. As discussed, the major advantages of the biological agents are that they increase the soil fertility and availability of nutrients in a more balanced and eco-friendly way. Moreover in addition to providing nutrients to the plant the biological agents which are presently used as biofertilizers are also responsible for providing multifarious benefits to the crop plants. These include protection from pathogens, synthesis of growth promoting substances and so on. Thus biological organisms or more appropriately microorganisms presently play a major role in mineral nutrition of the plants and also overall growth promotion. They require more popularization from agricultural point of view and should eventually replace the chemical fertilizer for a better eco-friendly approach.

## References


Aarabi F, Kusajima M, Tohge T, Konishi T, Gigolashvili T, Takamune M, Sasazaki Y, Watanabe M, Nakashita H, Fernie AR, Saito K, Takahashi H, Hubberten HM, Hoefgen R, Maruyama-Nakashita A (2016) Sulfur deficiency-induced repressor proteins optimize glucosinolate biosynthesis in plants. Sci Adv 2(10):e1601087

Abdolzadeh A, Wang X, Veneklaas EJ, Lambers H (2010) Effects of phosphorus supply on growth, phosphate concentration and cluster-root formation in three *Lupinus* species. Ann Bot 105 (3):365–374

Adams DG, Duggan PS (2008) Cyanobacteria-bryophyte symbioses. J Exp Bot 59(5):1047–1058

Aguirre G, Pilon M (2016) Copper delivery to chloroplast proteins and its regulation. Front Plant Sci 6:1250

Ahemad M, Kibret M (2014) Mechanisms and applications of plant growth promoting rhizobacteria: current perspective. J King Saudi Univ Sci 26(1):1–20

Ahmed E, Holmström SJ (2014) Siderophores in environmental research: roles and applications. Microb Biotechnol 7(3):196–208

Ahmed M, Rauf M, Mukhtar Z, Saeed NA (2017) Excessive use of nitrogenous fertilizers: an unawareness causing serious threats to environment and human health. Environ Sci Pollut Res Int 24(35):26983–26987

Alejandro S, Höller S, Meier B, Peiter E (2020) Manganese in plants: from acquisition to subcellular allocation. Front Plant Sci 11:300

Alvarez M, Huygens D, Olivares E, Saavedra I, Alberdi M, Valenzuela E (2009) Ectomycorrhizal fungi enhance nitrogen and phosphorus nutrition of *Nothofagus dombeyi* under drought conditions by regulating assimilative enzyme activities. Physiol Plant 136(4):426–436

Andreozzi A, Prieto P, Mercado-Blanco J, Monaco S, Zampieri E, Romano S, Valè G, Defez R, Bianco C (2019) Efficient colonization of the endophytes *Herbaspirillum huttiense* RCA24 and *Enterobacter cloacae* RCA25 influences the physiological parameters of *Oryza sativa* L. cv. Baldo rice. Environ Microbiol 21:3489–3504

Anonymous (n.d.) Food and Agriculture Organization, electronic files and web site. https://data.worldbank.org/indicator/AG.LND.AGRI.K2?end=2016&start=1961&view=chart

Anwar MS, Paliwal A, Firdous N, Verma A, Kumar A, Pande V (2019) Co-culture development and bioformulation efficacy of psychrotrophic PGPRs to promote growth and development of pea (Pisum sativum) plant. J Gen Appl Microbiol 65(2):88–95

Argiolas A, Puleo GL, Sinibaldi E, Mazzolai B (2016) Osmolyte cooperation affects turgor dynamics in plants. Sci Rep 6:30139

Ashford AE, Allaway WG (2002) The role of the motile tubular vacuole system in mycorrhizal fungi. In: Smith SE, Smith FA (eds) Diversity and integration in mycorrhizas. Developments in plant and soil sciences, vol 94. Springer, Dordrecht

Aziz A (2012) New method of large-scale production of *Azolla pinnata* var. *pinnata* R. Brown: a multipurpose crop. Int J Appl Agric Res 7(1):1–9

Aznar A, Chen NW, Rigault M, Riache N, Joseph D, Desmaële D, Mouille G, Boutet S, Soubigou-Taconnat L, Renou JP, Thomine S, Expert D, Dellagi A (2014) Scavenging iron: a novel mechanism of plant immunity activation by microbial siderophores. Plant Physiol 164 (4):2167–2183

Bai C, Reilly CC, Wood BW (2006) Nickel deficiency disrupts metabolism of ureides, amino acids, and organic acids of young pecan foliage. Plant Physiol 140(2):433–443

Bai G, Jenkins S, Yuan W, Graef GL, Ge Y (2018) Field-based scoring of soybean iron deficiency chlorosis using RGB imaging and statistical learning. Front Plant Sci 9:1002

Bak S, Beisson F, Bishop G, Hamberger B, Höfer R, Paquette S, Werck-Reichhart D (2011) Cytochromes p450. Arabidopsis Book. 9:e0144

Bashagaluke JB, Logah V, Opoku A, Sarkodie-Addo J, Quansah C (2018) Soil nutrient loss through erosion: impact of different cropping systems and soil amendments in Ghana. PLoS One 13(12):e0208250

Baslam M, Garmendia I, Goicoechea N (2013) The arbuscular mycorrhizal symbiosis can overcome reductions in yield and nutritional quality in greenhouse-lettuces cultivated at inappropriate growing seasons. Sci Hortic 164:145–154

Basyoni MM, Rizk EM (2016) Nematodes ultrastructure: complex systems and processes. J Parasit Dis 40(4):1130–1140

Beatty PH, Klein MS, Fischer JJ, Lewis IA, Muench DG, Good AG (2016) Understanding plant nitrogen metabolism through metabolomics and computational approaches. Plants (Basel) 5 (4):39

Beauzamy L, Nakayama N, Boudaoud A (2014) Flowers under pressure: ins and outs of turgor regulation in development. Ann Bot 114(7):1517–1533

Becquer A, Trap J, Irshad U, Ali MA, Claude P (2014) From soil to plant, the journey of P through trophic relationships and ectomycorrhizal association. Front Plant Sci 5:548

Behera BC, Yadav H, Singh SK, Mishra RR, Sethi BK, Dutta SK, Thatoi HN (2017) Phosphate solubilization and acid phosphatase activity of *Serratia sp.* isolated from mangrove soil of Mahanadi river delta, Odisha, India. J Genet Eng Biotechnol 15(1):169–178

Benedetto A, Magurno F, Bonfante P, Lanfranco L (2005) Expression profiles of phosphate transporter gene (GmosPT) from the endomycorrhizal fungus Glomus mosseae. Mycorrhiza 15(8):620–627

Berdeni D, Cotton TEA, Daniell TJ, Bidartondo MI, Cameron DD, Evans KL (2018) The effects of arbuscular mycorrhizal fungal colonisation on nutrient status, growth, productivity, and canker resistance of apple (*Malus pumila*). Front Microbiol 9:1461

Betti M, García-Calderón M, Pérez-Delgado CM, Credali A, Estivill G, Galván F, Vega JM, Márquez AJ (2012) Glutamine synthetase in legumes: recent advances in enzyme structure and functional genomics. Int J Mol Sci 13(7):7994–8024

Bhardwaj D, Ansari MW, Sahoo RK, Tuteja N (2014) Biofertilizers function as key player in sustainable agriculture by improving soil fertility, plant tolerance and crop productivity. Microb Cell Factories 13:66

Bhuvaneshwari K, Singh PK (2015) Response of nitrogen-fixing water fern *Azolla* biofertilization to rice crop. 3 Biotech 5(4):523–529

Billard V, Ourry A, Maillard A, Garnica M, Coquet L, Jouenne T, Cruz F, Garcia-Mina JM, Yvin JC, Etienne P (2014) Copper-deficiency in *Brassica napus* induces copper remobilization, molybdenum accumulation and modification of the expression of chloroplastic proteins. PLoS One 9(10):e109889

Birhane E, Sterck FJ, Fetene M, Bongers F, Kuyper TW (2012) Arbuscular mycorrhizal fungi enhance photosynthesis, water use efficiency, and growth of frankincense seedlings under pulsed water availability conditions. Oecologia 169(4):895–904

Bocchini M, Bartucca ML, Ciancaleoni S, Mimmo T, Cesco S, Pii Y, Albertini E, Del Buono D (2015) Iron deficiency in barley plants: phytosiderophore release, iron translocation, and DNA methylation. Front Plant Sci 6:514

Boiero L, Perrig D, Masciarelli O, Penna C, Cassán F, Luna V (2007) Phytohormone production by three strains of *Bradyrhizobium japonicum* and possible physiological and technological implications. Appl Microbiol Biotechnol 74(4):874–880

Borrelli P, Robinson DA, Fleischer LR, Lugato E, Ballabio C, Alewell C, Meusburger K, Modugno S, Schütt B, Ferro V, Bagarello V, Oost KV, Montanarella L, Panagos P (2017) An assessment of the global impact of 21st century land use change on soil erosion. Nat Commun 8 (1):2013

Böttinger L, Mårtensson CU, Song J, Zufall N, Wiedemann N, Becker T (2018) Respiratory chain supercomplexes associate with the cysteine desulfurase complex of the iron-sulfur cluster assembly machinery. Mol Biol Cell 29(7):776–785

Boutasknit A, Baslam M, Ait-El-Mokhtar M, Anli M, Ben-Laouane R, Douira A, El Modafar C, Mitsui T, Wahbi S, Meddich A (2020) Arbuscular mycorrhizal fungi mediate drought tolerance and recovery in two contrasting carob (*Ceratonia siliqua* L.) ecotypes by regulating stomatal, water relations, and (in) organic adjustments. Plants (Basel) 9(1):80

Bown HE, Watt MS, Clinton PW, Mason EG (2010) Influence of ammonium and nitrate supply on growth, dry matter partitioning, N uptake and photosynthetic capacity of *Pinus radiata* seedlings. Trees 24(6):1097–1107

Briccoli Bati C, Santilli E, Lombardo L (2015) Effect of arbuscular mycorrhizal fungi on growth and on micronutrient and macronutrient uptake and allocation in olive plantlets growing under high total Mn levels. Mycorrhiza 25(2):97–108

Brígido C, Glick BR, Oliveira S (2017) Survey of plant growth-promoting mechanisms in native Portuguese chickpea *Mesorhizobium isolates*. Microb Ecol 73(4):900–915

Brouwer P, Bräutigam A, Buijs VA, Tazelaar AO, van der Werf A, Schlüter U, Reichart GJ, Bolger A, Usadel B, Weber AP, Schluepmann H (2017) Metabolic adaptation, a specialized leaf organ structure and vascular responses to diurnal N2 fixation by *Nostoc azollae* sustain the astonishing productivity of *Azolla* ferns without nitrogen fertilizer. Front Plant Sci 8:442

Brouwer P, Schluepmann H, Nierop KG, Elderson J, Bijl PK, van der Meer I, de Visser W, Reichart GJ, Smeekens S, van der Werf A (2018) Growing Azolla to produce sustainable protein feed: the effect of differing species and CO2 concentrations on biomass productivity and chemical composition. J Sci Food Agric 98(12):4759–4768

Calabrese S, Pérez-Tienda J, Ellerbeck M, Arnould C, Chatagnier O, Boller T, Schüßler A, Brachmann A, Wipf D, Ferrol N, Courty PE (2016) GintAMT3 - a low-affinity ammonium transporter of the arbuscular mycorrhizal *Rhizophagus irregularis*. Front Plant Sci 7:679

Caldwell RB, Toque HA, Narayanan SP, Caldwell RW (2015) Arginase: an old enzyme with new tricks. Trends Pharmacol Sci 36(6):395–405

Calonne-Salmon M, Plouznikoff K, Declerck S (2018) The arbuscular mycorrhizal fungus *Rhizophagus irregularis* MUCL 41833 increases the phosphorus uptake and biomass of *Medicago truncatula*, a benzo[a]pyrene-tolerant plant species. Mycorrhiza 28(8):761–771

Camacho-Cristóbal JJ, Martín-Rejano EM, Herrera-Rodríguez MB, Navarro-Gochicoa MT, Rexach J, González-Fontes A (2015) Boron deficiency inhibits root cell elongation via an ethylene/auxin/ROS-dependent pathway in Arabidopsis seedlings. J Exp Bot 66 (13):3831–3840

Candido V, Campanelli G, D'Addabbo T, Castronuovo D, Perniola M, Camele I (2015) Growth and yield promoting effect of artificial mycorrhization on field tomato at different irrigation regimes. Sci Hortic 187:35–43

Carro L, Pujic P, Alloisio N, Fournier P, Boubakri H, Hay AE, Poly F, François P, Hocher V, Mergaert P, Balmand S, Rey M, Heddi A, Normand P (2015) Alnus peptides modify membrane porosity and induce the release of nitrogen-rich metabolites from nitrogen-fixing *Frankia*. ISME J 9(8):1723–1733

Carstensen A, Herdean A, Schmidt SB, Sharma A, Spetea C, Pribil M, Husted S (2018) The impacts of phosphorus deficiency on the photosynthetic electron transport chain. Plant Physiol 177 (1):271–284

Ceustermans A, Van Hemelrijck W, Van Campenhout J, Bylemans D (2018) Effect of arbuscular mycorrhizal fungi on *Pratylenchus penetrans* infestation in apple seedlings under greenhouse conditions. Pathogens 7(4):76

Chakdar H, Dastager SG, Khire JM, Rane D, Dharne MS (2018) Characterization of mineral phosphate solubilizing and plant growth promoting bacteria from termite soil of arid region. 3 Biotech 8(11):463

Chang ACG, Chen T, Li N, Duan J (2019) Perspectives on endosymbiosis in coralloid roots: association of cycads and cyanobacteria. Front Microbiol 10:1888

Chao L, Bofu P, Weiqian C, Yun L, Hao H, Liang C, Xiaoqing L, Xiao W, Fashui H (2008) Influences of calcium deficiency and cerium on growth of spinach plants. Biol Trace Elem Res 121(3):266–275

Charya LS, Garg S (2019) Advances in methods and practices of ectomycorrhizal research. In: Meena SM, Naik MM (eds) Advances in biological science research. Academic press, Cambridge, pp 303–325

Chen W, He ZL, Yang XE, Mishra S, Stoffella PJ (2013) Chlorine nutrition in higher plants: progress and perspectives. J Plant Nutr 33(7):943–952

Chen J, Huang M, Cao F, Pardha-Saradhi P, Zou Y (2017a) Urea application promotes amino acid metabolism and membrane lipid peroxidation in *Azolla*. PLoS One 12(9):e0185230

Chen S, Zhao H, Zou C, Li Y, Chen Y, Wang Z, Jiang Y, Liu A, Zhao P, Wang M, Ahammed GJ (2017b) Combined inoculation with multiple arbuscular mycorrhizal fungi improves growth, nutrient uptake and photosynthesis in cucumber seedlings. Front Microbiol 8:2516

Chen A, Husted S, Salt DE, Schjoerring JK, Persson DP (2019) The intensity of manganese deficiency strongly affects root endodermal suberization and ion homeostasis. Plant Physiol 181(2):729–742

Clawson ML, Caru M, Benson DR (1998) Diversity of *Frankia* strains in root nodules of plants from the families elaeagnaceae and rhamnaceae. Appl Environ Microbiol 64(9):3539–3543

Comadira G, Rasool B, Karpinska B, Morris J, Verrall SR, Hedley PE, Foyer CH, Hancock RD (2015) Nitrogen deficiency in barley (*Hordeum vulgare*) seedlings induces molecular and metabolic adjustments that trigger aphid resistance. J Exp Bot 66(12):3639–3655

Connorton JM, Balk J, Rodríguez-Celma J (2017) Iron homeostasis in plants - a brief overview. Metallomics 9(7):813–823

Costa JL, Paulsrud P, Rikkinen J, Lindblad P (2001) Genetic diversity of Nostoc symbionts endophytically associated with two bryophyte species. Appl Environ Microbiol 67 (9):4393–4396

Couturier J, Touraine B, Briat JF, Gaymard F, Rouhier N (2013) The iron-sulfur cluster assembly machineries in plants: current knowledge and open questions. Front Plant Sci 4:259

Desai S, Bagyaraj DJ, Ashwin R (2020) Inoculation with microbial consortium promotes growth of tomato and *Capsicum* seedlings raised in pro trays. Proc Natl Acad Sci India Sect B Biol Sci 90:21–28

D'Haeze W, Holsters M (2002) Nod factor structures, responses, and perception during initiation of nodule development. Glycobiology 12(6):79R–105R

Dong Z, Li W, Liu J, Li L, Pan S, Liu S, Gao J, Liu L, Liu X, Wang GL, Dai L (2019) The Rice phosphate transporter protein OsPT8 regulates disease resistance and plant growth. Sci Rep 9 (1):5408

Droux M (2004) Sulfur assimilation and the role of sulfur in plant metabolism: a survey. Photosynth Res 79(3):331–348

ELD (2015) Report for policy and decision makers: reaping economic and environmental benefits from sustainable land management. Economics of Land Degradation (ELD) Initiative, Bonn. https://repo.mel.cgiar.org/handle/20.500.11766/4881

Elhindi KM, El-Din AS, Elgorban AM (2017) The impact of arbuscular mycorrhizal fungi in mitigating salt-induced adverse effects in sweet basil (*Ocimum basilicum* L.). Saudi J Biol Sci 24(1):170–179

Evelin H, Devi TS, Gupta S, Kapoor R (2019) Mitigation of salinity stress in plants by arbuscular mycorrhizal symbiosis: current understanding and new challenges. Front Plant Sci 10:470

Ezawa T, Saito K (2018) How do arbuscular mycorrhizal fungi handle phosphate? New insight into fine-tuning of phosphate metabolism. New Phytol 220(4):1116–1121

Ezawa T, Smith SE, Smith FA (2002) P metabolism and transport in AM fungi. Plant Soil 244:221–230

Fabiano CC, Tezotto T, Favarin JL, Polacco JC, Mazzafera P (2015) Essentiality of nickel in plants: a role in plant stresses. Front Plant Sci 6:754

FAO and ITPS (2015) Status of the world's soils. Food and Agriculture Organization of the United Nations and Intergovernmental Technical Panel on Soils, Rome

Fernández H, Prandoni N, Fernández-Pascual M, Fajardo S, Morcillo C, Díaz E, Carmona M (2014) *Azoarcus* sp. CIB, an anaerobic biodegrader of aromatic compounds shows an endophytic lifestyle. PLoS One 9(10):e110771. Erratum in: PLoS One. 2014; 9(10):e110771

Ferrol N, Azcón-Aguilar C, Pérez-Tienda J (2019) Arbuscular mycorrhizas as key players in sustainable plant phosphorus acquisition: an overview on the mechanisms involved. Review. Plant Sci 280:441–447

Franco-Navarro JD, Brumós J, Rosales MA, Cubero-Font P, Talón M, Colmenero-Flores JM (2016) Chloride regulates leaf cell size and water relations in tobacco plants. J Exp Bot 67 (3):873–891

Gandhi UN, Chandra BS (2012) A comparative analysis of three classes of bacterial non specific acid phosphatases and archaeal phosphoesterases: evolutionary perspective. Acta Inform Med 20(3):167–173

Gao X, Guo H, Zhang Q, Guo H, Zhang L, Zhang C, Gou Z, Liu Y, Wei J, Chen A, Chu Z, Zeng F (2020) Arbuscular mycorrhizal fungi (AMF) enhanced the growth, yield, fiber quality and phosphorus regulation in upland cotton (*Gossypium hirsutum* L.). Sci Rep 10:2084

Garcia K, Chasman D, Roy S, Ané JM (2017) Physiological responses and gene co-expression network of mycorrhizal roots under K+ deprivation. Plant Physiol 173(3):1811–1823

Garcia-Gonzalez J, Sommerfeld M (2016) Biofertilizer and biostimulant properties of the microalga *Acutodesmus dimorphus*. J Appl Phycol 28:1051–1061

Gebremikael MT, Steel H, Buchan D, Bert W, De Neve S (2016) Nematodes enhance plant growth and nutrient uptake under C and N-rich conditions. Sci Rep 6:32862

Giovannetti M, Tolosano M, Volpe V, Kopriva S, Bonfante P (2014) Identification and functional characterization of a sulfate transporter induced by both sulfur starvation and mycorrhiza formation in *Lotus japonicus*. New Phytol 204(3):609–619

Giovannini L, Palla M, Agnolucci M, Avio L, Sbrana C, Turrini A, Giovannetti M (2020) Arbuscular mycorrhizal fungi and associated microbiota as plant biostimulants: research strategies for the selection of the best performing inocula. Agronomy 10(1):106

Goljanian-Tabrizi S, Amiri S, Nikaein D, Motesharrei Z (2016) The comparison of five low cost liquid formulations to preserve two phosphate solubilizing bacteria from the genera *Pseudomonas* and *Pantoea*. Iran J Microbiol 8(6):377–382

Gomez-Bellot MJ, Ortuño MF, Nortes PA, Vicente-Sánchez J, Bañón S, Sánchez Blanco MJ (2015) Mycorrhizal euonymus plants and reclaimed water: biomass, water status and nutritional responses. Sci Hortic 186:61–69

Gómez-Gallego T, Benabdellah K, Merlos MA, Jiménez-Jiménez AM, Alcon C, Berthomieu P, Ferrol N (2019) The *Rhizophagus irregularis* genome encodes two CTR copper transporters that mediate Cu import into the cytosol and a CTR-like protein likely involved in copper tolerance. Front Plant Sci 10:604

Gopal R, Sharma YK, Shukla AK (2016) Effect of molybdenum stress on growth, yield and seed quality in black gram. J Plant Nutr 39(4):463–469

Gopalakrishnan S, Srinivas V, Prakash B, Sathya A, Vijayabharathi R (2015a) Plant growth-promoting traits of *Pseudomonas geniculata* isolated from chickpea nodules. 3 Biotech 5 (5):653–661

Gopalakrishnan S, Srinivas V, Alekhya G, Prakash B (2015b) Effect of plant growth-promoting *Streptomyces sp.* on growth promotion and grain yield in chickpea (*Cicer arietinum* L). 3. Biotech 5(5):799–806

Govindarajan M, Balandreau J, Kwon SW, Weon HY, Lakshminarasimhan C (2008) Effects of the inoculation of *Burkholderia vietnamensis* and related endophytic diazotrophic bacteria on grain yield of rice. Microb Ecol 55(1):21–37

Grene R (2002) Oxidative stress and acclimation mechanisms in plants. Arabidopsis Book 1:e0036

Gruba P, Mulder J (2015) Tree species affect cation exchange capacity (CEC) and cation binding properties of organic matter in acid forest soils. Sci Total Environ 511:655–662

GSP (2017) Global soil partnership endorses guidelines on sustainable soil management. http://www.fao.org/global-soil-partnership/resources/highlights/detail/en/c/416516/

Guether M, Neuhäuser B, Balestrini R, Dynowski M, Ludewig U, Bonfante P (2009) A mycorrhizal-specific ammonium transporter from Lotus japonicus acquires nitrogen released by arbuscular mycorrhizal fungi. Plant Physiol 150(1):73–83

Guinel FC (2015) Ethylene, a hormone at the center-stage of nodulation. Front Plant Sci 6:1121

Guo W, Chen S, Hussain N, Cong Y, Liang Z, Chen K (2015) Magnesium stress signaling in plant: just a beginning. Plant Signal Behav 10(3):e992287

Hao Z, Xie W, Chen B (2019) Arbuscular mycorrhizal symbiosis affects plant immunity to viral infection and accumulation. Viruses 11(6):534

Harrison MJ, van Buuren ML (1995) A phosphate transporter from the mycorrhizal fungus *Glomus versiforme*. Nature 378(6557):626–629

Hebbern CA, Laursen KH, Ladegaard AH, Schmidt SB, Pedas P, Bruhn D, Schjoerring JK, Wulfsohn D, Husted S (2009) Latent manganese deficiency increases transpiration in barley (Hordeum vulgare). Physiol Plant 135(3):307–316

Hefferon K (2019) Biotechnological approaches for generating zinc-enriched crops to combat malnutrition. Nutrients 11(2):253

Helms K (1971) Calcium deficiency of dark-grown seedlings of Phaseolus vulgaris L. Plant Physiol 47(6):799–804

Hepler PK (2005) Calcium: a central regulator of plant growth and development. Plant Cell 17 (8):2142–2155

Hoffman BM, Lukoyanov D, Yang ZY, Dean DR, Seefeldt LC (2014) Mechanism of nitrogen fixation by nitrogenase: the next stage. Chem Rev 114(8):4041–4062

Høgh-Jensen H, Schjoerring J, Soussana JF (2002) The influence of phosphorus deficiency on growth and nitrogen fixation of white clover plants. Ann Bot 90(6):745–753

Holt S, Kankipati H, De Graeve S, Van Zeebroeck G, Foulquié-Moreno MR, Lindgreen S, Thevelein JM (2017) Major sulfonate transporter Soa1 in *Saccharomyces cerevisiae* and considerable substrate diversity in its fungal family. Nat Commun 8:14247

Hortal S, Plett KL, Plett JM, Cresswell T, Johansen M, Pendall E, Anderson IC (2017) Role of plant–fungal nutrient trading and host control in determining the competitive success of ectomycorrhizal fungi. ISME J 11:2666–2676

Hu Y, Ribbe MW (2015) Nitrogenase and homologs. J Biol Inorg Chem 20(2):435–445

Hu JL, Lin XG, Wang JH, Shen WS, Wu S, Peng SP, Mao TT (2010) Arbuscular mycorrhizal fungal inoculation enhances suppression of cucumber *Fusarium* wilt in greenhouse soils. Pedosphere 20(5):586–593

Hürlimann HC, Stadler-Waibel M, Werner TP, Freimoser FM (2007) Pho91 Is a vacuolar phosphate transporter that regulates phosphate and polyphosphate metabolism in *Saccharomyces cerevisiae*. Mol Biol Cell 18(11):4438–4445

Igiehon NO, Babalola OO, Aremu BR (2019) Genomic insights into plant growth promoting rhizobia capable of enhancing soybean germination under drought stress. BMC Microbiol 19 (1):159

Ingraffia R, Amato G, Frenda AS, Giambalvo D (2019) Impacts of arbuscular mycorrhizal fungi on nutrient uptake, N2 fixation, N transfer, and growth in a wheat/faba bean intercropping system. PLoS One 14(3):e0213672

Jentschke G, Brandes B, Kuhn AJ, Schröder WH, Godbold DL (2001) Interdependence of phosphorus, nitrogen, potassium and magnesium translocation by the ectomycorrhizal fungus *Paxillus involutus*. New Phytol 149(2):327–337

Jia YB, Yang XE, Feng Y, Jilani G (2008) Differential response of root morphology to potassium deficient stress among rice genotypes varying in potassium efficiency. J Zhejiang Univ Sci B 9 (5):427–434

Jiang J, Zhao J, Duan W, Tian S, Wang X, Zhuang H, Fu J, Kang Z (2019) TaAMT2;3a, a wheat AMT2-type ammonium transporter, facilitates the infection of stripe rust fungus on wheat. BMC Plant Biol 19(1):239

Jiménez-Gómez A, Flores-Félix JD, García-Fraile P, Mateos PF, Menéndez E, Velázquez E, Rivas R (2018) Probiotic activities of rhizobium laguerreae on growth and quality of spinach. Sci Rep 8(1):295

Jochum M, Moncayo LP, Jo YK (2018) Microalgal cultivation for biofertilization in rice plants using a vertical semi-closed airlift photobioreactor. PLoS One 13(9):e0203456

Johri AK, Oelmüller R, Dua M, Yadav V, Kumar M, Tuteja N, Varma A, Bonfante P, Persson BL, Stroud RM (2015) Fungal association and utilization of phosphate by plants: success, limitations, and future prospects. Front Microbiol 6:984

Ju M, Navarreto-Lugo M, Wickramasinghe S, Milbrandt NB, McWhorter A, Samia ACS (2019) Exploring the chelation-based plant strategy for iron oxide nanoparticle uptake in garden cress (*Lepidium sativum*) using magnetic particle spectrometry. Nanoscale 11(40):18582–18594

Kabir AH, Debnath T, Das U, Prity SA, Haque A, Rahman MM, Parvez MS (2020) Arbuscular mycorrhizal fungi alleviate Fe-deficiency symptoms in sunflower by increasing iron uptake and its availability along with antioxidant defense. Plant Physiol Biochem 150:254–262

Kaiser BN, Gridley KL, Ngaire Brady J, Phillips T, Tyerman SD (2005) The role of molybdenum in agricultural plant production. Ann Bot 96(5):745–754

Kang JP, Huo Y, Kim YJ, Ahn JC, Hurh J, Yang DU, Yang DC (2019) *Rhizobium panacihumi*sp. nov., an isolate from ginseng-cultivated soil, as a potential plant growth promoting bacterium. Arch Microbiol 201(1):99–105

Karthik C, Oves M, Thangabalu R, Sharma R, Santhosh SB, Indra Arulselvi P (2016) *Cellulosimicrobium funkei*-like enhances the growth of Phaseolus vulgaris by modulating oxidative damage under Chromium (VI) toxicity. J Adv Res 7(6):839–850

Khamar HJ, Breathwaite EK, Prasse CE, Fraley ER, Secor CR, Chibane FL, Elhai J, Chiu WL (2010) Multiple roles of soluble sugars in the establishment of *Gunnera-Nostoc* endosymbiosis. Plant Physiol 154(3):1381–1389

Kidd J, Manning P, Simkin J, Peacock S, Stockdale E (2017) Impacts of 120 years of fertilizer addition on a temperate grassland ecosystem. PLoS One 12(3):e0174632

Kim C, Kecskés ML, Deaker RJ, Gilchrist K, New PB, Kennedy IR, Kim S, Sa T (2005) Wheat root colonization and nitrogenase activity by *Azospirillum* isolates from crop plants in Korea. Can J Microbiol 51(11):948–956

Kiontke K, Fitch DH (2013) Nematodes. Curr Biol 23(19):R862–R864

Kisiala A, Laffont C, Emery RJ, Frugier F (2013) Bioactive cytokinins are selectively secreted by *Sinorhizobium meliloti* nodulating and nonnodulating strains. Mol Plant-Microbe Interact 26 (10):1225–1231

Kobae Y, Tamura Y, Takai S, Banba M, Hata S (2010) Localized expression of arbuscular mycorrhiza-inducible ammonium transporters in soybean. Plant Cell Physiol 51(9):1411–1415

Kobayashi NI, Saito T, Iwata N, Ohmae Y, Iwata R, Tanoi K, Nakanishi TM (2013) Leaf senescence in rice due to magnesium deficiency mediated defect in transpiration rate before sugar accumulation and chlorosis. Physiol Plant 148(4):490–501

Kobayashi T, Nozoye T, Nishizawa NK (2019) Iron transport and its regulation in plants. Free Radic Biol Med 133:11–20

Koegel S, Boller T, Lehmann MF, Wiemken A, Courty PE (2013) Rapid nitrogen transfer in the *Sorghum bicolor-Glomus mosseae* arbuscular mycorrhizal symbiosis. Plant Signal Behav 8(8): e25229

Kokkoris V, Hart M (2019) In vitro propagation of arbuscular mycorrhizal fungi may drive fungal evolution. Front Microbiol 10:2420

Kollah B, Patra AK, Mohanty SR (2016) Aquatic microphylla *Azolla*: a perspective paradigm for sustainable agriculture, environment and global climate change. Environ Sci Pollut Res Int 23 (5):4358–4369

Konečný J, Hršelová H, Bukovská P, Hujslová M, Jansa J (2019) Correlative evidence for co-regulation of phosphorus and carbon exchanges with symbiotic fungus in the arbuscular mycorrhizal *Medicago truncatula*. PLoS One 14(11):e0224938

Kopittke PM, Dalal RC, Finn D, Menzies NW (2017) Global changes in soil stocks of carbon, nitrogen, phosphorus, and Sulphur as influenced by long-term agricultural production. Glob Chang Biol 23(6):2509–2519

Kopittke PM, Menzies NW, Wang P, McKenna BA, Lombi E et al (2019) Environ Int 132:105078

Kraiser T, Gras DE, Gutiérrez AG, González B, Gutiérrez RA (2011) A holistic view of nitrogen acquisition in plants. J Exp Bot 62(4):1455–1466

Krewulak KD, Vogel HJ (2008) Structural biology of bacterial iron uptake. Biochim Biophys Acta 1778(9):1781–1804

Krogmeier MJ, McCarty GW, Shogren DR, Bremner JM (1991) Effect of nickel deficiency in soybeans on the phytotoxicity of foliar-applied urea. Plant Soil 135:283–286

Ku Y, Xu G, Tian X, Xie H, Yang X, Cao C, Chen Y (2018) Root colonization and growth promotion of soybean, wheat and Chinese cabbage by *Bacillus cereus* YL6. PLoS One 13(11): e0200181. Erratum in: PLoS One. 2018 Dec 27; 13(12):e0210035

Kumar D, Rampuria S, Singh NK, Kirti PB (2016) A novel zinc-binding alcohol dehydrogenase 2 from *Arachis diogoi*, expressed in resistance responses against late leaf spot pathogen, induces cell death when transexpressed in tobacco. FEBS Open Bio 6(3):200–210

Kurth C, Kage H, Nett M (2016) Siderophores as molecular tools in medical and environmental applications. Org Biomol Chem 14(35):8212–8227

Kuzyakov Y (2002) Factors affecting rhizosphere priming effects (review). J Plant Nutr Soil Sci 165:382–396

Le Pioufle O, Ganoudi M, Calonne-Salmon M, Ben Dhaou F, Declerck S (2019) Rhizophagus irregularis MUCL 41833 improves phosphorus uptake and water use efficiency in maize plants during recovery from drought stress. Front Plant Sci 10:897

Lemaître T, Gaufichon L, Boutet-Mercey S, Christ A, Masclaux-Daubresse C (2008) Enzymatic and metabolic diagnostic of nitrogen deficiency in Arabidopsis thaliana *Wassileskija accession*. Plant Cell Physiol 49(7):1056–1065

Li M, Zhao Z, Zhang Z, Zhang W, Zhou J, Xu F, Liu X (2017) Effect of boron deficiency on anatomical structure and chemical composition of petioles and photosynthesis of leaves in cotton (*Gossypium hirsutum* L.). Sci Rep 7(1):4420

Li J, Jia Y, Dong R, Huang R, Liu P, Li X, Wang Z, Liu G, Chen Z (2019a) Advances in the mechanisms of plant tolerance to manganese toxicity. Int J Mol Sci 20(20):5096

Li J, Meng B, Chai H, Yang X, Song W, Li S, Lu A, Zhang T, Sun W (2019b) Arbuscular mycorrhizal fungi alleviate drought stress in C3 (*Leymus chinensis*) and C4 (*Hemarthria altissima*) grasses via altering antioxidant enzyme activities and photosynthesis. Front Plant Sci 10:499

Lin BB, Egerer MH, Liere H, Jha S, Philpott SM (2018) Soil management is key to maintaining soil moisture in urban gardens facing changing climatic conditions. Sci Rep 8(1):17565

Lingua G, Bona E, Manassero P, Marsano F, Todeschini V, Cantamessa S, Copetta A, D'Agostino G, Gamalero E, Berta G (2013) Arbuscular mycorrhizal fungi and plant growth-promoting pseudomonads increases anthocyanin concentration in strawberry fruits (*Fragaria* x *ananassa* var. *Selva*) in conditions of reduced fertilization. Int J Mol Sci 14(8):16207–16225

Liu CH, Chao YY, Kao CH (2013) Effect of potassium deficiency on antioxidant status and cadmium toxicity in rice seedlings. Bot Stud 54(1):2

Liu A, Chen S, Chang R, Liu D, Chen H, Ahammed GJ, Lin X, He C (2014) Arbuscular mycorrhizae improve low temperature tolerance in cucumber via alterations in H2O2 accumulation and ATPase activity. J Plant Res 127(6):775–785

Liu Y, Pan Z, Zhuang Q, Miralles DG, Teuling AJ, Zhang T, An P, Dong Z, Zhang J, He D, Wang L, Pan X, Bai W, Niyogi D (2015) Agriculture intensifies soil moisture decline in northern China. Sci Rep 5:11261

Liu H, Wang X, Qi H, Wang Q, Chen Y, Li Q, Zhang Y, Qiu L, Fontana JE, Zhang B, Wang W, Xie Y (2017) The infection and impact of *Azorhizobium caulinodans* ORS571 on wheat (*Triticum aestivum* L.). PLoS One 12(11):e0187947

Liu G, Pfeifer J, de Brito Francisco R, Emonet A, Stirnemann M, Gübeli C, Hutter O, Sasse J, Mattheyer C, Stelzer E, Walter A, Martinoia E, Borghi L (2018) Changes in the allocation of endogenous strigolactone improve plant biomass production on phosphate-poor soils. New Phytol 217(2):784–798

Liu J, Liu J, Liu J, Cui M, Huang Y, Tian Y, Chen A, Xu G (2019) The potassium transporter SlHAK10 is involved in mycorrhizal potassium uptake. Plant Physiol 180(1):465–479

Llases ME, Morgada MN, Vila AJ (2019) Biochemistry of copper site assembly in heme-copper oxidases: a theme with variations. Int J Mol Sci 20(15):3830

Llopis-Torregrosa V, Vaz C, Monteoliva L, Ryman K, Engstrom Y, Gacser A, Gil C, Ljungdahl PO, Sychrová H (2019) Trk1-mediated potassium uptake contributes to cell-surface properties and virulence of *Candida glabrata*. Sci Rep 9(1):7529

Maathuis FJ, Diatloff E (2013) Roles and functions of plant mineral nutrients. Methods Mol Biol 953:1–21

Mahanty T, Bhattacharjee S, Goswami M, Bhattacharyya P, Das B, Ghosh A, Tribedi P (2017) Biofertilizers: a potential approach for sustainable agriculture development. Environ Sci Pollut Res Int 24(4):3315–3335

Mahmud K, Makaju S, Ibrahim R, Missaoui A (2020) Current progress in nitrogen fixing plants and microbiome research. Plants (Basel) 9(1):97

Malyan SK, Bhatia A, Kumar SS, Fagodiya RK, Pugazhendhi A, Duc PA (2019) Mitigation of greenhouse gas intensity by supplementing with *Azolla* and moderating the dose of nitrogen fertilizer. Biocatal Agric Biotechnol 20:101266

Mao D, Chen J, Tian L, Liu Z, Yang L, Tang R, Li J, Lu C, Yang Y, Shi J, Chen L, Li D, Luan S (2014) *Arabidopsis* transporter MGT6 mediates magnesium uptake and is required for growth under magnesium limitation. Plant Cell 26(5):2234–2248

Marques AT, Santos SP, Rosa MG, Rodrigues MA, Abreu IA, Frazão C, Romão CV (2014) Expression, purification and crystallization of MnSOD from *Arabidopsis thaliana*. Acta Crystallogr F Struct Biol Commun 70(Pt 5):669–672

Martha-Paz AM, Eide D, Mendoza-Cózatl D, Castro-Guerrero NA, Aréchiga-Carvajal ET (2019) Zinc uptake in the basidiomycota: characterization of zinc transporters in *Ustilago maydis*. Mol Membr Biol 35(1):39–50

Martínez-Cuenca MR, Martínez-Alcántara B, Quiñones A, Ruiz M, Iglesias DJ, Primo-Millo E, Forner-Giner MÁ (2015) Physiological and molecular responses to excess boron in *Citrus macrophylla* W. PLoS One 10(7):e0134372. Erratum in: PLoS One. 2015;10(9):e0137941. Erratum in: PLoS One. 2015; 10(11):e0142358

Maswada HF, Abd El-Razek UA, El-Sheshtawy AA, Mazrou YSA (2020) Effect of *Azolla filiculoides* on growth, physiological and yield attributes of maize grown under water and nitrogen deficiencies. J Plant Growth Regul. https://doi.org/10.1007/s00344-020-10120-5

Matthes MS, Robil JM, McSteen P (2020) From element to development: the power of the essential micronutrient boron to shape morphological processes in plants. J Exp Bot 71(5):1681–1693

Meeks JC, Elhai J (2002) Regulation of cellular differentiation in filamentous cyanobacteria in free-living and plant-associated symbiotic growth states. Microbiol Mol Biol Rev 66(1):94–121

Mendel RR, Hänsch R (2002) Molybdoenzymes and molybdenum cofactor in plants. J Exp Bot 53 (375):1689–1698

Mendoza-Cózatl D, Loza-Tavera H, Hernández-Navarro A, Moreno-Sánchez R (2005) Sulfur assimilation and glutathione metabolism under cadmium stress in yeast, protists and plants. FEMS Microbiol Rev 29(4):653–671

Mirza BS, Welsh A, Rasul G, Rieder JP, Paschke MW, Hahn D (2009) Variation in *Frankia* populations of the Elaeagnus host infection group in nodules of six host plant species after inoculation with soil. Microb Ecol 58(2):384–393

Mishra P, Dash D (2014) Rejuvenation of biofertilizer for sustainable agriculture and economic development. Consilience J Sustain Dev 11(1):41–61

Mishra AK, Mishra KN (2007) Use of Azolla pinnata as biofertilizer for the production of rice Pant-4 in Jaunpur District (U.P.), India. Plant Arch 7(1):313–316

Młodzińska E, Zboińska M (2016) Phosphate uptake and allocation - a closer look at *Arabidopsis thaliana* L. and *Oryza sativa* L. Front Plant Sci 7:1198

Moomaw EW, Hoffer E, Moussatche P, Salerno JC, Grant M, Immelman B, Uberto R, Ozarowski A, Angerhofer A (2013) Kinetic and spectroscopic studies of bicupin oxalate oxidase and putative active site mutants. PLoS One 8(3):e57933

Muthukumarasamy R, Kang UG, Park KD, Jeon WT, Park CY, Cho YS et al (2007) Enumeration, isolation and identification of diazotrophs from Korean wetland rice varieties grown with long-term application of N and compost and their short-term inoculation effect on rice plants. J Appl Microbiol 102:981–991

Nestler J, Wissuwa M (2016) Superior root hair formation confers root efficiency in some, but not all, rice genotypes upon P deficiency. Front Plant Sci 7:1935

Ngom M, Gray K, Diagne N, Oshone R, Fardoux J, Gherbi H, Hocher V, Svistoonoff S, Laplaze L, Tisa LS, Sy MO, Champion A (2016) Symbiotic performance of diverse *Frankia* strains on salt-stressed *Casuarina glauca* and *Casuarina equisetifolia* plants. Front Plant Sci 7:1331

Nieves-Cordones M, García-Sánchez F, Pérez-Pérez JG, Colmenero-Flores JM, Rubio F, Rosales MA (2019) Coping with water shortage: an update on the role of K+, Cl-, and water membrane transport mechanisms on drought resistance. Front Plant Sci 10:1619

Oakley B, North M, Franklin JF, Hedlund BP, Staley JT (2004) Diversity and distribution of *Frankia* strains symbiotic with *Ceanothus* in California. Appl Environ Microbiol 70 (11):6444–6452

Oertel C, Matschullat J, Zurba K, Zimmermann F, Erasmi S (2016) Greenhouse gas emissions from soils—a review. Geochemistry 76(3):327–352

Ogo S, Yamanaka T, Akama K, Nagakura J, Yamaji K (2018) Influence of ectomycorrhizal colonization on cesium uptake by *Pinus densiflora* seedlings. Mycobiology 46(4):388–395

Ohtomo R, Saito M (2005) Polyphosphate dynamics in mycorrhizal roots during colonization of an arbuscular mycorrhizal fungus. New Phytol 167(2):571–578

Olsson PA, van Aarle IM, Allaway WG, Ashford AE, Rouhier H (2002) Phosphorus effects on metabolic processes in monoxenic arbuscular mycorrhiza cultures. Plant Physiol 130 (3):1162–1171

Ordway EM, Naylor RL, Nkongho RN, Lambin EF (2019) Oil palm expansion and deforestation in Southwest Cameroon associated with proliferation of informal mills. Nat Commun 10(1):114

Ostaszewska M, Juszczuk IM, Kołodziejek I, Rychter AM (2014) Long-term Sulphur starvation of *Arabidopsis thaliana* modifies mitochondrial ultrastructure and activity and changes tissue energy and redox status. J Plant Physiol 171(7):549–558

Oyange WA, Chemining'wa GN, Kanya JI, Njiruh PN (2019) Effects of *Azolla* and inorganic nitrogen application on growth and yield of rice in mwea irrigation scheme. Int J Agron Agric Res 14(3):1–8

Pande A, Pandey P, Mehra S, Singh M, Kaushik S (2017) Phenotypic and genotypic characterization of phosphate solubilizing bacteria and their efficiency on the growth of maize. J Genet Eng Biotechnol 15(2):379–391

Pandey N, Sharma CP (1996) Copper effect on photosynthesis and transpiration in safflower. Indian J Exp Biol 34(8):821–822

Parada J, Valenzuela T, Gómez F, Tereucán G, García S, Cornejo P, Winterhalter P, Ruiz A (2019) Effect of fertilization and arbuscular mycorrhizal fungal inoculation on antioxidant profiles and activities in *Fragaria ananassa* fruit. J Sci Food Agric 99(3):1397–1404

Park YS, Dutta S, Ann M, Raaijmakers JM, Park K (2015) Promotion of plant growth by *Pseudomonas fluorescens* strain SS101 via novel volatile organic compounds. Biochem Biophys Res Commun 461(2):361–365

Peng HY, Qi YP, Lee J, Yang LT, Guo P, Jiang HX, Chen LS (2015) Proteomic analysis of *Citrus sinensis* roots and leaves in response to long-term magnesium-deficiency. BMC Genomics 16:253

Peng W, Sonne C, Lam SS, Ok YS, Alstrup AKO (2020) The ongoing cut-down of the Amazon rainforest threatens the climate and requires global tree planting projects: a short review. Environ Res 181:108887

Pereira AL, Vasconcelos V (2014) Classification and phylogeny of the cyanobiont *Anabaena azollae* Strasburger: an answered question? Int J Syst Evol Microbiol 64(Pt 6):1830–1840

Pilon-Smits EA, Quinn CF, Tapken W, Malagoli M, Schiavon M (2009) Physiological functions of beneficial elements. Curr Opin Plant Biol 12(3):267–274

Pohjanen J, Koskimäki JJ, Sutela S, Ardanov P, Suorsa M, Niemi K, Sarjala T, Häggman H, Pirttilä AM (2014) Interaction with ectomycorrhizal fungi and endophytic *Methylobacterium* affects nutrient uptake and growth of pine seedlings in vitro. Tree Physiol 34(9):993–1005

Polacco JC, Mazzafera P, Tezotto T (2013) Opinion: nickel and urease in plants: still many knowledge gaps. Plant Sci 199–200:79–90. Erratum in: Plant Sci. 2013 May; 205–06:1

Prasad D, Verma N, Bakshi M, Narayan OP, Singh AK, Dua M, Johri AK (2019) Functional characterization of a magnesium transporter of root endophytic fungus *Piriformospora indica*. Front Microbiol 9:3231

Printz B, Lutts S, Hausman JF, Sergeant K (2016) Copper trafficking in plants and its implication on cell wall dynamics. Front Plant Sci 7:601

Przybyla-Toscano J, Roland M, Gaymard F, Couturier J, Rouhier N (2018) Roles and maturation of iron-sulfur proteins in plastids. J Biol Inorg Chem 23(4):545–566

Qaswar M, Ahmed W, Jing H, Hongzhu F, Xiaojun S, Xianjun J, Kailou L, Yongmei X, Zhongqun H, Asghar W, Shah A, Zhang H (2019) Soil carbon (C), nitrogen (N) and phosphorus (P) stoichiometry drives phosphorus liability in paddy soil under long-term fertilization: a fractionation and path analysis study. PLoS One 14(6):e0218195

Qin SY, Sun XC, Hu CX, Tan QL, Zhao XH (2017) Uptake, transport and distribution of molybdenum in two oilseed rape (*Brassica napus* L.) cultivars under different nitrate/ammonium ratios. J Zhejiang Univ Sci B 18(6):512–521

Qiu YL, Yu J (2003) Azolla--a model organism for plant genomic studies. Genomics Proteomics Bioinformatics 1(1):15–25

Rai AN, Söderbäck E, Bergman B (2000) Tansley review no.116. Cyanobacterium–plant symbioses. New Phytol 147:449–481

Ramalho JC, Rebelo MC, Santos ME, Antunes ML, Nunes MA (1995) Effects of calcium deficiency on *Coffea arabica*. Nutrient changes and correlation of calcium levels with some photosynthetic parameters. Plant Soil 172:87–96

Rausch C, Daram P, Brunner S, Jansa J, Laloi M, Leggewie G, Amrhein N, Bucher M (2001) A phosphate transporter expressed in arbuscule-containing cells in potato. Nature 414 (6862):462–470

Razaq M, Zhang P, Shen HL, Salahuddin (2017) Influence of nitrogen and phosphorous on the growth and root morphology of *Acer mono*. PLoS One 12(2):e0171321

Razavipour T, Moghaddam SS, Doaei S, Noorhosseini SA, Damalas CA (2018) *Azolla* (*Azolla filiculoides*) compost improves grain yield of rice (*Oryza sativa* L.) under different irrigation regimes. Agric Water Manag 209:1–10

Riaz M, Yan L, Wu X, Hussain S, Aziz O, Jiang C (2018) Boron increases root elongation by reducing aluminum induced disorganized distribution of HG epitopes and alterations in subcellular cell wall structure of trifoliate orange roots. Ecotoxicol Environ Saf 165:202–210

Rimington WR, Pressel S, Duckett JG, Field KJ, Read DJ, Bidartondo MI (2018) Ancient plants with ancient fungi: liverworts associate with early-diverging arbuscular mycorrhizal fungi. Proc Biol Sci 285(1888):20181600

Riveras E, Alvarez JM, Vidal EA, Oses C, Vega A, Gutiérrez RA (2015) The calcium ion is a second messenger in the nitrate signaling pathway of arabidopsis. Plant Physiol 169 (2):1397–1404

Rosenblueth M, Ormeño-Orrillo E, López-López A, Rogel MA, Reyes-Hernández BJ, Martínez-Romero JC, Reddy PM, Martínez-Romero E (2018) Nitrogen fixation in cereals. Front Microbiol 9:1794

Roser M, Ritchie H, Ortiz-Ospina E (2020) World population growth. OurWorldInData.org. https://ourworldindata.org/world-population-growth

Ruuhola T, Keinänen M, Keski-Saari S, Lehto T (2011) Boron nutrition affects the carbon metabolism of silver birch seedlings. Tree Physiol 31(11):1251–1261

Ryden P, Sugimoto-Shirasu K, Smith AC, Findlay K, Reiter WD, McCann MC (2003) Tensile properties of *Arabidopsis* cell walls depend on both a xyloglucan cross-linked microfibrillar network and rhamnogalacturonan II-borate complexes. Plant Physiol 132(2):1033–1040

Saito S, Uozumi N (2019) Guard cell membrane anion transport systems and their regulatory components: an elaborate mechanism controlling stress-induced stomatal closure. Plants (Basel) 8(1):9

Sanderman J, Hengl T, Fiske GJ (2017) Soil carbon debt of 12,000 years of human land use. Proc Natl Acad Sci U S A 114(36):9575–9580. Erratum in: Proc Natl Acad Sci U S A. 2018 Feb 5

Sanders D, Pelloux J, Brownlee C, Harper JF (2002) Calcium at the crossroads of signaling. Plant Cell 14(Suppl):S401–S417

Santelia D, Lawson T (2016) Rethinking guard cell metabolism. Plant Physiol 172(3):1371–1392

Santi C, Bogusz D, Franche C (2013) Biological nitrogen fixation in non-legume plants. Ann Bot 111(5):743–767

Sapre S, Gontia-Mishra I, Tiwari S (2018) Klebsiella sp. confers enhanced tolerance to salinity and plant growth promotion in oat seedlings (*Avena sativa*). Microbiol Res 206:25–32

Sato T, Hachiya S, Inamura N, Ezawa T, Cheng W, Tawaraya K (2019) Secretion of acid phosphatase from extraradical hyphae of the arbuscular mycorrhizal fungus *Rhizophagus clarus* is regulated in response to phosphate availability. Mycorrhiza 29(6):599–605

Scharlemann JPW, Tanner EVJ, Hiederer R, Kapos V (2014) Global soil carbon: understanding and managing the largest terrestrial carbon pool. Carbon Manag 5(1):81–91

Schlüter U, Mascher M, Colmsee C, Scholz U, Bräutigam A, Fahnenstich H, Sonnewald U (2012) Maize source leaf adaptation to nitrogen deficiency affects not only nitrogen and carbon metabolism but also control of phosphate homeostasis. Plant Physiol 160(3):1384–1406

Schrettl M, Bignell E, Kragl C, Sabiha Y, Loss O, Eisendle M, Wallner A, Arst HN Jr, Haynes K, Haas H (2007) Distinct roles for intra- and extracellular siderophores during *Aspergillus fumigatus* infection. PLoS Pathog 3(9):1195–1207

Schröder I, Johnson E, de Vries S (2003) Microbial ferric iron reductases. FEMS Microbiol Rev 27 (2–3):427–447

Schwenke GD, Simpfendorfer SR, Collard BCY (2015) Confirmation of chloride deficiency as the cause of leaf spotting in durum wheat grown in the Australian northern grains region. Crop Pasture Sci 66:122–134

Selvakumar G, Yi PH, Lee SE, Shagol CC, Han SG, Sa T, Chung BN (2018) Effects of long-term subcultured arbuscular mycorrhizal fungi on red Pepper Plant growth and soil glomalin content. Mycobiology 46(2):122–128

Senovilla M, Castro-Rodríguez R, Abreu I, Escudero V, Kryvoruchko I, Udvardi MK, Imperial J, González-Guerrero M (2018) *Medicago truncatula* copper transporter 1 (MtCOPT1) delivers copper for symbiotic nitrogen fixation. New Phytol 218(2):696–709

Seto KC, Fragkias M, Güneralp B, Reilly MK (2011) A meta-analysis of global urban land expansion. PLoS One 6(8):e23777

Sharma A, Patni B, Shankhdhar D, Shankhdhar SC (2013) Zinc - an indispensable micronutrient. Physiol Mol Biol Plants 19(1):11–20

Shireen F, Nawaz MA, Chen C, Zhang Q, Zheng Z, Sohail H, Sun J, Cao H, Huang Y, Bie Z (2018) Boron: functions and approaches to enhance its availability in plants for sustainable agriculture. Int J Mol Sci 19(7):1856

Sim MY, Eom AH (2006) Effects of ectomycorrhizal fungi on growth of seedlings of *Pinus densiflora*. Mycobiology 34(4):191–195

Singh RP, Jha PN (2016) The multifarious PGPR *Serratia marcescens* CDP-13 augments induced systemic resistance and enhanced salinity tolerance of wheat (Triticum aestivum L.). PLoS One 11(6):e0155026

Singh T, Singh DK (2019) Rhizospheric *Microbacterium sp.* P27 showing potential of lindane degradation and plant growth promoting traits. Curr Microbiol 76(7):888–895

Singh JS, Pandey VC, Singh DP (2011) Effect of soil microorganisms: a new dimension of sustainable agriculture and environmental development. Agric Ecosyst Environ 140:339–353

Smirnoff N (2018) Ascorbic acid metabolism and functions: a comparison of plants and mammals. Free Radic Biol Med 122:116–129

Smith SE, Jakobsen I, Grønlund M, Smith FA (2011) Roles of arbuscular mycorrhizas in plant phosphorus nutrition: interactions between pathways of phosphorus uptake in arbuscular mycorrhizal roots have important implications for understanding and manipulating plant phosphorus acquisition. Plant Physiol 156(3):1050–1057

Smith P, House JI, Bustamante M, Sobocká J, Harper R, Pan G, West PC, Clark JM, Adhya T, Rumpel C, Paustian K, Kuikman P, Cotrufo MF, Elliott JA, Mcdowell R, Griffiths RI, Asakawa S, Bondeau A, Jain AK, Meersmans J, Pugh TA (2016) Global change pressures on soils from land use and management. Glob Chang Biol 22(3):1008–1028

Socha AL, Guerinot ML (2014) Mn-euvering manganese: the role of transporter gene family members in manganese uptake and mobilization in plants. Front Plant Sci 5:106

Solaiman MZ, Ezawa T, Kojima T, Saito M (1999) Polyphosphates in intraradical and extraradical hyphae of an arbuscular mycorrhizal fungus, *Gigaspora margarita*. Appl Environ Microbiol 65 (12):5604–5606

Sosa-Hernández MA, Leifheit EF, Ingraffia R, Rillig MC (2019) Subsoil arbuscular mycorrhizal fungi for sustainability and climate-smart agriculture: a solution right under our feet? Front Microbiol 10:744

Soumare A, Boubekri K, Lyamlouli K, Hafidi M, Ouhdouch Y, Kouisni L (2020) From isolation of phosphate solubilizing microbes to their formulation and use as biofertilizers: status and needs. Front Bioeng Biotechnol 7:425

Souza RD, Ambrosini A, Passaglia LM (2015) Plant growth-promoting bacteria as inoculants in agricultural soils. Genet Mol Biol 38(4):401–419

Srivastava S, Upadhyay MK, Srivastava AK, Abdelrahman M, Suprasanna P, Tran LP (2018) Cellular and subcellular phosphate transport machinery in plants. Int J Mol Sci 19(7):1914

Steffan JJ, Brevik EC, Burgess LC, Cerdà A (2018) The effect of soil on human health: an overview. Eur J Soil Sci 69(1):159–171

Sugawara M, Okazaki S, Nukui N, Ezura H, Mitsui H, Minamisawa K (2006) Rhizobitoxine modulates plant-microbe interactions by ethylene inhibition. Biotechnol Adv 24(4):382–388

Tagele SB, Kim SW, Lee HG, Kim HS, Lee YS (2018) Effectiveness of multi-trait *Burkholderia contaminans* KNU17BI1 in growth promotion and management of banded leaf and sheath blight in maize seedling. Microbiol Res 214:8–18

Taliman NA, Dong Q, Echigo K, Raboy V, Saneoka H (2019) Effect of phosphorus fertilization on the growth, photosynthesis, nitrogen fixation, mineral accumulation, seed yield, and seed quality of a soybean low-Phytate line. Plants (Basel) 8(5):119

Tamayo E, Gómez-Gallego T, Azcón-Aguilar C, Ferrol N (2014) Genome-wide analysis of copper, iron and zinc transporters in the arbuscular mycorrhizal fungus Rhizophagus irregularis. Front Plant Sci 5:547

Tittabutr P, Sripakdi S, Boonkerd N, Tanthanuch W, Minamisawa K, Teaumroong N (2015) Possible role of 1-aminocyclopropane-1-carboxylate (ACC) deaminase activity of *Sinorhizobium sp.* BL3 on symbiosis with mung bean and determinate nodule senescence. Microbes Environ 30(4):310–320

Tu C, Foster L, Alvarado A, McKenna R, Silverman DN, Frost SC (2012) Role of zinc in catalytic activity of carbonic anhydrase IX. Arch Biochem Biophys 521(1–2):90–94

United Nations, Convention to Combat Desertification (2011) Press release. https://www.unccd.int/news-events/world-business-leaders-launch-sustainable-land-management-business-forum

Valtanen K, Eissfeller V, Beyer F, Hertel D, Scheu S, Polle A (2014) Carbon and nitrogen fluxes between beech and their ectomycorrhizal assemblage. Mycorrhiza 24:645–650

van Velzen R, Holmer R, Bu F, Rutten L, van Zeijl A, Liu W, Santuari L, Cao Q, Sharma T, Shen D, Roswanjaya Y, Wardhani TAK, Kalhor MS, Jansen J, van den Hoogen J, Güngör B, Hartog M, Hontelez J, Verver J, Yang WC, Schijlen E, Repin R, Schilthuizen M, Schranz ME, Heidstra R, Miyata K, Fedorova E, Kohlen W, Bisseling T, Smit S, Geurts R (2018) Comparative genomics of the nonlegume *Parasponia* reveals insights into evolution of nitrogen-fixing rhizobium symbioses. Proc Natl Acad Sci U S A 115(20):E4700–E4709

Venkataraman S, Prasad BVLS, Selvarajan R (2018) RNA dependent RNA polymerases: insights from structure, function and evolution. Viruses 10(2):76

Villarreal AJC, Renzaglia KS (2006) Structure and development of Nostoc strands in *Leiosporoceros dussii* (Anthocerotophyta): a novel symbiosis in land plants. Am J Bot 93 (5):693–705

Volpe V, Giovannetti M, Sun XG, Fiorilli V, Bonfante P (2016) The phosphate transporters LjPT4 and MtPT4 mediate early root responses to phosphate status in non mycorrhizal roots. Plant Cell Environ 39(3):660–671

Walder F, Boller T, Wiemken A, Courty PE (2016) Regulation of plants' phosphate uptake in common mycorrhizal networks: role of intraradical fungal phosphate transporters. Plant Signal Behav 11(2):e1131372

Waldrop GL, Holden HM, St Maurice M (2012) The enzymes of biotin dependent $CO_2$ metabolism: what structures reveal about their reaction mechanisms. Protein Sci 21(11):1597–1619

Walker CD, Graham RD, Madison JT, Cary EE, Welch RM (1985) Effects of Ni deficiency on some nitrogen metabolites in cowpeas (*Vigna unguiculata* L. Walp). Plant Physiol 79 (2):474–479

Wang Q, Lu L, Wu X, Li Y, Lin J (2003) Boron influences pollen germination and pollen tube growth in *Picea meyeri*. Tree Physiol 23(5):345–351

Wang FY, Tong RJ, Shi ZY, Xu XF, He XH (2011) Inoculations with arbuscular mycorrhizal fungi increase vegetable yields and decrease phoxim concentrations in carrot and green onion and their soils. PLoS One 6(2):e16949

Wang M, Shi S, Lin F, Hao Z, Jiang P, Dai G (2012) Effects of soil water and nitrogen on growth and photosynthetic response of Manchurian ash (*Fraxinus mandshurica*) seedlings in northeastern China. PLoS One 7(2):e30754

Wang X, Zhang H, Gao Y, Zhang W (2016) Characterization of Cu/Zn-SOD enzyme activities and gene expression in soybean under low nitrogen stress. J Sci Food Agric 96(8):2692–2697

Wang Y, Lu J, Ren T, Hussain S, Guo C, Wang S, Cong R, Li X (2017) Effects of nitrogen and tiller type on grain yield and physiological responses in rice. AoB Plants 9(2):plx012

Wang Y, Wang M, Li Y, Wu A, Huang J (2018) Effects of arbuscular mycorrhizal fungi on growth and nitrogen uptake of *Chrysanthemum morifolium* under salt stress. PLoS One 13(4):e0196408

Wardle DA, Yeates GW, Williamson WM, Bonner KI, Barker GM (2004) Linking aboveground and belowground communities: the indirect influence of aphid species identity and diversity on a three trophic level soil food web. Oikos 107:283–294

Wei X, Shao M, Gale W, Li L (2014) Global pattern of soil carbon losses due to the conversion of forests to agricultural land. Sci Rep 4:4062

West NJ, Adams DG (1997) Phenotypic and genotypic comparison of symbiotic and free-living cyanobacteria from a single field site. Appl Environ Microbiol 63(11):4479–4484

Wiesmeier M, Poeplau C, Sierra CA, Maier H, Frühauf C, Hübner R, Kühnel A, Spörlein P, Geuß U, Hangen E, Schilling B, von Lützow M, Kögel-Knabner I (2016) Projected loss of soil organic carbon in temperate agricultural soils in the 21(st) century: effects of climate change and carbon input trends. Sci Rep 6:32525

Williams A, Hunter MC, Kammerer M, Kane DA, Jordan NR, Mortensen DA, Smith RG, Snapp S, Davis AS (2016) Soil water holding capacity mitigates downside risk and volatility in US rainfed maize: time to invest in soil organic matter? PLoS One 11(8):e0160974

Wilson BR, Bogdan AR, Miyazawa M, Hashimoto K, Tsuji Y (2016) Siderophores in iron metabolism: from mechanism to therapy potential. Trends Mol Med 22(12):1077–1090

Wissuwa M, Ismail AM, Yanagihara S (2006) Effects of zinc deficiency on rice growth and genetic factors contributing to tolerance. Plant Physiol 142(2):731–741

Xie X, Lin H, Peng X, Xu C, Sun Z, Jiang K, Huang A, Wu X, Tang N, Salvioli A, Bonfante P, Zhao B (2016) Arbuscular mycorrhizal symbiosis requires a phosphate transceptor in the *Gigaspora margarita* fungal symbiont. Mol Plant 9(12):1583–1608

Yadav V, Kumar M, Deep DK, Kumar H, Sharma R, Tripathi T, Tuteja N, Saxena AK, Johri AK (2010) A phosphate transporter from the root endophytic fungus Piriformospora indica plays a role in phosphate transport to the host plant. J Biol Chem 285(34):26532–26544

Yadav RK, Abraham G, Singh YV, Singh PK (2014) Advancements in the utilization of *Azolla-Anabaena* system in relation to sustainable agricultural practices. Proc Indian Natl Sci Acad 80 (2):301–316

Yao Y, Zhang M, TianY ZM, Zeng K, Zhang B, Zhao M, Yin B (2018) *Azolla* biofertilizer for improving low nitrogen use efficiency in an intensive rice cropping system. Field Crops Res 216:158–164

Ye L, Zhao X, Bao E, Cao K, Zou Z (2019) Effects of arbuscular mycorrhizal fungi on watermelon growth, elemental uptake, antioxidant, and photosystem II activities and stress response gene expressions under salinity-alkalinity stresses. Front Plant Sci 10:863

Yeates GW, Saggar S, Hedley CB, Mercer CF (1999) Increase in C-14-carbon translocation to the soil microbial biomass when five species of plant-parasitic nematodes infect roots of white clover. Nematology 1:295–300

Zabel F, Putzenlechner B, Mauser W (2014) Global agricultural land resources--a high resolution suitability evaluation and its perspectives until 2100 under climate change conditions. PLoS One 9(9):e107522. Erratum in: PLoS One. 2014; 9(12):e114980

Zeffa DM, Perini LJ, Silva MB, de Sousa NV, Scapim CA, Oliveira ALM, Amaral Júnior ATD, Azeredo Gonçalves LS (2019) *Azospirillum brasilense* promotes increases in growth and nitrogen use efficiency of maize genotypes. PLoS One 14(4):e0215332

Zhang X, Wang L, Ma F, Yang J, Su M (2017) Effects of arbuscular mycorrhizal fungi inoculation on carbon and nitrogen distribution and grain yield and nutritional quality in rice (Oryza sativa L.). J Sci Food Agric 97(9):2919–2925

Zhang D, Chang E, Yu X, Chen Y, Yang Q, Cao Y, Li X, Wang Y, Fu A, Xu M (2018a) Molecular characterization of magnesium chelatase in soybean [*Glycine max* (L.) Merr.]. front. Plant Sci 9:720

Zhang Y, Hu J, Bai J, Wang J, Yin R, Wang J, Lin X (2018b) Arbuscular mycorrhizal fungi alleviate the heavy metal toxicity on sunflower (*Helianthus annuus* L.) plants cultivated on a heavily contaminated field soil at a WEEE-recycling site. Sci Total Environ 628–629:282–290
Zhao K, Wu Y (2017) Effects of Zn deficiency and bicarbonate on the growth and photosynthetic characteristics of four plant species. PLoS One 12(1):e0169812
Zhao D, Kane M, Borders B, Harrison M (2008) Pine growth response to different site-preparation methods with or without post-plant herbaceous weed control on North Florida's Lower Coastal Plain. For Ecol Manag 255(7):2512–2523

# Biotechnological Approaches to Enhance Crop Quality for Iron and Zinc Nutrition

10

Shraddha Singh, Sudhakar Srivastava, and Penna Suprasanna

**Abstract**

Micronutrient deficiencies have assumed prominence as 'hidden hunger'; especially iron (Fe) and zinc (Zn) have received considerable attention. Deficiencies of iron and zinc in humans lead to a variety of health-related issues related to growth, metabolic, immune and disease and general fitness. Improving the nutritional quality of staple crops seems to be an effective and straightforward solution to the problem. Despite conventional breeding methods employed for this purpose, success has been limited to the existing diversity in the gene pool. Various other strategies have been used to combat these deficiencies including supplementation, food fortification and modification of food preparation and processing methods. Biofortification aims at either increasing accumulation of these minerals in edible parts, endosperm, or to increase their bioavailability. Another strategy is to use biotechnological tools to improve trace element nutrition in staple foods such as cereals and legumes. This may be achieved by the introduction of genes that code for trace element-binding proteins, overexpression of storage proteins already present and/or increased expression of proteins that are responsible for trace element uptake into plants. Alternative approach is also to use genetically modification to reduce, inhibitors of trace element absorption such as phytate. The present chapter outlines the progress made in the application of biotechnological approaches to enhance crop quality for Fe and Zn nutrition.

S. Singh (✉) · P. Suprasanna
Nuclear Agriculture and Biotechnology Division, Bhabha Atomic Research Centre, Mumbai, Maharashtra, India
e-mail: shradhas@barc.gov.in

S. Srivastava
Plant Stress Biology Laboratory, Institute of Environment and Sustainable Development, Banaras Hindu University, Varanasi, Uttar Pradesh, India



K. Mishra et al. (eds.), *Sustainable Solutions for Elemental Deficiency and Excess in Crop Plants*, https://doi.org/10.1007/978-981-15-8636-1_10

**Keywords**

Iron · Zinc · Fortification · Nutrients · Genetic engineering

## 10.1 Introduction

Hidden hunger refers to mineral nutrient deficiency, which affects millions of people worldwide especially the poor people in the developing and underdeveloped countries. However, depending on the diet composition and variety of foods, people may suffer from deficiency of one or more essential elements. Developing countries suffer more from mineral nutrient deficiency than the developed countries. Malnutrition affects the growth and development of children and impacts adults with lowered immunity and work efficiency. However, depending on the diet composition and variety of foods, all people may suffer from deficiency of one or more essential elements.

The basic source of nutrients for life on earth is plants. However, humans are mostly dependent on a few plant species and consume them as a staple food. These include most carbohydrate-rich grains such as wheat, corn and rice, which are typically nutrient-poor and often lack minerals like iron (Fe) and zinc (Zn). Rice is a common staple food that has low Zn concentration with poor bioavailability (Myers et al. 2014) and consumption of rice staple food causes huge problem of Zn deficiency among human populations (Impa and Johnson-Beebout 2012). Considering the limited bioavailability of Fe in rice grain, it has been estimated that approximately 540 g brown rice per day should be consumed to fulfil the daily Fe requirement; however, rice consumption varies greatly and is mostly below 500 g per day in most rice consuming countries (Huang et al. 2020). In case of Fe, the presence of phytate, an inhibitor of Fe resorption, also has been shown to reduce Fe bioavailability in rice grain (Stangoulis et al. 2007). The problem is also aggravated by the fact the most soils are heavily contaminated due to the presence of toxic elements such as cadmium (Cd) and arsenic (As) which inhibit plant growth, reduce nutrient availability and disrupt food chain. In such a scenario, it is imperative to boost the essential nutrient content and lower or eliminate toxic elements in the edible plant parts that contribute to human health.

The deficiency of Zn and Fe is common throughout the world causing people to suffer from Fe-induced anaemia and metabolic disorders leading to child mortality (WHO 2001, 2014; Maret and Sandstead 2006). Iron has several vital functions in the body including the role as a carrier of oxygen and hence its deficiency leads to anaemia, reduced mental development and growth in children, reduced strength and physical health of adults and leads to loss of life years (Christopher et al. 2013). Zinc deficiency causes loss of immunity, diminished learning ability, improper metabolism and digestion, dysfunction of the reproductive system, and physical growth (Krishnaswami 1998; Myers et al. 2014).

Zinc and Fe are also very important micronutrients for plant growth and development. Zinc is a constituent as structural, functional or regulatory factor of more than

300 enzymes including alcohol dehydrogenase, Cu-Zn superoxide dismutase, carbonic anhydrase, aldolases, enolase, isomerases, RNA and DNA polymerases, peptidases, etc. (Marschner 1995; Coleman 1998). The role of Zn in plants ranges from membrane function, photosynthesis, to protection against pathogens (Engelbrecht et al. 2004; Kramer and Clemens 2006). Zinc deficiency in plants causes several biochemical changes including membrane structural impairment, stunted growth and chlorosis on the leaves, reduced photosynthesis and sugar metabolism, short internodes and reduced transcription (Broadley et al. 2007; Marenco and Lopes 2007). Iron is an essential cofactor for several redox reactions including those of photosynthesis and respiration (Winterbourn 1995). Iron is a prosthetic group constituent of many enzymes and also plays a role in DNA synthesis. Iron deficiency results in chlorosis of leaves as Fe is also involved in the synthesis of chlorophyll and is essential for chloroplast structure and function.

It is important to note that breeding efforts in the past several decades have focused on increased yield, pest and disease resistance, while unknowingly, the aspect of mineral nutrition remained out of the priorities (Zhao et al. 2009). The low concentration of mineral elements in crop plants may also result from poor soils in which they grow. Further, the bioavailability of Zn and Fe in soil may be low as it is dependent on the physicochemical and biological properties of soil and also on the interactions amongst elements (Fageria et al. 2012; Hafeez et al. 2013). Temperature, available water in the soil, as well as climatic factors like atmospheric carbon dioxide concentration also affect nutrient bioavailability and accumulation in crop plants (Weih and Karlsson 2002; Fernando et al. 2014a, b). The regions with zinc-deficient soils are strongly correlated with the ones with high incidence of human Zn deficiency (Erenoglu et al. 2011). Iron is present in high amount on earth and it becomes a limiting factor for plant growth. This happens due to the interaction of Fe with oxygen to form insoluble oxides that are not available for uptake by plants (Aisen et al. 2001).

A better understanding of Zn and Fe acquisition (uptake, transport and homeostasis) will enable research efforts to enhance their levels in edible parts of crop plants. The aim of Zn and Fe biofortification of food grains is to increase both their concentration and bioavailability. While agronomic approaches of biofortification are used by the external application of micronutrient fertilizers or bioregulators, genetic methods use biotech or plant breeding based approaches.

## 10.2 Biofortification

Biofortification is considered as the most safe, easy, sustainable, and economical approach to address the problem of nutrient deficiency for human (Zhao and Steve 2009; Rawat et al. 2013; Kumar et al. 2019). Biofortification may be accomplished genetically through plant breeding and transgenic crop development and agronomically through mineral fertilizer application.

The major focus of agronomical method is to optimize the amount and timing of application of mineral nutrient fertilizers, and biological agents (plant growth promoting and elemental nutrition enhancing microorganisms). These approaches have been discussed in the other chapter of this book (Freire et al. 2020). Plant breeding and genetic engineering target the development of improved plant varieties that can take up and accumulate mineral elements in greater concentrations in the produce (grains) and also have them in higher bioavailability due to less accumulation of anti-nutrient compounds.

An important tool in breeding programmes has been the identification of appropriate germplasms with genetic variability for Zn and Fe concentration that can be utilized (Kotla et al. 2012). Plant breeding programmes include quantitative genetics, mutational breeding, marker-assisted breeding, quantitative trait loci (QTL) identification, etc. The HarvestPlus is an International programme aimed to mitigate the malnutrition problem through development of micronutrient enriched crop plants. As against the earlier focus, breeding programmes now pay more emphasis on nutritional aspects of varieties being developed along with high yields. This is true for elements like Zn, Cu and Fe (Bouis 2003; Welch and Graham 2004; Uauy et al. 2006; Garg et al. 2018). Micronutrient uptake and accumulation traits in plants are controlled by specific genes and are therefore heritable. There is significant genotypic variation for seed Fe and Zn accumulation in several staple crops, including rice, wheat, maize and bean (Glenn and Gregorio 2002; Genc et al. 2005; Mantovi et al. 2003). Thus, selective breeding for Fe and Zn enriched grains can be done to develop high yielding nutritive varieties. The germplasm needs to be exploited for QTL identification, and allocating specific genes to QTLs. Selection using molecular markers could be an efficient complementary breeding tool, especially when selection is performed under unfavourable conditions (Cakmak and Braun 2001). The quality of the phenotypic data is crucial for the success of gene/QTL analysis (Septiningsih et al. 2003; Li et al. 2007). After QTL identification, candidate loci, or genes can be determined through fine mapping and map-based cloning, and this information can be used for breeding. However, breeding approaches depend on the presence of sufficient genetic variability for the trait (Fe and Zn in grains) in the available germplasms. Since micronutrient uptake and accumulation are multi-gene controlled, it has been difficult to incorporate several genes in one background simultaneously to achieve desirable results. The success achieved by using this approach depends chiefly on natural variation that exists in the gene pool. Further, breeding for crop variety development is a lengthy, time-consuming process and the success of a variety may also depend on local climatic and soil factors and epigenetic modifications.

Detailed understanding of uptake, root to shoot translocation, xylem loading and unloading, cellular homeostasis, and grain loading might help in the development of suitable crop plants with enriched grain Zn and Fe concentration both through genetic approaches (Shehu and Jamala 2010; Ishimaru et al. 2011; Gao et al. 2012). Transporters play a crucial role in Zn and Fe uptake and transport in plants. However, transports lack specificity and therefore a transporter involved in uptake and transport of Zn would also function for Fe. Due to this, if there is a deficiency of other divalent elements like Fe, Cu and Mn then it might in turn lead to more Zn in plants owing to reduced competition (Alloway 2008; Hafeez et al. 2013). The lack of exclusive transporters in plants makes it difficult to target one gene or a few genes and achieve the biofortification. To take up Fe from the soil, plants have devised various mechanisms (Curie and Briat 2003; Hell and Stephan 2003). These include siderophore production and changes in pH and redox conditions in the rhizosphere through secretion of biomolecules (protons, acids, etc.) so as reduce iron from the insoluble ferric iron to the soluble ferrous form (Askwith and Kaplan 1998). These are referred to as strategy I (proton extrusion, reduction and uptake of Fe(II) and strategy II (phytosiderophores).

The genetic engineering techniques offer opportunity even in the absence of sufficient genotypic variation to incorporate the desired trait with in a crop. Transgenic technologies can improve genotypes by inserting genes in specific functions, making changes in focused metabolic pathways and that too at the desired location i.e. in grains. Further, this method offers the unique possibility of incorporating several genes together to achieve better results (Massot et al. 2013). However, genetically modified crop plants suffer presently from several regulations and are not permitted normally for cultivation. In the future, faster procedures to test the safety of genetically modified crops are needed to introduce such varieties for common use.

## 10.3 Genetic Engineering Studies for Biofortification of Fe and Zn

A number of studies have been performed to date with the purpose of biofortifying Fe and Zn in grains and produce of different crop plants without effect on yield. For Fe and Zn biofortification, the need exists to increase the uptake of Fe and Zn from the soil in combination with increased storage capacity. Promoters, genes, as well gene paralogs have been found to give good results in different crop species. The following discussion presents an overview of such studies to showcase the potential of genetic engineering in achieving measurable and exciting biofortification of Fe and Zn.

There has been significant progress in the genetic information on Fe accumulation which has helped research efforts aimed at manipulation of target plant species (Perrez-Massot et al. 2013). The transporter proteins involved in uptake and root-to-shoot and shoot-to-grain transport of Fe and Zn have been the preferred ones in several transgenic studies. The overexpression of a Zn transporter gene from *A. thaliana* (*AtZIP1*; Zinc regulated transporter (ZRT)-/Iron regulated transporter (IRT)-like Protein) in barley (*Hordeum vulgare*) increased the grain Zn concentration by twofold; however, it leads to smaller seeds (Ramesh et al. 2004). *AtCAX1* (Cation/Proton Exchanger), which is responsible for calcium (Ca) transport, has been found to transport Zn also and can be used to enhance Zn in transgenic plants (Shigaki et al. 2005). A *ZIP* family transporter, *AtIRT3* is known to be involved in the loading of Zn into the xylem (Lin et al. 2009). Lee et al. (2009) reported that mutation of *OsYSL15* (Yellow Stripe Like) in rice resulted in reduced Fe accumulation in rice seeds but Zn accumulation was not affected. The overexpression of

*AtIRT3* was shown to increase the accumulation of Zn in shoots. Overexpression of *IRT1* increased Fe concentration in rice leaves by 1.7-fold but only by 1.1-fold in grains (Lee and An 2009). However, the overexpression of *IRT1* with *PvFER1* (Ferritin) in the endosperm resulted in increase in Fe concentration up to fourfold in polished rice (Boonyaves et al. 2017). Once reached to the shoot, the mineral element has to be unloaded in the target organ, leaves. From, the leaves mineral elements are eventually transported to grains during grain filling time. The members of the YSL family are involved in Fe and Zn translocation. The expression level of *OsYSL4*, *OsYSL6*, and *OsYSL12* has been found to be correlated to Fe levels while that of *OsYSL2* and *OsYSL9* with Zn contents (Banerjee and Chandel 2011).

The overexpression of *AtHMA4* (Heavy Metal ATPase) in *A. thaliana* was found to result in twofold increase in Zn in leaves (Verret et al. 2004). Zinc concentrations in *Arabidopsis* leaves were also found to increase upon overexpression of *AtMTP3* (Metal Transporter Protein) (Arrivault et al. 2006) while upon reduced expression of *AtHMA2* (Eren and Argüello 2004) or *AtOPT3* (Oligopeptide Transporter) (Stacey et al. 2008). However, *HMA2/HMA4* are not specific for Zn and can also transport other divalent metals like Cd and therefore, the overexpression of these genes may result in negative effects due to Cd accumulation. As most of the transporters are non-specific, the need exists to identify specific domains in transporter proteins, which may be specific for nutritional elements like Zn and Fe and then engineer proteins to perform the specific function of Zn and Fe biofortification (Rouached 2012). Dixit et al. (2010) reported the expression of a *Neurospora crassa* zinc transporter gene (*tzn1*) in transgenic *Nicotiana tabacum* and showed increased zinc accumulation. The transporter was found to be specific for Zn as the Cd level was not found to increase. Wang et al. (2019) recently reported the expression of peanut (*Arachis hypogaea*) *AhNRAMP1* (Natural Resistance Associated Macrophage Proteins) in tobacco and rice and showed that Fe and Zn levels were enhanced. *AhNRAMP1* expression also led to increased biomass and yield rice.

Nicotianamine synthase (NAS) has also been a preferred target gene for the development of transgenics. This is because the biosynthesis of NA requires only one enzyme, NAS, which uses S-adenosyl methionine as a substrate. In a study by Masuda et al. (2009), transgenic rice was developed with overexpression of the NAS gene of *Hordeum vulgare* and it was found to accumulate up to threefold more Fe and Zn in polished rice grains (Masuda et al. 2009). Ferritin is an excellent Fe storage protein and can bind with 4500 Fe atoms. The Fe level of rice grains was observed to increase by twofold upon overexpression of NAS while up to sixfold upon overexpression ferritin and NAS genes. *OsNAS2* was suggested as a potential target to biofortify both Fe and Zn in rice (Johnson et al. 2011). However, another issue in biofortification efforts is the localization of mineral elements in grains. The maximum portion of the total mineral content of mineral elements is restricted in aleurone layers and embryo and these elements are removed during processing and polishing of rice seeds. For availing the benefits of biofortification, enriched mineral elements should be concentrated in the endosperm (Grennan 2011). The reductive Fe uptake can also be achieved through the secretion of small molecules, such as coumarin-derivatives and flavins (Connorton et al. 2017), which need to be evaluated in the future to evaluate the effects of their gene expression. The expression of *OsYSL2* with *HvNAS1* (*Hordeum vulgare*) and *SferH2* (soybean (*Glycine max*) ferritin gene) in rice was found to increase Fe and Zn levels in polished grains by 4.4-fold and 1.6-fold, respectively (Masuda et al. 2012). *Medicago falcata* ferritin gene was expressed into wheat and this resulted in increased grain Fe and Zn concentration by 73% and 44%, respectively in transgenic lines as compared to WT plants (Liu et al. 2016). In banana, Yadav et al. (2017) studied the overexpression of native ferritin (*MusaFer1*) in transgenic banana cv. Rasthali which accumulated more iron suggesting that this gene could become a candidate for future molecular breeding of banana.

Che et al. (2019) developed a knockout mutant for the gene OsVMT (Vacuolar Mugineic Acid Transporter), expressing a tonoplast transporter of 2-deoxymugineic acid (DMA). The mutant lines had a significant increase in Fe and Zn concentration in the polished rice grain due to reduced sequestration of DMA in the vacuoles and consequently higher solubility and availability of Fe and Zn in nodes for transport to the grains (Che et al. 2019). Díaz-Benito (2018) demonstrated in rice that the predominant method of seed loading of Fe is via Fe(III)-2-deoxymugineic acid (DMA) through the analysis of wild-type (WT) and transgenic rice lines over-expressing *OsNAS1* and/or barley nicotianamine amino transferase (*HvNAATb*).

Cereal grains have some anti-nutritional factors (ANF) like phytate, tannin and certain insoluble fibres which inhibit the absorption of elements like Fe and Zn from the food (Petry et al. 2012). Phytates are the most common ANF in grains and these have negative charges through which phytates make a complex with metals and make them unavailable. It has been postulated that if the concentration of phytates can be reduced, Zn and Fe nutrition quality may be enhanced. Few researchers have reduced the expression of genes coding for phytate to achieve a reduction in phytate in edible tissues (Lönnerdal 2003; Raboy 2002, 2003). Other researchers have targeted phytate reduction through the increase in phytase enzyme, which degrades phytate. The transgenics have been developed in wheat, rice, maize, and soybean (Lucca et al. 2001; Hong et al. 2004; Drakakaki et al. 2005; Brinch-Pedersen et al. 2006). Yamaji et al. (2017) developed transgenic lines for Pi transporter from rice, *OsSPDT* (SULTR-like Phosphorus Distribution Transporter), expressed in nodes and involved in P translocation to the grains. The knockout mutants of *OsSPDT* have 20–30% lower phytic acid in rice grains without a reduction in yield (Yamaji et al. 2017). The concentration of Fe and Zn in the grain was slightly increased in *OsSPDT* mutants along with their increased bioavailability due to reduced phytic acid levels.

The transcription factors have also been utilized in the transgenic approaches. The transcription factor *OsIRO2* (Iron Related Transcription Factor) is involved in the regulation of a number of genes responsible for Fe uptake. The overexpression of *OsIRO2* resulted in a three-fold increase in the Fe concentration in brown rice (Ogo et al. 2011). Rice mutants for Fe-binding hemerythrin RING ubiquitin ligases (named as *HRZ1* and *HRZ2*) have been identified which have a high expression of genes for Fe uptake and transport. The mutants have been found to have 1.7- to 3.5-fold more Fe in seeds than that in WT seeds (Kobayashi et al. 2013). Biofortifying approaches involving TFs as the master regulators will have a great scope in enhancing micronutrient accumulation and uptake.

## 10.4 Future Prospects

The increasing world population needs not only higher crop production but also more nutritive crop produce. This is to ensure management of both real hunger and hidden hunger. Some of the staple crops, and millets are now improved as nutrient-rich super crops and efforts are being made globally for bringing them into main stream cultivation. There has been significant progress in the development of transgenics for increased Fe and Zn nutrition of crop plants, however there is a need to develop deeper insights into the holistic genetic circuitry of Fe and Zn uptake, transport, and storage in grains. Till date, there has been a wealth of knowledge on the genes and promoters for target traits to improve the micronutrients in agronomically important crop plants, and it has become possible to develop transgenic lines with enhanced micronutrients. Future research should aim to integrate different genetic metabolic pathways to enhance a gamut of nutrients simultaneously, in a given crop. Such genetically enriched crops for micronutrients may help in alleviating the problems of malnutrition.

## References

Aisen P, Enns C, Wessling-Resnick M (2001) Chemistry and biology of eukaryotic iron metabolism. Int J Biochem Cell Biol 33:59

Alloway BJ (2008) Micronutrients and crop production: an introduction. In: Alloway BJ (ed) Micronutrient deficiencies in global crop production. Springer, Dordrecht

Arrivault S, Senger T, Krämer U (2006) The Arabidopsis metal tolerance protein AtMTP3 maintains metal homeostasis by mediating Zn exclusion from the shoot under Fe deficiency and Zn oversupply. Plant J 46(5):861–879

Askwith C, Kaplan J (1998) Iron and copper transport in yeast and its relevance to human disease. Trends Biochem Sci 23:135–138

Banerjee S, Chandel G (2011) Understanding the role of metal homeostasis related candidate genes in Fe/Zn uptake, transport and redistribution in rice using semi-quantitative RT-PCR. J Plant Molecular Biol Bioitechnol 2:33–46

Boonyaves K, Wu TY, Gruissem W, Bhullar NK (2017) Enhanced grain iron levels in rice expressing an iron regulated metal transporter, nicotianamine synthase and ferritin gene cassette. Front Plant Sci 8:130

Bouis H (2003) Micronutrient fortification of plants through plant breeding: can it improve nutrition in man at low cost? Proc Nutr Soc 62(2):403–411

Brinch-Pedersen H, Hatzack F, Stöger E, Arcalis E, Pontopidan K, Holm PB (2006) Heat-stable phytases in transgenic wheat (*Triticum aestivum* L.): disposition pattern, thermostability, and phytate hydrolysis. J Agric Food Chem 54:4624–4632

Broadley MR, White PJ, Hammond JP, Zelko I, Lux A (2007) Zinc in plants. New Phytol 173 (4):677–702

Cakmak I, Braun HJ (2001) Genotypic variation for zinc efficiency. In: Reynolds MP, Ortiz M, McNab A (eds) Application of physiology in wheat breeding. International Maize and Wheat Improvement Center, Mexico City, pp 183–199

Che J, Yokosho K, Yamaji N, Ma JF (2019) A vacuolarphytosiderophore transporter alters iron and zinc accumulation in polished rice grains. Plant Physiol 181:276–288

Christopher JL, Murray MD, Alan D, Lopez (2013) Global health measuring the global burden of disease. N Engl J Med 369:5

Coleman JE (1998) Zinc enzymes. Curr Opin Chem Biol 2(2):222–234

Connorton JM, Balk J, Rodríguez-Celma J (2017) Iron homeostasis in plants-a brief overview. Metallomics 9:813–823

Curie C, Briat JF (2003) Iron transport and signaling in plants. Annu Rev Plant Biol 54:183–206

Díaz-Benito P, Banakar R, Rodríguez-Menéndez S, Capell T, Pereiro R et al (2018) Iron and zinc in the embryo and endosperm of rice (*Oryza sativa* L.) seeds in contrasting 2′-deoxymugineic acid/nicotianamine scenarios front. Plant Sci 9:1–17

Dixit P, Singh S, Vamcheeswaran R, Patnala K, Eapen S (2010) Expression of a *Neurospora crassa* zinc transporter gene in transgenic *Nicotiana tabacum* enhances plant zinc accumulation without co-transport of cadmium. Plant Cell Environ 33:1697–1707

Drakakaki G, Marcell S, Glahn RP, Lund EK, Pariagh S, Fischer R, Christou P, Stoger E (2005) Endosperm-specific co-expression of recombinant soybean ferritin and Aspergillus phytase in maize results in significant increases in the levels of bioavailable iron. Plant Mol Biol 59:869–880

Englbrecht CC, Schoof H, Bohm S (2004) Conservation, diversification and expansion of $C_2H_2$ zinc finger proteins in the *Arabidopsis thaliana* genome. BMC Genomics 5:39

Eren E, Argüello JM (2004) Arabidopsis HMA2, a divalent heavy metal-transporting P(IB)-type ATPase, is involved in cytoplasmic Zn2 + homeostasis. Plant Physiol 136:3712–3723

Erenoglu B, Kutman UB, Ceylan Y, Yildiz B, Cakmak I (2011) Improved nitrogen nutrition enhances root uptake, root-to- shoot translocation and remobilization of zinc (65Zn) in wheat. New Phytol 189:438–448

Fageria NK, Moraes MF, Ferreira EPB, Knupp AM (2012) Biofortification of trace elements in food crops for human health. Commun Soil Sci Plant Anal 43:556–570

Fernando N, Panozzo J, Tausz M, Norton RM, Fitzgerald GJ, Myers S et al (2014a) Intra-specific variation of wheat grain quality in response to elevated [$CO_2$] at two sowing times under rain-fed and irrigation treatments. J Cereal Sci 59:137–144

Fernando N, Panozzo J, Tausz M, Norton RM, Neumann N, Fitzgerald GJ et al (2014b) Elevated $CO_2$ alters grain quality of two bread wheat cultivars grown under different environmental conditions. Agric Ecosyst Environ 185:24–33

Freire BM, Pereira RM, Lange CN, Batista BL (2020) Biofortification of crop plants: a practical approach to tackle elemental deficiency. In: Mishra K, Tandon PK, Srivastava S (eds) Sustainable solutions for elemental deficiency and excess in crop plants. Springer, Singapore

Gao H, Stomph C, Grant Z (2012) Improving zinc bioavailability in transition from flooded to aerobic rice. A review. Agron Sustain Dev 32(2):465–478

Garg M, Sharma N, Sharma S et al (2018) Biofortified crops generated by breeding, agronomy, and transgenic approaches are improving lives of millions of people around the world. Front Nutr 5:12

Genc Y, Humphries JM, Lyons GH, Graham RD (2005) Exploiting genotypic variation in plant nutrient accumulation to alleviate micronutrient deficiency in populations. J Trace Elem Med Biol 18:319–324

Glenn B, Gregorio (2002) Progress in breeding for trace minerals in staple crops. J Nutr 132:500S–502S

Grennan AK (2011) Metallothioneins, a diverse protein family. Plant Physiol 155(1750):1751

Hafeez B, Khanif YM, Saleem M (2013) Role of zinc in plant nutrition—a review. Am J Exp Agric 3:374–391

Hell R, Stephan UW (2003) Iron uptake, trafficking and homeostasis in plants. Planta 216:541–551

Hong CY, Cheng KJ, Tseng TH, Wang CS, Liu LF, Yu SM (2004) Production of two highly active bacterial phytases with broad pH optima in germinated transgenic rice seeds. Transgenic Res 13:29–39

Huang S, Wang P, Yamaji N, Ma JF (2020) Plant nutrition for human nutrition: hints from rice research and future perspectives. Mol Plant 13(6):825–835. https://doi.org/10.1016/j.molp.2020.05.007

Impa SM, Johnson-Beebout SE (2012) Mitigating zinc deficiency and achieving high grain Zn in rice through integration of soil chemistry and plant physiology research. Plant Soil 361:3–41

Ishimaru Y, Bashir K, Naoko K, Nishizawa (2011) Zn uptake and translocation in rice plants. Rice 4:21–27

Johnson AAT, Kyriacou B, Callahan DL, Carruthers L, Stangoulis J et al (2011) Constitutive overexpression of the OsNAS gene family reveals single-gene strategies for effective iron- and zinc-biofortification of rice endosperm. PLoS One 6:e24476

Kobayashi Y, Sano Y, Vannoni E, Goto E, Suzuki H et al (2013) Genetic dissection of medial habenula–interpeduncular nucleus pathway function in mice front. Behav Neurosci 12:445–457

Kotla A, Agarwal S, Batchu A, Prasad A, Mallikarjuna S, Thingnganing L, Sarla N (2012) Evaluating rice germplasm for iron and zinc concentration in brown rice and seed dimensions. J Phytol 4:19–25

Kramer U, Clemens S (2006) Functions and homeostasis of zinc, copper, and nickel in plants, molecule. In: Tamas MJ, Martinoia E (eds) Molecular biology of metal homeostasis and detoxification—from microbes to man. Springer, Berlin, pp 216–271

Krishnaswami K (1998) Country profile: India. Nutritional disorders–old and changing. Lancet 351:1268–1269

Kumar S, Palve A, Joshi C, Srivastava RK, Rukhsar (2019) Crop biofortification for iron (Fe), zinc (Zn) and vitamin A with transgenic approaches. Helion 5:12–23

Lee S, An G (2009) Over-expression of OsIRT1 leads to increased iron and zinc accumulations in rice. Plant Cell Environ 32:408–416

Lee S, Jeon US, Lee SJ, Kim YK, Persson DP et al (2009) Iron fortification of rice seeds through activation of the nicotianamine synthase gene. Proc Natl Acad Sci USA 106:22014–22019

Li S, Jia J, Wei X, Zhang X, Li L, Chen H, Fan Y, Sun H et al (2007) Aintervarietal genetic map and QTL analysis for yield traits in wheat. Mol Breed 20:167–178

Lin YF, Liang HM, Yang SY, Boch A, Clemens S, Chen CC et al (2009) Arabidopsis IRT3 is a zinc regulated and plasma membrane localized zinc/iron transporter. New Phytol 182:392–404

Liu DJ, Wang YB, Guo CH, Cong Q, Gong X, Zhang HJ (2016) Enhanced iron and zinc accumulation in genetically engineered wheat plants using sickle alfalfa (*Medicago falcata* L.) ferritin gene. Cereal Res Commun 44:44–59

Lönnerdal B (2003) Genetically modified plants for improved trace element nutrition. J Nutr 133:1490–1493

Lucca P, Hurrell R, Potrykus I (2001) Genetic engineering approaches to improve the bioavailability and the level of iron in rice grains. Theor Appl Genet 102:392–397

Mantovi P, Bonazzi G, Maestri E, Marmiroli N (2003) Accumulation of copper and zinc from liquid manure in agricultural soils and crop plant. Plant Soil 250:249–257

Marenco R, Lopes N (2007) Fisiologia vegetal, 2nd edn. UFV, Viçosa, MG

Maret W, Sandstead H (2006) Zinc requirements and the risks and benefits of zinc supplementation. J Trace Elem Med Biol 20:3–18

Marschner H (1995) Mineral nutrition of higher plants, 2nd edn. Academic Press, London

Massot EP, Banakar R, Galera SG, Lopez UZ, Sanahuja G et al (2013) The contribution of transgenic plants to better health through improved nutrition: opportunities and constraints. Genes Nutr 8:29–41

Masuda H, Usuda K, Kobayashi T et al (2009) Overexpression of the barley Nicotianamine synthase gene *HvNAS1* increases iron and zinc concentrations in rice grains. Rice 2:155–166

Masuda H, Ishimaru Y, Aung MS, Kobayashi T, Kakei Y, Takahashi M, Higuchi K, Nakanishi H, Nishizawa NK (2012) Iron biofortification in rice by the introduction of multiple genes involved in iron nutrition. Sci Rep 2:543–549

Myers SS, Zanobetti A, Kloog I, Huybers P, Leakey AD, Bloom AJ et al (2014) Increasing $CO_2$ threatens human nutrition. Nature 510:139–142

Ogo Y, Itai RN, Kobayashi T, Aung MS, Nakanishi H, Nishizawa NK (2011) OsIRO2 is responsible for iron utilization in rice and improves growth and yield in calcareous sol. Plant Mol Biol 75:593–605

Perez-Massot E, Banakar R, Gomez-Galera S, Zorrilla-Lopez U, Sanahuja G, Arjo G et al (2013) The contribution of transgenic plants to better health through improved nutrition: opportunities and constraints. Genes Nutr 8(1):29–41

Petry N, Egli I, Gahutu JB et al (2012) Stable iron isotope studies in Rwandese women indicate that the common bean has limited potential as a vehicle for iron biofortification. J Nutr 142:492–497

Raboy V (2002) Progress in breeding low phytate crops. J Nutr 132:503–505

Raboy V (2003) Molecules of interest: myo-inositol-1, 2, 3, 4, 5, 6-hexakisphosphate. Phytochemistry 64:1033–1043

Ramesh SA, Choimes S, Schachtman DP (2004) Over-expression of an *Arabidopsis* zinc transporter in *Hordeum vulgare* increases short-term zinc uptake after zinc deprivation and seed zinc content. Plant Mol Biol 54:373–385

Rawat N, Neelam K, Tiwari VK, Dhaliwal HS (2013) Biofortification of cereals to overcome hidden hunger. Plant Breed 132:437–445

Rouached H (2012) Recent developments in plant zinc homeostasis and the path toward improved biofortification and phytoremediation programs. Plant Signal Behav 8(1):2681–2689

Septiningsih EM, Prasetiyono J, Lubis E et al (2003) Identification of quantitative trait loci for yield and yield components in an advanced backcross population derived from the *Oryza sativa* variety IR64 and the wild relative *O. rufipogon*. Theor Appl Genet 107:1419–1432

Shehu HE, Jamala GY (2010) Available Zn distribution, response and uptake of Rice (*Oryza sativa*) to applied Zn along a Toposequence of Lake Gerio Fadama soils at Yola, North-Eastern Nigeria. J Am Sci 6(11):121–132

Shigaki T, Barkla BJ, Miranda-Vergara MC, Zhao J, Pantoja O, Hirschi KD (2005) Identification of a crucial histidine involved in metal transport activity in the *Arabidopsis* cation/H+ exchanger CAX1. J Biol Chem 280:30136–30142

Stacey MG, Patel A, McClain WE et al (2008) The Arabidopsis AtOPT3 protein functions in metal homeostasis and movement of iron to developing seeds. Plant Physiol 146(2):589–601

Stangoulis JCR, Huynh B, Welch RM et al (2007) Quantitative trait loci for phytate in rice grain and their relationship with grain micronutrient content. Euphytica 154:289–294

Uauy C, Distelfeld A, Fahima T, Blechl A, Dubcovsky J (2006) A NAC gene regulating senescence improves grain protein, *Zn*, and *Fe* content in wheat. Science 24:1298–1301

Verret F, Gravot A, Auroy P et al (2004) Overexpression of AtHMA4 enhances root-to-shoot translocation of zinc and cadmium and plant metal tolerance. FEBS Lett 576(3):306–312

Wang N, Qiu W, Dai J et al (2019) AhNRAMP1 enhances manganese and zinc uptake in plants. Front Plant Sci 12:373–380

Weih M, Karlsson PS (2002) Low winter soil temperature affects summertime nutrient uptake capacity and growth rate of mountain birch seedlings in the subarctic. Swedish lapland Arctic Antarct Alpine Res 34:434–439

Welch R, Graham R (2004) Breeding for micronutrients in staple food crops from a human nutrition perspective. J Exp Bot 55:353–364

Winterbourn CC (1995) Toxicity of iron and hydrogen peroxide: the Fenton reaction. Toxicol Lett 1995(82–83):969–974

World Health Organization (2001) Iron deficiency anaemia. Assessment, prevention and control. WHO/NHD/01.3

World Health Organization (2014) Global nutrition targets 2025: anaemia policy brief WHO/NMH/NHD/14.4

Yadav K, Patel P, Srivastava AK, Ganapathi TR (2017) Overexpression of native ferritin gene *MusaFer1* enhances iron content and oxidative stress tolerance in transgenic banana plants. PLoS One 12(11):e0188933

Yamaji N, Takemoto Y, Miyaji T, Mitani-Ueno N, Yoshida KT, Ma JF (2017) Reducing phosphorus accumulation inrice grains with an impaired transporter in the node. Nature 541:92–95

Zhao FJ, Steve M (2009) Biofortification and phytoremediation. Curr Opin Plant Biol 12:373–380

Zhao FJ, Su YH, Dunham SJ, Rakszegi M, Bedo Z, Mcgrath SP et al (2009) Variation in mineral micronutrient concentrations in grain of wheat lines of diverse origin. J Cereal Sci 49:290–295

# Part III

# Toxic Metals in Crop Plants

# 11 Toxic Metals in Crops: A Burgeoning Problem

Amit K. Mishra, Jaswant Singh, and Pratyush Pingita Mishra

**Abstract**

Food safety is a priority concern and the Sustainable Development Goals (Goal No. 2) have been established to overcome all forms of hunger and malnutrition by 2030, both qualitatively and quantitatively. In recent years, the quality of crop has been adversely impacted by the different contaminants which resulted in poor food quality and also endangered the human health. Heavy metals and metalloids like Cadmium, Chromium, Arsenic, Lead and Mercury are among the significant contaminants of the environment. Though, some metal elements are crucial (in micro concentrations) for the plant growth. Conversely, if the concentration of these metals (micronutrients) reaches the threshold level in soil, can obstructs growth of the underground and above ground plant parts. Some plants develop metal specific defence mechanisms against these contaminants, in addition there are several other methods also available to clean up the environment from these toxins. One of the effective, affordable and eco-friendly way out for the remediation of toxic metals is phytoremediation techniques. In this chapter, we have discussed the difference between nutrient and contaminant, their accumulation in crops. It also describes the health effects and some standard equation for risk assessment. Finally, we have discussed some eco-friendly options such as phytoremediation techniques, source reduction, role of genetic engineering and application of nanotechnology. These methods are proposed for the management of toxic heavy metals from soil.

**Keywords**

Agriculture · Ecological risk index · Hazard quotient · Incremental lifetime cancer risk (ILCR) · Phytoremediation · Toxic metals

A. K. Mishra · J. Singh (✉) · P. P. Mishra
Department of Environmental Sciences, Dr. Rammanohar Lohia Avadh University, Ayodhya, Uttar Pradesh, India





K. Mishra et al. (eds.), *Sustainable Solutions for Elemental Deficiency and Excess in Crop Plants*, https://doi.org/10.1007/978-981-15-8636-1_11

## 11.1 Introduction

Environment is defined as an organism and their surroundings, which is the external abiotic factor that affects and influences the development, survival and growth of organisms. The environment comprises five main components which are air, water, land, noise, flora and fauna. The physico-chemical or biological changes in any of the component considered as pollution of the environment. A pollutant is a substance in the environment, which causes intolerable effects, damaging the different dimensions of environment, life quality degradation and may even cause death. Presence of any such pollutant in the environment that may be contaminant or toxic and will cause destruction to living organisms in the environment is referred as pollution. Heavy metals are the pollutants which are present in the environment beyond the set standard limit (Tangahu et al. 2011).

"Heavy metals" refer to any metal elements (also some metalloids, includes transition metals, lanthanoids and actinoids) with density greater than 4 g/cm$^3$ or greater than water, and even at very low concentration can cause toxicity (Hawkes 1997; Duffus's 2002; Vamerali et al. 2010). In the environment the major sources of heavy metals are from natural, atmospheric, agricultural, industrial, domestic effluent, etc. Heavy metal toxicity in plants depends on various factors such as plant species, metal form specificity and concentration, soil pH, etc., because some metals are considered to be essential nutrients for plant. Some heavy metals neither break down nor easily metabolize; hence, they may bio-accumulate.

Such metals accumulate in ecological food chain through uptake and consumption by primary producer and different trophic level consumers, respectively. Plants are more susceptible to heavy metals as they are stationary; roots of the plants are key site for accumulation of heavy metal ions (Nagajyoti et al. 2010). Over the past two decades, the term "heavy metal" has been used increasingly in the literature and in legislation related to chemical hazards and the safe use of chemicals.

## 11.2 Metals: Nutrients or Contaminants

Metals are natural components in soil, essential for plants growth at very low concentrations, such as Copper (Cu), Iron (Fe), Cobalt (Co), Manganese (Mn), Zinc (Zn), Molybdenum (Mo) and Nickel (Ni) are termed as micronutrients (Grotz and Guerinot 2006; Hansch and Mendel 2009), whereas Antimony (Sb) and Chromium (Cr) are vital for animals (Misra and Mani 1991; Markert 1993; Tangahu et al. 2011).

Two major roles of essential heavy metals are (1) involvement in redox reaction, and (2) play a foremost role in several enzymatic activities (Nagajyoti et al. 2010) such as Mn and Cu in photosynthesis, Zn in DNA transcription, Ni in hydrolysis of urea into carbon dioxide and ammonia, and Co, Zn in legume nodulation and nitrogen fixation (Vamerali et al. 2010). Due to their greatly reactive nature, these metals can directly influence the growth, senescence and energy generating

**Table 11.1** Metals (Micronutrients) and their role

| Metals | Functional role | References |
|---|---|---|
| Fe | Central ion in heme proteins such as in nitrate reductase, in cytochromes; nitrite and sulphite reductase in Siroheme proteins; ferredoxin in iron-sulphur proteins; etc. | Hansch and Mendel (2009), Grotz and Guerinot (2006), Welch (1995), Campbell (1999), Prescott and John (1996), Siedow (1991) |
| Mn | Vital for evolution of oxygen in photosystem II and also as carboxykinase, and superoxide dismutase in chain of enzymatic reactions, etc. | Lidon et al. (2004), Lanquar et al. (2010), Filiz and Tombuloglu (2015) |
| Cu | Present in plastidial plastocyanin, Cu-Zn superoxide dismutase with other proteins also | Redinbo et al. (1994), Bueno et al. (1995), Yruela (2009, 2013) |
| Zn | Crucial for carbonic anhydrase, Cu-Zn superoxide dismutase, metalloproteinase, etc. | Mishra et al. (2014), Hacisalihoglu et al. (2003), Delorme et al. (2000), Takatsuji (1998) |
| Ni | Essential role in urease activity | Witte (2011), Sirko and Brodzik (2000), Psaras et al. (2000) |
| Mo | Act as part of the molybdenum cofactor, vital for nitrate reductase activity, etc. | Mendel and Schwarz (1999), Mendel (2007), Schwarz et al. (1997) |
| Co | Significant for the microorganisms in symbiotic nitrogen fixation | O'Hara (2001), Jayakumar et al. (2008) |

processes (Shah et al. 2010; Page and Feller 2015). Important essential micronutrients along with their functional roles have been described in Table 11.1.

These metals are toxic to plants if their concentration in soil reaches beyond acceptable limits, they may interrupt the uptake of essential nutrients and cause oxidative stress through free radicals. It is well recognized that heavy metals need to be removed or converted into less harmful compounds as it cannot be degraded (Henry 2000; Gaur and Adholeya 2004; Prasad 2008).

Heavy metals such as Cadmium, Chromium, Arsenic, Lead and Mercury are of special concern due to their potential toxicity to both plants and animals even at low concentrations (Sharma et al. 1995; Das et al. 1997; Shukla et al. 2007). Hence, these metals can get bio accumulated in living organism are distinguished from other pollutants. They can cause various health issues even at very low concentrations (Pehlivan et al. 2009).

### 11.2.1 Cadmium (Cd)

The toxicity of Cd is not new; it is well established that cadmium is highly toxic and due to readily water-soluble properties it is easily available to plants (Pinto et al. 2004). According to Zhang et al. (2002), Wu and Zhang (2002), Cd interferes with Fe, Zn, Cu and Mn during the absorption process in some plants. Hegedus et al. (2004) investigated that Cd also enhanced production of reactive oxygen species by inducing lipid peroxidation and breakdown of chlorophyll in plants. Along with this Cadmium also hinders the uptake of K, Ca, Mg and Fe as it also uses the similar transmembrane carriers. Accumulation of cadmium in plants may pose a very serious threat to human health and bio accumulated as it goes to higher trophic level of food chain; though, the direct consumption of Cd-contaminated soil also poses other risk to children (Rivetta et al. 1997; Nordberg 2003).

The great example of Cd poisoning is itai-itai disease that occurred in a town of Japan. A combination of symptoms of mass due to Cd toxicity such as Hematopoeisis, Osteomalacia, calcium malabsorption and renal tubular dysfunction is referred as itai-itai disease (Ogawa et al. 2004; Bernhoft 2013).

### 11.2.2 Chromium (Cr)

Cr is considered to be the seventh most abundant element on earth (Cervantes et al. 2001). Although Cr was found to be a stimulant for the growth of plant, however, several researchers around the globe reported its toxic effect on plants. Chromium has many oxidation states ranging from $Cr^{2-}$ to $Cr^{6+}$; though, Cr ($^{3+}$) and Cr ($^{6+}$) are of most concern (Krishnamurthy and Wilkens 1994). Becquer et al. (2003) reported that Cr ($^{6+}$), as chromate ($CrO_4{}^{2-}$) or dichromate ($Cr_2O_7{}^{2-}$), is more toxic because of its mobile nature in soil, while Cr ($^{3+}$) is less toxic, less mobile and is mainly combined with organic matter in soil. It is very difficult to distinguish the effect of trivalent and hexavalent form of Cr on plant, as both the species may interconvert in soil. Therefore, it is commonly referred as chromium toxicity in plants, instead of toxicity of trivalent or hexavalent. Chromium toxicity in plants can be identified by various visible symptoms such as leaf chlorosis, stunting and reduction in yield (Barcelo et al. 1993; Das et al. 1997; Boonyapookana et al. 2002).

### 11.2.3 Arsenic (As)

Arsenic is found in the environment which is extremely toxic to all the living organisms (Chutia et al. 2009). The IARC (International Agency for Research on Cancer) recognized Arsenic and As compounds as group 1 carcinogens. The contamination of the environment by As is attributed to its release via both natural and anthropogenic activities (States 2015; Shukla and Srivastava 2017). As is an extremely toxic element that exists in different form, and its toxicity depends on available form of Arsenic.

The inorganic species of arsenic are arsenite ($As^{3+}$), and arsenate ($As^{5+}$), generally considered the key species in the environment (Andrianisa et al. 2008). The pH, surrounding mineral composition and microbial activities, affects the oxidation state and forms (inorganic or organic) arsenic. The accumulation of arsenic in plants affects the growing mechanisms of crops, yield of crops, and may impact on health of lives (Rosas et al. 2016). There are many studies about the mobilization and accumulation of arsenic in soil. The $As^{3+}$ (arsenites) are more toxic (4–10 times more soluble in water) than the $As^{5+}$ (arsenates) (Ampiah-Bonney et al. 2007; Vaclavikova et al. 2008).

### 11.2.4 Lead (Pb)

In the environment, Pb is well-known toxic element for living organism. There are different forms of lead existing in the nature and at present one of the most common trace metal (European Commission 2002). Car exhaust, dust, gases from various industries, leaded fuels, old lead plumbing pipes are different sources of lead in the environment that contaminate the soil and plants. Lead is a long-term contaminant, highly immobile and holds in the upper 8 in. of the soil. In the environment Pb exists as an insoluble form, and poses various serious health issues, such as damage or retardation of brain (Cho-Ruk et al. 2006). The high soil lead concentrations need remedial action to overcome its adverse effect (Traunfeld and Clement 2001).

### 11.2.5 Mercury (Hg)

Mercury is highly persistent pollutant that occurs in different form in the environment with great bioaccumulation ability (Chang et al. 2009). The three soluble forms of mercury in the soil are $Hg^{\circ}$ metal, mercurous ion $Hg_2^{2+}$ and mercuric ion $Hg^{2+}$. The most poisonous form of mercury in the environment is Mercury salts and organomercury compounds. There are various sources of mercury contamination in the environment, such as mining, several industries, petrochemicals, medical instruments, batteries, fertilizer and fungicidal sprays and some domestic products (Resaee et al. 2005). However, various products are restricted or banned, but still there are some existing, older products in use (Musselman and QEP 2004). By interfering electron transport pathway in chloroplasts and mitochondria, mercury affects process of photosynthesis and oxidative metabolism in plants. Mercury obstructs the plant water uptake and reduces the action of aquaporins. The most extreme instance of Hg toxicity is Minamata disease, a great public health hazard happened in Japan and in Iraq (Bakir et al. 1973; Sas-Nowosielska et al. 2008; Reilly et al. 2012; Bernhoft 2012).

## 11.3 Heavy Metals in Agriculture

Atmospheric deposition, livestock compost, irrigation with wastewater, sewage sludge-based amendments, metal-based herbicides, pesticides and phosphate-based fertilizers are the sources of metals in environment and agriculture (Fig. 11.1) (Cai et al. 2009; Mansour 2009; Elgallal et al. 2016; Woldetsadik et al. 2017).

Some heavy metals like Fe, Cu, Cr, Pb, Ni and Mn are found in significantly greater concentrations in sludge from distilleries, electroplating and chemical industries, textile and leather (Chandra and Kumar 2017). Pandey (2006) showed

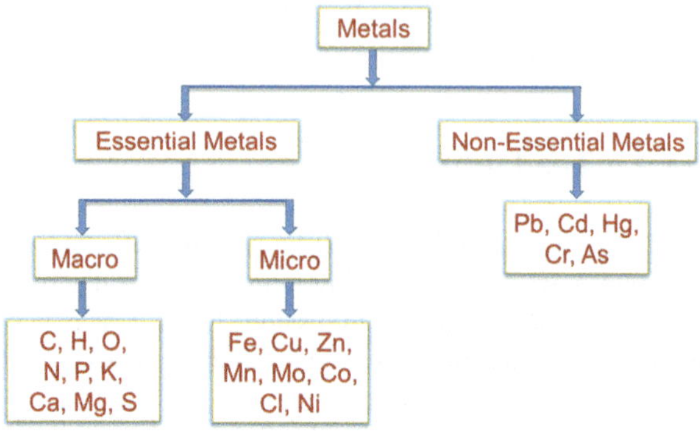


**Fig. 11.1** Metals in agriculture

that electroplating effluent that contains cadmium, nickel, chromium, copper and zinc could have severe effects such as inhibition of plant growth, chlorosis and necrosis in leaves and even can cause plant death. A study conducted in China shows that particulate matter adhere the metals released from the Pb-acid battery manufacture factories were deposited in soils and on leaves of crops. Phosphogypsum found in phosphate fertilizer waste can produce various metals in soil and crops. Another study shows that Cd and Cr were mostly present in soil, whereas Pb has been found highest in green peppers and tomatoes (Liu 2014; Al-Hawati and Al-Khashman 2015; Franca and Albuuerque 2017).

Air, Water, soil and sediments are the carriers of heavy metals in the environment where they could be immobilized. The physico-chemical properties of soil such as pH, cation exchange capacity, organic matter, redox potential, texture and clay content can influence the metal solubility and bioavailability (Logan and Chaney 1983; Verloo and Eeckhout 1990; Ozturk et al. 2008). Generally, the soil with high pH and higher clay or organic matter causes lesser bioavailability as the metals get bound firmly to soil increasing the residence time. Another important factor is soil temperature which causes the variations of metal accumulation by crops (Chang et al. 1987). Halim et al. (2003) found that organic matter forms metal complexes and whenever they get dissolved the soil solution increases the availability of metals.

## 11.4 Heavy Metals: Accumulation and Uptake

Several researchers have explored the mechanism behind the contaminants uptake by plants. Sinha et al. (2004) reported that the plants can act as excluders and accumulators. Accumulators can survive even after the toxins in their aerial tissues get concentrated. In their tissues they convert the toxins into inert forms by the process of biodegradation or biotransformation. And the excluders check the uptake of toxin into their biomass (Rai et al. 2019). As heavy metals mobilize in soil, root cells of the plants capture them, their movement depends primarily upon, (1) diffusion of metal along the concentration gradient; (2) roots interception, where root volume causes the displacement of soil volume and (3) flow of metal elements from bulk soil solution down the water potential gradient (Marschner 1995).

With increasing concentration of metals in the external medium the metal uptake increases both by roots and leaves. However, the uptake and increasing concentration have no linear relation. At low concentrations of metal in the external medium, the uptake efficiency of metals by the plants (or accumulation factor) is highest. Plant leaves can uptake both essential and non-essential metals, which can enter the leaves in the form of gases through the stomata, whereas the leaf cuticle allows the ionic form of metals (Lindberg et al. 1992).

Evapotranspiration is a process in which water evaporates from plant leaves, this results into the driven force behind the absorption of nutrients and other soil constituents by plant roots. In the meantime the contaminant is translocated from roots to shoots, which is harvested at the end and the contamination of soil reaches to consumer level.

On the basis of pollutant concentration in shoot biomass, Baker and Walker (1989) classified plants into three different groups, excluders, indicators and accumulators, with lower, similar and higher concentration of pollutants than that of the soil. Various defence mechanism gets activated when non-essential form of metals enters plants. Plants detoxification processes include reduced translocation compartmentalization, chelation with biotransformation and phytochelatins (Tomsett et al. 1992; Lasat 2002; Cobbet 2000; Angelova and Ivanov 2009).

## 11.5 Heavy Metals: Effects on Growth and Development

Roots are the main organ in plants that contact with toxic metals and accumulate more as compared to shoots (Salt et al. 1995; Wojcik and Tukiendorf 1999; Rout et al. 2001). The toxicity of metal differs with the type of metal elements. Heavy metals check either the entire plant growth or partially some plant parts (Shafiq and Iqbal 2005; Shanker et al. 2005). The parts of plant which establish direct contact with the soiled soils normally the roots, expresses rapid and sensitive changes in their growth pattern. The significant effects of a number of metals (Cu, Ni, Pb, Cd, Zn, Al, Hg, Cr, Fe) on the growth of shoots are well documented (Wong and Bradshaw 1982).

The generation of free radicals and reactive oxygen species from the heavy metals affect the growth of plants by deteriorating the important cellular components and posing constant oxidative damage (Pandey et al. 2005; Qureshi et al. 2005). Early seedling growth and germination of seed are sensitive to a certain extent towards fluctuating environmental conditions (Seregin and Ivanov 2001). The ability of plants towards tolerance of metal elements is often assessed by the germination performance and seedlings growth rate (Peralta et al. 2001). The higher concentrations of heavy metals obstruct seed germination and primary growth of rice, wheat and barley seedlings (Mahmood et al. 2007).

The metal also adversely affects the height and shoot growth of the plant (Rout et al. 1997). Reduction in plant height might be due to the nutrients and water check to the upper parts of the plant. Some morphological changes may also be caused by some metal like Cd such as necrosis and chlorosis of younger leaves, and drying of older leaves. Heavy metals can impact human health also along with plants. Dietary intake of these toxic metals can accumulate in fatty tissues and bones of humans leading to deteriorate immunological defences and reduction of essential nutrients (Iyengar and Nair 2000; Khan et al. 2010; Rai 2018).

The ingestion of contaminated crop causes serious health issues to all the living organism including humans, such as malnutrition, mental growth retardation, gastrointestinal cancer, fragile immunological mechanisms, etc. (Carrizales et al. 2006; Hu et al. 2013; Gress et al. 2015; Dickin et al. 2016). Most of the heavy metals are bio-accumulative and through the food chain can be transferred to the other media. Details of various health risks due to different heavy metals or metalloids have been described in Table 11.2.

## 11.6 Heavy Metals: Health Risk Assessment Indices

The human health risks that result from the ingestion of metals contaminated food crops have been assessed by different health hazard indices. Various indices are used to evaluate the metal transfer from soil to plant and associated health risk (Yang 2018). Study of Zhao (2014), on health risks, particularly heavy metal–induced cancer illustrate that arsenic, lead, mercury, chromium and cadmium had Target Hazard Quotient (THQ) values $>1$ in food crops, whereas the cancer of liver and gastric were caused by Pb and Hg, respectively.

Shaheen et al. (2016) conducted a study in 30 agro-ecological areas in a developing country on food crops intake to define the health indices. The results shown that the intake of heavy metals contaminated vegetables (mainly Cu) was more harmful to human health as compared to the ingestion of contaminated fruits. Similarly, the report of Obiora (2016) shows that lead can lead to Alzheimer's disease from the vegetables grown near a mine of lead and zinc. Further, Cui et al. (2005) found many people of a population suffer from renal dysfunction who consumed multiple metals contaminated foods.

Different indices which are being used widely by the researchers all around the globe to determine the ecotoxicological effects and health risks from the intake of contaminated food crops are as follows:

### 11.6.1 Bioconcentration Factor (BCF)

Bioconcentration factor somewhere also referred as Bioaccumulation factor (BAF) (Zhuang et al. 2009) is a process by which a metal element is accumulated into biota from its adjacent phases. The BCF is the proportion of the metal concentration of the crop/biota and the metal concentration of the soil/sediment (Mountouris et al. 2002; Sipter 2008). The BCF is a key tool used to calculate the transfer of hazardous toxins from soil or sediment to plants (Melgar et al. 2009).

**Table 11.2** Health risks from the consumption of toxic metal contaminated food

| S. no. | Heavy metal/ metalloid | Essential for plant | Essential for human | Health risks |
|---|---|---|---|---|
| 1 | Mercury | No | No | Damages lung and kidney, proteinuria, allergy, amalgam disease, hypersensitivity nephrotic syndrome |
| 2 | Cadmium | No | No | Adversely affects kidney functioning, inhibition of sex hormones, endocrine disruption, human carcinogen (group I) causing cancer of lung and breast, long-term exposure can result in itai-itai disease (Japan's incident) |
| 3 | Lead | No | No | Causes neurological, cardiovascular diseases in humans especially children, anaemia, abdominal pain, possible human carcinogen, and also affects mental growth |
| 4 | Copper | Yes | Yes | Higher concentration can induce liver damage and other gastric-related problems in humans, affect renal and metabolic functions |
| 5 | Chromium | No | Yes | Kidney/renal dysfunction/failure, haemolysis and gastrointestinal haemorrhage, collapse or dysfunction of respiratory system through lung cancer and pulmonary fibrosis |
| 6 | Nickel | Yes | Yes | Can affect renal functioning, integral component of urease enzyme in kidney, remarkable decrease in body and organ weights |
| 7 | Arsenic (metalloid) | No | No | Can cause cancer, dermal problems, respiratory complications, many diseases in cardiovascular, hepatic, Haematological gastrointestinal, renal, immune systems, neurological, reproductive developmental, and, multi-organ dysfunction, bone marrow depression, "rice-water" diarrhoea |
| 8 | Zinc | Yes | Yes | Respiratory problems, significant decrease (47%) in erythrocyte and concentration of superoxide dismutase in adult females |

Source: Rai et al. (2019) and Vamerali et al. (2010)

BCF can be calculated as follows:

$$\mathrm{BCF} = \frac{C_{\mathrm{crop/biota}}}{C_{\mathrm{soil/sediment}}}$$

where $C$ is the concentration of metal element in crop/biota corresponding to metal concentration in soil/sediment. All the values are in (mg kg$^{-1}$) (Chang et al. 2014; Zhao 2014).

A number of studies have used this tool and validated that the leafy vegetables show high value of BCF followed by tuberous crops, whereas the lowest values were in horticulture fruits and crops (Garcia et al. 2009; Liu et al. 2012; Yang et al. 2014).

### 11.6.2 Pollution Load Index (PLI)

Pollution Load Index, a quantitative assessment tool for soil pollution in terms of heavy metal, was developed by (Thomilson et al. 1980). PLI can be obtained as Contamination Factors (CF), which is calculated by deriving the nth root of the *n* factors. (nCFs). Therefore, the PLI for each sample can be calculated using:

$$\mathrm{PLI} = (\mathrm{CF}_1 \times \mathrm{CF}_2 \times \mathrm{CF}_3 \times \mathrm{CF}_n)^{1/n}$$

where CF is calculated as

$$\mathrm{CF} = \frac{\text{Concentration in Sample}}{\text{Concentration in Background}}$$

In other words it can be said as the ratio between metal in the soil and its background/baseline value (control or reference value) (Galan et al. 2002; Mmolawa et al. 2011).

The PL index is scored using a seven-point scale ranging from 0 to 6: 0 = none, 1 = none to medium, 2 = moderate, 3 = moderate to strong, 4 = strongly polluted, 5 = strong to very strong and 6 = very strong. This index provides an estimate of the metal concentration status, used in several studies by different researchers all around the globe for assessing heavy metal pollution (Rashed 2010; Sulaiman et al. 2019; Rai et al. 2019).

## 11.7 Ecological Risk Index (RI)

This terminology is used to recognize the Potential Ecological Risk Index developed by Hakanson (1980). This index have been used in many studies to assess the degree of pollution in sediment/soil and also risk factors for all hazardous heavy metals in a soil sample and the biotic responses (Effendia et al. 2016; Sulaiman et al. 2019). Ecological risk potential index of heavy metals can be calculated with the following formula:

$$C^i_f = \frac{C_n}{C_{\mathrm{nr}}}$$

$$E^i_f = T^i_r C^i_f$$

$$\mathrm{RI} = \sum E_r^i$$

where $C_f^i$ = single-metal pollution factor; $C_n$ = concentration of the given metal in sample; $C_{\mathrm{nr}}$ = reference value of metal; $E_r^i$ = potential ecological risk index of individual metal and $T_r^i$ = toxic factor of individual metal

The factor scores of different heavy metals according to Hakanson (1980) were As (10), Cd (30), Cr (2), Cu (5), Pb (5), Ni (5) and Zn (1). Some other studies have also given the individual metal toxic factor such as (5) Pb and Cu, (1) Zn, (30) Cd, (2) Cr and (6) Ni (Rai et al. 2019). $T_r^i$ is assumed to be 6 (Gan et al. 2000) and according to Chabukdhara et al. (2016) the value of RI is considered low (<50), moderate (50–100), considerable (100–200) and intense/high (>200).

## 11.8 Different Health Risk Assessment Indices

Health risk assessment is an evaluation of the probability of adverse health effects, for those who had exposure to contaminated environmental media (Health 1983; Wang et al. 2015). Heavy metals exposure depends mainly on the metal characteristics. Inhalation, Oral Ingestion and the Dermal Contact are the three major pathway by which humans get exposed to heavy metals (Li et al. 2017; Asaduzzaman et al. 2017).

Hazard Quotient (HQ), Daily Dietary Index (DDI), Health Risk Index (HRI), Hazard Index (HI), Daily Intake of Metals (DIM) are the different indices widely used in numerous study to evaluate the transfer of contaminants from food crops to humans (Hough et al. 2004; Pruvot et al. 2006; Khan et al. 2008; Zhuang et al. 2009; El-Kady and Abdel-Wahhab 2018).

## 11.9 Hazard Quotient

Non-carcinogenic hazards are generally characterized by HQ, a proportion of the chronic daily intake divided by the maximum acceptable dose of a toxic substance, which is a reference dose of a specific metal. Hazard quotient is characteristics of human health risk due to consumption of food crop contaminated with metal (Chien et al. 2002; Wang et al. 2005; Khan et al. 2009).

$$\mathrm{HQ} = \left[W_{\mathrm{plant}}\right] \times \frac{\left[M_{\mathrm{plant}}\right]}{R_f D \times B}$$

where $W_{\mathrm{plant}}$ = dry weight of contaminated plant part consumed (mg day$^{-1}$), $M_{\mathrm{plant}}$ = concentration of the metal in the vegetable(s) (mg kg$^{-1}$), $R_f D$ = food reference dose (toxicity threshold value of a substance) of the metal (mg day$^{-1}$) and $B$ = human body mass (kg)

HQ < 1 is safe, while HQ ≥ 1 may possibly pose a health risk (Al-Saleh et al. 2004; Cao et al. 2010; Wang et al. 2015).

## 11.10 Daily Dietary Index

Another index related to health risk is Daily Dietary Intake. As food crops are adulterated by heavy metals, therefore, the estimation of daily intake contaminated vegetable(s) needs to be assessed by the following formula:

$$\text{DDI} = \frac{\text{Metal in vegetable} \times \text{dry wt.of the vegetable} \times \text{approximate daily intake}}{\text{Average body mass}}$$

DDI provides the amount of a metal consumed by an individual on daily basis. This tool is also used in various study for the evaluation of metal consumed through dietary intake (Cui et al. 2004; Cui et al. 2005; Chary et al. 2008; Khan et al. 2009; Gall et al. 2015). The data required in this formula are generally collected by a questionnaire of the concerned area.

## 11.11 Daily Intake of Metals

DIM may also be used to access the individual metal concentration using following equation (Cui et al. 2004; Khan et al. 2010).

$$\text{DIM} = \frac{C_{\text{metal}} \times C_{\text{factor}} \times D_{\text{food}}}{B_{\text{average food}}}$$

where $C_{\text{metal}}$ = concentration of heavy metal in plants (mg kg$^{-1}$), $C_{\text{factor}}$ = conversion factor (fresh vegetable weight to dry vegetable weight), $D_{\text{food}}$ = daily intake of vegetables and $B_{\text{average food}}$ = average weight of the consumers.

The average daily intake of vegetable for adults 0.345 kg/person/day and for children 0.232 kg/person/day were considered, whereas the average body weights for adult 73 kg and children 32.7 kg, respectively (FAO 2000; Ge 1992; Wang et al. 2005). The proposed value of conversion factor is 0.085 used to convert the fresh vegetable weight to dry weight (Rattan et al. 2005).

## 11.12 Health Risk Index

The values of DIM and the reference dose (RfD) of food are put together to get the health risk index. HRI is assessed on the basis of intake of contaminated vegetables of the populations. The following equation is used for the calculation of HRI (USEPA 2002).

$$\mathrm{HRI} = \frac{\mathrm{DIM}}{\mathrm{RfD}}$$

From the literature the reference dose values for Zn, Ni and Cu were used as 0.3, 0.02 and 0.5 mg $kg^{-1}$/body weight $day^{-1}$, respectively (IRIS 2003; WHO 1982; Alam et al. 2003). In general, HRI value $<1$ indicates that the exposed population is considered safe, whereas HRI value $\geq 1$ could pose a health risk over multiple metals (IRIS 2003; Oves et al. 2012; Khan et al. 2013).

## 11.13 Hazard Index

To evaluate the overall non-carcinogenic health risk posed by more than one metal, a Hazard Index could be used. As there are many ways of exposure of multiple metals at a time, hence, a total exposure hazard index may possibly use to connect all the non-cancer risks caused by different pathways. The hazard index can be calculated as (Cao et al. 2010; Gall et al. 2015):

$$\mathrm{HI} = \mathrm{HQ1} + \mathrm{HQ2} + \ldots + \mathrm{HQn}$$

where HQn is the hazard quotient for one particular metal.

If the HI value is $<1$, the exposed inhabitants are suspected to experience adverse health effects, while greater than one HI value may possibly cause potential health effects from combined metals (Man et al. 2010; Gall et al. 2015).

Wilbur et al. (2004) explained that this method does not consider chemical interactions of the mixture metals. Therefore, the finding of HI may miscalculate or overestimate health hazard, if any kind of interferences is present. On the other hand a fair estimation of hazard requires same mechanism for all component (Cao et al. 2010).

## 11.14 Incremental Lifetime Cancer Risk (ILCR)

The heavy metal contamination exposure through different pathway, such as ingestion, inhalation, dermal and diet exposure, consists of carcinogenic as well as non-carcinogenic risk. Potential carcinogenic health risk of metal contaminant can be assessed by using ILCR value (Liu et al. 2013; Gomah et al. 2019).

$$\mathrm{ILCR} = \mathrm{CDI} \times \mathrm{CSF}$$

where CDI = chronic daily intake of a metal in (mg $kg\text{-}^{1}$) and CSF = cancer slope factor.

Chronic Daily Intake in (mg/kg/day) can be calculated from the following equation (WHO 2011; USEPA 2001):

$$CDI = {}^{(EDI \times EFr \times EDtot)}/_{AT}$$

where EDI = estimated daily intake; EFr = frequency of exposure; EDtot = total duration of exposure; AT = period of exposure.

Zhao (2014) and Rai et al. (2019) have reported that this tool is very useful for the assessment of cancer risk and associated other relevant factors. Li et al. (2011) have used this tool to understand the relation of arsenic exposure and cancer risk. ILCR helps to establish that the exposed population has greater cancer risk.

## 11.15 Heavy Metals: Management in Agriculture System

Human health and food safety are of a global importance which is at present threatened by heavy metals coming from irrigation of wastewater, sludge application and effluents from different industries. Therefore, the soil–crop system can be prevented by removal of heavy metal in soil. The soil to crops mechanisms are well understood for the heavy metals translocation. Reducing concentrations of heavy metals in soil to reduce the subsequent transfer to crops can be done directly by enforcing remediation efforts. Remediation technologies should be environment friendly, rapid and cost-effective. Different physical, biological, ecological and chemical approaches can also be conducted for the removal of heavy metal in soil (Rai et al. 2019).

## 11.16 Phytoremediation: A Green Solution

Ilya Raskin (1996) coined the term "phytoremediation" from Greek word "Phyto" (means plant) and Latin suffix "remedium" (means able to cure or remove an evil). Phytoremediation refers to reduction of the concentrations of contaminants using plants and associated soil microbes (Greipsson 2011). Phytoremediation can be used to mitigate the heavy metals, radionuclides and organic pollutants (Clemens 2001; Chehregani and Malayeri 2007; Vithanage et al. 2012). It is an eco-friendly, efficient and cost-effective remediation strategy (Turan and Esringu 2007; Lone et al. 2008; Saier and Trevors 2010; Sarma 2011; Kalve et al. 2011). The installation and maintenance costs of this techniques are also very low compared to other mitigation approaches (Van Aken 2009).

Phytoremediation is generally a group of numerous processes such as phytoextraction, phytostabilization, phytodegradation, rhizofiltration and phytovolatilization. Among all phytoextraction and phytostabilization are the most consistent for heavy metals (Adesodun et al. 2010; Shabani and Sayadi 2012).

### 11.16.1 Phytoextraction

Phytoextraction (also known as phytoabsorption or phytoaccumulation) is the uptake or absorption of contaminants by plant roots and translocation into the above ground biomass of the plants (Sekara et al. 2005). Finally, the plant can be harvested and burned to obtain energy and the metals get recycled from the ash (Yoon et al. 2006; Rafati et al. 2011). The translocation of metals to upper plant parts is a key biochemical process and is also essential for an effective phytoextraction (Zacchini et al. 2009; Tangahu et al. 2011). Plants like *Trifolium sp.* which gives many harvests in one growth period can have a great potential for the elimination of heavy metals using phytoextraction (Ali et al. 2012). Ali et al. (2013) also used Maize and Barley for the phytoextraction of heavy metals. Further Vamerali et al. (2010) suggested that the crops used for this purpose should not be considered for animal or human use.

### 11.16.2 Phytostabilization

Phytostabilization is a process where certain species of plants are used for immobilization of the contaminants at polluted sites. This process involves the accumulation of contaminants in plant tissues, absorption onto root surface and precipitation in rhizosphere of certain plant species which reduces the bioavailability of metals (Mendez and Maier 2008). This method can help to restore the contaminated sites with high metal concentrations where the natural vegetation is not possible. Phytostabilization also used to decrease leaching, run off and erosion as the plant roots fixes the soil (Berti and Cunningham 2000; Bandara et al. 2015). Several plants species have been specified in different studies for this techniques such as *Lemna minor, Helianthus annuus, Zea mays, Nicotiana tabacum, Trifolium repens* and the metal contaminants removed by them are Pb, Cu, Cd, Cr, Se, As, Zn, Ni and Fe (Redjala et al. 2011; Chen et al. 2012; Mukherjee et al. 2013; Udoka et al. 2014).

### 11.16.3 Rhizofiltration

Rhizofiltration is a process in which terrestrial and/or aquatic plants are used to eliminate the toxicants by the help of massive root. In this process the contaminants remain absorbed onto plants, adsorbed into roots, and sequesterize in the roots. Rhizofiltration is largely used to remediate wastewater, groundwater and surface water, with relatively lower concentrations (Ensley 2000). This technique can be used for remediation of metals such as lead, cadmium, copper, nickel, zinc and chromium, mainly retained within the roots (United States Protection Agency 2000). The ability to compatible for terrestrial as well as aquatic plants either in-situ or ex-situ applications makes rhizofiltration one of the most appropriate remediation method among other phytoremediation approaches (Jadia and Fulekar 2008).

For the removal of toxicants from water numerous plants species have been studied such as spinach, sunflower, tobacco, Indian mustard, rye, corn, etc. Study of Raskin and Ensley (2000) showed that, Indian mustard has great efficiency for the remediation of varied range of lead concentration.

### 11.16.4 Phytodegradation

Phytodegradation is the breakdown of complex organic pollutants into simpler and lesser toxic form by plants. Plants carry out this breakdown either by metabolic process or with the help of some enzymes like dehalogenase, reductases and oxygenase (Vishnoi and Srivastava 2008). The breakdown of organics in the soil through rhizospheric microbes is known as rhizodegradation; this process is much slower than phytodegradation (Ghosh and Singh 2005). The pollutants that are broken down into less complex products are utilized by the plant for quicker growth (Newman and Reynolds 2004). Many studies have reported that different plants have different phytodegrading abilities, for example, the use of genetically modified plants (Transgenic Poplars) has been reported for this purpose (Doty et al. 2007; Khandare and Govindwar 2015). There are numerous factors which affect the effectiveness of this practice such as concentration of pollutants in the soil and water, its uptake efficiency by plants, etc.

Various plants have been identified for phytodegradation method of removing toxic metals such as *Pteris vittata, Brassica juncea, Liriodendron tulipifera, Azolla caroliniana, Arundo donax, Cyperus papyrus and Typha angustifolia.* Mercury, methyl mercury and Arsenic can be removed by this technique (Bennicelli et al. 2004; Jomjun et al. 2010; Greipsson 2011).

### 11.16.5 Phytovolatilization

Uptake of a contaminant and transpiration by the plants are known as phytovolatilization process. As plants grow it releases the modified or less toxic form of the contaminant to the atmosphere (Moreno et al. 2004). Growing plants take up the water along with the contaminants, then pass the contaminants to the leaves and finally volatilize at relatively low concentrations (Muthusaravanan et al. 2018). Se, As and Hg are the toxic heavy metals which can be converted to dimethyl selenide and mercuric oxide and then evaporate.

Some plants species have been reported for this technique like *Typha latifolia, Sorghum bicolor, Picea abies, Populus tremula, Solanum nigra,* etc. and the metals removed by them are Zn, Mn, Co, Cd, Cr, Ni, Pb and As (Yang et al. 2014; Pastor et al. 2015; Zhao et al. 2016; Shackira and Puthur 2017; Radziemska et al. 2017).

## 11.17 Source Reduction

Reduction of heavy metals concentration at source is an effective approach for refining human well-being. The accumulation of heavy metals in food crops could be reduced significantly, if the use of inadequately treated effluent as well as sewage sludge for irrigation process is checked.

A comparative assessment of heavy metals in the soil samples and human hair (who consumed food grown in fields irrigated with wastewater and clean water) were conducted by Massaquoi (2015). They have reported that metal concentration in samples from wastewater-irrigated fields was considerably greater than the clean water irrigated fields. McLaughlin (1999) reported that concentration of Cd can be effectively decreased by reducing the contamination at sources, while plant advancement of breeding and agronomics can also control soil–food crop transfer.

Liao (2016) has found that road-side dust/particulate matter originated from urban traffic acts as a precursor of heavy metals in crops grown within 30 m of road-side. As the PM get deposited on the leaves of plant so, the road-side crops field are prone to take up the metal (e.g., lead in the leaves of *Amaranthus dubius*) (Nabulo 2006). The management of air-quality possibly reduces the concentration of PM, hence, decreases the deposition of these particles in the soil and crop field.

## 11.18 Application of Genetic Engineering

The aim of Genetic engineering technology is to alter the tolerance limit of plants, accumulation and metabolization of pollutants capacity. According to Ehrlich (1997), many genes of bacteria and yeasts have been identified and characterized for the acquisition, allocation and detoxification of metals. For the overproduction of recombinant proteins, the transgenic plants have been engineered to play significant role in transport of heavy metals along with chelation and assimilation. Metal chelating proteins such as citrate, metallothioneins, phytochelatins, phytosiderophores and ferritin have been produced to enhance the tolerance and accumulation.

Boosted aluminium tolerance can be achieved by increasing the activity of organic acid synthase. A full-length OsCS1 gene has been isolated for citrate synthase, highly induced by Aluminium toxicity in *Oryza sativa*. Increased citrate efflux conferred great tolerance to aluminium have been achieved by insertion of OsCS1 in numerous independent transgenic tobacco plants (Han et al. 2009). The overexpression of gamma-glutamylcysteine synthetase or glutathione synthetase in transgenic *Brassica juncea* (L.) Czern has been responsible for accumulation of metals such as Cadmium and Chromium (Reisinger et al. 2008).

Pavlikova et al. (2004) made an effort to improve the tolerance to cadmium, zinc and nickel by introducing a metallothionein gene in tobacco plant. Macek et al. (2002) also showed that accumulation of Cd has been increased significantly, by the tobacco plants containing transgene coding for the cluster of polyhistidine combined with yeast metallothionein.

## 11.19 Application of Nanoparticle Techniques

Nanoparticle techniques have been used remarkably by researchers for confirming soil security and for the reduction of metal's bioavailability as an important part of agro-nanotechnology (Shalaby 2016; Rai 2018). The recent advancement in molecular biology of plants through genetic engineering and phytosynthesis of nano-scale materials have also been emphasized for their controlled production in remediation (Kostal 2005).

Rai (2018) has stated that NP techniques and green chemicals are primarily used for the sustainable practices of agriculture and also for human health. Moreover, for assessing the level of toxicants in food crops and food safety analysis nano-sensors are applied (Kuswandi 2017). Nano-tools for remediation of metal contamination in soil might be cost-effective. The diverse nano-technologies must be demonstrated in case of formulation of varied pesticides that can make wastewater and sludge less toxic for food crops (Hazra et al. 2017).

Yousaf (2018) has observed that, wheat grown in polluted soil near industry has shown reduced bioavailability of carcinogenic metals by adsorption through biochar nanosheets. However, in rice, silica NPs help in inhibiting gene expression associated with the production of cadmium transporters (OsHMA3) that results in more Cadmium toxicity (Cui 2017). Therefore, clear understanding of the fate and adverse effects of NPs in the environment and food crops need urgent address.

## 11.20 Conclusion

Several heavy metals play a vital role for different biological functions such as nucleic acids and proteins biosynthesis, chlorophyll and secondary metabolites synthesis, etc. Some specific heavy metals also play an important physiological role such as stress tolerance, structural and functional integrity of various membranes and other cellular compounds. However, beyond tolerable limits, these metals become toxic depending upon the nature and species of metal and plants. In recent decades heavy metals concentration in the environment has increased significantly. Heavy metal transfer from soil to crop systems is complex and uses multi-faceted mechanisms.

Till now only few of the research work have properly used the epidemiological methods to define the human health risks globally. The existing remediation techniques are mainly used to mitigate heavy metal concentration in soil and crop to prevent the associated health risks. The phytoremediation technology, in which plants are used to uptake the metal, looks to be a successful way to mitigate heavy metals from the contaminated environment.

The phytoremediation technique has some advantages as compared with other conventional technologies. Even the phytoremediation technique seems to be one of the best alternatives, it also has some limitations. In-depth research needs to be conducted to minimize this limitation in order to apply this technique effectively.

To formulate appropriate remediation techniques and to prevent contaminant transfer into the food chain, rapid and precise mapping of soil contamination is required. For moderately contaminated soil, biological remediation, phytoremediation and PGPR can be an environment friendly and cost-effective options. To boost the local economies and livelihood with certain economical guarantees, some of the eco-feasible technologies like nanoparticle technology and awareness programs for farmer could be used.

## References

Adesodun JK, Atayese MO, Agbaje T, Osadiaye BA, Mafe O, Soretire AA (2010) Phytoremediation potentials of sunflowers (*Tithonia diversifolia* and *Helianthus annuus*) for metals in soils contaminated with zinc and lead nitrates. Water Air Soil Pollut 207:195–201

Alam MGM, Snow ET, Tanaka A (2003) Arsenic and heavy metal contamination of vegetables grown in Samta village. Bangladesh Sci Total Environ 308:83–96

Al-Hawati M, Al-Khashman O (2015) Health risk assessment of heavy metals contamination in tomato and green pepper plants grown in soils amended with phosphogypsum waste materials. Environ Geochem Health 37(2):287–304

Ali H, Naseer M, Sajad MA (2012) Phytoremediation of heavy metals by *Trifolium alexandrinum*. Int J Environ Sci 2:1459–1469

Ali H, Khan E, Sajad MA (2013) Phytoremediation of heavy metals—concepts and applications. Chemosphere 91:869–881

Al-Saleh I, Shinwari N, El-Doush I, Biuedo G, Al-Amodi M, Khogali F (2004) Comparison of mercury levels in various tissues of albino and pigmented mice treated with two different brands of mercury skin-lightening creams. Biometals 2:167–175

Ampiah-Bonney RJ, Tyson JF, Lanza GR (2007) Phytoextraction of arsenic from soil by *Leersia oryzoides*. Int J Phytoremediation 9(1):31–40

Andrianisa HA, Ito A, Sasaki A, Aizawa J, Umita T (2008) Biotransformation of arsenic species by activated sludge and removal of bio-oxidised arsenate from wastewater by coagulation with ferric chloride. Water Res 42(19):4809–4817

Angelova V, Ivanov K (2009) Bio-accumulation and distribution of heavy metals in black mustard (*Brassica nigra Koch*). Environ Monit Assess 153:449–459

Asaduzzaman K, Khandaker MU, Baharudin NAB, Amin YBM, Farook MS, Bradley DA, Mahmoud O (2017) Heavy metals in human teeth dentine: a bioindicator of metals exposure and environmental pollution. Chemosphere 176:221–230

Baker AJM, Walker PL (1989) Ecophysiology of metal uptake by tolerant plants. In: Shaw A (ed) Heavy metal tolerance in plants—evolutionary aspects. CRC Press, Boca Raton, pp 155–178

Bakir F, Damluji SF, Amin Zaki I (1973) Methylmercury poisoning in Iraq: an interuniversity report. Science 181(4096):230–241

Bandara T, Herath I, Kumarathilaka P (2015) Role of fungal bacterial co-inoculation and woody biochar on soil enzyme activity and heavy metal immobilization in serpentine soil. J Soil Sediment 17:665–673

Barcelo J, Poschenrieder C, Vazquez MD, Gunse B, Vernet JP (1993) Beneficial and toxic effects of chromium in plants: solution culture, pot and field studies. Studies in environmental science no. 55, paper presented at the 5th international conference on environmental contamination. Morges, Switzerland

Becquer T, Quantin C, Sicot M, Boudot JP (2003) Chromium availability in ultramafic soils from New Caledonia. Sci Total Environ 301:251–261

Bennicelli R, Stepniewska Z, Banach A (2004) The ability of *Azolla caroliniana* to remove heavy metals (hg(II), Cr(III), Cr(VI)) from municipal waste water. Chemosphere 55:141–146

Bernhoft RA (2012) Mercury toxicity and treatment: a review of the literature. J Environ Public Health 2012:460508

Bernhoft RA (2013) Cadmium toxicity and treatment. Scientific World J 2013:394652

Berti WR, Cunningham SD (2000) Phytostabilization of metals. In: Phytoremediation toxic met using plants to clean up environ. Wiley, London, pp 71–88

Boonyapookana B, Upatham ES, Kruatrachue M, Pokethitiyook P, Singhakaew S (2002) Phytoaccumulation and phytotoxicity of cadmium and chromium in duckweed *Wolffia globosa*. Int J Phytoremediation 4:87–100

Bueno P, Varela J, Gimenezgallego G, Delrio LA (1995) Peroxisomal copper, zinc-superoxide dismutase—characterization of the isoenzyme from watermelon cotyledons. Plant Physiol 108:1151–1160. https://doi.org/10.1104/pp.108.3.1151

Cai Q, Long ML, Zhu M, Zhou QZ, Zhang L, Liu J (2009) Food chain transfer of cadmium and lead to cattle in a lead-zinc smelter in Guizhou. China Environ Pollut 157:3078–3082

Campbell WH (1999) Nitrate reductase structure, function and regulation: bridging the gap between biochemistry and physiology. Annu Rev Plant Physiol Plant Mol Biol 50:277–303. https://doi.org/10.1146/annurev.arplant.50.1.277

Cao H, Chen J, Zhang J, Zhang H, Qiao L, Men Y (2010) Heavy metals in rice and garden vegetables and their potential health risks to inhabitants in the vicinity of an industrial zone in Jiangsu. Chin J Environ Sci 22:1792–1799

Carrizales L, Razo I, Tellez-Hernandez JI, Torres-Nerio R, Torres A, Batres LE, Cubillas AC, Dıaz-Barriga F (2006) Exposure to arsenic and lead of children living near a cu-smelter in San Luis Potosi, Mexico: importance of soil contamination for exposure of children. Environ Res 101:1–10

Cervantes C, Campos-Garcia J, Devars S, Gutierrez-Corona F, Loza-Tavera H, Torres-Guzman JC, Moreno-Sanchez R (2001) Interactions of chromium with microorganisms and plants. FEMS Microbiol Rev 25:335–347

Chabukdhara M, Munjal A, Nema AK, Gupta SK, Kaushal RK (2016) Heavy metal contamination in vegetables grown around peri-urban and urban-industrial clusters in Ghaziabad. India Hum Ecol Risk Assess 22(3):736–752

Chandra R, Kumar V (2017) Phytoextraction of heavy metals by potential native plants and their microscopic observation of root growing on stabilised distillery sludge as a prospective tool for in situ phytoremediation of industrial waste. Environ Sci Pollut Res 24(3):2605

Chang AC, Page AL, Warneke JE (1987) Long-term sludge application on cadmium and zinc accumulation in Swiss chard and radish. J Environ Qual 16:217–221

Chang TC, You SJ, Yu BS, Chen CM, Chiu YC (2009) Treating high-mercury-containing lamps using full scale thermal desorption technology. J Hazard Mater 162(2–3):967–972

Chang CY, Yu HY, Chen JJ, Li FB, Zhang HH (2014) Accumulation of heavy metals in leaf vegetables from agricultural soils and associated potential health risks in the Pearl River Delta, South China. Environ Monit Assess 186:1547–1560

Chary NS, Kamala CT, Raj DSS (2008) Assessing risk of heavy metals from consuming food grown on sewage irrigated soils and food chain transfer. Ecotoxicol Environ Saf 69:513–524

Chehregani A, Malayeri BE (2007) Removal of heavy metals by native accumulator plants. Int J Agri Biol 9:462–465

Chen GC, Liu Z, Zhang J, Owens G (2012) Phytoaccumulation of copper in willow seedlings under different hydrological regimes. Ecol Eng 44:285–289

Chien LC, Hung TC, Choang KY, Yeh CY, Meng PJ, Shieh MJ (2002) Daily intake of TBT, Cu, Zn, Cd and As for fishermen in Taiwan. Sci Total Environ 285:177–185

Cho-Ruk K, Kurukote J, Supprung P, Vetayasuporn S (2006) Perennial plants in the phytoremediation of lead contaminated soils. Australas Biotechnol 5(1):1–4

Chutia P, Kato S, Kojima T, Satokawa S (2009) Arsenic adsorption from aqueous solution on synthetic zeolites. J Hazard Mater 162(1):440–447

Clemens S (2001) Developing tools for phytoremediation: towards a molecular understanding of plant metal tolerance and accumulation. Int J Occup Med Environ Health 14:235–239
Cobbet CS (2000) Phytochelatins and their roles in heavy metal detoxification. Plant Physiol 123:825–832
Cui J (2017) Silica nanoparticles alleviate cadmium toxicity in rice cells: mechanisms and size effects. Environ Pollut 228:363–369
Cui YJ, Zhu YG, Zhai RH, Chen DY, Huang YZ, Qiu Y, Ling JZ (2004) Transfer of metals from soil to vegetables in an area near a smelter in Nanning, China. Environ Int 30:785–791
Cui Y, Zhu Y, Zhai R, Huang Y, Qiu Y, Liang J (2005) Exposure to metal mixtures and human health impacts in a contaminated area in Nanning, China. Environ Int 31:784–790
Das P, Samantaray S, Rout GR (1997) Studies on cadmium toxicity in plants: a review. Environ Pollut 98:29–36
Delorme VGR, McCabe PF, Kim DJ, Leaver CJ (2000) A matrix metalloproteinase gene is expressed at the boundary of senescence and programmed cell death in cucumber. Plant Physiol 123:917–927. https://doi.org/10.1104/pp.123.3.917
Dickin SK, Schuster-Wallace CJ, Qadir M, Pizzacalla K (2016) A review of health risks and pathways for exposure to wastewater use in agriculture. Environ Health Perspect 124 (7):900–909
Doty SL, Shang QT, Wilson AM, Moore AL, Newman LA, Strand SE (2007) Enhanced metabolism of halogenated hydrocarbons in transgenic plants containing mammalian P450 2E1. Proc Natl Acad Sci USA 97:6287–6291
Duffus JH (2002) "Heavy metals"—a meaningless term? Pure Appl Chem 74:793–807
Effendia H, Kawaroeb M, Mursalina S, Lestaria DF (2016) Ecological risk assessment of heavy metal pollution in surface sediment of Mahakam Delta, East Kalimantan. Procedia Environ Sci 33:574–582
Ehrlich HL (1997) Microbes and metals. Appl Microbiol Biotechnol 48:687–692
Elgallal M, Fletcher L, Evans B (2016) Assessment of potential risks associated with chemicals in wastewater used for irrigation in arid and semiarid zones: a review. Agric Water Manag 177:419–431
El-Kady AA, Abdel-Wahhab MA (2018) Occurrence of trace metals in foodstuffs and their health impact. Trends Food Sci Technol 75:36–45
Ensley BD (2000) Rationale for the use of phytoremediation. Phytoremediation of toxic metals: using plants to clean-up the environment. Wiley, New York
European Commission DG ENV (2002) Heavy metals in waste, Final report project ENV.E.3/ETU/2000/0058. http://ec.europa.eu/environment/waste/studies/pdf/heavymetalsreport.pdf
FAO (2000) Food balance sheet, Rome. National Institute of Health National Nutrition Survey, Islamabad
Filiz E, Tombuloglu H (2015) Genome-wide distribution of superoxide dismutase (SOD) gene families in *Sorghum bicolor*. Turk J Biol 39:49–59. https://doi.org/10.3906/biy-1403-9
Franca FCSS, Albuuerque AMA (2017) Heavy metals deposited in the culture of lettuce (*Lactuca sativa L.*) by the influence of vehicular traffic in Pernambuco, Brazil. Food Chem 215:171–176
Galan E, Gonzalez I, Fernandez-Caliani JC (2002) Residual pollution load of soils impacted by the Aznalcollar (Spain) mining spill after clean-up operations. Sci Total Environ 286:167–179
Gall JE, Boyd RS, Rajakaruna N (2015) Transfer of heavy metals through terrestrial food webs: a review. Environ Monit Assess 187:201
Gan J, Jia X, Qin L (2000) A primary study on ecological risk caused by the heavy metals in coastal sediments. J Fish China 24:533–538
Garcia MA, Alonso J, Melgar MJ (2009) Lead in edible mushrooms: levels and bioaccumulation factors. J Hazard Mater 167(1–3):777–783
Gaur A, Adholeya A (2004) Prospects of *arbuscular mycorrhizal* fungi in phytoremediation of heavy metal contaminated soils. Curr Sci 86(4):528–534
Ge KY (1992) The status of nutrient and meal of Chinese in the 1990s. Beijing People's Hygiene Press, Beijing, pp 415–434

Ghosh M, Singh SP (2005) A review on phytoremediation of heavy metals and utilization of it's by products. As J Energy Env 6(04):214–231
Gomah LG, Ngumbu RS, Voegborlo RB (2019) Dietary exposure to heavy metal contaminated Rice and Health risk to the population of Monrovia. J Environ Sci Public Health 3(3):474–482
Greipsson S (2011) Phytoremediation. Nat Educ Knowl 3:7
Gress J, de Oliveira LM, da Silva EB, Lessl JM, Wilson PC, Townsend T, Ma LQ (2015) Cleaning-induced arsenic mobilization and chromium oxidation from CCA-wood deck: potential risk to children. Environ Int 82:35–40
Grotz N, Guerinot ML (2006) Molecular aspects of Cu, Fe and Zn homeostasis in plants. Biochim Biophys Acta-Mol Cell Res 1763:595–608. https://doi.org/10.1016/j.bbamcr.2006.05.014
Hacisalihoglu G, Hart JJ, Wang YH, Cakmak I, Kochian LV (2003) Zinc efficiency is correlated with enhanced expression and activity of zinc-requiring enzymes in wheat. Plant Physiol 131:595–602. https://doi.org/10.1104/pp.011825
Hakanson L (1980) An ecological risk index for aquatic pollution control, a sediment logical approach. Water Res 14:975–1001
Halim M, Conte P, Piccolo A (2003) Potential availability of heavy metals to phytoextraction from contaminated soils induced by exogenous humic substances. Chemosphere 52:265–275
Han YY, Zhang WZ, Zhang BL, Zhang SS, Wang W, Ming F (2009) One novel mitochondrial citrate synthase from *Oryza sativa L.* can enhance aluminum tolerance in transgenic tobacco. Mol Biotechnol 42:299–305
Hansch R, Mendel RR (2009) Physiological functions of mineral micronutrients (cu, Zn, Mn, Fe, Ni, Mo, B, Cl). Curr Opin Plant Biol 12:259–266. https://doi.org/10.1016/j.pbi.2009.05.006
Hawkes JS (1997) Heavy metals. J Chem Educ 74:1369–1374
Hazra D, Karmakar R, Rajlakshmi P, Bhattacharya S, Mondal S (2017) Recent advances in pesticide formulations for eco-friendly and sustainable vegetable pest management: a review. Arch Agri Environ Sci 2(3):232–237
Health NRC, Committee on the Institutional Means for Assessment of Risks to Public Health (1983) Risk assessment in the federal government: managing the process. National Academy Press, Washington
Hegedüs A, Erdei S, Janda T, Toth E, Horvath G, Dubits D (2004) Transgenic tobacco plants over producing alfalfa aldose/aldehyde reductase show higher tolerance to low temperature and cadmium stress. Plant Sci 166:1329–1333
Henry JR (2000) In an overview of phytoremediation of lead and mercury. NNEMS report, Washington, pp 3–9
Hough RL, Breward N, Young SD, Crout NM, Tye AM, Moir AM, Thornton I (2004) Assessing potential risk of heavy metal exposure from consumption of home produced vegetables by urban populations. Environ Health Perspect 112:215–221
Hu J, Wu F, Wu S, Cao Z, Lin X, Wong MH (2013) Bioaccessibility, dietary exposure and human risk assessment of heavy metals from market vegetables in Hong Kong revealed with an in vitro gastrointestinal model. Chemosphere 91:455–461
IRIS (2003) Integrated risk information system-database. US Environmental Protection Agency, Washington
Iyengar V, Nair P (2000) Global outlook on nutrition and the environment: meeting the challenges of the next millennium. Sci Total Environ 249:331–346
Jadia CD, Fulekar MH (2008) Phytoremediation of heavy metals: recent techniques. Afr J Biotechnol 8(6):921–928
Jayakumar K, Vijayarengan P, Changxing Z, Gomathinayagam M, Jaleel CA (2008) Soil applied cobalt alters the nodulation, leg-haemoglobin content and antioxidant status of *Glycine max (L.)* Merr. Colloid Surf B-Biointerface 67:272–275. https://doi.org/10.1016/j.colsurfb.2008.08.012
Jomjun N, Siripen T, Maliwan S (2010) Phytoremediation of arsenic in submerged soil by wetland plants. Int J Phytoremediation 13:35–46

Kalve S, Sarangi BK, Pandey RA, Chakrabarti T (2011) Arsenic and chromium hyper accumulation by an ecotype of *Pteris vittata*-prospective for phytoextraction from contaminated water and soil. Curr Sci 100:888–894

Khan S, Aijun L, Zhang S, Hu Q, Zhu Y (2008) Accumulation of polycyclic aromatic hydrocarbons and heavy metals in lettuce grown in the soils contaminated with long term wastewater irrigation. J Hazard Mater 152:506–515

Khan S, Farooq R, Shahbaz S, Khan MA, Sadique M (2009) Health risk assessment of heavy metals for population via consumption of vegetables. World Appl Sci J 6(12):1602–1606

Khan S, Rehman S, Khan AZ, Khan MA, Shah MT (2010) Soil and vegetables enrichment with heavy metals from geological sources in Gilgit, northern Pakistan. Ecotoxicol Environ Saf 73:1820–1827

Khan MU, Malik RN, Muhammad S (2013) Human health risk from heavy metal via food crops consumption with wastewater irrigation practices in Pakistan. Chemosphere 93:2230–2238

Khandare RV, Govindwar SP (2015) Phytoremediation of textile dyes and effluents: current scenario and future prospects. Biotechnol Adv 33:1697–1714

Kostal J (2005) Customizable biopolymers for heavy metal remediation. J Nanopart Res 7 (4–5):517–523

Krishnamurthy S, Wilkens MM (1994) Environmental chemistry of Cr. Northeastern Geol 16 (1):14–17

Kuswandi B (2017) Nanosensors for the detection of food contaminants. In: Nanotechnology applications in food. Elsevier, Amsterdam, pp 307–333

Lanquar V, Ramos MS, Lelievre S, Barbier-Brygoo H, Krieger-Liszkay A, Kramer U, Thomine S (2010) Export of vacuolar manganese by AtNRAMP3 and AtNRAMP4 is required for optimal photosynthesis and growth under manganese deficiency. Plant Physiol 152:1986–1999. https://doi.org/10.1104/pp.109.150946

Lasat MM (2002) Phytoremediation of toxic metals: a review of biological mechanisms. J Environ Qual 31:109–120

Li G, Sun GX, Williams PN, Nunes L, Zhu YG (2011) Inorganic arsenic in Chinese food and its cancer risk. Environ Int 37:1219–1225

Li H-H, Chen L-J, Yu L, Guo Z-B, Shan C-Q, Lin J-Q, Gu Y-G, Yang Z-B, Yang Y-X, Shao J-R, Zhu X-M, Cheng Z (2017) Pollution characteristics and risk assessment of human exposure to oral bio-accessibility of heavy metals via urban street dusts from different functional areas in Chengdu, China. Sci Total Environ 586:1076–1084

Liao J (2016) Distribution and migration of heavy metals in soil and crops affected by acid mine drainage: public health implications in Guangdong Province, China. Ecotoxicol Environ Saf 124:460–469

Lidon FC, Barreiro MG, Ramalho JC (2004) Manganese accumulation in rice: implications for photosynthetic functioning. J Plant Physiol 161:1235–1244. https://doi.org/10.1016/j.jplph.2004.02.003

Lindberg SE, Meyers TP, Taylor GE Jr, Turner RR, Schroeder WH (1992) Atmosphere-surface exchange of mercury in a forest: results of modeling and gradient approached. J Geophys Res 97:2519–2528

Liu G (2014) An ecological risk assessment of heavy metal pollution of the agricultural ecosystem near a lead-acid battery factory. Ecol Indic 47:210–218

Liu L, Hu LL, Tang JJ, Li YF, Zhang Q, Chen X (2012) Food safety assessment of planting patterns of four vegetable type crops grown in soil contaminated by electronic waste activities. J Environ Manage 93(1):22–30

Liu X, Song Q, Tang Y, Li W, Xu J, Wu J, Wang F, Brookes PC (2013) Human health risk assessment of heavy metals in soil–vegetable system: a multi-medium analysis. Sci Total Environ 463–464:530–540

Logan TJ, Chaney RL (1983) Metals. In: Page AL (ed) Utilization of municipal wastewater and sludge on land. University of California, Riverside, pp 235–326

Lone MI, He Z, Stoffella PJ, Yang X (2008) Phytoremediation of heavy metal polluted soils and water: progresses and perspectives. J Zhejiang Univ Sci B 9:210–220

Macek T, Mackova M, Pavlikova D, Szakova J, Truksa M, Singh-Cundy A, Kotrba P, Yancey N, Scouten WH (2002) Accumulation of cadmium by transgenic tobacco. Acta Biotechnol 22:101–106

Mahmood T, Islam KR, Muhammad S (2007) Toxic effects of heavy metals on early growth and tolerance of cereal crops. Pak J Bot 39(2):451–462

Man YB, Sun XL, Zhao YG, Lopez BN, Chung SS, Wu SC (2010) Health risk assessment of abandoned agricultural soils based on heavy metal contents in Hong Kong, the world's most populated city. Environ Int 36:570–576

Mansour SA (2009) Monitoring of pesticides and heavy metals in cucumber fruits produced from different farming systems. Chemosphere 75:601–609

Markert B (1993) Plants as biomonitors-indicators of heavy metals in the terrestrial environment. VCH, Berlin, p 644

Marschner H (1995) Mineral nutrition of higher plants, 2nd edn. Academic Press, London, p 38

Massaquoi LD (2015) Heavy metal accumulation in soils, plants, and hair samples: an assessment of heavy metal exposure risks from the consumption of vegetables grown on soils previously irrigated with wastewater. Environ Sci Pollut Res 22(23):18456–18468

McLaughlin M (1999) Metals and micronutrients—food safety issues. Field Crop Res 60 (1–2):143–163

Melgar MJ, Alonso J, García MA (2009) Mercury in edible mushrooms and underlying soil: bioconcentration factors and toxicological risk. Sci Total Environ 407(20):5328–5334

Mendel RR (2007) Biology of the molybdenum cofactor. J Exp Bot 58:2289–2296. https://doi.org/10.1093/jxb/erm024

Mendel RR, Schwarz G (1999) Molybdoenzymes and molybdenum cofactor in plants. Crit Rev Plant Sci 18:33–69

Mendez MO, Maier RM (2008) Phytostabilization of mine tailings in arid and semiarid environments—an emerging remediation technology. Environ Health Perspect 116:278–283

Mishra P, Dixit A, Ray M, Sabat SC (2014) Mechanistic study of CuZn-SOD from Ipomoea carnea mutated at dimer interface: enhancement of peroxidase activity upon monomerization. Biochimie 97:181–193. https://doi.org/10.1016/j.biochi.2013.10.014

Misra SG, Mani D (1991) Soil pollution. Ashish Publishing House, Punjab

Mmolawa KB, Likuku AS, Gaboutloeloe GK (2011) Assessment of heavy metal pollution in soils along major roadside areas in Botswana. Afr J Environ Sci Technol 5(3):186–196

Moreno FN, Anderson CWN, Stewart RB (2004) Mercury phytoextraction and phytovolatilisation from hg-contaminated artisanal mine sites. Phytoremed Mercury Mine Wastes 136:147–159

Mountouris A, Voutsas E, Tassios D (2002) Bioconcentration of heavy metals in aquatic environments: the importance of bioavailability. Mar Pollut Bull 44:1136–1141

Mukherjee A, Bandyopadhyay A, Dutta S, Basu S (2013) Phytoaccumulation of iron by callus tissue of *Clerodendrum indicum* (*L*). Chem Ecol 29:564–571

Musselman JF, QEP (2004) Sources of mercury in wastewater, Pretreatment corner. http://www.cetinc.com/cmsdocuments//7%20%20Sources%20of%20in%20Wastewater%20(0204).pdf

Muthusaravanan S, Sivarajasekar N, Vivek JS, Paramasivan T, Naushad M, Prakashmaran J, Gayathri V, Al-Duaij OK (2018) Phytoremediation of heavy metals: mechanisms, methods and enhancements. Environ Chem Lett 16:1339–1359

Nabulo G (2006) Assessment of lead, cadmium, and zinc contamination of roadside soils, surface films, and vegetables in Kampala City. Uganda Environ Res 101(1):42–52

Nagajyoti PC, Lee KD, Sreekanth TVM (2010) Heavy metals, occurrence and toxicity for plants: a review. Environ Chem Lett 8:199–216

Newman LA, Reynolds CM (2004) Phytodegradation of organic compounds. Curr Opin Biotechnol 15:225–230

Nordberg G (2003) Cadmium and human health: a perspective based on recent studies in China. J Trace Elem Exp Med 16:307–319

O'Hara GW (2001) Nutritional constraints on root nodule bacteria affecting symbiotic nitrogen fixation: a review. Aust J Exp Agric 41:417–433

Obiora SC (2016) Heavy metals and health risk assessment of arable soils and food crops around Pb–Zn mining localities in Enyigba, southeastern Nigeria. J African Earth Sci 116:182–189

Ogawa T, Kobayashi E, Okubo Y, Suwazono Y, Kido T, Nogawa K (2004) Relationship among prevalence of patients with Itai-itai disease, prevalence of abnormal urinary findings, and cadmium concentrations in rice of individual hamlets in the Jinzu River basin, Toyama prefecture of Japan. Int J Environ Health Res 14(4):243–252

Oves M, Khan MS, Zaidi A, Ahmad E (2012) Soil contamination, nutritive value, and human health risk assessment of heavy metals: an overview. In: Zaidi A, Wani PA, Khan MS (eds) Toxicity of heavy metals to legumes and bioremediation. Springer, New York, pp 1–27

Ozturk M, Yucel E, Gucel S, Sakcali S, Aksoy A (2008) Plants as biomonitors of trace elements pollution in soil. In: Prasad MNV (ed) Trace elements: environmental contamination, nutritional benefits and health implications. Wiley, New York, pp 723–744

Page V, Feller U (2015) Heavy metals in crop plants: transport and redistribution processes on the whole plant level. Agronomy 2015(5):447–463

Pandey SN (2006) Accumulation of heavy metals (cd, Cr, cu, Ni and Zn) in *Raphanus sativus L.* and *Spinacia oleracea L.* plants irrigated with industrial effluent. J Environ Biol 27(2):381–384

Pandey V, Dixit V, Shyam R (2005) Antioxidative responses in relation to growth of mustard (*Brassica juncea* cv. Pusa JaiKisan) plants exposed to hexavalent chromium. Chemosphere 61:40–47

Pastor J, GutiErrez-ginEs MJ, HernAndez AJ (2015) Heavy-metal phytostabilizing potential of *Agrostis castellana* Boiss and reuter. Int J Phytoremediation 17:988–998

Pavlikova D, Macek T, Mackova M, Sura M, Szakova J, Tlustos P (2004) The evaluation of cadmium, zinc and nickel accumulation ability of transgenic tobacco bearing different transgenes. Plant Soil Environ 50:513–517

Pehlivan E, Ozkan AM, Dinc S, Parlayici S (2009) Adsorption of Cu2+ and Pb2+ ion on dolomite powder. J Hazard Mater 167(1–3):1044–1049

Peralta JR, Torresdey JLG, Tiemann KJ, Gomez E, Arteaga S, Rascon E (2001) Uptake and effects of five heavy metals on seed germination and plant growth in alfalfa (*Medicago sativa*) *L.* Environ Contam Toxicol 66:727–734

Pinto AP, Mota AM, de Varennes A, Pinto FC (2004) Influence of organic matter on the uptake of cadmium, zinc, copper and iron by sorghum plants. Sci Total Environ 326:239–247

Prasad MNV (2008) Trace elements as contaminants and nutrients: consequences in ecosystems and human health. Wiley, New York

Prescott AG, John P (1996) Dioxygenases: molecular structure and role in plant metabolism. Annu Rev Plant Physiol Plant Mol Biol 47:245–271. https://doi.org/10.1146/annurev.arplant.47.1.245

Pruvot C, Douay F, Herve F, Waterlot C (2006) Heavy metals in soil, crops and grass as a source of human exposure in the former mining areas. J Soil Sediment 6:215–220

Psaras GK, Constantinidis T, Cotsopoulos B, Manetas Y (2000) Relative abundance of nickel in the leaf epidermis of eight hyper accumulators: evidence that the metal is excluded from both guard cells and trichomes. Ann Bot 86:73–78. https://doi.org/10.1006/anbo.2000.1161

Qureshi MI, Israr M, Abdin MZ, Iqbal M (2005) Responses of *Artemisia annua L.* to lead and salt induced oxidative stress. Environ Exp Bot 53:185–193

Radziemska M, Vaverkova MD, Baryla A (2017) Phytostabilisation management strategy for stabilizing trace elements in contaminated soils. Int J Environ Res Public Health 14:958

Rafati M, Khorasani N, Moattar F, Shirvany A, Moraghebi F, Hosseinzadeh S (2011) Phytoremediation potential of *Populus alba* and *Morus alba* for cadmium, chromium and nickel absorption from polluted soil. Int J Environ Res 5:961–970

Rai PK (2018) Phytoremediation of emerging contaminants in wetlands. CRC Press, Boca Raton, p 248

Rai PK, Leeb SS, Zhangc M, Tsangd YF, Kime K (2019) Heavy metals in food crops: health risks, fate, mechanisms, and management. Environ Int 125:365–385

Rashed MN (2010) Monitoring of contaminated toxic and heavy metals, from mine tailings through age accumulation, in soil and some wild plants at Southeast Egypt. J Hazard Mater 178:739–746
Raskin I (1996) Plant genetic engineering may help with environmental cleanup. Proc Natl Acad Sci USA 93:3164–3166
Raskin I, Ensley BD (2000) Phytoremediation of toxic metals: using plants to clean up the environment. Wiley, New York
Rattan RK, Datta SP, Chhonkar PK, Suribabu K, Singh AK (2005) Long-term impact of irrigation with sewage effluents on heavy metal content in soils, crops and groundwater—a case study. Agric Ecosyst Environ 109:310–322
Redinbo MR, Yeates TO, Merchant S (1994) Plastocyanin—structural and functional analysis. J Bioenerg Biomembr 26:49–66. https://doi.org/10.1007/BF00763219
Redjala T, Zelko I, Sterckeman T (2011) Relationship between root structure and root cadmium uptake in maize. Environ Exp Bot 71:241–248
Reilly SB, McCarty KM, Steckling N, Lettmeier B (2012) Mercury exposure and children's health. Curr Probl Pediatr Adolesc Health Care 40:186–215
Reisinger S, Schiavon M, Terry N, Pilon-Smits EAH (2008) Heavy metal tolerance and accumulation in Indian mustard (*Brassica juncea L.*) expressing bacterial gamma-glutamylcysteine synthetase or glutathione synthetase. Int J Phytoremediation 10:440–454
Resaee A, Derayat J, Mortazavi SB, Yamini Y, Jafarzadeh MT (2005) Removal of mercury from chlor-alkali industry wastewater using *Acetobacter xylinum* cellulose. Am J Environ Sci 1 (2):102–105
Rivetta A, Negrini N, Cocucci M (1997) Involvement of Ca+– calmodulin in Cd2+ toxicity during the early phases of radish (*Raphanus sativus L.*) seed germination. Plant Cell Environ 20:600–608
Rosas CJ, Portugal L, Ferrer L, Hinojosa RL, Guzmán MJL, Hernández RA, Cerd V (2016) An evaluation of the bioaccessibility of arsenic in corn and rice samples based on cloud point extraction and hydride generation coupled to atomic fluorescence spectrometry. Food Chem 204:475–482
Rout GR, Samantaray S, Das P (1997) Differential chromium tolerance among eight mungbean cultivars grown in nutrient culture. J Plant Nutr 20:473–483
Rout GR, Samantaray S, Das P (2001) Differential lead tolerance of rice and black gram genotypes in hydroponic culture. Rost Vyroba (Praha) 47:541–548
Saier MH, Trevors JT (2010) Phytoremediation. Water Air Soil Pollut 205:61–63
Salt DE, Prince RC, Pickering IJ, Raskin I (1995) Mechanisms of cadmium mobility and accumulation in Indian mustard. Plant Physiol 109:1427–1433
Sarma H (2011) Metal hyperaccumulation in plants: a review focusing on phytoremediation technology. J Environ Sci Technol 4:118–138
Sas-Nowosielska A, Galimska-Stypa R, Kucharski R, Zielonka U, Małkowski E, Gray L (2008) Remediation aspect of microbial changes of plant rhizosphere in mercury contaminated soil. Environ Monit Assess 137(1–3):101–109
Schwarz G, Boxer DH, Mendel RR (1997) Molybdenum cofactor biosynthesis—the plant protein Cnx1 binds molybdopterin with high affinity. J Biol Chem 272:26811–26814. https://doi.org/10.1074/jbc.272.43.26811
Sekara A, Poniedzialeek M, Ciura J, Jedrszczyk E (2005) Cadmium and lead accumulation and distribution in the organs of nine crops: implications for phytoremediation. Pol J Environ Stud 14:509–516
Seregin IV, Ivanov VB (2001) Physiological aspects of cadmium and lead toxic effects on higher plants. Russian J Plant Physiol 4:523–544
Shabani N, Sayadi MH (2012) Evaluation of heavy metals accumulation by two emergent macrophytes from the polluted soil: an experimental study. Environmentalist 32:91–98
Shackira AM, Puthur JT (2017) Enhanced phytostabilization of cadmium by a halophyte—*Acanthus ilicifolius L.* Int J Phytoremediation 19:319–326

Shafiq M, Iqbal MZ (2005) Tolerance of Peltophorum pterocarpum D. C Baker Ex K Heyne seedlings to lead and cadmium treatment. J New Seeds 7:83–94

Shah FUR, Ahmad N, Masood RK, Videa JRP, Ahmad FD (2010) Heavy metal toxicity in plants. In: Plant adaptation and phytoremediation. Springer, London

Shaheen N, Irfan N, Khan IN, Islam S, Islam M, Ahmed M (2016) Presence of heavy metals in fruits and vegetables: health risk implications in Bangladesh. Chemosphere 152:431–438

Shalaby TA (2016) Nanoparticles, soils, plants and sustainable agriculture. In: Nanoscience in food and agriculture, vol 10. Springer, Cham, pp 283–312

Shanker AK, Cervantes C, Loza-Tavera H, Avudainayagam S (2005) Chromium toxicity in plants. Environ Int 31:739–751

Sharma DC, Chaterjee C, Sharma CP (1995) Chromium accumulation and its effects on wheat (*Triticum aestivum L.* cv. DH220) metabolism. Plant Sci 111:145–151

Shukla A, Srivastava S (2017) Emerging aspects of bioremediation of arsenic. In: Singh R, Kumar S (eds) Green technologies and environmental sustainability. Springer, Cham, pp 395–407

Shukla OP, Dubey S, Rai UN (2007) Preferential accumulation of cadmium and chromium: toxicity in *Bacopa monnieri L.* under mixed metal treatments. Environ Contam Toxicol 78:252–257

Siedow JN (1991) Plant lipoxygenase—structure and function. Annu Rev Plant Physiol Plant Mol Biol 42:145–188. https://doi.org/10.1146/annurev.arplant.42.1.145

Sinha RK, Herat S, Tandon PK (2004) Phytoremediation: role of plants in contaminated site management. In: Book of environmental bioremediation technologies. Springer, Berlin, pp 315–330

Sipter E (2008) Site-specific risk assessment in contaminated vegetable gardens. Chemosphere 71 (7):1301–1307

Sirko A, Brodzik R (2000) Plant ureases: roles and regulation. Acta Biochim Pol 47:1189–1195

States JC (2015) Arsenic: exposure sources, health risks, and mechanisms of toxicity. Wiley, Hoboken

Sulaiman MB, Salawu K, Barambu AU (2019) Assessment of concentrations and ecological risk of heavy metals at resident and remediated soils of uncontrolled mining site at Dareta Village, Zamfara, Nigeria. J Appl Sci Environ Manag 23(1):187–193

Takatsuji H (1998) Zinc-finger transcription factors in plants. Cell Mol Life Sci 54:582–596. https://doi.org/10.1007/s000180050186

Tangahu BV, Abdullah SRS, Basri H, Idris M, Anuar N, Mukhlisin M (2011) A review on heavy metals (As, Pb, and Hg) uptake by plants through phytoremediation. Int J Chem Eng 2011:939161. https://doi.org/10.1155/2011/939161

Thomilson DC, Wilson DJ, Harris CR, Jeffrey DW (1980) Problem in heavy metals in estuarics and the formation of pollution index. Helgol Wiss Meeresunlter 33(1–4):566–575

Tomsett AB, Sewell AK, Jones SJ, Miranda JR, de Thurman DA (1992) Metal-binding proteins and metal-regulated gene expression in higher plants. In: Wray JL (ed) Inducible plant proteins: their biochemistry and molecular biology. Cambridge University Press, Cambridge, pp 1–24

Traunfeld JH, Clement DL (2001) Lead in garden soils. Home and garden. Maryland Cooperative Extension, University of Maryland, College Park. http://www.hgic.umd.edu/media/documents/hg18.pdf

Turan M, Esringu A (2007) Phytoremediation based on canola (*Brassica napus L.*) and Indian mustard (*Brassica juncea L.*) planted on spiked soil by aliquot amount of cd, cu, Pb, and Zn. Plant Soil Environ 53:7–15

Udoka OC, Ekanem EO, Harami MA, Tafawa A (2014) Phytoaccumulation potentials of *Tamarindus indica*. Int J Innov Scientific Res 11(1):72–78

United States Protection Agency (2000) Introduction to phytoremediation. EPA 600/R-99/107. U.S. Environmental Protection Agency, Office of Research and Development, Cincinnati

USEPA (2001) Risk-based concentration table. United States Environmental Protection Agency, Washington

USEPA (2002) Region 9, preliminary remediation goals. http://www.epa.gov/region09/waste/sfind/prg

Vaclavikova M, Gallios GP, Hredzak S, Jakabsky S (2008) Removal of arsenic from water streams: an overview of available techniques. Clean Techn Environ Policy 10(1):89–95
Vamerali T, Bandiera M, Mosca G (2010) Field crops for phytoremediation of metal contaminated land. A review. Environ Chem Lett 8:1–17
Van Aken B (2009) Transgenic plants for enhanced phytoremediation of toxic explosives. Curr Opin Biotechnol 20:231–236
Verloo M, Eeckhout M (1990) Metal species transformations in soil: an analytical approach. Int J Environ Anal Chem 39:170–186
Vishnoi SR, Srivastava PN (2008) Phytoremediation-green for environmental clean. In: The 12th world lake conference, pp 1016–1021
Vithanage M, Dabrowska BB, Mukherjee B, Sandhi A, Bhattacharya P (2012) Arsenic uptake by plants and possible phytoremediation applications: a brief overview. Environ Chem Lett 10:217–224
Wang X, Sato T, Xing B, Tao S (2005) Health risks of heavy metals to the general public in Tianjin, China via consumption of vegetables and fish. Sci Total Environ 350:28–37
Wang L, Cui X, Cheng H, Chen F, Wang J, Zhao X, Lin C, Pu X (2015) A review of soil cadmium contamination in China including a health risk assessment. Environ Sci Pollut Res 22 (21):16441–16452
Welch RM (1995) Micronutrient nutrition of plants. Crit Rev Plant Sci 14:49–82. https://doi.org/10.1080/713608066
WHO (World Health Organization) (1982) Toxicological evaluation of certain food additives. Joint FAO/WHO expert committee on food additives. WHO Food Additive Series No. 683. World Health Organization, Geneva
Wilbur SB, Hansen H, Pohl H, Colman J, McClure P (2004) Using the ATSDR guidance manual for the assessment of joint toxic action of chemical mixtures. Environ Toxicol Pharmacol 18:223–230
Witte CP (2011) Urea metabolism in plants. Plant Sci 180:431–438. https://doi.org/10.1016/j.plantsci.2010.11.010
Wojcik M, Tukiendorf A (1999) Cd-tolerance of maize, rye and wheat seedlings. Acta Physiol Plant 21:99–107
Woldetsadik D, Drechsel P, Keraita B, Itanna F, Gebrekidan H (2017) Heavy metal accumulation and health risk assessment in wastewater-irrigated urban vegetable farming sites of Addis Ababa, Ethiopia. Int J Food Contam 4:9
Wong MH, Bradshaw AD (1982) A comparison of the toxicity of heavy metals, using root elongation of rye grass, *Lolium perenne*. New Phytol 91:255–261
World Health Organization (2011) Adverse health effect of heavy metals in children. World Health Organization, Geneva
Wu FB, Zhang GP (2002) Genotypic variation in kernel heavy metal concentrations in barley and as affected by soil factors. J Plant Nutr 25:1163–1173
Yang Y (2018) Assessing cadmium exposure risks of vegetables with plant uptake factor and soil property. Environ Pollut 238:263–269
Yang S, Liang S, Yi L (2014) Heavy metal accumulation and phytostabilization potential of dominant plant species growing on manganese mine tailings. Front Environ Sci Eng 8:394–404
Yoon J, Cao X, Zhou Q, Ma LQ (2006) Accumulation of Pb, Cu, and Zn in native plants growing on a contaminated Florida site. Sci Total Environ 368:456–464
Yousaf B (2018) Comparative effects of biochar-nanosheets and conventional organic-amendments on health risks abatement of potentially toxic elements via consumption of wheat grown on industrially contaminated-soil. Chemosphere 192:161–170
Yruela I (2009) Copper in plants: acquisition, transport and interactions. Funct Plant Biol 36:409–430. https://doi.org/10.1071/FP08288
Yruela I (2013) Transition metals in plant photosynthesis. Metallomics 5:1090–1109. https://doi.org/10.1039/c3mt00086a

Zacchini M, Pietrini F, Mugnozza GS, Iori V, Pietrosanti L, Massacci A (2009) Metal tolerance, accumulation and translocation in poplar and willow clones treated with cadmium in hydroponics. Water Air Soil Pollut 197:23–34
Zhang GP, Fukami M, Sekimoto H (2002) Influence of cadmium on mineral concentration and yield components in wheat genotypes differing in cd tolerance at seedling stage. Field Crop Res 4079:1–7
Zhao Q (2014) Potential health risks of heavy metals in cultivated topsoil and grain, including correlations with human primary liver, lung and gastric cancer, in Anhui province, eastern China. Sci Total Environ 470-471:340–347
Zhao L, Li T, Zhang X (2016) Pb uptake and phytostabilization potential of the mining ecotype of *Athyrium wardii* (Hook.) grown in Pb-contaminated soil. Clean Soil Air Water 44:1184–1190
Zhuang P, Murray B, McBride XH, Lia N, Li Z (2009) Health risk from heavy metals via consumption of food crops in the vicinity of Dabaoshan mine. South China Sci Total Environ 407:1551–1561

# Heavy Metal Contamination of Environment and Crop Plants

12

Anuradha Patel, Sanjesh Tiwari, Amandeep Raju, Neeraj Pandey, Madhulika Singh, and Sheo Mohan Prasad

**Abstract**

Heavy metal contamination is a prime environmental concern that threatens plants productivity, animal, and human health. Heavy metals occur in the earth's crust, but due to their persistent and stable character, they cannot be degraded or destroyed. Various natural and anthropogenic processes release these heavy metals into the environment and contaminate the agricultural soils which is further associated with health hazard due to dietary intake of contaminated vegetables. The modern agricultural practices i.e. prolonged application of fertilizers, pesticides has resulted in the degradation of the ecosystem and accumulation of heavy metals especially in crop plants. Vegetable crops are vital for the human diet and in particular provide the nutrients to maintain normal health. The exposure of plants to heavy metals results in accumulation of metal content inside the plant cell via specific transport mechanisms. This alters the physiological and biochemical process that negatively affects the growth and development of plants. Heavy metals consequently alter the electron transport chain and generate ROS; singlet oxygen ($_1O_2$), hydroxyl radical ($HO^•$), superoxide radical ($O_2^{•-}$), and hydrogen peroxide ($H_2O_2$). Albeit, ROS at low concentrations serves the role as important signalling molecules that are generally in equilibrium with antioxidant molecules but at higher concentrations impose adverse effect by interacting with macromolecules. Besides this, heavy metals also down-regulate the specific genes that are involved in the stress responses. Thus introduction of heavy metals in food chain and consumption of contaminated vegetables and its

Anuradha Patel and Sanjesh Tiwari contributed equally with all other contributors.

A. Patel · S. Tiwari · A. Raju · N. Pandey · M. Singh · S. M. Prasad (✉)
Ranjan Plant Physiology and Biochemistry Laboratory, Department of Botany, University of Allahabad, Prayagraj, Uttar Pradesh, India



K. Mishra et al. (eds.), *Sustainable Solutions for Elemental Deficiency and Excess in Crop Plants*, https://doi.org/10.1007/978-981-15-8636-1_12

toxicity is a serious concern. Furthermore, strategy and policy are required to control the limits of accumulation in crop plants and vegetables.

**Keywords**

Heavy metals · Accumulation · Crop plants · Toxicity in plants

## 12.1 Introduction

Confronted through present environmental conventions, heavy metals are the striking environmental contaminants; and its toxicity is a delinquent for growing concern to ecological and environmental motives (Arif et al. 2019). The "heavy metals" denotes to category of metals and metalloids that have relatively high density and become toxic even at low concentrations (Lenntech Water Treatment and Air Purification 2004). In general heavy metals are defined as the metallic element having density greater than 4 $g/cm^3$, or fivefold, more than water. These include essential elements such as copper (Cu), zinc (Zn), manganese (Mn), and molybdenum (Mo) and toxic elements as: vanadium (V), arsenic (As), boron (B), aluminium (Al), cadmium (Cd), lead (Pb), mercury (Hg), and chromium (Cr) (Gupta et al. 2002). Apart from the natural activities such as weathering of rocks and volcanic eruption, indiscriminate use of pesticides and insecticides in agricultural fields and increased industrialization (mining and gasoline industry, fossil fuel burning, paint and leather industry, sewage and sludge's) are the chief foundations for introducing heavy metal into the environment (Zhang et al. 2010; Yadav et al. 2016). Heavy metals such as Fe, Zn, Mn, and Cu at lesser concentration also function as the microelements due to their key requirement for growth and development in plants and also acts as co-factor for several enzymes involved in metabolic processes that are needed for the proper growth and functioning. The enlistment of metal pollutant from outside to inside; the cell wall of the root has various transporter proteins that facilitate the uptake of heavy metal across plasma membrane due to higher affinity for metals (Hall and Williams 2003). Plant cell owes numerous carrier/transporter such as HMAs (heavy metal ATPases), Nramps (natural resistance-associated macrophage proteins), the ZIP (ZRT, IRT-like proteins) family, CDF (cation diffusion) facilitator, and the ABC transporters (ATP-binding cassette), the IRT1, NRAMP3/4 and ZIP4 family commonly transports $Fe^{2+}$, $Zn^{2+}$, $Mn^{2+}$, $Cd^{2+}$, and HMA1 transports, $Cu^{2+}$, $Cd^{2+}$, $Zn^{2+}$, and $Co^{2+}$ (Puig and Peñarrubia 2009; Yadav et al. 2016).

The acquisition and elevated concentration of HM leads to the accumulation of HMs in plants causing physiological changes, like reduction in growth, disturbance in redox homeostasis, amplified senescence, distorted photosynthetic apparatus, and alteration of enzyme activity and biosynthesis of biomolecules thereby disturbing the assembly of cell membrane, mitochondria, lysosome, and endoplasmic reticulum

(Wang and Shi 2001; Hu et al. 2016). The HMs accumulation also causes cell dehydration which proves to be a critical state for the enzyme functioning and maintenance of organelles, and membranes (Duan and Cai 2012). Apart from physiological and molecular responses, autoxidation of HMs leads to the generation of Reactive Oxygen Species (ROS). Heavy metal such as As, Cd, Pb, and Hg are class I carcinogen and hazardous to plants, animals, and humans (Jallad 2015). Reactive oxygen species (ROS) generation is common endpoint of aerobic life under stress condition due to impairment in the electron transport chain. Albeit ROS also function as vital signalling molecules under low levels and modulate the activity of specific defence proteins; however, high levels of ROS are extremely toxic to plants which result into oxidative stress. Oxidative stress occurs due to the overproduction of ROS or due to the decrease of cellular antioxidative levels. Accumulation of ROS disrupts the membrane structure and damages the macromolecule like DNA, amino acid, lipids, and carbohydrate (Wysocki and Tamas 2010; Sharma et al. 2011; Hu et al. 2016). In order to overcome the stress condition, plants are endowed with multiple survival strategies and they utilize a number of physiological and molecular mechanisms to minimize possible damage (Mahajan and Tuteja 2005; Liang et al. 2013). These strategies include regulation of gene expression of different proteins or enzymes, modulating the amount of metals to exclude non-essential forms through regulating their uptake, accumulation, compartmentation, chelation and sequestration as well as extrusion (Sharma et al. 2008) and also plants have inbuilt defence system involving enzymatic and non-enzymatic antioxidant. Enzymatic antioxidants include superoxide dismutase (SOD) that convert $O_2^{\bullet-}$ in to $H_2O_2$, peroxidase (POD), catalase (CAT), ascorbate peroxidase (APX) which further detoxified by glutathione-s-transferase (GST) that form conjugation with the metal and lead to its removal and also involve in synthesis of phytochelatins. Non-enzymatic antioxidants such as ascorbic acid, flavonoids, glutathionc, and proline directly scavenge the free radicals. Thus, heavy metal toxicity and their transportation is a strategy adopted by plants to overcome heavy metal stress. This chapter summarizes the numerous studies focused on the stress effects of HMs, and plant adaptations and molecular approaches involved.

## 12.2 Heavy Metals in Environment

Heavy metals (HMs) are among most baring and priority environmental pollutants. They enter into the environment through different sources as discussed in Fig. 12.1 and tend to equally affect the terrestrial and aquatic ecosystem. Even at low trace levels their toxicity is more prominent in plants. Unlike organic contaminants, heavy metals are non-biodegradable and tend to accumulate and are often toxic to plants, animals, and humans. Heavy metals such as Cu, Fe, Mn, Zn, Ni, Co, Cd, As, etc. hamper the growth and productivity while some (Fe, Zn, Mn, and Cu) are needed for the proper growth and functioning. Essential heavy metal such as Cu is predominantly involved in photosynthesis, acts as a co-factor in enzymes amine oxidases,

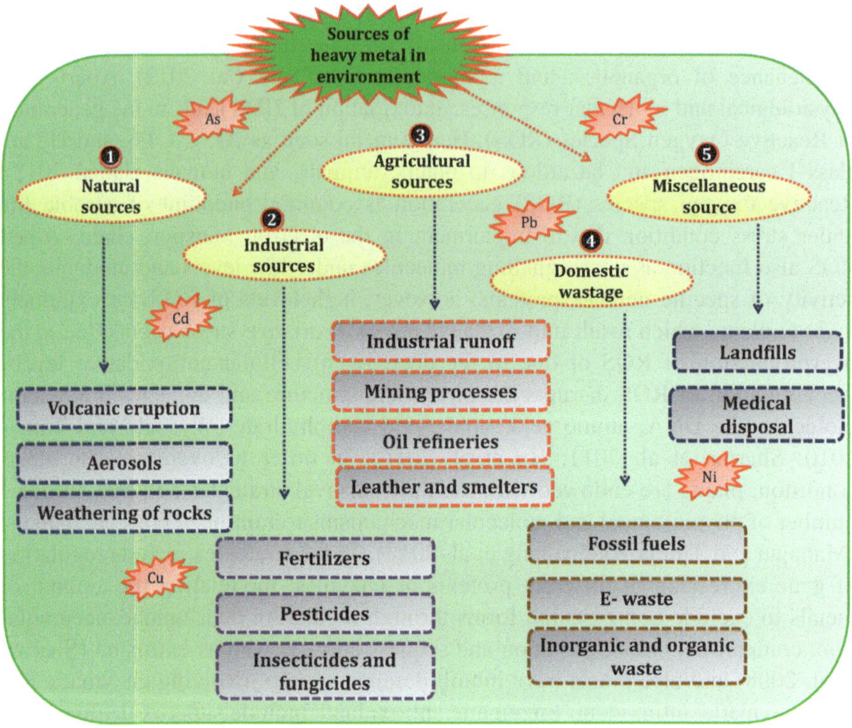


**Fig. 12.1** Various sources of heavy metals in the environment

lysyl oxidase, also acts as an electron donor in PS I (Chatterjee et al. 2006), and is an integral component of SOD and AOX. By the same token, Zn acts as a co-factor for enzyme carbonic anhydrase (CA), and SOD (Nagajyoti et al. 2010) and Ni is the essential component of enzyme urease while Mg and Mn participates in enzymes malic dehydrogenase and oxalo-succinic decarboxylase. Mn is also needed for water splitting at photosystem II and for SOD (Nagajyoti et al. 2010). In addition to the essential metal the sources for the toxic heavy metals in the environment have been convened into two groups, i.e., natural and anthropogenic origin. The natural sources such as weathering of geologic parent material, volcanic eruption, and rock outcroppings are the reason for the alleviated concentration of HMs in the environment. Sedimentary rock adds for higher concentration of Cr, Mn, Co, Ni, Cu, Zn, Cd, Sn, Hg, and Pb and volcanic eruption emits high amounts of Al, Zn, Mn, Pb, Ni, Cu, and Hg along with toxic and harmful gases (Seaward and Richardson 1990). Besides this, fly ashes and dust particle also add to the enhanced level of Mn, Zn, Cr, Ni, and Pb (Ross 1994). Other natural interferences such as marine aerosols and forest fires also exert an influential increase in HMs concentration. Carbonaceous matter produced during the fire contributes to the volatile heavy metals such as Hg and Se. Through the leaching from leaves and stems, decomposition, and volatilization processes, the emitted heavy metals enter the soil system. Several toxic HMs have been perceived in inland coastal areas payable by sea sprays and aerosols bent in oceanic events. Additional to the natural interventions the basis of anthropogenic origin such as use of inorganic and organic fertilizers liming, sewage sludge, irrigation waters, and pesticides serve as key sources of heavy metals to agricultural soil. Predominantly fungicides, inorganic fertilizers, and phosphate fertilizers have variable levels of Cd, Cr, Ni, Pb, and Zn depending on their sources. Although the levels of heavy metals in agricultural soil are very small, recurrent usage of phosphate fertilizer and their long persistence leads to high accumulation of toxic metals (Verkleji 1993). Moreover industrial sources including mining, refineries (spoil heaps, ores, smelters and metal finish and recycling of metals) emit diverse HMs depending on the category of mining. For example, coalmines are cradles of As, Cd, Fe, etc., which directly or indirectly enrich the soil around the coalfield. The consumption of Hg in gold mining mobilizes high concentration of Hg from gold mines that have become a significant source of this pollutant to the environment (Lacerda 1997). Finally the runoff from the industrial sources serves as a key source for elevated HMs concentration in rivers and lakes and thus contaminating the water bodies. Domestic effluents entail untreated waste waters, substances that escape through filters of biological treatmnets and all these are discharged to water bodies thus contaminating aquatic ecosystems. The indiscriminate use of detergents is a chronic pollution hazard, as it affects the water quality and causes eutrophication. Most enzyme detergents contain trace amounts of the elements such as Fe, Mn, Cr, Co, Zn, Sr, and B (Angino et al. 1970). Increased natural and man-made interventions have given away to airborne particulates. Reliant on prevalent climatic conditions, these particulates turn to be windly blown over great distances; however, they returned to the lithosphere as precipitations by rain or snowfall. Geothermal sources, such as volcanic eruptions, have caused significant atmospheric pollution (Eshleman et al. 1971). The elevated concentration of toxic metals enter in plant cell via various transporter involved and are responsible for the toxic responses in plant as discussed in following section.

## 12.3 Transport of Heavy Metals in Crop Plants

Due to increased anthropogenic activity and human interventions, the concentration of heavy metals has increased in the environment and contaminating soil and water system. The augmented level of heavy metal in the soil interferes with the growth and productivity of the plants. Metals are as actively taken up by plants *via* the root cells through specific transportation mechanisms and tend to accumulate in plant body. The increased concentration of heavy metal inside plant body significantly alters the growth and development processes. Metals enter the cell through secondary transporters, such as channel proteins and/or proton carrier proteins and bind easily due to negative charge on plasma membrane of plant cell (Mishra and Dubey 2006; Tangahu et al. 2011). Several pathways and mechanisms operate for the transportation of heavy metals, within the lipid bilayer membrane such as transporter proteins that are embedded to form either channel, or pore that facilitates the transport of metal ions inside plant system. Trans-membrane helices is a pore like structure either cysteine-rich or histidine-rich that mediates the transport of metal ions.

The metal transporters involved in the uptake of metal ions are categorized as

1. NRAMP (Natural Resistance-Associated Macrophage Protein),
2. CTR/COPT (Copper Transporter)
3. ZIP (ZTR/IRT related Proteins)

Beside this some aquaporins (AQP) also seem to be involved in transportation of certain metalloids such as arsenic (As) (Ali et al. 2009). The NRAMP transporters are mainly expressed in roots and are involved in the transportation of metals across the plasma membrane and the tonoplast (Maser et al. 2001). The NRAMP family plays major role in general metal ion transporters and can transport $Mn^{2+}$, $Zn^{2+}$, $Cu^{2+}$, $Fe^{2+}$, $Cd^{2+}$, $Ni^{2+}$, and $Co^{2+}$across the plant cell (Nevo and Nelson 2004). The NRAMP family was first discovered in rice where only three NRAMP (OsNramp 1–3) were reported (Belouchi et al. 1997). The plant NRAMPs are classified into two sub-families: one includes AtNramps 1 and 6 and the other AtNRAMPs 2–5 (Maser et al. 2001). The NRAMP proteins are highly conserved and have 12 trans-membrane domains that have a transport motif similar to animal shaker-type $K^+$ channel (Williams et al. 2000). The Fe deficient condition is up-regulated by AtNRAMP 1, 3, and 4 in *A. thaliana* roots; AtNRAMP3 is involved in Cd uptake and sensitivity (Thomine et al. 2000) and expressed in the vascular bundles of roots, stems, and leaves therefore also function in long-distance metal transport (Thomine et al. 2003). LeNRAMP1 is up-regulated by Fe deficiency and localizes to the root vascular parenchyma in roots of tomato (Bereczky et al. 2003). In barley, when enough amount of nitrogen is available then in presence of Cd, Nramp transcript is down-regulated but in nitrogen deficient condition, it is up-regulated (Finkemeier et al. 2003).

In eukaryotes, CTR/COPT transporters have been found, and are known as COPT transporters in plants (Sancenon et al. 2004) and Ctr in animals and fungi (Kim et al. 2013). These proteins have metal binding domain that possess three trans-membrane regions where they perform as homo or heterotrimer and form channel that specifically transport $Cu_2^+$ ions (Sinani et al. 2007; De Feo et al. 2010). Expression of COPT1 transporters occurred maximum in leaves and was also present in stems and flowers but absent in roots (Puig and Thiele 2002). In *Arabidopsis* the expression of additional five members COPT1 transporter family has been reported (Sancenon et al. 2004).

Further, divalent cations are transported by ZIP transporters across membranes, i.e., Fe, Mn, Zn, and Cd. There are about 85 ZIP transporters that have now been identified from bacteria, archaea, and all types of eukaryotes, including 15 genes in *Arabidopsis.* The predicted amino acid sequences alignment suggests that ZIPs can be grouped into four sub-families, while genes from the higher plants fall into a single group (Maser et al. 2001). The ZIP proteins have eight trans-membrane domains with the amino- and carboxyl-terminal ends are on the outer surface of the plasma membrane site (Guerinot 2000). Under the influence of metals as Zn or Fe there is expression of ZIP transporter proteins in *Arabidopsis thaliana* which is a stress-responsive protein (Nishida et al. 2008). In plants, transport of Ni is mediated by the expression of IRT1, while in yeast there are two types of ZIP *viz* ZRT1 and ZRT2 depending on the similarity with IRT1 (Thomine et al. 2000). In general two systems occur for Zn transport; one is activated under Zn deficient condition while second is activated when there is enough Zn (Claus and Chavarría-Krauser 2012) and former is encoded by gene ZRT1 while later by ZRT2 gene (Manara 2012). For transport of Fe and Zn there is expression of four major genes viz., IRT1, ZIP1, ZIP2, and ZIP3 as reported in *A. thaliana* (Lin and Aarts 2012). Apart from this, another Zn transporter named ZIP4 is identified by analysis of the genomic sequence in *A. thaliana* and expressed under Zn starvation (Kupper and Kochian 2010). Similarly, rice plant expresses OsZIP8 gene that encodes a membrane ZIP transporter in roots and shoots under Zn limiting conditions (Lee et al. 2010). Further, maize (*Zea mays*) expresses nine ZIP-encoding genes and all ZmZIP proteins are present on ER's plasma membrane. The up-regulation of these ZIPs in maize depends on developmental stages; as late developmental stages express ZmZIP6 and ZmIRT1 while early stages express ZmZIP4 and expression of ZmZIP5 occurs during development of endosperm and embryo (Li et al. 2013). Similarly in *Medicago truncatula*, the Zn homeostasis is mediated by up-regulation of three ZIP transporters such as MtZIP1, MtZIP6, and MtZIP5 (Stephens et al. 2011). Several other transporters involved in facilitating the entry of toxic metals and metalloid and their toxic effects have been discussed in Table 12.1.

Besides this several factors that affect the transportation of metal from soil to plant are soil pH, organic acids and bio-surfactants, chelators such as siderophores, redox reactions, and adsorption and desorption of metal ions (Rajkumar et al. 2012). The siderophores are the low-molecular-weight compounds (400–1000 kDa) that form complexes with metals such as Fe, Zn, Pb, Al, and Cd and solubilize metals present in unavailable forms and make them available to the plants (Schalk et al. 2011). By binding with chelators such as amino acids, nicotinamine, or organic acid, in root cells, the metal ions are loaded into xylem and then transported to shoots (Kramer et al. 1996). The transport protein mediates the transportation of metals from xylem to various part of plant and these are grouped into families, the P1B-type ATPases, MATEs, and OPTs. The P1B-type ATPases form a phosphorylated intermediate and are also called as HMAs (Heavy Metal-transporting ATPases). These have eight trans-membrane domains that consist of Cys-Pro-Cys/His/Ser motif regarded as involved in metal translocation (Ashrafi et al. 2011). MATE (Multidrug and Toxic Compounds Extrusion) proteins are membrane-bound transporters that expel out drugs and toxic compounds from the cell (Durrett et al. 2007).

**Table 12.1** Showing the transporter involved in metal uptake and the toxicity symptom in plants

| S. N. | Family of transporter | Plants | Gene | Metal transported | Expressed in plant part | Toxicity | References |
|---|---|---|---|---|---|---|---|
| 1. | Zinc regulated transporter (ZRT) | *Arabidopsis thaliana* | zip 1-12 | Zinc (Zn) | Roots | Stunted growth, leaf curling, and decease of leaf tip | Weber et al. (2004), Lopez-Millan et al. (2004) |
| | | *Oryza sativa* | zip 4 | | | | |
| | | *Medicago truncatula* | zip1 | | | | |
| 2. | P-type ATPase transporter | *Arabidopsis thaliana* | AtHMA1-8 | Copper (Cu) | Roots and Shoots | suppressed growth, reduced plant biomass, reduction in germination percentage decrease in plant chlorophyll and protein content | Xing et al. (2008), Talke et al. (2006) |
| | | *Arabidopsis halleri* | AhHMA3-4 | Zinc (Zn) | | | |
| | | *Lycopersicum esculentum* | TcHMA4, GmHMA8, OsHMA9 | Cadmium (Cd), Cobalt (Co), Lead (Pb) | | | |
| 3. | Natural resistance-associated microphage proteins (NRAMP) | *Lycopersicon esculentum* | Narmp1-3 | Iron (Fe) | Shoots and Roots | stunted top, thickened root growth and spotted leaf | Bereczky et al. (2003), Mizuno et al. (2005), Xiao et al. (2008) |
| | | *Thalspi japonocum* | Narmp4 | Iron (Fe) | | | |
| | | *Malus baccata* | Narmp1 | Iron (Fe) | | | |
| 4. | Copper transporter | *Arabidopsis thaliana* | Copt1 | Copper (Cu) | Plasma membrane and cytoplasm | Interveinal chlorosis reduced growth | Sancenon et al. (2004) |
| 5. | Yellow strip like (YSL) | *Arabidopsis thaliana* | Ysl2 | Iron I (Fe), Copper (Cu) | Roots and leaves | Reduces leaves size | DiDonato et al. (2004), Gendre et al. (2006) |
| | | *T. caerulescens* | Ysl3 | Iron (Fe), Nickel (Ni) | | | |

| | | | | | | | |
|---|---|---|---|---|---|---|---|
| 6. | Al active malate transporter (ALMT) | *Triticum* sp.<br>*Secale cereale* | Almt1 | Aluminium (Al) | Roots | Suppressed root development<br>Short root | Sasaki et al. (2004), Collins et al. (2008) |
| 7. | Cation diffusion facilitator (CDF) | *Arabidopsis thaliana*<br>*A. halleri* | Mtp1 | Zinc (Zn) | Leaves and Roots | Stunted growth, curling of leaf, reduction in growth of leaf tip and reduction in photosynthetic pigment content | Kawachi et al. (2008), Willems et al. (2007), Kim et al. (2004), Shingu et al. (2005) |
| | | *Thlaspi goesingense* | Mtp1 | Zinc (Zn), Nickel (Ni) | | | |
| | | *Nicotiana tabacum* | Mtp1 | Zinc (Zn), Cobalt (Co) | | | |
| 8. | Nicotianamine synthase (NAS) | *Arabidopsis halleri* | nas2, nas3 | Zinc (Zn) | Roots | Early leaf chlorosis and had sterile flowers | Talke et al. (2006) |
| 9. | Multi drug resistance efflux (MATE) | *Arabidopsis thaliana* | AtDTX1 | Cadmium (Cd) | Plasma membrane | Chlorosis, leaf rolls, and stunting include bronzing stippling of leaves discoloration of leaves | Li et al. (2002), Rogers and Guerinot (2002) |
| | | | FRD | Iron (Fe) | | | |
| 10. | Low affinity cation transporter | *Triticum aestivum* | LCT1 | Calcium(Ca), Cadmium(Cd) | Plasma membrane | Distorted or chlorotic with irregular margins, spotting or necrotic areas. | Schachtman et al. (1997), Clemens et al. (1998) |
| 11. | Magnesium transporter | *Arabidopsis thaliana* | AtMHX | Magnesium (Mg), Zinc (Zn) | Vascular cylinder | Reduced leaf size<br>Stunted growth | Li et al. (2001) |
| 12. | Cation/H+ antiporter | *Arabidopsis thaliana* | CAX1, CAX2 | Calcium(Ca), Cadmium (Cd), and Manganese (Mn) | Root tonoplast vesicle and vacuole | Reduced leaf size, distorted, discoloration<br>Stunted growth | Hirschi et al. (2000), Shigaki et al. (2003) |

## 12.4 Effects of Heavy Metals on Crop Plants

Some heavy metals (HMs) present in the soil like Cu, Zn, Mn, Mo, and Ni are essential for carrying out the physiological and biochemical processes (Shahid et al. 2015). But other HMs such as As, Pb, Hg, Cr, Al, Cd, etc. when present above their permissible limit cause great amount of loss in crop productivity around the globe (Pierart et al. 2015). These toxic metals cause abnormalities in normal functioning of plant cells and incur metabolic disorders leading to reduced growth in plants (Amari et al. 2017). After accumulating in the plants *via* roots, these heavy metals cause plants to produce massive amount of reactive oxygen species (ROS) such as superoxide radical ($O_2^{\bullet-}$), $H_2O_2$, and hydroxyl radical ($OH^-$) due to which redox status of the cells is severely affected and peroxidation of lipids occurs causing membrane damage (Singh et al. 2020) (Fig. 12.2).

### 12.4.1 Effect on Growth and Pigments

Heavy metals are taken up by the plant in a soluble state *via* root (Blaylock and Huang 2000). Many of the heavy metals are needed by the plants for maintenance but excessive levels can be toxic. The mechanism through which plants take essential heavy metals also paves the way for uptake of non-essential metals (Djingova and Kuleff 2000). These toxic heavy metals affect the growth and activities of soil microbes negatively. Many of the useful soil microbes have shown to inhibit their activities due to interference of heavy metals and thus plant growth is affected which can further lead to its death (Schaller and Diez 1991). For metals such as As, Hg, Pb, Cr, and Cd, even low concentrations have adverse effects on growth of the plants. Kibra (2008) reported that 1 mgHg/kg soil caused reduction in growth, plant height, and panicle formation in rice plant. Similarly reduced shoot and root growth were also seen in wheat plants with 5 mg/L Cd in soil solution (Ahmad et al. 2012). Even the essential heavy metals at their higher concentrations cause reduction in growth. In tomato plants, 50 mgCo/kg showed improvement in the overall growth but 100–250 mgCo/kg caused severe decline in growth parameters of the plant (Jayakumar et al. 2013). In *Lythrum salicaria,* Pb and Cu at 500 and 1000 mg/kg caused complete death of the plant which indicates in field areas where different heavy metals can co-exist could further instigate the damage caused to the plants (Nicholls and Mal 2003). In maize seedlings, effects of six heavy metals were seen (Cd, Cr, Co, Mn, and Pb) and plant showed reduction in overall growth and protein content and toxicity of these metals were seen as: Cd > Co > Hg > Mn > Pb > Cr.

Light absorption is the critical step of photosynthesis through two major classes of photosynthetic pigments, i.e., chlorophylls and carotenoids. Under heavy metal stress conditions due to decrease of photosynthetic pigment the photosynthesis also decreased as reported in variety of plant species (Kaur et al. 2018). Reduced levels of chlorophylls in response to various toxic heavy metals are majorly seen due to inhibition of chlorophyll biosynthesis with impairment of several reactions of its

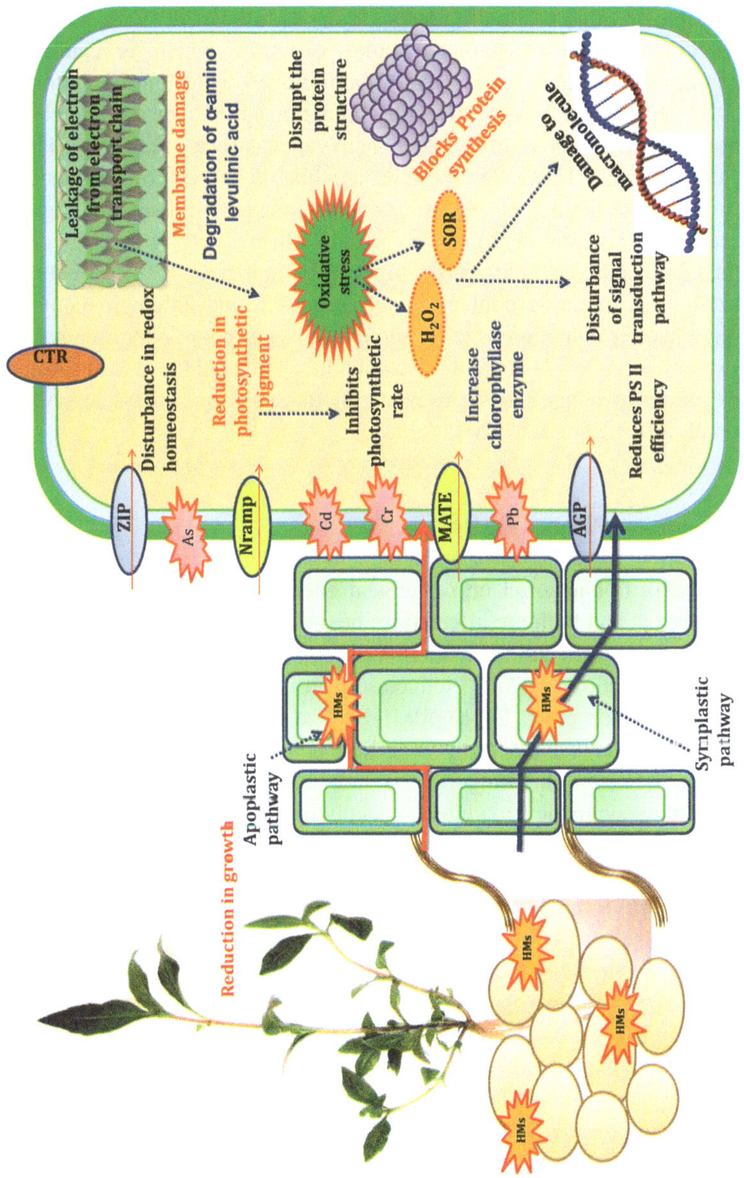


**Fig. 12.2** Pathway of metal transportation and toxicity mediated by heavy metals

synthesis pathways (Xue et al. 2013). Chlorophyll biosynthesis depends on its precursor δ-aminolevulinic acid (ALA) in the plastids. ALA synthesis is the regulatory step of chlorophyll biosynthesis and it occurs by glutamate reduction (Parmar et al. 2013). Many of the heavy metals such as Mn, Ni, Co, Cd, and Pb have negatively affected this rate limiting step. Heavy metals (Al, Cd, Hg, Pb, Se) also affect the condensation of two molecules of ALA which is catalysed by δ-aminolevulinate dehydratase, to form porphobilinogen, a precursor for tetrapyrrole synthesis (Skrebsky et al. 2008). Chlorophyll function can also be inhibited by $Mg^{2+}$ substitution of the chlorophyll molecule by heavy metals such as Zn, Cu, Cd, and Hg (Kupper et al. 1998). These chlorophylls are unable to perform photosynthesis because of their interference of resonance energy transfer from antenna pigment to reaction centres (Karukstis 1991). Heavy metals influence on carotenoid content is less examined compared to chlorophyll molecules (Kaur et al. 2018). Effect of these heavy metals on carotenoid content is not consistent and although many of the studies have shown reduction of carotenoid levels but some experimental works have shown enhanced levels of carotenoid content which could be due to carotenoids also work as antioxidants and show plant tolerance towards these heavy metals (Parmar et al. 2013; Kaur et al. 2018).

### 12.4.2 Effect on Photosynthesis

The solar energy is converted into chemical energy by PSI and PSII membrane protein complexes and both are affected by heavy metal toxicity. PSI catalyses the oxidation of plastocyanin and the reduction of ferredoxin under light conditions and reaction centre, ferredoxin-$NADP^+$ reductase, and iron-sulphur centres are its main inhibitory sites (Kaur et al. 2018). PSII is as a multi-subunit pigment–protein complex with the enzymatic activity of light-dependent water-oxidizing plastoquinone reductase (Nelson and Junge 2015) and susceptible to various stresses and damaged the photosynthetic apparatus especially in case of heavy metal stress (Kaur et al. 2018). Generally, these heavy metals disrupt the functioning of oxygen-evolving complexes (OECs) and harm the proteins on the PSII donor and acceptor side, quinone B (QB) and the pheophytin-iron-quinone A (Pheo-Fe-QA) which finally leads to decline in PSII quantum yield (Karimi et al. 2013). The reduction seen in the photochemical activities of PSII was observed through a change of thylakoid membranes (Tanyolac et al. 2007). Photoinhibition occurs on the PSII acceptor and donor side and is triggered by Al (Jiang et al. 2008; Moustaka et al. 2016). Al stressed plants of *Citrus grandis* L. showed photoinhibition at PSII sites and decline in quantum yield of electron transport chain from QA to PSI (Jiang et al. 2008). In *Triticum aestivum*, Al stress brought closure of reaction centre of PSII which further led to decline in the ratio of oxidized to reduced QA and diminished PSII excitation pressure and quantum yield (Moustaka et al. 2016). Arsenic (As) stress in mustard seedlings diminished the quantum yield of primary photochemistry, electron transport, yield of electron transport per trapped exciton and performance index of PSII (Singh et al. 2020). Quantum yield of PSII in As stressed soybean seedlings was also found to be reduced (Piršelova et al. 2016). But in contrast, As-hyperaccumulators spp. *Pteris cretica* and *Isatis cappadocica* showed no change in quantum efficiency of PSII (Hong Bin et al. 2012; Karimi et al. 2013). Electron transport is inhibited by Cd through the reducing side of PSI. Electron transfer was interfered by Cd in maize seedlings from redox-active tyrosine (Tyr) residues D1-161 (Ishikita and Knapp 2001). Further, Cd was seen to decline maximum quantum yield in soybean seedlings and PSI and PSII activities weakened in *Schima superba* Gardner and Champ (Piršelova et al. 2016). Cr by large affects the OEC and QA reduction as it was related to alteration in D1 protein turnover and change in 24 and 33 kDa proteins of the OEC (Ali et al. 2006). Stress of Cr also propagates reduction in the quantity of active reaction centres of PSII, lower rate of electron transport, and thus alteration in antenna size heterogeneity of PSII (Mathur et al. 2016). Pb stress mainly inhibits electron transport in PSII than PSI. In *Spirodela polyrrhiza* seedlings, Pb got accumulated in PSII and disrupted its secondary structure, lowering the visible light absorbance, and the energy transfer among amino acids in PSII protein–pigment complex and energy transport from tyrosine residue to chlorophyll were reduced (Qufei and Fashui 2009). Pb was also seen to block the electron transport between plastocyanin and P700, and further led to restricting electron transport to the donor site of PSI (Belatik et al. 2013). Deng et al. (2013) also saw reduction in quantum yield and electron transport where restriction was enhanced in the donor site of PSI and PSII in *Microsorum pteropus* with increasing levels of Hg.

### 12.4.3 Effect on Oxidative Stress

As heavy metal stress affects the photosynthetic apparatus of plants, it also leads to generation of reactive oxygen species (ROS). Accumulation of ROS in plants is depended on generation and scavenging balance of ROS (Gill and Tuteja 2010), which further depends on conditions of growth like temperature, humidity, heavy metal intake and light intensity, etc. Therefore, excess uptake of heavy metals becomes a limiting factor of $CO_2$ fixation in chloroplasts and combined with further decline of electron transport chain of photosynthetic apparatus paves way for major ROS generation site (Gill and Tuteja 2010). In mitochondria, huge reduction of electron transport chains also serves a key site to production of ROS (Keunen et al. 2011). About 1–5% of oxygen consumed by isolated mitochondria is converted into ROS (Moller et al. 2007). Hydrogen peroxide ($H_2O_2$) is generated in the peroxisomes during photorespiration in which oxidation of glycolate to glyoxylic acid takes place. Thus, different ROS like singlet oxygen ($1O_2$), superoxide anion ($O^{\bullet-2}$), $H_2O_2$, and hydroxyl radicals ($^{\bullet}OH$) are generated in these organelles due to spin inversion and different electron transfer reactions to $O_2$ in the electron transport chains (Gill and Tuteja 2010). Heavy metals namely Cu, Cd, Cr, Hg, Fe, Pb, Al, As, and Zn induce ROS production in plants (Marquez-Garcia et al. 2011; Shahid et al. 2015; Singh et al. 2016). But the occurrence of ROS generation differs for redox-active and redox-inactive heavy metals (Opdenakker et al. 2012). Redox-active heavy metals such as Fe and Cu participate in Haber–Weiss/Fenton reactions where at neutral pH, $H_2O_2$ breaks down to $OH^{\bullet}$ (Keunen et al. 2011). But in case of redox-inactive metals, namely Cd, Pb, Hg, As, Zn, and Ni, these inhibit enzymatic activities due to affinity for –SH groups on the enzyme (Cuypers et al. 2011). The redox-inactive heavy metals due to affinity for electron sharing form covalent bonds with protein sulfhydryl groups. These metals after interacting with enzymes make them inactive mainly at the catalytic site. Enzymes are inactivated by heavy metals such as Pb where it binds to functional groups (−COOH) present in proteins. Further, it is seen that ROS balance in plant cells is disrupted by heavy metals by the displacement of essential cations which finally leads to overproduction of ROS. For example, toxic heavy metals can replace Zn which acts as co-factor for enzymes causing its inhibition and oxidative stress. Toxic heavy metals are also found to deplete GSH levels inside cells (Lee et al. 2012; Pourrut et al. 2013). This way, major antioxidant activity is greatly reduced due to which ROS imbalance occurs. These metals by binding and reducing GSH and its derivatives further cease the ability of stressed plants to scavenge the ROS levels inside the cell (Lee et al. 2012). Oxidative stress is induced by heavy metals involving plasma membrane-bound NADPH oxidase. These NADPH oxidases utilize the cytosolic NADPH causing production of superoxide radical ($O2^{\bullet-}$) which by SOD is dismutated to $H_2O_2$ (Pourrut et al. 2013). ROS production induced by toxic heavy metals through NADPH oxidase was studied in Cd stressed pea, Ni stressed *Triticum duram*, and Pb stressed seedlings of *Vicia faba* (Rodriguez-Serrano et al. 2006; Hao et al. 2006).

### 12.4.4 Effect on Macromolecule

ROS induced by toxic heavy metals can provoke extensive destruction to plants such as lipid peroxidation, enzyme inhibition, protein oxidation, and damage to the RNA and DNA (Cuypers et al. 2011). Pourrut et al. (2011) have reported that ROS induced indirect effect of these heavy metals are more toxic in comparison with the direct effects. Therefore, the initiation of heavy metal toxicity is done by the ROS (Shahid et al. 2015). Inside the plant, primary target site of toxic heavy metals is plasma cell membranes (Cuypers et al. 2011). Like various other abiotic stresses, lipid peroxidation is caused through ROS generation by the toxic heavy metals (Shahid et al. 2015; Raju et al. 2020; Singh et al. 2020). Damage of cell membranes by lipid peroxidation due to exposure of heavy metals is one of the most vicious effects seen in plants (Pourrut et al. 2013). Enhanced lipoxygenase activity can initiate oxylipins formation which finally leads to lipid peroxidation (Gill and Tuteja 2010). Heavy metal induced oxidative stress has been related to increased lipoxygenase activity in *A. thaliana* and *Lessonia nigrescens* (Vanhoudt et al. 2011; Kumar et al. 2012). PSII inhibition is generally seen with heavy metal exposure to plants which causes increased $O2^{\bullet-}$ formation in leaves due to which peroxidation of lipids occurs (Farmer and Mueller 2013). In plant tissues where chlorophyll is not present, lipid peroxidation occurs due to $OH^{\bullet}$, a radical formed by Fe- or Cu-catalysed degradation of $H_2O_2$ (Farmer and Mueller 2013). Although $O_2^{\bullet-}$

and $H_2O_2$ are apt in starting the reactions responsible for lipid peroxidation, only $OH^{\bullet}$ is sufficiently reactive, mostly in the presence of transition metals such as Cu or Fe (Pourrut et al. 2013). Among ROS, most damage of the DNA components is done by $OH^{\bullet}$ (Jones et al. 2011). ROS interactions with DNA molecules result in base deletions, strand breaks, and damage to cross-links and pyrimidine dimers (Gill and Tuteja 2010; Gastaldo et al. 2008). Among various DNA sites which can be affected, base deletion is mostly induced by heavy metals (Gastaldo et al. 2008). The base ring nitrogens, exocyclic base keto groups, ribose hydroxyls, and negatively charged phosphate oxygen atoms are the four potential sites of DNA on which these toxic heavy metals can bind (Oliveira et al. 2008). Heavy metals Cu, Hg, Pb, Cd, and Cr bind with DNA at sulfhydryl groups and the phosphate backbone (Sheng et al. 2008). Heavy metals form chromosomal breaks or mitotic anomalies which result in micronuclei formation and cause extensive damage to the DNA (Marcato-Romain et al. 2009). Spindle apparatus of the dividing cells are interfered with heavy metals to cause DNA damage (Gastaldo et al. 2008). Genotoxicity was seen in *Phaseolus vulgaris* seedlings exposed to different heavy metals such as Hg, Cr, B, and Zn with concentrations of 150 and 350 ppm. Genotoxicity was also seen along with relation to chloroplast damage in Pb stressed *Brassica rapa* with use of RAPD profile in which lead nitrate concentration ranged from 0.5 to 5 mM (Cenkci et al. 2010). In *Vicia faba* exposed to Pb stress, micronuclei were formed in root tips through formation of ROS (Shahid et al. 2011). This study was further confirmed by Pourrut et al. (2011) in which oxidative stress was closely linked with DNA strand breaks and micronuclei production *Vicia faba* root tips exposed to Pb stress.

Another major target of heavy metals induced ROS damage is protein synthesis (Singh et al. 2020). The alteration in protein structure and content due to heavy metals can occur through various mechanisms like replacement of Mg and other essential metal ions with free heavy metal ions in the various metal dependent proteins, binding of heavy metal ions. Wherever the ROS is induced by heavy metals, they interact with sulphur containing amino acids and thiol groups of proteins (Shahid et al. 2015). Heavy metals ions by binding with –COOH, $-NH_2$, and –SH groups form complexes with proteins (Tan et al. 2010). Therefore, these altered molecules are not able to function properly because of the structural modifications and cannot attach itself to substrate side and thus causes malfunction of the cell. When heavy metals bind to these active groups of proteins, they inactivate different enzyme systems, or alter protein structure, which is related to the catalytic properties of enzymes. Amino acid side groups like Trp, Met, His, Pro, Lys, Arg, Tyr, and Cys are oxidized by ROS induced heavy metals (Shahid et al. 2015). Carbohydrates are the universal energy sources and key macromolecules for plant metabolism and function (Guan-fu 2011). They are the major products of photosynthesis and wholly involved in growth, storage, and development of plants. They are also responsible for maintaining integrity of plasma membrane (Guan-fu 2011). Carbohydrate content of plants is affected due to heavy metals stress induced by ROS. Expression of transporter genes of carbohydrates which is induced by toxic heavy metals can decline the oxidative damage caused by ROS overproduction (Nguyen et al. 2010). Glucose is also found to increase the chances of plant survival against H2O2 toxicity by speeding the cellular defences (Averill-Bates and Przybytkowski 1994). Carbohydrate structure is altered due to ROS and these energy sources are unable to sustain metabolic activities of plant (Zadak et al. 2009). All of this leads to weakening of plant defence systems and reduced food supply and storage make these plants more vulnerable to heavy metal toxicity.

## 12.5 Role of ROS as Signalling Molecule

Appearance of first oxygen-evolving photo-autotrophs i.e. cyanobacteria around 2.7 billion years ago is responsible for the acquaintance of oxygen in the environment and initiation of aerobic metabolism. During the process of photosynthesis, autotrophs generate ROS which is an unavoidable phenomenon. The ROS are highly reactive and excessive accumulation of ROS inside the cell turns to be toxic as it directly damages the macromolecules like DNA, RNA, protein, and lipids and induces mutation (Sharma et al. 2012). However, under low concentrations, the ROS behaves as signalling molecule that regulates the various physiological and biochemical processes.

Among ROS, $H_2O_2$ which is a potent toxicant is commonly produced in cellular organelles such as mitochondria, chloroplast, and peroxisome. Several studies have been focused illuminating that $H_2O_2$ plays a substantial role as a signalling molecule and directs several physiological and developmental pathways such as induction of programmed cell death (Vavilala et al. 2015), seed germination (Barba-Espín et al. 2011), development of roots (Ma et al. 2014), regulation of stomatal movement (Ge et al. 2015), and induction of flower by the expression of specific genes and proteins. The high stability and mobility across the membrane and having half-life in milliseconds suggest its potent role as signalling molecule. It is present in range of μmol per gram fresh weight in the leaf extracts and the concentration increases from base to tip of leaf. The $H_2O_2$ via simple diffusion or presence of specific $H_2O_2$ transporter or channels is transported from its site of production towards the specific cell (Stocker et al. 2018). A special class of aquaporins (AQP) named peroxiporins is majorly involved in the transport of $H_2O_2$ due to its similarity with water molecule as reported in *Arabidopsis* (Tian et al. 2016). In the study of Hooijmaijers et al. (2012), it was observed that exogenous treatment of $H_2O_2$ can alter aquaporin expression, indicating a feedback loop between $H_2O_2$ concentration and peroxiporin expression.

Role of $H_2O_2$ as a signalling molecule was first identified by the study carried by (Desikan et al. 2005) which reports that the functions of ethylene under exogenous $H_2O_2$ supplementation and $H_2O_2$ was found to itself bind with the receptor of ethylene, RET1. Similarly in kimchi variety of cabbage (*Brassica rapa*) the cell death triggered by nitrosative stress is increased by $H_2O_2$ (Kim et al. 2015). Further, the enhancement of thermo-tolerance in maize seedlings is arbitrated by application of $H_2O_2$ that explained the cross talk between NO and hydrogen sulphide ($H_2S$) (Li et al. 2015). In study of Gonzalez et al. (2012), exogenous $H_2O_2$ diminished the Cu stress in *Ulva compressa* by regulating the NO and $Ca^{2+}$ level. Stomatal movement in plants under the influence of brassinosteroids was regulated by endogenous level of $H_2O_2$ at every stage of plant life and under various abiotic stress conditions (Shi et al. 2015). $H_2O_2$ also triggers signalling molecules such as hormones and protein kinase. The interaction of NO and $H_2O_2$ regulates root growth under zinc stress and assuaged zinc stress through enhancing antioxidant system, declining lipid peroxidation as well as up-regulating resistance gene expression as reported in wheat seedlings (Duan et al. 2015). Moreover, $H_2O_2$ and $Ca_2^+$ signalling also operates in ABA responses to drought stress in *A. thaliana* through $Ca_2^+$-dependent protein kinase 8 (CPK8) which regulate catalase 3 (CAT3) activity arbitrating stomatal movement (Zou et al. 2015). Parallelly, Qiao et al. (2015) reported that a $Ca_2^+$-binding protein (rice annexin OsANN1) possibly enhances heat stress tolerance by modulating $H_2O_2$ production. Overproduction of $H_2O_2$ induced by heat stress increased *OsANN*1 expression and up-regulated the level of SOD and CAT expression, which constructed a signalling mechanism for stress defence in plants (Qiao et al. 2015).

Further, senescence, an important strategy to recover and recycle nutrients, is regulated by ambient concentrations of $H_2O_2$ and was first discovered by Bieker et al. (2012). Concentrations of $H_2O_2$-scavanging enzymes inside the senescence part/leaf are regulated timely. $H_2O_2$ concentration also regulates the programmed cell death (PCD) correlated with activation of hormones like ethylene (Wang et al. 2017) or salicylic acid (Liu et al. 2015a, b). It was found that concentration of $H_2O_2$ enhanced at initial point of leaf senescence and peaked again during the terminal stage (Jajić et al. 2015). Furthermore, transgenic lines with lower $H_2O_2$ levels display delayed senescence (Bieker et al. 2012). In brief, $H_2O_2$ was once considered as a poisonous molecule in plants but based on current studies, $H_2O_2$ may be a vital signalling molecule which controls plant growth and development.

The developmental processes in plants are governed by two key steps (1) cell division and (2) cellular differentiation and both steps are initially regulated by ROS homeostasis (Tsukagoshi et al. 2010). In *A. thaliana* the root development is controlled by ROS as oxygen radicals ($O_2^{⍰}$ ) get accumulated in the meristematic region while $H_2O_2$ accumulates in the elongation region and homeostasis is kept under control of transcription factor named UPBEAT1 (UPB1) that found to be up-regulated in the transition root zone (Dunand et al. 2007; Tsukagoshi et al. 2010). Transcription factor named PFT1/MED25 (phytochrome and flowering time 1) and MED8 are required for the differentiation of root hairs and expressed at the sites where PFT1 positively regulates the synthesis of ROS by activating class II peroxidases (Xu and Li 2011). Apart from cell division, seed germination or breaking of the seed dormancy is also regulated by ROS levels as it plays an important role during seed germination; ROS are known to play role in corrosion of endosperm, mobilization of reserve food of seeds, and pathogen defence (El-Maarouf-Bouteau et al. 2014). Further, exogenous $H_2O_2$ supplementation reduces the level of ABA associated with increase in the GA concentration that helps in breaking the seed dormancy (Oracz and Karpinski 2016). Although ROS are largely perceived as either negative by-products of aerobic metabolism or makers for plant stress, but their functions as a signalling molecule and vital role in growth and development still need to be explored.

## 12.6 Molecular Responses Against Heavy Metals

Plants are unceasingly facing adverse environmental condition and due to their sessile nature they exhibit altered physiology and biochemical processes. Among abiotic components the heavy metal toxicity is the commonly bared issue that affects the growth and productivity of plants. Furthermore, some heavy metals such as Fe, Cu, Co, and Zn act as a co-factor in enzymatic reactions; however, excessive concentration of these metals generates oxidative stress and are damaging for both plants and humans and are very toxic at low concentrations (Shi et al. 2019). The HMs are present in the soils in form of ions or combined with oxygen to form oxides and actively taken up by roots and travel via xylem to different parts. The uptake, translocation, detoxification, and accumulation of HMs are key processes for metal accumulation in plants. These processes are controlled by various physiological and molecular regulatory mechanisms. Plants that experience excessive concentration of heavy metals up-regulate the expression of various proteins/genes under the molecular level to minimize the toxicity.

Under unfavourable environmental stresses, a common stress-inducible protein first identified in *E. coli* by using matrix-assisted laser desorption/ionization (MALDI) analysis has molecular weight of 13.5 kDa and named universal stress protein (USP) and ubiquitously present in plants (Kvint et al. 2003). The USPs are shown to involve in proper protein folding, preventing the denaturation of macromolecules, and enhance the protein trafficking inside the cell (Vollmer and Bark 2018) and also exhibit DNA binding, repairing, and refolding activities that can support organisms to protect their nucleic acids from external stresses (Drumm et al. 2009). All USPs have a metal binding domain and other cytosolic motifs and their expression is tissue specific or at various developmental stages under stress conditions (Wang et al. 2017). Study demonstrates that *Brassica napus* have large number of USPs i.e. have 142 USPs in plant kingdom followed by *Triticum aestivum*, *Brassica rapa*, *Solanum lycopersicum*, *Solanum tuberosum* that have 123, 71, 42, 41 USPs, respectively. Catalytic motifs are various types such as serine/threonine kinase, tyrosine kinase, U-box, SWI2/snf2 and Mudr (SWIM)-zinc finger, HomeoDomain leucine zipper (HDzip), and cation exchanger C1 motif that enhances the protection strategies from foreign stresses (Kerk et al. 2003). The fusion of a cation exchange motif with the USP domain endows plants with protection from sodium toxicity. Treatment with various stress-inducing factors including heavy metals mainly increases the activity of cotton USP promoters. Activity of a 949 bp fragment of the cotton USP promoter is significantly increased in transgenic tobaccos during the stress treatment, as shown in USP mRNA levels (Zahur et al. 2009).

Besides role of USPs, various genes and proteins actively participate in the detoxification and sequestration of heavy metals in the plant cells. These proteins either excrete the heavy metals from the cell (in case of arsenic) or enhance the accumulation of metals (in case of other metals). These genes mainly belong to the gene families such as NRAMP family, HMA family, ATP-binding cassette (ABC) family, and the Cation Diffusion Facilitator (CDF) family (Cailliatte et al. 2009). For example in *A. thaliana* homeostasis of heavy metals as Cd is mainly governed by the action of AtNRAMP6 protein and transport of $Co_2^+$, $Zn_2^+$, $Cd_2^+$, and $Pb_2^+$ into vacuoles is mediated by the actions of AtHMA3 (Morel et al. 2009). In *S. plumbizincicola* the Cd and Zn hyperaccumulator, SpHMA3 is a tonoplast-localized transporter and plays a pivotal role in the vacuolar storage of HMs (Liu et al. 2017). AtABCC1 and AtABCC2 are responsible for the transport of Cd-PCs and Hg-PCs into vacuoles for HM storage in *A. thaliana* (Park et al. 2012). Proteins of CDF family are integral membrane component having six trans-membrane domains (TMDs) that perform transport of metal ions out of the cytoplasm into extracellular space or into the vacuoles (Montanini et al. 2007). In plants the CDF family proteins are well known as metal tolerance proteins (MTPs) and are mainly three types Zn-CDF, Zn/Fe-CDF, and Mn-CDF and are seven types as MTP1, MTP5, MTP6, MTP7, MTP8, MTP9, and MTP12 that are differentially expressed in different plants (Gustin et al. 2011). Transport for HM ions, including $Mn_2^+$, $Fe_2^+$, $Zn_2^+$, $Cd_2^+$, and $Ni_2^+$, from the cytosol to the outside of the cell or into intracellular compartments is exclusively mediated by CDF family members (Podar et al. 2012; Ricachenevsky et al. 2013). A detailed study was performed by Ram et al. (2019) to show the expression of these MTPs in *Oryza sativa* under different time intervals and in different plant parts. After clear observations, it was found that expression of MTPs increased with increase in time under the Co, Cd, and Ni exposure and after one week of treatment expression of most MTP genes was down-regulated in both root and shoot tissues. Apart of transport proteins, secondary signalling pathway also plays a major role in tolerance against metal stress mediated by mitogen-activated protein kinase (MAPK) family. In *Oryza sativa* there is up-regulation of two different types of MAPKs as *OsMPK3* in leaves, *OsMPK3* and *OsMPK4* in roots under increasing As concentration in leaves and roots (Rao et al. 2011). Similarly, 50 and 150 μM concentration of As significantly up-regulate the expression of 46 kDa MAPK in *Brassica juncea* (Gupta et al. 2009). Besides its major role in growth and development, regulation of plant hormones and its related genes also mediate the stress tolerance. After treated with arsenate, *Oryza sativa* significantly up-regulated the genes needed for synthesis of jasmonic acid (JA) (*OsDAD*, *OsLOX*, *OsAOS*, and *OsPCL*), abscisic acid (ABA) (*OsNCED*), ethylene (*OsACS*), cytokinin (*OsLOG*), and gibberellins (GA) (*GA2ox3*) (Huang et al. 2012; Yu et al. 2012). Concentration of plant hormones works in efficient manner under the control of several miRNAs. Under As-stress there is accumulation of miR159 and miR319 noticed that enhance the stress hormones; ABA and JA (Srivastava et al. 2012).

To combat negative effects induced by metals, plants have an array of enzymatic and non-enzymatic antioxidant defence systems and among them ascorbate-glutathione cycle plays prominent role in scavenging ROS. Metal detoxification is primarily mediated by glutathione (GSH) and enzyme glutathione S-transferases (GSTs) (Zhang and Ge 2008). Basic structure of GST includes the presence of N-terminal thioredoxin-like domain of GSH binding site and C-terminal helical bundle domain with the hydrophobic substrate binding site and plant GSTs appear to be grouped into six classes and mainly are tau (GSTU) and phi (GSTF) (Cummins et al. 2011). Up-regulation of gene AtGSTF8 was noticed under high concentration of Cd in *Arabidopsis* which encodes Phi8 (type of GST) and involve in metal detoxification (Hou et al. 2019).

In order to minimize the negative effects induced by heavy metals, plants produce phytochelatins (PCs) that are small cysteine-rich glutathione (GSH) oligomers, that is responsible for the metal chelation and sequestration inside the vacuole and stimulated by ABC transporters (Pochodylo and Aristilde 2017). Formation of PC is mediated by some specific genes and enzyme majorly phytochelatin syntheses (PCs) (Emamverdian et al. 2015). The level of PCs is diverse among different plant species and it is a post-translational event regulated by concentrations of metalloids. In some cases, up-regulation of phytochelatin synthase (PCs) gene in plants does not ensure the tolerance behaviour of plants against metal stress. For example, overexpression of AtPCS1 in *A. thaliana* induced hypersensitivity against Cd and Zn without imparting metal tolerance (Lee et al. 2003). For metal sequestration the heavy metal transporter gene (HMT1) participates in the synthesis of ABC type transporters and is also responsible for the translocation of low-molecular-weight PC complex (LMW-PC) with heavy metals into the vacuoles (Cobbett and Goldsbrough 2002). Further, for the detoxification of As and Cd *Arabidopsis* plant expresses the transporter proteins ABCC1 and ABCC2, respectively, and also some metal ions such as Zn, Cu, Mn, and Fe increased the PCs transport efficiency in barley vacuoles, attesting that PCs might contribute to both the homeostasis of important metals and detoxification of non-essential toxic metal(loid)s in plants (Song et al. 2014). In number of plants such as *Avicennia marina*, *Kandelia obovata*, and *Cicer arietinum* the PCs synthesis participates in metal detoxification. (Al-Huqail et al. 2017; Dai et al. 2017)

Beside PCs, metallothioneins (MTs) are metal binding cysteine-rich protein actively mobilizes and sequestrate the HMs (Pirzadeh and Shahpiri 2016). The distinct role of MTs comprises (a) homeostasis of essential metal ions, (b) confiscation of the toxic heavy metals, and (c) minimizing the oxidative stress induced by stress factors (Liu et al. 2015a, b). Further, homeostasis of essential metal ions are also important as Cu directly regulate growth of plants by acting as co-factor in photosynthesis and ambient Cu concentration is maintained by the expression of MTs. For example MTs deficient plants accumulate less Cu as compared to wild type plants that have MTs proteins (Benatti et al. 2014). In plants, metal sequestration takes place inside the vacuoles after formation of MT-HMs complex but the exact mechanism that mediates the transport of this complex towards the vacuoles from cytoplasm is not so clear (Yang et al. 2011). Besides this, MTs expression is tissues specifics as phloem express MT1a and MT2b, mesophyll cell and root tips express MT2a and MT3 (Hassinen et al. 2011). Similarly, two OsMT genes (OsMT1b and OsMT2c) have been reported that bind to metal ions and scavenge ROS accumulation (Liu et al. 2015a, b; Pirzadeh and Shahpiri 2016) under Cr stress. Overall, it was demonstrated that the MTs in plants stimulate metal chelation and decontamination *via* scavenging ROS. Together with the PCs and MTs in protecting plants under damaging conditions heat shock proteins (HSPs) and heat shock factors (HSFs) also play prominent role under challenging conditions (Al-Whaibi 2011). HSFs are transcriptional activators that recognize the stress signals and result in the transcription of HSP-encoding genes (Reddy et al. 2014). These HSPs are also known as molecular chaperones or stress-responsive proteins that accumulate at higher levels to protect other proteins from being dysfunctional. The rapid increase in the level of HSPs upon exposure to various stresses indicates their importance in stress response. The HSPs belong to six major classes on the basis of the approximate molecular mass and nature of function: small heat shock proteins (sHSPs), HSP40 (J-proteins), HSP60 (chaperonins), HSP70 (DnaK), HSP90, and HSP100 (Clp proteins). Controversy with the expression of these HSPs is that some workers stated that they are only being expressed when plants face heat stress but a number of studies in the modern era describe its emerging role in mitigating the various kind of stress including heavy metals (Singh et al. 2016). This indicates that HSP chaperones are integral members of stress-responsive cascade and therefore, considering their importance, genome-wide studies have been performed in several model crops such as *Arabidopsis*, and foxtail millet (Sarkar et al. 2009; Singh et al. 2016). In brief, it is possible that the HSPs family of proteins might play diverse roles in modulating signal transduction pathways under normal growth condition as well as abiotic stress; the overall response is discussed in Fig. 12.3.

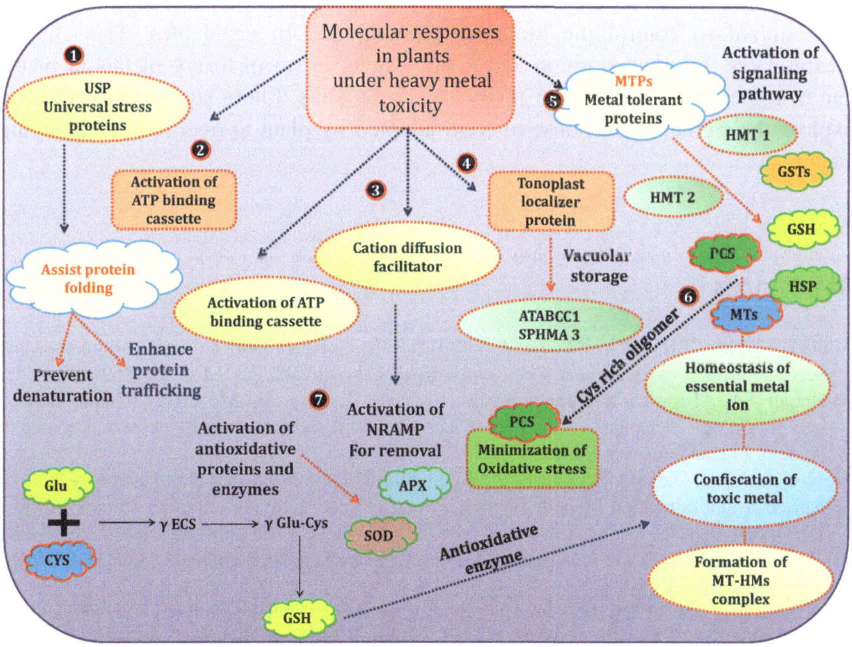


**Fig. 12.3** Molecular responses in plants under heavy metal toxicity

## 12.7 Conclusion and Future Perspective

In the recent scenario, heavy metal toxicity is one of the major problems acknowledged globally and a serious concern for the plant, animal, and human health. It also deteriorates the quality of environment. Various natural and anthropogenic processes release these heavy metals into the ambient environment and contaminate the agricultural soils associated with health hazard due to dietary intake of contaminated vegetables. The increased industrialization and human interventions has ultimately resulted into the heavy metal contamination of soil that disturbs the uptake and transportation of essential elements and thereby reduces the plant growth and causes nutrient scarcity, ROS generation, and reduction in photosynthetic activity. Plants exposed to heavy metals stress show accumulation via specific transportation that modifies the physiological and biochemical process and directly affects plant growth and yield. The disturbance in the electron transport chain generates ROS. In general, ROS are in equilibrium with the help of antioxidants and participated in cell signalling by acts as signalling molecules while imposed adverse effect by interacting with macromolecules under high concentrations. However, activation of molecular stress responses in plants against heavy metals regulates various physiological, biochemical, and molecular reactions. Thus intrusion of heavy metals in food chain and its toxicity is a serious concern. Furthermore, strategy and policy are required to control the limits of accumulation in vegetables. This chapter recapitulates the transportation and toxicity mechanism of heavy metals in plants and also their regulation under molecular level. Thus, future study thus needed to explore the signalling response of ROS adopted by plant to overawed challenging condition.

## References

Ahmad I, Akhtar MJ, Zahir ZA, Jamil A (2012) Effect of cadmium on seed germination and seedling growth of four wheat (*Triticum aestivum* L.) cultivars. Pak J Bot 44:1569–1574

Al-Huqail AA, Al-Rashed SA, Ibrahim MM, El-Gaaly GA, Qureshi MI (2017) Arsenic induced eco-physiological changes in Chickpea (Cicer arietinum) and protection by gypsum, a source of sulphur and calcium. Sci Hortic 217:226–233

Ali NA, Dewez D, Didur O, Popovic R (2006) Inhibition of photosystem II photochemistry by Cr is caused by the alteration of both D1 protein and oxygen evolving complex. Photosynth Res 89:81–87

Ali W, Isayenkov SV, Zhao FJ, Maathuis FJM (2009) Arsenite transport in plants. Cell Mol Life Sci 66:2329–2339

Al-Whaibi MH (2011) Plant heat-shock proteins: a mini review. J King Saud Univ Sci 23:139–115

Amari T, Ghnaya T, Abdelly C (2017) Nickel, cadmium and lead phytotoxicity and potential of halophytic plants in heavy metal extraction. S Afr J Bot 111:99–110

Angino EE, Magnuson LM, Waugh TC, Galle OK, Bredfeldt J (1970) Arsenic in detergents-possible danger and pollution hazard. Science 168:389–392

Arif N, Sharma NC, Yadav V, Ramawat N, Dubey NK, Tripathi DK, Chauhan DK, Sahi S (2019) Understanding heavy metal stress in a rice crop: toxicity tolerance mechanism, and amelioration strategies. J Plant Biol 62:239–253

Ashrafi E, Alemzadeh A, Ebrahimi M, Ebrahimie E, Dadkhodaei N, Ebrahimi M (2011) Amino acid features of P1B-ATPase heavy metal transporters enabling small numbers of organisms to cope with heavy metal pollution. Bioinf Biol Insights 5:59–82

Averill-Bates DA, Przybytkowski E (1994) The role of glucose in cellular defences against cytotoxicity of hydrogen peroxide in Chinese hamster ovary cells. Arch Biochem Biophys 312:52–58

Barba-Espín G, Diaz-Vivancos P, Job D, Belghazi M, Job C, Hernández JA (2011) Understanding the role of H2O2 during pea seed germination: a combined proteomic and hormone profiling approach. Plant Cell Environ 34:1907–1919

Belatik A, Hotchandani S, Tajmir-Riahi HA, Carpentier R (2013) Alteration of the structure and function of photosystem I by $Pb^{2+}$. J Photochem Photobiol B 123:41–47

Belouchi A, Kwan T, Gros P (1997) Cloning and characterization of the OsNramp family from *Oryza sativa*, a new family of membrane proteins possibly implicated in the transport of metal ions. Plant Mol Biol 33:1085–1092

Benatti RM, Yookongkaew N, Meetam M, Guo WJ, Punyasuk N, AbuQamar S et al (2014) Metallothionein deficiency impacts copper accumulation and redistribution in leaves and seeds of Arabidopsis. New Phytol 202:940–951

Bereczky Z, Wang HY, Schubert V, Ganal M, Bauer P (2003) Differential regulation of nramp and irt metal transporter genes in wild type and iron uptake mutants of tomato. J Biol Chem 278:24697–24704

Bieker S, Riester L, Stahl M, Franzaring J, Zentgraf U (2012) Senescence-specific alteration of hydrogen peroxide levels in *Arabidopsis thaliana* and oilseed rape spring variety *Brassica napus* L. cv. Mozart F. J Integr Plant Biol 54:540–554

Blaylock MJ, Huang JW (2000) Phytoextraction of metals. In: Raskin I, Ensley BD (eds) Phytoremediation of toxic metals: using plants to clean up the environment. Wiley, New York, pp 53–70

Cailliatte R, Lapeyre B, Briat JF, Mari S, Curie C (2009) The NRAMP6 metal transporter contributes to cadmium toxicity. Biochem J 422:217–228

Cenkci S, Cigerci IH, Yildiz M, Özay C, Bozdag A, Terzi H (2010) Lead contamination reduces chlorophyll biosynthesis and genomic template stability in *Brassica rapa* L. Environ Exp Bot 67:467–473

Chatterjee C, Gopal R, Dube BK (2006) Physiological and biochemical responses of French bean to excess cobalt. J Plant Nutr 29:127–136

Claus J, Chavarría-Krauser A (2012) Modeling regulation of zinc uptake via ZIP transporters in yeast and plant roots. PLoS One 7(6):e37193

Clemens S, Antosiewicz DM, Ward JM, Schachtman DP, Schroeder JI (1998) The plant cDNA LCT1 mediates the uptake of calcium and cadmium in yeast. Proc Natl Acad Sci U S A 95:12043–12048

Cobbett C, Goldsbrough P (2002) Phytochelatins and metallothioneins: roles in heavy metal detoxification and homeostasis. Annu Rev Plant Biol 53:159–182

Collins NC, Shirley NJ, Saeed M, Pallotta M, Gustafson JP (2008) An ALMT1 gene cluster controlling aluminum tolerance at the Alt4 locus of rye (Secale cereale L.). Genetics 179:669–682

Cummins I, Dixon DP, Freitag-Pohl S, Skipsey M, Edwards R (2011) Multiple roles for plant glutathione transferases in xenobiotic detoxification. Drug Metab Rev 43:266–280

Cuypers A, Smeets K, Ruytinx J (2011) The cellular redox state as a modulator in cadmium and copper responses in *Arabidopsis thaliana* seedlings. J Plant Physiol 168:309–316

Dai M, Lu H, Liu W, Jia H, Hong H, Liu J, Yan C (2017) Phosphorus mediation of cadmium stress in two mangrove seedlings *Avicennia marina* and *Kandelia obovata* differing in cadmium accumulation. Ecotoxicol Environ Saf 139:272–279

De Feo C, Mootien S, Unger V (2010) Tryptophan scanning analysis of the membrane domain of CTR-copper transporters. J Membr Biol 234:113–123

Deng C, Zhang D, Pan X (2013) Toxic effects of mercury on PSI and PSII activities, membrane potential and trans thylakoid proton gradient in *Microsorum pteropus*. J Photochem Photobiol B 127:1–7

Desikan R, Hancock JT, Bright J, Harrison J, Weir I, Hooley R, Neill SJ (2005) A role for ETR1 in hydrogen peroxide signaling in stomatal guard cells. Plant Physiol 137:831–834

DiDonato RJ, Roberts LA, Sanderson T, Eisley RB, Walker EL (2004) Arabidopsis Yellow Stripe-Like2 (YSL2): a metal-regulated gene encoding a plasma membrane transporter of nicotianamine metal complexes. Plant J 39:403–414

Djingova R, Kuleff I (2000) Instrumental techniques for trace analysis. In: Vernet JP (ed) Trace elements: their distribution and effects in the environment. Elsevier, London

Drumm JE, Mi K, Bilder P, Sun M, Lim J, Bielefeldt-Ohmann H (2009) Mycobacterium tuberculosis universal stress protein Rv2623 regulates bacillary growth by ATP-Binding: requirement for establishing chronic persistent infection. PLoS Pathog 5:1000460

Duan J, Cai W (2012) OsLEA3-2, an abiotic stress induced gene of rice plays a key role in salt and drought tolerance. PLoS One 7:45117

Duan X, Li X, Ding F, Zhao J, Guo A, Zhang L (2015) Interaction of nitric oxide and reactive oxygen species and associate deregulation of root growth in wheat seedlings under zinc stress. Ecotox Environ Safe 113:95–102

Dunand C, Crevecoeur M, Penel C (2007) Distribution of superoxide and hydrogenperoxide in Arabidopsis root and their influence on root development: possible interaction with peroxidases. New Phytol 174:332–341

Durrett TP, Gassmann W, Rogers EE (2007) The FRD3- mediated efflux of citrate into the root vasculature is necessary for efficient iron translocation. Plant Physiol 144:197–205

El-Maarouf-Bouteau H, Sajjad Y, Bazin J, Langlade N, Cristescu SM, Balzergue S, Baudouin E, Bailly C (2014) Reactive oxygen species, abscisic acid and ethylene interact to regulate sunflower seed germination. Plant Cell Environ. https://doi.org/10.1111/pce.12371

Emamverdian A, Ding Y, Mokhberdoran F, Xie Y (2015) Heavy metal stress and some mechanisms of plant defense response. Sci World J 2015:756120

Eshleman A, Siegel SM, Siegel BZ (1971) Is mercury from Hawaiian volcanoes a natural source of pollution? Nature 223:471–475

Farmer EE, Mueller MJ (2013) ROS-Mediated lipid peroxidation and RES-activated signaling. Annu Rev Plant Biol 64:429–450

Finkemeier I, Kluge C, Metwally A, Georgi M, Grotjohann N, Dietz KJ (2003) Alterations in Cd-induced gene expression under nitrogen deficiency in *Hordeum vulgare*. Plant Cell Environ 26:821–833

Gastaldo J, Viau M, Bouchot M, Joubert A, Charvet A-M, Foray N (2008) Induction and repair rate of DNA damage: a unified model for describing effects of external and internal irradiation and contamination with heavy metals. J Theor Biol 251:68–81

Ge XM, Cai HL, Lei X, Zou X, Yue M, He JM (2015) Hetero trimeric G protein mediates ethylene-induced stomatal closure via hydrogenperoxide synthesis in *Arabidopsis*. Plant J 82:138–150

Gendre D, Czernic P, Conéjéro G, Pianelli K, Briat JF, Lebrun M, Mari S (2006) TcYSL3, a member of the YSL gene family from the hyperaccumulator Thlaspi caerulescens, encodes a nicotianamine- Ni/Fe transporter. Plant J 49:1–15

Gill SS, Tuteja N (2010) Reactive oxygen species and antioxidant machinery in abiotic stress tolerance in crop plants. Plant Physiol Biochem 48:909–930

Gonzalez A, de los Ángeles Cabrera M, Henríquez MJ, Contreras RA, Morales B, Moenne A (2012) Cross talk among calcium, hydrogenperoxide, and nitric oxide and activation of gene expression involving calmodulins and calcium-dependent protein kinases in Ulva compress a exposed to copper excess. Plant Physiol 158:1451–1462

Guan-fu F (2011) Changes of oxidative stress and soluble sugar in anthers involve in rice pollen abortion under drought stress. Agric Sci China 10:1016–1025

Guerinot ML (2000) The ZIP family of metal transporters. Biochim Biophys Acta 1465:190–198

Gupta DK, Rai UN, Tripathi RD, Inouhe M (2002) Impacts of fly-ash on soil and plant responses. J Plant Res 115:401–409

Gupta M, Sharma P, Sarin NB, Sinha AK (2009) Differential response of arsenic stress in two varieties of *Brassica juncea* L. Chemosphere 74:1201–1208

Gustin JL, Zanis MJ, Salt DE (2011) Structure and evolution of the plant cation diffusion facilitator family of ion transporters. BMC Evol Biol 11:76

Hall JL, Williams LE (2003) Transition metal transporters in plants. J Exp Bot 54(2601):2613

Hao F, Wang X, Chen J (2006) Involvement of plasma-membrane NADPH oxidase in nickel-induced oxidative stress in roots of wheat seedlings. Plant Sci 170:151–158

Hassinen VH, Tervahauta AI, Schat H, Kärenlampi SO (2011) Plant metallothioneins–metal chelators with ROS scavenging activity? Plant Biol 13:225–232

Hirschi KD, Korenkov VD, Wilganowski NL, Wagner GJ (2000) Expression of Arabidopsis CAX2 in tobacco: altered metal accumulation and increased manganese tolerance. Plant Physiol 124:125–134

Hong Bin W, Xie F, Yan-Zhuo Y (2012) The effects of arsenic and induced phytoextraction methods on photosynthesis in Pteris species with different arsenic accumulating abilities. Environ Exp Bot 75:298–306

Hooijmaijers C, Rhee JY, Kwak KJ, Chung GC, Horie T, Katsuhara M, Kang H (2012) Hydrogen peroxide permeability of plasma membrane aquaporins of Arabidopsis thaliana. J Plant Res 125:147–153

Hou X, Tan L, Tang SF (2019) Molecular mechanism study on the interactions of cadmium (II) ions with Arabidopsis thaliana glutathione transferase Phi8. Spectrochim Acta A Mol Biomol Spectrosc 216:411–417

Hu T, Zhu S, Tan L, Qi W, He S, Wang G (2016) Overexpression of OsLEA4 enhances drought, high salt and heavy metal stress tolerance in transgenic rice (Oryza sativa L.). Environ Exp Bot 123:68–77

Huang T-L, Nguyen QTT, Fu S-F, Lin C-Y, Chen Y-C, Huang H-J (2012) Transcriptomic changes and signalling pathways induced by arsenic stress in rice roots. Plant Mol Biol 80:587–608

Ishikita H, Knapp EW (2001) Function of Redox-Active Tyrosine in Photosystem II. Biophys J 90 (11):3886–3896

Jajić I, Sarna T, Szewczyk G, Strzałka K (2015) Changes in production of reactive oxygen species in illuminated thylakoids isolated during development and senescence of barley. J Plant Physiol 184:49–56

Jallad KN (2015) Heavy metal exposure from ingesting rice and its related potential hazardous health risks to humans. Environ Sci Pollut Res 22:15449–15458

Jayakumar K, Rajesh M, Baskaran L, Vijayarengan P (2013) Changes in nutritional metabolism of tomato (*Lycopersicon esculantum* Mill.) plants exposed to increasing concentration of cobalt chloride. Int J Food Nutr Saf 4:62–69

Jiang HX, Chen LS, Zheng JG et al (2008) Aluminum-induced effects on photosystem II photochemistry in citrus leaves assessed by the chlorophyll a fluorescence transient. Tree Physiol 28:1863–1871

Jones GC, Corin KC, Van Hille RP, Harrison STL (2011) The generation of toxic reactive oxygen species (ROS) from mechanically activated sulphide concentrates and its effect on thermophilic bioleaching. Miner Eng 24:1198–1208

Karimi N, Siyahat Shayesteh L, Ghasmpour H, Alavi M (2013) Effects of arsenic on growth, photosynthetic activity and accumulation in two new hyperaccumulating populations of *Isatis cappadocica* Desv. J Plant Growth Regul 32:823–830

Karukstis KK (1991) Chlorophyll fluorescence as a physiological probe of the photosynthetic apparatus. In: Scheer H (ed) Chlorophylls. CRC Press, Boca Raton, pp 770–797

Kaur P, Bali S, Verma V (2018) Responses of photosynthesis under metal stress. In: Singh VP, Singh SSR, Srivastava PK, Prasad SM (eds) Environment and photosynthesis: a future prospect. Studium Press India, pp 241–260

Kawachi M, Kobae Y, Mimura T, Maeshima M (2008) Deletion of a histidine-rich loop of AtMTP1, a vacuolar Zn2þ/Hþ antiporter of Arabidopsis thaliana, stimulates the transport activity. J Biol Chem 283:8374–8383

Kerk D, Bulgrien J, Smith DW, Gribskov M (2003) Arabidopsis proteins containing similarity to the universal stress protein domain of bacteria. Plant Physiol 131:1209–1219

Keunen E, Remans T, Bohler S, Vangronsveld J, Cuypers A (2011) Metal- induced oxidative stress and plant mitochondria. Int J Mol Sci 12:6894–6918

Kibra MG (2008) Effects of mercury on some growth parameters of rice (Oryza sativa L.). Soil Environ 27:23–28

Kim D, Gustin JL, Lahner B, Persans MW, Baek D, Yun DJ, Salt DE (2004) The plant CDF family member TgMTP1 from the Ni/Zn hyperaccumulator Thlaspi goesingense acts to enhance efflux of Zn at the plasma membrane when expressed in Saccharomyces cerevisiae. Plant J 39:237–251

Kim H, Wu X, Lee J (2013) SLC31 (CTR) family of copper transporters in health and disease. Mol Asp Med 34:561–570

Kim YJ, Lee YH, Lee HJ, Jung H, Hong JK (2015) H2O2 production and gene expression of antioxidant enzymes in kimchi cabbage (*Brassica rapa var. glabra* Regel) seedlings regulated by plant development and nitrosative stress-triggered cell death. Plant Biotechnol Rep 9:67–78

Kramer U, Cotter-Howells JD, Charnock JMM, Baker AJ, Smith JAC (1996) Free histidine as a metal chelator in plants that accumulate nickel. Nature 379:635–638

Kumar A, Prasad MNV, Sytar O (2012) Lead toxicity, defense strategies and associated indicative biomarkers in *Talinum triangulare* grown hydroponically. Chemosphere 89:1056–1165

Kupper H, Kochian LV (2010) Transcriptional regulation of metal transport genes and mineral nutrition during acclimatization to cadmium and zinc in the Cd/Zn hyperaccumulator, *Thlaspi caerulescens* (Ganges population). New Phytol 185(1):114–129

Kupper H, Kupper F, Spiller M (1998) In situ detection of heavy metal substituted chlorophylls in water plants. Photosynth Res 58:123–133

Kvint K, Nachin L, Diez A, Nyström T (2003) The bacterial universal stress protein: function and regulation. Curr Opin Microbiol 6:140–145

Lacerda LD (1997) Global mercury emissions from gold and silver mining. Water Air Soil Pollut 97:209–221

Lee S, Moon JS, Ko T-S, Petros D, Goldsbrough PB, Korban SS (2003) Overexpression of Arabidopsis phytochelatin synthase paradoxically leads to hypersensitivity to cadmium stress. Plant Physiol 131:656–663

Lee S, Kim SA, Lee J, Guerinot ML, An G (2010) Zinc deficiency-inducible OsZIP8 encodes a plasma membrane-localized zinc transporter in rice. Mol Cell 29(6):551–558

Lee J-C, Son Y-O, Pratheeshkumar P, Shi X (2012) Oxidative stress and metal carcinogenesis. Free Radic Biol Med 53:742–757

Lenntech Water Treatment and Air Purification (2004) Water treatment. Lenntech, Rotterdamseweg. http://www.excelwater.com/thp/filters/Water-Purification.htm

Li L, Tutone AF, Drummond RSM, Gardner RC, Luan S (2001) A novel family of magnesium transport genes in Arabidopsis. Plant Cell 13:2761–2775

Li L, He Z, Pandey GK, Tsuchiya T, Luan S (2002) Functional cloning and characterization of a plant efflux carrier for multidrug and heavy metal detoxification. J Biol Chem 277:5360–5368

Li S, Zhou X, Huang Y, Zhu L, Zhang S, Zhao Y, Guo J, Chen J, Chen R (2013) Identification and characterization of the zinc-regulated transporters, iron-regulated transporter-like protein (ZIP) gene family in maize. BMC Plant Biol 13(1):1–14

Li ZG, Luo LJ, Sun YF (2015) Signal crosstalk between nitric oxide and hydrogen sulphide may be involved in hydrogen peroxide – induced thermo-tolerance in maize seedlings. Russ J Plant Physiol 62:507–514

Liang J, Zhou M, Zhou X, Jin Y, Xu M, Lin J (2013) JcLEA, a novel LEA-like protein from Jatropha curcas, confers a high level of tolerance to dehydration and salinity in Arabidopsis thaliana. PLoS One 8:e83056

Lin YF, Aarts MG (2012) The molecular mechanism of zinc and cadmium stress response in plants. Cell Mol Life Sci 69(19):3187–3206
Liu J, Shi X, Qian M, Zheng L, Lian C, Xia Y, Shen Z (2015a) Copper-induced hydrogen peroxide upregulation of a metallothionein gene, OsMT2c, from Oryza sativa L. confers copper tolerance in Arabidopsis thaliana. J Hazard Mater 294:99–108
Liu J, Xu Y, Zhang Z, Wei J (2015b) Hydrogen peroxide promotes programmed cell death and salicylic acid accumulation during the induced production of sesquiterpenes in cultured cell suspensions of Aquilaria sinensis. Funct Plant Biol 42:337–346
Liu H, Zhao H, Wu L, Liu A, Zhao FJ, Xu W (2017) Heavy metal ATPase 3 (HMA3) confers cadmium hypertolerance on the cadmium/zinc hyperaccumulator *Sedum plumbizincicola*. New Phytol 215:687–698
Lopez-Millan AF, Ellis DR, Grusak MA (2004) Identification and characterization of several new members of the ZIP family of metal ion transporters in Medicago truncatula. Plant Mol Biol 54:583–596
Ma F, Wang LJ, Li JL, Samma MK, Xie YJ, Wang R (2014) Interaction between HY1 and H2O2 in auxin-induced lateral root formation in *Arabidopsis*. Plant Mol Biol 85:49–61
Maser P, Thomine S, Schroeder JI, Ward JM, Hirschi K, Sze H (2001) Phylogenetic relationships within cation transporter families of *Arabidopsis*. Plant Physiol 126:1646–1667
Mahajan S, Tuteja N (2005) Cold, salinity and drought stresses: an overview. Arch Biochem Biophys 444:139–158
Manara A (2012) Plant responses to heavy metal toxicity. In: Furini A (ed) Plants and heavy metals. Springer, New York, pp 27–53
Marcato-Romain C-E, Pinelli E, Pourrut B, Silvestre J, Guiresse M (2009) Assessment of the genotoxicity of Cu and Zn in raw and anaerobically digested slurry with the *Vicia faba* micronucleus test. Mutat Res 672:113–118
Marquez-Garcia B, Horemans N, Cuypers A, Guisez Y, Cordoba F (2011) Antioxidants in *Erica andevalensis*: a comparative study between wild plants and cadmium-exposed plants under controlled conditions. Plant Physiol Biochem 49:110–115
Mathur S, Kalaji HM, Jajoo A (2016) Investigation of deleterious effects of chromium phytotoxicity and photosynthesis in wheat plant. Photosynthetica 54:185–192
Mishra S, Dubey RS (2006) Heavy metal uptake and detoxification mechanisms in plants. Int JAgric Res 1:122–141
Mizuno T, Usui K, Horie K, Nosaka S, Mizuno N, Obata H (2005) Cloning of three ZIP/NRAMP transporter genes from a Ni hyperaccumulator plant Thlaspi japonicum and their Ni 2þ-transport abilities. Plant Physiol Biochem 43:793–801
Moller IM, Jensen PE, Hansson A (2007) Oxidative modifications to cellular components in plants. Annu Rev Plant Biol 58:459–481
Montanini B, Blaudez D, Jeandroz S, Sanders D, Chalot M (2007) Phylogenetic and functional analysis of the Cation Diffusion Facilitator (CDF) family: improved signature and prediction of substrate specificity. BMC Genomics 8:107
Morel M, Crouzet J, Gravot A, Auroy P, Leonhardt N, Vavasseur A, Richaud P (2009) AtHMA3, a P1B-ATPase allowing Cd/Zn/Co/Pb vacuolar storage in Arabidopsis. Plant Physiol 149:894–904
Moustaka J, Ouzounidou G, Baycu G, Moustakas M (2016) Aluminum resistance in wheat involves maintenance of leaf $Ca^{2+}$ and $Mg^{2+}$ content, decreased lipid peroxidation and Al accumulation, and low photosystem II excitation pressure. Biometals 29:611–623
Nagajyoti PC, Lee KD, Sreekanth TVM (2010) Heavy metals, occurrence and toxicity for plants: a review. Environ Chem Lett 8:199–216
Nelson N, Junge W (2015) Structure and energy transfer in photosystems of oxygenic photosynthesis. Annu Rev Biochem 84:659–683
Nevo Y, Nelson N (2004) The mutation F227I increases the coupling of metal ion transport in DCT1. J Biol Chem 27951:53056–53061

Nguyen GN, Hailstones DL, Wilkes M, Sutton BG (2010) Drought stress: role of carbohydrate metabolism in drought-induced male sterility in rice anthers. J Agron Crop Sci 196:346–357
Nicholls AM, Mal TK (2003) Effects of lead and copper exposure on growth of an invasive weed, *Lythrum salicaria* L. (Purple Loosestrife). Ohio J Sci 103:129–133
Nishida S, Mizuno T, Obata H (2008) Involvement of histidine-rich domain of ZIP family transporter TjZNT1 in metal ion specificity. Plant Physiol Biochem 46:601–606
Oliveira SCB, Corduneanu O, Oliveira-Brett AM (2008) In situ evaluation of heavy metal-DNA interactions using an electrochemical DNA biosensor. Bioelectrochemistry 72:53–58
Opdenakker K, Remans T, Keunen E, Vangronsveld J, Cuypers A (2012) Exposure of *Arabidopsis thaliana* to Cd or Cu excess leads to oxidative stress mediated alterations in MAPKinase transcript levels. Environ Exp Bot 83:53–61
Oracz K, Karpinski S (2016) Phytohormones signaling pathways and ROS involvement in seed germination. Front Plant Sci 7:864
Park J, Song WY, Ko D, Eom Y, Hansen TH, Schiller M, Lee TG, Martinoia E, Lee Y (2012) The phytochelatin transporters AtABCC1 and AtABCC2 mediate tolerance to cadmium and mercury. Plant J 69:278–288
Parmar P, Kumari N, Sharma V (2013) Structural and functional alterations in photosynthetic apparatus of plants under cadmium stress. Bot Stud 54:45
Pierart A, Shahid M, Séjalon-Delmas N, Dumat C (2015) Antimony bioavailability: knowledge and research perspectives for sustainable agricultures. J Hazard Mater 289:219–234
Piršelova B, Boleček P, Gálusová T (2016) Effect of cadmium and arsenic on chlorophyll fluorescence of selected soybean cultivars. Russ J Plant Physiol 63:499–503
Pirzadeh S, Shahpiri A (2016) Functional characterization of a type 2 metallothionein isoform (OsMITI-2b) from rice. Int J Biol Macromol 88:491–496
Pochodylo AL, Aristilde L (2017) Molecular dynamics of stability and structures in phytochelatin complexes with Zn, Cu, Fe, Mg, and Ca: implications for metal detoxification. Environ Chem Lett 15:495–500
Podar D, Scherer J, Noordally Z, Herzyk P, Nies D, Sanders D (2012) Metal selectivity determinants in a family of transition metal transporters. J Biol Chem 287:3185–3196
Pourrut B, Jean S, Silvestre J, Pinelli E (2011) Lead-induced DNA damage in *Vicia faba* root cells: potential involvement of oxidative stress. Mutat Res 726:123–128
Pourrut B, Shahid M, Douay F, Dumat C, Pinelli E (2013) Molecular mechanisms involved in lead uptake, toxicity and detoxification in higher plants. In: Corpas FJ, Palma JM, Gupta DK (eds) Heavy metal stress in plants. Springer, Berlin, pp 121–147
Puig S, Peñarrubia L (2009) Placing metal micronutrients in context: transport and distribution in plants. Curr Opin Plant Biol 12:299–306
Puig S, Thiele DJ (2002) Molecular mechanisms of copper uptake and distribution. Curr Opin Chem Biol 6:171–180
Qiao B, Zhang Q, Liu D, Wang H, Yin J, Wang R (2015) A calcium-binding protein, rice annex in OsANN1, enhances heat stress tolerance by modulating the production of H2O2. J Exp Bot 66:5853–5866
Qufei L, Fashui H (2009) Effects of $Pb^{2+}$ on the structure and function of photosystem II of *Spirodela polyrrhiza*. Biol Trace Elem Res 129:251
Rajkumar M, Sandhya S, Prasad MNV, Freitas H (2012) Perspectives of plant-associated microbes in heavy metal phytoremediation. Biotechnol Adv 30:1562–1574
Raju AD, Parihar P, Singh R (2020) Synergistic action of indole acetic acid with homobrassinolide in easing the NaCl-induced toxicity in Solanum melongena L. seedlings. Acta Physiol Plant 42:68
Ram H, Kaur A, Gandass N, Singh S, Deshmukh R, Sonah H, Sharma TR (2019) Molecular characterization and expression dynamics of MTP genes under various spatio-temporal stages and metal stress conditions in rice. PLoS One 14(5):e0217360
Rao KP, Vani G, Kumar K, Wankhede DP, Misra M, Gupta M, Sinha AK (2011) Arsenic stress activates MAP kinase in rice roots and leaves. Arch Biochem Biophys 506:73–82

Reddy PS, Kavi Kishor PB, Seiler C (2014) Unraveling regulation of the small heat shock proteins by the heat shock factor HvHsfB2c in barley: its implications in drought stress response and seed development. PLoS One 9:e89125

Ricachenevsky FK, Menguer PK, Sperotto RA, Williams LE, Fett JP (2013) Roles of plant metal tolerance proteins (MTP) in metal storage and potential use in biofortification strategies. Front Plant Sci 4:144

Rodriguez-Serrano M, Romero-Puertas MC, Zabalza A, Corpas FJ, Gomez M, Del Rio LA, Sandalio LM (2006) Cadmium effect on oxidative metabolism of pea (*Pisum sativum* L.) roots. Imaging of reactive oxygen species and nitric oxide accumulation in vivo. Plant Cell. Environment 29:1532–1544

Rogers EE, Guerinot ML (2002) FRD3, a member of the multidrug and toxin efflux family, controls iron deficiency responses in Arabidopsis. Plant Cell 14:1787–1799

Ross SM (1994) Toxic metals in soil–plant systems. Wiley, Chichester, p 469

Sancenon V, Puig S, Mira H, Thiele DJ, Pen Äarrubia L (2004) Identification of a copper transporter family in *Arabidopsis thaliana*. Plant Mol Biol 51:577–587

Sarkar NK, Kim YK, Grover A (2009) Rice sHsp genes: genomic organization and expression profiling under stress and development. BMC Genomics 10:393

Sasaki T, Yamamoto Y, Ezaki B, Katsuhara M, Ahn SJ, Ryan PR, Delhaize E, Matsumoto H (2004) A wheat gene encoding an aluminum- activated malate transporter. Plant J 37:645–653

Schachtman DP, Kumar R, Schroeder JI, Marsh EL (1997) Molecular and functional characterization of a novel low-affinity cation transporter (LCT1) in higher plants. Proc Natl Acad Sci U S A 94:11079–11084

Schalk IJ, Hannauer M, Braud A (2011) New roles for bacterial siderophores in metal transport and tolerance. Environ Microbiol 13:2844–2854

Schaller A, Diez T (1991) Plant specific aspects of heavy metal uptake and comparison with quality standards for food and forage crops. In: Sauerbeck D, Lubben S (eds) Der Einfluß von festen Abfallen auf Boden, Pflanzen. KFA, Julich, pp 92–125. (German)

Seaward MRD, Richardson DHS (1990) Atmospheric sources of metal pollution and effects on vegetation. In: Shaw AJ (ed) Heavy metal tolerance in plants evolutionary aspects. CRC Press, Boca Raton, pp 75–94

Shahid M, Pinelli E, Pourrut B, Silvestre J, Dumat C (2011) Lead-induced genotoxicity to *Viciafaba* L. roots in relation with metal cell uptake and initial speciation. Ecotoxicol Environ Saf 74:78–84

Shahid M, Khalid S, Abbas G, Shahid N, Nadeem M, Sabir M (2015) Heavy metal stress and crop productivity. In: Hakeem KR (ed) Crop production and global environmental issues. Springer, Cham, pp 1–25

Sharma SK, Goloubinoff P, Christen P (2008) Heavy metal ions are potent inhibitors of protein folding. Biochem Bioph Res Co 372:341–345

Sharma SK, Goloubinoff P, Christen P (2011) Non-native proteins as newly-identified targets of heavy metals and metalloids. In: Banfalvi G (ed) Cellular effects of heavy metals. Springer, Heidelberg, pp 263–274

Sharma P, Jha AB, Dubey RS, Pessarakli M (2012) Reactive oxygen species, oxidative damage, and antioxidative defense mechanism in plants under stressful conditions. J Bot 2012:217037. https://doi.org/10.1155/2012/217037

Sheng Z, Chaohai W, Chaodeng L, Haizhen W (2008) Damage to DNA of effective microorganisms by heavy metals: impact on wastewater treatment. J Environ Sci 20:1514–1518

Shi C, Qi C, Ren HY, Huang AX, Hei SM, She XP (2015) Ethylene mediates brassinosteroid-induced stomatal closure via Ga protein-activated hydrogen peroxide and nitricoxide production in *Arabidopsis*. Plant J 82:280–301

Shi WG, Liu W, Yu W, Zhang Y, Ding S, Li H, Mrak T, Kraigher H, Luo ZB (2019) Abscisic acid enhances lead translocation from the roots to the leaves and alleviates its toxicity in *Populus×canescens*. J Hazard Mater 362:275–285

Shigaki T, Pittman JK, Hirschi KD (2003) Manganese specificity determinants in the Arabidopsis metal/H+ antiporter CAX2. J Biol Chem 278:6610–6617

Shingu Y, Kudo T, Ohsato S, Kimura M, Ono Y, Yamaguchi I, Hamamoto H (2005) Characterization of genes encoding metal tolerance proteins isolated from Nicotiana glauca and Nicotiana tabacum. Biochem Biophys Res Commun 331:675–680

Sinani D, Adle DJ, Kim H, Lee J (2007) Distinct mechanisms for Ctr 1- mediated copper and cisplatin transport. J Biol Chem 282:26775–26785

Singh RK, Jaishankar J, Muthamilarasan M (2016) Genome-wide analysis of heat shock proteins in C4 model, foxtail millet identifies potential candidates for crop improvement under abiotic stress. Sci Rep 6:32641

Singh R, Parihar P, Prasad SM (2020) Interplay of Calcium and Nitric Oxide in improvement of Growth and Arsenic-induced Toxicity in Mustard Seedlings. Sci Rep 10:6900

Skrebsky EC, Tabaldi LA, Pereira LB (2008) Effect of cadmium on growth, micronutrient concentration, and δ-aminolevulinic acid dehydratase and acid phosphatase activities in plants of *Pfaffia glomerata*. Braz J Plant Physiol 20:285–294

Song WY, Mendoza-Cozatl DG, Lee Y, Schroeder JI, Ahn SN, Lee HS (2014) Phytochelatin–metal (loid) transport into vacuoles shows different substrate preferences in barley and Arabidopsis. Plant Cell Environ 37:1192–1201

Srivastava S, Srivastava AK, Suprasanna P, D'souza S (2012) Identification and profiling of arsenic stress-induced microRNAs in Brassica juncea. J Exp Bot 64:303–315

Stephens BW, Cook DR, Grusak MA (2011) Characterization of zinc transport by divalent metal transporters of the ZIP family from the model legume *Medicago truncatula*. Biometals 24 (1):51–58

Stocker S, Van Laer K, Mijuskovic A, Dick TP (2018) The Conundrum of Hydrogen Peroxide Signaling and the Emerging Role of Peroxiredoxins as Redox Relay Hubs. Antioxid Redox Signal 28:558–573

Talke IN, Kramer U, Hanikenne M (2006) Zinc-dependent global transcriptional control, transcriptional deregulation, and higher gene copy number for genes in metal homeostasis of the hyperaccumulator Arabidopsis halleri. Plant Physiol 142:148–167

Tan Y-F, O'Toole N, Taylor NL, Millar AH (2010) Divalent metal ions in plant mitochondria and their role in interactions with proteins and oxidative stress-induced damage to respiratory function. Plant Physiol 152:747–761

Tangahu B, Abdullah S, Basri H, Idris M, Anur N, Mukhlisin M (2011) A review on heavy metals (As, Pb, and Hg) uptake by plants through phytoremediation. Int J Chem Eng 1155(10):939161

Tanyolac D, Ekmekci Y, Unalan S (2007) Changes in photochemical and antioxidant enzyme activities in maize (*Zea mays* L.) leaves exposed to excess copper. Chemosphere 67:89–98

Thomine S, Wang R, Ward JM, Crawford NM, Schroeder JI (2000) Cadmium and iron transport by members of a plant metal transporter family in *Arabidopsis* with homology to Nramp genes. Proc Natl Acad Sci U S A 97(9):4991–4996

Thomine S, Lelie Ávre F, Debarbieux E, Schroeder JI, Barbier Brygoo H (2003) AtNRAMP3, a multispecific vacuolar metal transporter involved in plant responses to iron deficiency. Plant J 34:685–695

Tian S, Wang X, Li P, Wang H, Ji H, Xie J, Qiu Q, Shen D, Dong H (2016) Plant aquaporin AtPIP1; 4 links apoplastic H2O2 induction to disease immunity pathways. Plant Physiol 171:1635–1650

Tsukagoshi H, Busch W, Benfey PN (2010) Transcriptional regulation of ROS controls transition from proliferation to differentiation in the root. Cell 143:606–616

Vanhoudt N, Vandenhove H, Horemans N, Remans T, Opdenakker K, Smeets K, Bello DM, Wannijn J, Van Hees M, Vangronsveld J (2011) Unraveling uranium induced oxidative stress related responses in *Arabidopsis thaliana* seedlings. Part I: responses in the roots. J Environ Radioact 102:630–637

Vavilala SL, Gawde KK, Sinha M, D'Souza JS (2015) Programmed cell death is induced by hydrogenperoxide but not by excessive ionic stress of sodium chloride in the unicellular green alga *Chlamydomonas reinhardtii*. Eur J Phycol 50:422–438

Verkleji JAS (1993) The effects of heavy metals stress on higher plants and their use as bio monitors. In: Markert B (ed) Plant as bioindicators: indicators of heavy metals in the terrestrial environment. VCH, New York, pp 415–424

Vollmer AC, Bark SJ (2018) Twenty-five years of investigating the universal stress protein: function, structure, and applications. Adv Appl Microbiol 102:1–36

Wang S, Shi X (2001) Molecular mechanisms of metal toxicity and carcinogenesis. Mol Cell Biochem 222:3–9

Wang H, Lin J, Chang Y, Jiang C-Z (2017) Comparative transcriptomic analysis reveals that ethylene/$H_2O_2$-mediated hypersensitive response and programmed cell death determine the compatible interaction of sand pear and Alternaria alternata. Front Plant Sci 8:195

Weber M, Harada E, Vess C, von Roepenack-Lahaye E, Clemens S (2004) Comparative microarray analysis of Arabidopsis thaliana and Arabidopsis halleri roots identifies nicotianamine synthase, a ZIP transporter and other genes as potential metal hyperaccumulation factors. Plant J 37:269–281

Willems G, Drager DB, Courbot M, Gode C, Verbruggen N, Saumitou Laprade P (2007) The genetic basis of Zn tolerance in the metallophyte Arabidopsis halleri ssp. halleri (Brassicaceae): an analysis of quantitative trait loci. Genetics 176:659–674

Williams LE, Pittman JK, Hall JL (2000) Emerging mechanisms for heavy metal transport in plants. Biochim Biophys Acta 1465:104–126

Wysocki R, Tamas MJ (2010) How Saccharomyces cerevisiae copes with toxic metals and metalloids. FEMS Microbiol Rev 34:925–951

Xiao H, Yin L, Xu X, Li T, Han Z (2008) The Iron-regulated transporter, MbNRAMP1, isolated from Malus baccata is involved in Fe, Mn and Cd trafficking. Ann Bot 102:881–889

Xing JP, Jiang RF, Ueno D, Ma JF, Schat H, McGrath SP, Zhao FJ (2008) Variation in root-to-shoot translocation of cadmium and zinc among different accessions of the hyperaccumulators Thlaspi caerulescens and Thlaspi praecox. New Phytol 178:315–325

Xu R, Li Y (2011) Control of final organ size by mediator complex subunit 25 in Arabidopsis thaliana. Development 138:4545–4554

Xue XC, Gao HY, Zhang LT (2013) Effects of cadmium on growth, photosynthetic rate and chlorophyll content in leaves of soybean seedlings. Biol Plant 57:587–590

Yadav V, Arif N, Singh S, Srivastava PK, Sharma S, Tripathi DK, Dubey NK, Chauhan DK (2016) Exogenous mineral regulation under heavy metal stress: advances and prospects. Biochem Pharmacol 5:2167–0501

Yang J, Wang Y, Liu G, Yang C, Li C (2011) Tamarix hispida metallothionein-like ThMT3, a reactive oxygen species scavenger, increases tolerance against Cd2+, Zn2+, Cu2+, and NaCl in transgenic yeast. Mol Biol Rep 38:567–1574

Yu LJ, Luo YF, Liao B, Xie LJ, Chen L, Xiao S, Li JT, Hu SN, Shu WS (2012) Comparative transcriptome analysis of transporters, phytohormone and lipid metabolism pathways in response to arsenic stress in rice (*Oryza sativa*). New Phytol 195:97–112

Zadak Z, Hyspler R, Ticha A, Hronek M, Fikrova P, Rathouska J, Hrnciarikova D, Stetina R (2009) Antioxidants and vitamins in clinical conditions. Physiol Res 58(Suppl 1):S13–S17

Zahur M, Maqbool A, Ifran M, Barozai MY, Rashid B, Riazuddin S (2009) Isolation and functional analysis of cotton universal stress protein promoter in response to phytohormones and abiotic stresses. Mol Biol 43:628–635

Zhang CH, Ge Y (2008) Response of glutathione and glutathione S-transferase in rice seedlings exposed to cadmium stress. Rice Sci 15:73–76

Zhang MK, Liu ZY, Wang H (2010) Use of single extraction methods to predict bioavailability of heavy metals in polluted soils to rice. Commun Soil Sci Plan 41:820–831

Zou JJ, Li XD, Ratnasekera D, Wang C, Liu WX, Song LF (2015) *Arabidopsis* calcium-dependent protein kinase 8 and catalase 3 function in abscisic acid-mediated signalling and H2O2 homeostasis in stomatal guard cell sunder drought stress. Plant Cell 27:1445–1460

# Mechanism of Toxic Metal Uptake and Transport in Plants

# 13

Jyoti Mathur and Priti Chauhan

**Abstract**

Heavy metals such as zinc, iron, nickel, and copper are essential micronutrients for the plant growth and development, while other metals like lead, cadmium, arsenic are toxic. Plants have developed various mechanisms for metal uptake and transfer from soil to root cells. Various transporters are used for uptake and detoxification of metal ions. As a first line of defense, plants that are often exposed to concentrations of heavy metal ions prevent or minimize their absorption into root cells by trapping metal ions into apoplast, binding to the cell wall or preventing their transfer into the cell. The plants utilize metal transport detoxification techniques, chelation, and metal sequestration into the vacuole. If nonessential metals exceed concentrations, then an active metabolic mechanism results in the production of chelating compounds such as metallothioneins and phytochelatins. The complexes of metals are then sequestered to vacuoles via specific transporters.

**Keywords**

Metals · Phytochelatins · Vacuolar sequestration · Transporters

## Abbreviations

| | |
|---|---|
| ABC | ATP binding cassette |
| AtIRT | *Arabidopsis* iron-resistance transporter |
| CTR | Copper transporters |
| EIN2 | Ethylene-insensitive 2 |

J. Mathur (✉) · P. Chauhan
Department of Bioscience and Biotechnology, Banasthali Vidyapith, Banasthali, Rajasthan, India



K. Mishra et al. (eds.), *Sustainable Solutions for Elemental Deficiency and Excess in Crop Plants*, https://doi.org/10.1007/978-981-15-8636-1_13

| | |
|---|---|
| HMA | Heavy metal-transporting P-type ATPase |
| IRT | Iron-resistance transporter |
| MRP | Multidrug resistance-associated protein |
| MT | Metallothioneins |
| NRAMPs | Natural resistance-associated macrophage proteins |
| PC | Phytochelatins |
| PDR | Pleiotropic drug resistance |
| YSL | Yellow stripe 1-like protein |
| ZIP | Zinc resistance transporter and iron-resistance transporter-like proteins |

## 13.1 Introduction

Plant mineral nutrition is important for agriculture and human. Heavy metal ions such as $Fe^{2+}$, $Zn^{2+}$, $Cu^{2+}$, $Mn^{2+}$, $Co^{2+}$, and $Ni^{2+}$ are vital micronutrients for metabolism of the plant but toxic when these essential and nonessential metals such as Pb, Cd, Hg, and Ag are in high concentrations (Pilon 2011; Olsen and Palmgren 2014). Such metals can become highly toxic and cause symptoms like stunting, chlorosis and necrosis, leaf discoloration, and reduction in root length. Plants uptake metals from root and transport to shoot and leave, where metals are stored in the tissues (Yang and Ye 2015; Kalaivanan and Ganeshamurthy 2016). Plants selectively acquire the elements. The distribution and concentration of metals in the plants depend on species of plants and species of elements and also on soil properties like metal bioavailability, redox, pH, dissolved oxygen, and cation exchange efficiency, etc.

Tolerance against high concentrations of these metal ions in plants can potentially be achieved by upregulation of the antioxidant defense mechanism or by effluxing or compartmentalization and sequestration. Plants act both ways as "accumulators" and "excluders." Accumulators sustain even after high concentrations of metals in above ground parts. Such accumulators can biodegrade and biotransform pollutants inside their tissues into inert types. The excluders prevent the absorption of toxins into their above ground tissues (Maestri et al. 2010). Plants have developed various specific and effective environmental systems for acquiring essential nutrients, at even low concentrations. Further plants have developed very specific mechanisms for translocation and storing micronutrients. Translocation of metals into shoots is necessary after accumulation of metals in roots although root biomass harvesting is usually not feasible (Masood et al. 2012).

Hyperaccumulator plants define the various plants which belong to distant families but have the capacity to grow on metal polluted soil and start accumulating large concentrations of HMs in aerial parts, lead to phytotoxicity (Rascio and Navari-Izzo 2011). There are three hall marks of a good hyperaccumulator: (1) Significantly increase rate of the metal uptake, (2) Faster translocation of root to shoot, and (3) Greater potential for detoxification and sequestration of metals in the leaves.

## 13.2 Metal Binding to Extracellular Exudates and Cell Wall

The plant releases phytosiderophores such as avenic and mugineic acids due to shortages of essential metal ions (Kinnersley 1993). As the first line of protection against metals, the roots of the plant release exudates into the soil matrix. These exudates have significant role in chelating metals and regulating their uptake within cells. Interaction of heavy metals in the apoplasts such as Cu and Zn helps in regulation of metal concentration in root cells (Zhao et al. 2011). Cell wall also performs an important function in metal stabilization by supplying pectic sites, histidyl groups, extracellular callose, and mucilage, thereby preventing metal from being absorbed into the cytosol. Studies show that metal absorption and tolerance against metal can be modulated by chemical properties of cell wall (Verbruggen et al. 2009). The resistant plant species *Silene vulgaris* accumulates different metal ions by interacting with proteins in the epidermal cell walls (Revathi and Subhashree 2015).

## 13.3 Metal Ions Transport Through Plasma Membrane in the Roots

Plant plasma membranes can be considered the first "living" system to counteract the toxicity of heavy metals. Metals can rapidly affect the role of plasma membrane, as evidenced by increased cell leakage in the presence of high amount of the metals, especially Cu. $K^+$ efflux from roots of *Agrostis capillaries* is triggered by Cu ions (Wainwright and Woolhouse 1977). Plant damage can result due to a variety of mechanisms, which include cross-linking and oxidation of protein thiols, degradation of essential membrane proteins like $H^+$-ATPase and the membrane lipids fluidity. Thus tolerance could involve in protecting the integrity of the plasma membrane from damage to heavy metal which could lead to increase the leakage of solutes from cells. In addition to tolerance, cell membranes act as an important function in heavy metal homeostasis. The plant uses a variety of transporters for uptake of metals. Metal transporters sustain physiological concentrations of these metals on plasma membrane and tonoplast. These carriers belong to different families (Williams et al. 2000; Guerinot 2010).

### 13.3.1 ZIP Family (Zinc Resistance Transporter and Iron-Resistance Transporter-Like Proteins)

The ZIP family is a ubiquitous family of divalent metal transporters that help transport Fe, Zn, and Mn and also allows the transport of toxic metals such as Cd. Members of the ZIP family differ in their selection and uniqueness of substrates (Guerinot 2000; Mäser et al. 2001; Mizuno et al. 2005; Lin et al. 2007). Although no ZIP transporter structure is available, computational modeling alone suggests that they act as homodimer, with eight transmembrane domains in each monomer

(Antala et al. 2015). In *Arabidopsis* about 85 members of the family were described from archaea, bacteria, and plants with 15 genes (Mäser et al. 2001).

The first ZIP transporter, IRT1 was discovered in *A. thaliana*. AtIRT1, OsIRT1, and HvIRT1 are IRT1 homologs which are present on plasma membrane and have major role in the absorption of $Fe^{2+}/Fe^{3+}$ and $Mn^{2+}$ (Vert et al. 2002). Additionally, IRT1 is involved in absorption of $Zn^{2+}$ and $Cd^{2+}$ in the plants (Cohen et al. 1998). In high concentration of Ni, AtIRT1 is involved in the $Ni^{2+}$ transportation and induced accumulation. The significant number of ZIP genes, 15 putative ZIP proteins in *Arabidopsis,* and 12 putative ZIP proteins in rice may suggest some redundancy in their functions. ZIP transporters need to increase metal ion accumulation and metal accumulation potential in hyperaccumulator species (Kramer et al. 2007). In *A. thaliana,* roots and shoots are part of the stress mechanism in response to Fe or Zn loading. These include 8 transmembrane domains and also have a similar topology with the N and C-termini exposed between a transmembrane domain III and IV and between a histidine-enriched domain of the apoplast and a variable cytoplasmic loop, consist for binding and specificity of metals (Guerinot 2000; Nishida et al. 2008). ZRT3 is another transporter identified as a functional complement in *Saccharomyces cerevisiae*, but it seems to be a result of Zn mobilization from vacuole.

### 13.3.2 NRAMPs (Natural Resistance-Associated Macrophage Proteins) Family

The Nramps family of integral membrane proteins includes metal ions ($Fe^{2+}$, $Zn^{2+}$, $Cd^{2+}$, $Cu^{2+}$, $Ni^{2+}$, and $Mn^{2+}$) transportation through the membranes of bacteria, fungi, plants, and animals. The Nramps family was usually confined to corn, *Oryza sativa*, in higher plants. Three Nramps (OsNramp 1–3) were first identified in the rice plant (Belouchi et al. 1995, 1997). This family of genes has now been recognized in a number of the higher plants (Williams et al. 2000; Mäser et al. 2001). One of those genes, AtNramp2, is extremely related to the Nramps in mammals and rice. Another gene named EIN2 (Ethylene-Insensitive 2) is a part of the ethylene signaling pathway, shows relatively less similarity than the Nramp gene. However, the hydrophobic region of these proteins shows significant similarity in both EIN2 and Nramp proteins. Six Nramps were shown in *Arabidopsis* for transport of Cd, Fe, and Mn. In plants, the Nramps tend to cluster into two subfamilies: first subfamily comprises AtNramps 1–6 and second one with Nramps 2–5 (Mäser et al. 2001). Overexpression of AtNramp1 in Transgenic *A. thaliana* increased resistance of the plants to Fe (Curie et al. 2000). Similar patterns of gene expression and Fe regulation were observed in AtNramp3 and AtNramp4, found in vacuolar membrane (Cailliatte et al. 2010). Cd can also be taken up through AtNramp3 (Thomine et al. 2000). AtNramp3 is present in vascular bundles of root, shoot and leaves can therefore be also used long-distance metals transport (Thomine et al. 2003). LeNramp1, that is present in the tomato root vascular parenchyma, is up-regulated with Fe deficiency (Bereczky et al. 2003).

Such results suggest that distinct members of this family play physiological roles and some Nramps contribute to absorption and homeostasis of Fe, Mn, and Cd (Hall and Williams 2003; Cailliatte et al. 2010). OsNramp5 accumulates Mn, translocate to the roots and also allows uptake of Cd in rice plant (Yang et al. 2004; Oomen et al. 2009; Ishikawa et al. 2012).

### 13.3.3 Ctr/COPT Transporters (Copper Transporter)

COPT1 was first described in *Arabidopsis* and this was found to complement a yeast mutant defect in the uptake and transport of Cu. Although all members are ubiquitous, variations in the relative abundance have been found in the roots, shoots, leaves, and flowers (Nuria et al. 2010). In comparison, COPT1 exhibits a high affinity in CTR2, potentially low affinity Cu transporter in the yeast and both are members of broad eukaryotic Cu transporter family. Yeast also has two Cu-membrane-located high affinity plasma transporters, CTR1 and CTR3 (Van Ho et al. 2002). Other research found another related gene in the database for genome sequencing (on the BAC clone AC005623). It is 71% similar to COPT1 and predicted to encode proteins with similar size to N-terminus enriched in methionine, histidine, and serine, similar to COPT1. COPT1 has been reported to show high expression in the leaves, lower expression in the shoots and flowers but absent in the roots. Therefore, other homolog may be accountable for Cu uptake in roots. Cd induced the accumulation of Cu in the roots of *A. thaliana* and also increases the synthesis of mRNA by 3 plasma membrane-bound Cu uptake transporters, COPT1, COPT2, and COPT6 (Gayomba et al. 2013)

## 13.4 Efflux Pumping at Plasma Membrane and Reduced Metals Uptake

In plant response to heavy metals, plasma membrane plays a significant function in preventing or decreasing metal uptake into cells and by pumping them outside cells via active efflux. It also appears that metabolic penalty for more specific metal absorption mechanisms, thus preventing the entry of toxic metal ions that is higher than that of efflux systems (Silver 1996). There are only some plasma membrane efflux transporters such as transporter family P1B-ATPases and these are involved in heavy metal plant responses.

### 13.4.1 P1B-ATPases

It is a big superfamily of transporters and their role is in pumping of various cations via cell membranes. These include the plant and fungal $H^+$-ATPases, animal $Na^+/K^+$ ATPases, and $Ca^{2+}$ ATPases present in different species (Axelsen and Palmgren 2001). Usually, it contains 8–12 transmembrane domains with a big cytoplasmic loop. However, eight transmembrane domains are estimated in the heavy metal ATPases (Williams et al. 2000; Mills et al. 2003, 2005). Subfamily P1B shows a significant role in metal transport. These transporters pump out cytoplasmic heavy metal in apoplast. *Arabidopsis* comprises eight ATPases of the type P1B (Mills et al. 2003). Heavy metal ATPase (HMA) cluster is divided into 2 phylogenetic research subclasses which are referred as Zn/Co/Pb/Cd and Cu/Ag group (Wang et al. 2011). First P1B type ATPase was AtHMA4 from *Arabidopsis* that was categorized in group Zn/Co/Cd/Pb. Metal uptake by HMAs is more selective than carriers such as HMA2, HMA3, and HMA4 which transport Cd and Zn (Kramer et al. 2007).

## 13.5 Movement of Metal Root to Shoot

Most heavy metals are retained in the root cells after being taken up from soil. In root cells, they may be subjected to detoxification by chelation in the cytoplasm. Further, metals may be translocated efficiently to shoot through xylem (Fig. 13.1). This includes the availability of heavy metal for xylem transport resulting in low sequestration in roots.

Xylem cell wall has high ability to exchange cations; metal cation movement is significantly inhibited because these metals are not chelated with ligands. Constitutively, high concentrations of small organic molecules can be contained in the hyperaccumulator roots, which can serve like ligands binding metals. The function of organic acids as ligands in root cells, mainly malate and citrate, is especially controversial because of their low interaction constants with the metals which forms a complex at the cytosolic pH values negligible (Haydon and Cobbett 2007). Histidine and nicotinamine form complexes with bivalent metal cations and take part in hyperaccumulation of metals. A range of facts indicate that effective transfer of major metal concentration in the plants is based on increased xylem loading through overexpression of genes contrary to non-hyperaccumulator plants. The greater movement of arsenic to the shoots in hyperaccumulator *Pteris vittata* in comparison to non-hyperaccumulator happens mostly as arsenite, which is present in the xylem sap (Su et al. 2008). Most arsenate is decreased rapidly to arsenite through the action of glutathione-dependent arsenate reductase in roots of hyperaccumulating ferns. (Duan et al. 2005)

### 13.5.1 HMA Family of Transporters

HMAs transporters act as pumps to eliminate metal from cell and serve the same as carriers to move Cd and Zn into the xylem from nearby tissues (Mills et al. 2003). AtHMA4 is a divalent ion which is present in plasma membrane. It acts as transporter for homeostasis of Zn and detoxification of Cd while it requires efflux and plant metal movement (Mills et al. 2003; Verret et al. 2004). Overexpression of AtHMA4 protein enhances tolerance against Zn and Cd and increases translocation of metals in plant. AtHMA5 is found in roots and other plant tissues, and is activated

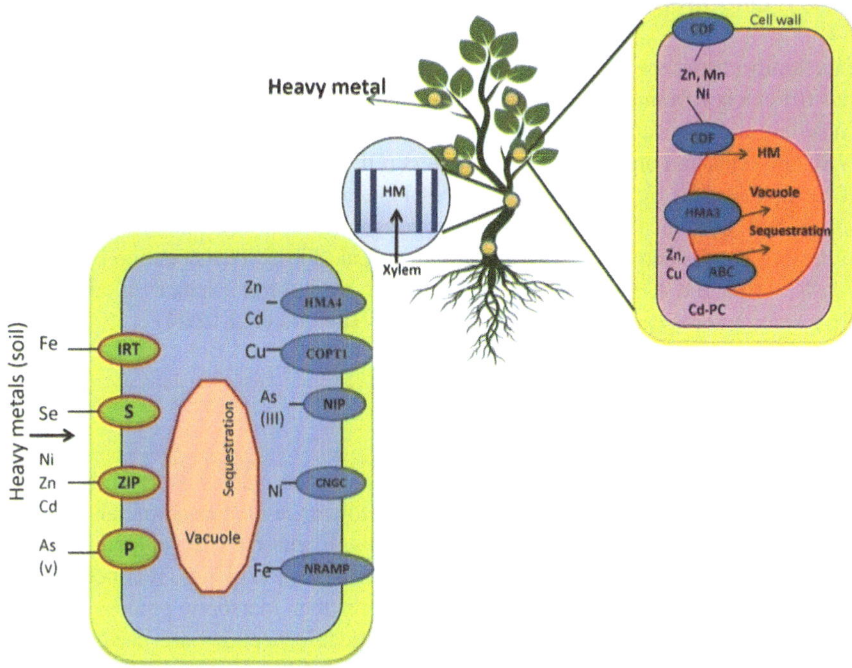


**Fig. 13.1** A schematic transport system of heavy metal involved in metal uptake

by Cu. AtHMA5 is expressed at high levels in the roots and stimulated by Cu in other tissues, but HMA5 mutants are extremely sensitive to Cu and absorb this metal more in the roots (Andrés-Colás et al. 2006).

### 13.5.2 Multidrug and Toxic Compound Extrusion Family (MATE)

This family is membrane-bound efflux protein that carries cell borne toxic metals in extrusion. AtALF5 (*A. thaliana* aberrant lateral root formation 5), the first plant MATE transporter was identified and found to be active in multidrug resistance (Li et al. 2002). These transporters mediate secondary transport, using either the $Na^+$ or the $H^+$ electrochemical gradient. Such transporters transport metal ions in opposite side to that of $Na^+$ or $H^+$, some MATE acts as substrate efflux transporters from cytoplasm to apoplasts. Such proteins are occupied in the transfer of citrate mandatory to translocate iron or to detoxify aluminum. In Arabidopsis, FRD3, a MATE protein citrate transporter is required for the movement of iron (Durrett et al. 2007). These transporters located in roots and cells (Durrett et al. 2007).

### 13.5.3 Oligopeptide Transporter Family

This family carries membrane proteins which help in translocation of their substrates into Cd cytosol from extracellular environment or organelles (Schaaf et al. 2004). These are a super family which includes two subfamilies: the yellow stripe-like (YSL) proteins and oligopeptide transporters (OPTs) YSL transporters are involved in the homeostasis by translocation of metal chelates, implying a degree of protection of both biological and substrates functions (DiDonato et al. 2004). Complexes of Fe(III)- phytosiderophore absorb Fe (Curie et al. 2001). Translocation of Fe, Zn, Cu, Ni, Mn, and Cd is done by the ZmYS1 transporter (Schaaf et al. 2004). Phytosiderophores can chelate these metal ions (Roberts et al. 2004).

## 13.6 Chelation in the Cytosol

Chelation of metals by ligands with high affinity in the cytosol is possibly a very significant mechanism for detoxification and tolerance to heavy metals. Heavy metals are not essential and are harmful to cells. Plants have developed different methods to accumulate excess metals and escape their toxic effects. If concentration of toxic metals exceeds a certain level within cells, then metabolic process can lead to chelating compound formation. In cytosol specific peptides are used as metal chelating agents, such as metallothioneins (MTs), phytochelatins (PCs), amino acids, and organic acid. Huge numbers of small molecules including amino acids and phosphate compounds are involved in metal chelation (Andra et al. 2010).

### 13.6.1 Phytochelatins

Plants adopted a detoxification method for HMs that is the development of low weight peptides of short-thiol rich for metal chelation. The PCs are repetitions of glutathione (GSH), synthesized with the enzyme phytochelatin synthase (PCS). PCs have a high binding affinity to toxic HMs (Hassinen et al. 2011). PCs protect plants from metal and other stresses such as excess salt, heat, and UV. PCs are also used as biomarkers to detect HM early on in the plants (Manara 2012; Saba et al. 2013; Song et al. 2014). These are processed in cytosol and effectively transferred to vacuole as their final aim to form a metal phytochelatin complex of high molecular weight (Fig. 13.2).

PCs are attached to certain metal anions and cations such as Pb, Cu, Zn, Cd, Ag, and Ni, through sulfhydryl and carboxyl groups (Rauser 1995; Cobbett 2000; Gupta et al. 2013). PCs can be developed and stored within plant roots and tissues. PC level in roots is recorded twice as high as the leaves when sunflower was treated with Cd (Yurekli and Kucukbay 2003). Among hyperaccumulators, it appears that they implement processes that include translocation of PCs from the root to shoot over long distances (Inouhe 2005). The lengths of the PCs chain indicate variability within and between plant species and HM types. In legumes, PCs with longer chains

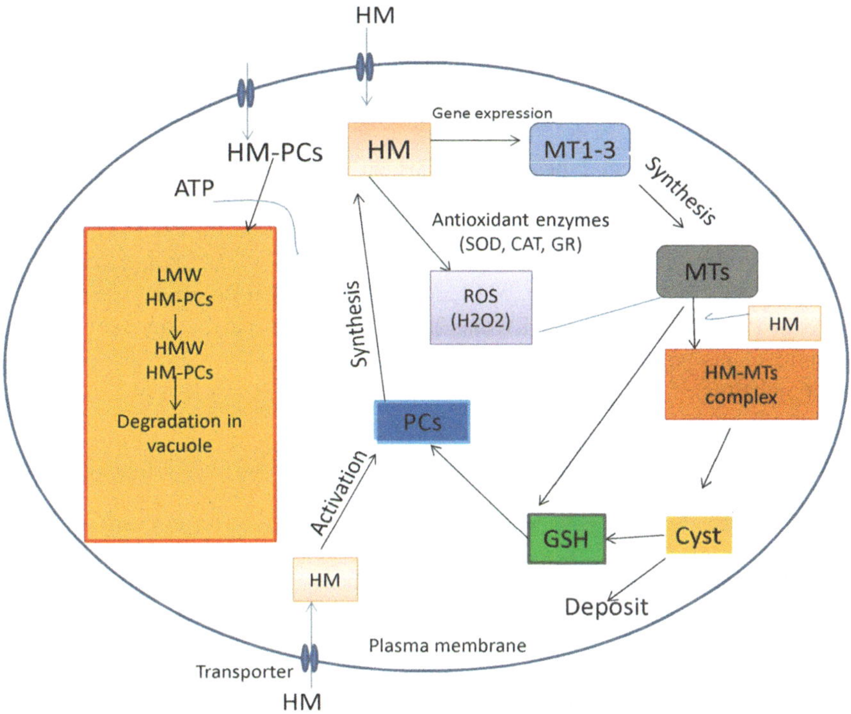


**Fig. 13.2** Representation of the cellular mechanism of phytochelatins and metallothioneins in heavy metal detoxification

are reported to be more strongly binding to Pb compared to shorter PCs (Piechalak et al. 2002). An artificial synthesis of the phytochelatin synthase gene increased its susceptibility in transgenic tobacco plants to differing amounts of Cd (Postrigan et al. 2012). Using *Brassica juncea,* it was shown that Cd accumulation is induced by a rapid induction of PC biosynthesis and that content of PCs is sufficient to chelate all Cd contamination.

### 13.6.2 Metallothioneins (MTs)

Metallothioneins are a super family of low weight proteins with very high cysteine levels that allow the combination of specific metal ions with the $d^{10}$ electron structure in the form of the metal thiolate clusters (Bhogavalli 2007). The MT gene reported in a number of plants that include *Arabidopsis*, where MT3 and MT4 variants have been recognized in addition to class 1 and 2. While MTs can be stimulated by metal treatments and evidence of increased metallothionein levels in response to Cd & Zn heavy metal tolerance is available. It plays a definite role in transporting as well as storing few fundamental ions such as Zn and Cu. MTs also contribute to detoxifying metal ions such as Cd and Hg (Fig. 13.2). Furthermore, these high thiol proteins and their incorporation by the stress conditions of the cells indicate a factor or action in direct search for cell damaging ROS, i.e., $HO^{\bullet}$ and $O_2^{\bullet}$ resulting in the development of disulfide bridges of Cys residues (Cobbett and Goldsbrough 2002).

Plants have different types of MTs which are classified into four different subgroups according to the Cys residue arrangement (1) MT type 1 is shown in the roots, (2) MT type 2 exists predominantly in the shoots, (3) MT type 3 is produced in leaves and at the stage in fruit maturation, and (4) MT type 4 is abundant in seed production. In soybean, MT1, MT2, and MT3 were mainly occupied in detoxification of Cd (Huang and Wang 2009). As an alternative, they may function as antioxidants but there is lack of research although another role in plasma membrane repairing.

### 13.6.3 Amino Acids and Organic Acids

Organic acids and amino acids can also chelate heavy metals and detoxification by chelation in plants can be attained by acid secretion. Carboxylic and amino acids like citric, malic, and histidine are alternative ligands for contaminants and have a role in detoxification. Citrate is well known for its high chelation affinity of metals such as Cd, Fe, Ni, Zn, and Co. Furthermore, the correlation between exudation of citric acid and resistance to aluminum was shown in *Phaseolus vulgaris.* In Ni-hyperaccumulating plant *Alyssum lesbiacum,* a 36-fold increment was reported in histidine content of xylem sap (Hall 2002).

## 13.7 Sequestration of Metal into Vacuole by Tonoplast Transporters

Heavy metals accumulate in cytosol in high concentrations; plants need to remove them to reduce their adverse effects. In response with elevated amount of ions, plants use these pumps to transfer to apoplast of the ions. Major ion storage compartment is the vacuole that constitutes up to 90% of total cell volume (Vögeli-Lange and Wagner 1990). Cd stimulates PC synthesis and then forms a Cd–PC complex, which is transported by a Cd/H antiport and an ATP-dependent PC-transporter into the vacuole. MTP1 (Metal transport protein), a gene that encodes a protein located at tonoplast, is highly overexpressed in Zn/Ni hyperaccumulator leaves (Kim et al. 2004; Gustin et al. 2009). MTP1 plays a significant role in the accumulation and tolerance of Zn. Genes of MTP members also stimulate Ni storage in vacuole of *Thlaspi goesingense* shoots. Therefore, MTP1 is found in both vacuoles and plasma membranes, indicating its role in cytoplasm-to-cell wall action of Zn and Ni efflux (Kim et al. 2004).

### 13.7.1 ABC Transporters

These carry heavy metals in the vacuoles. The multidrug resistance-associated protein (MRP) and pleiotropic drug resistance (PDR) are especially involved in chelated metal sequestration. The vacuoles of plant cell are also key location for PC–Cd complexes accumulation and stabilization. Appropriate Cd and sulfide are added in the vacuoles to form the primary source of Cd storage, HMW complexes. First vacuolar HMT1 transporter was known for its efficiency for complement *S. pombe* mutant. AtMRP1 and AtMRP2, two ABC family transporters are reported for carrying complexes in vacuole of *A. Thaliana* (Kolaj-Robin et al. 2015). The substrates are in conjunction with an antiport $H^+$, such as $Fe^{2+}$, $Zn^{2+}$, or $Mn^{2+}$ (Rubio-Sanz et al. 2013; Gupta et al. 2014; Raimunda et al. 2012). The protein metal binding domains are:

1. present in the transmembrane region,
2. is located in membrane cytosol interface, and
3. in C-terminal domain, but transportation appears to include only (1) and (2) directly (Montanini et al. 2007).

### 13.7.2 HMA Transporters

These are reported in a broad diversity of flora and fauna. They are powered by ATP hydrolysis and transported heavy metals (Pb, Cu, Cd, and Co) from cytoplasm into plasma membranes. AtHMA3 leads by sequestration into the vacuole to homeostasis Cd and Zn (Kramer et al. 2007). AtHMA3 can also act in detoxifying a wide variety of the heavy metals by retaining them in vacuoles, as overexpression induces resistance to heavy metals (Morel et al. 2009). HMA2, HMA3, and HMA4 transport metal ions such as Zn and Cd. HMA4 protein that plays a significant role in transport of Zn via root to shoot also protects plants from Cd through its efflux. Expression of the C-type ABC transporter (OsABCC1), a member of ATP-binding cassette in *Oryza sativa* has been identified for detoxification and reduction As accumulation in the rice grains.

## References

Andra SS, Datta R, Sarkar D, Makris KC, Mullens CP, Sahi SV et al (2010) Synthesis of phytochelatins in vetiver grass upon lead exposure in the presence of phosphorus. Plant Soil 326(1–2):171–185

Andrés-Colás N, Sancenón V, Rodríguez-Navarro S, Mayo S, Thiele DJ, Ecker JR, Peñarrubia L (2006) The *Arabidopsis* heavy metal P-type ATPase HMA5 interacts with metallochaperones and functions in copper detoxification of roots. Plant J 45(2):225–236

Antala S, Ovchinnikov S, Kamisetty H, Baker D, Dempski RE (2015) Computation and functional studies provide a model for the structure of the zinc transporter hZIP4. J Biol Chem 290 (29):17796–17805

Axelsen KB, Palmgren MG (2001) Inventory of the superfamily of P-type ion pumps in *Arabidopsis*. Plant Physiol 126:696–706
Belouchi A, Cellier M, Kwan T, Saini HS, Leroux G, Gros P (1995) The macrophage- specific membrane protein Nramp controlling natural resistance to infections in mice has homologues expressed in the root system of plants. Plant Mol Biol 29:1181–1196
Belouchi A, Kwan T, Gros P (1997) Cloning and characterization of OsNramp family from Oryza sativa, a new family of membrane proteins possibly implicated in the transport of metal ions. Plant Mol Biol 33(6):1085–1092
Bereczky Z, Wang H-Y, Schubert V, Ganal M, Bauer P (2003) Differential regulation of Nramp and IRT metal transporter genes in wild type and iron uptake mutants of tomato. J Biol Chem 278:24697–24704
Bhogavalli S (2007) Characterization of a Type II metallothionein from *Helianthus annuus* using recombinant DNA techniques. ETD Archive, p 329
Cailliatte R, Schikora A, Briat JF, Mari S, Curie C (2010) High- affinity manganese uptake by the metal transporter NRAMP1 is essential for *Arabidopsis* growth in low manganese conditions. Plant Cell 22:904–917
Cobbett CS (2000) Phytochelatins and their roles in heavy metal detoxification. Plant Physiol 123 (3):825–832
Cobbett C, Goldsbrough P (2002) Phytochelatins and metallothioneins: roles in heavy metal detoxification and homeostasis. Annu Rev Plant Biol 53:159–182
Cohen CK, Fox TC, Garvin DF, Kochian LV (1998) The role of Iron-deficiency stress responses in stimulating heavy-metal transport in plants. Plant Physiol 116:1063–1072
Curie C, Alonso JM, Le Jean M, Ecker JR, Briat JF (2000) Involvement of NRAMP1 from *Arabidopsis thaliana* in iron transport. Biochem J 347:749–755
Curie C, Panaviene Z, Loulergue C, Dellaporta SL, Briat JF, Walker EL (2001) Maize yellow stripe1 encodes a membrane protein directly involved in Fe(III) uptake. Nature 409:346–349
DiDonato JRJ, Roberts LA, Sanderson T, Eisley RB, Walker EL (2004) Arabidopsis Yellow Stripe-Like2 (YSL2): a metal-regulated gene encoding a plasma membrane transporter of nicotianamine–metal complexes. Plant J 39(3):403–414
Duan GL, Zhu YG, Tong YP, Cai C, Kneer R (2005) Characterization of arsenate reductase in the extract of roots and fronds of Chinese brake fern, an arsenic hyperaccumulator. Plant Physiol 138(1):461–469
Durrett TP, Gassmann W, Rogers EE (2007) The FRD3-mediated efflux of citrate into the root vasculature is necessary for efficient iron translocation. Plant Physiol 144:197–205
Gayomba SR, Jung HI, Yan J, Danku J, Rutzke MA, Bernal M et al (2013) The CTR/COPT-dependent copper uptake and SPL7-dependent copper deficiency responses are required for basal cadmium tolerance in *A. thaliana*. Metallomics 5(9):1262–1275
Guerinot ML (2000) The ZIP family of metal transporters. Biochem Biophys Acta 1465:190–198
Guerinot ML (2010) Iron in cell biology of metals and nutrients. In: Hell R, Mendel R (eds) Plant cell monographs, vol 17. Springer, Berlin/Heidelberg, pp 75–94
Gupta D, Vandenhove H, Inouhe M (2013) Heavy metal stress in plants. role of phytochelatins in heavy metal stress and detoxification mechanisms in plants. Springer, Berlin, pp 73–94
Gupta S, Chai J, Cheng J, D'Mello R, Chance MR, Fu D (2014) Visualizing the kinetic power stroke that drives proton-coupled Zn(II) transport. Nature 512:101–104
Gustin JL, Loureiro ME, Kim D, Na G, Tikhonova M, Salt DE (2009) MTP1-dependent Zn sequestration into shoot vacuoles suggests dual roles in Zn tolerance and accumulation in Zn-hyperaccumulating plants. Plant J 57(6):1116–1127
Hall JL (2002) Cellular mechanisms for heavy metal detoxification and tolerance. J Exp Bot 53 (366):1–11
Hall JL, Williams LE (2003) Transition metal transporters in plants. J Exp Bot 54(393):2601–2613
Hassinen VH, Tervahauta AI, Schat H, Kärenlampi SO (2011) Plant metallothioneins–metal chelators with ROS scavenging activity? Plant Biol 13:225–232

Haydon MJ, Cobbett CS (2007) Transporters of ligands for essential metal ions in plants. New Phytol 174(3):499–506

Huang GY, Wang YS (2009) Expression analysis of type 2 metallothionein gene in mangrove species (*Bruguiera gymnorrhiza*) under heavy metal stress. Chemosphere 77(7):1026–1029

Inouhe M (2005) Phytochelatins. Braz J Plant Physiol 17(1):65–78

Ishikawa S, Ishimaru Y, Igura M, Kuramata M, Abe T, Senoura T et al (2012) Ion-beam irradiation, gene identification, and marker-assisted breeding in the development of low-cadmium rice. Proc Natl Acad Sci U S A 109:19166–19171

Kalaivanan D, Ganeshamurthy AN (2016) Mechanisms of heavy metal toxicity in plants. In: Srinivasa Rao NK, Shivashankara KS, Laxman RH (eds) Abiotic stress physiology of horticultural crops. Springer, New Delhi, pp 85–102

Kim D, Gustin JL, Lahner B, Persans MW, Baek D, Yun DJ, Salt DE (2004) The plant CDF family member TgMTP1 from the Ni/Zn hyperaccumulator *Thlaspi goesingense* acts to enhance efflux of Zn at the plasma membrane when expressed in *Saccharomyces cerevisiae*. Plant J 39 (2):237–251

Kinnersley AM (1993) The role of phytochelates in plant growth and productivity. Plant Growth Regul 12(3):207–218

Kolaj-Robin O, Russell D, Hayes KA, Pembroke JT, Soulimane T (2015) Cation diffusion facilitator family: structure and function. FEBS Lett 589:1283–1295

Kramer U, Talke IN, Hanikenne M (2007) Transition metal transport. FEBS Lett 581:2263–2272

Li L, He Z, Pandey GK, Tsuchiya T, Luan S (2002) Functional cloning and characterization of a plant efflux carrier for multidrug and heavy metal detoxification. J Biol Chem 277:5360–5368

Lin R, Wang X, Luo Y, Du W, Guo H, Yin D (2007) Effects of soil cadmium on growth, oxidative stress and antioxidant system in wheat seedlings (*Triticum aestivum* L.). Chemosphere 69:89–98

Maestri E, Marmiroli M, Visioli G, Marmiroli N (2010) Metal tolerance and hyperaccumulation: costs and trade-offs between traits and environment. Environ Exp Bot 68(1):1–13

Manara A (2012) Plant responses to heavy metal toxicity. In: Furini A (ed) Plants and heavy metals. Springer, Dordrecht, pp 27–53

Mäser P, Thomine S, Schroeder JI, Ward JM, Hirschi K, Sze H, Talke IN, Amtmann A, Maathuis FJM, Sanders D, Harper JF, Tchieu J, Gribskov M, Persans MW, Salt DE, Kim SA, Guerinot ML (2001) Phylogenetic relationships within cation transporter families of *Arabidopsis*. Plant Physiol 126:1646–1667

Masood A, Iqbal N, Khan NA (2012) Role of ethylene in alleviation of cadmium-induced photosynthetic capacity inhibition by Sulphur in mustard. Plant Cell Environ 35:524–533

Mills RF, Krijger GC, Baccarini PJ, Hall JL, Williams LE (2003) Functional expression of AtHMA4, a P1B-type ATPase of the Zn/Co/Cd/Pb subclass. Plant J 35:164–176

Mills RF, Francini A, Ferreira da Rocha PSC, Baccarini PJ, Aylett M, Krijger GC, Williams LE (2005) The plant P1B-type ATPase AtHMA4 transports Zn and Cd and plays a role in detoxification of transition metals supplied at elevated levels. FEBS Lett 579:783–791

Mizuno T, Usui K, Horie K, Nosaka S, Mizuno N, Obata H (2005) Cloning of three ZIP/Nramp transporter genes from a Ni hyperaccumulator plant *Thlaspi japonicum* and their $Ni^{2+}$-transport abilities. Plant Physiol Biochem 43:793–801

Montanini B, Blaudez D, Jeandroz S, Sanders D, Chalot M (2007) Phylogenetic and functional analysis of the cation diffusion facilitator (CDF) family: improved signature and prediction of substrate specificity. BMC Genomics 8:107

Morel M, Crouzet J, Gravot A, Auroy P, Leonhardt N, Vavasseur A, Richaud P (2009) AtHMA3, a P1B-ATPase allowing Cd/Zn/Co/Pb vacuolar storage in *Arabidopsis*. Plant Physiol 149:894–904

Nishida S, Mizuno T, Obata H (2008) Involvement of histidine-rich domain of ZIP family transporter TjZNT1 in metal ion specificity. Plant Physiol Biochem 46(5-6):601–606

Nuria AC, Ana PG, Sergi P, Lola PA (2010) Deregulated copper transport affects *Arabidopsis* development especially in the absence of environmental cycles. Plant Physiol 153:170–184

Olsen LI, Palmgren MG (2014) Many rivers to cross: the journey of zinc from soil to seed. Front Plant Sci 5:30

Oomen RJ, Wu J, Lelievre F, Blanchet S, Richaud P, Barbier-Brygoo H, Aarts MG, Thomine S (2009) Functional characterization of NRAMP3 and NRAMP4 from the metal hyperaccumulator *Thlaspi caerulescens*. New Phytol 181:637–650

Piechalak A, Tomaszewska B, Baralkiewicz D, Malecka A (2002) Accumulation and detoxification of lead ions in legumes. Phytochemistry 60(2):153–162

Pilon M (2011) Moving copper in plants. New Phytol 192:305–307

Postrigan BN, Knyazev AB, Kuluev BR, Yakhin OI, Chemeris AV (2012) Expression of the synthetic phytochelatin gene in tobacco. Russ J Plant Physiol 59(2):275–280

Raimunda D, Subramanian P, Stemmler T, Argüello JM (2012) A tetrahedral coordination of zinc during transmembrane transport by P-type $Zn^{2+}$-ATPases. Biochem Biophys Acta 1818:1374–1377

Rascio N, Navari-Izzo F (2011) Heavy metal hyperaccumulating plants: how and why do they do it? And what makes them so interesting? Plant Sci 180(2):169–181

Rauser WE (1995) Phytochelatins and related peptides. Structure, biosynthesis, and function. Plant Physiol 109(4):1141–1149

Revathi S, Subhashree V (2015) Physiological and biochemical mechanisms of heavy metal tolerance. Int J Environ Sci 3:1339–1354

Roberts LA, Pierson AJ, Panaviene Z, Walker EL (2004) Yellow Stripe1. Expanded roles for the maize iron-phytosiderophore transporter. Plant Physiol 135:112–120

Rubio-Sanz L, Prieto RI, Imperial J, Palacios JM, Brito B (2013) Functional and expression analysis of the metal-inducible dmeRF system from *Rhizobium leguminosarum* bv. viciae. Appl Environ Microbiol 79(20):6414–6422

Saba H, Jyoti P, Neha S (2013) Mycorrhizae and phytochelators as remedy in heavy metal contaminated land remediation. Int Res J Environ Sci 2(1):74–78

Schaaf G, Ludewig U, Erenoglu BE, Mori S, Kitahara T, von Wirén N (2004) ZmYS1 functions as a proton-coupled symporter for phytosiderophore- and nicotianamine-chelated metals. J Biol Chem 279:9091–9096

Silver S (1996) Bacterial resistance to toxic metal ions—a review. Gene 179:9–19

Song WY, Mendoza-Cózatl DG, Lee Y et al (2014) Phytochelatin-metal(loid) transport into vacuoles shows different substrate preferences in barley and *Arabidopsis*. Plant Cell Environ 37(5):1192–1201

Su YH, McGrath SP, Zhu YG, Zhao FJ (2008) Highly efficient xylem transport of arsenite in the arsenic hyperaccumulator *Pteris vittata*. New Phytol 180(2):434–441

Thomine S, Wang R, Ward JM, Crawford NM, Schroeder JI (2000) Cadmium and iron transport by members of a plant metal transporter family in *Arabidopsis* with homology to Nramp genes. Proc Natl Acad Sci U S A 97:4991–4996

Thomine S, LelieÁvre F, Debarbieux E, Schroeder JI, Barbier-Brygoo H (2003) AtNRAMP3, a multispecific vacuolar metal transporter involved in plant responses to iron deficiency. Plant J 34:685–695

Van Ho A, Ward DM, Kaplan J (2002) Transition metal transport in yeast. Annu Rev Microbiol 56 (1):237–261

Verbruggen N, Hermans C, Schat H (2009) Molecular mechanisms of metal hyperaccumulation in plants. New Phytol 181(4):759–776

Verret F, Gravot A, Auroy P, Leonhardt N, David P, Nussaume L, Vavasseur A, Richaud P (2004) Overexpression of AtHMA4 enhances root-to-shoot translocation of zinc and cadmium and plant metal tolerance. FEBS Lett 576:306–312

Vert G, Grotz N, Dédaldéchamp F, Gaymard F, Guerinot ML, Briat J-F, Curie C (2002) IRT1, an *Arabidops*is transporter essential for iron uptake from the soil and for plant growth. Plant Cell 14:1223–1233

Vögeli-Lange R, Wagner GJ (1990) Subcellular localization of cadmium and cadmium-binding peptides in tobacco leaves. Implication of a transport function for cadmium-binding peptides. Plant Physiol 92:1086–1093

Wainwright SJ, Woolhouse HW (1977) Some physiological aspects of copper and zinc tolerance in *Agrostis tenuis* Sibth.: cell elongation and membrane damage. J Exp Bot 28(4):1029–1036

Wang X, Winter D, Ashrafi G, Schlehe J, Wong YL, Selkoe D, Rice S, Steen J, LaVoie MJ, Schwarz TL (2011) PINK1 and Parkin target Miro for phosphorylation and degradation to arrest mitochondrial motility. Cell 147(4):893–906

Williams LE, Pittman JK, Hall JL (2000) Emerging mechanisms for heavy metal transport in plants. Biochim Biophys Acta 77803:1–23

Yang J, Ye Z (2015) Antioxidant enzymes and proteins of wetland plants: their relation to Pb tolerance and accumulation. Environ Sci Pollut Res 22(3):1931–1939

Yang XE, Long XX, Ye HH, He ZL, Calvert DV, Stoffella PJ (2004) Cadmium tolerance and hyperaccumulation in a new Zn-hyperaccumulating plant species (*Sedum alfredii* Hance). Plant Soil 259(1–2):181–189

Yurekli F, Kucukbay Z (2003) Synthesis of phytochelatins in *Helianthus annuus* is enhanced by cadmium nitrate. Acta Bot Croat 62(1):21–25

Zhao S, Lian F, Duo L (2011) EDTA-assisted phytoextraction of heavy metals by turfgrass from municipal solid waste compost using permeable barriers and associated potential leaching risk. Bioresour Technol 102:621–626

# Cadmium: Bioavailability in Soils and Phytotoxicity

# 14

Harmanjit Kaur and Sofi Javed Hussain

**Abstract**

Contamination of soils with heavy metals is a major environmental problem which causes threat to the production of crops globally as well as compromises human health due to soil-to-plant transfer of metals. In this context, increasing levels of cadmium (Cd) in farming soils are a reality these days. Cd is the most prevalent non-essential element for plants and is one of the most toxic heavy metals due to its deleterious effects at very low concentrations and high water solubility. Cd uptake from soil occurs through plant roots, depending upon its availability and concentration. There are several factors which can modulate the Cd availability in soils. Cd causes multifaceted alterations in plants at morphological and physiological levels, resulting in phytotoxicity. Accumulation of Cd in plants induces oxidative stress due to overproduction of reactive oxygen species (ROS), which results in severe damage at the cellular level, eventually impairing plant growth and productivity. In response to Cd toxicity, plants employ different strategies including enzymatic and non-enzymatic antioxidant defense mechanisms which contribute in alleviating cellular damage and balancing redox homeostasis. This chapter aims to summarize various factors influencing the bioavailability of Cd in soils along with our present understanding of Cd-mediated toxic responses as well as tolerance mechanisms in plants and also highlights the potential targets of Cd toxicity for improved plant metal tolerance.

H. Kaur (✉)
Department of Botany, Akal University, Talwandi Sabo, Bathinda, Punjab, India

S. J. Hussain
Department of Botany, Aligarh Muslim University, Aligarh, Uttar Pradesh, India



K. Mishra et al. (eds.), *Sustainable Solutions for Elemental Deficiency and Excess in Crop Plants*, https://doi.org/10.1007/978-981-15-8636-1_14

**Keywords**

Antioxidants · Bioavailability · Cadmium · Morphological · Physiological · Oxidative stress

## 14.1 Introduction

Cadmium ($^{112.41}Cd_{48}$), belonging to group IIB transition elements, was discovered in 1817 by a German chemist Friedrich Strohmeyer. Out of eight naturally occurring isotopic forms of Cd, two are radioactive (Sebastian et al. 2019). Cd mostly occurs in +2 oxidation state, though +1 oxidation state has been reported in some diatomic compounds. Greenockite is an essential ore of Cd (CdS); however, the metal is largely purified from Zn ores (Sebastian et al. 2019). Cd exists in all types of soils, mostly as a minor component, however, it can attain elevated levels in farming soils. Among the various elements causing heavy metal contamination in soil, Cd has been registered as the most harmful element globally by the United Nations Environment Program (UNEP) (Wang et al. 2015) because of prolonged biological half-life of 10–30 years, extremely mobile nature, great solubility in water, readily taken up by plant roots, and the low concentration required to exhibit its toxic effects on plants (Chou et al. 2011). It is regarded as a non-nutrient element, as it has no role in plant growth and life as a whole, except Cd-carbonic anhydrase in marine diatoms (Morant-Manceau et al. 2007). Cd concentrations in unpolluted soils are normally below 0.5 mg kg$^{-1}$ dry soil; however, they can reach up to 3.0 mg kg$^{-1}$ dry soil or more than 100 mg dm$^{-3}$ in the top 25 cm layer of soil due to several human activities (Tezotto et al. 2012; Szolnoki et al. 2013). Agricultural soils are polluted with Cd owing to (1) use of phosphatic manures, fertilizers, pesticides, and sewage sludge, (2) irrigation with industrial effluents, and (3) urban and industrial activities (Song et al. 2009; Cabala et al. 2011). In phosphatic fertilizers, Cd content varies from less than 3.6 to 527 mg kg$^{-1}$ (Satarug et al. 2003). Besides these human-based emissions, Cd is also discharged via natural sources, for example, continental dusts, volcanic emissions, erosion of underlying rock layer, and forest fires (Grobelak et al. 2019). The major application of Cd at industrial level is in Ni-Cd batteries, pigment formulations, surface coverings and plating, stabilizing agent for plastics and synthetics, non-ferrous alloys, and electro-optics (Sebastian et al. 2019).

Cd strongly binds with various soil structures and only a small fraction of Cd in soil is phytoavailable (Sun et al. 2009). Cd is effortlessly absorbed by plant roots and transported to aerial organs, causing morphological, physiological, biochemical, cellular, and molecular alterations, ultimately affecting plant growth and productivity (Shanmugaraj et al. 2013; Song et al. 2017). The daily dietary intake of Cd has been restricted to 60–70 μg (Chunhabundit 2016). The permitted quantity of Cd in cereal grains is 100 μg kg$^{-1}$. Green manure crops, for instance, *Avena strig*osa Schreb. and *Crotalaria juncea* L. accumulate more than 100 mg kg$^{-1}$ dry weight of Cd in their leaves (Uraguchi et al. 2006). Cd concentrations in mature leaves are categorized as normal (0.05–0.2 mg kg$^{-1}$), excessive/toxic (5–30 mg kg$^{-1}$), and tolerable (0.05–0.5 mg $kg^{-1}$) (Kabata-Pendias and Pendias 2010). Cd hyperaccumulators can tolerate and accumulate Cd up to 100 mg Cd $kg^{-1}$ (DW) in shoots (0.01% of shoot DW), which is 100-fold higher than that accumulated by non-hyperaccumulator plants (Kramer 2010). Under Cd stress of moderate to high toxicity, most of the Cd hyperaccumulators do not exhibit toxic symptoms. On the contrary, non-accumulators experience toxicity symptoms when exposed to Cd. Therefore, it is imperative to comprehend the deleterious effects of Cd on plants, as the main route of Cd intake in humans is via plant-based food products. With this background, the chapter attempts to present current knowledge regarding several soil characteristics associated with Cd phytoavailability and response of plants to Cd stress at various levels, with an aim to provide a better understanding of the detoxification mechanisms in plants and to improve plant tolerance towards Cd stress.

## 14.2 Cd Bioavailability at Soil–Plant Interface

The risk connected with Cd accretion in soils is more strongly linked to the Cd bioavailability rather than to the total Cd concentration in soil (Carrillo-Gonzalez et al. 2006). Cd sequestration in phosphates or carbonates present in soils may decrease the availability of Cd to plants. On the contrary, Carrillo-Gonzalez et al. (2006) proposed that since the free activity of $Cd^{2+}$ in the soil solution is much less than the chemical equilibrium of established solid phases, the mobility as well as solubility of Cd in soils is possibly regulated by co-precipitation or sorption processes. Physico-chemical attributes of soil as well as soil solution, for instance, cation exchange capacity (CEC), organic matter, pH, presence of other ions, redox potential (Eh), and temperature control the fraction of Cd which is bioavailable (Basar 2009; Parrotta et al. 2015). Fischerová et al. (2006) and Tlustoš et al. (2006a) examined remediation prospects of soils polluted with heavy metals, viz. As, Cd, and Zn and noticed that the metals accumulated in the plant biomass depended considerably on their availability in the soil. Black et al. (2011) concluded that metal bioavailability in soils by and large depends on the particular metal as well as developmental stage of the plant.

### 14.2.1 Organic Matter

Organic matter content is a critical soil feature that influences availability of Cd in soils because of its ability to retain metal cation (Sarwar et al. 2010). Elevated contents of organic matter in soil augment the CEC and accordingly can enhance Cd adsorption. Soils having higher CEC hold more Cd in the negative sites, resulting in reduced Cd accumulation (Sebastian and Prasad 2014b). Organic matter changes the soluble/exchangeable form of Cd into an organically bound fraction, thereby decreasing its bioavailability (Chen et al. 2010). Organic materials generally mount up on the outer layers of soil profile as either bulky mass or as a covering on particular matter (Kirkham 2006). Organic matter and pH of soil may act together in their effects on Cd bioavailability. Sauve (2003) observed that total metal concentrations as well as soil pH were not constant markers of bioavailable Cd and that the organic matter increased the sorption affinity of soil for Cd to 30 times than mineral soil. Ge and Hendershot (2005) expressed that organic matter was the chief sorbent for heavy metals in organic forest soils. Soils having excessive quantities of organic matter, for example, those containing biosolids amendments or manure have large concentrations of dissolved organic carbon which can form soluble complexes with Cd (Sauve et al. 2000a). Therefore, in Cd-polluted soils, application of several organic amendments, for instance, composts, farmyard manure (FYM), and biosolids can efficiently decrease Cd phytoavailability.

### 14.2.2 pH

Among soil parameters influencing availability of Cd, soil pH is generally recognized as the most important factor (Basta et al. 2005). Alteration in pH changes the Cd adsorption by soils and thus, its concentration in the soil solution. Since speciation differs with pH, the concentration of unbounded metal ion causing a certain degree of noxious effect can be stated as a function of pH only (Lofts et al. 2004). According to Kabata-Pendias and Pendias (2001), Cd bioavailability gets altered from very high at pH 2 to very low at pH 7. Combined data from 70 studies revealed that 50% of the differences in soil Cd Kd (soil distribution coefficient) could be explicated by pH, nevertheless, taking soil organic matter as a subsequent autonomous variable considerably enhanced the regression (Sauve et al. 2000b). The Kd rates for Cd were greater with higher soil organic matter contents. Multiple regression analysis revealed that soil Cd and pH were the major parameters affecting Cd concentrations in a meta-analysis of 162 wheat (*Triticum aestivum*) and 215 barley (*Hordeum vulgare*) grain samples (Adams et al. 2004). pH of the soil significantly influenced the Cd accumulation in barley (Wu and Zhang 2002) and there was a considerable negative correlation between Cd content in seeds and soil pH under low Cd levels. Jauert et al. (2002) studied the ability of strawberry clover roots to alter pH of rhizosphere and absorb Cd and confirmed negative correlation between rhizosphere pH and Cd uptake. McLaren et al. (2004) added Cd-spiked sludge to five diverse soils from New Zealand: two pasture soils and three forest soils and examined leaching of Cd from soils constantly for 3 years. They observed that pH of the leachate from the soils devoid of sludge was greater (6.0–6.5) than the pH of the leachate from the soils amended with sludge (5.0 or below) and established increased metal leaching with decrease in pH ensuing from application of sludge. Bergkvist and Jarvis (2004) described a model to determine ambiguity over the long-term outcomes of sludge application to agricultural land and proposed that the most vital factors associated with leaching and crop uptake were Cd loading, the partition coefficient for inorganic materials derived from sludge, and the factor controlling the effect of pH on sorption. Li et al. (2005) explained the relation between types of soil and rice genotypes on Cd bioavailability by growing the two genotypes on two different soils: one having pH of 4.95 and other with pH of 6.54 and witnessed that grain Cd contents surpassed the Chinese food guidelines and recommended that the raising of cultivars on acidic soils should be avoided. Nursita et al. (2009) noticed maximum soluble-exchangeable Cd concentration in Australian soil which had the least pH, CEC, clay, organic carbon, and very low Fe content while soil having the highest pH, CEC, clay, organic carbon, and Fe content contained maximum Cd in the form of organic and oxide fraction. Studies by Liu et al. (2015) and He et al. (2017) in wheat and rice, respectively, established the existence of an inverse correlation between soil pH and Cd bioavailability, i.e. with decline in soil pH, plant Cd uptake increases. Hence, in Cd-polluted acidic soils, Cd availability is larger than alkaline and neutral soils. Addition of lime to slightly acidic soils increased the soil pH and reduced the concentration of Cd in rice plants (Yang et al. 2018). A recent study suggested that blend of lime and gypsum in contaminated soil is efficient in decreasing the phytoavailability of Cd, Pb, and As in medicinal plants (Kim et al. 2018).

### 14.2.3 Presence of Other Ions

Occurrence of other ions in soil affects Cd availability as a result of complexation, ionic potency, and competition for soil/root surface exchange sites (Tlustos et al. 2006b). Gothberg et al. (2004) suggested that concentration of the external nutrient solution was of significance for the accretion and toxicity of Cd in *Ipomoea aquatica*. Cd forms Cd-chloro complexes in the presence of $Cl^-$, which decrease soil sorption of Cd, thereby enhancing solubility as well as Cd availability (Grant and Sheppard 2008). Numerous cations like $Ca^{2+}$, $Mn^{2+}$, $Mg^{2+}$, and $Zn^{2+}$ compete with Cd for uptake by plants (Tlustos et al. 2006b).

### 14.2.4 Redox Potential

Rhizospheric redox potential (Eh) around plant roots differs from that of greater part of soil. For instance, rice rhizosphere Eh is always higher as compared to non-rhizosphere Eh due to emission of $O_2$ by rice roots. Nonetheless, rhizosphere Eh is mostly lesser than that of bulk soil (Sarwar et al. 2010) since roots use oxygen and secrete reducing compounds, e.g. phenolics. Heavy metals affect the rhizosphere pH which successively may influence Eh. Additionally, reactions of metals, particularly redox reactions, may be strongly influenced by rhizosphere Eh. Eh changes the electric charge of some minerals that may modify their CEC. Hydrogen sulphide ($H_2S$) is generated in soil when Eh is low. Cd combines with $S^{2-}$ and forms insoluble CdS which is not readily absorbed by plants. Su et al. (2000) observed that upon enhancing the redox potential from $-150$ to 200 mV, the exchangeable form of Cd declined whereas the reductive Cd increased. Bolan and Duriasamy (2003) demonstrated that supplementation of phosphate as $KH_2PO_4$ enhanced the soil pH, Eh, exchangeable and soluble Cd-fraction, leading to its poor availability.

Larger Eh decreases Cd and causes precipitation of the Cd ion (Sebastian and Prasad 2014a).

### 14.2.5 Speciation

The chemical speciation of a heavy metal in soil includes fractionation of its entire content into acid-extractable, exchangeable, reducible, oxidizable, and residual forms (Christine et al. 2002). The acid-extractable and exchangeable forms are mobile that are believed to be easily available. The reducible and oxidizable fractions are leached out under severe situations, whereas the residual form is almost inert (Wong and Selvam 2006). Instead of total Cd content, Cd speciation plays an essential function in ascertaining the mobility and ultimately its bioavailability (Hu et al. 2011). The $Cd^{2+}$ form is mostly regarded as the most bioavailable form, though ligand-increased Cd uptake has also been demonstrated (Berkelaar and Hale 2003). The particulate-bound Cd availability in a soil–plant scheme declines in the order: exchangeable < carbonates < metal-organic complexes < organics < Fe and Mn oxides < mineral lattices (Krishnamurti et al. 1995). The most copious Cd species in 35 soil samples from Czech Republic and carbonate-bound in arable soils was oxide-bound (Szakova et al. 1999; Banat et al. 2007). Few studies have described Cd forms in soil via sequential extraction procedures (Kim and Owens 2009).

### 14.2.6 Aging

Additional factor that influences availability of Cd is aging. Because of long lasting chemical processes, the bio-accessibility of metals declines with time and with the period of its contact with soil (Lock and Janssen 2003a), with slight or no diminution in total metal concentration. According to Lock and Janssen (2003b), effect of aging on availability of a metal for a specific soil can be assessed by analyzing the difference between the proportion of metal sorbed instantly after spiking and the proportion expected to be sorbed after aging on pH basis. Granato et al. (2004) quantified Cd at long-term sludge dumping site immediately after the termination of a 10-year use of sludge at the land and then 10 years afterwards and observed that Cd contents in roots and leaves of maize decreased to about half suggesting that the available Cd in the soil declined with the passage of time, although the total amount did not change much.

### 14.2.7 Root Exudations

Roots of plants also alter bioavailability of metals because of exudation of protons ($H^+$), amino acids, organic acids (OAs), phytosiderophores, and enzymes (Yang et al. 2005). $H^+$ secretion by roots, both actively or passively, makes the rhizosphere acidic and enhances dissolution of metal. Legume plants have a tendency to acidify their rhizosphere as they absorb more cations (ammonium) than anions (nitrate), which is balanced via secretion of $H^+$. Extrusion of $H^+$ by the roots is carried out by plasma membrane $H^+$ pumps and $H^+$-ATPases. P-deficiency stimulated exudation of citrate in Lupin by increasing the action of plasma membrane $H^+$ pumps and $H^+$-ATPases (Ligaba et al. 2004). Such alterations in rhizosphere pH mediated by roots have a key function both in the bioavailability of several pH dependent nutrients and noxious metals as well as various trace metals (Hinsinger et al. 2006). Root exudates are components of plant photosynthates (about 12–40%) which are secreted into rhizosphere throughout plant development via roots into the soil, comprising amino acids, sugars, polysaccharides, proteins, peptides, and few OAs (Dong et al. 2007; Sarwar et al. 2010). Root exudations contain both organic ligands (e.g. amino acids, carbohydrates, humic acids, nucleic acids, OAs, proteins, etc.) and inorganic ligands (e.g. $Cl^-$, $CO_3^{2-}$, $NH_4^+$, $SO_4^{2-}$, $PO_4^{3-}$, etc.). These ligands form complexes with metal ions and affect pH as well as Eh along with chemical attributes in the rhizosphere. OAs (having low molecular weight) commonly occur in soils as natural component of root and microbial exudates in addition to decomposed remains of plants and animals (Jiang et al. 2011) and assist in dissolving the metals in the soil, making them available for plant absorption (Sidhu et al. 2019). OAs modulate soil pH, influence metal solubility in the soil, and ensure their uptake by plants (Javed et al. 2013). OAs with higher number of hydroxyl and carboxylic groups have larger effect on metal desorption because of their greater complexing potential (Jing et al. 2007). Among all the carboxylates secreted in the rhizosphere, citrate, malate, and oxalate play a major role in forming complexes with metals (Mucha et al. 2005). Nonetheless, studies have suggested that carboxylates form less stable complexes than phytosiderophores with Cu, Fe, Cd, and Zn (Xu et al. 2005). Researches have revealed interactions between OAs secreted as root exudates and Cd-tolerant and Cd-nontolerant ecotypes of hyperaccumulators. For instance, roots of Cd-hyperaccumulating ecotype of *Sedum alfredii* released roughly a two-fold larger oxalate than non-hyperaccumulating ecotype (Tao et al. 2016). Under different Cd, Ni, and Cu concentrations, roots of *Phragmites australis* liberated oxalic, citric, and malic acids, whereas *Halimione portulacoides* secreted oxalic and malic acids (Rocha et al. 2016). On the contrary, few researchers have also reported metal sequestration capability of OAs in soil, thus reducing Cd uptake and defending roots from deleterious effects of heavy metals (Adeleke et al. 2017; Kaur et al. 2017).

## 14.3 Cadmium Toxicity in Plants

Cd causes numerous morphological, physiological, biochemical, and structural changes in plants, ultimately leading to phytotoxicity (Table 14.1). Nevertheless, the threshold of phytotoxic concentration of Cd differs depending on species, ecotypes, cultivars, etc. The potential mechanisms responsible for Cd-induced responses in plants have been discussed in the following sub-sections:

**Table 14.1** Effect of cadmium toxicity on different physiological, biochemical, and molecular traits in some agriculturally important plants

| Plant | Cd treatments | Effects | References |
|---|---|---|---|
| *Cajanus cajan* | 25 and 50 mg $kg^{-1}$ of soil | Decline in nodulation, nitrogen fixation, chlorophyll, yield; nutrient imbalance, oxidative stress; boost in antioxidant enzymatic and non-enzymatic defense system | Garg and Kaur (2012, 2013a, b), Kaur and Garg (2017) |
| *Glycine max* | 4, 8, and 16 mg $kg^{-1}$ sand | Reduction in root and shoot length, plant biomass, nodulation, $N_2$ fixation | Sheirdil et al. (2012) |
| *Hordeum vulgare* | 14, 28, and 42 mg Cd $kg^{-1}$ sand | Reduced root dry mass, leaf gas exchange, photosynthetic pigments content, and electron transport activity; enhanced membrane lipid peroxidation, $K^+$ leakage | Vassilev and Lidon (2011) |
| *Oryza sativa* | 0.05, 0.10, 0.15, and 0.20 mM | Decreased root growth; increased $H_2O_2$ contents; no effect on SOD, APX, and GOPX activities but inhibited activity of CAT in roots | Cho et al. (2012) |
| *Phaseolus vulgaris* | 1 mM | Reduction in growth traits, pod yield and pod protein; increased electrolyte leakage, lipid peroxidation; decreased MSI, RWC | Rady (2011) |
| *Pisum sativum* | 5 mM | Inhibition in activities of mitochondrial cytochrome c oxidase, NCCR, SCCR; stimulation in the activities of G-6PDH, 6-PGDH, and ADH | Smiri et al. (2010a) |
| | 5 mM | Delayed seed germination capacity; decrease in GR activity in peroxisomal fraction and mitochondria of cotyledons as well as embryonic axis while increase in GRX and GPX | Smiri et al. (2010b) |
| | 10, 20, 50, and 100 μM | Growth abnormalities such as root browning, leaf chlorosis, wilting, leaf abscission; decline in protein contents | Bavi et al. (2011) |
| *Solanum lycopersicum* | 100 μM | Decrease in plant biomass; abortion of all flowers at immature flower bud stage; decreased carotenoid and chlorophyll contents; α-tocopherol, asparagine, and tyrosine accumulation increased | Hediji et al. (2010) |
| | 30 and 300 μM | Decline in plant growth; increase in soluble sugars and amino acid contents | Zoghlami et al. (2011) |
| *Triticum aestivum* | 5, 20, 50, 100, and 200 mg $L^{-1}$ | Decreased seed germination percentage, mean germination time, seedling dry weight, root length, | Asgharipour et al. (2011) |

(continued)

**Table 14.1** (continued)

| Plant | Cd treatments | Effects | References |
|---|---|---|---|
| | | shoot height; increase in proline and carbohydrates | |
| | 10 mg $L^{-1}$ | Reduced photosynthesis and chlorophyll contents | Li et al. (2015) |
| *Triticum urartu* | 10 μM | Increased proline and GSH contents in shoots and roots; slight damage to leaf tips; severe damage to root tips and increased root secretions | Qiao et al. (2019) |
| *Vicia faba* | 50, 100, and 200 μM | Strong brownishing, thickening growth, and branching inhibition in roots; reduced mitotic activity whereas increased micronucleus formation in root tips; higher $H_2O_2$ and MDA; loss of plasma membrane integrity | Souguir et al. (2011) |
| *Vigna radiata* | 3, 6, 9, and 12 mg $kg^{-1}$ soil | Yellowing and tangible chlorosed area among the leaf veins; decrease in net photosynthesis; increase in substomatal $CO_2$ level, nutrient imbalance | Wahid et al. (2008) |
| | 100 mg $kg^{-1}$ soil | Reduction in shoot and root dry weights, depression in pod and seed characteristics; nutrient imbalance | Ghani (2010) |
| *Vigna unguiculata* | 25, 50, and 100 mg $kg^{-1}$ soil | Reduction in plant growth, leaf area; characteristic reddish brown spots on primary and secondary leaves and blackening, browning, burning, and weakening of stems and leaves; oxidative stress; increased GR; reduced AsA and GSH pools as well as AsA/DHA and GSH/GSSG ratio | Anjum et al. (2011) |
| *Zea mays* | 10, 30, and 50 mg $kg^{-1}$ | Decrease in germination percentage, growth parameters such as root and shoot length, fresh and dry weight of root and shoot, total leaf area, and photosynthetic pigments | Vijayaragavan et al. (2011) |
| | 5, 10, 30, and 50 mg $kg^{-1}$ | Decline in net photosynthetic rate, whereas transpiration rate, stomatal conductance, and intercellular $CO_2$ concentration enhanced in heading period and then decreased in maturity period | Zhang et al. (2012) |

(continued)

**Table 14.1** (continued)

| Plant | Cd treatments | Effects | References |
|---|---|---|---|
| | 100 μM | Inhibited growth, yield, photosynthetic attributes and chlorophyll contents | Anjum et al. (2017) |
| | 1, 3 and 5 ppm | Decrease in total phenolics and GSH content | Akinyemi et al. (2017) |

*ADH* alcohol dehydrogenase, *APX* ascorbate peroxidase, *AsA* ascorbic acid, *CAT* catalase, $CO_2$ carbon dioxide, *DHA* dehydroascorbic acid, *G-6PDH* glucose-6 phosphate dehydrogenase, *GB* glycine betaine, *GOPX* guaiacol peroxidase, *GPX* glutathione peroxidase, *GR* glutathione reductase, *GRX* glutaredoxin, *GSH/GSSG* reduced glutathione/oxidized glutathione, *GSH* reduced glutathione, $H_2O_2$ hydrogen peroxide, *MDA* malondialdehyde, *MSI* membrane stability index, *NCCR* *NADH* cytochrome C reductase, *6-PGDH* 6-phosphogluconate-dehydrogenases, *RWC* relative water content, *SCCR* succinate-cytochrome C reductase, *SOD* superoxide dismutase

### 14.3.1 Morphology, Growth and Yield Responses

Cd inhibits seed germination and development of seedlings (Table 14.1) in plants through altered water potential (Ci et al. 2010). Disturbed redox metabolism could be another reason for negative effect of Cd on seed germination (Smiri et al. 2010c). Seedling root growth is considered to be comparatively more sensitive to Cd stress as compared to seed germination (Kiran and Sahin 2006). Cd did not affect seed germination but interfered with seedling growth in rice, indicating differential response of the two events towards Cd toxicity (Rascio et al. 2008). Uptake and accumulation of Cd in plants cause disturbed plant growth homeostasis, general symptoms being chlorosis, necrosis, wilting, and stunted height (Zoghlami et al. 2011; Polle et al. 2013). Leaf chlorosis as a result of Cd toxicity is considered to be connected directly or indirectly with Fe or Zn (Jiang et al. 2004). Even number and surface area of leaves decrease in response to Cd stress (e.g. lettuce and *Amaranthus hypochondriacus*) (Xie et al. 2019; Kolahi et al. 2020). Browning of roots is observed in plants exposed to Cd (Lux et al. 2010; Chang et al. 2013). Cd stress enhances diameter of roots owing to increase in the size of parenchymatous and cortical cells, which have a role in boosting the plant resistance towards radial flow of water (Maksimovic et al. 2007). Cd toxicity affects the total and specific root length, surface area, number of roots as well as structure of root system in different plants (Wei et al. 2012; Huang et al. 2015). Reduction in root elongation is one of the most sensitive responses to Cd stress and it manifests earlier than most of the other physiological processes. The root apex and in particular, the distal portion of transition region within the apex is the prime target of Cd toxicity (Qiao et al. 2019). Initiation of adventitious roots, rather than persistence of primary root was witnessed by Molina et al. (2008) in *Vigna mungo* under Cd stress. Cd impairs lateral root formation, whereas the principal root turns brown, rigid, and twisted (Rascio and Navari-Izzo 2011). Roots are the earliest place of contact and toxicity to Cd and therefore root biomass is usually more negatively affected than aerial organs of plants (Gopal and Nautiyal 2011). Cd influences ultrastructure of meristematic cells modifying r-RNA precursor biosynthesis and hence affects root growth (Marcano et al. 2002). Cd-induced reduction in growth could be attributable to hindered cell division and elongation that largely takes place by an irreversible obstruction of proton pump responsible for the process (Liu et al. 2004). Moreover, Cd impedes plant growth due to decrease in mitotic activity, chromosomal aberrations, and toxicity of nucleoli in the apical meristems (He et al. 2010). Cd also decreases the biosynthesis of cell wall components, damages Golgi apparatus, inactivates proteins, and alters polysaccharide metabolism (Heidari and Sarani 2011; Muneer et al. 2011). Toxic effects of Cd on plant growth depend on plant genotype/cultivar as well as on intensity of stress (Zhang et al. 2009; Anjum et al. 2011). Plant growth stage with regard to Cd toxicity also determines the overall health of plants (Street et al. 2010).

Reduction in plant growth and development due to Cd toxicity results in decreased yield which corresponds to the combined effects of hindered physiological and metabolic processes (Garg and Kaur 2013a; Barman et al. 2020). Cd toxicity affects fruit set, size, and quality which are essential features for marketability, particularly for agriculturally important crops (Shekhar et al. 2011). Cd also negatively affects plant reproduction by restricting pollen germination and tube growth (Sabrine et al. 2010). Wahid and Ghani (2008) demonstrated that loss/or decreased chlorophyll synthesis under Cd stress affected growth, biomass, and yield during the ontogeny of *Vigna radiata* plants. Decrease in biomass and yield under Cd exposure could be due to lesser carbon assimilation because of reduced photosynthesis (Ghani 2011).

### 14.3.2 Nodulation and Nitrogen Fixation

Among the several symbiotic stages, nodulation is considered to be more sensitive to Cd as compared to root/shoot growth of legumes. It has been established that Cd triggers oxidative stress in nodules, thereby affecting nitrogen fixation as well as assimilation (Balestrasse et al. 2003). Moreover, Cd hampers nitrogen metabolism in many crop plants by decreasing the activities of nitrogen assimilating enzymes, for instance, nitrate reductase, nitrite reductase, glutamate dehydrogenase, and glutamine synthetase (Dhingra and Priefer 2006; Maarouf et al. 2009). Cd is known to exert considerable toxic effects on survivability and nitrogen-fixing efficiency of symbiotic rhizobia (Younis 2007). Likewise, Cd negatively influences composition and metabolic activity of rhizobial strains (Pereira et al. 2006). Cd reduces nitrogenase activity and number as well as biomass of nodules, disrupts nodular ultrastructure and causes nodule senescence (Wani et al. 2006; Noriega et al. 2007). Harmful effects of Cd on structure and function of nodules (Table 14.1), for instance, occurrence of glycoprotein in intercellular spaces of nodule cortex, changes in symbiosomes, deposition of Cd in cell walls and oxidative stress have been reported (Carpena et al. 2003). Even $N_2$-fixing area and cells decrease under Cd stress (Chen et al. 2003). Manier et al. (2009) suggested nodulation index as a bioindicator of augmented Cd in soil. Reduction in nodulation under Cd stress could also be ascribed to diminution in root biomass (Garg and Aggarwal 2011). Furthermore, brown colored nodules develop under Cd excess owing to loss of leghemoglobin pigment which finally reduces the $N_2$-fixing potential of nodules, i.e. acetylene reduction assay (ARA) (Saraswat and Rai 2011).

### 14.3.3 Photosynthesis and Carbon Assimilation

Diminution in growth and biomass under high levels of Cd has been mostly ascribed to disturbed photosynthesis. Cd perturbs photosynthesis by damaging chlorophyll biosynthesis, assembly of pigment protein complexes and thylakoids, chloroplast structure and replication, functionality of photosystems, enzymes of dark reactions, utilization of sugar and transport and oxidative stress (Hattab et al. 2009; Malec et al. 2010). Additionally, stomatal conductance and index, transpiration, and net carbon dioxide uptake are significantly decreased with increased Cd levels in the growth media (Balakhnina et al. 2005; Wahid et al. 2008). Such alterations could be attributed to abnormal stomatal behavior like high stomatal resistance, low stoma density, and stomatal closure arising from Cd-induced water deficit (Poschenreider and Barcelo 2004; Ying et al. 2010). Because of similar chemical nature with $Ca^{2+}$ ions, Cd permeates guard cells via voltage-dependent $Ca^{2+}$ channels and interrupts regulation of stomata by interfering with Ca signaling (Perfus-Barbeoch et al. 2002). Yellowing of leaves is one of the primary evident symptoms of Cd toxicity which is directly related with loss and/or hindered chlorophyll synthesis (Liu et al. 2011a) because of substitution of Mg with Cd in chlorophyll molecules (Parmar et al. 2013). Moreover, reduction in carotenoid and chlorophyll contents under Cd exposure (Table 14.1) is correlated with decrease in the expression of genes regulating isoprenoid biosynthetic pathway (Hediji et al. 2010). In general, carotenoids are less affected by Cd than chlorophylls; therefore, the carotenoid to chlorophyll ratio is larger in Cd-stressed plants (Dobrikova et al. 2017). Enhancement in carotenoid: chlorophyll ratio has also been linked to Cd-induced oxidative stress that hampers photosynthetic light use efficiency (Martins et al. 2011). Damaged structure of chloroplast in terms of distorted shape and enlarged thylakoid membranes under Cd stress is often observed in plants (Gratao et al. 2009). Furthermore, high Cd concentrations reduce the size and number of chloroplasts as well as the extent of stacking of thylakoid membranes, resulting in thylakoid swelling (Moussa and El-Gamal 2010). Cd treatments reduce the quantity of lipids in chloroplast membrane (necessary for regulating the structure of PSII complexes) together with alterations in their composition and fatty acid content, thereby leading to lesser fatty acid unsaturation (Nouairi et al. 2006). Basa et al. (2014) have also reported distinct changes in the architecture of photosynthetic apparatus, comprising reduction in the amount of PSII supercomplexes and enhancement in the monomeric form of the LHCII antennae under Cd stress. $Cd^{2+}$ is a competitive inhibitor of $Ca^{2+}$ in the $Mn_4Ca$ cluster of oxygen evolving complex (OEC), leading to inhibition of the oxygen evolution (Faller et al. 2005). Sigfridsson et al. (2004) have also suggested that Cd quickly impairs the activity of PSII on the donor side connecting OEC and the initial electron donor of P680 (Tyr $Y_Z$) by two mechanisms: (1) it obstructs the oxidation of $Y_Z$ in a $Ca^{2+}$-independent way and (2) impedes the OEC in a $Ca^{2+}$-dependent manner. Cd brings about alterations in the $Mn_4Ca$ group of the OEC, leading to higher number of PSII centers in the primary $S_0$ state, as well as in the hindered $S_B$ state (Dobrikova et al. 2017). Moreover, Cd toxicity leads to some modifications in the secondary quinone acceptor, $Q_B$, on the PSII acceptor side, thereby interrupting electron flow from $Q_A$ to $Q_B$ (Faller et al. 2005). Cd reduces Mn uptake which is involved in photolysis of $H_2O$ by PSII at donor site of photosynthesis, thereby inhibiting the photosynthetic electron flow (Pagliano et al. 2006). Herbette et al. (2006) observed Cd-induced downregulation of genes encoding enzymes of chlorophyll synthesis, proteins of photosystem I and II, $H^+$-ATPases, electron transporters, and enzymes of Calvin cycle in *Arabidopsis*. Cd affects numerous sites of PSII, particularly the number of active reaction centers (Gonzalez-Mendoza et al. 2007). In contrast, Delperee and Lutts (2008) demonstrated that photochemical efficiency of PSII did not decline despite reduction in number of reaction centers and inhibition of electron transfer to PSII acceptors in tomato under Cd stress. In some plants exposed to Cd stress, PSI is more sensitive than PSII (Fagioni et al. 2009), while in others vice versa is true (Dobrikova et al. 2017). Cd toxicity reduces effective quantum yield of photochemical energy conversion of PSII, photosynthetic active radiation, coefficient of photochemical quenching, and chlorophyll fluorescence decrease ratio (Wan et al. 2011) (Fig. 14.1). Some reports suggest that Cd hampers photosynthetic light reactions more than Calvin–Benson cycle (Kupper et al. 2007), whereas others report Calvin cycle to be the main target of Cd toxicity rather than PSII (Wang et al. 2009). It has been suggested that Cd firstly damages the photosynthetic carbon reduction cycle and afterwards affects the photosynthetic electron transport and photosynthetic rate (Mobin and Khan 2007; Ahammed et al. 2013). Decrease in $CO_2$ reduction reactions (Fig. 14.1) is correlated with reduced activities of phosphoenolpyruvate carboxylase (PEPCase) and ribulose 1,5-bis-phosphate carboxylase (RuBPcase) (Krantev et al. 2008). Cd-induced oxidative alteration of thiol groups of rubisco has been reported to lower the photosynthetic rate in plants such as tomato (Liu et al. 2008a). Thus, it can be concluded that the Cd-induced inhibition of photosynthesis can be ascribed to: (1) stomatal and nonstomatal limitations, (2) decrease in photosynthetic pigments, and (3) damage to the photosynthetic apparatus.

### 14.3.4 Respiration

Mitochondria can be regarded as a target site of Cd owing to its accumulation in these organelles of plant cells (Smiri et al. 2010c). Plant mitochondrial electron transport chain (ETC) is susceptible to Cd stress as Cd obstructs ETC by hampering electron flow via cytochrome complex, mainly insensitive towards cyanide (Delperee and Lutts 2008). Mitochondria are sensitive to redox disturbances stimulated by Cd, as their redox status recuperates slower from an oxidative explosion than cytoplasm or chloroplasts (Bi et al. 2009; Schwarzländer et al. 2009). The recognized sites for ROS synthesis are complexes I and III of the mitochondrial ETC where $O_2^{\bullet-}$ generated is subsequently changed to $H_2O_2$ (Vanlerberghe 2013). Additionally, Cd damages functioning of mitochondria to a certain extent by altering the redox balance (Smiri et al. 2010c). Cd-induced respiratory damage causes reduced ATP production. Once mitochondrial respiration capacity declines, pentose phosphate pathway and alternative alcoholic fermentation get activated to

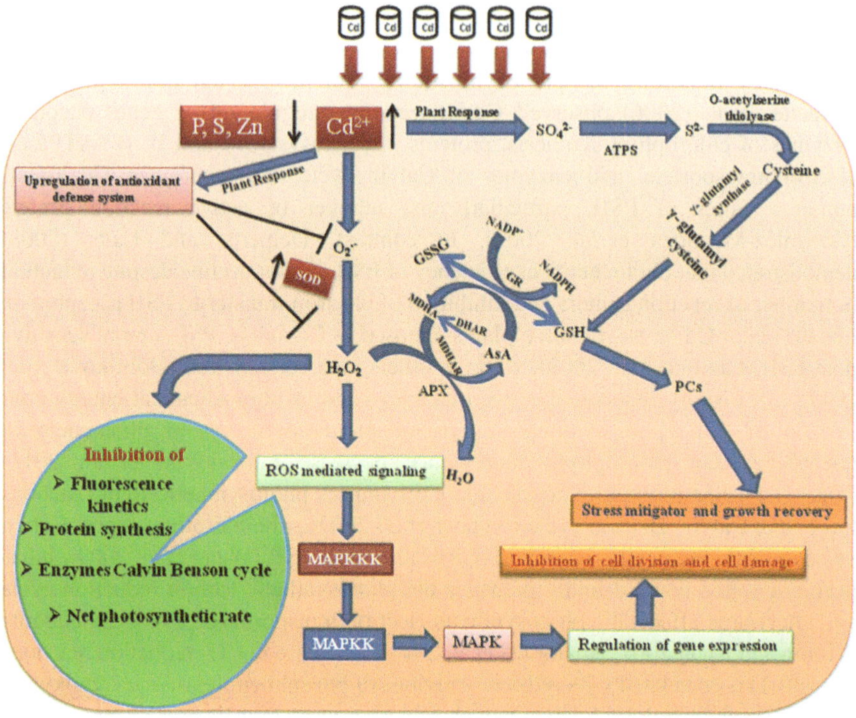


**Fig. 14.1** Diagrammatic representation of cadmium mediated detrimental effects and the response of plants. *APX* ascorbate peroxidase, *AsA* ascorbic acid, *ATPS* adenosine triphosphate sulfurylase, *DHAR* dehydroascorbate reductase, *GR* glutathione reductase, *GSH* reduced glutathione, *GSSG* oxidized glutathione, *MAPKKK* mitogen activated protein kinase kinase kinase, *MDHA* monodehydroascorbic acid, *MDHAR* monodehydroascorbate reductase, *P* phosphorus, *PCs* phytochelatins; *ROS* reactive oxygen species, *SOD* superoxide dismutase, *S* sulfur, *Zn* zinc

counterbalance the loss in energy supply (Shetty et al. 2002). Furthermore, Cd induces diminution in root respiration that results in decreased metabolic processes, eventually leading to hampered growth and yield (Ci et al. 2010). However, increase in plant respiration rate under Cd stress has also been reported by a few researchers (Zhou et al. 2020). Respiration is regarded as one of the principal events during initial phases of seed germination. Cd not only impedes seed germination but also enzymatic activities involved in respiratory metabolism of germinating seeds (Smiri et al. 2010c). In addition to this, mitochondrial proteome consists of subunits of respiratory complexes and supercomplexes which are hampered by Cd in germinating seeds (Smiri et al. 2009), perhaps through degradation of proteins or interference in their synthesis. Cd impedes the activities of several respiratory enzymes, such as hexokinase, glucose-6-phosphate dehydrogenase (G6PDH), 6-phosphogluconate dehydrogenase (6PGDH), and alcohol dehydrogenase (Chugh and Sawhney 1999), though contrary results have also been reported (Smiri et al. 2010a) (Table 14.1). Cd damages ultrastructure of mitochondria through increase in its volume, strong reduction of ATPase activity, rupturing of mitochondrial outer membrane with extrusion of cytochrome c (Yang et al. 2011). Irregular clusters of mitochondria around the chloroplasts or aggregation in the cytoplasm with their simultaneous blocked movement play an essential role in Cd-induced cell death (Bi et al. 2009).

### 14.3.5 Plant–Water Relationships and Nutrient Uptake

Cd enters into cytosol via Ca channels present in the plasma membrane and alters the plant–water relations (Perfus-Barbeoch et al. 2002; Lutts et al. 2004) owing to the reduction in absorptive surface of root hairs. Cd disturbs the uptake of water as well as nutrient metabolism (Mn, Fe, Cu, Ca, Mg, K, Zn, and P), causing severe nutrient deficiencies in various plant parts (Wahid et al. 2008; DalCorso et al. 2008). Cd-induced alteration in nutrient balances of plants (Table 14.1) results in physiological disorders as well as reduced growth and yield, ultimately affecting their productivity (Dong et al. 2006). Heavy metals hamper water transport activity of plant vacuolar membrane (Maurel and Chrispeels 2001) along with plasma membrane aquaporins (Martinez-Ballesta et al. 2003). In this context, Cd alters permeability of plasma membrane which results in decrease in the water content of plants (Fernandez et al. 2013). Elevated Cd levels impair the mechanisms of water and nutrient uptake due to disintegration of root-tip cells, resulting in decreased relative water content (RWC), cytokinin transport, and increase in the abscisic acid content which causes closing of stomata (Zoghlami et al. 2011). Decreased leaf water potential and RWC were observed in Cd-stressed date palm plants by Zouari et al. (2016).

Plants take up nutrients by an active transport mechanism in the form of divalent cations that compete with $Cd^{2+}$ due to similar routes for uptake and transport (Fig. 14.1). Cd-induced leaf chlorosis occurs as a result of antagonistic effect of Cd on the uptake of Cu, Fe, Mn, P, and Zn, causing their deficiencies particularly in Cd-sensitive varieties (Chen et al. 2007). Furthermore, $Fe^{2+}$ deficiency causes reduction in ferredoxins and chlorophyll content, thereby affecting photosynthesis (Kosakowska et al. 2004). Loading of Fe into xylem elements and its subsequent translocation to shoots is directly correlated with occurrence of citrate in the xylem sap and the citrate transporter FRD3 is believed to be downregulated by Cd (Yamaguchi et al. 2009). Fe deficiency stimulated expression of certain genes which enhanced Cd accumulation in peanut roots (Chen et al. 2019). Moreover, phosphate or sulfate deficiency reduces Cd uptake by competitively augmenting Fe uptake and accumulation in plants, authenticating competitive interactions between Fe and Cd (Astolfi et al. 2012; Yang et al. 2016). Correspondingly, depleted levels of $K^+$ under Cd stress lead to an imbalance in plant water transport. Effect of Cd on nutrient contents in plants is related to applied Cd concentration, elements, growth medium, plant organ, exposure time, and genotype (Cheng et al. 2009; Goncalves et al. 2009). Cd-induced decrease in nutrient composition has been reported in various crop plants (Lopez-Millan et al. 2009; Ahmad et al. 2011). On the contrary, enhancement in the contents of mineral nutrients (Zn, Fe, Mg, K, Ca, and P) due to Cd application has also been witnessed which might be due to the synergistic effect between Cd and the mineral nutrients (Li et al. 2016). Thus, reduced growth could be the consequence of direct effects (Cd accumulation in tissues) and/or indirect effects (restricted mineral and water uptake).

### 14.3.6 Cd-Induced Reactive Oxygen Species (ROS): Impact on Membranes, Lipids, Proteins, and Nucleic Acids

Cd does not directly produce ROS [superoxide radicals ($O_2^{\bullet -}$), singlet oxygen ($_1O^2$), hydrogen peroxide ($H_2O_2$), and hydroxyl radical ($OH^{\bullet}$)] through biological redox reactions, i.e. Cd ions do not change their oxidation state as they are not involved in Fenton and Haber–Weiss reactions (Clemens 2006) because of its redox potential (-820 mV) (Souguir et al. 2011). It can indirectly trigger generation of different ROS (Fig. 14.1) due to (1) reduced activities of antioxidative enzymes; (2) stimulation of different peroxidases and oxidases; (3) disruption of electron transport chain; or (4) lipid peroxidation linked to Cd-mediated activation of lipoxgenases (Azpilicueta et al. 2007; Rodríguez-Serrano et al. 2009). Cd-induced ROS react with biomolecules and cause oxidative injuries, for instance, peroxidation of membrane lipids, oxidation of proteins, inactivation of enzymes as well as damage to DNA and RNA, eventually leading to cell death (Singh et al. 2016). Generation of $H_2O_2$ in Cd-treated cells could be a synchronized event with calmodulin and Ca-dependent protein kinases (CDPK) involved in its signal transduction cycle (Olmos et al. 2003). Romero-Puertas et al. (2004) detected localization of Cd-induced $H_2O_2$ on the inner side of plasma membrane as well as tonoplast, while $O_2^{\bullet -}$ largely in tonoplasts of bundle sheath cells adjacent to the vascular tissue of pea leaves. Cd-induced accumulation of $H_2O_2$ has also been reported to initiate from interior of root cells, chiefly from mitochondrial electron transport (Heyno et al. 2008). Roots modulate Cd toxicity as they sense Cd in the rhizosphere and then pass suitable signals to the shoots. Tamas et al. (2009) studied changes in ROS along the longitudinal axis of barley root tips and observed initiation of Cd-induced $O_2^{\bullet -}$ and $H_2O_2$ generation in the elongation zone which further enhanced to the differentiation zone. Further, there was reduction in $OH^{\bullet}$ production along the whole root tip. These researchers suggested that not only the enhanced levels of ROS but also their distribution was implicated in the Cd-induced stress responses of barley root tip. Both temporal and spatial generation of ROS function in controlling root growth, differentiation, and plant development (Gapper and Dolan 2006). $O_2^{\bullet -}$ and $H_2O_2$ had particular accretion zones in the root apoplast of *Arabidopsis* (Dunand et al. 2007). In rice seedlings, Cd-induced $H_2O_2$ production was found mainly in epidermis, exodermis, sclerenchyma, and vascular cylinder (Cho et al. 2012). Bulk of Cd-induced ROS generation might be through plasma membrane NADPH oxidase; however, apoplastic peroxidases might also play a role in production of ROS (Pérez-Chaca et al. 2014). NADPH oxidases utilize cytosolic NADPH and produce $O_2^{\bullet -}$ which is rapidly dismutated to $H_2O_2$ by superoxide dismutase (SOD). It has been reported that Cd instigates a chain of three waves of ROS production: NADPH oxidase-dependent accretion of $H_2O_2$ followed by $O_2^{\bullet -}$ accumulation in mitochondria and lastly fatty acid hydroperoxides (Garnier et al. 2006). Accrual of δ-aminolevulinic acid (ALA) because of hampered activity of ALA dehydratase by Cd also enhances ROS levels in plants (Noriega et al. 2007).

Accumulation of malondialdehyde (MDA)/thiobarbituric acid reactive substances (TBARS) is a sign of severe lipid peroxidation in cells (Younis et al. 2016). In plants, free fatty acid peroxidation can take place via enzymatic as well as non-enzymatic means with lipid hydroperoxides (LOO) generated as by-products. Enzymatic lipid peroxidation is catalyzed by lipoxygenases (LOXs; EC 1.13.11.12), whereas in non-enzymatic, ROS/metal ions are involved (Skorzynska-Polit 2007). Few reports exist regarding the effect of Cd on LOX activity in plants (Roychoudhury et al. 2012). Lipid peroxidation-induced aldehydes are mutagenic molecules which produce MDA-guanine adducts (Roldan-Arjona and Ariza 2009). Ann et al. (2011) performed gene expression studies in *A. thaliana* seedlings and observed that in roots, cytosolic $LOX_1$ was strongly stimulated under Cu/Cd while $LOX_2$ gene expression decreased. Nevertheless, both LOX mRNAs increased in leaves under metal stress, with $LOX_2$ being more augmented. Additionally, Cd caused considerable gene expression of ROS producing enzymes (*RBOHD*, *RBOHC*, and *RBOHF*), suggesting involvement of NADPH oxidases in ROS production. Cd also modifies the functionality of membranes by changing their lipid composition which in turn can affect some enzymatic activities connected with membranes like $H^+$-ATPase (Belkhadi et al. 2010). Some features of lipid biochemistry alter following Cd stress, for example, qualitative and quantitative changes in lipids, hampered lipid biosynthetic pathways and decrease in unsaturated fatty acids (Djebali et al. 2005). Increase in MDA/TBARS, $H_2O_2$ content, and electrolyte leakage (EL) has been reported in several Cd-stressed plants (Srivastava et al. 2011; Farooq et al. 2016).

Degeneration of proteins has been suggested as an indicator of Cd-induced oxidative stress. Oxidative alterations in proteins are due to the formation of carbonyl derivatives on side chains of histidine, lysine, and proline residues. Protein carbonylation is an irreparable oxidation process which inhibits multiple enzymes. Enhanced carbonyl content has been reported in several plant species exposed to Cd (Rellan-Alvarez et al. 2006). Some target proteins of Cd-induced oxidative stress have also been categorized such as CAT, GR, Rubisco, and Mn-SOD (Romero-Puertas et al. 2004). Cd toxicity decreases protein content due to $H_2O_2$-induced oxidation of proteins and enhanced proteolytic activity resulting in lowered biomass (Gopal and Nautiyal 2011).

Several studies have suggested that Cd impairs the structure of nucleolus, DNA, and RNA in plants (Souguir et al. 2011). Cd damages structure and function of DNA via modifying bases, single and double strand break, abasic sites, DNA–protein cross links, bulky adducts, point mutations, ploidy changes, etc. (Liu et al. 2005, 2012). Cd causes a range of abnormalities in plant chromosomes, such as sticky chromosomes, double micronucleus (MCN), anaphase bridges, chromosome breaks, and laggards (Souguir et al. 2011; Nefic et al. 2013). Enhanced levels of ROS often result in DNA mutations and chromosomal aberrations. DNA is oxidized largely by $OH^{\bullet}$ and $O_2^{\bullet-}$ radicals which react with guanine (G), while $H_2O_2$ has less effect. Number of $OH^{\bullet}$ radicals can enhance up to 10,000 under Cd stress with 100,000 alterations in DNA bases in a single cell every day (Kranner and Colville 2011). The molecular means by which Cd induces genotoxicity in plants are not clear. It is believed that Cd interacts directly with DNA via binding with A, T, and G bases (Hossain and Huq 2002). Cd functions as a mutagen mainly through direct inhibition of DNA mismatch repair (MMR), leading to genetic instability (Jin et al. 2003). Study carried out by Liu et al. (2008b) revealed that Cd stress modified the expression of genes related to DNA MMR system in *Arabidopsis*, indicating that Cd had damaging effect on mismatch identification process of *Mut Sα* and *Mut Sβ* complex as well as nucleotide excision process of MMR system. In a subsequent study, Liu et al. (2009) reported modification in the expression of DNA MMR genes in *Arabidopsis* seedlings using RT-PCR method and suggested enhanced expression of MMR genes as an alteration in signal transduction pathways via increase in protein phosphorylation as well as stimulation of transcription factors because of enhanced ROS production. Qiao et al. (2019) reported impairment of two DNA repair mechanisms in Cd-stressed wheat seedlings; base excision repair (BER) and MMR as well as reduced expression of genes encoding various constituents of the DNA replication pathway (helicases, minichromosome maintenance protein, proliferating cell nuclear antigen, and DNA ligases). Moreover, in response to DNA damage, diverse ribonucleotide reductase (RNR) genes (necessary for DNA replication and repair) are distinctively activated. Mediouni et al. (2008) observed biphasic induction of two RNR genes in tomato plants (*AtRNR* 2, 3 bis, and *AtRNR* 2, 5 genes) under Cd exposure, suggesting that expression of RNR was stimulated by Cd-induced ROS. Furthermore, low levels of Cd impair the activities of phosphatases and kinases which are engaged in DNA repair (Whiteside et al. 2010).

### 14.3.7 Metabolic Antioxidative Defense Mechanisms: Limiting Cd-Induced ROS-Mediated Damage

Plants have developed a complex system of antioxidative defense mechanisms which detoxify Cd-induced ROS in a process called redox homeostasis. ROS-scavenging antioxidant defense machinery comprises enzymatic as well as non-enzymatic antioxidants (Table 14.1) that differ at several cellular and subcellular levels in many plants (Gill and Tuteja 2011). Though an active antioxidative metabolism does not symbolize a Cd tolerance mechanism in an exact sense, it is imperative for functioning of plants under metal stress conditions.

#### 14.3.7.1 Enzymatic Antioxidative Defense System

Plants have a diverse range of enzymes such as superoxide dismutase (SOD), catalase (CAT), ascorbate peroxidase (APX), glutathione reductase (GR), monodehydroascorbate reductase (MDHAR), dehydroascorbate reductase (DHAR), glutathione peroxidase (GPX), guaiacol peroxidase (GOPX), and glutathione sulfotransferases (GST) to lessen the damaging effects arising from Cd-induced oxidative stress (Fig. 14.1), thus shielding cells against metabolic injury and dysfunction (Gill and Tuteja 2011). The collaborative activities of all the antioxidant enzymes help in tolerance, adaptation, and survival of plants under metal stress conditions. Nevertheless, response of antioxidant enzymes to Cd is divisive, as both increase and decrease in the activities of several antioxidant enzymes have been observed in Cd-treated plants (Rodriguez-Serrano et al. 2006; Tian et al. 2011), indicating that antioxidant systems, besides being implicated in detoxification process, could be susceptible to Cd toxicity. Low Cd concentrations and brief periods of treatment usually result in increase in antioxidative enzymes (Liu et al. 2011b), whereas prolonged treatment periods cause decrease in enzymatic activities (Rodriguez-Serrano et al. 2006). At low Cd levels antioxidative enzymes scavenge ROS so as to regulate normal functioning of cells, however, when Cd stress surpasses the protective capacity of antioxidant enzymes, their activities decrease rapidly, exhibiting an inhibitory effect (Chen et al. 2007). In addition to this, the response of antioxidant enzymes to Cd varies with Cd concentration, plant age, growing conditions as well as among species and plant organ/tissues (root, stem, leaf, fruits) (Gratao et al. 2008; Sharma and Dietz 2006). Variations in stress tolerance among diverse plant species and even among genotypes within a species are usually correlated with the differential activation of enzymatic antioxidant machinery under stressful conditions (Chen et al. 2007). Even differential responses of antioxidant enzymes could be ascribed to diverse levels of ROS production in functionally distinct plant organs (Yannarelli et al. 2007). The activities of SOD and nitric oxide (NO) metabolism get reduced together with enhanced generation of ROS due to Cd-induced Ca deficiency (Rodríguez-Serrano et al. 2009). Root proteome analysis of Cd-exposed *Brassica juncea* revealed upregulation of FeSOD, whereas Cu/ZnSOD were downregulated (Alvarez et al. 2009). $H_2O_2$ and $O_2$ are required in the signaling events that manage differences in transcript levels of CuZn-SOD, CAT, and GR in plants under Cd stress (Romero-Puertas et al. 2004). Modifications in transcript levels of many genes involved in antioxidative defense (*CSD2*, *FSD1*, *MSD1*, *CAT1*, *APX1*, *DHAR*, *GR1*, *GPX2*) under Cd stress have been reported (Smeets et al. 2008; Cuypers et al. 2011). Using bioinformatics, Kolahi et al. (2020) identified 14 and 19 genes coding for APX and SOD in Cd-stressed lettuce. Furthermore, overexpression of genes, for instance, AtTSB1, OXS3 (encoding oxidative stress 3), and ThVHAc1 also enhances plant tolerance to Cd stress (Hsiao et al. 2008; Wang et al. 2016; Yang et al. 2016).

#### 14.3.7.2 Non-enzymatic Antioxidative Defense System

Apart from antioxidant enzymes, plants possess some non-enzymatic antioxidant compounds, for instance, α-tocopherol, phenolics, flavonoids, and ascorbic acid, which have capacity to quench ROS without themselves changing to destructive radicals (Iqbal et al. 2010). Glutathione (GSH), metallothioneins (MTs), and phytochelatins (PCs) are involved in detoxification of heavy metals and are members of a group of biogenic ligands/thiols produced by plants. Legume nodules contain additional compounds with antioxidant properties like polyamines, uric acid, flavonoids, and other phenolics (Becana et al. 2000).

### α-tocopherols, Phenolics, and Flavonoids

α-tocopherols (vitamin E) are regarded as recyclable chain reaction terminators of polyunsaturated fatty acid (PUFA) radicals produced as a result of oxidation of lipids. They shield lipids and other components of membrane by physically quenching and chemically combining with $O_2$ in chloroplasts, consequently protecting structure as well as function of PSII. $O_2^{\bullet-}$ quenching by tocopherols is extremely efficient, and it is believed that a single α-tocopherol molecule can defuse up to 120 $O_2^{\bullet-}$ molecules in vitro before being disintegrated (Wu and Tang 2004). Enzymes associated with vitamin E biosynthesis were upregulated in response to Cu and Cd in *Arabidopsis* plants, and vitamin E-deficient mutants (*vte1*) exhibited increased oxidative stress and sensitivity to both the metals (Collin et al. 2008). Under metal stress, phenolics can be formed from cinnamic acid which in turn is synthesized from phenylalanine with the help of phenylalanine ammonia-lyase (PAL) (Kovacik et al. 2009). The antioxidant function of phenolic compounds is because of their high affinity to chelate metals (Agati et al. 2012), which perhaps may be linked to the high nucleophilic nature of the aromatic rings, rather than to precise chelating groups within the molecule. Additionally, flavonoids are effective in scavenging free radicals and contribute to plant metal tolerance (including Cd) via chelating complexes in the cell wall (Kostyuk et al. 2007).

### Ascorbic Acid

Ascorbic acid (AsA, vitamin C) is the main, perhaps the only, antioxidant present in the apoplast in concentration of 1–2 mM in nodules, 5–25 mM in leaves, and 25–50 mM in chloroplasts (Becana et al. 2011). It is a water-soluble and the most important antioxidant that directly quenches $O_2$, $O_2^{\bullet-}$, and $OH^{\bullet}$ and restore tocopherol from tocopheroxyl radical, thereby providing protection to membranes (Ahmad et al. 2009). In nodules, AsA can reduce ferric LHb and $LHb^{IV}$. AsA is the substrate of cAPX and the corresponding organeller isoforms, which are essential constituents of AsA-GSH cycle (Fig. 14.1). Besides this, it functions as a co-factor of violaxanthin de-epoxidase, hence maintains dissipation of overloaded excitation energy (Shao et al. 2007). Increase in AsA levels has been observed under Cd stress in plants (Marquez-Garcia et al. 2012). Liu et al. (2007) observed that AsA contents enhanced in *Bechmeria nivea* plants under low Cd stress, while they got reduced under high Cd. Likewise, short-term (24 h) exposure of *Arabidopsis* seedlings to Cd toxicity resulted in increased AsA levels in roots (Smeets et al. 2009) while 1 week exposure to Cd decreased AsA contents in *Ceratophyllum demersum* (Aravind and Prasad 2005). The biosynthetic ability of AsA is weakened under severe stress conditions as AsA pool is usually ascertained not only by its rate of regeneration but also synthesis. It has been reported that restoration of AsA pool under extreme Cd stress is inadequate or that AsA synthesis is lesser than AsA breakdown (Chao et al. 2010). Therefore, the loss of AsA content under excessive Cd stress is most probably due to greater degradation than synthesis and the oxidation of AsA to DHA.

**Glutathione**

The tripeptide glutathione (GSH, γ-glutamyl-cysteinyl-glycine) is an essential, copious, and ubiquitous, sulfhydryl containing non-protein thiol (NP-SH) which functions in controlling $H_2O_2$ levels and restoration of AsA via AsA-GSH cycle (Foyer and Noctor 2005a). It directly plays a role as radical scavenger by combining with metal induced ROS, for example, $^1O_2$, $O_2^{\bullet-}$, and $OH^{\bullet}$ (Gill and Tuteja 2010). It also confers protection to the membranes by keeping α-tocopherol and zeaxanthin in reduced states (Gill and Tuteja 2010). It functions as a disulfide reductant to protect thiol (-SH) group of enzymes and as a substrate for GPX and GSH-transferases during Cd exposure. Furthermore, it protects cells from oxidative damage and alters metal toxicity by modifying metal uptake rates. GSH exists in two forms, reduced (GSH) or oxidized (GSSG) (Fig. 14.1). The reduction capacity of glutathione is dependent on intracellular GSH/GSSG ratio. Foyer and Noctor (2005b) are of the view that GSH/GSSG ratio is an index of cellular redox balance and it may be involved in ROS assessment. GSH1 is solely located in plastids, whereas bulk of GSH2 is restricted to the cytosol (Wachter and Rausch 2005). In addition to the common occurrence of GSH, homoglutathione (hGSH, γ-glutamylcysteinyl-β alanine) is found in Fabaceae family (Rouhier et al. 2008) and hydroxymethyl glutathione (hmGSH, γ-glutamylcysteinyl-serine) is present in Poaceae family (Okumura et al. 2003). Enhanced GSH content is generally correlated with the ability of plants to tolerate metal induced oxidative stress (Garg and Kaur 2012). mRNA levels of genes involved in GSH synthesis (*gsh1* and *gsh2*) and PCS (*pcs1*) increased in *A. thaliana* under Cd stress (Semane et al. 2007). Contrarily, low level of GSH linked with Cd sensitivity may be due to reduced synthesis of PCs. Decreased GSH pool has also been attributed to higher utilization of GSH during AsA regeneration from DHA or due to the direct interaction of GSH with Cd (Pietrini et al. 2003). Moreover, reduction in GSH pool at low Cd concentrations while increase in GSH levels at high Cd has also been reported (Anjum et al. 2014; Buono et al. 2014). Anjum et al. (2008) reported plant phenological stage-dependent modulation of GSH pool in *B. campestris*. Wojcik and Tukiendorf (2011) demonstrated that decline in GSH levels led to decreased Cd tolerance in *A. thaliana* plants, nevertheless, enhanced GSH levels did not increase Cd tolerance rather it was toxic for plants.

**Metallothioneins**

MTs (products of mRNA translation) are gene-encoded, low-molecular-weight (8-14 kDa) cysteine (cys)-rich polypeptides, activated in response to presence of metals (Cobbett and Goldsbrough 2002). Cysteine residues in MTs combine with metal ions via mercaptide bonds (DalCorso et al. 2008). MTs are categorized into two groups: class I MTs-extensively present in vertebrates and class II prevalent in plants, fungi as well as non-vertebrate animals (Cobbett and Goldsbrough 2002). Depending on the arrangement of cysteine moieties, plant MTs are categorized into four subfamilies: MT1, MT2, MT3, and MT4, with differential expressions in plant tissues and probably performing diverse functions (Cobbett and Goldsbrough 2002). Overexpression of *B. juncea MT2* in *A. thaliana* conferred enhanced tolerance to Cd and Cu (Zhigang et al. 2006). Likewise, induction of MT gene expression was observed in barley roots under Cd stress (Finkemeier et al. 2003). Overexpression of *AtMT2b* increased the plant tolerance to Cd and decreased Cd translocation from root to shoot in tobacco under Cd stress (Grispen et al. 2011). When MT1 was knocked down in *Arabidopsis*, all the knock-down lines exhibited hypersensitivity to Cd and accumulated many-fold lesser levels of Cd (Zimeri et al. 2005). It is believed that Cd-induced $H_2O_2$ generation plays a significant role in stimulating MT synthesis in plants (Zhang et al. 2017).

**Phytochelatins**

GSH is not only involved in direct Cd complexation and its sequestration in vacuole but also participates indirectly in Cd detoxification by functioning as a substrate for phytochelatin (PC) synthesis (DalCorso et al. 2008). PCs are non-translationally synthesized from GSH in a transpeptidation reaction catalyzed by γ-glutamyl cysteine dipeptidyl transpeptidase (phytochelatin synthase: PCS, EC 2.3.2.15) enzyme, which gets activated in the presence of heavy metals including Cd (Gill and Tuteja 2011) (Fig. 14.1). There are a few compounds alike but not identical to PCs, collectively termed as iso-PCs, especially, hydroxymethyl-PCs (hm PCs) found in some Poaceae members, for example, oat, wheat, rice, and rye (Klapheck et al. 1994). hm PCs contain Ser in place of Gly [i.e. $(\gamma\text{-Glu-Cys})_n$-Ser], therefore they are also named as iso PC-Ser and are generally present together with common PCs (Klapheck et al. 1994). Other iso-PCs are also reported with different terminal amino acids (for instance, alanine, glutamine, and glutamic acid) and possess a C-terminal altered residue excluding glycine (Shah 2011). In legumes, (γ-Glu-Cys)-Ala peptides, termed as homo-phytochelatins (hPCs) are present which are synthesized from hGSH (Becana et al. 2011). PCs are found in several plants, including dicots, monocots, gymnosperms, and algae, suggesting that PCs pathway evolved quite early for imparting tolerance to plants against adverse environmental conditions (Shukla et al. 2016). *Lotus japonicus* plants exposed to Cd accumulated PCs and hPCs in roots as well as nodules (Ramos et al. 2008). Cd-exposed root extracts of bread wheat contained hmGSH, GSH, $PC_2$, $hmPC_2$, $desGlyPC_2$, $PC_3$, $hmPC_3$, $desGlyPC_3$, and $PC_4$ while only hmGSH, GSH, $PC_2$, $PC_3$, and $PC_4$ were detected in leaf extracts, suggesting high contents of Cd in roots (Ranieri et al. 2005). Similarly, Vazquez et al. (2006) witnessed high PC levels in roots of Cd-stressed lupin plants, with $PC_3$ being the most copious followed by $PC_4$. Lindberg et al. (2007) demonstrated increased PCs in root and leaf protoplasts of 5–7 day old wheat seedlings grown under Cd stress, with PCs in both shoot and root protoplasts being $(\text{Glu-Cys})_n$ Gly [where $n = 2$ or 3], whereas in leaf protoplasts $(\text{Glu-Cys})_n$-Ser [Ser-PC where $n = 2$] were present. Cabala et al. (2011) demonstrated elevated synthesis of $PC_2$ and $PC_3$ in tolerant broad bean cultivar Giza 40 than sensitive cultivar Giza 2 while reverse was true for GSH content. In *Oenothera odorata*, $PC_2$ and $PC_3$ were twice as high as $PC_4$ in response to elevated levels of Cd (Son et al.

2012). On the contrary, Xie et al. (2019) observed $PC_2$ and $PC_4$ contents to be 3–4 times higher than $PC_3$ in the leaves of *Amaranthus hypochondriacus*. PCs produced in cytoplasm bind with $Cd^{2+}$ to form low molecular weight (LMW) complexes which are compartmentalized into the vacuole via specific ABC-type transporter. Two ABCC-type transporters from *A. thaliana*, ABCC1 and ABCC2 have been reported to increase Cd tolerance and accumulation through sequestration of PC-Cd complexes in vacuoles (Park et al. 2012). Inside vacuole, more $Cd^{2+}$ (transported by $Cd^{2+}/H^+$ antiporter) together with sulfide and other PC chains binds to LMW complexes to generate high molecular weight (HMW) complexes (Zhang et al. 2010). The latter are believed to be made up of CdS crystallite core surrounded with PCs and are more stable with higher affinity for Cd. HMW complexes dissociate inside vacuole because of its acidic pH. Cd then binds with amino acids or organic acids, whereas apo-PCs are either disintegrated by hydrolases or return to cytosol. This method confers plants with a moderate level of tolerance against Cd. PCs can undergo long-distance translocation either from shoot to root or vice versa (Pal and Rai 2010). Experimental evidences indicate phloem as the main vascular route for long-distance transportation of Cd in the form of Cd-PC as well as Cd-GSH complexes (Mendoza-Cózatl et al. 2008). Although PCs do have a role in Cd detoxification and compartmentalization, their involvement in Cd tolerance is unclear. Gallego et al. (2005) demonstrated involvement of PC synthesis in detoxification of Cd in sunflower calluses and advocated that plant ability to grow under Cd toxicity was associated with the capacity to sustain high intracellular GSH levels. Similarly, Fenik et al. (2007) while working on two cell lines of *Nicotiana plumbaginifolia* [without inhibition of PC synthesis (Cd-R line) and with inhibition of butionine sulfoximine (Cd-Ri line)] reported fivefold and threefold enhancement in PC levels, respectively, and suggested that PCs contributed to Cd resistance in callus lines. High expression of *PCS1* gene boosts PC content which confers increased level of tolerance to plants grown on Cd-polluted soils (Liu et al. 2012). Nevertheless, researchers have reported that PCS overexpression in plants surprisingly led to hypersensitivity to metals or in other words PCs did not contribute towards metal tolerance and detoxification (Uraguchi et al. 2009). Possible explanation for this could be that overproduction of PCs reduces GSH levels and therefore can cause oxidative stress. Application of buthionine sulfoximine (BSO) decreased PC accumulation without concurrent enhancement in *Thlaspi caerulescens* plants sensitivity to Cd (Wojcik et al. 2005). Accumulation of Cd in *Pisum sativum* plants resulted in decrease in mono-thiol levels which in turn limited PC synthesis (Lima et al. 2006). Mishra et al. (2006) reported enhanced cysteine, NP-SH, GSH, and PCs levels in *Bacopa monnieri* under Cd stress which decreased with increase in applied Cd concentration and duration. A cad1-3 mutant of *A. thaliana* (knockout mutant for PCS1) exhibited defective callose deposition upon Cd exposure, suggesting role of AtPCS1 not only in PCs synthesis, but also in deposition of callose (De Benedictis et al. 2018). PCs are synthesized and accumulated in roots as well as leaves. Some reports suggest higher production of PCs in the roots (Rahman et al. 2016; Yan et al. 2017), whereas others indicate their higher synthesis in leaves (Mohamed et al. 2012; Szalai et al. 2013).

**Cysteine and Chaperones**
Cysteine is an essential component for the synthesis of GSH, PCs, and thiolate peptides engaged in Cd detoxification and transport to vacuoles (Fig. 14.1). Genetic alteration in *A. thaliana* using *Atcys-3A* construct expressing cytosolic O-acetyl serine (thiol) lyase (OASTL) revealed considerably increased cysteine biosynthesis in transformant under Cd stress than the wild type, which was responsible for enhanced Cd tolerance (Dominguez-Soils et al. 2004), Similarly, *Arabidopsis* mutants deficient in OASTL exhibited higher sensitivity to Cd (Lopez-Martin et al. 2008). Pre-treatment of cysteine increased Cd tolerance in rice seedlings by regulating cellular redox status with the assistance of GSH (Wang et al. 2017). Mitigation of Cd toxicity is also found to be connected with chaperones. Chaperones function in repairing misfolded proteins under Cd stress. An endoplasmic reticulum chaperone binding protein called LcBiP conferred Cd tolerance in tobacco plants (Guan et al. 2015).

Overall, these multiple reports indicate that under Cd stress plants adapt to a new metabolic equilibrium through co-ordinated synthesis and utilization of various antioxidant enzymes as well as non-enzymatic antioxidants to combat Cd toxicity.

### 14.3.8 Accumulation of Compatible Solutes/Osmolytes

Cd toxicity triggers the accretion of a group of varied "stress metabolites" such as soluble sugars, free amino acids, proline, and glycine betaine (GB), which corresponds to a general defense response of plants against metal stress. Compatible organic solutes decrease water potential of the cells below external values, forcing water intake by the cell, hence supporting turgor pressure high enough to maintain plant growth (Trovato et al. 2008). Therefore, osmolytes help in adaptation and ultimate survival of plants during heavy metal stress such as Cd. Soluble sugars play a key role in the pro-oxidant and antioxidant balance, thereby controlling oxidative stress induced by metals like Cd. Sugar accrual probably regulates internal osmolarity and protects cellular biomolecules and membranes. In addition to this, soluble sugars assist the plants in maximizing enough carbohydrate storage reserves to sustain basal metabolism under stressful conditions (Dubey and Singh 1999). Carbohydrate accumulation in roots and leaves has been reported for various plant species under Cd stress (Zoghlami et al. 2011; Zhou et al. 2020). Under conditions of Cd toxicity, certain amino acids can form complexes with toxic ions inside plant cells and contribute towards metal homeostasis. In *A. thaliana* cell cultures and leaves, various vital enzymes engaged either in proteasome pathway or in amino acid biosynthesis were upregulated by Cd treatments (Polge et al. 2009). Elevated free amino acid concentrations under Cd toxicity are reported to be concentration, organ, and genotype dependent (Zhu et al. 2018). Enhanced levels of proline and GB in plants under heavy metal stress are directly linked with increased stress tolerance (Sharma and Dietz 2006). However, proline confers more efficient protection against Cd stress than GB (Islam et al. 2009). Imperative functions of proline in plants against metal toxicity comprise osmoregulation, antioxidative defense, and Cd chelator forming non-toxic complexes with Cd (Ashraf and Foolad 2007). Because of its role in quenching $_1O^2$ and scavenging $OH^•$ radicals, proline assists in stabilizing proteins, DNA, and membranes. Accumulation of proline under stressful environments can act as a sink for surplus reductants providing $NAD^+$ and $NADP^+$ which are essential for maintaining respiratory and photosynthetic processes (Kishor et al. 2005). It has also been suggested as a means of storing excess nitrogen and carbon, a method to buffer cytosolic pH and a component of stress signal inducing adaptive responses (Verbruggen and Hermans 2008). Enhanced proline levels following Cd exposure have been reported in crop plants (Srivastava et al. 2011) as well as hyperaccumulator plants (Bao et al. 2011). Moreover, differential accumulation of proline in various plant organs and even cultivars has also been reported (Nedjimi and Daoud 2009). In pea, genotypic variation in proline accumulation was demonstrated by Metwally et al. (2005), indicating that it had genetic basis. Augmentation in proline levels in plants exposed to Cd could be due to water stress or physiological drought caused by Cd (Srivastava et al. 2011). Proline induces stomatal closure which suppresses the uptake and translocation of Cd in several plant species due to inhibition in transpiration (Sharma and Dietz 2006), indicating indirect role of proline in conferring metal tolerance to plants. However, Vassilev and Lidon (2011) considered higher levels of proline in two barley cultivars to be a resultant of Cd-induced proteolysis and not Cd-imposed osmotic adjustment. Muneer et al. (2011) advocated possible assimilation of proline in PC and MT synthesis to sequester and alleviate Cd stress. He et al. (2011) reported moderate increase in proline levels in roots and barks of *Populus* × *canescens* after $CdSO_4$ exposure for 10 and 20 days and in leaves after 20 days, not supporting an essential role of proline in poplar for scavenging Cd-induced ROS. GB does not scavenge ROS directly but alleviates the deleterious effects of oxidative stress by activating ROS-scavenging enzymes and/or limiting the production of ROS (Chen and Murata 2008). Potential mechanisms for GB-mediated increased tolerance of plants to stress include (1) protection of photosynthetic machinery, (2) involvement as a chaperon assisting in refolding of enzymes, (3) reduction in the membrane lipid peroxidation, (4) activation of particular genes whose products are implicated in stress tolerance, and (5) protection of electron transport chain in mitochondria (Sakamoto and Murata 2000; Hamilton and Heckathorn 2001). Literature also revealed that proline and GB pre-treatment mitigates Cd toxicity via ROS scavenging (Duman et al. 2011).

## 14.4 Conclusions

Pollution of agricultural soils by heavy metals through extensive industrialization together with nonjudicious use of chemical fertilizers has grown into a matter of great concern for researchers. Cd is extremely toxic non-essential metal and its occurrence in the arable soils is of importance in terms of health for plants, man as well as biosphere. The soil attributes influencing the phytoavailability of Cd are those which regulate the amount of Cd in soil solution and its sorption–desorption from/into the soil solution. Several studies demonstrated that Cd affects the plants by reducing their growth, biomass, and yield. Cd lowers the seed germination, respiration, photosynthesis, mineral nutrition and disturbs the nitrogen fixing capability of legumes. In addition, Cd toxicity induces oxidative stress in plants which triggers the antioxidant defense machinery to combat the deleterious effects of Cd. In spite of this, large knowledge gaps still exist in terms of genes, regulatory gene networks, and various proteins and metabolites produced in response to Cd stress in plants. Lastly, it can be concluded that further investigations using genomics, transcriptomics, proteomics, and metabolomics should be carried out to explore the detailed mechanisms of Cd toxicity as well as tolerance in plants.

**Acknowledgement** Harmanjit Kaur is grateful to University Grants Commission (UGC), New Delhi, India for providing financial assistance for carrying out the related research.

## References

Adams ML, Zhao FJ, McGrath SP, Nicholson FA, Chambers BJ (2004) Predicting cadmium concentrations in wheat and barley grain using soil properties. J Environ Qual 33:532–541

Adeleke R, Nwangburuka C, Oboirien B (2017) Origins, roles and fate of organic acids in soils: a review. South Afr J Bot 108:393–406

Agati G, Azzarello E, Pollastri S, Tattini M (2012) Flavonoids as antioxidants in plants: location and functional significance. Plant Sci 196:67–76

Ahammed GJ, Choudhary SP, Chen S, Xia X, Shi K, Zhou Y, Yu J (2013) Role of brassinosteroids in alleviation of phenanthrene-cadmium co-contamination-induced photosynthetic inhibition and oxidative stress in tomato. J Exp Bot 64(1):199–213

Ahmad P, Jaleel CA, Azooz MM, Nabi G (2009) Generation of ROS and non-enzymatic antioxidants during abiotic stress in plants. Bot Res Int 2:11–20

Ahmad P, Nabi G, Ashraf M (2011) Cadmium-induced oxidative damage in mustard [*Brassica juncea* (L.) Czern. & Coss.] plants can be alleviated by salicylic acid. South Afr J Bot 77:36–44

Akinyemi AJ, Faboya OL, Olayide I, Faboya OA, Ijabadeniyi T (2017) Effect of cadmium stress on non-enzymatic antioxidant and nitric oxide levels in two varieties of maize (*Zea mays*). Bull Environ Contam Toxicol 98(6):845–849

Alvarez S, Berla BM, Sheffield J, Cahoon RE, Jez JM, Hicks LM (2009) Comprehensive analysis of the *Brassica juncea* root proteome in response to cadmium exposure by complementary proteomic approaches. Proteomics 9:2419–2431

Anjum NA, Umar S, Iqbal M, Khan NA (2008) Ontogenic variation in response of *Brassica campestris* L. to cadmium toxicity. J Plant Interact 3(3):189–198

Anjum NA, Umar S, Iqbal M, Khan NA (2011) Cadmium causes oxidative stress in mungbean by affecting the antioxidant enzyme system and ascorbate-glutathione cycle metabolism. Russ J Plant Physiol 58(1):92–99

Anjum NA, Umar S, Iqbal M (2014) Assessment of cadmium accumulation, toxicity, and tolerance in Brassicaceae and Fabaceae plants- implications for phytoremediation. Environ Sci Pollut Res 21:10286–10293

Anjum SA, Tanveer M, Hussain S, Ashraf U, Khan I, Wang L (2017) Alteration in growth, leaf gas exchange, and photosynthetic pigments of maize plants under combined cadmium and arsenic stress. Water Air Soil Pollut 228(1):13

Ann C, Karen S, Jos R, Kelly O, Els K, Tony R, Nele H, Nathalie V, Suzy VS, Frank VB, Yves G, Jan C, Jaco V (2011) The cellular redox state as a modulator in cadmium and copper responses in *Arabidopsis thaliana* seedlings. J Plant Physiol 168:309–316

Aravind P, Prasad MNV (2005) Modulation of cadmium induced oxidative stress in *Ceratophyllum demersum* by zinc involves ascorbate-glutathione cycle and glutathione metabolism. Plant Physiol Biochem 43:107–116

Asgharipour MR, Khatamipour M, Razavi-Omrani M (2011) Phytotoxicity of cadmium on seed germination, early growth, proline and carbohydrate content in two wheat varieties. Adv Environ Biol 5(4):559–565

Ashraf M, Foolad MR (2007) Roles of glycine betaine and proline in improving plant abiotic stress resistance. Environ Exp Bot 59:206–216

Astolfi S, Zuchi S, Neumann G, Cesco S, di Toppi LS, Pinton R (2012) Response of barley plants to Fe deficiency and Cd contamination as affected by S starvation. J Exp Bot 63:1241–1250

Azpilicueta CE, Benavides MP, Tomaro ML, Gallego SM (2007) Mechanism of CATA3 induction by cadmium in sunflower leaves. Plant Physiol Biochem 45:589–595

Balakhnina T, Kosobryukhov A, Ivanov A, Kreslavskii V (2005) The effect of cadmium on $CO_2$ exchange, variable fluorescence of chlorophyll, and the level of antioxidant enzymes in pea leaves. Russ J Plant Physiol 52:15–20

Balestrasse KB, Benavides MP, Gallego S, Tomaro ML (2003) Effect of cadmium stress on nitrogen metabolism in nodules and roots of soybean plants. Funct Plant Biol 30:57–64

Banat KM, Howari FM, Tomah MM (2007) Chemical fractionation and heavy metal distribution in agricultural soils, north of Jordan valley. Soil Sediment Contam 16(1):89–107

Bao T, Sun T, Sun L (2011) Effect of cadmium hyperaccumulation on antioxidative defense and proline accumulation of *Solanum nigrum* L. Afr J Biotechnol 10(37):7198–7206

Barman F, Majumdar S, Arzoo SH, Kundu R (2020) Genotypic variation among 20 rice cultivars/ landraces in response to cadmium stress grown locally in West Bengal, India. Plant Physiol Biochem 148:193–206

Basa B, Lattanzio G, Solti A, Toth B, Abadia J, Fodor F, Sarvari E (2014) Changes induced by cadmium stress and iron deficiency in the composition and organization of thylakoid complexes in sugar beet (*Beta vulgaris* L.). Environ Exp Bot 101:1–11

Basar H (2009) Methods for estimating phytoavailable metals in soils. Commun Soil Sci Plant Anal 40:1087–1105

Basta NT, Ryan JA, Chaney RL (2005) Trace element chemistry in residual-treated soils: key concepts and metal bioavailability. J Environ Qual 34:49–63

Bavi K, Kholdebarin B, Moradshahi A (2011) Effect of cadmium on growth, protein content and peroxidase activity in pea plants. Pak J Bot 43(3):1467–1470

Becana M, Dalton DA, Moran JF, Iturbe-Ormaetxe I, Matamoros MA, Rubio MC (2000) Reactive oxygen species and antioxidants in legume nodules. Physiol Plant 109:372–381

Becana M, Matamoros M, Udvardi M, Dalton DA (2011) Recent insights into antioxidant defenses of legume root nodules. New Phytol 188:960–976

Belkhadi A, Hediji H, Abbes Z, Nouairi I, Barhoumi Z, Zarrouk M, Chaibi W, Djebali W (2010) Effects of exogenous salicylic acid pre-treatment on cadmium toxicity and leaf lipid content in *Linum usitatissimum* L. Ecotoxicol Environ Saf 73:1004–1011

Bergkvist P, Jarvis N (2004) Modeling organic carbon dynamics and cadmium fate in long-term sludge-amended soils. J Environ Qual 33:181–191

Berkelaar EJ, Hale BA (2003) Accumulation of cadmium by durum wheat roots: bases for citrate-mediated exceptions to the free ion model? Environ Toxicol Chem 22:1155–1161

Bi YH, Chen WL, Zhang WN, Zhou Q, Yun LJ, Xing D (2009) Production of reactive oxygen species, impairment of photosynthetic function and dynamic changes in mitochondria are early events in cadmium-induced cell death in *Arabidopsis thaliana*. Biol Cell 101:629–643

Black A, McLaren RG, Reichman SM, Speir TW, Condron LM (2011) Evaluation of soil metal bioavailability estimates using two plant species (*L. perenne* and *T. aestivum*) grown in a range of agricultural soils treated with biosolids and metal salts. Environ Pollut 159:1523–1535

Bolan NS, Duriasamy VP (2003) Role of inorganic and organic soil amendments on immobilisation and phytoavailability of heavy metals: a review involving specific case studies. Aust J Soil Res 41:533–555

Buono DD, Mimmo T, Terzano R, Tomasi N, Cesco S (2014) Effect of cadmium on antioxidative enzymes, glutathione content, and glutathionylation in tall Fescue. Biol Plant 58:773–777
Cabala R, Slovakova L, Zohri ME, Frank H (2011) Accumulation and translocation of Cd metal and the Cd-induced production of glutathione and phytochelatins in *Vicia faba*. L. Acta Physiol Plant 33:1239–1248
Carpena RO, Vázquez S, Esteban E, Fernández-Pascual M, de Felipe MR, Zornoza P (2003) Cadmium-stress in white lupin: effects on nodule structure and functioning. Plant Physiol Biochem 41:911–919
Carrillo-Gonzalez R, Simunek J, Sauve S, Adriano D (2006) Mechanisms and pathways of trace element mobility in soils. Adv Agron 91:111–178
Chang YS, Chang YJ, Lin CT, Lee MC, Wu CW, Lai YH (2013) Nitrogen fertilization promotes the phytoremediation of cadmium in *Pentas lanceolata*. Int Biodeterior Biodegrad 85:709–714
Chao YY, Hong CY, Kao CH (2010) The decline in ascorbic acid is associated with cadmium toxicity of rice seedlings. Plant Physiol Biochem 48:374–381
Chen THH, Murata N (2008) Glycinebetaine: An effective protectant against abiotic stress in plants. Trends Plant Sci 13(9):499–505
Chen YX, He YF, Yang Y, Yu YL, Zheng SJ, Tian GM, Luo YM, Wong MH (2003) Effect of cadmium on nodulation and $N_2$-fixation of soybean in contaminated soils. Chemosphere 50:781–787
Chen J, Zhu C, Lin D, Sun ZX (2007) The effects of Cd on lipid peroxidation, hydrogen peroxide content and antioxidant enzyme activities in Cd-sensitive mutant rice seedlings. Can J Plant Sci 87:49–57
Chen HS, Huang QY, Liu LN, Cai P, LiangW LM (2010) Poultry manure compost alleviates the phytotoxicity of soil cadmium: influence on growth of pakchoi (*Brassica chinensis* L.). Pedosphere 20:63–70
Chen C, Cao Q, Jiang Q, Li J, Yu R, Shi G (2019) Comparative transcriptome analysis reveals gene network regulating cadmium uptake and translocation in peanut roots under iron deficiency. BMC Plant Biol 19:35
Cheng WD, Yao HG, Zhang HM, Tao XG (2009) Influences of cadmium on grain mineral nutrient contents of two rice genotypes differing in grain cadmium accumulation. Rice Sci 16 (2):151–156
Cho SC, Chao YY, Hong CY, Kao CH (2012) The role of hydrogen peroxide in cadmium-inhibited root growth of rice seedlings. Plant Growth Regul 66:27–35
Chou TS, Chao YY, Huang WD, Hong CY, Kao CH (2011) Effect of magnesium deficiency on antioxidant status and cadmium toxicity in rice seedlings. J Plant Physiol 168(10):1021–1030
Christine G, Sylvaine T, Michel A (2002) Fractionation studies of trace elements in contaminated soils and sediments: a review of sequential extraction procedures. Trends Anal Chem 21:451–467
Chugh LK, Sawhney SK (1999) Effect of cadmium on activities of some enzymes of glycolysis and pentose phosphate pathway in pea. Biol Plant 42(3):401–407
Chunhabundit R (2016) Cadmium exposure and potential health risk from foods in contaminated area, Thailand. Toxicol Res 32(1):65–72
Ci D, Jiang D, Wollenweber B, Dai T, Jing Q, Cao W (2010) Cadmium stress in wheat seedlings: growth, cadmium accumulation and photosynthesis. Acta Physiol Plant 32:365–373
Clemens S (2006) Toxic metal accumulation, responses to exposure and mechanisms of tolerance in plants. Biochimie 88:1707–1719
Cobbett C, Goldsbrough P (2002) Phytochelatins and metallothioneins: roles in heavy metal detoxification and homeostasis. Annu Rev Plant Biol 53:159–182
Collin V, Eymery F, Genty B, Rey P, Havaux M (2008) Vitamin E is essential for the tolerance of *Arabidopsis thaliana* to metal-induced oxidative stress. Plant Cell Environ 31:244–257
Cuypers A, Smeets K, Ruytinx J, Opdenakker K, Keunen E, Remans T, Horemans N, Vanhoudt N, Van Sanden S, Van Belleghem F, Guisez Y, Colpaert J, Vangronsveld J (2011) The cellular

redox state as a modulator in cadmium and copper responses in *Arabidopsis thaliana* seedlings. J Plant Physiol 168(4):309–316

DalCorso G, Farinati S, Maistri S, Furini A (2008) How plants cope with cadmium: staking all on metabolism and gene expression. J Integr Plant Biol 50:1268–1280

De Benedictis M, Brunetti C, Brauer EK, Andreucci A, Popescu SC, Commisso M, Guzzo F, Sofo A, Ruffini Castiglione M, Vatamaniuk OK, Sanita di Toppi L (2018) The *Arabidopsis thaliana* knockout mutant for *Phytochelatin Synthase 1* (*cad1-3*) is defective in callose deposition, bacterial pathogen defense and auxin content. Front Plant Sci 9:19

Delperee C, Lutts S (2008) Growth inhibition occurs independently of cell mortality in tomato (*Solanum lycopersicum*) exposed to high cadmium concentrations. J Integr Plant Biol 50 (3):300–310

Dhingra HR, Priefer UR (2006) Impact of cadmium on structural and functional aspect of pea (*Pisum sativum* L.) root nodules. J Plant Biol 33:201–208

Djebali W, Zarrouk M, Brouquisse R, El Kahoui S, Limam F, Ghorbel MH, Chaibi W (2005) Ultrastructure and lipid alterations induced by cadmium in tomato (*Lycopersicon esculentum*) chloroplast membranes. Plant Boil 7:258–268

Dobrikova AG, Yotsova EK, Borner A, Landjeva SP, Apostolova EL (2017) The wheat mutant DELLA-encoding gene (Rht-B1c) affects plant photosynthetic responses to cadmium stress. Plant Physiol Biochem 114:10–18

Dominguez-Soils JR, Lopez-Martin MC, Ager FJ, Ynsa MD, Romero LC, Gotor C (2004) Increased cysteine availability is essential for cadmium tolerance and accumulation in *Arabidopsis thaliana*. Plant Biotechnol J 2:469–476

Dong J, Wu F, Zhang G (2006) Influence of cadmium on antioxidant capacity and four microelement concentrations in tomato seedlings (*Lycopersicon esculentum*). Chemosphere 64 (10):1659–1666

Dong J, Mao WH, Zhang GP, Wu FB, Cai Y (2007) Root excretion and plant tolerance to cadmium toxicity – a review. Plant Soil Environ 53:193–200

Dubey RS, Singh AK (1999) Salinity induces accumulation of soluble sugars and alters the activity of sugar metabolising enzymes in rice plants. Biol Plant 42:233–239

Duman F, Aksoy A, Aydin Z, Temizgul R (2011) Effects of exogenous glycinebetaine and trehalose on cadmium accumulation and biological responses of an aquatic plant (*Lemna gibba* L.). Water Air Soil Pollut 217:545–556

Dunand C, Crèvecoeur M, Penel C (2007) Distribution of superoxide and hydrogen peroxide in *Arabidopsis* root and their influence on root development: possible interaction with peroxidases. New Phytol 174(2):332–341

Fagioni M, D'Amici GM, Timperio AM, Zolla L (2009) Proteomic analysis of multiprotein complexes in the thylakoid membrane upon cadmium treatment. J Proteome Res 8(1):310–326

Faller P, Kienzler K, Krieger-Liszkay A (2005) Mechanism of $Cd^{2+}$ toxicity: $Cd^{2+}$ inhibits photoactivation of photosystem II by competitive binding to the essential $Ca^{2+}$ site. Biochim Biophys Acta 1706:158–164

Farooq MA, Ali S, Hameed A, Bharwana SA, Rizwan M, Ishaque W, Faridc M, Mahmood K, Iqbal Z (2016) Cadmium stress in cotton seedlings: Physiological, photosynthesis and oxidative damages alleviated by glycinebetaine. South Afr J Bot 104:61–68

Fenik SI, Solodushko VG, Kalinyak TB, Blume AB (2007) The role of Cd-binding proteins and phytochelatins in the formation of resistance to cadmium in the cell lines of *Nicotiana plumbaginifolia*. Cytol Genet 41(1):6–11

Fernandez R, Bertrand A, Reis R, Mourato MP, Martins LL, Gonzalez A (2013) Growth and physiological responses to cadmium stress of two populations of *Dittrichia viscosa* (L.) Greuter. J Hazard Mater 244:555–562

Finkemeier I, Kluge C, Metwally A, Georgi M, Grotjohann N, Dietz KJ (2003) Alteration in Cd-induced gene expression under nitrogen deficiency in *Hordeum vulgare*. Plant Cell Environ 26:821–833

Fischerová Z, Tlustoš P, Száková J, Šichorová K (2006) A comparison of phytoremediation capability of selected plant species for given trace elements. Environ Pollut 144:93–100
Foyer CH, Noctor G (2005a) Redox homeostasis and antioxidant signaling: a metabolic interface between stress perception and physiological responses. Plant Cell 17:1866–1875
Foyer CH, Noctor G (2005b) Oxidant and antioxidant signalling in plants: a re-evaluation of the concept of oxidative stress in a physiological context. Plant Cell Environ 28:1056–1071
Gallego SM, Kogan MJ, Azpilicueta CE, Pena C, Tomaro ML (2005) Glutathione-mediated antioxidative mechanisms in sunflower (*Helianthus annuus* L.) cells in response to cadmium stress. Plant Growth Regul 46:267–276
Gapper C, Dolan L (2006) Control of plant development by reactive oxygen species. Plant Physiol 141(2):341–345
Garg N, Aggarwal N (2011) Effects of interactions between cadmium and lead on growth, nitrogen fixation, phytochelatin, and glutathione production in mycorrhizal *Cajanus cajan* (L.) Millsp. J Plant Growth Regul 30(3):286–300
Garg N, Kaur H (2012) Influence of zinc on cadmium-induced toxicity in nodules of pigeonpea (*Cajanus cajan* L. Millsp.) inoculated with arbuscular mycorrhizal (AM) fungi. Acta Physiol Plant 34:1363–1380
Garg N, Kaur H (2013a) Impact of Cd-Zn interactions on metal uptake, translocation and yield in *Cajanus cajan* (L.) Millsp. genotypes colonized by arbuscular mycorrhizal (AM) fungi. J Plant Nutr 36:67–90
Garg N, Kaur H (2013b) Response of antioxidant enzymes, phytochelatins and glutathione production towards Cd and Zn stresses in *Cajanus cajan* (L.) Millsp. genotypes colonized by arbuscular mycorrhizal fungi. J Agron Crop Sci 199:118–133
Garnier L, Simon-Plas F, Thuleau P, Angel JP (2006) Cadmium affects tobacco cells by a series of three waves of reactive oxygen species that contribute to cytotoxicity. Plant Cell Environ 29:1956–1969
Ge Y, Hendershot W (2005) Modeling sorption of Cd, Hg and Pb in soils by the NICA (non-ideal competitive adsorption)-Donnan model. Soil Sediment Contam 14:53–69
Ghani A (2010) Effect of cadmium toxicity on the growth and yield components of mungbean [*Vigna radiate* (L.) Wilczek]. World Appl Sci J 8:26–29
Ghani A (2011) Varietal differences in Canola (*Brassica napus* L.) for the growth, yield and yield components exposed to cadmium stress. J Animal Plant Sci 21(1):57–59
Gill SS, Tuteja N (2010) Reactive oxygen species and antioxidant machinery in abiotic stress tolerance in crop plants. Plant Physiol Biochem 48:909–930
Gill SS, Tuteja N (2011) Cadmium stress in crop plants. Plant Signal Behav 6(2):215–222
Goncalves JF, Antes FG, Maldaner J, Pereira LB, Tabaldi LA, Rauber R, Rossato LV, Bisognin DA, Dressler VL, de Moraes Flores EM, Nicoloso FT (2009) Cadmium and mineral nutrient accumulation in potato plantlets grown under cadmium stress in two different experimental culture conditions. Plant Physiol Biochem 47:814–821
Gonzalez-Mendoza D, Espadas Y, Gil F, Santamaria JM, Zapata-Perez O (2007) Multiple effects of cadmium on the photosynthetic apparatus of *Avicennia germinans* L. as probed by OJIP chlorophyll fluorescence measurements. Z Naturforsch C 62:265–272
Gopal R, Nautiyal N (2011) Phytotoxic effects of cadmium exposure and metal accumulation in sunflower. J Plant Nutr 34:1616–1624
Gothberg A, Greger M, Holm K, Bengtsson BE (2004) Influence of nutrient levels on uptake and effects of mercury, cadmium, and lead in water spinach. J Environ Qual 33:1247–1255
Granato TC, Pietz RI, Knafl GJ, Carlson CR Jr, Tata P, Lue-Hing C (2004) Trace element concentrations in soil, corn leaves, and grain after cessation of biosolids applications. J Environ Qual 33:2078–2089
Grant CA, Sheppard SC (2008) Fertilizer impacts on cadmium availability in agricultural soils and crops. Hum Ecol Risk Assess 14:210–228

Gratao PL, Monteiro CC, Antunes AM, Peres LEP, Azevedo RA (2008) Acquired tolerance of tomato (*Lycopersicon esculentum* cv. Micro-Tom) plants to cadmium-induced stress. Ann Appl Biol 153:321–333

Gratao PL, Monteiro CC, Rossi ML, Martinelli AP, Peres LEP, Medici LO, Lea PJ, Azevedo RA (2009) Differential ultrastructural changes in tomato hormonal mutants exposed to cadmium. Environ Exp Bot 67:387–394

Grispen VM, Hakvoort HW, Bliek T, Verkleij JA, Schat H (2011) Combined expression of the *Arabidopsis* metallothionein MT2b and the heavy metal transporting ATPase HMA4 enhances cadmium tolerance and the root to shoot translocation of cadmium and zinc in tobacco. Environ Exp Bot 72:71–76

Grobelak A, Świątek J, Murtaś A, Jaskulak M (2019) Cadmium-induced oxidative stress in plants, cadmium toxicity, and tolerance in plants: from physiology to remediation. In: Hasanuzzaman M, Prasad MNV, Fujita M (eds) Cadmium toxicity and tolerance in plants. Academic, San Diego, pp 213–231

Guan C, Jin C, Ji J, Wang G, Li X (2015) LcBiP, a endoplasmic reticulum chaperone binding protein gene from *Lycium chinense*, confers cadmium tolerance in transgenic tobacco. Biotechnol Prog 31(2):358–368

Hamilton EW, Heckathorn SA (2001) Mitochondrial adaptations to NaCl. Complex I is protected by anti-oxidants and small heat-shock proteins, whereas complex II is protected by proline and betaine. Plant Physiol 126:1266–1274

Hattab S, Dridi B, Chouba L, Kheder MB, Bousetta H (2009) Photosynthesis and growth responses of pea *Pisum sativum* L. under heavy metals stress. J Environ Sci 21:1552–1556

He J, Ren Y, Pan X, Yan Y, Zhu C, Jiang D (2010) Salicylic acid alleviates the toxicity effect of cadmium on germination, seedling growth, and amylase activity of rice. J Plant Nutr Soil Sci 173:300–305

He J, Qina J, Longa L, Maa Y, Li H, Li K, Jiang X, Liu T, Polle A, Liang Z, Luo ZB (2011) Net cadmium flux and accumulation reveal tissue-specific oxidative stress and detoxification in *Populus* × *canescens*. Physiol Plant 143:50–63

He H, Tam NF, Yao A, Qiu R, Li WC, Ye Z (2017) Growth and Cd uptake by rice (*Oryza sativa*) in acidic and Cd-contaminated paddy soils amended with steel slag. Chemosphere 189:247–254

Hediji H, Djebali W, Cabasson C, Maucourt M, Baldet P, Bertrand A, Zoghlami LB, Deborde C, Moing A, Brouquisse R, Chaibi W, Gallusci P (2010) Effects of long-term cadmium exposure on growth and metabolomic profile of tomato plants. Ecotoxicol Environ Saf 73:1965–1974

Heidari M, Sarani S (2011) Effects of lead and cadmium on seed germination, seedling growth and antioxidant enzymes activities of mustard (*Sinapis arvensis* L.). J Agric Biol Sci 6(1):44–47

Herbette S, Taconnat L, Hugouvieux V, Piette L, Magniette MLM, Cuine S, Auroy P, Richaud P, Forestier C, Bourguignon J, Renou JP, Vavasseur A, Leonhardt N (2006) Genome-wide transcriptome profiling of the early cadmium response of *Arabidopsis* roots and shoots. Biochimie 88:1751–1765

Heyno E, Klose C, Krieger-Liszkay A (2008) Origin of cadmium induced reactive oxygen species production: mitochondrial electron transfer versus plasma membrane NADPH oxidase. New Phytol 179:687–699

Hinsinger P, Plassard C, Jaillard B (2006) Rhizosphere: a new frontier for soil biogeochemistry. J Geochem Explor 88:210–213

Hossain Z, Huq F (2002) Studies on the interaction between $Cd^{2+}$ ions and nucleobases and nucleotides. J Inorg Biochem 90:97–105

Hsiao PY, Su RC, Ko SS, Tong CG, Yang RY, Chan MT (2008) Overexpression of *Arabidopsis thaliana* tryptophan synthase beta 1 (*AtTSB1*) in *Arabidopsis* and tomato confers tolerance to cadmium stress. Plant Cell Environ 31:1074–1085

Hu L, McBride MB, Cheng H, Wu J, Shi J, Xu J, Wu L (2011) Root-induced changes to cadmium speciation in the rhizosphere of two rice (*Oryza sativa* L.) genotypes. Environ Res 111:356–361

Huang B, Xin J, Dai H, Liu A, Zhou W, Yi Y, Liao K (2015) Root morphological responses of three hot pepper cultivars to Cd exposure and their correlations with Cd accumulation. Environ Sci Pollut Res 22(2):1151–1159

Iqbal N, Masood A, Nazar R, Syeed S, Khan NA (2010) Photosynthesis, growth and antioxidant metabolism in mustard (*Brassica juncea* L.) cultivars differing in cadmium tolerance. Agric Sci China 9(4):519–527

Islam MM, Hoque MA, Okuma E, Jannat R, Banu MNA, Jahan MS, Nakamura Y, Murata Y (2009) Proline and glycinebetaine confer cadmium tolerance on tobacco bright yellow-2 cells by increasing ascorbate-glutathione cycle enzyme activities. Biosci Biotechnol Biochem 73 (10):2320–2323

Jauert P, Schumacher TE, Boe A, Reese RN (2002) Rhizosphere acidification and cadmium uptake by strawberry clover. J Environ Qual 31:627–633

Javed MT, Stoltz E, Lindberg S, Greger M (2013) Changes in pH and organic acids in mucilage of *Eriophorum angustifolium* roots after exposure to elevated concentrations of toxic elements. Environ Sci Pollut Res 20:1876–1880

Jiang XJ, Luo YM, Liu Q, Liu SL, Zhao QG (2004) Effects of cadmium on nutrient uptake and translocation by Indian mustard. Environ Geochem Health 26:319–324

Jiang H, Li T, Han X, Yang X, He Z (2011) Effects of pH and low molecular weight organic acids on competitive adsorption and desorption of cadmium and lead in paddy soils. Environ Monit Assess 184(10):6325–6335

Jin YH, Clark AB, Slebos RJC, Al-Refai H, Taylor JA, Kunkel TA, Resnick MA, Gordenin DA (2003) Cadmium is a mutagen that acts by inhibiting mismatch repair. Nat Genet 34:326–329

Jing YD, He ZL, Yang XE (2007) Effects of pH, organic acids and competitive cations on mercury desorption in soils. Chemosphere 69:1662–1669

Kabata-Pendias A, Pendias H (2001) Trace elements in soils and plants. CRC Press, Boca Raton

Kabata-Pendias A, Pendias H (2010) Trace elements in soils and plants, 3rd edn. CRC Press, Boca Raton

Kaur H, Garg N (2017) Zinc-arbuscular mycorrhizal interactions: effect on nutrient pool, enzymatic antioxidants, and osmolyte synthesis in pigeonpea nodules subjected to Cd stress. Commun Soil Sci Plant Anal 48:1684–1700

Kaur R, Yadav P, Sharma A, Thukral AK, Kumar V, Kohli SK, Bhardwaj R (2017) Castasterone and citric acid treatment restores photosynthetic attributes in *Brassica juncea* L. under Cd (II) toxicity. Ecotoxicol Environ Saf 145:466–475

Kim KR, Owens G (2009) Chemodynamics of heavy metals in long-term contaminated soils: metal speciation in soil solution. J Environ Sci 21(11):1532–1540

Kim HS, Seo BH, Kuppusamy S, Lee YB, Lee JH, Yang JE, Owens G, Kim KR (2018) A DOC coagulant, gypsum treatment can simultaneously reduce As, Cd and Pb uptake by medicinal plants grown in contaminated soil. Ecotoxicol Environ Saf 148:615–619

Kiran Y, Sahin A (2006) The effects of cadmium on seed germination, root development and mitotic of root tip cells of lentil (*Lens culinaris* Medik). World J Agric Sci 2(2):196–200

Kirkham MB (2006) Cadmium in plants on polluted soils: Effects of soil factors hyperaccumulation, and amendments. Geoderma 137:19–32

Kishor PBK, Sangam S, Amrutha RN, Sri Laxmi P, Naidu KR, Rao KRSS, Rao S, Reddy KJ, Theriappan P, Sreenivasulu N (2005) Regulation of proline biosynthesis, degradation uptake and transport in higher plants: Its implications in plant growth and abiotic stress tolerance. Curr Sci 88(3):424–438

Klapheck S, Fliegner W, Zimmer I (1994) Hydroxymethyl-phytochelatins [gamma-glutamylcysteine-serine] are metal-induced peptides of the Poaceae. Plant Physiol 104 (4):1325–1332

Kolahi M, Kazemi EM, Yazdi M, Goldson-Barnaby A (2020) Oxidative stress induced by cadmium in lettuce (*Lactuca sativa* Linn.): Oxidative stress indicators and prediction of their genes. Plant Physiol Biochem 146:71–89

Kosakowska A, Lewandowska J, Ston J, Burkiewicz A (2004) Qualitative and quantitative composition of pigments in *Phaeodactylum tricornutum* (Bacillariophyceae) stressed by iron. Biometals 17:45–52

Kostyuk VA, Potapovich AI, Kostyuk TV, Cherian MG (2007) Metal complexes of dietary flavonoids: evaluation of radical scavenger properties and protective activity against oxidative stress in vivo. Cell Mol Biol 53:61–68

Kovacik J, Klejdus B, Hedbavny J, Stork F, Backor M (2009) Comparison of cadmium and copper effect on phenolic metabolism, mineral nutrients and stress-related parameters in *Matricaria chamomilla* plants. Plant Soil 3(20):231–242

Kramer U (2010) Metal hyperaccumulation in plants. Annu Rev Plant Biol 61:517–534

Kranner I, Colville L (2011) Metals and seeds: biochemical and molecular implications and their significance for seed germination. Environ Exp Bot 72(1):93–105

Krantev A, Yordanova R, Janda T, Szalai G, Popova L (2008) Treatment with salicylic acid decreases the effect of cadmium on photosynthesis in maize plants. J Plant Physiol 165:920–930

Krishnamurti GSR, Huang PM, Van Rees KCJ, Kozak LM, Rostad HPW (1995) Speciation of particulate-bound cadmium in soils and its bioavailability. Analyst 120:659–665

Kupper H, Parameswaran A, Leitenmaier B, Trtilek M, Setlik I (2007) Cadmium-induced inhibition of photosynthesis and long-term acclimation to cadmium stress in the hyperaccumulator *Thlaspi caerulescens*. New Phytol 175:655–674

Li Z, Li L, Pan G, Chen J (2005) Bioavailability of Cd in a soil–rice system in China: soil type versus genotype effects. Plant Soil 271:165–173

Li Y, Chen Z, Xu S, Zhang L, Hou W, Yu N (2015) Effect of combined pollution of Cd and B [a] P on photosynthesis and chlorophyll fluorescence characteristics of wheat. Pol J Environ Stud 24 (1):121–131

Li X, Zhou Q, Sun X, Ren W (2016) Effects of cadmium on uptake and translocation of nutrient elements in different welsh onion (*Allium fistulosum* L.) cultivars. Food Chem 194:101–110

Ligaba A, Yamaguchi M, Shen H, Sasaki T, Yamamoto Y, Matsumoto H (2004) Phosphorus deficiency enhances plasma membrane $H^+$-ATPase activity and citrate exudation in greater purple lupin (*Lupinus pilosus*). Funct Plant Biol 31:1075–1083

Lima AIG, Pereira SIA, Figueira EMP, Caldeira GCN, Caldeira HDQ (2006) Cadmium detoxification in roots of *Pisum sativum* seedlings: relationship between toxicity levels, thiol pool alterations and growth. Environ Exp Bot 55:149–162

Lindberg S, Landberg T, Greger M (2007) Cadmium uptake and interaction with phytochelatins in wheat protoplasts. Plant Physiol Biochem 45:47–53

Liu D, Jiang W, Gao X (2004) Effects of cadmium on root growth, cell division and nucleoli in root tip cells of garlic. Biol Plant 47:79–83

Liu W, Li PJ, Qi XM, Zhou QX, Zheng L, Sun TH, Yang YS (2005) DNA changes in barley (*Hordeum vulgare*) seedlings induced by cadmium pollution using RAPD analysis. Chemosphere 61:158–167

Liu Y, Wang X, Zeng G, Qu D, Gu J, Zhou M, Chai L (2007) Cadmium-induced oxidative stress and response of the ascorbate–glutathione cycle in *Bechmeria nivea* (L.) Gaud. Chemosphere 69:99–107

Liu KL, Shen L, Wang JQ, Sheng JP (2008a) Rapid inactivation of chloroplastic ascorbate peroxidase is responsible for oxidative modification to rubisco in tomato (*Lycopersicon esculentum*) under cadmium stress. J Integr Plant Biol 50(4):415–426

Liu W, Yang YS, Francis D, Rogers HJ, Li P, Zhang Q (2008b) Cadmium stress alters gene expression of DNA mismatch repair related genes in *Arabidopsis* seedlings. Chemosphere 73:1138–1144

Liu W, Zhou Q, Li P, Gao H, Han YP, Li XJ, Yang YS, Li Y (2009) DNA mismatch repair related gene expression as potential biomarkers to assess cadmium exposure in *Arabidopsis* seedlings. J Hazard Mater 167:1007–1013

Liu C, Guo J, Cui Y, Lu T, Zhang X, Shi G (2011a) Effects of cadmium and salicylic acid on growth, spectral reflectance and photosynthesis of castor bean seedlings. Plant Soil 344:131–141

Liu Z, Chen W, He X (2011b) Cadmium-induced changes in growth and antioxidative mechanisms of a medicine plant (*Lonicera japonica* Thunb.). J Med Plant Res 5(8):1411–1417

Liu Z, Gu C, Chen F, Yang D, Wu K, Chen S, Jiang J, Zhang Z (2012) Heterologous expression of a *Nelumbo nucifera* phytochelatin synthase gene enhances cadmium tolerance in *Arabidopsis thaliana*. Appl Biochem Biotechnol 166(3):722–734

Liu K, Lv J, He W, Zhang H, Cao Y, Dai Y (2015) Major factors influencing cadmium uptake from the soil into wheat plants. Ecotoxicol Environ Saf 113:207–213

Lock K, Janssen CR (2003a) Influence of ageing on zinc bioavailability in soils. Environ Pollut 126:371–374

Lock K, Janssen CR (2003b) Influence of aging on metal availability in soils. Rev Environ Contam Toxicol 178:1–21

Lofts S, Spurgeon DJ, Svendsen C, Tipping E (2004) Deriving soil critical limits for Cu, Zn, Cd, and Pb: a method based on free ion concentrations. Environ Sci Technol 38:3623–3631

Lopez-Martin MC, Becana M, Romero LC, Gotor C (2008) Knocking out cytosolic cysteine synthesis compromises the antioxidant capacity of the cytosol to maintain discrete concentrations of hydrogen peroxide in *Arabidopsis*. Plant Physiol 147:562–572

Lopez-Millan AF, Sagardoy R, Solanas M, Abadia A, Abadia J (2009) Cadmium toxicity in tomato (*Lycopersicon esculentum*) plants grown in hydroponics. Environ Exp Bot 65:376–385

Lutts S, Lefevre I, Delperee C, Kivits S, Dechamps C, Robledo A, Correal E (2004) Heavy metal accumulation by the halophyte species Mediterranean Saltbush. J Environ Qual 33:1271–1279

Lux A, Martinka M, Vaculik M, White PJ (2010) Root responses to cadmium in the rhizosphere: a review. J Exp Bot 62(1):21–37

Maarouf DH, Debouba M, Ghorbel MH, Gouia H (2009) Tissue-specific cadmium accumulation and its effects on nitrogen metabolism in tobacco (*Nicotiana tabacum*, Bureley v. Fb9). C R Biol 332:58–68

Maksimovic I, Kastori R, Krstic L, Lukovic J (2007) Steady presence of cadmium and nickel affects root anatomy, accumulation and distribution of essential ions in maize seedlings. Biol Plant 51 (3):589–592

Malec P, Maleva MG, Prasad MNV, Strzalka K (2010) Responses of *Lemna trisulca* L. (Duckweed) exposed to low doses of cadmium: thiols, metal binding complexes, and photosynthetic pigments as sensitive biomarkers of ecotoxicity. Protoplasma 240:69–74

Manier N, Deram A, Broos K, Denayer FO, Haluwyn CV (2009) White clover nodulation index in heavy metal contaminated soils- a potential bioindicator. J Environ Qual 38:685–692

Marcano L, Carruyo I, Del Campo A, Monteil X (2002) Effect of cadmium on the nucleoli of meristematic cells of onion *Allium cepa* L: an ultrastructural study. Environ Res 88:30–35

Marquez-Garcia B, Horemans N, Torronteras R, Cordoba F (2012) Glutathione depletion in healthy cadmium-exposed *Erica andevalensis*. Environ Exp Bot 75:159–166

Martinez-Ballesta MC, Martinez V, Charvajal M (2003) Aquaporin functionality in relation to H+-ATPase activity in root cells of *Capsicum annuum* grown under salinity. Physiol Plant 117:413–420

Martins LL, Mourato MP, Cardoso AI, Pinto AP, Mota AM, Goncalves MLS, Varennes A (2011) Oxidative stress induced by cadmium in *Nicotiana tabacum* L: effects on growth parameters, oxidative damage and antioxidant responses in different plant parts. Acta Physiol Plant 33 (4):1375–1383

Maurel C, Chrispeels MJ (2001) Aquaporins. A molecular entry into plant water relations. Plant Physiol 125:135–138

McLaren RG, Clucas LM, Taylor MD, Henry T (2004) Leaching of macronutrients and metals from undisturbed soils treated with metal-spiked sewage sludge. 2. Leaching of metals. Aust J Soil Res 42:459–471

Mediouni C, Houlne G, Chaboute ME, Ghorbel MH, Jemal F (2008) Cadmium and copper genotoxicity in plants. In: Abdelly C, Ozturk M, Ashraf M, Grignon C (eds) Biosaline agriculture high salinity tolerance. Birkhauser Verlag, Cham, pp 325–333

Mendoza-Cózatl DG, Butko E, Springer F, Torpey JW, Komives EA, Kehr J, Schroeder JI (2008) Identification of high levels of phytochelatins, glutathione and cadmium in the phloem sap of *Brassica napus*. A role for thiol-peptides in the long-distance transport of cadmium and the effect of cadmium on iron translocation. Plant J 54:249–259

Metwally A, Safronova VI, Belimov AA, Dietz KJ (2005) Genotypic variation of the response to cadmium toxicity in *Pisum sativum* L. J Exp Bot 56:167–178

Mishra S, Srivastava S, Tripathi RD, Govindaranjan R, Kuriakose SV, Prasad MNV (2006) Phytochelatin synthesis and response of antioxidants during cadmium stress in *Bacopa monnieri* L. Plant Physiol Biochem 44:25–37

Mobin M, Khan NA (2007) Photosynthetic activity, pigment composition and antioxidative response of two mustard (*Brassica juncea*) cultivars differing in photosynthetic capacity subjected to cadmium stress. J Plant Physiol 164(5):601–610

Mohamed AA, Castagna A, Ranieri A, Sanità di Toppi L (2012) Cadmium tolerance in *Brassica juncea* roots and shoots is affected by antioxidant status and phytochelatin biosynthesis. Plant Physiol Biochem 57:15–22

Molina AS, Nievas C, Chaca MVP, Garibotto F, Gonzalez U, Marsa SM, Luna C, Gimenez MS, Zirulnik F (2008) Cadmium-induced oxidative damage and antioxidative defense mechanisms in *Vigna mungo* L. Plant Growth Regul 56:285–295

Morant-Manceau A, Nguyen TLN, Pradier E, Tremblin G (2007) Carbonic anhydrase activity and photosynthesis in marine diatoms. Eur J Phycol 42:263–270

Moussa H, El-Gamal S (2010) Effect of salicylic acid pretreatment on cadmium toxicity in wheat. Biol Plant 54:315–320

Mucha AP, Almeida CMR, Bordalo AA, Vasconcelos MTSD (2005) Exudation of organic acids by a marsh plant and implications on trace metal availability in the rhizosphere of estuarine sediments. Estuar Coastal Shelf Sci 65:191–198

Muneer S, Qadri TN, Mahmooduzaffar A, Siddiqi TO (2011) Cytogenetic and biochemical investigations to study the response of *Vigna radiate* to cadmium stress. Afr J Plant Sci 5 (3):183–192

Nedjimi B, Daoud Y (2009) Cadmium accumulation in *Atriplex halimus* subsp. Schweinfurthii and its influence on growth, proline, root hydraulic conductivity and nutrient uptake. Flora 204:316–324

Nefic H, Musanovic J, Metovic A, Kurteshi K (2013) Chromosomal and nuclear alterations in root tip cells of *Allium cepa* L. induced by Alprazolam. Mediev Archaeol 67(6):388–392

Noriega GO, Balestrasse KB, Batlle A, Tomaro ML (2007) Cadmium induced oxidative stress in soybean plants also by the accumulation of δ- aminolevulinic acid. Biometals 20:841–851

Nouairi I, Ammar WB, Youssef NB, Daoud DBM, Ghorbal MH, Zarrouk M (2006) Comparative study of cadmium effects on membrane lipid composition of *Brassica juncea* and *Brassica napus* leaves. Plant Sci 170:511–519

Nursita AI, Singh B, Lees E (2009) Cadmium bioaccumulation in *Proisotoma minuta* in relation to bioavailability in soils. Ecotoxicol Environ Saf 72:1767–1773

Okumura R, Koizumi Y, Sekiya J (2003) Synthesis of hydroxymethylglutathione from glutathione and L-serine catalysed by carboxypeptidase. Biosci Biotechnol Biochem 67:434–437

Olmos E, Martinez-Solano JR, Piqueras A, Hellin E (2003) Early steps in the oxidative burst induced by cadmium in cultured tobacco cells (BY-2 line). J Exp Bot 54:291–301

Pagliano C, Ravioli M, Dalla VF, Gabbrielli R, Gonnelli C, Rascio N, Barbato R, La Rocca N (2006) Evidence for PS II donor-side type photoinhibition induced by cadmium treatment of rice (*Oryza sativa*). J Phtotchem Photobiol B Biol 84:70–78

Pal R, Rai JPN (2010) Phytochelatins: peptides involved in heavy metal detoxification. Appl Biochem Biotechnol 160:945–963

Park J, Song WY, Ko D, Eom Y, Hansen TH, Stokholm M, Lee TG, Martinoia E, Lee Y (2012) The phytochelatin transporters AtABCC1 and AtABCC2 mediate tolerance to cadmium and mercury. Plant J 69:278–288

Parmar P, Kumari N, Sharma V (2013) Structural and functional alterations in photosynthetic apparatus of plants under cadmium stress. Bot Stud 54:45–51

Parrotta L, Guerriero G, Sergeant K, Cai G, Hausman JF (2015) Target or barrier? The cell wall of early- and later-diverging plants vs cadmium toxicity: differences in the response mechanisms. Front Plant Sci 6:133

Pereira SIA, Lima AIG, Figueira EMAP (2006) Screening possible mechanisms mediating cadmium resistance in *Rhizobium leguminosarum* bv. Viciae isolated from contaminated Portuguese soils. Microb Ecol 52:176–186

Pérez-Chaca MV, Rodríguez-Serrano M, Molina AS, Pedranzani HE, Zirulnik F, Sandalio LM, Romero-Puertas MC (2014) Cadmium induces two waves of reactive oxygen species in *Glycine max* (L.) roots. Plant Cell Environ 37:1672–1687

Perfus-Barbeoch L, Leonhardt N, Vavasseur A, Forestier C (2002) Heavy metal toxicity: cadmium permeates through calcium channels and disturbs the plant water status. Plant J 32:539–548

Pietrini F, Innelli MA, Pasqualini S, Massacci A (2003) Interaction of cadmium with glutathione and photosynthesis in developing leaves and chloroplast of *Phragmites australis* (Cav.)Trin. ex. Steudel. Plant Physiol 133:829–837

Polge C, Jaquinod M, Holzer F, Bourguignon J, Walling L, Brouquisse R (2009) Evidence for the existence in *Arabidopsis thaliana* of the proteasome proteolytic pathway-activation in response to cadmium. J Biol Chem 284:35412–35424

Polle A, Klein T, Kettner C (2013) Impact of cadmium on young plants of *Populus euphratica* and *P. canescens*, two poplar species that differ in stress tolerance. New For 44(1):13–22

Poschenreider C, Barcelo J (2004) Water relations in heavy metal stressed plants. In: Prasad MNV (ed) Heavy metal stress in plants: from biomolecules to ecosystems, 2nd edn. Springer, Berlin, pp 249–270

Qiao K, Liang S, Wang F, Wang H, Hu Z, Chai T (2019) Effects of cadmium toxicity on diploid wheat (*Triticum urartu*) and the molecular mechanism of the cadmium response. J Hazard Mater 374:1–10

Rady MM (2011) Effect of 24-epibrassinolide on growth, yield, antioxidant system and cadmium content of bean (*Phaseolus vulgaris* L.) plants under salinity and cadmium stress. Sci Hortic 129:232–237

Rahman M, Haq N, Williams I (2016) Phytoaccumulation of arsenic, cadmium and lead by *Brassica juncea* parents and their F1 hybrids. J Environ Prot 7:613–622

Ramos J, Naya L, Gay M, Abian J, Becana M (2008) Functional characterization of an unusual phytochelatin synthase, LjPCS3, of *Lotus japonicus*. Plant Physiol 148:536–545

Ranieri A, Castagna A, Scebba F, Careri M, Zagnoni I, Predieri G, Pagliari M, Sanita di Toppi L (2005) Oxidative stress and phytochelatin characterization in bread wheat exposed to cadmium stress. Plant Physiol Biochem 43:45–54

Rascio N, Navari-Izzo F (2011) Heavy metal hyperaccumulating plants: how and why do they do it? And what makes them so interesting? Plant Sci 180:169–181

Rascio N, Vecchia FD, Rocca NL, Barbato R, Pagliano C, Raviolo M, Gonnelli C, Gabbrielli R (2008) Metal accumulation and damage in rice (cv. Vialone nano) seedlings exposed to cadmium. Environ Exp Bot 62:267–278

Rellan-Alvarez R, Ortega-Villasante C, Alvarez-Fernandez A, del Campo FF, Hernandez LE (2006) Stress responses of *Zea mays* to cadmium and mercury. Plant Soil 279:41–50

Rocha ACS, Almeida CMR, Basto MCP, Vasconcelos MTS (2016) Marsh plant response to metals: exudation of aliphatic low molecular weight organic acids (ALMWOAs). Estuar Coast Shelf Sci 171:77–84

Rodriguez-Serrano M, Romero-Puertas MC, Zabalza A, Corpas FJ, Gomez M, del Rio LA, Sandalio LM (2006) Cadmium effect on oxidative metabolism of pea (*Pisum sativum* L.)

roots. Imaging of reactive oxygen species and nitric oxide accumulation in vivo. Plant Cell Environ 29:1532–1544

Rodríguez-Serrano M, Romero-Puertas MC, Pazmiño DM, Testillano PS, Risueño MC, Del Río LA, Sandalio LM (2009) Cellular response of pea plants to cadmium toxicity: cross talk between reactive oxygen species, nitric oxide, and calcium. Plant Physiol 150:229–243

Roldan-Arjona T, Ariza RR (2009) Repair and tolerance of oxidative DNA damage in plants. Mutat Res 681:169–179

Romero-Puertas MC, Rodriguez-Serrano M, Corpas FJ, Gomez M, del Rio LA, Sandalio LM (2004) Cadmium-induced subcellular accumulation of $O_2^{\bullet-}$ and $H_2O_2$ in pea leaves. Plant Cell Environ 27:1122–1134

Rouhier N, Lemaire SD, Jacquot JP (2008) The role of glutathione in photosynthetic organisms: emerging functions for glutaredoxins and glutathionylation. Annu Rev Plant Biol 59:143–166

Roychoudhury A, Basu S, Sengupta DN (2012) Antioxidants and stress-related metabolites in the seedlings of two indica rice varieties exposed to cadmium chloride toxicity. Acta Physiol Plant 34(3):835–847

Sabrine H, Afif H, Mohamed B, Hamadi B, Maria H (2010) Effects of cadmium and copper on pollen germination and fruit set in pea (*Pisum sativum* L.). Sci Hortic 125:551–555

Sakamoto A, Murata N (2000) Genetic engineering of glycine betaine synthesis in plants: current status and implications for enhancement of stress tolerance. J Exp Bot 51:81–88

Saraswat S, Rai JPN (2011) Prospective application of *Leucaena leucocephala* for phytoextraction of Cd and Zn and nitrogen fixation in metal polluted soils. Int J Phytoremediation 13:271–288

Sarwar N, Saifullah MSS, Zia MH, Naeem A, Bibi S, Farid G (2010) Role of mineral nutrition in minimizing cadmium accumulation by plants. J Sci Food Agric 90:925–937

Satarug S, Baker JR, Urbenjapol S, Haswell-Elkins M, Reilly PEB, Williams DJ, Moore MR (2003) A global perspective on cadmium pollution and toxicity in non-occupationally exposed population. Toxicol Lett 137:65–83

Sauve S (2003) The role of chemical speciation in bioavailability. In: Naidu R, Gupta VVSR, Rogers S, Kookana RS, Bolan NS, Adriano DC (eds) Bioavailability: toxicity and risk relationships in ecosystems. India Book House Pvt. Ltd., Chennai, pp 59–82

Sauve S, Norvell WA, McBride M, Hendershot W (2000a) Speciation and complexation of cadmium in extracted soil solutions. Environ Sci Technol 34:291–296

Sauve S, Hendershot W, Allen HE (2000b) Solid-solution partitioning of metals in contaminated soils: dependence on pH, total metal burden, and organic matter. Crit Rev Environ Sci Technol 34:1125–1131

Schwarzländer M, Fricker MD, Sweetlove LJ (2009) Monitoring the in vivo redox state of plant mitochondria: effect of respiratory inhibitors, abiotic stress and assessment of recovery from oxidative challenge. Biochim Biophys Acta 1787:468–475

Sebastian A, Prasad MNV (2014a) Cadmium minimization in rice. A review. Agron Sustain Dev 34:155–173

Sebastian A, Prasad MNV (2014b) Vertisol prevents cadmium accumulation in rice: analysis by ecophysiological toxicity markers. Chemosphere 108:85–92

Sebastian A, Nangia A, Prasad MNV, Rattanapolsan L, Nakbanpote W (2019) Cadmium toxicity and tolerance in micro- and phytobiomes. In: Hasanuzzaman M, Prasad MNV, Fujita M (eds) Cadmium toxicity and tolerance in plants. Academic, San Diego, pp 19–46

Semane B, Cuypers A, Smeets K, Belleghem FV, Horemans N, Schat H, Vangronsveld J (2007) Cadmium responses in *Arabidopsis thaliana*: glutathione metabolism and antioxidative defence system. Physiol Plant 129:519–528

Shah K (2011) Cadmium metal detoxification and hyperaccumulators. In: Sherameti I, Varma A (eds) Detoxification of heavy metals, soil biology. Springer, Berlin, pp 181–203

Shanmugaraj BM, Harish MC, Balamurugan S, Sathishkumar R (2013) Cadmium induced physio-biochemical and molecular response in *Brassica juncea*. Int J Phytoremediation 15:206–218

Shao HB, Jiang SY, Li FM, Chu LY, Zhao CX, Shao MA, Zhao XN, Li F (2007) Some advances in plant stress physiology and their implications in the systems biology era. Colloid Surf B 54 (1):33–36

Sharma SS, Dietz KJ (2006) The significance of amino acids and amino acid-derived molecules in plant responses and adaptation to heavy metal stress. J Exp Bot 57:711–726

Sheirdil RA, Bashir K, Hayat R, Akhtar MS (2012) Effect of cadmium on soybean (*Glycine max* L.) growth and nitrogen fixation. Afr J Biotechnol 11(8):1886–1891

Shekhar CC, Sammaiah D, Rambabu M, Reddy KJ (2011) Effect of cadmium on tomato growth and yield attributes. J Microbiol Biotechnol Res 1(3):109–112

Shetty PJ, Atallah MT, Shetty K (2002) Effect of UV treatment on proline-linked pentose phosphate pathway for phenolic and L-DOPA synthesis in dark germinated *Vicia faba*. Process Biochem 37:1285–1295

Shukla D, Trivedi PK, Nath P, Tuteja N (2016) Metallothioneins and phytochelatins: role and perspectives in heavy metal(loid)s stress tolerance in crop plants. In: Tuteja N, Gill SS (eds) Abiotic stress response in plants. Wiley, Weinheim, pp 237–264

Sidhu GPS, Bali AS, Bhardwaj R (2019) Role of organic acids in mitigating cadmium toxicity in plants. In: Hasanuzzaman M, Prasad MNV, Nahar K (eds) Cadmium tolerance in plants: agronomic, molecular, signaling, and omic approaches. Elsevier, London, pp 255–279

Sigfridsson KGV, Bernat G, Mamedov F, Styring S (2004) Molecular interference of $Cd^{2+}$ with photosystem II. Biochim Biophys Acta 1659:19–31

Singh S, Parihar P, Singh R, Singh VP, Prasad SM (2016) Heavy metal tolerance in plants: role of transcriptomics, proteomics, metabolomics, and ionomics. Front Plant Sci 6:1143

Skorzynska-Polit E (2007) Lipid peroxidation in plant cells, its physiological role and changes under heavy metal stress. Acta Soc Bot Pol 76(1):49–54

Smeets K, Ruytinx J, Semane B, Belleghem FV, Remans T, Sanden SV, Vangronsveld J, Cuypers A (2008) Cadmium-induced transcriptional and enzymatic alterations related to oxidative stress. Environ Exp Bot 63:1–8

Smeets K, Opdenakker K, Remans T, Sanden SV, Belleghem FV, Semane B, Horemans N, Guisez Y, Vangronsveld J, Cuypers A (2009) Oxidative stress-related responses at transcriptional and enzymatic levels after exposure to Cd or Cu in a multipollution context. J Plant Physiol 166:1982–1992

Smiri M, Chaoui A, El Ferjani E (2009) Respiratory metabolism in the embryonic axis of germinating pea seed exposed to cadmium. J Plant Physiol 166:259–269

Smiri M, Chaoui A, Rouhier N, Gelhaye E, Jacquot JP, El Ferjani E (2010a) Effect of cadmium on resumption of respiration in cotyledons of germinating pea seeds. Ecotoxicol Environ Saf 73:1246–1254

Smiri M, Chaoui A, Rouhier N, Gelhaye E, Jacquot JP, Ferjani EE (2010b) Redox regulation of the glutathione reductase/iso-glutaredoxin system in germinating pea seed exposed to cadmium. Plant Sci 179:423–436

Smiri M, Choui A, Rouhier N, Kamel C, Gelhaye E, Jacquot JP, El-Ferjani E (2010c) Cadmium induced mitochondrial redox changes in germinating pea seed. Biometals 23:973–984

Son KH, Kim DY, Koo N, Kim KR, Kim JG, Owens G (2012) Detoxification through phytochelatin synthesis in *Oenothera odorata* exposed to Cd solutions. Environ Exp Bot 75:9–15

Song A, Li Z, Zhang J, Xue G, Fan F, Liang Y (2009) Silicon-enhanced resistance to cadmium toxicity in *Brassica chinensis* L. is attributed to Si-suppressed cadmium uptake and transport and Si-enhanced antioxidant defense capacity. J Hazard Mater 172:74–83

Song Y, Jin L, Wang X (2017) Cadmium absorption and transportation pathways in plants. Int J Phytoremediation 19:133–141

Souguir D, Erjani E, Ledoigt G, Goupil P (2011) Sequential effects of cadmium on genotoxicity and lipoperoxidation in *Vicia faba* roots. Ecotoxicology 20:329–336

Srivastava R, Khan R, Manzoor N, Mahmooduzzafar A (2011) Responses of cadmium exposures on growth, physio-biochemical characteristics and the antioxidative defence system of soybean (*Glycine max* L.). J Phytology 3(10):20–25

Street RA, Kulkarni MG, Stirk WA, Southway C, van Staden J (2010) Effect of cadmium on growth and micronutrient distribution in wild garlic (*Tulbaghia violacea*). South Afr J Bot 76:332–336

Su L, Zhang Y, Lin X, Luo A (2000) Cadmium toxicity and tolerance in vascular plants. Plant Nutr Fertiliz Sci 6(1):106–112

Sun Y, Zhou Q, Wang L, Liu W (2009) The influence of different growth stages and dosage of EDTA on Cd uptake and accumulation in Cd-hyperaccumulator (*Solanum nigrum* L.). Bull Environ Contam Toxicol 82:348–353

Szakova J, Tlustos P, Balik J, Pavlikova D, Vanek V (1999) The sequential analytical procedure as a tool for evaluation of As, Cd and Zn mobility in soil. Fresenius J Anal Chem 363:594–595

Szalai G, Krantev A, Yordanova R, Popova LP, Janda T (2013) Influence of salicylic acid on phytochelatin synthesis in *Zea mays* during Cd stress. Turk J Bot 37(4):708–714

Szolnoki Z, Farsang A, Puskás I (2013) Cumulative impacts of human activities on urban garden soils: origin and accumulation of metals. Environ Pollut 177:106–115

Tamas L, Valentovicova K, Haluskova L, Huttova J, Mistrik I (2009) Effect of cadmium on the distribution of hydroxyl radical, superoxide and hydrogen peroxide in barley root tip. Protoplasma 236:67–72

Tao Q, Hou D, Yang X, Li T (2016) Oxalate secretion from the root apex of *Sedum alfredii* contributes to hyperaccumulation of Cd. Plant Soil 398:139–152

Tezotto T, Favarin JL, Azevedo RA, Alleoni LRF, Mazzafera P (2012) Coffee is highly tolerant to cadmium, nickel and zinc: plant and soil nutritional status, metal distribution and bean yield. Field Crop Res 125:25–34

Tian S, Lua L, Zhang J, Wang K, Brown P, He Z, Liang J, Yang X (2011) Calcium protects roots of *Sedum alfredii* H. against cadmium-induced oxidative stress. Chemosphere 84(1):63–69

Tlustoš P, Száková J, Hrubý J, Hartman I, Najmanová J, Nedělník J, Pavlíková D, Batysta M (2006a) Removal of As, Cd, Pb, and Zn from contaminated soil by high biomass producing plants. Plant Soil Environ 52:413–423

Tlustos P, Szakova J, Korinek K, Pavlikova D, Hanc A, Balik J (2006b) The effect of liming on cadmium, lead and zinc uptake reduction by spring wheat grown in contaminated soil. Plant Soil Environ 52:16–24

Trovato M, Mattioli R, Costantino P (2008) Multiple roles of proline in plant stress tolerance and development. Rendiconti Lincei 19:325–346

Uraguchi S, Watanabe I, Yoshitomi A, Kiyono M, Kuno K (2006) Characteristics of cadmium accumulation and tolerance in novel Cd-accumulating crops, *Avena strigosa* and *Crotalaria juncea*. J Exp Bot 57:2955–2965

Uraguchi S, Kiyono M, Sakamoto T, Watanabe I, Kuno K (2009) Contributions of apoplasmic cadmium accumulation, antioxidative enzymes and induction of phytochelatins in cadmium tolerance of the cadmium-accumulating cultivar of black oat (*Avena strigosa* Schreb.). Planta 230:267–276

Vanlerberghe GC (2013) Alternative oxidase: a mitochondrial respiratory pathway to maintain metabolic and signalling homeostasis during abiotic and biotic stress in plants. Int J Mol Sci 14:6805–6847

Vassilev A, Lidon F (2011) Cd-induced membrane damages and changes in soluble protein and free amino acid contents in young barley plants. Emir J Food Agric 23(2):130–136

Vazquez S, Goldsbrough P, Carpena RO (2006) Assessing the relative contributions of phytochelatins and the cell wall to cadmium resistance in white lupin. Physiol Plant 128:487–495

Verbruggen N, Hermans C (2008) Proline accumulation in plants: a review. Amino Acids 35:753–759

Vijayaragavan M, Prabhahar C, Sureshkumar J, Natarajan A, Vijayarengan P, Sharavanan S (2011) Toxic effect of cadmium on seed germination, growth and biochemical contents of cowpea (*Vigna unguiculata* L.) plants. Int Multidisciplinary Res J 1(5):1–6
Wachter A, Rausch T (2005) Regulation of glutathione (GSH) synthesis in plants: novel insight from *Arabidopsis*. FAL Agric Res 283:149–155
Wahid A, Ghani A (2008) Varietal differences in mungbean (*Vigna radiata*) for growth, yield, toxicity symptoms and cadmium accumulation. Ann Appl Biol 152:59–69
Wahid A, Ghani A, Javed F (2008) Effect of cadmium on photosynthesis, nutrition and growth of mungbean. Agron Sustain Dev 28:273–280
Wan G, Najeeb U, Jilani G, Naeem MS, Zhou W (2011) Calcium invigorates the cadmium-induced *Brassica napus* L. plants by strengthening their photosynthetic system. Environ Sci Pollut Res Int 18(9):1478–1486
Wang H, Zhao SC, Liu RL, Zhou W, Jin JY (2009) Changes of photosynthetic activities of maize (*Zea mays* L.) seedlings in response to cadmium stress. Photosynthetica 47(2):277–283
Wang L, Cui X, Cheng H, Chen F, Wang J, Zhao X, Lin C, Pu X (2015) A review of soil cadmium contamination in China including a health risk assessment. Environ Sci Pollut Res 22(21):1–12
Wang AY, Wang MY, Liao Q, He XQ (2016) Characterization of Cd translocation and accumulation in 19 maize cultivars grown on Cd contaminated soil: Implication of maize cultivar selection for minimal risk to human health and for phytoremediation. Environ Sci Pollut Res 23:5410–5419
Wang W, Cang L, Zhou D, Yu Y (2017) Exogenous amino acids increase antioxidant enzyme activities and tolerance of rice seedlings to cadmium stress. Environ Prog Sustain Energy 36:155–161
Wani PA, Khan MS, Zaidi A (2006) An evaluation of the effects of heavy metals on the growth, seed yield, and grain protein of lentil in pots. Ann Appl Biol 27:23–24
Wei S, Li Y, Zhan J, Wang S, Zhu J (2012) Tolerant mechanisms of *Rorippa globosa* (Turcz.) Thell. hyperaccumulating Cd explored from root morphology. Bioresour Technol 118:455–459
Whiteside JR, Box CL, McMillan TJ, Allinson SL (2010) Cadmium and copper inhibit both DNA repair activities of polynucleotide kinase. DNA Repair 9:83–89
Wojcik M, Tukiendorf A (2011) Glutathione in adaptation of *Arabidopsis thaliana* to cadmium stress. Biol Plant 55(1):125–132
Wojcik M, Vangronsveld J, Tukiendorf A (2005) Cadmium tolerance in *Thlaspi caerulescens* I. Growth parameters, metal accumulation and phytochelatin synthesis in response to cadmium. Environ Exp Bot 53:151–161
Wong JWC, Selvam A (2006) Speciation of heavy metals during co-composting of sewage sludge with lime. Chemosphere 63(6):980–986
Wu YS, Tang KX (2004) MAP kinases cascades responding to environmental stress in plants. Acta Bot Sin 46(2):127–136
Wu FB, Zhang GP (2002) Genotypic variation in kernel heavy metal concentrations in barley and as affected by soil factors. J Plant Nutr 25:1163–1173
Xie M, Chen W, Lai X, Dai H, Sun H, Zhou X, Chen T (2019) Metabolic responses and their correlations with phytochelatins in *Amaranthus hypochondriacus* under cadmium stress. Environ Pollut 252:1791–1800
Xu JK, Yang LX, Wang YL, Wang ZQ (2005) Advances in the study uptake and accumulation of heavy metal in rice (*Oryza sativa*) and its mechanisms. Chin Bull Bot 22:614–622
Yamaguchi H, Fukuoka H, Arao T, Ohyama A, Nunome T, Miyatake K, Negoro S (2009) Gene expression analysis in cadmium-stressed roots of a low cadmium-accumulating solanaceous plant, *Solanum torvum*. J Exp Bot 61:423–437
Yan S, Hui Y, Zebin W, Xuhui K, Qi-Tang W (2017) Root cell wall and phytochelatins in low-cadmium cultivar of *Brassica parachinensis*. Pedosphere 2017:1–12
Yang X, Feng Y, He Z, Stoffella PJ (2005) Molecular mechanisms of heavy metal hyperaccumulation and phytoremediation. J Trace Elem Med Biol 18:339–353

Yang HY, Shi GX, Xu QS, Wang HX (2011) Cadmium effects on mineral nutrition and stress-related indices in *Potamogeton crispus*. Russ J Plant Physiol 58(2):253–260

Yang GY, Wang C, Wang YC, Guo YC, Zhao YL, Yang CP, Gao CQ (2016) Overexpression of ThVHAc1 and its potential upstream regulator, ThWRKY7, improved plant tolerance of cadmium stress. Sci Rep 6:18752

Yang Y, Chen J, Huang Q, Tang S, Wang J, Hu P, Shao G (2018) Can liming reduce cadmium (Cd) accumulation in rice (*Oryza sativa*) in slightly acidic soils? A contradictory dynamic equilibrium between Cd uptake capacity of roots and Cd immobilisation in soils. Chemosphere 193:547–556

Yannarelli GG, Fernandez-Alvarez AJ, Santa-Cruz DM, Tomaro ML (2007) Glutathione reductase activity and isoforms in leaves and roots of wheat plants subjected to cadmium stress. Phytochemistry 68:505–512

Ying RR, Qiu RL, Tang YT, Hu PJ, Qiu H, Chen HR, Shi TH, Morel JL (2010) Cadmium tolerance of carbon assimilation enzymes and chloroplast in Zn/Cd hyperaccumulator *Picris divaricata*. J Plant Physiol 167:81–87

Younis M (2007) Responses of *Lablab purpureus-rhizobium symbiosis* to heavy metals in pot and field experiments. World J Agric Sci 3(1):111–122

Younis U, Malik SA, Rizwan M, Qayyum MF, Ok YS, Shah MHR, Rehman RA, Ahmad N (2016) Biochar enhances the cadmium tolerance in spinach (*Spinacia oleracea*) through modification of Cd uptake and physiological and biochemical attributes. Environ Sci Pollut Res 23:21385–21394

Zhang S, Zhang H, Qin R, Jiang W, Liu D (2009) Cadmium induction of lipid peroxidation and effects on root tip cells and antioxidant enzyme activities in *Vicia faba* L. Ecotoxicology 18:814–823

Zhang ZC, Chen BX, Qiu BS (2010) Phytochelatin synthesis plays a similar role in shoots of the cadmium hyperaccumulator *Sedum alfredii* as in non-resistant plants. Plant Cell Environ 33 (8):1248–1255

Zhang L, Zhang H, Guo W, Tian Y, Chen Z, Wei X (2012) Photosynthetic responses of energy plant maize under cadmium contamination stress. Adv Mater Res 356-360:283–286

Zhang H, Lv S, Xu H, Hou D, Li Y, Wang F (2017) $H_2O_2$ is involved in the metallothionein-mediated rice tolerance to copper and cadmium toxicity. Int J Mol Sci 18:1–12

Zhigang A, Cuijie L, Yuangang Z, Yejie D, Wachter A, Gromes R, Rausch T (2006) Expression of BjMT2, a metallothionein 2 from *Brassica juncea*, increases copper and cadmium tolerance in *Escherichia coli* and *Arabidopsis thaliana*, but inhibits root elongation in *Arabidopsis thaliana* seedlings. J Exp Bot 57:3575–3582

Zhou J, Cheng K, Huang G, Chen G, Zhou S, Huang Y, Zhang J, Duan H, Fan H (2020) Effects of exogenous 3-indoleacetic acid and cadmium stress on the physiological and biochemical characteristics of *Cinnamomum camphora*. Ecotoxicol Environ Saf. 191:109998

Zhu G, Xiao H, Guo Q, Zhang Z, Zhao J, Yang D (2018) Effects of cadmium stress on growth and amino acid metabolism in two Compositae plants. Ecotoxicol Environ Saf 158:300–308

Zimeri AM, Dhankher OP, McCaig B, Meagher RB (2005) The plant MT1 metallothioneins are stabilized by binding cadmiums and are required for cadmium tolerance and accumulation. Plant Mol Biol 58:839–855

Zoghlami LB, Djebali W, Abbes Z, Hediji H, Maucourt M, Moing A, Brouquisse R, Chaibi W (2011) Metabolite modifications in *Solanum lycopersicum* roots and leaves under cadmium stress. Afr J Biotechnol 10(4):567–579

Zouari M, Ahmed CB, Zorrig W, Elloumi N, Rabhi M, Delmail D, Rouina BB, Labrousse P, Abdallah FB (2016) Exogenous proline mediates alleviation of cadmium stress by promoting photosynthetic activity, water status and antioxidative enzymes activities of young date palm (*Phoenix dactylifera* L.). Ecotoxicol Environ Saf 128:100–108

# Cadmium: Uptake in Plants and Its Alleviation Via Crosstalk Between Phytohormones and Sulfur

15

Harmanjit Kaur and Sofi Javed Hussain

**Abstract**

Addition of heavy metals to arable soils is becoming a leading issue worldwide owing to broad array of unobstructed and nonstop anthropogenic activities, particularly in developing countries. Among the various heavy metals, cadmium (Cd) nowadays is becoming a daunting challenge for agricultural production and human health globally. Enhanced level of Cd in soils causes deleterious effects on flora as well as fauna due to its access into the food web via uptake and accumulation in plants, especially agricultural crops. Various transporters are engaged in Cd uptake in plants, for instance, Fe-regulated transporters (IRT) like proteins, Zn-regulated transporters (ZRT), ATP-binding cassette transporters (ABC), natural resistance-associated macrophage proteins (NRAMP), P-type ATPase, to name a few. Phytohormones regulate various signalling pathways and play an essential role in various processes associated with growth and development of plants, hence assist in mitigating Cd-stress. Additionally, sulfur (S) being a component of various antioxidant metabolic compounds functions in increasing plant Cd tolerance. Alleviation of Cd-mediated noxious effects in plants through up-regulation of different enzymatic as well as non-enzymatic antioxidants is well established. Apart from the self-defence mechanisms operative in plants, exogenous application of phytohormones and S is an essential and sustainable approach in alleviating Cd toxicity. In the present chapter, we focus on the routes and transporters via which Cd is taken up by plants and mechanisms of crosstalk between different phytohormones and S in ameliorating detrimental effects of Cd on plants are also highlighted.

H. Kaur (✉)
Department of Botany, Akal University, Talwandi Sabo, Bathinda, Punjab, India

S. J. Hussain
Department of Botany, Aligarh Muslim University, Aligarh, Uttar Pradesh, India



K. Mishra et al. (eds.), *Sustainable Solutions for Elemental Deficiency and Excess in Crop Plants*, https://doi.org/10.1007/978-981-15-8636-1_15

**Keywords**

Antioxidants · Cadmium · Phytohormones · Sulfur · Transporters

## 15.1 Introduction

Agricultural plants are confronted with several abiotic and biotic stresses, which limit their productivity. Among various abiotic stresses, heavy metal toxicity has been rising at an alarming pace and has developed into a universal concern in the modern age (Ahmad et al. 2015). Heavy metals are primarily noxious, either at all levels (e.g., Hg, Pb, Cd, and As) or above specific threshold levels (e.g., Mg, Fe, Co, Cu, Zn, Ca, Mn, and K) and damage the natural ecosystems, by negatively influencing crop growth and productivity as well as soil richness (Mudgal et al. 2010). Nearly 60% of the worldwide farming soils have mineral problems (Cakmak 2002), for instance, metal toxicities together with nutritional deficiencies (Hajiboland and Farhanghi 2010; Khan et al. 2014a). Soils contaminated with Cd are particularly detrimental for human health as Cd significantly accumulates in the food chain (Gall et al. 2015) via agricultural crops which are the crucial cause of poisoning in humans (Gill and Tuteja 2011). It is estimated that out of 30,000 tonnes yearly addition of Cd to environment, 13,000 is added through several anthropogenic actions (Gallego et al. 2012). At the expense of similar ionic radii with other essential inorganic elements, Cd gains entry into the plant cells, i.e. Cd competes with other essential elements (Sharma and Dietz 2009; Lin and Aarts 2012). Cd is quickly taken up by roots of plants and transferred to the aerial parts, with the assistance of many transporters (Verbruggen et al. 2009; Lux et al. 2011), resulting in cellular, molecular, biochemical, and physiological alterations that affect plant growth and morphology (Song et al. 2017). The up-coming molecular knowledge of Cd uptake and retention in root as well as translocation from root-to-shoot would facilitate the development of crops with low Cd-accumulating ability (Clemens et al. 2013).

Several plans of action have been designed for alleviation of Cd toxicity in plants, among which plant growth regulators (PGRs, viz. salicylic acid, ethylene, brassinosteroids, jasmonic acid, and nitric oxide) are the key macroorganic molecules which play a significant role in mitigation of Cd stress in plants. PGRs are innate chemical substances needed in small amounts, and exert invigorating or inhibitory influence on plants (Bali et al. 2019). These plant hormones are considered to take part in signal transduction during stress conditions (Ke et al. 2015; Fujita et al. 2011). The primary essential mineral nutrients, including nitrogen (N), phosphorus (P), and potassium (K), are needed for growth, development, and productivity of plants under normal conditions. N-, P-, and K-mediated enhanced abiotic stress tolerance has been described in different plants (Anjum et al. 2010). Among the various essential mineral nutrients required by the plants, sulfur (S) comes fourth after N, P, and K. S in plant kingdom is an important constituent of many biologically active compounds including vitamins, coenzymes, phytohormones, antioxidants, thiols (sulfolipids), secondary S-compounds (alliins, glucosinolates, phytochelatins), etc. (Khan et al. 2008; Anjum et al. 2014). Plants acquire S in the form of sulfates, which is reduced to sulfide in the chloroplasts of the plant cells to synthesize cysteine (reduced S-donor). S-assimilation benefits plants through the generation of reduced S-compounds (such as glutathione, GSH), which impart tolerance against abiotic stress factors (Khan et al. 2013). It has been suggested that different PGRs crosstalk with other essential nutrients like S, hence shows a positive response towards Cd toxicity (Masood et al. 2012; Asgher et al. 2014). Having acknowledged the significance of phytohormones and S and the negative influence of Cd on plants, it becomes necessary to comprehend the interplay between PGRs and S in Cd stress mitigation. With this backdrop, the present chapter summarizes the mechanism of Cd uptake and translocation in conjunction with transporter proteins along with its alleviation via crosstalk between various PGRs and S in plants, with a goal intended to improve plant resistance to Cd stress.

## 15.2 Cd Uptake, Translocation, and Accumulation: Role of Transporters

Uptake of Cd by roots and its subsequent translocation to the aerial parts are the two main physiological processes correlated with accumulation of Cd in plants (Clemens 2006). Cd absorption from soil is largely in the form of $Cd^{2+}$ ions. Cd uptake can also occur as inorganic ($CdSO_4$, $CdCl_2$, $CdCl^+$) or organic complexes (phytometallophore) (Irfan et al. 2013). Absorption of Cd through the root cell membrane is regulated by electro-chemical gradient variation between the activity of Cd in cytoplasm and in apoplasts of root. The huge negative potential of membrane, which might go above −200 mV in epidermal cells of roots, is responsible for providing a strong driving force for the uptake of cations even at small concentrations (Benavides et al. 2005; Hasan et al. 2009). Entry of Cd into the roots largely takes place via root apical cell plasma membrane which is believed to be concentration-dependent process showing saturable kinetics linked to a linear factor in wheat (Hart et al. 2002; Harris and Taylor 2004), *Noccaea caerulescens* (Zhao et al. 2002), maize (Han et al. 2006), and *Arabidopsis halleri* (Zhao et al. 2006). Entry of $Cd^{2+}$ into the epidermal layer of root mainly involves three pathways: (i) exchange of $H^+$ (released from dissociation of $H_2CO_3$) with $Cd^{2+}$ and its subsequent entry via apoplast pathway (Yamaguchi et al. 2011); (ii) via ion channels ($Ca^{2+}$, $Zn^{2+}$, $Fe^{2+}$) and through symplast pathway (Sadana et al. 2003); and (iii) by means of yellow-stripe 1-like (YSL) proteins as Cd–ligand complexes (Curie et al. 2009). Cd can be transported via roots, stems, and leaves through both apoplast and symplast pathways (Song et al. 2017). $Cd^{2+}$ builds up in the root cells or can pass through the cells of cortex (Yamaguchi et al. 2011). Cortical cell walls and casparian strips of the endodermal and exodermal layers play a vital role in regulating Cd transport in apoplast (Redjala et al. 2011). Small quantity of $Cd^{2+}$ traverses the tonoplast and accumulates in the vacuole, whereas remaining progress in the symplasm from one cell to another through plasmodesmata until they enter pericycle cells and are loaded into the xylem (Salt et al. 1995). Radial symplastic transfer and fast loading into xylem are needed for effective uptake of Cd as well as its translocation in plants (Verbruggen et al. 2009). Absorption of Cd and its subsequent loading into xylem are considered to be metabolically active and energy-dependent processes (Mori et al. 2009). Majority of $Cd^{2+}$ is held inside the roots with small amount being translocated to the shoots (Benzarti et al. 2010). Moreover, uptake of Cd into roots is believed to be comparatively fast, while its translocation to aerial parts is slower (Zhang et al. 2005). Alfalfa, chicory, pumpkin, and red beets build up more Cd in leaves than roots, whereas barley, common beans, maize, and wheat accumulate more Cd in roots compared to leaves (Sękara et al. 2005; Lysenko et al. 2015; Dobrikova et al. 2017). Moreover, broad leaved plants accumulate higher Cd than thin leaved ones (Wang et al. 2014).

Root-to-shoot Cd translocation by and large takes place through xylem and driving force is transpiration from the leaves (Lu et al. 2009). Not much is known about impregnation of cell wall with suberin in the course of development of exodermis and endodermis, which influences plasticity and limits Cd movement into the stele (Lux et al. 2011). Subsequent to the entry of Cd into the stele, it can re-enter the apoplast prior to xylem loading (Hu et al. 2009). In this regard, Akhter et al. (2014) found elevated Cd signals in walls of the xylem vessels of barley. Cd can also be reallocated all through the plant body by phloem. Even Cd transportation into seeds or grains occurs via phloem (Tanaka et al. 2007). Position-emitting tracer imaging system (PETIS) tests have revealed the significance of phloem in Cd allocation in rice and translocation to the grains (Fujimaki et al. 2010). Mendoza-Cózatl et al. (2008) detected increased phytochelatins (PCs), glutathione (GSH), and Cd levels in phloem sap of *Brassica napus* and proposed that besides xylem, phloem is also a pathway for the long-distance transport of Cd–PC and Cd–GSH complexes. Even cysteine levels in the phloem have been found to increase after Cd exposure, indicating its function in the long-distance transport of Cd (Jozefczak et al. 2012). Various researchers have also advocated the role of organic acids in enhancing Cd translocation to plant shoots (Ehsan et al. 2014; Mnasri et al. 2015; Al Mahmud et al. 2018).

Cd allocation inside plant is affected by its translocation from roots to aerial organs by xylem, unloading from xylem to phloem loading and transfer via phloem from sources to sinks (Riesen and Feller 2005). Cd uptake and accumulation differ not only between plant species but also among cultivars/genotypes (Barman et al. 2020). Since Cd uptake via roots is the primary stage of Cd accumulation, root uptake attributes explain variations in Cd-accumulating ability between plant species, cultivars, and ecotypes (Redjala et al. 2009; Xin et al. 2017). Among varied plant parts, Cd content diminishes in the order roots > leaves > fruits > seeds (Sharma et al. 2006; Zhang et al. 2008). Although Cd is easily taken up, its limited translocation from roots to stems and to leaves as well as fruits causes lesser contents in tubers, fruits, and grains as compared to roots or foliage (Harris and Taylor 2004; Arao and Ishikawa 2006). Variations among plant species with respect to seed Cd content may be partially linked to the variations in abilities of plants to control transfer of Cd from xylem to phloem and from phloem to kernels (Tanaka et al.

2007). Higher accrual of Cd in grains could be correlated with larger total Cd uptake by plant (Stolt et al. 2003). However, according to Greger and Lofstedt (2004), wheat grain Cd content correlated with shoot Cd content as well as with root-to-shoot translocation during vegetative period. Conversely, Rodda et al. (2011) observed uptake of Cd from root to rice grain at post-flowering phase and suggested it as a chief source of grain Cd.

Reallocation of Cd in developing cereals is believed to be a vital physiological process affecting the quality of grain yield (Chen et al. 2007). In this context, Sankaran and Ebbs (2008) expressed that concentration of Cd provided to the roots of Indian mustard during seed maturation was most important in influencing Cd content in seeds. However, Kashiwagi et al. (2009) advocated distinct routes of Cd translocation and accumulation prior and subsequent to heading in rice as one of the deciding parameters of Cd contents in leaves or grains. Furthermore, Uraguchi et al. (2009) proposed root-to-shoot transfer of Cd through xylem as the main physiological process influencing accumulation of Cd in leaves as well as grains of rice plants. It seems that Cd absorption and translocation processes are part of a multifaceted "firewall" system which functions in restricting translocation of Cd from root to aerial parts, attaining different equilibriums depending on external Cd concentration (Nocito et al. 2011).

Several transition metal transporters have been recognized in plants, of which six families are engaged in Cd uptake and translocation, viz. ZRT (Zn-regulated transporter) Fe- regulated transporter (IRT)-like proteins (ZIP), ABC (ATP-binding cassette transporter), NRAMP (natural resistance-associated macrophage proteins), CDF (cation diffusion facilitator), P-type ATPase, and CAX (cation/$H^+$ exchangers) (Lin and Aarts 2012). The contribution of these transporters in Cd uptake has been reported mostly in rice and *Arabidopsis*.

### 15.2.1 ZIPs

ZIPs are regarded as one of the most important group of proteins engaged in uptake of Zn in plants and few of them also transport Cd (Grotz and Guerinot 2006). ZIPs have been discovered in both dicots as well as monocots, for example, *Arabidopsis*, barley, maize, *Medicago,* and rice (Zheng et al. 2018). Besides ZIP1; ZIP3, ZIP4, ZIP5, ZIP9, ZIP10, ZIP12, and IRT3 genes are engaged in Zn/Cd uptake (Lin and Aarts 2012). Yeast *ZRT1* gene encodes a high affinity transport system which can transport Cd (Gomes et al. 2002). By complementing a yeast mutant (having non-functional Zn transportation system) with *Thlaspi caerulescens* cDNA library, Lasat et al. (2000) cloned the ZNT1 cDNA responsible for encoding a high affinity Zn transporter. Nevertheless, ZNT1/Tc ZNT1 can also assist low affinity Cd transport (Lasat et al. 2000; Pence et al. 2000). IRT1 was accountable for considerable enhancement in Cd uptake by ecotype Ganges of *T. caerulescens*, but not in Prayon ecotype, which had less ability to hyperaccumulate Cd (Lombi et al. 2002). Furthermore, IRT-1 transgenic *Arabidopsis* accumulated higher Cd than wild-type plants (Connolly et al. 2002). In rice genome, 17 ZIPs have been identified. OsIRT1 and

OsIRT2 operate in Cd uptake from soil (Nakanishi et al. 2006) and over-expression of OsIRT1 enhances Cd accumulation (Lee and An 2009). A recent comparative study revealed differences in gene expression of ZIPs between rice and *Arabidopsis* in response to Cd. In rice, mostly the shoot ZIPs were up-regulated, whereas in *Arabidopsis* it was largely the root ZIP genes (Zheng et al. 2018).

### 15.2.2 ABCs

ATP-binding cassette (ABC) transporters are Mg ATP-dependent pumps which are independent of vacuolar proton gradient (Rea 2007). The main groups of these transporters are the multidrug resistance-related proteins (MRP), multidrug-resistance proteins (MDR) or P-glycoproteins (PGP), and products of the pleiotropic drug resistance genes (PDR) (Martinoia et al. 2002). ABC transporters have minimum one transmembrane subunit and a nucleotide binding domain which assists in hydrolysis of ATP for Cd uptake. Yeast Cd factor (YCF1), belonging to MRP family, can mediate Cd resistance. It is a MgATP-invigorated vacuolar transporter which sequester compounds subsequent to their S-conjugation with GSH. The hydrophobic N-terminal enlargement of YCF1p includes a linker region in cytosol which is crucial for Cd resistance (Mason and Michaelis 2002). The yeast Bpt1p (YCF1p homologue) seems to participate to a little extent in Cd transport (Sharma et al. 2002). Additionally, YCF1 gene over-expression in *A. thaliana* enhanced Cd-GSH uptake in YCF1 transgenics in comparison to wild plants (Song et al. 2003). The examination of microarray having genes encoding prospective ABC transporters revealed increased expression of AtPDR8 gene which conferred tolerance to Cd stressed *A. thaliana*. Increased tolerance was associated with diminished level of $Cd^{2+}$ suggesting that this gene function as an imperative efflux transporter of Cd (Kim et al. 2007). AtATM3 (ATP-binding transporter of *Arabidopsis*) is a protein present in mitochondria and is up-regulated in roots exposed to Pb(II) or Cd(II) (Kim et al. 2006). AtMRP1–14 are expressed in roots as well as leaves and AtMRP3, 4, 6, 7, and 14 are up-regulated in response to Cd (Bovet et al. 2003). In tobacco, heterologous expression of AtMRP7 caused higher Cd retention in roots, indicating its role in regulating root-to-shoot translocation (Wojas et al. 2009). Furthermore, another member of ABC transporter family, multidrug and toxic compound extrusion (MATE) has been described to control detoxification of Cd through its exportation outside the cytoplasm (Huang et al. 2016).

### 15.2.3 NRAMPs

NRAMPs comprise a huge family of intrinsic membrane proteins (also recognized as DCT1—divalent cation transporter 1 or DMT1—divalent metal transporter 1), and NRAMP genes have been categorized in many plant species (Wei et al. 2009). Yeast expressing AtNRAMP1, AtNRAMP3, AtNRAMP4, and AtNRAMP6 revealed enhanced Cd sensitivity and AtNRAMP3, AtNRAMP4, and AtNRAMP6 overexpression in *Arabidopsis* bestowed hypersensitivity to Cd (Lanquar et al. 2005; Cailliatte et al. 2009). In rice, OsNRAMP1 and OsNRAMP5 are the main transporters involved in Cd uptake (Takahashi et al. 2011; Sasaki et al. 2012). AtNRAMP6 is an intracellular metal transporter which influences distribution of Cd within the cell (Cailliatte et al. 2009).

### 15.2.4 CDFs

CDFs (renamed as MTPs-metal tolerance proteins) have six transmembrane domains and the C-terminal is involved in binding of Cd. CDFs from *Thlaspi goesingense* function in sequestering Cd into the vacuoles (Maser et al. 2001). Shingu et al. (2005) separated NgMTP1 from *Nicotiana glauca* along with NtMTP1a and NtMTP1b from *N. tabacum* which was expressed in mutants of yeast and conferred Zn/Cd-tolerance.

### 15.2.5 P-type ATPases

P-type ATPases family utilizes energy derived from hydrolysis of ATP to transport several ions across the membrane and are categorized into numerous subfamilies, consisting of heavy metal transporting P1B-ATPases. In *Arabidopsis*, P1B-ATPases subfamily includes eight members, four of which (HMA1–4) are engaged in Zn/Co/Cd/Pb transport across plasma lemma (Mills et al. 2003). They are also termed as CPx-ATPase. Some reports indicate contribution of AtHMA2 and AtHMA4 in Cd transport (Verret et al. 2005; Wong and Cobbett 2009). HMA2 and HMA4 are present in cellular membrane and function in xylem transport of both Zn and Cd from roots to shoots of plants (Kum et al. 2009). Xylem loading of Cd is carried out by AtHMA4 in roots of *A. thaliana* as well as its homologues in *A. halleri* (Hanikenne et al. 2008). OsHMA5, OsHMA6, and OsHMA9 were up-regulated under enhanced Zn and Cd levels in rice (Lee et al. 2007). Morel et al. (2009) observed that plants over-expressing HMA3 gene exhibited increased Cd tolerance in *Arabidopsis*. Furthermore, OsHMA3 is involved in Cd transportation across root vacuoles (Miyadate et al. 2011), whereas TcHMA3 is responsible for sequestrating Cd into leaf vacuoles (Ueno et al. 2011).

### 15.2.6 CAXs

CAX2 (calcium exchanger 2) transporter identified from *Arabidopsis* is believed to be a high affinity $Cd^{2+}/H^{+}$ antiporter (Korenkov et al. 2007a). *Arabidopsis* has 6 CAXs, i.e. CAX1-CAX6 (Shigaki and Hirschi 2006) and all of them have the potential to transport Cd, but CAX2 and CAX4 have highest Cd transport capabilities (Korenkov et al. 2007b). Expression of CAX1 increased the accumulation as well as tolerance of Cd in *Petunia* (Wu et al. 2011).

### 15.2.7 Other Transporters

Oligopeptide transporter (OPT) members, for example, OPT6 transport Cd-GSH complexes as well as derivatives of GSH (Cagnac et al. 2004). Additionally, Cd can enter into roots as Cd-chelates via yellow stripe 1-like (YSL) proteins belonging to the OPT family of transporters (Curie et al. 2009). Expression of TaVP1 (vacuolar proton pump of V-PPase type in wheat) affected Cd translocation from roots to aerial parts in transgenics of tobacco (Khoudi et al. 2012). Expression of an additional transporter LCT1 (low affinity cation transporter) increased the uptake of Cd and simultaneously enhanced the defensive action of Ca against Cd toxicity in tobacco (Antosiewicz and Henning 2004). LCT1 is expressed in the vascular bundles of the stem as well as in cells encircling vascular bundles, predominantly in the reproductive period in wheat. LCT1 inactivation via RNA interference (RNAi) decreased Cd levels in phloem sap of rice plants raised in soil spiked with low Cd levels (Uraguchi et al. 2011). Decreased Cd contents in kernels of the *lcd* mutant indicated involvement of OsLCD in transporting Cd within rice (Shimo et al. 2011). Furthermore, Cd uptake into root cells can occur through non-selective cation channels, for instance, voltage-insensitive cation channels (VICC), hyperpolarization-activated Ca channels (HACC), and depolarization-activated Ca channels (DACC) (Qin et al. 2020). Additionally, a Ca transport route could also be engaged in uptake of Cd (Perfus-Barbeoch et al. 2002). In *Thlaspi caerulescens*, Cd uptake was suggested to occur either by a high affinity uptake system for Fe (Vert et al. 2002) or a low affinity uptake system for Ca or Zn (Molitor et al. 2005). Moreover, over-expression of ATPase ZntA conferred decreased Cd accumulation and increased tolerance in *Arabidopsis* (Lee et al. 2003). PCR1 (plasma membrane-localized Cd efflux transporter) activity increases plant tolerance to Cd by exporting it outside the cell (Lochlainn et al. 2011).

## 15.3 Alleviation of Cd Toxicity Via Crosstalk Between Phytohormones and Sulfur

Numerous attempts have been carried out by the researchers to impede Cd toxicity in plants. Phytohormones are known to play multifaceted roles in regulating several plant processes. Recently, many studies have validated the role of plant hormones in mitigating negative effects of various abiotic stresses including metal stress. Sulfur (S) is one of the vital mineral nutrients needed for normal well-being of plants. Additionally, it homogenizes metabolism of plants under favorable as well as stressful conditions (Iqbal et al. 2013a; Nazar et al. 2014). This section underlines the mechanisms instigated by concomitant application of phytohormones (salicylic acid, ethylene, brassinosteroids, jasmonic acid, nitric oxide) and S in increasing plant tolerance under Cd-stressed conditions.

Positive role of some other plant hormones (auxins, gibberellins, cytokinins, abscicic acid) in imparting Cd tolerance to plants has been summarized in Table 15.1.

**Table 15.1** Alleviation of toxic effects of cadmium on some plants via crosstalk with phytohormones (IAA, GA, cytokinins, ABA)

| Plant | Phytohormone | Role under Cd stress | References |
|---|---|---|---|
| *Oryza sativa* (Rice) | Auxin (IAA) | Regulation of root growth via enhanced expression of MAPK related genes and repression of ROS<br>Increased plant height, tiller number, panicles, grain weight; decreased translocation of Cd to grains | Zhao et al. (2012, 2013)<br>Farooq et al. (2015) |
| *Camellia sinensis* (Tea) | IAA | Reduced accumulation of Cd in plant parts, enhanced growth and development | Zhang et al. (2020) |
| *Triticum aestivum* (Wheat) | IAA | Increased growth, RWC, photosynthetic pigments, enzymatic antioxidant defence system (SOD, CAT, POX) | Agami and Mohamed (2013) |
| *Hordeum vulgare* (Barley) | IAA | Enhanced activity of GST | Bočová et al. (2013) |
| *Arabidopsis thaliana* | IAA | Increased transcriptome level of nitrilase (AtNIT) gene in roots | Vitti et al. (2013) |
| *Solanum melongena* (Eggplant) | IAA | Elevated pigment content and rate of photosynthesis | Singh and Prasad (2016) |
| *Solanum lycopersicum* (Tomato) | IAA | Improved endogenous level of NO led to the increased activity of enzymes regulating AsA-GSH cycle which in turn neutralized the ROS in roots and increased the growth and photosynthetic parameters | Khan et al. (2019) |
| *Pisum sativum* (Pea) | IAA-Aspartate | Enhanced level of carbonylated protein, increased activities of CAT and POX, decreased $H_2O_2$ levels | Ostrowski et al. (2016) |
| *Trigonella foenum-graecum* (Fenugreek) | IAA | Low IAA doses augmented the activities of SOD, POD, CAT, and GST, maintained the redox status (AsA/DHA and GSH/GSSG) while high IAA doses suppressed the activity of DHAR and GR and disturbed the redox status | Bashri and Prasad (2016) |
| *Arabidopsis thaliana* | Gibberellic acid (GA) | Reduced ROS levels, lipid peroxidation, NO; suppressed IRT1 gene expression meant for Cd uptake | Zhu et al. (2012) |
| *Hordeum vulgare* (Barley) | $GA_3$ | Enhanced activity of hydrolytic enzymes in germinating seeds, increased sugar and amino acid content in endosperm, promotion of mobilization of protein and starch reserves from endosperm to seedling roots | Amri et al. (2016) |
| *Parthenium hysterophorus* (Carrot grass) | $GA_3$ | Improved plant growth and biomass, enhanced accumulation of Cd in roots, increased phytoremediation capability of plants | Hadi et al. (2014) |
| *Vicia faba* (Broad bean) | $GA_3$ | Stimulation of mitotic activity, increased protein content and biochemical constituents in seeds | Mansour and Kamel (2005) |

(continued)

**Table 15.1** (continued)

| Plant | Phytohormone | Role under Cd stress | References |
|---|---|---|---|
| *Solanum lycopersicum* (Tomato) | Cytokinin (Kinetin) | Reduced levels of ROS, increased integrity of photosystem II and activities of enzymes of AsA-GSH cycle | Singh et al. (2018) |
| *Pisum sativum* (Pea) | Cytokinin | Enhanced photosynthetic parameters and growth | Al-Hakimi (2007) |
| *Solanum tuberosum* (Potato) | Abscisic acid (ABA) | Regulation of phytochelatin and cysteine levels | Stroiński et al. (2010) |
| *Sedum alfredii* | ABA | Enhanced transcriptome level of Cd stress responsive genes (i.e. *HsfA4c, HMA4* expression in roots and *HMA2, HMA3, CAD, NAS* expression in shoots), accumulation of Cd in shoots | Lu et al. (2020) |
| *Oryza sativa* (Rice) | ABA | Decreased rate of transpiration, Cd accumulation and increased Cd tolerance in rice seedlings | Hsu and Kao (2003) |
| *Solanum photeinocarpum* | ABA | Increased biomass, chlorophyll, Cd content in shoots and Cd extraction ability (effects were dose dependent) | Wang et al. (2016) |

*AsA* ascorbic acid, *CAD* cinnamyl alcohol dehydrogenase, *CAT* catalase, *DHA* dehydroascorbic acid, *DHAR* dehydroascorbate reductase, *GR* glutathione reductase, *GSH* reduced glutathione, *GSSG* oxidized glutathione, *GST* glutathione sulfotransferase, *HMA* heavy metal ATPase, *IRT* iron regulated transporter, *MAPK* mitogen activated protein kinase, *NAS* nicotianamine synthase, *NO* nitric oxide, *POX/POD* peroxidase, *ROS* reactive oxygen species, *RWC* relative water content, *SOD* superoxide dismutase

### 15.3.1 Crosstalk Between Salicylic Acid, Sulfur, and Cd

Salicylic acid (SA) is a natural endogenous phenolic signalling molecule synthesized from t-cinnamic acid and plays a significant defensive role under stressful conditions by regulating several plant physiological and biochemical processes (Gunes et al. 2007). Many reports have suggested that SA plays a pivotal role in ameliorating metal stress in plants (Zhou et al. 2009). Various analogues of SA (2,6 dichloro-isonicotinic acid) as well as its precursors (benzoic acid or O-coumaric acid) increase the plant tolerance against metals by elevating antioxidative capacity, suggesting their involvement in alleviation of metal stress (Horvath et al. 2007). It has been reported that SA ameliorates Cd toxicity by reducing its uptake, enhancing photosynthetic capacity, and up-regulating antioxidative enzymes in *Cannabis sativa* (Shi et al. 2009). Exogenous supplementation of 3 mM SA enhanced the endogenous level of SA in Cd-exposed *Oryza sativa* due to increased expression of transcription factor OsWRKY45, thereby overcoming membrane damage by lowering the superoxides like $H_2O_2$ (Chao et al. 2010). Moreover, exogenous treatment of SA under Cd stress up-regulates the activity of phenylalanine ammonium lyase (PAL), which acts as an accumulating agent of soluble phenolics in the roots and these phenolic metabolites are involved in plant defence by scavenging reactive oxygen species (ROS) and chelating heavy metals (Kováčik et al. 2009; Sgherri et al. 2004). Popova et al. (2009) observed SA-induced diminution in accumulation of Cd in pea plants. It has been established that SA acts as an accumulating agent of various antioxidant enzymes under Cd stress in plants (Wang et al. 2006). Exogenous application of SA in barley played a significant role in suppressing Cd-induced generation of superoxides through up-regulation of antioxidant enzymes like catalase (CAT) and ascorbate peroxidase (APX) (Metwally et al. 2003). Pre-treatment of seeds with SA under Cd stress enhanced the growth of radical and plumule in alfalfa plants (Dražić et al. 2006). Similarly, seed priming with SA under Cd stress enhanced the activities of PEP carboxylase, RUBP carboxylase, antioxidant enzymes, for instance, APX and SOD and suppressed CAT activity in maize plants (Krantev et al. 2008). SA played an important role in lowering $H_2O_2$, thiobarbituric acid reactive substances (TBARS) and $O_2$ in rice, thereby increasing tolerance level of plants against Cd (Panda and Patra 2007). Various reports have suggested that SA enhances the growth and development of plants by modulating the metabolism of nutrients under abiotic stresses (Nazar et al. 2015; Tufail et al. 2013; Wang et al. 2011). Supplementation of mineral nutrients like sulfur (S) plays a vital role in reducing Cd-induced toxicity, as it is a key constituent of amino acids like methionine and cysteine, antioxidants (GSH), phytochelatins (PCs), and secondary metabolites (Khan et al. 2014b). It is believed that crop yield and resistance to Cd stress are directly proportional to sulfur assimilation (Gallego et al. 2012; Koprivova and Kopriva 2014). It has been reported that high concentrations of Cd enhance the high affinity sulfate transport system (HATs), thereby increasing sulfate uptake capacity (Nocito et al. 2007). Moreover, Cd induces activation of certain genes involved in the synthesis of enzymes involved in sulfate assimilatory pathway (Herbette et al. 2006; Khan et al. 2007) (Fig. 15.1). The periodic treatment of Cd (2 or 6 h) as well as depletion of sulfate (12–24 h) led to the up-regulation of two sulfate transporter encoding genes *Sultr1;1* and *Sultr2;1* in *Arabidopsis* (Herbette et al. 2006; Sarry et al. 2006). In most of the plants, sulfur is present in the reduced glutathione (GSH) form, which acts as a key precursor of PCs under Cd stress (Inouhe 2005). Application of GSH with sulfur under Cd exposure protected the photosynthetic apparatus in *Brassica campestris* (Anjum et al. 2008). Furthermore, application of SA enhanced some sulfur containing secondary metabolites like glucosinolates, defensins, and thionins under abiotic stress in *B. napus* (Irchhaiya et al. 2015).

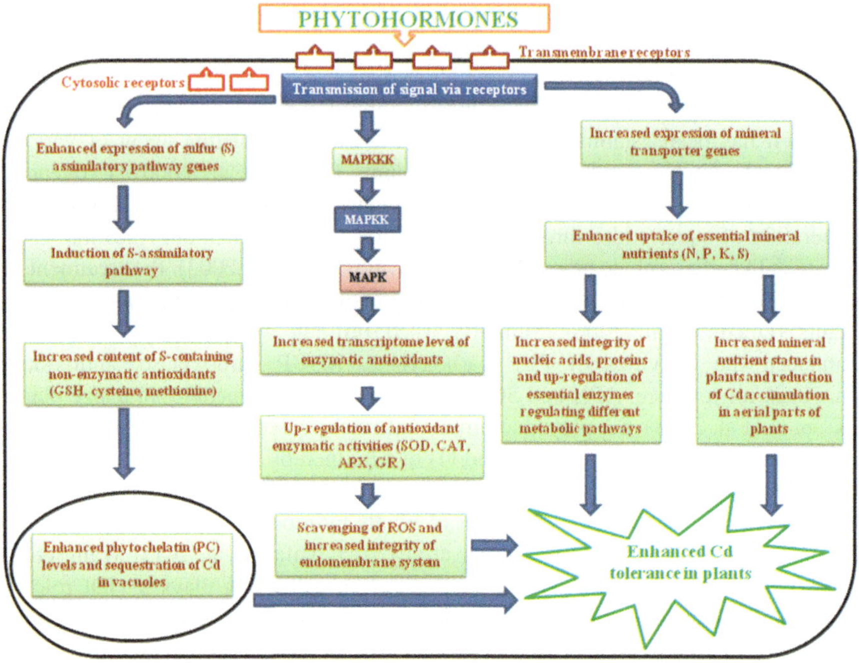


**Fig. 15.1** Schematic representation of role of phytohormones in alleviating the detrimental effects of Cd in plants (adopted from Asgher et al. 2014). *APX* ascorbate perodixase, *CAT* catalase, *Cd* cadmium, *GR* glutathione reductase, *GSH* reduced glutathione, *K* potassium, *MAPKKK* mitogen activated protein kinase kinase kinase, *N* nitrogen, *P* phosphorus, *ROS* reactive oxygen species, *SOD* superoxide dismutase

### 15.3.2 Crosstalk Between Ethylene, Sulfur, and Cd

Methionine is a precursor of phytohormone ethylene, and S-adenosyl methionine synthetase converts methionine into S-adenosyl methionine (SAM), which in turn is converted into 1-amino cyclopropane carboxylic acid (ACC) as well as 5-deoxy-5-methylthioadenosine (MTA) with the help of enzyme ACC synthase (ACS). ACC is finally changed to ethylene by ACC-oxidase (ACO). In this ethylene biosynthetic pathway, ACO and ACS are the key enzymes which function in oxidation as well as synthesis of ACC, respectively. It has been reported that abiotic stresses, such as, nutrient deficiency (P and Fe deficiency) leads to higher production of ethylene in plants (Romera et al. 1999). Ethylene plays a significant role in combination with S in regulating various plant processes under abiotic stresses (Iqbal et al. 2013b). Under Cd stress, concomitant application of S and ethephon to *B. juncea* enhanced cysteine, methionine, and GSH levels in comparison to control plants since both the amendments synergistically up-regulated the activities of serine acetyl transferase (SAT) as well as ATP-sulfurylase (ATPS) and enhanced intracellular S contents (Khan et al. 2016). Furthermore, Cd stressed plants exhibited higher accumulation of superoxides, $H_2O_2$ and experienced lipid peroxidation via up-regulation of ACC-synthase (ACS), increased ethylene production and reduction in net photosynthesis and ultimately growth. However, combined treatments of ethephon and S under Cd stress enhanced the growth and net photosynthesis in *B. juncea* by reducing ROS, glucose and ethylene production (Khan et al. 2016). The authors further demonstrated that individual applications of S and ethephon under Cd stress reduced almost the same amount of Cd content in roots as well as leaves of plants; however, their combined supplementation reduced much higher Cd content. S and ethylene synergistically regulate GSH synthesis under ozone and Cd stress (Masood et al. 2012; Yoshida et al. 2009). In *B. juncea* and *Triticum aestivum*, application of ethylene and S constitutively mitigated Cd-mediated oxidative stress (Khan et al. 2015; Masood et al. 2012). Individual application of S and ethephon to Cd stressed *B. juncea* decreased the glucose content to the same extent (11.3%), whereas their combined application reduced the glucose levels up to 16% in Cd-treated plants over controls (Khan et al. 2016; Masood et al. 2012). Tholen et al. (2007, 2008) demonstrated that endogenous glucose level suppressed the expression of Rubisco in ethylene insensitive *Nicotiana tabacum* genotypes (lacking functional ethylene receptors), thereby reducing the photosynthetic ability. Under Cd stress, increase in glucose content inhibited Rubisco activity, resulting in reduced photosynthesis in plants. However, supplementation of S along with ethephon to Cd-treated plants reduced the glucose content, leading to enhanced Rubisco activity and better photosynthetic capacity. Similarly, application of S to Cd stressed tobacco plants increased the cysteine content as compared to control plants (Ning et al. 2010). S, being an essential component of cysteine, up-regulated the synthesis of O-acetylserine lyase-encoding gene GmOASTL4 in tobacco, resulting in enhanced cysteine contents (Ning et al. 2010). Cysteine in turn helps in synthesizing various intracellular ligands like PCs, cysteine rich polypeptides, short gene encoding polypeptides, and GSH which are engaged in detoxification of heavy metals (Anjum et al. 2008). Khan et al. (2015) evaluated the interactive effects of ethylene along with antioxidant and proline metabolism in enhancing Cd resistance in wheat plants. Authors reported that application of S/Se under Cd stress resulted in down-regulation of proline oxidase (PROX) and up-regulation of glutamyl kinase (GK), leading to higher accumulation of proline. Moreover, Se or S enhanced the activity of glutathione peroxidase (GPOX) and glutathione reductase (GR) and decreased the ethylene levels, thus improving photosynthesis and growth of plants. It has been proved via application of ethylene biosynthesis inhibitor, aminoethoxyvinylglycine (AVG) that S and ethylene synergistically mitigate Cd stress in plants (Asgher et al. 2014). Rodríguez-Serrano et al. (2009) reported that ROS mediated over-synthesis of ethylene along with methyl jasmonate in pea played a crucial role in regulating pathogenesis related proteins which protected the cellular proteins from Cd injury. These two hormones are involved in increasing tolerance in *Arabidopsis* against metalloid selenite (Tamaoki et al. 2008).

### 15.3.3 Crosstalk Between Brassinosteroids, Sulfur, and Cd

Plant steroidal hormone called brassinosteroid (BR) plays a major role in improving plant growth, photosynthesis and mitigates the oxidative stress through enhancement of antioxidative system under various abiotic stresses (Ahammed et al. 2013; Bajguz and Piotrowska-Niczyporuk 2014). Exogenous treatment of BRs alleviates Cd toxicity in plants by stimulating the activities of antioxidative enzymes, ultimately leading to improved growth, photosynthesis and yield (Hasan et al. 2008; Hayat et al. 2012; Vázquez et al. 2013). In this context, exogenous application of 24-epibrassinolide to *Phaseolus vulgaris* improved Cd-resistance (Rady 2011). Supplementation of 24-epibrassinolide to Cd stressed *Raphanus sativus* increased the activities of polyphenol oxidase (PPO) and glutathione S-transferase (GST) enzymes (Sharma et al. 2012). Cao et al. (2013) revealed that BRs function in the amplification of Cd phytoextraction in rice seedlings. Kaur et al. (2017) demonstrated that exogenous supplementation of castasterone (6-ketone type active BRs) along with CA (citric acid) to Cd-treated *B. juncea* plants decreased the accumulation of oxidized products of membrane lipids called malondialdehyde (MDA) and $H_2O_2$, hence protecting the cells against Cd-induced membrane damage. Similarly, co-application of 28-homobrassinolide or 24-epibrassinolide decreased Cd contents in rice seedlings (Xu et al. 2018).

To the best of our knowledge, no report is available in literature which supports the crosstalk between BRs and S for the alleviation of toxic effects of Cd in plants.

### 15.3.4 Crosstalk Between Jasmonic Acid, Sulfur, and Cd

Significance of jasmonic acid (JA) in alleviating phytotoxic effects of heavy metals has been discussed recently (Chen et al. 2014; Yan et al. 2015) (Fig. 15.1). Application of inhibitors of JA biosynthetic pathway revealed the role of JA in increasing growth and photosynthetic parameters in Cd stressed *Arabidopsis thaliana* (Maksymiec and Krupa 2002). Exogenous supplementation of methyl jasmonate (MeJA) to Cd-exposed *O. sativa* modulated the activities of glutathione reductase (GR), superoxide dismutase (SOD), peroxidase (POD), and catalase (CAT) as well as enhanced the accumulation of S-containing redox compound, i.e. GSH (Singh and Shah 2014). Similarly, MeJA application has been reported to increase the production of S-containing phytochelatins (PCs), for instance, in *A. thaliana* (Maksymiec et al. 2007) and restored *KOMT2* gene expression in *Kandelia obovata* leaves under Cd stress (Chen et al. 2014). Study of Sasaki-Sekimoto et al. (2005) indicated significant role of JA in up-regulating the various genetic factors controlling nine metabolic pathways in *Arabidopsis,* including AsA-GSH pathway. JA enhanced the contents of cysteine, ascorbic acid (AsA) and activity of key enzyme of AsA recycling pathway, i.e. dehydroascorbate reductase (DHAR) along with the genes responsible for the synthesis of GSH (γ-glutamylcysteine synthetase and glutathione synthetase) in Cd-stressed *A. thaliana* (Xiang and Oliver 1998). JA is engaged in assimilation of sulfate ($SO_4^{2-}$) through its regulatory role in primary as well as secondary S metabolism (Fig. 15.1). On the contrary, it has been reported that deficiency of $SO_4^{2-}$ results in synthesis of JA in plants (Jost et al. 2005; Nikiforova et al. 2003). Reports indicate that JA up-regulates the expression of genes responsible for reduction of $SO_4^{2-}$, synthesis of GSH, and metabolism of glucosinolate (GS) (Harada et al. 2000; Jost et al. 2005). However, JA regulated genes are believed to be independent of coronatine insensitive 1(COI1) JA receptor. Instead, a CYP20-3 (a plastidic cyclophilin JA binding protein) has been found whose physiological ligand is 12-oxo-phytodienoic acid (OPDA) (Park et al. 2013). Binding of OPDA to the CYP20-3 accelerates the interaction between cyclophilin and serine acetyltransferase (SAT) leading to increased production of GSH and cysteine which in turn modulates the cellular redox potential, thereby activating the TGA transcription factors (Park et al. 2013).

### 15.3.5 Crosstalk Between Nitric Oxide, Sulfur, and Cd

Exogenous supplementation of nitric oxide (NO) can alleviate the harmful effects of Cd in plants by up-regulating the antioxidant enzyme system or directly eradicating ROS (Fig. 15.1). Reports in literature have highlighted the beneficial role of exogenous application of NO as well as the mechanism through which it alleviates the negative effects of Cd stress in plants. Reduced activity of nitric oxide synthase (NOS) decreased the endogenous level of NO in *O. sativa* under Cd stress (Xiong et al. 2009). Authors further demonstrated that addition of sodium nitroprusside (SNP) alleviated the toxic effects of Cd in rice by reducing its accumulation in cell wall as well as in the leaf cells, validating the magnificent role of NO in increasing Cd tolerance. Microarray analysis further confirmed the importance of NO in regulating the genes related to $Fe/Cd^{2+}$ transporters (Besson-Bard et al. 2009). It has been suggested that application of NO ameliorates oxidative stress via activation of antioxidant enzyme system, $H^+$-ATPase and increment in the mineral absorption ability along with maintenance of hormonal balance (Liu et al. 2015). Application of SNP attenuated the detrimental effects of Cd in *O. sativa* by enhancing the chlorophyll and protein synthesis (Hsu and Kao 2004). In the same study, SNP reduced the levels of $H_2O_2$, MDA, GSH, AsA and up-regulated the specific antioxidant enzyme activities. Pre-treatment with NO in *Helianthus annuus* reduced the Cd-mediated oxidative stress and increased the growth, synthesis of chlorophyll, S-containing GSH, AsA and activities of CAT and APX (Laspina et al. 2005). Additionally, NO acts as signalling molecule in a cascade of events and modulates the transcriptome level of various genes as well as S-nitrosylation, such as, post-transcriptional modification under Cd stress (Arasimowicz-Jelonek et al. 2011; De Michele et al. 2009; Rodríguez-Serrano et al. 2009). Innocenti et al. (2007) demonstrated that exogenous application of NO modulated GSH synthesis by regulating the transcriptomics of *c-ecs* and *gshs* genes in *Medicago truncatula*. Cd caused reduction in intracellular levels of NO and GSH in *M. truncatula* roots and exogenous supplementation of NO enhanced the NO levels and GSH contents (Xu et al. 2010), suggesting interplay between increased levels of NO and S-containing compound GSH in alleviating Cd toxicity. Wang et al. (2015) reported increased S requirement by the plants grown under stressed conditions, which was further confirmed from the formation of GSNO (S-nitrosoglutathione) from GSH and NO. Various reports suggest that NO enhances the plant tolerance against oxidative stress generated by metals via up-regulating the key enzymes involved in GSH biosynthetic pathway (Xiong et al. 2010; Wang et al. 2015), thereby indirectly increasing S assimilation.

## 15.4 Conclusions

Cd is a heavy metal of foremost health and environmental concern which has provoked interest in global research due to its non-biodegradability and high accumulation in plants. There are various transporters via which Cd can be taken up and translocated to different parts of plants. Several studies have inspected the effects of Cd on physiology and metabolism of plants. For instance, Cd competes with other essential elements for binding to the same transporters and leads to their deficiency in plants. Moreover, Cd mediated over generation of ROS increases protein denaturation, lipid peroxidation and oxidation of nucleic acids. Crosstalk between plant hormones like SA, ethylene, BR, JA, NO, and S plays a crucial role in protecting plants from Cd toxicity by regulating antioxidant systems, osmolyte synthesis, improving photosynthesis and yield, thereby imparting stress tolerance. PGRs and S application to plants, thus represents a novel strategy to reduce Cd toxicity which could be utilized as a physiological tool in sustainable agriculture. Nevertheless, more detailed research is necessitated to understand the relationship between phytohormones and mineral nutrients, underlining Cd-toxicity. Therefore, future investigations should be focused on identification of genes as well as new potential targets of phytohormones and nutrients for the development of mechanisms involved in Cd-stress tolerance.

## References

Agami RA, Mohamed GF (2013) Exogenous treatment with indole-3-acetic acid and salicylic acid alleviates cadmium toxicity in wheat seedlings. Ecotoxicol Environ Saf 94:164–171

Ahammed GJ, Choudhary SP, Chen S, Xia X, Shi K, Zhou Y, Yu J (2013) Role of brassinosteroids in alleviation of phenanthrene–cadmium co-contamination-induced photosynthetic inhibition and oxidative stress in tomato. J Exp Bot 64(1):199–213

Ahmad P, Sarwat M, Bhat NA, Wani MR, Kazi AG, Tran LS (2015) Alleviation of cadmium toxicity in *Brassica juncea* L. (Czern. & Coss.) by calcium application involves various physiological and biochemical strategies. PLos One 10(1):e0114571

Akhter MF, Omelon CR, Gordon RA, Moser D, Macfie SM (2014) Localization and chemical speciation of cadmium in the roots of barley and lettuce. Environ Exp Bot 100:10–19

Al Mahmud J, Hasanuzzaman M, Nahar K, Bhuyan MB, Fujita M (2018) Insights into citric acid-induced cadmium tolerance and phytoremediation in *Brassica juncea* L.: coordinated functions of metal chelation, antioxidant defense and glyoxalase systems. Ecotoxicol Environ Saf 147:990–1001

Al-Hakimi AMA (2007) Modification of cadmium toxicity in pea seedlings by kinetin. Plant Soil Environ 53(3):129–135

Amri B, Khamassi K, Ali MB, da Silva JAT, Kaab LBB (2016) Effects of gibberellic acid on the process of organic reserve mobilization in barley grains germinated in the presence of cadmium and molybdenum. S Afr J Bot 106:35–40

Anjum NA, Umar S, Ahmad A, Iqbal M, Khan NA (2008) Sulphur protects mustard (*Brassica campestris* L.) from cadmium toxicity by improving leaf ascorbate and glutathione. Plant Growth Regul 54(3):271–279

Anjum NA, Chan MT, Umar S (2010) In: Anjum NA, Umar S, Chan M-T (eds) Ascorbate-glutathione pathway and stress tolerance in plants. Springer Science & Business Media, pp 1–429

Anjum NA, Gill SS, Gill R (2014) Cysteine–jack of all glutathione-based plant stress defense trades. In: Anjum NA, Gill SS, Gill R (eds) Plant adaptation to environmental change-significance of amino acids and their derivatives. CAB International, UK, pp 35–52

Antosiewicz DM, Henning J (2004) Over-expression of LCT1 in tobacco enhances the protective action of calcium against cadmium toxicity. Environ Pollut 129:237–245

Arao T, Ishikawa S (2006) Genotypic differences in cadmium concentration and distribution of soybeans and rice. Jpn Agric Res Q 40:21–30

Arasimowicz-Jelonek M, Floryszak-Wieczorek J, Gwóźdź EA (2011) The message of nitric oxide in cadmium challenged plants. Plant Sci 181:612–620

Asgher M, Khan NA, Khan MIR, Fatma M, Masood A (2014) Ethylene production is associated with alleviation of cadmium-induced oxidative stress by sulphur in mustard types differing in ethylene sensitivity. Ecotoxicol Environ Saf 106:54–61

Bajguz A, Piotrowska-Niczyporuk A (2014) Brassinosteroids implicated in growth and stress responses. In: Tran LS, Pal S (eds) Phytohormones: a window to metabolism, signaling and biotechnological applications. Springer, New York, pp 163–190

Bali AS, Sidhu GPS, Kumar V, Bhardwaj R (2019) Mitigating cadmium toxicity in plants by phytohormones. In: Hasanuzzaman M, Prasad MNV, Fujita M (eds) Cadmium toxicity and tolerance in plants. Academic Press, Elsevier, pp 375–396

Barman F, Majumdar S, Arzoo SH, Kundu R (2020) Genotypic variation among 20 rice cultivars/ landraces in response to cadmium stress grown locally in West Bengal, India. Plant Physiol Biochem 148:193–206

Bashri G, Prasad S (2016) Exogenous IAA differentially affects growth, oxidative stress and antioxidants system in Cd stressed *Trigonella foenum-graecum* L. seedlings: toxicity alleviation by up-regulation of ascorbate-glutathione cycle. Ecotoxicol Environ Saf 132:329–338

Benavides MP, Gallego SM, Tomaro ML (2005) Cadmium toxicity in plants. Braz J Plant Physiol 17:21–34

Benzarti S, Hamdi H, Mohri S, Ono Y (2010) Response of antioxidative enzymes and apoplastic bypass transport in *Thlaspi caerulescens* and *Raphanus sativus* to cadmium stress. Int J Phytoremed 12:733–744

Besson-Bard A, Astier J, Rasul S, Wawer I, Dubreuil-Maurizi C, Jeandroz S, Wendehenne D (2009) Current view of nitric oxide-responsive genes in plants. Plant Sci 177:302–309

Bočová B, Huttová J, Mistrík I, Tamás L (2013) Auxin signalling is involved in cadmium induced glutathione-S-transferase activity in barley root. Acta Physiol Plant 35(9):2685–2690

Bovet L, Eggmann T, Meylan-Bettex M, Polier J, Kammer P, Marin E, Feller U, Martinoia E (2003) Transcript levels of MRFs: induction of *ATMRP3* after cadmium treatments. Plant Cell Environ 26:371–381

Cagnac O, Bourbouloux A, Chakrabarty D, Zhang MY, Delrot S (2004) AtOPT6 transports glutathione derivatives and is induced by primisulfuron. Plant Physiol 135:1378–1387

Cailliatte R, Lapeyre B, Briat JF, Mari S, Curie C (2009) The NRAMP6 metal transporter contributes to cadmium toxicity. Biochem J 422:217–228

Cakmak I (2002) Plant nutrition research: priorities to meet human needs for food in sustainable ways. Plant Soil 247:3–24

Cao F, Liu L, Ibrahim W, Cai Y, Wu F (2013) Alleviating effects of exogenous glutathione, glycinebetaine, brassinosteroids and salicylic acid on cadmium toxicity in rice seedlings (*Oryza sativa*). Agrotechnol 2(1):107–112

Chao YY, Chen CY, Huang WD, Kao CH (2010) Salicylic acid-mediated hydrogen peroxide accumulation and protection against Cd toxicity in rice leaves. Plant Soil 329:327–337

Chen F, Wu F, Dong J, Vincze E, Zhang G, Wang F, Huang Y, Wei K (2007) Cadmium translocation and accumulation in developing barley grains. Planta 227:223–232

Chen J, Yan Z, Li X (2014) Effect of methyl jasmonate on cadmium uptake and antioxidative capacity in *Kandelia obovata* seedlings under cadmium stress. Ecotoxicol Environ Saf 104:349–356

Clemens S (2006) Toxic metal accumulation, responses to exposure and mechanisms of tolerance in plants. Biochimie 88:1707–1719

Clemens S, Aarts MGM, Thomine S, Verbruggen N (2013) Plant science: the key to preventing slow cadmium poisoning. Trends Plant Sci 18(2):92–99

Connolly E, Fett J, Guerinot M (2002) Expression of the IRT1 metal transporter is controlled by metals at the levels of transcript and protein accumulation. Plant Cell 14:1347–1357

Curie C, Cassin G, Couch D, Divol F, Higuchi K, Le Jean M, Mari S (2009) Metal movement within the plant: contribution of nicotianamine and yellow stripe 1-like transporters. Ann Bot 103(1):1–11

De Michele R, Vurro E, Rigo C, Costa A, Elviri L, Di Valentin M, Careri M, Zottini M, di Toppi LS, Schiavo FL (2009) Nitric oxide is involved in cadmium induced programmed cell death in *Arabidopsis* suspension cultures. Plant Physiol 150:217–228

Dobrikova AG, Yotsova EK, Borner A, Landjeva SP, Apostolova EL (2017) The wheat mutant DELLA-encoding gene (Rht-B1c) affects plant photosynthetic responses to cadmium stress. Plant Physiol Biochem 114:10–18

Dražić G, Mihailović N, Lojić M (2006) Cadmium accumulation in *Medicago sativa* seedlings treated with salicylic acid. Biol Plant 50(2):239–244

Ehsan S, Ali S, Noureen S, Mahmood K, Farid M, Ishaque W, Shakoor MB, Rizwan M (2014) Citric acid assisted phytoremediation of cadmium by *Brassica napus* L. Ecotoxicol Environ Saf 106:164–172

Farooq H, Asghar HN, Khan MY, Saleem M, Zahir ZA (2015) Auxin-mediated growth of rice in cadmium-contaminated soil. Turk J Agric For 39(2):272–276

Fujimaki S, Suzui N, Ishioka NS, Kawachi N, Ito S, Chino M, Nakamura S (2010) Tracing cadmium from culture to spikelet: noninvasive imaging and quantitative characterization of absorption, transport, and accumulation of cadmium in an intact rice plant. Plant Physiol 152:1796–1806

Fujita Y, Fujita M, Shinozaki K, Yamaguchi-Shinozaki K (2011) ABA-mediated transcriptional regulation in response to osmotic stress in plants. J Plant Res 124:509–525

Gall JE, Boyd RS, Rajakaruna N (2015) Transfer of heavy metals through terrestrial food webs: a review. Environ Monit Assess 187(4):201

Gallego SM, Pena LB, Barcia RA, Azpilicueta CE, Iannone MF, Rosales EP, Benavides MP (2012) Unravelling cadmium toxicity and tolerance in plants: insight into regulatory mechanisms. Environ Exp Bot 83:33–46

Gill SS, Tuteja N (2011) Cadmium stress tolerance in crop plants: probing the role of sulfur. Plant Signal Behav 6(2):215–222

Gomes DS, Fragoso LC, Riger CJ, Panek AD, Eleutherio ECA (2002) Regulation of cadmium uptake by *Saccharomyces cerevisiae*. Biochim Biophys Acta 1573:21–25

Greger M, Lofstedt M (2004) Comparison of uptake and distribution of cadmium in different cultivars of bread and durum wheat. Crop Sci 44:501–507

Grotz N, Guerinot ML (2006) Molecular aspects of Cu, Fe and Zn homeostasis in plants. Biochim Biophys Acta 1763(7):595–608

Gunes A, Inal A, Alpaslan M, Eraslan F, Bagci EG, Cicek N (2007) Salicylic acid induced changes on some physiological parameters symptomatic for oxidative stress and mineral nutrition in maize (*Zea mays* L.) grown under salinity. J Plant Physiol 164:728–736

Hadi F, Ali NZ, Ahmad A (2014) Enhanced phytoremediation of cadmium contaminated soil by *Parthenium hysterophorus* plant: effect of gibberellic acid (GA3) and synthetic chelator, alone and in combinations. Bioremed J 18(1):46–55

Hajiboland R, Farhanghi F (2010) Remobilization of boron, photosynthesis, phenolic metabolism and anti-oxidant defense capacity in boron-deficient turnip (Brassica *rapa* L.) plants. Soil Sci Plant Nutr 56:427–437

Han F, Shan X, Zhang S, Wen B, Owens G (2006) Enhanced cadmium accumulation in maize roots-the impact of organic acids. Plant Soil 289:355–368

Hanikenne M, Talke IN, Haydon MJ, Lanz C, Nolte A, Motte P, Kroymann J, Weigel D, Kramer U (2008) Evolution of metal hyperaccumulation required cis-regulatory changes and triplication of HMA4. Nature 453:391–395
Harada E, Kusano T, Sano H (2000) Differential expression of genes encoding enzymes involved in sulfur assimilation pathways in response to wounding and jasmonate in *Arabidopsis thaliana*. J Plant Physiol 156:272–276
Harris N, Taylor G (2004) Cadmium uptake and translocation in seedlings of near isogenic lines of durum wheat that differ in grain cadmium accumulation. BMC Plant Biol 4(4). https://doi.org/10.1186/1471-2229-4-4
Hart JJ, Welch RM, Norvell WA, Kochian LV (2002) Transport interactions between cadmium and zinc in roots of bread and durum wheat cultivars. Physiol Plant 116:73–78
Hasan SA, Hayat S, Ali B, Ahmad A (2008) 28-Homobrassinolide protects chickpea (*Cicer arietinum*) from cadmium toxicity by stimulating antioxidants. Environ Pollut 151:60–66
Hasan SA, Fanduddin Q, Ali B, Hayat S, Ahmad A (2009) Cadmium: toxicity and tolerance in plants. J Environ Biol 30(2):165–174
Hayat S, Alyemeni MN, Hasan SA (2012) Foliar spray of brassinosteroid enhances yield and quality of *Solanum lycopersicum* under cadmium stress. Saudi J Biol Sci 19:325–335
Herbette S, Taconnat L, Hugouvieux V, Piette L, Magniette ML, Cuine S, Renou JP (2006) Genome-wide transcriptome profiling of the early cadmium response of *Arabidopsis* roots and shoots. Biochimie 88(11):1751–1765
Horvath E, Szalai G, Janda T (2007) Induction of abiotic stress tolerance by salicylic acid signalling. J Plant Growth Regul 26:290–300
Hsu YT, Kao CH (2003) Role of abscisic acid in cadmium tolerance of rice (*Oryza sativa* L.) seedlings. Plant Cell Environ 26(6):867–874
Hsu YT, Kao CH (2004) Cadmium toxicity is reduced by nitric oxide in rice leaves. Plant Growth Regul 42:227–238
Hu PJ, Qiu RL, Senthilkumar P, Jiang D, Chen ZW, Tang YT, Liu FJ (2009) Tolerance accumulation and distribution of zinc and cadmium in hyperaccumulator *Potentilla griffithii*. Environ Exp Bot 66(2):317–325
Huang YY, Shen C, Chen JX, He CT, Zhou Q, Tan X, Yuan JG, Yang ZY (2016) Comparative transcriptome analysis of two *Ipomoea aquatica* Forsk. cultivars targeted to explore possible mechanism of genotype-dependent accumulation of cadmium. J Agric Food Chem 64:5241–5250
Innocenti G, Pucciariello C, Le Gleuher M, Hopkins J, de Stefano M, Delledonne M, Puppo A, Baudouin E, Frendo P (2007) Glutathione synthesis is regulated by nitric oxide in *Medicago truncatula* roots. Planta 225:1597–1602
Inouhe M (2005) Phytochelatins. Braz J Plant Physiol 17(1):65–78
Iqbal N, Masood A, Khan MI, Asgher M, Fatma M, Khan NA (2013a) Cross-talk between sulfur assimilation and ethylene signaling in plants. Plant Signal Behav 8:1–9
Iqbal N, Trivellini A, Masood A, Ferrante A, Khan NA (2013b) Current understanding on ethylene signaling in plants: the influence of nutrient availability. Plant Physiol Biochem 73:128–138
Irchhaiya R, Kumar A, Yadav A, Gupta N, Kumar S, Gupta N (2015) Metabolites in plants and its classification. World J Pharm Pharmaceut Sci 4:287–305
Irfan M, Hayat S, Ahmad A, Alyemeni MN (2013) Soil cadmium enrichment: allocation and plant physiological manifestations. Saudi J Biol Sci 20:1–10
Jost R, Altschmied L, Bloem E, Bogs J, Gershenzon J, Hähnel U, Hänsch R, Hartmann T, Kopriva S, Kruse C, Mendel RR (2005) Expression profiling of metabolic genes in response to methyl jasmonate reveals regulation of genes of primary and secondary sulfur-related pathways in *Arabidopsis thaliana*. Photosynth Res 86:491–508
Jozefczak M, Remans T, Vangronsveld J, Cuypers A (2012) Glutathione is a key player in metal-induced oxidative stress defenses. Int J Mol Sci 13(3):3145–3175

Kashiwagi T, Shindoh K, Hirotsu N, Ishimaru K (2009) Evidence for separate translocation pathways in determining cadmium accumulation in grain and aerial plant parts in rice. BMC Plant Biol 9(8). https://doi.org/10.1186/1471-2229-9-8

Kaur R, Yadav P, Sharma A, Thukral AK, Kumar V, Kohli SK, Bhardwaj R (2017) Castasterone and citric acid treatment restores photosynthetic attributes in *Brassica juncea* L. under Cd (II) toxicity. Ecotoxicol Environ Saf 145:466–475

Ke Q, Wang Z, Ji CY, Jeong JC, Lee HS, Li H, Xu B, Deng X, Kwak SS (2015) Transgenic poplar expressing Arabidopsis YUCCA6 exhibits auxin-overproduction phenotypes and increased tolerance to abiotic stress. Plant Physiol Biochem 94:19–27

Khan NA, Samiullah Singh S, Nazar R (2007) Activities of antioxidative enzymes, sulphur assimilation, photosynthetic activity and growth of wheat (*Triticum aestivum*) cultivars differing in yield potential under cadmium stress. J Agron Crop Sci 193(6):435–444

Khan NA, Singh S, Umar S (2008) Sulfur assimilation and abiotic stress in plants. Springer, Berlin

Khan MIR, Asgher M, Iqbal N, Khan NA (2013) Potentiality of sulphur-containing compounds in salt stress tolerance. In: Ahmad P, Azooz M, Prasad M (eds) Ecophysiology and responses of plants under salt stress. Springer, New York, pp 443–472

Khan GA, Bouraine S, Wege S, Li Y, de Carbonnel M, Berthomieu P, Poirier Y, Rouached H (2014a) Coordination between zinc and phosphate homeostasis involves the transcription factor PHR1, the phosphate exporter PHO1, and its homologue PHO1;H3 in *Arabidopsis*. J Exp Bot 65(3):871–884

Khan MA, Castro-Guerrero NA, Mendoza-Cozatl D (2014b) Moving toward a precise nutrition: preferential loading of seeds with essential nutrients over non-essential toxic elements. Front Plant Sci 5:51

Khan MIR, Nazir F, Asgher M, Per TS, Khan NA (2015) Selenium and sulfur influence ethylene formation and alleviate cadmium-induced oxidative stress by improving proline and glutathione production in wheat. J Plant Physiol 173:9–18

Khan NA, Asgher M, Per TS, Masood A, Fatma M, Khan MIR (2016) Ethylene potentiates sulfur-mediated reversal of cadmium inhibited photosynthetic responses in mustard. Front Plant Sci 7:1628

Khan MY, Prakash V, Yadav V, Chauhan DK, Prasad SM, Ramawat N, Sharma S (2019) Regulation of cadmium toxicity in roots of tomato by indole acetic acid with special emphasis on reactive oxygen species production and their scavenging. Plant Physiol Biochem 142:193–201

Khoudi H, Maatar Y, Gouiaa S, Masmoudi K (2012) Transgenic tobacco plants expressing ectopically wheat $H^+$-pyrophosphatase ($H^+$-PPase) gene *TaVP1* show enhanced accumulation and tolerance to cadmium. J Plant Physiol 169(1):98–103

Kim DY, Bovet L, Kushnir S, Noh EW, Martinoia E, Lee Y (2006) AtATM3 is involved in heavy metal resistance in *Arabidopsis*. Plant Physiol 140:1–11

Kim DY, Bovet L, Maeshima M, Martinoia E, Lee Y (2007) The ABC transporter AtPDR8 is a cadmium extrusion pump conferring heavy metal resistance. Plant J 50:207–218

Koprivova A, Kopriva S (2014) Molecular mechanisms of regulation of sulfate assimilation: first steps on a long road. Front Plant Sci 5:589

Korenkov V, Hirschi K, CrutchWeld JD, Wagner GJ (2007a) Enhancing tonoplast Cd/H antiport activity increases Cd, Zn, and Mn tolerance, and impacts root/shoot Cd partitioning in *Nicotiana tabacum* L. Planta 226:1379–1387

Korenkov V, Park S, Cheng NH, Sreevidya C, Lachmansingh J, Morris J, Hirschi K, Wagner GJ (2007b) Enhanced $Cd^{2+}$-selective root tonoplast-transport in tobaccos expressing *Arabidopsis* cation exchangers. Planta 225(2):403–411

Kováčik J, Klejdus B, Hedbavny J, Štork F, Bačkor M (2009) Comparison of cadmium and copper effect on phenolic metabolism, mineral nutrients and stress-related parameters in *Matricaria chamomilla* plants. Plant Soil 320(1–2):231–242

Krantev A, Yordanova R, Janda T, Szalai G, Popova L (2008) Treatment with salicylic acid decreases the effect of cadmium on photosynthesis in maize plants. J Plant Physiol 165 (9):920–931

Kum C, Wong E, Cobbett C (2009) HMA P-type ATPases are the major mechanism or root-to-shoot Cd translocation in *Arabidopsis thaliana*. New Phytol 181:71–78

Lanquar V, Lelièvre F, Bolte S, Hamès C, Alcon C, Neumann D, Vansuyt G, Curie C, Schröder A, Krämer U, Barbier-Brygoo H, Thomine S (2005) Mobilization of vacuolar iron by AtNRAMP3 and AtNRAMP4 is essential for seed germination on low iron. EMBO J 24(23):4041–4051

Lasat MM, Pence NS, Garvin DF, Ebbs SD, Kochian LV (2000) Molecular physiology of zinc transport in Zn hyperaccumulator *Thlaspi caerulescens*. J Exp Bot 51:71–79

Laspina NV, Groppa MD, Tomaro ML, Benavides MP (2005) Nitric oxide protects sunflower leaves against Cd-induced oxidative stress. Plant Sci 169:323–330

Lee S, An G (2009) Over-expression of OsIRT1 leads to increased iron and zinc accumulations in rice. Plant Cell Environ 32:408–416

Lee J, Bae H, Jeong J, Lee JY, Yang YY, Hwang I, Martinoia E, Lee Y (2003) Functional expression of a bacterial heavy metal transporter in *Arabidopsis* enhances resistance to and decreases uptake of heavy metals. Plant Physiol 133:589–596

Lee S, Ki YY, Lee Y, An G (2007) Rice P1B-type heavy-metal ATPase, OsHMA9, is a metal efflux protein. Plant Physiol 145:831–842

Lin YF, Aarts M (2012) The molecular mechanism of zinc and cadmium stress response in plants. Cell Mol Life Sci 69:3187–3206

Liu S, Yang R, Pan Y, Ma M, Pan J, Zhao Y, Cheng Q, Wu M, Wang M, Zhang L (2015) Nitric oxide contributes to minerals absorption, proton pumps and hormone equilibrium under cadmium excess in *Trifolium repens* L. plants. Ecotoxicol Environ Saf 119:35–46

Lochlainn SO, Bowen HC, Fray RG, Hammond JP, King GJ, White PJ, Graham NS, Broadley MR (2011) Tandem quadruplication of HMA4 in the zinc (Zn) and cadmium (Cd) hyperaccumulator *Noccaea caerulescens*. PLoS One 6:e17814

Lombi E, Tearall KL, Howarth JR, Zhao FJ, Hawkesford MJ, McGrath SP (2002) Influence of iron status on cadmium and zinc uptake by different ecotypes of the hyperaccumulator *Thlaspi caerulescens*. Plant Physiol 128:1359–1367

Lu LL, Tian SK, Yang X, Li TQ, He ZL (2009) Cadmium uptake and xylem loading are active processes in the hyperaccumulator *Sedum alfredii*. J Plant Physiol 166:579–587

Lu Q, Chen S, Li Y, Zheng F, He B, Gu M (2020) Exogenous abscisic acid (ABA) promotes cadmium (Cd) accumulation in *Sedum alfredii* Hance by regulating the expression of Cd stress response genes. Environ Sci Pollut Res 27:8719–8731

Lux A, Martinka M, Vaculik M, White PJ (2011) Root responses to cadmium in the rhizosphere: a review. J Exp Bot 62(1):21–37

Lysenko EA, Klaus AA, Pshybytko NL, Kusnetsov VV (2015) Cadmium accumulation in chloroplasts and its impact on chloroplastic processes in barley and maize. Photosynth Res 125:291–303

Maksymiec W, Krupa Z (2002) The in vivo and in vitro influence of methyl jasmonate on oxidative processes in *Arabidopsis thaliana* leaves. Acta Physiol Plant 24:351–357

Maksymiec W, Wojcik M, Krupa Z (2007) Variation in oxidative stress and photochemical activity in *Arabidopsis thaliana* leaves subjected to cadmium and excess copper in the presence or absence of jasmonate and ascorbate. Chemosphere 66:421–427

Mansour MM, Kamel EAR (2005) Interactive effect of heavy metals and gibberellic acid on mitotic activity and some metabolic changes of *Vicia faba* L. plants. Cytologia 70(3):275–282

Martinoia E, Klein M, Geisler M, Bovet L, Forestier C, Kolukisaoglu U, Muller-Rober B, Schulz B (2002) Multifunctionality of plant ABC transporters-more than just detoxifiers. Planta 214:345–355

Maser P, Thomine S, Schroeder JI, Ward JM, Hirschi K, Sze H, Talke IN, Amtmann A, Maathius FJM, Sanders D, Harper JF, Tchieu J, Gribskov M, Persans MW, Salt DE, Kim SA, Guerinot

ML (2001) Phylogenetic relationships within cation transporter families of *Arabidopsis*. Plant Physiol 126:1646–1667
Mason DL, Michaelis S (2002) Requirement of the N-terminal extension for vacuolar trafficking and transport activity of yeast YCF1p, an ATP-binding cassette transporter. Mol Biol Cell 13:4443–4455
Masood A, Iqbal N, Khan NA (2012) Role of ethylene in alleviation of cadmium-induced photosynthetic capacity inhibition by sulphur in mustard. Plant Cell Environ 35(3):524–533
Mendoza-Cózatl DG, Butko E, Springer F, Torpey JW, Komives EA, Kehr J, Schroeder JI (2008) Identification of high levels of phytochelatins, glutathione and cadmium in the phloem sap of *Brassica napus*. A role for thiol-peptides in the long-distance transport of cadmium and the effect of cadmium on iron translocation. Plant J 54:249–259
Metwally A, Finkemeier I, Georgi M, Dietz KJ (2003) Salicylic acid alleviates the cadmium toxicity in barley seedlings. Plant Physiol 132(1):272–281
Mills RF, Krijger GC, Baccarini PJ, Hall JL, Williams LE (2003) Functional expression of AtHMA4, a P1B-type ATPase of the Zn/Co/Cd/Pb subclass. Plant J 35:164–176
Miyadate H, Adachi S, Hiraizumi A, Tezuka K, Nakazawa N, Kawamoto T, Katou K, Kodama I, Sakurai K, Takahashi H, Satoh-Nagasawa N, Watanabe A, Fujimura T, Akagi H (2011) OsHMA3, a P1B-type of ATPase affects root-to shoot cadmium translocation in rice by mediating efflux into vacuoles. New Phytol 189:190–199
Mnasri M, Ghabriche R, Fourati E, Zaier H, Sabally K, Barrington S, Lutts S, Abdelly C, Ghnaya T (2015) Cd and Ni transport and accumulation in the halophyte *Sesuvium portulacastrum*: implication of organic acids in these processes. Front Plant Sci 6:156
Molitor M, Dechamps C, Gruber W, Meerts P (2005) *Thlaspi caerulescens* on non-metalliferous soil in Luxembourg: ecological niche and genetic variation in mineral element composition. New Phytol 165:503–512
Morel M, Crouzet J, Gravot A (2009) AtHMA3, a P1B-ATPase allowing Cd/Zn/Co/Pb vacuolar storage in *Arabidopsis*. Plant Physiol 149:894–904
Mori S, Uraguchi S, Ishikawa S, Arao T (2009) Xylem loading process is a critical factor in determining Cd accumulation in the shoots of *Solanum melongena* and *Solanum torvum*. Environ Exp Bot 67:127–132
Mudgal V, Madaan N, Mudgal A (2010) Heavy metals in plants: phytoremediation: plants used to remediate heavy metal pollution. Agric Biol J N Am 1:40–46
Nakanishi H, Ogawa I, Ishimaru Y, Mori S, Nishizawa NK (2006) Iron deficiency enhances cadmium uptake and translocation mediated by the $Fe^{2+}$ transporters OsIRT1 and OsIRT2 in rice. Soil Sci Plant Nutr 52:464–469
Nazar R, Khan MIR, Iqbal N, Masood A, Khan NA (2014) Involvement of ethylene in reversal of salt-inhibited photosynthesis by sulfur in mustard. Physiol Plant 152(2):331–344
Nazar R, Umar S, Khan NA (2015) Exogenous salicylic acid improves photosynthesis and growth through increase in ascorbate-glutathione metabolism and S assimilation in mustard under salt stress. Plant Signal Behav 10:e1003751
Nikiforova V, Freitag J, Kempa S, Adamik M, Hesse H, Hoefgen R (2003) Transcriptome analysis of sulfur depletion in *Arabidopsis thaliana*: interlacing of biosynthetic pathways provides response specificity. Plant J 33:633–650
Ning H, Zhang C, Yao Y, Yu D (2010) Overexpression of a soybean O-acetylserine (thiol) lyase-encoding gene GmOASTL4 in tobacco increases cysteine levels and enhances tolerance to cadmium stress. Biotechnol Lett 32(4):557–564
Nocito FF, Lancilli C, Giacomini B, Sacchi GA (2007) Sulfur metabolism and cadmium stress in higher plants. Plant Stress 1(2):142–156
Nocito FF, Lancilli C, Dendena B, Lucchini G, Sacchi GA (2011) Cadmium retention in rice roots is influenced by cadmium availability, chelation and translocation. Plant Cell Environ 34:994–1008
Ostrowski M, Ciarkowska A, Jakubowska A (2016) The auxin conjugate indole-3-acetyl aspartate affects responses to cadmium and salt stress in *Pisum sativum* L. J Plant Physiol 191:63–72

Panda SK, Patra HK (2007) Effect of salicylic acid potentiates cadmium-induced oxidative damage in *Oryza sativa* L. leaves. Acta Physiol Plant 29(6):567–575

Park SW, Li W, Viehhauser A, He B, Kim S, Nilsson AK, Andersson MX, Kittle JD, Ambavaram MM, Luan S, Esker AR (2013) Cyclophilin 20-3 relays a 12 oxophytodienoic acid signal during stress responsive regulation of cellular redox homeostasis. Proc Natl Acad Sci 110:9559–9564

Pence NS, Larsen PB, Ebbs SD, Letham DL, Lasat MM, Garvin DF, Eide D, Kochian LV (2000) The molecular physiology of heavy metal transport in the Zn/Cd hyperaccumulator *Thlaspi caerulescens*. Proc Natl Acad Sci U S A 97:4956–4960

Perfus-Barbeoch L, Leonhardt N, Vavasseur A, Forestier C (2002) Heavy metal toxicity: cadmium permeates through calcium channels and disturbs the plant water status. Plant J 32:539–548

Popova LP, Maslenkova LT, Yordanova RY, Ivanova AP, Krantev AP, Szalai G, Janda T (2009) Exogenous treatment with salicylic acid attenuates cadmium toxicity in pea seedlings. Plant Physiol Biochem 47(3):224–231

Qin S, Liu H, Nie Z, Rengel Z, Gao W, Li C, Zhao P (2020) Toxicity of cadmium and its competition with mineral nutrients for uptake by plants: a review. Pedosphere 30(2):168–180

Rady MM (2011) Effect of 24-epibrassinolide on growth, yield, antioxidant system and cadmium content of bean (*Phaseolus vulgaris* L.) plants under salinity and cadmium stress. Sci Hortic (Amsterdam) 129:232–237

Rea PA (2007) Plant ATP-binding cassette transporters. Annu Rev Plant Biol 58:347–375

Redjala T, Sterckeman T, Morel JL (2009) Cadmium uptake by roots: contribution of apoplast and of high- and low-affinity membrane transport systems. Environ Exp Bot 67:235–242

Redjala T, Zelko I, Streckeman T, Legue V, Lux A (2011) Relationship between root structure and root cadmium uptake in maize. Environ Exp Bot 71:241–248

Riesen O, Feller U (2005) Redistribution of nickel, cobalt, manganese, zinc, and cadmium via the phloem in young and maturing wheat. J Plant Nutr 28:421–430

Rodda MS, Li G, Reid RJ (2011) The timing of grain Cd accumulation in rice plants: the relative importance of remobilisation within the plant and root Cd uptake post-flowering. Plant Soil 347:105–114

Rodríguez-Serrano M, Romero-Puertas MC, Pazmino DM, Testillano PS, Risueño MC, Luis A, Sandalio LM (2009) Cellular response of pea plants to cadmium toxicity: cross talk between reactive oxygen species, nitric oxide, and calcium. Plant Physiol 150(1):229–243

Romera FJ, Alcantara E, De La Guardia MD (1999) Ethylene production by Fe deficient roots and its involvement in the regulation of Fe deficiency stress responses by strategy I plants. Ann Bot 83(1):51–55

Sadana US, Samal D, Claassen N (2003) Differences in manganese efficiency of wheat (*Triticum aestivum* L.) and raya (*Brassica juncea* L.) as related to root-shoot relations and manganese influx. J Plant Nutr Soil Sci 166(3):385–389

Salt DE, Prince RC, Pickering IJ, Raskin I (1995) Mechanisms of cadmium mobility and accumulation in Indian mustard. Plant Physiol 109:1427–1433

Sankaran RP, Ebbs SD (2008) Transport of Cd and Zn to seeds of Indian mustard (*Brassica juncea*) during specific stages of plant growth and development. Physiol Plant 132:69–78

Sarry JE, Kuhn L, Ducruix C, Lafaye A, Junot C, Hugouvieux V, Amekraz B (2006) The early responses of *Arabidopsis thaliana* cells to cadmium exposure explored by protein and metabolite profiling analyses. Proteomics 6(7):2180–2198

Sasaki A, Yamaji N, Yokosho K, Ma JF (2012) Nramp5 is a major transporter responsible for manganese and cadmium uptake in rice. Plant Cell 24:2155–2167

Sasaki-Sekimoto Y, Taki N, Obayashi T, Aono M, Matsumoto F, Sakurai N, Suzuki H, Hirai MY, Noji M, Saito K, Masuda T (2005) Coordinated activation of metabolic pathways for antioxidants and defence compounds by jasmonates and their roles in stress tolerance in *Arabidopsis*. Plant J 44:653–668

Sękara A, Poniedzialek M, Ciura J, Jędrszczyk E (2005) Cadmium and lead accumulation and distribution in the organs of nine crops: implications for phytoremediation. Polish J Environ Stud 14:509–516

Sgherri C, Stevanovic B, Navari Izzo F (2004) Role of phenolics in the antioxidative status of the resurrection plant *Ramonda serbica* during dehydration and rehydration. Physiol Plant 122 (4):478–485

Sharma SS, Dietz KJ (2009) The relationship between metal toxicity and cellular redox imbalance. Trends Plant Sci 14(1):43–50

Sharma KG, Mason DL, Liu G, Rea PA, Bachhawat AK, Michaelis S (2002) Localization, regulation, and substrate transport properties of Bpt1p, a *Saccharomyces cerevisiae* MRP-type ABC transporter. Eukaryot Cell 1:391–400

Sharma M, Gautam KH, Handique AK (2006) Toxic heavy metal stress in paddy: metal accumulation profile and development of a novel stress protein in seed. Indian J Plant Physiol 11:227–233

Sharma N, Hundal GS, Sharma I, Bharadwaj R (2012) Effect of 24-epibrassinolide on protein content and activities of glutathione-S-transferase and polyphenoloxidase in *Raphanus sativus* L. plants under cadmium and mercury metal stress. Terr Aquat Environ Toxicol 6:1–7

Shi GR, Cai QS, Liu QQ, Wu L (2009) Salicylic acid-mediated alleviation of cadmium toxicity in hemp plants in relation to cadmium uptake, photosynthesis, and antioxidant enzymes. Acta Physiol Plant 31(5):969–977

Shigaki T, Hirschi KD (2006) Diverse functions and molecular properties emerging for CAX cation/$H^+$ exchangers in plants. Plant Biol 8:419–429

Shimo H, Ishimaru Y, An G, Yamakawa T, Nakanishi H, Nishizawa NK (2011) Low cadmium (LCD), a novel gene related to cadmium tolerance and accumulation in rice. J Exp Bot 62:5727–5734

Shingu Y, Kudo T, Ohsato S, Kimura M, Ono Y, Yamaguchi I, Hamamoto H (2005) Characterization of genes encoding metal tolerance proteins isolated from *Nicotiana glauca* and *Nicotiana tabacum*. Biochem Biophys Res Commun 331:675–680

Singh S, Prasad SM (2016) Kinetin ameliorates cadmium induced toxicity on growth, pigments and photosynthesis by regulating antioxidant potential in tomato seedlings. Int J Sci Eng Appl Sci 2:1–10

Singh I, Shah K (2014) Exogenous application of methyl jasmonate lowers the effect of cadmium-induced oxidative injury in rice seedlings. Phytochemistry 108:57–66

Singh S, Singh A, Srivastava PK, Prasad SM (2018) Cadmium toxicity and its amelioration by kinetin in tomato seedlings vis-à-vis ascorbate-glutathione cycle. J Photochem Photobiol B: Biol 178:76–84

Song WY, Sohn EJ, Martinoia E, Lee YJ, Yang YY, Jasinski M, Forestier C, Hwang I, Lee Y (2003) Engineering tolerance and accumulation of lead and cadmium in transgenic plants. Nat Biotechnol 21:914–919

Song Y, Jin L, Wang X (2017) Cadmium absorption and transportation pathways in plants. Int J Phytoremed 19:133–141

Stolt JP, Sneller FEC, Bryngelsson T, Lundborg T, Schat H (2003) Phytochelatin and cadmium accumulation in wheat. Environ Exp Bot 49:21–28

Stroiński A, Chadzinikolau T, Giżewska K, Zielezińska M (2010) ABA or cadmium induced phytochelatin synthesis in potato tubers. Biol Plant 54(1):117–120

Takahashi R, Ishimaru Y, Senoura T, Shimo H, Ishikawa S, Arao T, Nakanishi H, Nishizawa NK (2011) The OsNRAMP1 iron transporter is involved in Cd accumulation in rice. J Exp Bot 62:4843–4850

Tamaoki M, Freeman JL, Pilon-Smits EA (2008) Co-operative ethylene and jasmonic acid signaling regulates selenite resistance in *Arabidopsis*. Plant Physiol 146(3):1219–1230

Tanaka K, Fujimaki S, Fujiwara T, Yoneyama T, Hayashi H (2007) Quantitative estimation of the contribution of the phloem in cadmium transport to grains in rice plant (*Oryza sativa* L.). Soil Sci Plant Nutr 53:72–77

Tholen D, Pons TL, Voesenek LA, Poorter H (2007) Ethylene insensitivity results in down-regulation of Rubisco expression and photosynthetic capacity in tobacco. Plant Physiol 144 (3):1305–1315

Tholen D, Pons TL, Voesenek LA, Poorter H (2008) The role of ethylene perception in the control of photosynthesis. Plant Signal Behav 3(2):108–109

Tufail A, Arfan M, Gurmani AR, Khan A, Bano A (2013) Salicylic acid induced salinity tolerance in maize (*Zea mays*). Pak J Bot 45:75–82

Ueno D, Milner MJ, Yamaji N, Yokosho K, Koyama E, Zambrano MC, Kaskie M, Ebbs S, Kochian LV, Ma JF (2011) Elevated expression of TcHMA3 plays a key role in the extreme Cd tolerance in a Cd-hyperaccumulating ecotype of *Thlaspi caerulescens*. Plant J 66:852–862

Uraguchi S, Kiyono M, Sakamoto T, Watanabe I, Kuno K (2009) Contributions of apoplasmic cadmium accumulation, antioxidative enzymes and induction of phytochelatins in cadmium tolerance of the cadmium-accumulating cultivar of black oat (*Avena strigosa* Schreb.). Planta 230:267–276

Uraguchi S, Kamiya T, Sakamoto T, Kasai K, Sato Y, Nagamura Y, Yoshida A, Kyozuka J, Ishikawa S, Fujiwara T (2011) Low-affinity cation transporter (OsLCT1) regulates cadmium transport into rice grains. Proc Natl Acad Sci U S A 108:20959–20964

Vázquez MN, Guerrero YR, González LM, de la Noval WT (2013) Brassinosteroids and plant responses to heavy metal stress. An overview. Open J Metal 3:34–41

Verbruggen N, Hermans C, Schat H (2009) Molecular mechanisms of metal hyperaccumulation in plants. New Phytol 181:759–776

Verret F, Gravot A, Auroy P, Preveral S, Forestier C, Vavasseur A, Richaud P (2005) Heavy metal transport by AtHMA4 involves the N-terminal degenerated metal binding domain and the C-terminal His11 stretch. FEBS Lett 579:1515–1522

Vert G, Grotz N, Dedaldechamp F, Gaymard F, Guerinot ML, Briata JF, Curie C (2002) IRT1, an *Arabidopsis* transporter essential for iron uptake from the soil for plant growth. Plant Cell 14:1223–1233

Vitti A, Nuzzaci M, Scopa A, Tataranni G, Remans T, Vangronsveld J, Sofo A (2013) Auxin and cytokinin metabolism and root morphological modifications in *Arabidopsis thaliana* seedlings infected with Cucumber mosaic virus (CMV) or exposed to cadmium. Int J Mol Sci 14 (4):6889–6902

Wang L, Chen S, Kong W, Li S, Archbold DD (2006) Salicylic acid pre-treatment alleviates chilling injury and affects the antioxidant system and heat shock proteins of peaches during cold storage. Postharvest Biol Technol 41(3):244–251

Wang C, Zhang S, Wang P, Hou J, Qian J, Ao Y, Li L (2011) Salicylic acid is involved in the regulation of nutrient elements uptake and oxidative stress in *Vallisneria natans* (Lour.) Hara under Pb stress. Chemosphere 84:136–142

Wang L, Xu Y, Sun Y, Liang X, Lin D (2014) Identification of pakchoi cultivars with low cadmium accumulation and soil factors that affect their cadmium uptake and translocation. Front Environ Sci Eng 8(6):877–887

Wang D, Liu Y, Tan X, Liu H, Zeng G, Hu X, Jian H, Gu Y (2015) Effect of exogenous nitric oxide on antioxidative system and S-nitrosylation in leaves of *Boehmeria nivea* (L.) Gaud under cadmium stress. Environ Sci Pollut Res 22:3489–3497

Wang J, Lin L, Luo L, Liao MA, Lv X, Wang Z, Tang Y (2016) The effects of abscisic acid (ABA) addition on cadmium accumulation of two ecotypes of *Solanum photeinocarpum*. Environ Monit Assess 188(3):182

Wei W, Chai T, Zhang Y, Han L, Xu J, Guan Z (2009) The *Thlaspi caerulescens* NRAMP homologue TcNRAMP3 is capable of divalent cation transport. Mol Biotechnol 41:15–21

Wojas S, Hennig J, Plaza S, Geisler M, Siemianowski O, Sklodowska A, Ruszczynska A, Bulska E, Antosiewicz DM (2009) Ectopic expression of *Arabidopsis* ABC transporter MRP7 modifies cadmium root-to-shoot transport and accumulation. Environ Pollut 157(10):2781–2789

Wong CKE, Cobbett CS (2009) HMA P-type ATPases are the major mechanism for root-to-shoot Cd translocation in *Arabidopsis thaliana*. New Phytol 181:71–78

Wu Q, Shigaki T, Williams KA, Han JS, Kim CK, Hirschi KD, Park S (2011) Expression of an *Arabidopsis* $Ca^{2+}/H^+$ antiporter CAX1 variant in *Petunia* enhances cadmium tolerance and accumulation. J Plant Physiol 168:167–173

Xiang C, Oliver DJ (1998) Glutathione metabolic genes co-ordinately respond to heavy metals and jasmonic acid in *Arabidopsis*. Plant Cell 10:1539–1550

Xin J, Zhao X, Tan Q, Sun X, Hu C (2017) Comparison of cadmium absorption, translocation, subcellular distribution and chemical forms between two radish cultivars (*Raphanus sativus* L.). Ecotoxicol Environ Saf 145:258–265

Xiong J, An L, Lu H, Zhu C (2009) Exogenous nitric oxide enhances cadmium tolerance of rice by increasing pectin and hemicellulose contents in root cell wall. Planta 230:755–765

Xiong J, Fu G, Tao L, Zhu C (2010) Roles of nitric oxide in alleviating heavy metal toxicity in plants. Arch Biochem Biophys 497:13–20

Xu J, Wang W, Yin H, Liu X, Sun H, Mi Q (2010) Exogenous nitric oxide improves antioxidative capacity and reduces auxin degradation in roots of *Medicago truncatula* seedlings under cadmium stress. Plant Soil 326:321–330

Xu B, Yu JY, Xie T, Li YL, Liu MJ, Guo JX, Li HL, Yu Y, Zheng CY, Chen YH, Wang G (2018) Brassinosteroids and iron plaque affect arsenic and cadmium uptake by rice seedlings grown in hydroponic solution. Biol Plant:1–7

Yamaguchi N, Mori S, Baba K, Kaburagi-Yada S, Arao T, Kitajima N, Terada Y (2011) Cadmium distribution in the root tissues of solanaceous plants with contrasting root-to-shoot Cd translocation efficiencies. Environ Exp Bot 71(2):198–206

Yan Z, Zhang W, Chen J, Li X (2015) Methyl jasmonate alleviates cadmium toxicity in *Solanum nigrum* by regulating metal uptake and antioxidative capacity. Biol Plant 59:373–381

Yoshida S, Tamaoki M, Ioki M, Ogawa D, Sato Y, Aono M, Nakajima N (2009) Ethylene and salicylic acid control glutathione biosynthesis in ozone-exposed *Arabidopsis thaliana*. Physiol Plant 136(3):284–298

Zhang H, Jiang Y, He Z, Ma M (2005) Cadmium accumulation and oxidative burst in garlic (*Allium sativum*). J Plant Physiol 162:977–984

Zhang L, Zhang L, Song F (2008) Cadmium uptake and distribution by different maize genotypes in maturing stage. Commun Soil Sci Plant Anal 39:1517–1531

Zhang C, He Q, Wang M, Gao X, Chen J, Shen C (2020) Exogenous indole acetic acid alleviates Cd toxicity in tea (*Camellia sinensis*). Ecotoxicol Environ Saf 190:110090

Zhao FJ, Hamon RE, Lombi E, McLaughlin MJ, McGrath SP (2002) Characteristics of cadmium uptake in two contrasting ecotypes of the hyperaccumulator *Thlaspi caerulescens*. J Exp Bot 53:535–543

Zhao FJ, Jiang RF, Dunham SJ, McGrath SP (2006) Cadmium uptake, translocation and tolerance in the hyperaccumulator *Arabidopsis halleri*. New Phytol 172:646–654

Zhao FY, Han MM, Zhang SY, Wang K, Zhang CR, Liu T, Liu W (2012) Hydrogen peroxide-mediated growth of the root system occurs via auxin signalling modification and variations in the expression of cell-cycle genes in rice seedlings exposed to cadmium stress. J Integr Plant Biol 54(12):991–1006

Zhao FY, Hu F, Zhang SY, Wang K, Zhang CR, Liu T (2013) MAPKs regulate root growth by influencing auxin signaling and cell cycle-related gene expression in cadmium stressed rice. Environ Sci Pollut Res 20(8):5449–5460

Zheng X, Chen L, Li X (2018) *Arabidopsis* and rice showed a distinct pattern in ZIPs genes expression profile in response to Cd stress. Bot Stud 59:22

Zhou ZS, Guo K, Elbaz AA, Yang ZM (2009) Salicylic acid alleviates mercury toxicity by preventing oxidative stress in roots of *Medicago sativa*. Environ Exp Bot 65(1):27–34

Zhu XF, Jiang T, Wang ZW, Lei GJ, Shi YZ, Li GX, Zheng SJ (2012) Gibberellic acid alleviates cadmium toxicity by reducing nitric oxide accumulation and expression of IRT1 in *Arabidopsis thaliana*. J Hazard Mater 239:302–307

# Agronomic Management Practices to Tackle Toxic Metal Entry into Crop Plants

# 16

Tatiana Pedron, Vitória Aparecida Procópio, Bruno Alves Rocha, and Bruno Lemos Batista

**Abstract**

Contamination of soil, water, and food crops by the main toxic elements (arsenic, cadmium, lead, and mercury) is a ubiquitous environmental problem that has resulted from natural sources and mainly human activities (industrialization, urbanization, and intensive agricultural practices). Since contaminated soil and water can affect the environment, agricultural safety, and human health, the remediation of these toxic metals is a challenging task. However, it is essential to ensure that remediation processes are safe and consider the limitation of each process. In this chapter, we explore soil contamination with arsenic, cadmium, lead, and mercury and summarize information about the methods employed to remediate each of these toxic elements. Conventional methods of remediation possess disadvantages in the form of environmental and financial burdens. This fact leads to the search for alternative approaches of remediation for contaminated sites. Compared with other technologies, phytoremediation is a cost-effective and environmentally friendly approach for the remediation of chemical elements contaminated soil. Current knowledge gaps and future research needs are identified to facilitate a better understanding in relation to capture, mobility, translocation, accumulation, and tolerance strategies for these chemical toxic elements by plants, to make remediation techniques more and more effective

**Keywords**

Remediation · Mitigation · Arsenic · Mercury · Lead · Cadmium · Food safety

T. Pedron · V. A. Procópio · B. L. Batista (✉)
Centro de Ciências Naturais e Humanas, Universidade Federal do ABC, São Paulo, Brazil
e-mail: bruno.lemos@ufabc.edu.br

B. A. Rocha
Instituto de Ciências Ambientais, Químicas e Farmacêuticas, Universidade Federal de São Paulo, Diadema, Brazil



K. Mishra et al. (eds.), *Sustainable Solutions for Elemental Deficiency and Excess in Crop Plants*, https://doi.org/10.1007/978-981-15-8636-1_16

## 16.1 Introduction

The presence of contaminants in the environment shows a need for solutions. Sustainable and economically viable solutions for detoxification and removal of toxic elements such as arsenic (As), cadmium (Cd), lead (Pb), and mercury (Hg) are welcome. These elements are released in the environment by anthropogenic activities such as mining, industrial, and agricultural. The environmental pollution by metals and metalloids in soils, waters, and sediments brings risks to human health since the cultivation of food in the contaminated environment causes direct contamination of the edible parts of vegetables and, consequently, all forms of life in a specific food chain (LeDuc and Terry 2005; Dikilitas et al. 2016).

Exposure to these elements through human feeding can cause several health damages, even at low concentrations. Currently, there is no minimum safe concentration of intake for As. Ingestion of As between 0.3 and 8 μg $kg^{-1}$ by body weight per day can increase the risk of cancers (lung, skin, and bladder). Exposure to Pb, Cd, and Hg can cause neurological, renal, and cardiovascular/reproductive problems (Flora et al. 2012; EFSA 2014; Kim et al. 2016; Jiang et al. 2017).

These elements enter the food chain usually through the soil, where they are present in the form of free ions, soluble complexes, and insoluble chemical species causing serious damage to plants. Besides, these inorganic compounds are not easily eliminated or converted into innocuous species by chemical or biological processes and, therefore, due to low solubility, persist in soils and waters (Dikilitas et al. 2016).

However, there are techniques used to remedy contaminated soils. Solidification, vitrification, thermal treatment, and others are examples of expensive techniques that present higher risks of contamination (they sometimes involve manipulation and transporting of contaminated soils to specific treatment sites) compared to other techniques. Remediation techniques with plants, management of soils and water, in most cases, are cheaper and more efficient. It is still possible to use simultaneously microorganisms, fertilizers, compounds containing phosphorus (P), silicon (Si), iron (Fe), and nanoparticles (Favas et al. 2014; Silva et al. 2020; Wang et al. 2019).

Phytoremediation is an interesting and advantageous process that uses vegetation to treat superficially contaminated soils, sediments, and waters (Schnoor 1997; LeDuc and Terry 2005). This mitigation mechanism is an alternative for physical and chemical corrective approaches. Moreover, the use of plants and microorganisms to minimize the possible impacts of environmental contamination by metals/metalloids is an eco-friendly tool (Sarwar et al. 2017).

The use of plants (*Leersia oryzoides, Brassica juncea, Salix* sp.) in phytoremediation has the fundamental role of decreasing the mobility/availability of elements in soils, through mechanisms such as adsorption, precipitation, exudation, and complexation. These immobilization mechanisms make it possible to decrease the leaching of these elements in soils and, consequently, to minimize groundwater contamination, decreasing possible contaminations of other crop plants that use that groundwater for irrigation (Tangahu et al. 2011; Sarwar et al. 2017).

There are several advantages in phytoremediation for crops and soils, such as increased biomass production, hyperaccumulation, wide distribution, tolerance to

**Table 16.1** Advantages and limitations of phytoremediation

| Technique | Contaminants | Plants | Advantages | Limitations |
|---|---|---|---|---|
| Phytoextraction | Cd, Ni, Cu, Zn, Pb, Se, As, and organic compounds | *Elsholtzia splendens*, *Alyssum bertolonii*, *Thlaspi caerulescens*, and *Pteris vittata* | Abundant biomass production in a short time | Slow growing and bioproductivity[a] |
| Phytostabilization | Metals precipitated in insoluble forms and compound inorganic and organic | *Haumaniastrum*, *Eragrostis*, *Ascolepis*, *Gladiolus*, and *Alyssum* | Less disruptive and enhances ecosystem restoration | Extensive fertilization or soil modification |
| Phytovolatilization | Hg, Se, and As | *Astragalus bisulcatus*, *Nicotiana tabacum*, *Liriodendron tulipifera*, or *Brassica napus* | Contaminant transformed into less toxic forms | Contaminant might accumulate in vegetation and fruits |

[a]Bio productivity is rather small and shallow root systems (Adapted: Gratão et al. 2005; Favas et al. 2014; Farraji et al. 2016)

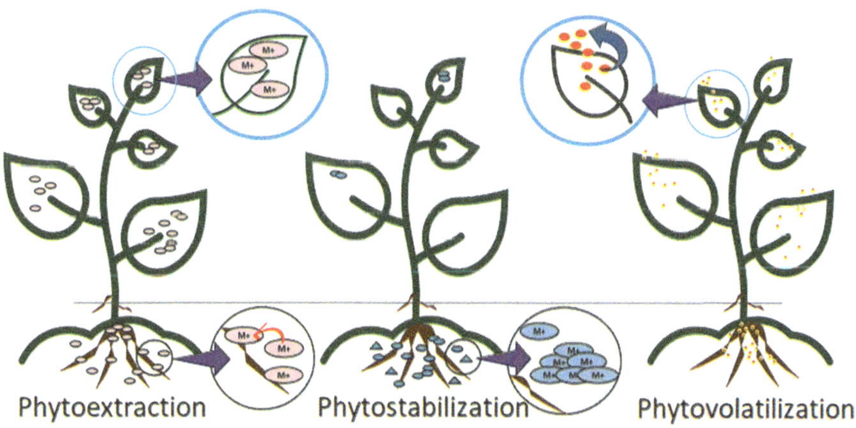


**Fig. 16.1** Mechanisms of phytoremediation. (Adapted: Favas et al. 2014)

the toxic effects for several elements, easier cultivation and harvesting. However, there are limitations in the long term, needing continuous monitoring of soils and water. Table 16.1 presents the advantages and limitations of phytoremediation (Tangahu et al. 2011; Sarwar et al. 2017).

Among phytoremediation are phytoextraction, phytovolatilization, and phytostabilization (Fig. 16.1). Phytoextraction promotes soil remediation through the absorption, translocation, and accumulation of elements by roots in which shoots can be harvested. Concerning phytovolatilization, it consists of the absorption and transpiration of elements by plants. The element is absorbed by the plant and then metabolized/modified and released by stomata into the atmosphere (Schmidt 2003; Tangahu et al. 2011). Phytostabilization is the decrease of mobility/availability of metals in the soil. This method of phytoremediation is complemented by the application of alkalizing agents, organic material, phosphorus (P), and others (Bolan et al. 2011; Favas et al. 2014). Therefore, in a polluted environment by potentially toxic elements, phytoremediation has an important role in food safety and economic issues (best solutions in relation to cost/beneficial soil decontamination).

## 16.2 Arsenic

### 16.2.1 Arsenic in Water and Soils

Arsenic is found in the environment in the form of minerals such as arsenopyrite, orpiment, realgar, arsenides, among others. Weathering and anthropogenic actions promote As-availability in soils, where As can be adsorbed and/or desorbed, precipitated, dissolved in different environmental compartments (Zhang and Selim 2008).

In natural environments, the primary concentrations of As in soils are variable and directly associated with the geological formation of each region, varying from $<1$ to 40 mg $kg^{-1}$ (Gomez-Caminero et al. 2001). However, concentrations in contaminated areas by anthropogenic sources vary from $<1$ to 250,000 mg $kg^{-1}$ (Pitten et al. 1999).

In general, the natural contamination of groundwater and soils occurs through the dissolution and desorption of rocks containing As (Khalid et al. 2017). Arsenic concentrations in water may vary according to the body of water and the location. In oceans, concentrations vary from 0.15–8 μg $L^{-1}$. In surface waters, the average concentrations are $<10$ μg $L^{-1}$. The concentrations in groundwater vary according to the mineralogical characteristics of the aquifer (Gomez-Caminero et al. 2001; Sarkar and Paul 2016).

Several factors can influence the availability of As in groundwater and soil, such as organic matter, ions, dissolution and precipitation, adsorption/desorption (Sarkar and Paul 2016). In the environment, As has relative mobility due to pH and Eh. In reducing conditions (anaerobic environments), with low Eh (0–200 mV), $As^{3+}$ (arsenite) is predominant, having higher solubility. In oxidizing conditions (aerobic environment) with high Eh (200–500 mV), $As^{5+}$ (arsenate) is prevalent (Heikens et al. 2007). This indicates that the As-form in soils is strongly dependent on pH and Eh. Arsenite is generally less adsorbed compared to $As^{5+}$ for a variety of solvents. Predominantly in aerobic soils, $As^{5+}$ is adsorbed in amorphous Fe (hydr)oxides, which favor its immobilization. In anoxic conditions, in which there is a decrease in Eh (pH $> 7$) compared with aerobic soils, $As^{5+}$ is more available/launched in the soil and converted to $As^{3+}$ (Goldberg 2002; McLaughlin et al. 2011; Ma et al. 2017).

Arsenate and $As^{3+}$ (inorganic chemical species of As) in soils can be methylated by microorganisms, converting to organic As chemical species such as monomethylarsonic acid (MMA) and dimethylarsinic acid (DMA). However, inorganic As are the most common forms found in soils compared to organic As. In soils, methylated species not only occur through the methylation of inorganic As, but also through external addition, such as the use of fertilizers/pesticides containing methylated compounds of As (Turpeinen et al. 1999; Mandal and Suzuki 2002; Lomax et al. 2012; Cai et al. 2015). Concisely, As, which has a high oxygen binding capacity, transforms in several chemical species in soils, especially anionic As ($[AsO_3]^{-3}$ and $[AsO_4]^{-3}$), that are easily available for absorption by plants (McLaughlin et al. 2011).

### 16.2.2 Arsenic in Plants

Several deleterious effects can be caused to the plant under exposure to As, such as low growth, productivity, and fertility, decrease of biomass, inhibition of root extension/proliferation, and impacts in morphology and physiological processes (Abbas et al. 2018). Plant exposure to inorganic As ($As^{3+}$ and $As^{5+}$) is more deleterious than to organic species (MMA and DMA). Physical, chemical, biological characteristics of soils have played an important role in the availability of As and the interconversion of inorganic species ($As^{3+} \leftrightarrow As^{5+}$) for plant uptake (Farooq et al. 2016). Arsenic uptake (inorganic and organic chemical species) by plants is dependent on several factors such as plant species/cultivar, irrigation, nitrogen (N) and phosphorous (P) availability, soil conditions, and others (Abedin et al. 2002; Lomax et al. 2012, Silva et al. 2020).

Abedin et al. (2002) did an experiment with *Oryza sativa* L. covering two growing seasons, rice varieties Boro (dry season) and rice varieties Aman (wet season). They showed that $As^{3+}$ and $As^{5+}$ uptake by Boro varieties was lower than that for Aman varieties under the same conditions. In plants, $As^{5+}$can be reduced to $As^{3+}$ by HAC1 (High Arsenic Content 1 enzyme). This is an enzyme responsible for $As^{5+}$ reductase activity. The absence of HAC1 in roots promotes an increase of As accumulation up to 300 times (Chao et al. 2014). Besides the influence of plant varieties, irrigation and P/N are important factors that influence As accumulation in plants. According to Silva et al. (2020), the concentration of As decreased as a function of the increased water tension (decreased irrigation) in three evaluated rice varieties (BRS Sinuelo, Puitá INTA CL, BRS Pampa). This is because in more oxygenated environments As becomes less soluble. Lomax et al. (2012) demonstrated that under induced deficiency of N and P, tomatoes (*S. lycopersicum* L.) significantly decreased the concentration of As in the roots and shoot compared to controls.

However, there are specific genetic characteristics of each plant conditioning the uptake, absorption, and transport of As. Examples are genes involved in the efflux of As in plants such as arsA and arsB ($As^{3+}$ efflux), arsC ($As^{5+}$ reduction), aoxB ($As^{3+}$ oxidation), and arsM ($As^{3+}$ methylation). Moreover, the plants have specific transporters as *AtNIP5*; *1*, *AtNIP6*; *1*, *AtNIP7*; *1*, *OsPht1*; *8*, *OsNIP2*; *1*, and others. Aquaporin channels such as *AtNIP5*; *1*, *AtNIP6*; *1*, *AtNIP7*; *1* are responsible for transporting As in rice via roots. Specifically, transporters such as *OsPht1*; *8* are responsible for the efficient uptake of $As^{5+}$, once $As^{5+}$ has a chemical structure analogous to inorganic P. Arsenite is readily absorbed by root cells passively due to the concentration gradient which has a direct relationship with *OsNIP2; 1* (Lomax et al. 2012; Li et al. 2016a; Chen et al. 2017; Hasanuzzaman et al. 2018; Roychowdhury et al. 2018; Cao et al. 2020). Other channels with Lsi1 (Low silicon rice 1) and Lsi2 (Low silicon rice 2) belonging to the aquaporin family are important transporters of Si. However, they are the main entry route for the transport of $As^{3+}$ (Khan and Gupta 2018).

Regarding organic As (MMA and DMA) in rice, the transport occurs by Si channels (OsLsi1). Arsenic-methylation occurs in soils, by microorganisms, since the plants cannot methylate the inorganic species of As. Lomax et al. (2012) proved this observation by treating rice (*Oryza sativa* L.) and tomato (*S. lycopersicum* L.) with $As^{3+}$ (1859 mg $kg^{-1}$) or DMA (1859 mg $kg^{-1}$). The plants presented >97% of the As-specie used in the treatment during all exposure periods (tillering, the extension of shoots, and flowering).

### 16.2.3 Arsenic Remediation by Agronomic Management

Contamination by As in soils creates not only serious problems for plantations but also economic and health public issues. There are currently several techniques for removing As from soils such as excavation, vitrification, and soil flushing (Table 16.2).

In the late twentieth century, contaminated soils by potentially toxic elements were recovered using soil excavation and burial at the average cost of US$ 247.1 per $m^2$, accounting in the USA alone for a total expenditure of about US$ 300 billion (Raskin et al. 1997; Gonzaga et al. 2006). However, these expensive techniques are hard to use in areas and cause an environmental disturbance. Phytoremediation techniques using As hyperaccumulators such as *P. vittata* are cheaper (around US $ 11.6 per $m^2$). *Pteris vittata* accumulates high rates of As at concentration proportional to As concentrations in soil (Wan et al. 2016).

Kumar et al. (2017) conducted a study in which uniformly sized seedlings of *Pteris pinnata* with fully developed the first two leaves were exposed to various concentrations of As (37, 92, 184, 276, and 368 mg $kg^{-1}$) for 7 days. They demonstrated an accumulation of As ranging from 1129 to 3322 mg $kg^{-1}$ dry weight in roots and 345 to 3662 mg $kg^{-1}$ dry weigh in shoots. However, other plants have higher efficiency in accumulating As in their roots. *P. australis* accumulates about 32 mg $kg^{-1}$ probably due to factors such as higher surface area per unit mass and the tubular extension of specialized epidermal cells from and root hairs, favoring higher uptake of ions (Ghassemzadeh et al. 2008). Yang et al. (2018) conducted experiments in pots to evaluate the effects of monoammonium phosphate ($NH_4H_2PO_4$) and citric acid on the absorption of As by *P. vittata* in contaminated

**Table 16.2** Techniques for removal of arsenic and other elements from contaminated soils

| Techniques | Description |
|---|---|
| Excavation | The method consists of physically removing the contaminated soil to designated landfills. Requires special transportation and destination requirements |
| Vitrification | A durable chemical method, resistant to leaching, in which As is chemically bound within a matrix of vitreous mass, forming silico-arsenates |
| Solidification | The method consisting of leveling or physically protecting the soil within a stabilized mass |
| Soil flushing | The method has the purpose of mobilizing the contaminant through dissolution or emulsification using water, chemical, or organic products without excavating the contamination material itself |
| Soil venting | Combines steam extraction well with blowers or vacuum pumps to remove vapors (volatile compounds) of soil and thus reduce the levels of contaminant residues |
| Soil washing | A washing solution is applied to the unexcavated contaminated zone by flooding or sprinkling to extract pollutants from the soil |
| Soil extracting | Single stage chemical extractors are used for soil decontamination, in which neutral salt-based extracts can be applied; chelator-based extracts; acid-based extracts; or synthetic root exudate-based extracts |

(Adapted: Johnson et al. (1990), Abumaizar and Smith (1999), USEPA (2002), Gonzaga et al. (2006), van der Ent et al. (2019))

soils containing between 26 (fluvo-aquic soil) and 210 (brown soil) mg $kg^{-1}$ of As. The use of $NH_4H_2PO_4$ and citric acid improved the bioavailability of As in the soil and consequently the absorption of As by the plant up to six times. The use of $NH_4H_2PO_4$ was more efficient compared to citric acid. Leather leaves *P. vittata* cultivated in fluvo-aquic soil had an average concentration of 515 mg $kg^{-1}$, while those cultivated in brown soil had an average concentration of 2340 mg $kg^{-1}$.

Studies have been carried out to decrease the absorption of As through irrigation management. Silva et al. (2020) observed that As concentrations decreased due to the increase in water tension (decreased irrigation) for three rice varieties (BRS Pampa, BRS Sinuelo, and Puitá INTA CL). This water management was carried out in two cycles, in which it was 121–135 days for BRS Sinuelo CL and Puitá INTA CL and 106–120 days for BRS Pampa. The plants were grown under different water management systems (water tensions varying from 10, 15, and 20 kPa) considering the phenological phases of the crop cycle (vegetative, reproductive, and maturation). Significant decrease in As concentration, approximately between 200 and 400 μg $kg^{-1}$, occurred during the cycle reproductive 1 (R1) of the plant, underwater tension of 20 kPa. Another study performed by Rahaman et al. (2011) showed an As decrease around 24% in the grains under an intermittent irrigation system. This regime consisted of irrigating the plants from the 15th to 40th days after the transplant, decreasing the water levels and waiting 2 to 4 days to irrigate again.

The use of stabilizing and oxidizing agents, such as Fe, Al, and Mn oxides is another way for As mitigation in soils (Li et al. 2019a; Mikkonen et al. 2019; Wang et al. 2019). Wang et al. (2020a) used the addition of biogenic manganese oxide (BMO) in the soil for As-immobilization. BMO was synthesized by *P. putida* strain

MnB1, a typical $Mn^{2+}$ oxidizing bacterium, and other compounds (ammonium iron sulfate, sodium citrate, and sodium pyrophosphate). Biogenic manganese oxide was added in concentrations ranging from 0.013% to 0.067% in soils containing As (24.9 mg $kg^{-1}$) during different periods (15, 30, 45, and 60 days). After treatment, $As^{3+}$ was fully converted to $As^{5+}$ on the BMO surface. Moreover, it was observed that a significant decrease in leaching of As with the addition of BMO, in which the solubility of As decreased between 49% and 67% according to the treatment.

Li et al. (2019a) developed their experiments with soils contaminated by As (73 mg $kg^{-1}$) and used Mn for soil remediation. The study used α-$MnO_2$ nanorods at rates of 0.2 to 2.0% (w/w). α-$MnO_2$ was added to pots containing 15 kg of soil treated with fertilizers containing N, P, and K and irrigated with water until saturation, followed by drainage. After 7 days, germinated rice seeds were transplanted into pots and cultivated under irrigation until 105 days, when they were collected. In general, the application of α-$MnO_2$ improved the residual fraction of As and decreased the effectiveness/reactivity of As forms in the soil. The treatment using 1% α-$MnO_2$ increased the residual As to 57%, indicating thus a decrease in As forms in the soil. Concerning husked rice, the total concentration of As decreases between 43% and 77% (compared to control: 0.96 mg $kg^{-1}$), depending on the rate of α-$MnO_2$ added to the soil. The lowest concentration of As (0.22 mg $kg^{-1}$) in rice was obtained from the 2% α-$MnO_2$.

Other studies have been carried out to decrease the contamination of As in plants through interaction with other chemical elements, such as Si. Arsenite, for example, shares the same transporter of silicic acid, in which Si is always combined with oxides or silicates, being absorbed as silicic acid for plants and converted to amorphous Si (Ma et al. 2008). Gang et al. (2018) cultivated rice (*Oryza sativa* L.) in the presence of Si (20 g $SiO_2$ gel/kg of soil) under flood conditions. The fertilization with Si significantly increased the concentration of As in the roots of the rice, about 61%; however, in the shoots, leaf, husk, and grain the concentrations decreased from 24% to 52%. The treatment with Si increased to 59% the concentration of inorganic As in the roots. Conversely, inorganic As decreased in a range of 20–52% in the shoots, leaf, bark, and grain. In the roots, the results indicated that the highest concentration of inorganic As was approximately 80% for $As^{5+}$. Comparing the treatments with Si, the percentages of $As^{3+}$ and $As^{5+}$ were decreased from approximately 40–17%, respectively. Khan and Gupta (2018) showed that growth parameters such as seed germination and root length of the aerial part of *Oryza sativa* L. treated only with $As^{3+}$ decreased seed twinning by 42%. However, compared to the combination $As^{3+}$ + Si, they found increased seed germination by 40%. After treatment with $As^{3+}$, the length of the aerial part and the root decreased.

Phosphorous has a similar chemical structure of $As^{5+}$, entering cells through the same transporter (Zhao et al. 2009). Therefore, several studies were carried out to use P to mitigate the absorption of As. However, these studies only demonstrate a decrease in the toxic effects of As when the plant is treated with P. The results do not demonstrate effectiveness for the inhibition of As uptake since by competition by the same cell transporter. Only at much higher levels of P than the plant requirements, As uptake was decreased due to the increased adsorption of As in the Fe plaques present in roots (Hossain 2006; Zhao et al. 2009; Lee et al. 2016; Li et al. 2016b).

Iron plaques have a high affinity for $As^{5+}$ and $HPO_4^{2-}$, functioning as a deposit or source of these compounds depending on the physical–chemical conditions of the soil. Besides, its presence in the roots increases the absorption of $As^{3+}$ by roots (Chen et al. 2005; Hossain 2006; Zhao et al. 2009). According to Voegelin et al. (2007), $As^{5+}$is sequestered to the maximum extent (between 70% and 80%) by Fe plaques. Moreover, Fe and Fe plaques influence the accumulation of various micro- and macronutrients by the plant (Panthri and Gupta 2019).

There is another plant defense for As after its uptake. Arsenite can be bounded by phytochelatins (PCs) and stored into vacuoles of cell roots. Phytochelatins are peptides containing thiol group (-SH) which bounds $As^{3+}$.In plants, PCs have a specific role in the translocation and accumulation of As, in which the production of PCs and the formation of As-PCs in roots can decrease the translocation of As to the shoots and grains of rice (Batista et al. 2014; Zhang et al. 2009). Therefore, selecting/breeding cultivars focusing on PCs may play also an important role for As mitigation in plants. Finally, the use of microorganisms is also used for As mitigation in rice grains. Segura et al. (2018) applied two fungi (*Aspergillus* sp. and *Penicillium* sp.) isolated from paddy rice rhizosphere in soils (background and spiked with 5 mg $kg^{-1}$ $As^{5+}$) during rice cultivation. The authors reported an increase in plant growth and higher levels of the less toxic As chemical specie (organic As) in grains treated with microorganisms, decreasing the risk of As ingestion by rice consumption.

## 16.3 Cadmium

### 16.3.1 Cadmium in Water and Soils

Cadmium (Cd) is a naturally occurring ubiquitous element of environmental concern. This metal is found in association with Zn, Pb, and Cu sulfidic ore deposits, as well as some phosphate rock formations. Uses of Cd include batteries, pigments, stabilizers, pesticides, coatings, and some alloys (Faroon et al. 2012).

Geogenic and anthropogenic sources can increase the concentrations of Cd in soils and water. The major sources of natural Cd emissions are weathering of rocks or airborne soil particles launched from volcanoes (ATSDR 2008; Faroon et al. 2012). However, the main source of Cd in the environment occurs through anthropogenic activities. The major emissions of Cd are through industrial processes and product use and energy use in industry (EEA 2019).

The WHO Guidelines for Drinking Water Quality recommend a guideline value for Cd of 3 μg $L^{-1}$. It is necessary to point that even in low doses the Cd is carcinogenic for humans (IARC 1993; WHO 2011b; USEPA 2009). Cadmium occurs usually up to 1 μg $L^{-1}$ in natural waters. In uncontaminated soils worldwide, the average abundance of Cd ranges from 0.02 to 25 mg $kg^{-1}$ and they are associated with the rock formation in each location (Kubier et al. 2019).

Several physical and chemical properties (Fig. 16.2) influence the availability of Cd in the environment. For example, the increase in pH from 4.5 to 6.5 results in

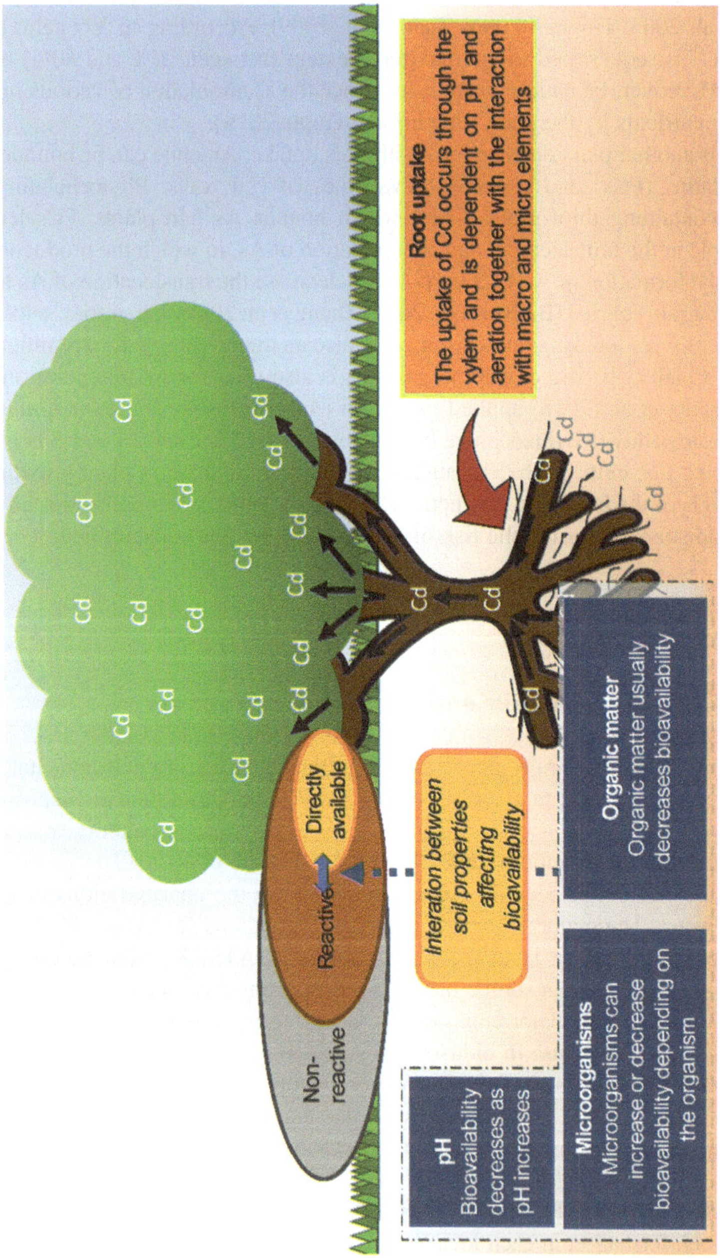


**Fig. 16.2** Properties that affect plant uptake of Cd. (Adapted: Meter et al. 2019; Shahid et al. 2019)

higher absorption of Cd in soil particles, which consequently decrease the Cd extractability. Concerning Eh, the increases in Eh increase the soluble portion of Cd. Nevertheless, other factors can affect the availability of Cd, such as the presence of organic material in the environment that favors the complexation of Cd with inorganic carbon and possibly the precipitation of $CdCO_3$. This is explained because, during the degassing of organic matter, there is the formation of dissolved carbon, decreasing the solubility of Cd (He and Shing 1993; Pinto et al. 2004; Pierangeli et al. 2005; Esfandbod 2010; Khan et al. 2017; Kubier et al. 2019; Meng et al. 2019).

Moreover, biotic factors play an important role in the solubility and mobility of Cd in soil. Microorganisms associated with plants can modulate the expression of defense-related proteins and the modulatory expression of transport of non-essential elements. In addition, bacteria have an important role in the solubility of Cd. These bacteria are important for the degradation of organic matter under anoxic conditions (Meng et al. 2019; Shahid et al. 2019).

### 16.3.2 Cadmium in Plants

The exposure to Cd inhibits many enzymes such as superoxide dismutase, catalase, ascorbate peroxidase, besides alter cell division, leading to chromosomal aberration, low growth, decreased photosynthesis, and productivity in plants. The toxicity of Cd in plants is generally noticeable through stunting and chlorosis, in which exposure of the plant to high concentrations of Cd can suppress the uptake of Fe, Mn, Ca, Mg, P, K, and water (Das et al. 1997; Benavides et al. 2005; Gill et al. 2011). However, there is a relationship between Cd and Zn, in which both elements share the same carrier protein in most plants, where the enhancement between Cd/Zn is governed by the relative concentration of Cd in the soil. A low Cd/Zn ratio allows the plant to quickly absorb Zn and limit Cd entry (Meter et al. 2019).

One of the first entry points of such elements into the plants is the rhizosphere. Here, the elements are transported through the root hairs and epidermal cells from endoderm to parenchymal cells, where they can enter the xylem conducting vessels. In general, it is in this region of rhizosphere that the first limitation of Cd entry in the plant occurs, through diffusion, selectivity, and uptake control processes. However, processes of reduction of $Fe^{2+}$ in $Fe^{3+}$ or $Mn^{5+}$ and $Mn^{3+}$ in $Mn^{2+}$ can favor the mobilization of Cd (Welch and Norvell 1999; Song et al. 2017).

Cadmium has no specific pathway of entry into the plant cells. However, there are transporters and proteins that favor the uptake of Cd by roots, such as ZIP, OsIRTs, Yellow-STripe 1-Like, NRAMP, and LCT1 (Shahid et al. 2016). Through biochemical pathways controlled by different types of genes, the physiological tolerance mechanism of Cd is different in each plant according to variety and species, as *Triticum aestivum* and *Triticum turgidum* var. *durum* (Das et al. 1997; Benavides et al. 2005). Experiments carried out with the mustard species *Brassica juncea*

L. showed different tolerance in different cultivars, in which the cultivar Alankar showed higher tolerance to Cd than the cultivar RH30. Alankar presented a lower concentration of Cd and stress, higher dry mass, and antioxidant activity (Gill et al. 2011).

### 16.3.3 Cadmium Remediation by Agronomic Management

Among various technologies to remediate Cd contaminated sites, phytoremediation is a low cost and environmentally sustainable alternative that presents efficiency in mobilizing, capturing, and absorbing Cd from soils. *Lupinus* species have been used in contaminated soils containing Cd concentrations of 9, 18, and 27 mg $kg^{-1}$. This plant showed tolerance of 88%, 82%, and 49%, respectively. Fifty-eight percent of Cd was retained in the roots regardless of treatment (Ehsan et al. 2009).

Lan et al. (2020) used the variety of ramie (*Boehmeria nivea* L.) together with organic agents such as coconut shell, mushroom residue, peanut bran, and bone meal, in addition to inorganic agents such as Fe–Si–Ca. In soils, the application of both organic and inorganic agents significantly increased the pH, decreasing the soil concentration of Cd extractable with $CaCl_2$ by more than 90% and consequently increasing the availability of other elements such as P and S.

In addition to the use of plants for Cd phytoremediation, phytostabilizers agents can be used. Chen et al. (2020) used plant growth regulators [indole-3-acetic acid, gibberellin A3 (GA3), 6-Benzylaminopurine (6-BA), and 24-epibrassinolide (EBL)] to improve stress tolerance and phytoremediation of Cd in mustard (*Brassica juncea* L.). The use of these regulators enabled less formation of lipid peroxidation and active oxygen in the plant and increased activities of antioxidant enzymes such as superoxide dismutase, peroxidase, catalase, and ascorbate peroxidase. The results suggested that plant growth regulators might improve stress tolerance and phytoremediation efficiency using *Brassica juncea* in soils contaminated by Cd.

It is also possible to control plant exposure to Cd through irrigation management (e.g., *Oryza sativa* L). According to Silva et al. (2020), increasing soil water tension (water tensions ranging from 10, 15, and 20 kPa) significantly increases the accumulation of Cd in all phenological phases of *Oryza sativa* L. Cadmium concentrations increased by more than 50 mg $kg^{-1}$ in relation to the vegetative phase in the three cultivars (BRS Pampa, BRS Sinuelo, and Puitá INTA CL) compared to the reproductive phase. Another study performed in rice fields in Hangzhou (China) by Hu et al. (2015), showed that the Cd concentration in rice decreased from 0.21 (aerobic) to 0.02 mg $kg^{-1}$ (flooding) with increased irrigation, varying aerobic conditions, and constant flooding in soils containing 0.48 mg $kg^{-1}$.

Additionally, nanoparticles have shown great potential for the treatment of contaminated soil due to the small particle size, large surface area, high reactivity, and controllable soil deliverability. For instance, Adrees et al. (2020) used Fe-nanoparticles (Fe-NPs) for remediation of soil contaminated with wastewater from the city of Multan in Pakistan. In pots containing 5 $kg^{-1}$ contaminated soil with Cd, germinated *Triticum aestivum* (wheat) seedlings were planted and then Fe-Nps were added in concentrations of 25, 50, and 100 mg $kg^{-1}$. These seedlings were cultivated during the Rabi season (crops are sown in the winter and harvested in the spring in regions of India and Pakistan) under two irrigation regimes (normal conditions at 70% water holding capacity, WHC, and under stress water condition of 35% WHC). Cadmium concentrations were 4.2, 8.8, and 0.21 mg $kg^{-1}$ in shoots, roots, and grains, respectively, in soils treated with 100 mg $kg^{-1}$ of Fe-NPs under normal irrigation conditions. The hydraulic stress regime and without treatment with Fe-NPs, Cd concentrations were twofold higher for shoots (7.44 mg $kg^{-1}$) and roots (17.3 mg $kg^{-1}$) and were sixfold higher in grains (1.26 mg $kg^{-1}$). Qiao et al. (2017) used $Fe_3$ $(PO_4)_2$-NPs and biochar (10:1 ratio) for soil remediation by Cd-contamination (5 mg $kg^{-1}$) during germination and development of cabbage mustard once. The results indicated an effective immobilization of Cd in the soil (81% after 28 days of cultivation) and a decreased bioaccessibility, indicating that the biochar compound + $Fe_3(PO_4)_2$-NPs may inhibit the Cd uptake (from 45% to 70%) for belowground and aboveground parts of cabbage mustard.

The use of biofertilizers can also influence the plant's uptake of Cd. Wang et al. (2019) used biofertilizers in soil contaminated with Cd (1.32 mg $kg^{-1}$), obtained from the province of Hunan (China) in the cultivation of *Oryza sativa* L. Pots containing 500 g of soil were used in which the biofertilizers (3%) added to the soil consisted of different mixtures such as bovine manure supplemented with micronutrients, organic substrates, oil and pig manure, and bagasse and straw fermentation. Cultivation was performed for 5 weeks and irrigation maintained at 75% (WHC) until collection. The treatment that showed the highest result consisted of using the biofertilizer containing bovine manure, in which it showed an increase in biomass of 85% and 48% for root and shoot, respectively, compared to control. Besides, the use of the same biofertilizer resulted in 72% of Cd-reduction in roots compared to controls. Therefore, they demonstrated that the remediation of Cd is associated with the type of biofertilizer used once there is a direct relationship with microorganisms (Bacteroidetes, Gemmatimonadetes, and Proteobacteria) in the rhizosphere that stabilize Cd in soils.

Regarding the use of microorganisms, Jiang et al. (2009) used bacteria of the genus *B. subtilis* 38 (mutated) together with biofertilizer (NovoGro) in soils containing 10 mg $kg^{-1}$ of Cd. The concomitant use of bacteria and fertilizer under ideal conditions (temperature, pH, and humidity) allowed a better immobilization capacity of Cd in soil, about 0.87 mg $kg^{-1}$ compared to the controls (3.34 mg $kg^{-1}$). In addition, they observed a significant decrease in the extractable fraction of Cd in soils (about 8.7%).

Other studies have used sulfate (S) to mitigate Cd in *Triticum aestivum*. Shi et al. (2020) cultivated *Triticum aestivum* in contaminated soils (mining activities) with Cd (3.0 mg $kg^{-1}$) from the Shantou region (China). Two cultivars (Jinmai 85 and Mianmai) from *Triticum aestivum* were grown in 6 kg pots during the wheat growing season (October to May) at three levels of S concentration, at 0, 60, and 120 mg $kg^{-1}$. The water irrigation was between 65 and 70% of WHC. The samples were collected at the flowering and maturation phase. The application of S affected the concentration of Cd in the root and decreases the concentration in the grains.

Compared to the controls, the two cultivars under S exposure (120 mg kg$^{-1}$) showed 39% of Cd increase in roots, while in the grains they showed a decrease of 24%. Moreover, they reported an increase in the concentration of N, P, and S in both cultivars, N = 25 and 33%, P = 30 and 37%, and S = 57 and 68% when applied S at 60 and 120 mg kg$^{-1}$, respectively.

There are several remediation techniques for Cd in the soil, and these can be combined for a better efficiency of Cd capturing and mobilization in soils. Among the different techniques, Li et al. 2019b used soil conditioners to remedy and fertilize a weak alkaline agricultural land contaminated with Cd (2.93 mg kg$^{-1}$) in situ field trials. The soil was divided into five equal plots and conditioned separately for treatments and control. The treatments consisted of MFA (microbial agent with fulvic acid), biochar (pyrogenic organic carbon material of strong polarity), sepiolite (a mineral with a porous structure, with high ion exchange capacity), and finally a mixture of the three treatments (the dose of each component was 1/3 of single application). In these soils, wheat was grown under normal production and management conditions. The treatments showed a decrease in the soil available Cd by 39–71% and in the *Triticum aestivum* of 42–87%. The treatment with sepiolite showed a higher decrease in Cd availability in soils (0.41 mg kg$^{-1}$, 71%). However, the yielding with sepiolite (6%) was lower in relation to the treatment with MFA (9%) and the mixture of treatments (17%). The mixture of treatments provided a *Triticum aestivum* production of 7590 kg/hm$^2$ against 6495 kg/hm$^2$ of the control soil.

## 16.4 Lead

### 16.4.1 Lead in Water and Soils

Lead is found in mineral rocks such as galena, sphalerite (ZnS), and chalcopyrite ($CuFeS_2$), in which concentrations can vary from 7 to 150 mg kg$^{-1}$ depending on the type of rock. In soils, concentrations can vary from 10 to 700 mg kg$^{-1}$, where Pb is associated with sulfur and oxygen, such as PbS, $PbSO_4$, and $PbCO_3$ (Ruby et al. 1999; Tabelin et al. 2018). Soils contaminated with Pb have about 400–800 mg kg$^{-1}$ of soil and in industrial areas this value can reach 1000 mg kg$^{-1}$(Sharma and Dubey 2005).

Lead is widely distributed in the environment and the main source of contamination comes from anthropogenic actions (about 95%). Lead has been used for centuries, mainly during the Roman Empire and is still widely used in the twenty-first century for the manufacture of pigments, ammunition, cables, solders, and batteries. Among the items manufactured with Pb, the battery has a greater prominence, since about 85% of the global consumption of Pb is destined for this equipment. The production of batteries has gained space in the international market due to high demand as purposes for the use in vehicles, storage of energy generated by photovoltaic, wind-cell turbines, and backup power supplies (ILA 2012; WHO 2017).

The emission of compounds such as $PbSO_4$, PbO, and Pb contributes 71% to the total with high importance for ecological/human risks (Smith and Flegal 1995; Klee and Graedel 2004; Liang and Mao 2016). Among the largest sectors of Pb emission in the environment are the industry and product use (38%), road transport (19%), and commercial, institutional, and households (13%) (EEA 2019).

The presence of mineral fertilizers with regular application in the soil increases the absorption and availability of mobile Pb species in the soil, confirmed by a 41-year field trial in barley or sunflower crops followed by oat-winter rye in rotation (Sharma and Dubey 2005). Several studies estimated the Pb persistence in soils varying from 150 to 5000 years. However, the Pb persistence in soils is associated with factors such as soil type, soil particle size, cation exchange capacity, and pH (between 3.0 and 8.5 absorption tends to increase with increasing pH) (Shaw 1989; Sharma and Dubey 2005). There are reports that soils with pH values between 5.5 and 7.5 decrease Pb availability for plants (even those with an affinity for the element) because the presence of phosphate or carbonate precipitates minimizes the Pb solubility in soils (Sharma and Dubey 2005).

In soils, Pb tends to accumulate on the surface (3–5 cm), decreasing the concentration with depth. However, Pb mobility, transport, and leaching in soils are affected by high concentrations of available Pb, which may eventually pollute underground water systems (Yeboah et al. 2019). In natural waters, concentrations of Pb can vary from 1 to 3 $\mu g\ L^{-1}$ (Harrison and Laxen 1981). According to the World Health Organization the guideline value is 10 $\mu g\ L^{-1}$ of Pb in drinking water (WHO 2017).

Most of the Pb in soils is in a precipitated form or intensely bound to organic or colloidal. Therefore, Pb adsorption can easily occur in soils constituted by higher levels of clay, organic matter, and mainly humic acid that contributes to minimizing the absorption of Pb in roots (Werther et al. 1986; Peters and Shem 1992; Sharma and Dubey 2005). Soil carbonates play an important role in controlling the behavior of Pb in soils, in which Pb carbonate compounds are predominantly insoluble ($pH > 6$) as well as phosphates, oxides, and hydroxides (Bradl 2004; Wuana and Okieimen 2011).

The increase in pH due to an increased dissociation of functional carboxylic groups favors the stability of metal-organic complexes, arising from the strong bond between Pb and organic matter. Lead in the cationic chemical specie ($Pb^{2+}$) forms strong complexes with the organic matter in soils (Schulin et al. 1995). In general, the organic matter is included in the low-density fraction of the soil (3.32 g/cc) which can favor a higher fixation of the ionic Pb by the organic matter, due to the dissolution of the particulate Pb which tend to increase the concentration close to the soil surface (Zimdahl and Skogerboe 1977).

Lead is the lowest bioavailable metal in soil. Therefore, based on plant nutrition studies, the chelating agents are necessary to increase the absorption of Pb by high biomass plants used in phytoremediation (Leštan et al. 2008). The presence of chelating agents such as EDTA, NTA, and [S,S]-EDDS results in less Pb adsorption in the soil, due to the strong complexing capacity of these chelating agents, favoring plant uptake (Peters and Shem 1992; Leštan et al. 2008).

Studies have shown that the combination of EDTA with acetic acid increases two times the absorption of Pb in Indian mustard shoots. The lower pH of the rhizosphere decreases the Pb-retention in the cell wall in the form of Pb carbonate; moreover, the combination of more than one chelating agent such as EDTA and [S, S] -EDDS increases the efficiency of phytoextraction of Pb. Chelating agents present different levels of efficiency in the extraction of metals from soils. The combination of chelating agents favors the solubility of one specific metal, while other chelates and solubilized-cations in the soil, such as Ca, decrease the uptake by competition (Sharma and Dubey 2005; Leštan et al. 2008). However, the formation of complexes is controlled by several factors such as adsorption, mineral dissolution, and degradation of chelating agents that need to be considered during a remediation treatment.

## 16.4.2 Lead in Plants

In contact with the root (apoplast), Pb is captured passively by the endoderm. The translocation of Pb in the plant is lower than Cd, for example. Even at low contaminating concentrations, Pb can directly affect the plant, such as decrease in photosynthesis, affecting the production of chlorophyll/carotene and the mitotic activity due to the interruption in the repairing and replication of DNA (Dikilitas et al. 2016). However, the extension is dependent on the physiology of the plant. Lead contamination influences the germination, seedling development, interfering in length and dry weight of root and aerial part of plants. These biochemical problems directly impair the growth, physiology, and yielding of crop plants (Saleem et al. 2018; Silva et al. 2020) (Table 16.3). The highest concentrations are probably found in the lower (root) parts of plants compared with the aerial parts since the first defense strategy of the plant is the exclusion of entry of Pb ions by the root system, thus forming barriers in the cell walls (calluses and lignificants) (Koeppe 1977; Sharma and Dubey 2005; Dikilitas et al. 2016; Mitra et al. 2020).

Lead at high concentrations in soils in affected areas is hardly absorbed by plants due to its low solubility and strong interaction with the soil, high pH, and availability of organic matter (Koeppe 1977). Exposure of rice seedlings (*Oryza sativa* L.) for 10 and 20 days of cultivation with nutrient medium fortified with 165 mg $kg^{-1}$ and

**Table 16.3** General effects of physiological disturbance caused by Pb toxicity in plants

| Physiological processes | Effects of Pb |
|---|---|
| Enzyme activity | Negative[a]/positive[b] |
| Membrane structure | Negative[a] |
| Water potential | Negative[a] |
| Electron transport | Negative[a] |
| Hormonal status | Negative[a] |

[a]Inhibition of enzyme activity, alteration in membrane permeability, decrease of water potential, hormone status, and electron transport activities
[b]Increase of enzyme activity (Adapted: Sharma and Dubey 2005)

331 mg $kg^{-1}$of Pb $(NO_3)_2$ presented decreased growth of roots (between 22% and 42%) and decreasing in aerial parts (approximately 25%). Concerning the Pb-levels found in the plant parts, the accumulation was 1.7 and 3.3 higher in roots than aerial parts, explained by the lower transport due to the root endoderm (Sharma and Dubey 2005).

Although there are several mechanisms to restrict the translocation of Pb in plants (immobilization, precipitation, and sequestration), Pb is still translocated from the root cortex to apoplastic spaces and then to sympathetic spaces of the endoderm through passive uptake mechanisms. The PCs are polypeptides of low molecular weight rich in cysteine, commonly synthesized by plants exposed to toxic metals such as Pb. Phytochelatins work as an important detoxification mechanism in plants by sequestering metal ions. The interaction of PC-metal varies according to the element. For example, PC-Pb interaction is less pronounced compared to PC-Cd, due to the high number of coordination of Pb (6–8) and the larger ionic radius size (Pb, octahedral) (Koeppe 1977; Sharma and Dubey 2005; Dikilitas et al. 2016).

### 16.4.3 Lead Remediation by Agronomic Management

Phytoremediation techniques for Pb can be about 43 times less expensive compared to other soil treatment techniques (Table 16.2) such as soil venting, soil washing, and solvent extraction (Schnoor 1997).

Liu et al. (2018) used *Trifolium repens* L. for soil phytoremediation. They added to uncontaminated soils 100 and 500 mg kg of $Pb^{2+}$. After stabilization of the Pb in the soils, the seeds of *Trifolium repens* L. were cultivated for 14 weeks. The tests showed that the accumulated Pb in *Trifolium repens* L. were 55.81 and 90.3 mg $kg^{-1}$, respectively. During phytoremediation, major changes in microbial diversity and community structure were affected, both in relation to the rhizosphere and the soil. However, after phytoremediation, the microbial community structure has become more abundant. They also observed a decrease of the Pb/acid-soluble fractions in soils after cultivation.

Li et al. (2016b) used the bacterium *Rhodobacter sphaeroides* isolated from oil field to remedy soils contaminated by Pb (21 mg $kg^{-1}$) and spiked (0–1500 mg $kg^{-1}$ $PbNO_3$). After 30 days of inoculation of the bacteria in the soil, wheat seeds were then planted and cultivated for 14 days and then removed from the soil. The results indicated that *Rhodobacter sphaeroides* did not remove Pb from the soil; however, its chemical species change. The most available fractions were converted into less accessible and inert Pb, where the exchangeable part decreased considerably while the residual phase increased. The authors point that the main mechanism of *Rhodobacter sphaeroides* is the formation of precipitates in inert compounds ($PbSO_4$ and PbS), despite the low root (14.7%) and leaf (24%) remediation.

*Nicotiana glauca* R. Graham (shrub tobacco) is a plant that grows quickly with high biomass, is repulsive by herbivores, and has a wide geographic distribution depending on the location. Gisbert et al. (2003) used *Nicotine glauca* for Pb remediation due to its phytoremediation properties. The authors tested a genetically modified plant: the wheat TaPCS1 gene related to the synthesis of PC was inserted to increase the tolerance of *Nicotine glauca* to metals such as Pb. *Nicotiana glauca* was planted in soils containing 1572 mg $kg^{-1}$ of Pb. The concentration was increased by about 50% (52 mg $kg^{-1}$) in the aerial part of genetically modified plants. Besides, Pb absorption reached up by about 85% (82 mg $kg^{-1}$) increment in the roots of the plants that contained the TaPCS1 gene.

Due to its immobilizing capacity and the decreased bioavailable fraction of Pb in sediments and aqueous solutions, nano-hydroxyapatite (nHAP) has been used to mitigate Pb in soils (Zhang et al. 2010). In this sense, several studies have been carried out. Yang et al. (2016) used nHAP (nano-hydroxyapatite) and nHAP@BC (nHAP supported on biochar) in soils containing 640 mg $kg^{-1}$ Pb in which cabbage mustard were planted. Lead-immobilization rate was 72% (nHAP) and 57% (nHAP@BC) after a 28 days remediation using 8% nHAP and nHAP@BC materials. Ding et al. (2017) used nHAP in the cultivation of ryegrass. They showed decreasing mobility and availability of Pb in soils. Cultivation was carried out on background soils containing 58 mg $kg^{-1}$ of Pb with an additional ranging from 400 to 1200 mg $kg^{-1}$ of Pb. Compared to conditions without nHAP, the addition of nHAP decreased the Pb content in ryegrass shoots and roots by 13.19–20.3% and 2.86–21.1%, respectively. Finally, a non-conventional and environmentally technique using clean energy was also tested for Pb-removal from soils. Hussein and Alatabe (2019) carried out two experiments that consisted of evaluating the removal of Pb from contaminated soils (1500 mg $kg^{-1}$) by applying continuous and non-continuous voltages. The continuous voltage was generated by two solar panels, each with a maximum voltage of 17 volts. The soils used for this experiment were sandy loam, silty, and sand. Remediation lasted 7 days with a potential gradient of about 1.2 V/cm. The experimental results showed that the overall removal efficiencies of Pb were 90.7% (sand), 63.3% (sandy), and 42.8% (silty loam soils).

## 16.5 Mercury

Mercury (Hg) is a dense, silvery-white metal. Naturally, it exists in elemental, inorganic, and organic chemical species. Elemental Hg is liquid at standard conditions of temperature and pressure. However, because of its high vapor pressure it can also be found as vapor in the air. Mercury contamination in soil, water, and air is associated with potential toxicity to humans and ecosystems (Aghaei et al. 2019; Wang et al. 2020b). Mercury is ranked third by the US Government Agency for Toxic Substances and Disease Registry, as the most toxic elements or substances that continues to be disposed into our water sources and soils, released to the atmosphere, and consumed by humans and animals (Rice et al. 2014; Huang et al. 2020). Mercury is released into the environment by both natural sources and human activities, and today it is a ubiquitous pollutant. The anthropogenic Hg emissions greatly exceed natural sources. These sources inevitably lead to the release of Hg into the atmosphere, the accumulation of Hg in the aquatic environment, and the deposition of Hg in soil (Ranieri et al. 2020; Gong et al. 2019; Luo et al. 2019, 2020).

### 16.5.1 Mercury in Water and Soils

Mercury contamination can be much more widespread than that of other potentially toxic elements due to its high mobility (Selin 2009; Xie et al. 2020). Mercury may undergo complex physical, chemical, and biological transformations in the environment, e.g., the atmospheric transport of Hg, the photochemical oxidation and subsequent deposition of mercury in water and soil (Xie et al. 2020). After air emissions, Hg has transported some distance within the atmosphere, returning to Earth's surface through wet and dry deposition. In this way, more than 90% of the emitted Hg ends up in terrestrial ecosystems, with soils being the largest recipient (Ranieri et al. 2020). Once deposited, $Hg^{2+}$ interacts with soil through a variety of chemical and biological pathways. Therefore, soil is by far the largest reservoir of terrestrial Hg and plays a major role in its global biogeochemical cycle (Beckers and Rinklebe 2017; Huang et al. 2020).

The threat posed by Hg contaminated soils to humans is persistent, with an estimated 90 tons of anthropogenic Hg pollution accumulated in surface soils worldwide (O'Connor et al. 2019). The average background contents of Hg in different types of soils from all over the world range between 0.58 and 1.8 mg $kg^{-1}$, and the worldwide mean content is estimated at 1.1 mg $kg^{-1}$ (Ranieri et al. 2020).

Mercury exists in soils in many different states and species. This element has three oxidation states: (1) $Hg^0$ (elemental mercury), (2) $Hg^+$ (mercurous), or (3) $Hg^{2+}$ (mercuric). Of these, the mercurous form is not stable under typical environmental conditions and, therefore, is rarely observed. Because of high volatility and susceptibility to oxidation, metallic $Hg^0$ is typically only present in the atmosphere. It can, however, be found as a free phase in soils heavily polluted by anthropogenic activities involving this form of Hg. The deposition of atmospheric $Hg^0$ to soils occurs over large spatial and temporal scales whereas atmospheric $Hg^{2+}$ is nearly always quickly disposed to the pedosphere by wet or dry deposition. Once introduced in soils, Hg in the oxidized mercuric $Hg^{2+}$ state readily forms inorganic Hg salts and minerals, such as $HgCl_2$, HgO, or HgS, or, under appropriate conditions, organo-Hg compounds (O'Connor et al. 2019; Raj and Maiti 2019; Natasha et al. 2020). In addition to the deposition of Hg to the soil from air, another form of Hg deposition to the soil is the dropping of Hg-laden plant leaves on the ground (Raj and Maiti 2019).

The major Hg species in waters are elemental Hg, which is present as a dissolved gas in the water, $Hg^{2+}$ ions referred to as inorganic Hg (iHg), and methylated Hg species (mono- and dimethylmercury). Mono- and dimethylmercury (MMeHg and DMeHg, respectively) occur due to biomethylation processes. Both species are derived from divalent Hg, and hence MMeHg is present as $CH_3Hg^+$, which forms stable compounds with present anions, whereas DMeHg is a highly volatile gas dissolved in water. The dissolved total Hg contains between 3% and 50% of $Hg^0$, 30–90% of iHg; MMeHg varies between <1% and up to 40%. DMeHg (<2%) is dissolved in natural waters to a much lower extent (Fitzgerald et al. 2007; Leopold et al. 2010; Huber and Leopold 2016). More than 90% of the Hg loading of surface waters results from atmospheric deposition. The total concentrations of Hg in natural waters range between sub-ng $L^{-1}$ levels for seawater up to about 10 ng $L^{-1}$ for coastal waters and freshwaters (Huber and Leopold 2016). In contaminated waters, Hg concentrations can reach the μg $L^{-1}$ range and the predominant species depends on the emitting source (Huber and Leopold 2016). The maximum concentration of Hg in drinking water was recommended at 6 μg $L^{-1}$ for inorganic Hg by the World Health Organization (WHO 2011a). The US Environmental Protection Agency specifies as 2 μg $L^{-1}$ the maximum contaminant level of inorganic Hg in drinking water (USEPA 2002).

It is important to highlight that Hg is one of the most hazardous pollutants in aquatic ecosystems because it is readily transformed to MMeHg, which bioaccumulates in aquatic organisms and therefore, fish consumption is currently the main source for non-occupational human Hg exposure (Huber and Leopold 2016; Yan et al. 2019).

### 16.5.2 Mercury in Plants

Estimates of gaseous $Hg^0$ emissions to the atmosphere over land, long considered a critical Hg source, have been revised downward, and most terrestrial environments now are considered net sinks of atmospheric Hg due to substantial Hg uptake by plants (Obrist et al. 2018). The three principal routes of Hg uptake by vascular plants are mainly through: (1) the roots from soil solution as ionic Hg; (2) the stomata from the atmosphere as volatile Hg and; (3) foliar adsorption of divalent, reactive gaseous Hg, and particulate Hg. They efficiently absorb Hg from root hairs as Hg is highly water-soluble and easily converted into gaseous phase (Ranieri et al. 2020; Natasha et al. 2020).

Mercury is stored in roots as a mechanism of plant defense, thus making phytoextraction difficult. It is suggested that vascular plants accumulate Hg through root uptake from soils via the transpiration stream, through the stoma from the atmosphere, and through foliar adsorption of wet and dry deposited Hg. Several studies indicated that Hg is not taken into roots in significant amounts relative to the amount of Hg available in the root zone. More than 95–99% of Hg that is taken up by the roots remains in them and is not translocated to the leaves (Ericksen and Gustin 2004; Du et al. 2005; Moreno-Jiménez et al. 2009; Shiyab et al. 2009; Lomonte et al. 2010; Ranieri et al. 2020).

The toxic effects of Hg on plants include the following: growth retardation and yield production, alterations in nutrient uptake, genotoxicity, generation of reactive oxygen species and membrane lipid peroxidation, changes in the integrity of bio-membranes, inhibition of photosynthesis, and perturbation of almost any function in which critical or non-protected proteins are involved due to mercury's high affinity for sulfhydryl groups (Patra and Sharma 2000; Ali et al. 2002; Patra et al. 2004; Israr et al. 2006; Elbaz et al. 2010; Malar et al. 2015a, b; Wang et al. 2015; Lv et al. 2018; Natasha et al. 2020). The defense mechanisms against Hg-induced oxidative stress include enzymatic antioxidants and some non-enzymatic antioxidants which include glutathione, PCs, salicylic acids, ascorbic acid, selenium, proline, and tocopherols (Stahl and Sies 2003; Rellán-Álvarez et al. 2006; Shao et al. 2008; Sobrino-Plata et al. 2009; Meng et al. 2011; Zhang et al. 2017; Natasha et al. 2020).

Environmental Hg contamination can result in food chain contamination, especially due to its accumulation in edible plant parts. Consumption of Hg-rich edible plants is one of the key sources of Hg exposure to humans and animals. Since Hg does not possess any biological role and has genotoxic and carcinogenic potential, it is critical to monitor its biogeochemical behavior in the soil–plant system and its influence in terms of possible food chain contamination and human exposure (Natasha et al. 2020).

### 16.5.3 Mercury Remediation by Agronomic Management

Mercury-contaminated soil affects the quality of agricultural products, damaging ecosystems directly and compromising environmental and human health via the food chain (Liu et al. 2017, 2018; Wang et al. 2017). Like other potentially toxic elements, Hg cannot be degraded in ecosystems, and therefore remediation should be based on removal or immobilization processes (Wang et al. 2020b). Remediation of Hg in contaminated sites includes the removal or containment of contaminated materials, the application of amendments to reduce mobility and bioavailability, landscape/water body manipulations to reduce MeHg production, and food web manipulations to reduce MeHg accumulated in desired fish species. These approaches can utilize physical, chemical, thermal, and biological methods (Eckley et al. 2020; Shahid et al. 2019). The four major technologies used for remediating of Hg contaminated soil are soil washing, stabilization, thermal treatment, and phytoremediation (Xu et al. 2015).

Mercury soil washing is a physical separation process in which water is used to decrease the concentration of Hg in soil. The principle of the soil washing technique is based on the concept that most metal elements bind with the fine particles in soil and that these contaminants are removed along with the particles by washing with water (USEPA 2007). When chemicals are applied to remove Hg from soil, the process of removal is called chemical extraction, and this method may be used with physical separation (USEPA 2007; Dermont et al. 2008; Xu et al. 2015; Raj and Maiti 2019).

The stabilization process involves a physical–chemical process in which Hg is altered into stable chemical forms and highly insoluble in various pH ranges of soil (Alloway 2012; Xu et al. 2015; Raj and Maiti 2019). Therefore, the availability of Hg for uptake in plants is decreased by stabilization (USEPA 2007; Raj and Maiti 2019). This method involves physically binding or enclosing contaminants within a stabilized mass (solidification) or inducing chemical reactions between the stabilizing agent and the contaminants to decrease their mobility (stabilization). The stabilization procedure involves mixing soil with chemical binders such as cement, sulfide, and phosphate binders, polyester resins, or polysiloxane compounds to create a slurry or paste which is allowed time to cure into a solid form. For Hg, powder reactivated carbon and thiol-functionalized zeolite have been used as binders (Zhang and Bishop 2002; Zhang et al. 2009; Wang et al. 2012).

In thermal treatment, high temperatures (320–700 °C) are used along with low pressure to volatilize elemental Hg and Hg compounds, removing Hg from the soil. This is followed by a condensation of the Hg vapors into liquid elemental Hg. In the thermal treatment process, the efficiency removal of Hg from the soil is around 40%, and high-temperature and low pressure treatment can remove high concentrations of Hg (ca. 30,000 mg/kg) from the soil (Vik and Bardos 2003; USEPA 2007; Xu et al. 2015; Raj and Maiti 2019).

Phytoremediation is a new in situ remediation process that uses plants to remove potentially toxic elements from soil. It is gaining widespread public acceptance and has a low cost compared with traditional technologies. This process is used to remove or decrease the level of contaminants in the soil through the process of translocation, in which the contaminants are translocated from the soil to plant tissues (Raj and Maiti 2019; Singh and Kumar 2019).

Phytoremediation involving phytostabilization, phytoextraction and phytovolatilization relies on selected plants to clean up Hg contamination in the soil (Xu et al. 2015; Ranieri et al. 2020). Phytostabilization immobilizes Hg in the soil through absorption and accumulation of Hg in roots or through Hg precipitation in the root area; this process will prevent the migration of Hg by soil erosion. Aspirant plants for phytostabilization need to have extensive root systems, sensitivity to Hg toxicity and high plant survival rate (Wang et al. 2005; Xu et al. 2015; Ranieri et al. 2020). For instance, it was demonstrated that willow (*Salix viminalis* and *S. schwerinii*) can accumulate bioavailable Hg in the root system, thereby reducing the concentration of bioavailable Hg in the rhizosphere while leaving total Hg in the soil relatively unchanged (Wang et al. 2005).

Phytoextraction is the use of plants to remove pollutants and contaminants from soil. It is the process of Hg uptake/absorption and translocation by plant roots into the aboveground parts that can then be harvested and burned (burning for Hg is not considered due to its low boiling temperature). Chemical additives to soil, such as ammonium thiosulfate, potassium iodide, and halogen salts, can be used to assist in phytoextraction by promoting Hg bioavailability to make plant accumulation much more effective from plants that are not hyperaccumulators (Pant et al. 2011; Pedron et al. 2011; Petruzzelli et al. 2013; Wang et al. 2014; Xu et al. 2015; Ranieri et al. 2020).

Phytovolatilization refers to the uptake and transpiration of such elements by plants. The element is taken up by plant roots, transported through the xylem, and is finally released to the atmosphere from cellular tissues (evaporates or vaporizes). Few studies have reported the use of plants for the phytovolatilization of Hg, because mercury emission from leaf tissues is affected by environmental parameters such as light intensity and air temperature (Leonard et al. 1998; Wang et al. 2012; Xu et al. 2015; Ranieri et al. 2020).

Occasionally microorganisms play an important role in the phytoremediation process, and among the microorganisms involved in phytoremediation, rhizosphere bacteria has received attention. The rhizosphere bacteria can increase the metal bioavailability in various ways, including altering soil pH, releasing chelators and oxidation/reduction reactions (Franchi et al. 2017; Wang et al. 2020b). Mercury resistant bacteria are used in this microbial reduction ($Hg^{2+}$ to $Hg^0$) and volatilization process because they possess the *mer operon*. The *mer operon* are several genes related to the mercury resistance and reduction mechanism of these bacteria (Kumari et al. 2020; Wang et al. 2020b). However, up to date, no bioremediation studies using microorganisms for Hg remediation on a large scale have been conducted.

## 16.6 Final Considerations and Perspectives

Elements such as As, Cd, Pb, and Hg are potentially toxic, even at low concentrations, for humans and plants. There are several sources of these elements in the environment, natural or anthropogenic, in which the main negative influences for agricultural soil and crops are decreased productions and contaminations. The contamination of crop plants has a direct impact on the food chain. Therefore, it is evident the need for monitoring, control, and remediation of these elements, ensuring efficient production and safer food. In this sense, several studies and strategies such as phytoremediation, use of microorganisms, water management, use of nanotechnology aimed at the decrease of contamination associated with reduction of effective costs and improvement of crop yielding. However, there is a need for more knowledge concerning the capture, mobility, translocation, accumulation, and tolerance strategies for these elements by plants, to make remediation techniques more and more effective.

## References


Abbas G, Murtaza B, Bibi I, Shahid M, Niazi NK, Khan MI, Hussain M (2018) Arsenic uptake, toxicity, detoxification, and speciation in plants: physiological, biochemical, and molecular aspects. Int J Environ Res Public Health 15(1):59. https://doi.org/10.3390/ijerph15010059

Abedin MJ, Feldmann J, Meharg AA (2002) Uptake kinetics of arsenic species in rice plants. Plant Physiol 128(3):1120–1128. https://doi.org/10.1104/pp.010733

Abumaizar RJ, Smith EH (1999) Heavy metal contaminants removal by soil washing. J Hazard Mater 70(1-2):71–86. https://doi.org/10.1016/S0304-3894(99)00149-1

Adrees M, Khan ZS, Ali S, Hafeez M, Khalid S, urRehman MZ, Rizwan M (2020) Simultaneous mitigation of cadmium and drought stress in wheat by soil application of iron nanoparticles. Chemosphere 238:124681. https://doi.org/10.1016/j.chemosphere.2019.124681

Agency for Toxic Substances and Disease Registry (ATSDR) (2008) Case studies in environmental medicine (CSEM). Cadmium toxicity. Atlanta, GA, USA. https://www.atsdr.cdc.gov/csem/cadmium/docs/cad. Accessed 11 Apr 2020

Aghaei E, Alorro RD, Tadesse B, Browner R (2019) A review on current practices and emerging technologies for sustainable management, sequestration and stabilization of mercury from gold processing streams. J Environ Manag 249:109367

Ali MB, Chun HS, Lee CB (2002) Response of antioxidant enzymes in rice (Oryza sauva L. cv. Dongjin) under mercury stress. J Plant Biol 45(3):141–147

Alloway BJ (ed) (2012) Heavy metals in soils: trace metals and metalloids in soils and their bioavailability. Springer, Netherlands

Batista BL, Nigar M, Mestrot A, Rocha BA, Barbosa Junior F, Price AH, Feldmann J (2014) Identification and quantification of phytochelatins in roots of rice to long-term exposure: evidence of individual role on arsenic accumulation and translocation. J Exp Bot 65 (6):1467–1479. https://doi.org/10.1093/jxb/eru018

Beckers F, Rinklebe J (2017) Cycling of mercury in the environment: Sources, fate, and human health implications: A review. Crit Rev Environ Sci Technol 47(9):693–794

Benavides MP, Gallego SM, Tomaro ML (2005) Cadmium toxicity in plants. Braz J Plant Physiol 17(1):21–34. https://doi.org/10.1590/S1677-04202005000100003

Bolan NS, Park JH, Robinson B, Naidu R, Huh K (2011) Phytostabilization: a green approach to contaminant containment. In: Advances in agronomy, vol 112. Academic Press, Cambridge, pp 145–204

Bradl HB (2004) Adsorption of heavy metal ions on soils and soils constituents. J Colloid Interface Sci 277(1):1–18. https://doi.org/10.1016/j.jcis.2004.04.005

Cai L, Xu Z, Bao P, He M, Dou L, Chen L, Zhu YG (2015) Multivariate and geostatistical analyses of the spatial distribution and source of arsenic and heavy metals in the agricultural soils in Shunde, Southeast China. J Geochem Explor 148:189–195. https://doi.org/10.1016/j.gexplo.2014.09.010

Cao GH, Li ZD, Wang XF, Zhang X, Zhao RH, Gu W, He S (2020) Phosphate transporters, PnPht1; 1 and PnPht1; 2 from Panax notoginseng enhance phosphate and arsenate acquisition. BMC Plant Biol 20(1):1–14. https://doi.org/10.1186/s12870-020-2316-7

Chao DY, Chen Y, Chen J, Shi S, Chen Z, Wang C, Salt DE (2014) Genome-wide association mapping identifies a new arsenate reductase enzyme critical for limiting arsenic accumulation in plants. PLoS Biol 12(12):e1002009

Chen Z, Zhu YG, Liu WJ, Meharg AA (2005) Direct evidence showing the effect of root surface iron plaque on arsenite and arsenate uptake into rice (Oryza sativa) roots. New Phytol 165 (1):91–97. https://doi.org/10.1111/j.1469-8137.2004.01241.x

Chen Y, Han YH, Cao Y, Zhu YG, Rathinasabapathi B, Ma LQ (2017) Arsenic transport in rice and biological solutions to reduce arsenic risk from rice. Front Plant Sci 8:268. https://doi.org/10.3389/fpls.2017.00268

Chen L, Long C, Wang D, Yang J (2020) Phytoremediation of cadmium (Cd) and uranium (U) contaminated soils by Brassica juncea L. enhanced with exogenous application of plant growth regulators. Chemosphere 242:125112. https://doi.org/10.1016/j.chemosphere.2019.125112

Das P, Samantaray S, Rout GR (1997) Studies on cadmium toxicity in plants: a review. Environ Pollut 98(1):29–36

Dermont G, Bergeron M, Mercier G, Richer-Laflèche M (2008) Soil washing for metal removal: a review of physical/chemical technologies and field applications. J Hazard Mater 152(1):1–31

Dikilitas M, Karakas S, Ahmad P (2016) Effect of lead on plant and human DNA damages and its impact on the environment. In: Plant metal interaction. Elsevier, Amsterdam, pp 41–67. https://doi.org/10.1016/B978-0-12-803158-2.00003-5

Ding L, Li J, Liu W, Zuo Q, Liang SX (2017) Influence of nano-hydroxyapatite on the metal bioavailability, plant metal accumulation and root exudates of ryegrass for phytoremediation in lead-polluted soil. Int J Environ Res Public Health 14(5):532. https://doi.org/10.3390/ijerph14050532

Du X, Zhu YG, Liu WJ, Zhao XS (2005) Uptake of mercury (Hg) by seedlings of rice (Oryza sativa L.) grown in solution culture and interactions with arsenate uptake. Environ Exp But 54:1–7. https://doi.org/10.1016/j.envexpbot.2004.05.001

Elbaz A, Wei YY, Meng Q, Zheng Q, Yang ZM (2010) Mercury-induced oxidative stress and impact on antioxidant enzymes in Chlamydomonas reinhardtii. Ecotoxicology 19 (7):1285–1293

Eckley CS, Gilmour CC, Janssen S, Luxton TP, Randall PM, Whalin L, Austin C (2020) The assessment and remediation of mercury contaminated sites: A review of current approaches. Sci Total Enviro 707:136031

Ehsan M, Delgado KS, Alarcón AV, Chavez AA, Dela Cruz Landero N (2009) Phytostabilization of cadmium contaminated soils by" Lupinus uncinatus" Schldl. Span J Agric Res 7(2):390–397. https://doi.org/10.5424/sjar/2009072-430

Environmental Protection Agency (USEPA) (2007) Treatment technologies for mercury in soil, waste, and water. US EPA, Office of Superfund Remediation and Technology Innovation Washington, DC 20460, EPA-542-R-07-003.

Ericksen JA, Gustin MS (2004) Foliar exchange of mercury as a function of soil and air mercury concentrations. Sci Total Environ 324:271–279. https://doi.org/10.1016/j.scitotenv.2003.10.034

Esfandbod AM (2010) Cadmium adsorption behavior of some soils from northern of Iran. In: Proceedings of the 19th World Congress of Soil Science: Soil solutions for a changing world, Brisbane, Australia, 1–6 August 2010. Symposium 3.5. 1 Heavy metal contaminated soils. International Union of Soil Sciences (IUSS), c/o InstitutfürBodenforschung, UniversitätfürBodenkultur, pp 20–22

European Environment Agency (EEA) (2019) Heavy metal emission. In: Indicator assessment data and maps. https://www.eea.europa.eu/data-and-maps/indicators/eea32-heavy-metal-hm-emissions-1/assessment-10. Accessed 9 Apr 2020

European Food Safety Authority (EFSA) (2014) Dietary exposure to inorganic arsenic in the European population. EFSA J 12(3):3597

Faroon O, Ashizawa A, Wright S, Tucker P, Jenkins K, Ingerman L, Rudisill C (2012) Toxicological profile for cadmium. Agency for Toxic Substances and Disease Registry (US), Atlanta, GA

Farooq MA, Islam F, Ali B, Najeeb U, Mao B, Gill RA, Zhou W (2016) Arsenic toxicity in plants: cellular and molecular mechanisms of its transport and metabolism. Environ Exp Bot 132:42–52. https://doi.org/10.1016/j.envexpbot.2016.08.004

Farraji H, Zaman NQ, Tajuddin R, Faraji H (2016) Advantages and disadvantages of phytoremediation: a concise review. Int J Env Tech Sci 2:69–75

Favas PJ, Pratas J, Varun M, D'Souza R, Paul MS (2014) Phytoremediation of soils contaminated with metals and metalloids at mining areas: potential of native flora. Environ Risk Assess Soil Contamin 3:485–516

Fitzgerald WF, Lamborg CH, Hammerschmidt CR (2007) Marine biogeochemical cycling of mercury. Chem Rev 107(2):641–662

Flora G, Gupta D, Tiwari A (2012) Toxicity of lead: a review with recent updates. Interdiscip Toxicol 5(2):47–58. https://doi.org/10.2478/v10102-012-0009-2

Franchi E, Rolli E, Marasco R, Agazzi G, Borin S, Cosmina P, Petruzzelli G (2017) Phytoremediation of a multi contaminated soil: mercury and arsenic phytoextraction assisted by mobilizing agent and plant growth promoting bacteria. J Soils Sediment 17(5):1224–1236

Gang LI, Zheng M, Jianfeng TANG, Hojae SHIM, Chao CAI (2018) Effect of silicon on arsenic concentration and speciation in different rice tissues. Pedosphere 28(3):511–520

Ghassemzadeh F, Yousefzadeh H, Arbab-Zavar MH (2008) Arsenic phytoremediation by Phragmites australis: green technology. Int J Environ Stud 65(4):587–594. https://doi.org/10.1080/00207230802273387

Gill SS, Khan NA, Tuteja N (2011) Differential cadmium stress tolerance in five Indian mustard (Brassica juncea L.) cultivars: an evaluation of the role of antioxidant machinery. Plant Signal Behav 6(2):293–300. https://doi.org/10.4161/psb.6.2.15049

Gisbert C, Ros R, De Haro A, Walker DJ, Bernal MP, Serrano R, Navarro-Aviñó J (2003) A plant genetically modified that accumulates Pb is especially promising for phytoremediation. Biochem Biophys Res Commun 303(2):440–445. https://doi.org/10.1016/S0006-291X(03)00349-8

Goldberg S (2002) Competitive adsorption of arsenate and arsenite on oxides and clay minerals. Soil Sci Soc Am J 66(2):413–421. https://doi.org/10.2136/sssaj2002.4130

Gomez-Caminero A, Howe PD, Hughes M, Kenyon E, Lewis DR, Moore M, Ng J (2001) Arsenic and arsenic compounds. World Health Organization, Geneva

Gong Y, Huang Y, Wang M, Liu F, Zhang T (2019) Application of iron-based materials for remediation of mercury in water and soil. Bull Environ Contam Toxicol 102(5):721–729

Gonzaga MIS, Santos JAG, Ma LQ (2006) Arsenic phytoextraction and hyperaccumulation by fern species. Sci Agric 63(1):90–101. https://doi.org/10.1590/S0103-90162006000100015

Gratão PL, Prasad MNV, Cardoso PF, Lea PJ, Azevedo RA (2005) Phytoremediation: green technology for the clean up of toxic metals in the environment. Braz J Plant Physiol 17 (1):53–64. https://doi.org/10.1590/S1677-04202005000100005

Harrison RM, Laxen DPH (1981) Lead pollution. In: Causes and control. Chapman and Hall, London

Hasanuzzaman M, Nahar K, Fujita M (eds) (2018) Mechanisms of arsenic toxicity and tolerance in plants. Springer, Singapore

He QB, Singh BR (1993) Effect of organic matter on the distribution, extractability and uptake of cadmium in soils. J Soil Sci 44(4):641–650. https://doi.org/10.1111/j.1365-2389.1993.tb02329.x

Heikens A, Panaullah GM, Meharg AA (2007) Arsenic behaviour from groundwater and soil to crops: impacts on agriculture and food safety. In: Reviews of environmental contamination and toxicology. Springer, New York, pp 43–87

Hossain MF (2006) Arsenic contamination in Bangladesh—an overview. Agric Ecosyst Environ 113(1-4):1–16. https://doi.org/10.1016/j.agee.2005.08.034

Hu P, Ouyang Y, Wu L, Shen L, Luo Y, Christie P (2015) Effects of water management on arsenic and cadmium speciation and accumulation in an upland rice cultivar. J Environ Sci 27:225–231. https://doi.org/10.1016/j.jes.2014.05.048

Huang JH, Shetaya WH, Osterwalder S (2020) Determination of (Bio)-available mercury in soils: a review. Environ Pollut 263:114323

Huber J, Leopold K (2016) Nanomaterial-based strategies for enhanced mercury trace analysis in environmental and drinking waters. TrAC Trends Anal Chem 80:280–292

Hussein AA, Alatabe MJA (2019) Remediation of lead-contaminated soil, using clean energy in combination with electro-kinetic methods. Pollution 5(4):859–869. https://doi.org/10.22059/POLL.2019.275250.579

International Agency for Research on Cancer, International Agency for Research on Cancer (IARC) (1993) Beryllium, cadmium, mercury, and exposures in the glass manufacturing industry, vol 58. International Agency for Research on Cancer, Lyon, France

International Lead Association (ILA) (2012) Lead use statistics. Lead facts. http://www.ila-lead.org/lead-facts/lead-uses--statistics. Accessed 5 Apr 2020

Israr M, Sahi S, Datta R, Sarkar D (2006) Bioaccumulation and physiological effects of mercury in Sesbania drummondii. Chemosphere 65(4):591–598

Jiang C, Sun H, Sun T, Zhang Q, Zhang Y (2009) Immobilization of cadmium in soils by UV-mutated Bacillus subtilis 38 bioaugmentation and NovoGro amendment. J Hazard Mater 167(1–3):1170–1177. https://doi.org/10.1016/j.jhazmat.2009.01.107

Jiang Y, Chao S, Liu J, Yang Y, Chen Y, Zhang A, Cao H (2017) Source apportionment and health risk assessment of heavy metals in soil for a township in Jiangsu Province, China. Chemosphere 168:1658–1668. https://doi.org/10.1016/j.chemosphere.2016.11.088

Johnson PC, Stanley CC, Kemblowski MW, Byers DL, Colthart JD (1990) A practical approach to the design, operation, and monitoring of in situ soil-venting systems. Groundwater Monit Rem 10(2):159–178. https://doi.org/10.1111/j.1745-6592.1990.tb00347.x

Khalid S, Shahid M, Niazi NK, Rafiq M, Bakhat HF, Imran M, Dumat C (2017) Arsenic behaviour in soil-plant system: biogeochemical reactions and chemical speciation influences. In: Enhancing cleanup of environmental pollutants. Springer, Cham, pp 97–140

Khan E, Gupta M (2018) Arsenic–silicon priming of rice (Oryza sativa L.) seeds influence mineral nutrient uptake and biochemical responses through modulation of Lsi-1, Lsi-2, Lsi-6 and nutrient transporter genes. Sci Rep 8(1):1–16. https://doi.org/10.1038/s41598-018-28712-3

Khan MA, Khan S, Khan A, Alam M (2017) Soil contamination with cadmium, consequences and remediation using organic amendments. Sci Total Environ 601:1591–1605. https://doi.org/10.1016/j.scitotenv.2017.06.030
Kim KH, Kabir E, Jahan SA (2016) A review on the distribution of Hg in the environment and its human health impacts. J Hazard Mater 306:376–385. https://doi.org/10.1016/j.jhazmat.2015.11.031
Klee RJ, Graedel TE (2004) Elemental cycles: a status report on human or natural dominance. Annu Rev Environ Resour 29:69–107. https://doi.org/10.1146/annurev.energy.29.042203.104034
Koeppe DE (1977) The uptake, distribution, and effect of cadmium and lead in plants. Sci Total Environ 7(3):197–206. https://doi.org/10.1016/0048-9697(77)90043-2
Kubier A, Wilkin RT, Pichler T (2019) Cadmium in soils and groundwater: A review. Appl Geochem 108:104388. https://doi.org/10.1016/j.apgeochem.2019.104388
Kumar D, Tripathi DK, Liu S, Singh VK, Sharma S, Dubey NK, Chauhan DK (2017) Pongamia pinnata (L.) Pierre tree seedlings offer a model species for arsenic phytoremediation. Plant Gene 11:238–246. https://doi.org/10.1016/j.plgene.2017.06.002
Kumari S, Jamwal R, Mishra N, Singh DK (2020) Recent developments in environmental mercury bioremediation and its toxicity: a review. Environ Nanotechnol Monit Manag 13:100283
Lan MM, Liu C, Liu SJ, Qiu RL, Tang YT (2020) Phytostabilization of Cd and Pb in highly polluted farmland soils using ramie and amendments. Int J Environ Res Public Health 17 (5):1661. https://doi.org/10.3390/ijerph17051661
LeDuc DL, Terry N (2005) Phytoremediation of toxic trace elements in soil and water. J Ind Microbiol Biotechnol 32(11–12):514–520. https://doi.org/10.1007/s10295-005-0227-0
Lee CH, Wu CH, Syu CH, Jiang PY, Huang CC, Lee DY (2016) Effects of phosphorous application on arsenic toxicity to and uptake by rice seedlings in As-contaminated paddy soils. Geoderma 270:60–67. https://doi.org/10.1016/j.geoderma.2016.01.003
Leopold K, Foulkes M, Worsfold P (2010) Methods for the determination and speciation of mercury in natural Waters - A review. Anal Chim Acta 663:127–138. https://doi.org/10.1016/j.aca.2010.01.048
Leštan D, Luo CL, Li XD (2008) The use of chelating agents in the remediation of metal-contaminated soils: a review. Environ Pollut 153(1):3–13. https://doi.org/10.1016/j.envpol.2007.11.015
Li N, Wang J, Song WY (2016a) Arsenic uptake and translocation in plants. Plant Cell Physiol 57 (1):4–13. https://doi.org/10.1093/pcp/pcv143
Li X, Peng W, Jia Y, Lu L, Fan W (2016b) Bioremediation of lead contaminated soil with Rhodobacter sphaeroides. Chemosphere 156:228–235. https://doi.org/10.1016/j.chemosphere.2016.04.098
Li B, Zhou S, Wei D, Long J, Peng L, Tie B, Lei M (2019a) Mitigating arsenic accumulation in rice (Oryza sativa L.) from typical arsenic contaminated paddy soil of southern China using nanostructured α-MnO2: pot experiment and field application. Sci Total Environ 650:546–556. https://doi.org/10.1016/j.scitotenv.2018.08.436
Li J, Zhang P, Ye J, Zhang G, Cai Y (2019b) Simultaneous in-situ remediation and fertilization of Cd-contaminated weak-alkaline farmland for wheat production. J Environ Manag 250:109528. https://doi.org/10.1016/j.jenvman.2019.109528
Liang J, Mao JS (2016) Risk assessment of lead emissions from anthropogenic cycle. Trans Nonferrous Metals Soc China 26(1):248–255. https://doi.org/10.1016/S1003-6326(16)64148-1
Liu Z, Wang LA, Xu J, Ding S, Feng X, Xiao H (2017) Effects of different concentrations of mercury on accumulation of mercury by five plant species. Ecol Eng 106:273–278
Liu C, Lin H, Dong Y, Li B, Liu Y (2018) Investigation on microbial community in remediation of lead-contaminated soil by Trifolium repens L. Ecotoxicol Environ Saf 165:52–60. https://doi.org/10.1016/j.ecoenv.2018.08.054
Lomax C, Liu WJ, Wu L, Xue K, Xiong J, Zhou J, Zhao FJ (2012) Methylated arsenic species in plants originate from soil microorganisms. New Phytol 193(3):665–672. https://doi.org/10.1111/j.1469-8137.2011.03956.x

Lomonte C, Doronila AI, Gregory D, Baker AJM, Kolev SD (2010) Phytotoxicity of biosolids and screening of selected plant species with potential for mercury phytoextraction. J Hazard Mater 173:494–501. https://doi.org/10.1016/j.jhazmat.2009.08.112

Leonard TL, Taylor GE, Gustin MS, Fernandez GC (1998) Mercury and plants in contaminated soils: 1. Uptake, partitioning, and emission to the atmosphere. Environ Toxicol Chem 17:2063–2071. https://doi.org/10.1002/etc.5620171024

Luo H, Cheng Y, He D, Yang EH (2019) Review of leaching behavior of municipal solid waste incineration (MSWI) ash. Sci Total Environ 668:90–103

Luo H, Cheng Q, Pan X (2020) Photochemical behaviors of mercury (Hg) species in aquatic systems: a systematic review on reaction process, mechanism, and influencing factor. Sci Total Environ 720:137540

Lv S, Yang B, Kou Y, Zeng J, Wang R, Xiao Y, Zhao C (2018) Assessing the difference of tolerance and phytoremediation potential in mercury contaminated soil of a non-food energy crop, Helianthus tuberosus L.(Jerusalem artichoke). PeerJ 6:e4325

Ma JF, Yamaji N, Mitani N, Xu XY, Su YH, McGrath SP, Zhao FJ (2008) Transporters of arsenite in rice and their role in arsenic accumulation in rice grain. Proc Natl Acad Sci 105 (29):9931–9935. https://doi.org/10.1073/pnas.0802361105

Ma L, Wang L, Jia Y, Yang Z (2017) Accumulation, translocation and conversion of six arsenic species in rice plants grown near a mine impacted city. Chemosphere 183:44–52. https://doi.org/10.1016/j.chemosphere.2017.05.089

Malar S, Sahi SV, Favas PJC, Venkatachalam P (2015a) Assessment of mercury heavy metal toxicity-induced physiochemical and molecular changes in Sesbania grandiflora L. Int J Environ Sci Technol 12(10):3273–3282

Malar S, Sahi SV, Favas PJ, Venkatachalam P (2015b) Mercury heavy-metal-induced physiochemical changes and genotoxic alterations in water hyacinths [Eichhornia crassipes (Mart.)]. Environ Sci Pollut Res 22(6):4597–4608

Mandal BK, Suzuki KT (2002) Arsenic round the world: a review. Talanta 58(1):201–235. https://doi.org/10.1016/S0039-9140(02)00268-0

McLaughlin MJ, Smolders E, Degryse F, Rietra R (2011) Uptake of metals from soil into vegetables. In: Dealing with contaminated sites. Springer, Dordrecht, pp 325–367

Meng de K, Chen J, Yang ZM (2011) Enhancement of tolerance of Indian mustard (Brassica juncea) to mercury by carbon monoxide. J Hazard Mater 186:1823–1829. https://doi.org/10.1016/j.jhazmat.2010.12.062

Meng D, Li J, Liu T, Liu Y, Yan M, Hu J, Yin H (2019) Effects of redox potential on soil cadmium solubility: insight into microbial community. J Environ Sci 75:224–232. https://doi.org/10.1016/j.jes.2018.03.032

Meter A, Atkinson RJ, Laliberte B (2019) Cadmium in cacao from latin America and the Caribbean. A review of research and potential mitigation solutions. Bioversity International, Rome

Mikkonen HG, Van De Graaff R, Collins RN, Dasika R, Wallis CJ, Howard DL, Reichman SM (2019) Immobilisation of geogenic arsenic and vanadium in iron-rich sediments and iron stone deposits. Sci Total Environ 654:1072–1081. https://doi.org/10.1016/j.scitotenv.2018.10.427

Mitra A, Chatterjee S, Voronina AV, Walther C, Gupta DK (2020) Lead toxicity in plants: a review. Lead in plants and the environment. Springer, Cham, pp. 99–116

Moreno-Jiménez E, Esteban E, Carpena-Ruiz RO, Peñalosa JM (2009) Arsenic- and mercuryinduced phytotoxicity in the Mediterranean shrubs Pistacia lentiscus and Tamarix gallica grown in hydroponic culture. Ecotoxicol Environ Saf 72:1781–1789. https://doi.org/10.1016/j.ecoenv.2009.04.022

Natasha SM, Khalid S, Bibi I, Bundschuh J, Niazi NK, Dumat C (2020) A critical review of mercury speciation, bioavailability, toxicity and detoxification in soil-plant environment: ecotoxicology and health risk assessment. Sci Total Environ 711:134749

O'Connor D, Hou D, Ok YS, Mulder J, Duan L, Wu Q, Rinklebe J (2019) Mercury speciation, transformation, and transportation in soils, atmospheric flux, and implications for risk management: a critical review. Environ Int 126:747–761

Obrist D, Kirk JL, Zhang L, Sunderland EM, Jiskra M, Selin NE (2018) A review of global environmental mercury processes in response to human and natural perturbations: Changes of emissions, climate, and land use. Ambio 47(2):116–140

Pant P, Allen M, Tansel B (2011) Mercury contamination in the riparian zones along the East Fork Poplar Creek at Oak Ridge. Ecotoxicol Environ Saf 74(3):467–472

Panthri M, Gupta M (2019) Facets of iron in arsenic exposed Oryza sativa varieties: A manifestation of plant's adjustment at morpho-biochemical and enzymatic levels☆. Environ Pollut 255:113289. https://doi.org/10.1016/j.envpol.2019.113289

Patra M, Sharma A (2000) Mercury toxicity in plants. Bot Rev 66(3):379–422

Patra M, Bhowmik N, Bandopadhyay B, Sharma A (2004) Comparison of mercury, lead and arsenic with respect to genotoxic effects on plant systems and the development of genetic tolerance. Environ Exp Bot 52(3):199–223

Pedron F, Petruzzelli G, Barbafieri M, Tassi E, Ambrosini P, Patata L (2011) Mercury mobilization in a contaminated industrial soil for phytoremediation. Commun Soil Sci Plant Anal 42 (22):2767–2777

Peters RW, Shem L (1992) Adsorption/desorption characteristics of lead on various types of soil. Environ Prog 11(3):234–240. https://doi.org/10.1002/ep.670110322

Petruzzelli G, Pedron F, Rosellini I, Barbafieri M (2013) Phytoremediation towards the future: focus on bioavailable contaminants. In: Plant-based remediation processes. Springer, Berlin, Heidelberg, pp 273–289

Pierangeli MAP, Guilherme LRG, Curi N, Silva MLN, Lima JMD, Costa ETDS (2005) Efeito do pH na adsorção e dessorção de cádmio em Latossolos brasileiros. Revista Brasileira de Ciência do Solo 29(4):523–532. https://doi.org/10.1590/S0100-06832005000400005

Pinto AP, Mota AD, De Varennes A, Pinto FC (2004) Influence of organic matter on the uptake of cadmium, zinc, copper and iron by sorghum plants. Sci Total Environ 326(1–3):239–247

Pitten FA, Müller G, König P, Schmidt D, Thurow K, Kramer A (1999) Risk assessment of a former military base contaminated with organoarsenic-based warfare agents: uptake of arsenic by terrestrial plants. Sci Total Environ 226(2-3):237–245. https://doi.org/10.1016/S0048-9697(98)00400-8

Qiao Y, Wu J, Xu Y, Fang Z, Zheng L, Cheng W, Zhao D (2017) Remediation of cadmium in soil by biochar-supported iron phosphate nanoparticles. Ecol Eng 106:515–522. https://doi.org/10.1016/j.ecoleng.2017.06.023

Rahaman S, Sinha AC, Mukhopadhyay D (2011) Effect of water regimes and organic matters on transport of arsenic in summer rice (Oryza sativa L.). J Environ Sci 23(4):633–639

Raj D, Maiti SK (2019) Sources, toxicity, and remediation of mercury: an essence review. Environ Monit Assess 191(9):566

Ranieri E, Moustakas K, Barbafieri M, Ranieri AC, Herrera-Melián JA, Petrella A, Tommasi F (2020) Phytoextraction technologies for mercury-and chromium-contaminated soil: a review. J Chem Technol Biotechnol 95(2):317–327

Raskin I, Smith RD, Salt DE (1997) Phytoremediation of metals: using plants to remove pollutants from the environment. Curr Opin Biotechnol 8(2):221–226. https://doi.org/10.1016/S0958-1669(97)80106-1

Rellán-Álvarez R, Ortega-Villasante C, Álvarez-Fernández A, Del Campo FF, Hernández LE (2006) Stress responses of Zea mays to cadmium and mercury. Plant Soil 279(1-2):41–50

Rice KM, Walker EM Jr, Wu M, Gillette C, Blough ER (2014) Environmental mercury and its toxic effects. J Prev Med Public Health 47(2):74

Roychowdhury R, Khan MH, Choudhury S (2018) Arsenic in rice: an overview on stress implications, tolerance and mitigation strategies. In: Plants under metal and metalloid stress. Springer, Singapore, pp 401–415

Ruby MV, Schoof R, Brattin W, Goldade M, Post G, Harnois M, Edwards D (1999) Advances in evaluating the oral bioavailability of inorganics in soil for use in human health risk assessment. Environ Sci Technol 33(21):3697–3705. https://doi.org/10.1021/es990479z

Saleem M, Asghar HN, Zahir ZA, Shahid M (2018) Impact of lead tolerant plant growth promoting rhizobacteria on growth, physiology, antioxidant activities, yield and lead content in sunflower in lead contaminated soil. Chemosphere 195:606–614. https://doi.org/10.1016/j.chemosphere.2017.12.117

Sarkar A, Paul B (2016) The global menace of arsenic and its conventional remediation-a critical review. Chemosphere 158:37–49. https://doi.org/10.1016/j.chemosphere.2016.05.043

Sarwar N, Imran M, Shaheen MR, Ishaque W, Kamran MA, Matloob A, Hussain S (2017) Phytoremediation strategies for soils contaminated with heavy metals: modifications and future perspectives. Chemosphere 171:710–721. https://doi.org/10.1016/j.chemosphere.2016.12.116

Schmidt U (2003) Enhancing phytoextraction. J Environ Qual 32(6):1939–1954. https://doi.org/10.2134/jeq2003.1939

Schnoor JL (1997) Phytoremediation. Ground-Water Remediation Technologies Analysis Center (GWRTAC) and Series: TE-98-01. https://clu-in.org/download/toolkit/phyto_e.pdf. Accessed 03 May 2020

Schulin R, Geiger G, Furrer G (1995) Heavy metal retention by soil organic matter under changing environmental conditions. In: Biogeodynamics of pollutants in soils and sediments. Springer, Berlin, Heidelberg, pp 53–85

Segura FR, Paulelli ACC, Braga GÚL, Dos Reis Pedreira Filho W, Silva FF, Batista BL (2018) Promising filamentous native fungi isolated from paddy soils for arsenic mitigation in rice grains cultivated under flooded conditions. J Environ Chem Eng 6(4):3926–3932. https://doi.org/10.1016/j.jece.2018.05.036

Selin NE (2009) Global biogeochemical cycling of mercury: a review. Annu Rev Environ Resour 34:43–63

Shahid M, Dumat C, Khalid S, Niazi NK, Antunes PM (2016) Cadmium bioavailability, uptake, toxicity and detoxification in soil-plant system. In: Reviews of environmental contamination and toxicology, vol 241. Springer, Cham, pp 73–137. https://doi.org/10.1007/398_2016_8

Shahid M, Javed MT, Mushtaq A, Akram MS, Mahmood F, Ahmed T, Azeem M (2019) Microbe-mediated mitigation of cadmium toxicity in plants. In: Cadmium toxicity and tolerance in plants. Academic Press, Cambridge, MA, pp 427–449. https://doi.org/10.1016/B978-0-12-814864-8.00017-6

Shao HB, Chu LY, Shao MA, Jaleel CA, Hong-mei M (2008) Higher plant antioxidants and redox signaling under environmental stresses. Comptes Rendus Biologies 331(6):433–441

Sharma P, Dubey RS (2005) Toxicidade de chumbo em plantas. Braz J Plant Physiol 17(1):35–52. https://doi.org/10.1590/S1677-04202005000100004

Shaw J (1989) Heavy metal tolerance in plants: evolutionary aspects. CRC Press, Boca Raton

Shi G, Lu H, Liu H, Lou L, Zhang P, Song G, Ma H (2020) Sulfate application decreases translocation of arsenic and cadmium within wheat (Triticum aestivum L.) plant. Sci Total Environ 713:136665. https://doi.org/10.1016/j.scitotenv.2020.136665

Shiyab S, Chen J, Han FX, Monts DL, Matta FB, Gu M, Su Y (2009) Phytotoxicity of mercury in Indian mustard (Brassica juncea L.). Ecotoxicol Environ Saf 72:619–625. https://doi.org/10.1016/j.ecoenv.2008.06.002

Silva JT, Paniz FP, Pedron T, Torres DP, da Rocha Concenço FIG, Parfitt JMB, Batista BL (2020) Selected soil water tensions at phenological phases and mineral content of trace elements in rice grains–mitigating arsenic by water management. Agric Water Manag 228:105884. https://doi.org/10.1016/j.agwat.2019.105884

Singh S, Kumar V (2019) Mercury detoxification by absorption, mercuric ion reductase, and exopolysaccharides: a comprehensive study. Environ Sci Pollut Res 27(22):27181–27201

Smith DR, Flegal AR (1995) Lead in the biosphere: recent trends. Ambio 24(1):21–23

Sobrino-Plata J, Ortega-Villasante C, Flores-Cáceres ML, Escobar C, Del Campo FF, Hernández LE (2009) Differential alterations of antioxidant defenses as bioindicators of mercury and cadmium toxicity in alfalfa. Chemosphere 77(7):946–954

Song Y, Jin L, Wang X (2017) Cadmium absorption and transportation pathways in plants. Int J Phytorem 19(2):133–141. https://doi.org/10.1080/15226514.2016.1207598

Stahl W, Sies H (2003) Antioxidant activity of carotenoids. Mol Asp Med 24(6):345–351
Tabelin CB, Igarashi T, Villacorte-Tabelin M, Park I, Opiso EM, Ito M, Hiroyoshi N (2018) Arsenic, selenium, boron, lead, cadmium, copper, and zinc in naturally contaminated rocks: a review of their sources, modes of enrichment, mechanisms of release, and mitigation strategies. Sci Total Environ 645:1522–1553
Tangahu BV, Abdullah S, Rozaimah S, Basri H, Idris M, Anuar N, Mukhlisin M (2011) A review on heavy metals (As, Pb, and Hg) uptake by plants through phytoremediation. Int J Chem Eng 2011:939161. https://doi.org/10.1155/2011/939161
Turpeinen R, Pantsar-Kallio M, Häggblom M, Kairesalo T (1999) Influence of microbes on the mobilization, toxicity and biomethylation of arsenic in soil. Sci Total Environ 236 (1-3):173–180. https://doi.org/10.1016/S0048-9697(99)00269-7
United States Environmental Protection Agency (USEPA) (2002) Office of Solid Waste, and Emergency Response. In: Arsenic treatment technologies for soil, waste, and water. Diane Publishing, Darby, PA
United States Environmental Protection Agency (USEPA) (2009) National primary drinking water regulations. https://www.epa.gov/ground-water-and-drinkingwater/national-primary-drinkingwater-egulations. Accessed 9 Apr 2020
Van der Ent A, Nkrumah PN, Tibbett M, Echevarria G (2019) Evaluating soil extraction methods for chemical characterization of ultramafic soils in Kinabalu Park (Malaysia). J Geochem Expl 196:235–246. https://doi.org/10.1016/j.gexplo.2018.10.004
Vik EA, Bardos P (2003) Remediation of contaminated land technology implementation in Europe: [prepared by working group "Remediation technologies" of the concerted action "Contaminated Land Rehabilitation Network for Environmental Technologies"(CLARINET)...]. Umweltbundesamt
Voegelin A, Weber FA, Kretzschmar R (2007) Distribution and speciation of arsenic around roots in a contaminated riparian floodplain soil: Micro-XRF element mapping and EXAFS spectroscopy. Geochim Cosmochim Acta 71(23):5804–5820. https://doi.org/10.1016/j.gca.2007.05.030
Wan X, Lei M, Chen T (2016) Cost–benefit calculation of phytoremediation technology for heavy-metal-contaminated soil. Sci Total Environ 563:796–802. https://doi.org/10.1016/j.scitotenv.2015.12.080
Wang Y, Stauffer C, Keller C, Greger M (2005) Changes in Hg fractionation in soil induced by willow. Plant Soil 275(1-2):67–75
Wang J, Feng X, Anderson CW, Xing Y, Shang L (2012) Remediation of mercury contaminated sites–a review. J Hazard Mater 221:1–18
Wang J, Feng X, Anderson CW, Wang H, Wang L (2014) Thiosulphate-induced mercury accumu lation by plants: metal uptake and transformation of mercury fractionation in soil-results from a field study. Plant Soil 375(1-2):21–33
Wang ZQ, Huang H, Deng JM, Liu JQ (2015) Scaling the respiratory metabolism to phosphorus relationship in plant seedlings. Sci Rep 5:16377
Wang J, Xia J, Feng X (2017) Screening of chelating ligands to enhance mercury accumulation from historically mercury-contaminated soils for phytoextraction. J Environ Manag 186:233–239
Wang M, Li S, Chen S, Meng N, Li X, Zheng H, Wang D (2019) Manipulation of the rhizosphere bacterial community by biofertilizers is associated with mitigation of cadmium phytotoxicity. Sci Total Environ 649:413–421. https://doi.org/10.1016/j.scitotenv.2018.08.174
Wang YN, Tsang YF, Wang H, Sun Y, Song Y, Pan X, Luo S (2020a) Effective stabilization of arsenic in contaminated soils with biogenic manganese oxide (BMO) materials. Environ Pollut 258:113481. https://doi.org/10.1016/j.envpol.2019.113481
Wang L, Hou D, Cao Y, Ok YS, Tack FM, Rinklebe J, O'Connor D (2020b) Remediation of mercury contaminated soil, water, and air: A review of emerging materials and innovative technologies. Environ Int 134:105281

Welch RM, Norvell WA (1999) Mechanisms of cadmium uptake, translocation and deposition in plants. In: Cadmium in soils and plants. Springer, Dordrecht, pp 125–150. https://doi.org/10.1007/978-94-011-4473-5_6

Werther J, Hilligardt R, Kroning H (1986) Sand from dredge sludge-development of processes for the mechanical treatment of dredged material. In Contaminated soil. International TNO conference. pp 887–889

World Health Organization (WHO) (2011a) Guidelines for drinking-water quality, 4th edn

World Health Organization (WHO) (2011b) Cadmium in drinking-water: background document for development of WHO guidelines for drinking-water quality (No. WHO/SDE/WSH/03.04/80/Rev/1), WHO, Geneva

World Health Organization (WHO) (2017) Guidelines for drinking-water quality: first addendum to the fourth edition. WHO, Geneva

Wuana RA, Okieimen FE (2011) Heavy metals in contaminated soils: a review of sources, chemistry, risks and best available strategies for remediation. ISRN Ecol 2011:402647

Xie F, Dong K, Wang W, Asselin E (2020) Leaching of mercury from contaminated solid waste: a mini-review. Miner Process Extr Metall Rev 41(3):187–197

Xu J, Bravo AG, Lagerkvist A, Bertilsson S, Sjöblom R, Kumpiene J (2015) Sources and remediation techniques for mercury contaminated soil. Environ Int 74:42–53

Yan H, Li Q, Yuan Z, Jin S, Jing M (2019) Research progress of mercury bioaccumulation in the aquatic food chain, china: a review. Bull Environ Contam Toxicol 102(5):612–620

Yang Z, Fang Z, Zheng L, Cheng W, Tsang PE, Fang J, Zhao D (2016) Remediation of lead contaminated soil by biochar-supported nano-hydroxyapatite. Ecotoxicol Environ Saf 132:224–230. https://doi.org/10.1016/j.ecoenv.2016.06.008

Yang J, Yang SS, Lei M, Yang JX, Wan XM, Chen TB, Liu SQ (2018) Comparison among soil additives for enhancing Pteris vittata L.: Phytoremediation of As-contaminated soil. Int J Phytorem 20(13):1300–1306. https://doi.org/10.1080/15226514.2017.1319325

Yeboah IB, Tuffour HO, Abubakari A, Melenya C, Bonsu M, Quansah C, Adjei-Gyapong T (2019) Mobility and transport behavior of lead in agricultural soils. Sci African 5:e00117. https://doi.org/10.1016/j.sciaf.2019.e00117

Zhang J, Bishop PL (2002) Stabilization/solidification (S/S) of mercury-containing wastes using reactivated carbon and Portland cement. J Hazard Mater 92:199–212. https://doi.org/10.1016/s0304-3894(02)00019-5

Zhang H, Selim HM (2008) Reaction and transport of arsenic in soils: equilibrium and kinetic modeling. Adv Agron 98:45–115. https://doi.org/10.1016/S0065-2113(08)00202-2

Zhang XY, Wang QC, Zhang SQ, Sun XJ, Zhang ZS (2009) Stabilization/ solidification (S/S) of mercury-contaminated hazardous wastes using thiol-functionalized zeolite and Portland cement. J Hazard Mater 168(2-3):1575–1580

Zhang Z, Li M, Chen W, Zhu S, Liu N, Zhu L (2010) Immobilization of lead and cadmium from aqueous solution and contaminated sediment using nano-hydroxyapatite. Environ Pollut 158 (2):514–519. https://doi.org/10.1016/j.envpol.2009.08.024

Zhang T, Lu Q, Su C, Yang Y, Hu D, Xu Q (2017) Mercury induced oxidative stress, DNA damage, and activation of antioxidative system and Hsp70 induction in duckweed (*Lemna minor*). Ecotoxicol Environ Saf 143:46–56

Zhao FJ, Ma JF, Meharg AA, McGrath SP (2009) Arsenic uptake and metabolism in plants. New Phytol 181(4):777–794. https://doi.org/10.1111/j.1469-8137.2008.02716.x

Zimdahl RL, Skogerboe RK (1977) Behavior of lead in soil. Environ Sci Technol 11 (13):1202–1207. https://doi.org/10.1021/es60136a004

# Microbial Inoculation to Alleviate the Metal Toxicity in Crop Plants and Subsequent Growth Promotion

17

Fathima Afsal, Arnab Majumdar, Jisha Suresh Kumar, and Sutapa Bose

**Abstract**

With increasing anthropogenic activities, soil pollution by heavy metals and metalloids is causing serious quality issues in variable crops irrespective of irrigation pattern or seasonal impact. Rice, wheat, all kinds of lentils and leafy vegetables are contaminated with such metal(loid)s, and soil microbiota has been proven to be a vital biomanagement agent in remediation of such pollution. Rhizospheric bacteria, fungi, along with the mycorrhizal association, and algae are experimentally proven by many researchers over the years that these bioagents have the potential to mitigate metal toxicity at high level and can be applied at fields with proper implementation processes for alleviating the toxic metal(loid) stress on crops. This chapter has summarized the role of soil microbial communities in mitigation of soil metal(loid)s from being phyto-available and compromising crop's quality.

F. Afsal
Department of Earth Sciences, Indian Institute of Science Education and Research (IISER) Kolkata, Mohanpur, West Bengal, India

Department of Biological Sciences, Indian Institute of Science Education and Research (IISER) Bhopal, Bhauri, Madhya Pradesh, India

A. Majumdar · S. Bose (✉)
Department of Earth Sciences, Indian Institute of Science Education and Research (IISER) Kolkata, Mohanpur, West Bengal, India

J. S. Kumar
Department of Earth Sciences, Indian Institute of Science Education and Research (IISER) Kolkata, Mohanpur, West Bengal, India

The Faculty of Biology, Ludwig-Maximilians-University of Munich, Planegg, Germany





K. Mishra et al. (eds.), *Sustainable Solutions for Elemental Deficiency and Excess in Crop Plants*, https://doi.org/10.1007/978-981-15-8636-1_17

**Keywords**

Heavy metals and metalloids · Plant growth promoting rhizobacteria · Algal and cyanobacterial inoculation · Mycorrhizal association

## 17.1 Introduction

Soil is the storehouse of numerous beneficial macro- and micro-nutrients for plants. Despite elemental abundance in Earth's crust, the bioavailability and subsequent uptake of nutrients are dependent on various factors such as the soil pH, organic carbon content, bioavailability and the surrounding microbial community (Majumdar et al. 2019). However, the lack of selectivity in plants to uptake these essential elements leads to the uptake of non-essential elements as well when they are bioavailable (Clemens and Ma 2016). Several studies have thus previously reported the uptake of Zn, Cu, Fe, Ni, Cd and Pb by plants (Rattan et al. 2005; Singh et al. 2010; Bai et al. 2011). Heavy metals (HM) are indeed required by plants as micro-nutrients to act as cofactors to be a part of the prosthetic groups of enzymes that are involved in their metabolic pathways (Burd et al. 2000). However, studies have reported the detrimental effects of HM on plants due to their uptake beyond a limit, as a higher concentration of nutritional elements more than the requirement can cause damages (Sarkar et al. 2017). HM accumulation has been rising as a major concern especially in agriculture. The routes of heavy metal entry into plants could be because of various reasons such as: (a) Dependence on contaminated wastelands for agriculture: Rising population and increasing land usage have left farmers to use heavy metal-contaminated sites for agriculture. (b) Irrigation with the sewage water: The decline in the availability of freshwater has led to the exploration of wastewaters for irrigation, particularly in peri-urban areas of developing countries. (c) Another most common way of metal entry into crops is via leaching of contaminated water into agricultural lands and groundwater. A case study by Rattan et al. (2005) has shown that irrigation for 20 years with sewage water led to the accumulation of Zn (208%), Cu (170%), Fe (170%), Ni (63%) and Pb (29%) in the soil, compared to tube-well water-irrigated soils. Although the crops studied exhibited enhanced productivity, it could partly be due to the alleviation of the already existed Fe-deficiency syndrome. Similar long-term investigations are required regarding the usage of effluent water for irrigation.

However, there are also plants that can tolerate the accumulation of higher levels of toxic metals. They are called hyperaccumulators, and studies have discovered about 300 plants that hyperaccumulate nickel, 26 for cobalt, 24 for copper, 16 for zinc and 11 for manganese, respectively, in addition to several others (ul Islam et al. 2007). In the current scenario of rapid industrialization and hazardous heavy metal (loid) accumulation at an exponential rate, phytoremediation using hyperaccumulators has begun to be considered as a feasible option (Majumdar et al. 2018). Most of the studies employ plants to remove heavy metal(loid)s via phytoextraction (usage of plants to extract metals from the soil, transport and translocate them to harvestable above-ground mass), phytodegradation (usage of plant enzymes to degrade metal complexes, both internally and via secreted enzymes), rhizodegradation (usage of plants to stimulate the microbial community living around it to degrade complexes), phytostabilization (usage of plants to reduce mobility and hence bioavailability of metals via their root exudates) and phytovolatilization (usage of plants to take up contaminants and convert them into their volatile forms) (Kokyo et al. 2013; Soni and Jain 2014). The success of phytoremediation depends on the ability of plants to hyperaccumulate metals and produce large amounts of biomass (Burd et al. 2000). However, since the hyperaccumulators possess very high amounts of metals, their disposal has always been a problem. Incineration year after year has been increasing the burden and hence reducing the feasibility of conventional phytoremediation. Hence, lately, the usage of biofuel crops is receiving more attention because of their dual usage as a phytoremediator and as valuable biomass (Kokyo et al. 2013). This implies a need for exploring the potential of crops for the same and investigating ways to increase metal accumulation, in order to alleviate its toxic effects. Thus, the demand to enhance hyperaccumulation capacities of phytoremediator crops and the dire need to reduce metal toxicity in food crops share the common motivation to explore the potential of microbial inoculation to alleviate metal toxicity and thereby subsequently promote the growth of crop plants.

The role of the rhizosphere is crucial in HM and metalloids mobility and bioavailability due to the active microbial interactions taking place within a plant root system that involves pH fluctuations, redox transformations of soil elements and breaking down of complex carbon sources. Metalloids like arsenic in the soil can inhibit microbial population variably depending on the species richness (Majumdar et al. 2020). Release of metabolites from plant roots triggers microbial colonization around the root surface, and subsequent microbial activities act on the HM biotransformation or uptake depending on community richness (Seshadri et al. 2015). Rhizospheric pH tends to be acidic due to several microbial attributes like release of organic acids, the exchange between cations and anions, microbial substrate-induced respiration and changes in the oxidative state of essential elements like Fe, Mn and N (Hinsinger et al. 2003). Although reports suggest that continuous accumulation of heavy metal(loid)s in the soil can reduce the functional diversity of soil microbial ecosystem (Crowley 2008), along with accumulation of such toxic elements, resistance towards HM in microbes is also growing, and this can be used for specific bioremoval processes (Ellis et al. 2003). Figure 17.1 briefly represents the schematic diagram of microbial processes involved in metal(loid) uptake, the transformation processes and the mechanisms for plant's growth enhancement. Table 17.1 shows a diversified involvement of microbes in response to soil HM pollution and thereby leading to the betterment of associated crops.

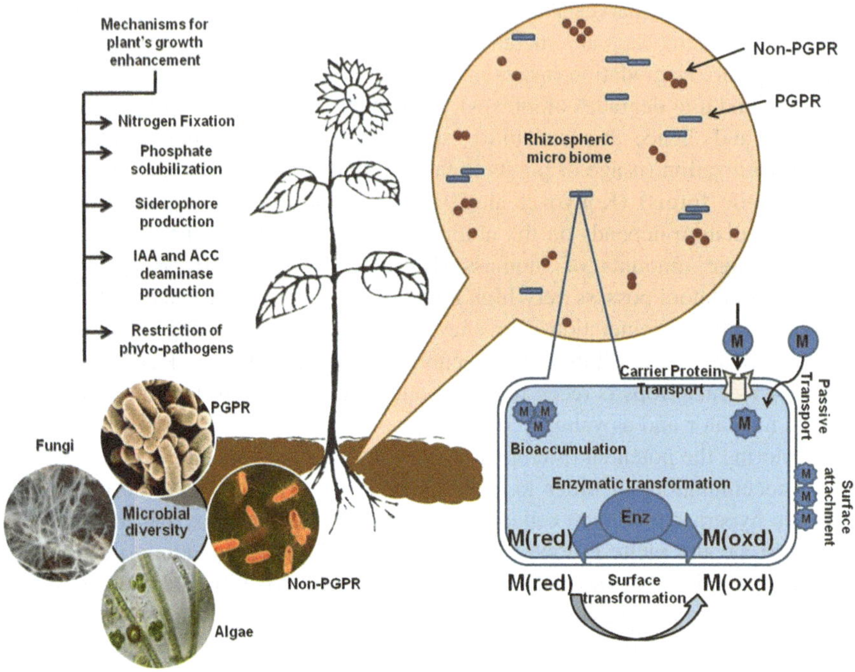


**Fig. 17.1** Mechanisms of metal(loid)s management by soil microbes and plants growth promotion in rhizosphere

## 17.2 Microbial Resistance Towards Different Toxic Elements

### 17.2.1 Tolerance Towards Nutrient Over Richness

Nutrients are undoubtedly playing vital roles in plant and microbial health. The sources of nutrients can vary from natural (mainly as a result of weathering of rocks) to human-related inputs (runoff from agricultural lands, households and industries). However, the presence of excess nutrients in water bodies has been a topic of discussion in the last decade in particular, as a result of the increase in industrial and agricultural practices leading to nutrient leach off from lands into groundwater and freshwater bodies. Increased algal growth, biological and chemical oxygen demands and culmination in eutrophication have been the described effects. One evolutionary strategy adopted by microbes, especially those possessing the ability to multiply faster compared to others, to confront over-richness of nutrients is to utilize the resources available to grow and increase in number. Godwin and Cotner (2018) reported the presence of more strains from phyla with higher $\mu_{max}$ (growth rate as a fraction of the maximum growth rate), particularly the phyla gamma-proteobacteria

**Table 17.1** Involvement of soil microbial community in heavy metal(loid)s uptake with enhancement of plant's growth

| Type of crop/cereal | Microorganism | Metal stress on plants | Microbial remediation on metal stressed plants | Reference |
|---|---|---|---|---|
| *Oryza sativa* (Rice) | *Ochrobactrum* sp., *Bacillus* sp. | Oxidative stress indicated by high levels of SOD and MDA, in addition to reduction in biomass | Overall biomass and root/shoot ratio were enhanced by bacterial inoculation with the aid of ACC deaminase, protease and catalase activities | Pandey et al. (2013) |
| | *K. pneumoniae* strain K5 | Cd-induced toxicity reduced the activity of α-amylase that resulted in metabolizable sugar deprivation, which ultimately decreased the germination percentage as well as the reduction of root and shoot growth | Via IAA, phosphate solubilization, ACC deaminase and N fixation ability, it increased seed germination, root and shoot length, root and shoot fresh weight. Improved antioxidant enzyme activity and several biochemical parameters (proline, α-amylase, protease, total sugar, total protein, chlorophyll content) decreased stress-ethylene level, reduced Cd accumulation in seedling | Pramanik et al. (2017) |
| | *Bacillus flexus* | 280 mM As(V), 32 mM As(III) | In the presence As (20 and 80 mg $kg^{-1}$), inoculated plants performed well as compared to un-inoculated plants. The grain yield (g $pot^{-1}$) of inoculated plants were 7.7 [As(20)] and 5.2 [As(80)] whereas un-inoculated plants had 6.6 and 4.8 (g $pot^{-1}$) grain yield | Das et al. (2016) |
| | *Kocuria flava* AB402, *Bacillus vietnamensis* AB403 | As-induced reduction in root and shoot length, dry and wet weight, and chlorophyll content | Siderophore and IAA-mediated alleviation of As toxicity increased chlorophyll and biomass content | Mallick et al. (2018) |
| *Triticum aestivum* (Wheat) | *Azotobacter* sp. | High Cd and Cr content in the food crop | Restricted the uptake of Cd and Cr by wheat plants; the underlying mechanism is uncertain | Joshi and Juwarkar (2009) |

(continued)

**Table 17.1** (continued)

| Type of crop/ cereal | Microorganism | Metal stress on plants | Microbial remediation on metal stressed plants | Reference |
|---|---|---|---|---|
| | *Piriformospora indica* | Cd stress reduced shoot length and shoot and root dry weights and chlorophyll content | The plants became more resistant to Cd stress, which included a reduction in Cd content of shoot and enhanced growth | Shahabivand et al. (2012) |
| | *Enterobacter ludwigii*, *E. cloacae*, *Klebsiella pneumoniae* | Significant reduction in growth parameters (shoot length, root length, fresh and dry weight of roots and shoots); and desiccation due to Hg | $NH_3$, siderophore, IAA and EPS production; ACC deaminase activity; Zn, P and K solubilization | Gontia-Mishra et al. (2016) |
| | *Bacillus cereus*, *Pseudomonas moraviensis* | Reduced biomass, chlorophyll content and germination due to stress induced by Cu, Cr, Co, Cd, Ni, Mn and Pb | Improved seed germination and increased fresh weight; plant height and chlorophyll contents also significantly increased | Hassan et al. (2017) |
| *Zea mays* (Maize) | *Pseudomonas aeruginosa*, *Pseudomonas fluorescens*, *Ralstonia metallidurans* | High uptake of Cr and Pb, leading to reduced biomass | Increased Cr and Pb shoot concentration, shoot biomass, as well as translocation factor; reduced total metal uptake without change in metal mobility in the soil | Braud et al. (2009) |
| | *Rhizobium leguminosarum* TAL-102, *Azotobacter chroococcum* | Pd toxicity reduced plant height and overall biomass | Assisted root growth and branching, increased root and shoot biomass; although it parallely had increased root metal content | Hadi and Bano (2010) |
| | *Ralstonia eutropha*, *Chryseobacterium humi* | Phytotoxic Cd levels reduced plant biomass | Increased root and shoot biomass | Moreira et al. (2014) |
| | *Bacillus*, *Klebsiella*, *Leifsonia*, *Enterobacter* sp. | Reduction in relative seedling growth of maize cultivars upon Cd exposure | Overall catalase, oxidase, phosphate solubilization, exopolysaccharide (EPS) and auxin production increased shoot/root growth and dry biomass | Ahmad et al. (2016) |

| | | | | |
|---|---|---|---|---|
| *Solanum lycopersicum* (Tomato) | *Kluyvera ascorbata* SUD165 | Ni toxicity decreased protein content (except for Indian mustard), plant dry weight and chlorophyll contents | Increase in dry weight, and protein and chlorophyll contents | Burd et al. (2000) |
| | *Pseudomonas aeruginosa*, *Burkholderia gladioli* | Cd stress triggered oxidative stress by accumulation of superoxide anions, MDA and $H_2O_2$ | Inoculation enhanced stress tolerance in the seedlings via expression of antioxidants | Khanna et al. (2019) |
| | *Methylobacterium oryza* CBMB20, *Burkholderia* sp. CBMB40 | Reduced root elongation, and biomass production | Reduction in the accumulations of Ni (II) and Cd(II) in roots and shoots | Madhaiyan et al. (2007) |
| | *Bacillus* sp. J119 | No significant reduction in growth | Promoted the growth of tomato, consequently increasing total Cd accumulation in the plants | Sheng et al. (2008b) |
| *Glycine max* (Soybean) | *Glomus etunicatum*, *G. macrocarpum* | Mn toxicity caused leaf puckering, necrotic spots, shorter petioles and browning of main veins and petioles | Enhanced plant growth and P uptake; stimulated the ATP-dependent sequestration of Mn or Mn-chelates in the vacuoles, formed low-solubility P–Mn complexes | Nogueira et al. (2007) |
| | *Bradyrhizobium* sp. | Cd stress reduced biomass and chlorophyll content | Increased photosynthesis pigments and mineral nutrients (Fe or Mg) in plant leaves, shoot dry weights, reduced Cd concentration in root and shoot | Guo and Chi (2014) |
| *Brassica juncea* (Indian mustard) | *Variovorax paradoxus*, *Rhodococcus* sp. | Cd induced plant stress and ethylene biosynthesis | ACC deaminase-dependent root elongation | Belimov et al. (2005) |
| | *Kluyvera ascorbata* SUD165 | Ni toxicity decreased protein content (except for Indian mustard), plant dry weight and chlorophyll contents | Increase in dry weight, and protein and chlorophyll contents | Burd et al. (2000) |
| | *Pseudomonas* sp., *Bacillus* sp. | Higher amount of Cr decreased root length, shoot length, plant fresh and dry weight | Protected the plants against the inhibitory effects of chromium, probably due to the production of IAA, siderophores and solubilization of phosphate | Rajkumar et al. (2006) |

(continued)

**Table 17.1** (continued)

| Type of crop/ cereal | Microorganism | Metal stress on plants | Microbial remediation on metal stressed plants | Reference |
|---|---|---|---|---|
| | *Azotobacter chroococcum*, *Bacillus megaterium*, *Bacillus mucilaginosus* | Cu- Pb- and Zn-associated toxicity reduced shoot biomass | Produced larger above-ground biomass and altered metal bioavailability in the soil | Wu et al. (2006) |
| | *Sinorhizobium* sp. Pb002 | Reduced plant biomass due to Pb toxicity | Increase in plant biomass, possibly dependent on ACC deaminase | Di Gregorio et al. (2006) |
| | *Bacillus subtilis* | Ni induced severe growth inhibition in plants | Plant growth promotion via IAA production and phosphate solubilization | Zaidi et al. (2006) |
| | *Bacillus edaphicus* NBT | Pb toxicity reduced biomass | Increased biomass production with plant growth promoting features were accompanied with high levels of Pb uptake | Sheng et al. (2008c) |
| | *Achromobacter xylosoxidans* strain Ax10 | Decrease in root length, shoot length, plant fresh and dry weight due to Cu-induced toxicity | Promoted plant growth via production of IAA and solubilization of phosphate, concurrently with Cu accumulation | Ma et al. (2009) |
| | *Enterobacter aerogenes*, *Rahnella aquatilis* | Ni and Cr decreased growth, as well as the protein and chlorophyll content | Through siderophore, IAA and ACC deaminase production, stimulated plant biomass and enhanced phytoextraction of metals | Kumar et al. (2009) |
| | *Pseudomonas* sp. | Ni stress reduced fresh and dry biomass production | Improved biomass production via ACC deaminase, siderophores, IAA and solubilizing P | Ma et al. (2011) |
| | *Ralstonia* sp. J1-22-2, *Pantoea agglomerans* Jp3-3, *Pseudomonas thivervalensis* Y1-3-9 | Reduction in biomass due to presence of Cu | ACC deaminase-dependent protection from metal toxicity, while increasing metal uptake | Zhang et al. (2011) |
| | *Staphylococcus arlettae* | Reduced chlorophyll content, P uptake and biomass due to As-related toxicity | Via production of IAA, siderophores and ACC deaminase, inoculation significantly increased biomass, protein, chlorophyll and carotenoids contents | Srivastava et al. (2013) |

| | | | | |
|---|---|---|---|---|
| | *Pseudomonas* sp., *Bacillus megaterium* | Inhibition in the growth and biomass production due to Ni toxicity | Protected the plants against the inhibitory effects of Ni, probably due to the production of IAA, siderophore and solubilization of phosphate | Rajkumar and Freitas (2008b) |
| *Brassica napus* (Rapeseed) | *Pseudomonas fluorescens*, *Pseudomonas putida* | Germination percent, length and fresh weight of seedlings significantly decreased by increasing Ni concentration | *Pseudomonas* strains had a promoting effect on germination percent of high Ni level-treated seeds; also increased adult plant biomass | Ashour et al. (2006) |
| | *Pseudomonas* sp., *Bacillus* sp., *Xanthomonas* sp. | Cd inhibited root elongation | Increased root and shoot biomass, in addition to Cd uptake | Sheng and Xia (2006) |
| | *Pseudomonas fluorescens* G10, *Microbacterium* sp. G16 | Pb uptake reduced the growth and biomass production | Stimulated plant growth by synthesis of IAA, siderophore or ACC deaminase activity | Sheng et al. (2008a) |
| | *Bacteroidetes bacterium*, *Pseudomonas fluorescens*, *Variovorax* sp. | Decrease in the root growth because of Cd and Zn toxicity | Improved plant growth and chlorophyll content | Dąbrowska et al. (2017) |
| | *Pseudomonas tolaasii* ACC23, *Pseudomonas fluorescens* ACC9, *Mycobacterium* sp. ACC14 | Cd reduced shoot and root elongation, and chlorophyll content | Increased plant biomass (via IAA and siderophore production in ACC9 and ACC23, while ACC14 produced only IAA) | Dell'Amico et al. (2008) |
| *Brassica nigra* (Black mustard) | *Microbacterium* sp. CE3R2, *Microbacterium* sp. NE1R5, *Curtobacterium* sp. NM1R1, *Microbacterium* sp. NM3E9 | Cu, Zn, As and Pb toxicity decreased root elongation and seed germination | Via siderophores, IAA and solubilization of phosphate, they improved seed germination and root growth in the presence of heavy metals | Roman-Ponce et al. (2017) |
| | *Kocuria* sp. CRB15 | Cu-dependent reduction in root and shoot elongation | The production of IAA, ammonia, HCN and P solubilization helped increase root and shoot elongation | Hansda and Kumar (2017) |

(continued)

**Table 17.1** (continued)

| Type of crop/ cereal | Microorganism | Metal stress on plants | Microbial remediation on metal stressed plants | Reference |
|---|---|---|---|---|
| *Cajanus cajan* (Pigeon pea) | *Proteus vulgaris* KNP3 | Reduction in biomass, root length, shoot growth, chlorophyll content and germination | Increased biomass, root length, shoot growth and chlorophyll content, besides reducing the Cu load of the soil | Rani et al. (2008) |
| | *Glomus mosseae* | Cd and Pb induced stress (indicated by GSH production) decreased root and shoot dry weights, nodule formation, leghaemoglobin content and nitrogenase activity | Attenuated the toxic effects of metal stress, and increased biomass by biosorption (minimizing metal translocation to the shoots) and dilution effects | Garg and Aggarwal (2011) |
| *Pisum sativum* (Common pea) | *Pseudomonas putida* KNP9 | Cd and Pb accumulation caused reduction in biomass and chlorophyll content | Enhanced biomass and reduced Cd and Pb accumulation in root and shoot | Tripathi et al. (2005) |
| | *Rhizobium* sp. RP5 | Ni- and Zn-associated toxicity decreased the dry matter, nodule numbers and nodule mass | With abilities of N fixation, growth promotion and reduction of toxicity of Ni and Zn, it increased the dry matter, nodule numbers, seed yield, grain protein, leghaemoglobin, total N and glutathione reductase activity | Wani et al. (2008) |
| | *Pseudomonas brassicacearum*, *Pseudomonas marginalis* | Root growth was inhibited by Cd toxicity; also decreased shoot content of P, K, Ca and Fe | Counteracted the Cd-induced inhibition of nutrient uptake by plants probably through stimulation of root growth and enhancement of nutrient active processes | Safronova et al. (2006) |
| *Vigna unguiculata* (Cow pea) | *Streptomyces acidiscabies* E13 (derived cell-free supernatant containing siderophores and auxins) | Significant lipid peroxidation occurred in plants under metal stress conditions, a marker for cellular oxidative damage | Better root growth and higher uptake of Al, Cu, Fe, Mn, Ni and U | Dimkpa et al. (2009) |

| | | | | |
|---|---|---|---|---|
| *Cicer arietinum* (Chickpea) | *Pseudomonas* sp. | Ni at high concentrations reduced biomass | Siderophore-mediated chelation of Ni led to enhanced plant growth and reduced Ni uptake | Tank and Saraf (2009) |
| *Vigna radiate* (Green gram) | *Bradyrhizobium* sp. (vigna) RM8 | Phytotoxic effects of Ni and Zn such as decrease in the dry matter accumulation, nodule numbers, leghaemoglobin and nodule mass due to biomolecular damage | Growth promotion and attenuation of the toxic effects of Ni and Zn; increase in nodules, leghaemoglobin and N content | Wani et al. (2007) |
| | *Pseudomonas putida*, *Comamonas* sp. | Reduction in biomass, and root and shoot length, all associated with Cd toxicity | Significant increase in root length, shoot length, wet weight and dry weight | Saluja and Sharma (2014) |
| *Vigna mungo* (Black gram) | *Pseudomonas aeruginosa* | Cd was toxic, causing stunting, poor rooting and shrunken leaves, affected germination dose-dependently | Overcame Cd toxicity via lessened Cd accumulation, extensive rooting and enhanced plant growth | Ganesan (2008) |
| *Ricinus communis* (Castor bean) | *Pseudomonas* sp., *Pseudomonas jessenii* | Ni, Cu and Zn toxicity reduced root and shoot biomass | Both strains increased the shoot and root biomass via plant growth promoting activities | Rajkumar and Freitas (2008a) |
| | *Psychrobacter* sp. | Ni toxicity reduced amounts of Fe, P, chlorophyll and protein content, and catalase activity | Increased Fe, P, chlorophyll and protein contents and catalase activity in the tissues | Ma et al. (2010) |
| *Capsicum annuum* (Green chilli) | *Cellulosimicrobium cellulans* KUCr3 | Shoot length, root length and the production of chlorophyll were severely hampered by Cr | Improved the growth parameters of plants due to its IAA production and phosphate mineralization; also decreased the Cr uptake in plants | Chatterjee et al. (2009) |
| *Coffea Arabica* (Arabian coffee) | *Glomus clarum*, *Gigaspora margarita*, *Acaulospora* sp. | Cu and Zn stress reduced biomass accumulation and P content | Increased plant growth, P uptake, reduced metal translocation to the shoots (uptake by extra-radical hyphae) | Andrade et al. (2010) |

(continued)

**Table 17.1** (continued)

| Type of crop/ cereal | Microorganism | Metal stress on plants | Microbial remediation on metal stressed plants | Reference |
|---|---|---|---|---|
| *Sorghum bicolor* (Sweet sorghum) | *Bacillus* sp. SLS18 | Cd stress decreased the aerial part dry weight | Increased the aerial parts and root dry weights via production of IAA, siderophores and ACC deaminase | Luo et al. (2011) |
| | *Bacillus cereus* TCR17, *Providencia rettgeri* TCR21, *Myroides odoratimimus* TCR22 | Cr toxicity decreased root length, shoot length, fresh and dry weight, chlorophyll content | Increased plant growth, antioxidant status (SOD, catalase and ascorbate peroxidase) and decreased proline and MDA, helping plants to reduce stress-induced oxidative damage | Bruno et al. (2020) |
| *Calopogonium mucunoides* (Ground nut) | *G. etunicatum* | Reduced the plant growth, nutrient absorption and ability to produce root nodules | Conferred Pb tolerance to plants by reducing ROS generation; promoted biomass production nutrient uptake (P, S and Fe) and nodulation of plants | de Souza et al. (2012) |
| *Vigna unguiculata* (Yardlong bean) | *Photobacterium* spp. | Hg-induced phytotoxicity in growth parameters | Significant increase in biomass, root length, seed number and increased mercury uptake limited to roots | Mathew et al. (2015) |
| *Eruca sativa* (Miller Plant) | *Pseudomonas putida* | Cd toxicity led to decrease in biomass, chlorophyll content and increase in proline content | Via ACC deaminase activity, increased shoot length, root length, fresh and dry weight, and chlorophyll content | Kamran et al. (2015) |

and bacteroidetes in eutrophic lakes because of their nutrient- and energy-rich conditions.

Several microbes belonging to the genus *Pseudomonas* and *Bacillus* have the ability to sequester and accumulate excess nutrients, especially nitrogen and phosphorous, which are the two emerging contaminants in agricultural lands because of the excessive use of fertilizers. Two discrete enriched microbial consortia containing nitrogen and phosphorous accumulators, when applied, were able to restrict the nutrients within the plant root zone, ensuring prevention of eutrophication through leaching while making it available for uptake by plants (Chaudhuri et al. 2017). Another method adopted by microbes to meet this crisis of excess nutrients is by secreting chelating molecules that bind nutrients and metals. These chelating agents can either make nutrients available for plants or sequester toxic heavy metals, thereby reducing their mobility.

### 17.2.2 Resistance Towards Heavy Metal(loid)s Pollution

Essential or non-essential metals, depending on the type of metals and their speciation and ionic potential, can exhibit toxicity when present above their threshold concentrations (Sullivan and Gadd 2019). Root plaque has been described to be a characteristic feature of wetland plants. It performs dual functions—by being a source when it functions as a facilitator for a sequestered chemical, and a sink when it functions as a barrier for that chemical. In addition to the presence of iron oxyhydroxides and manganese oxides, in some cases, bacterial biofilms have also played role in sequestration and/or accumulation of Al, As, Cd, Cu, Cr, Mn, Ni, P, Pb, Se and/or Zn (Khan et al. 2016). Siderophores are chelating agents released by microbes and have the potential to sequester iron from the surrounding region, thereby making it available to the plants. However, Sullivan and Gadd (2019) have identified several reports that establish another evidential role of siderophores in the complexation of several metals, including Cr, Co, Pb, Mo, Mn and Zn. For instance, *P. aeruginosa* produces a siderophore named pyochelin, which can form complexes with a variety of non-ferrous metals (Braud et al. 2009; Manoj et al. 2020).

Other possible mechanisms for microbial resistance include internal compartmentalization, outer membrane vesicle formation and secretion of substances that may precipitate, adsorb or detoxify metal(loid)s (Sullivan and Gadd 2019).

## 17.3 Bacterial Inoculation and Metal Uptake

### 17.3.1 Application of PGPR Microbes

With alarming rates of pollution, the concentration of metals in soil has increased over the past decade. This has drastically affected crops in terms of toxicity because of the capability of heavy metals to inactivate plant enzymes either by binding strongly the oxygen, nitrogen and sulphur atoms in plant cells or through substitution of one metal ion with another one in the molecule (Shilev et al. 2007). One way to alleviate the toxicity caused by heavy metals will be by inoculating the crops with plant growth-promoting rhizobacteria (PGPR). They are a group of bacteria that can promote the growth of plants and, in addition, can help reduce the metal-associated toxicity. The well-known PGPR include members of the genera Arthrobacter, Azospirillum, Azoarcus, Bacillus, Burkholderia, Enterobacter, Gluconacetobacter, Herbaspirillum, Klebsiella, Paenibacillus, Pseudomonas and Serratia (Dardanelli et al. 2010).

In general, the rhizosphere refers to the narrow region of soil that is immediate to plant roots and hosts a rich microbial flora, which plays important roles in several plant processes and in maintaining plant health. The microbial community flourishes due to the presence of plant root exudates that can pose as a nutrient source to promote microbial growth. In response, microbial populations in the rhizosphere are known to modify the chemical properties of the rhizospheric soil via the release of organic compounds, redox changes and soil acidification, thereby resulting in changes in the environmental mobility and availability of metal contaminants (Usman and Mohamed 2009).

PGPR are usually free-living bacteria associated with roots of plants, and nitrogen fixation is only a minor component of the benefit they offer to plants, unlike their symbiotic nitrogen-fixing counterparts. These PGPR have the potential to enhance the plant's growth via either direct or indirect measures. The direct promotion of plant growth by PGPR is mainly by secreting beneficial compounds to plants, or by facilitating the uptake of certain nutrients from the environment. The indirect promotion of plant growth is via reducing or preventing the harmful effects of phytopathogenic organisms (Glick 1995).

Some of the mechanisms adopted by PGPR to enhance plant growth include (Manoj et al. 2020) (i) Phytohormone production: Plant growth regulators such as auxins, cytokines, polyamines etc. can help regulate various physiological and metabolic functions of the plant under metal stress, even if present in fewer amounts. (ii) Nitrogen fixation: Nitrogen is one of the essential components of plant metabolic reactions. It has been reported that PGPR inoculation in Ni-contaminated soil can stimulate the nitrogen uptake and protect plants from reactive oxygen species (ROS) due to the ROS scavenging defence system possessed by the nitrogen-fixing bacteria (such as via glutathione reductase activity) (Khan et al. 2009). (iii) Phosphate solubilization: Phosphate solubilizing rhizobacteria help convert insoluble P material into soluble forms available for plant growth, thereby leading to increased photosynthesis and yield (Han and Lee 2005). Nutrient deficiency (Fe, P, Mg, Ca) is a major concern associated with plants grown in Ni-contaminated soils, and inoculation with PGPR has been shown to significantly increase the plant P content (Ma et al. 2010). (iv) Siderophores and ammonia production: Heavy metal contamination can often be correlated with iron deficiency of plants (Burd et al. 2000). Siderophores, as mentioned previously, are chelating agents released by microbes and have the potential to sequester iron from the surrounding region, thereby making it available to the plants. They can convert $Fe^{3+}$ (complex, insoluble form) into $Fe^{2+}$

form (soluble form) for efficient plant uptake (Manoj et al. 2020). Ammonia is an alternative nitrogen source for plants. Enhancement of phytoremediation efficiency of plant was observed after inoculation with Cr(VI) tolerant rhizobacterial strain with ammonia production (Oves et al. 2013; Manoj et al. 2020) and excess production of ammonia can also serve as a triggering factor for the virulence of opportunistic plant pathogens (Panhwar et al. 2013; Manoj et al. 2020). In addition to these mechanisms, PGPR secrete various other metabolites also, which are not directly involved in plant growth but can reduce the plant stress caused by biotic (phytopathogens) and abiotic (heavy metal-induced phytotoxicity) factors (Manoj et al. 2020).

Once identified, the heavy metal remediating microbes could be further screened for their potential to promote plant growth based on its ability, (i) to produce IAA, (ii) to utilize ACC as the sole N source, (iii) to solubilize phosphorus, and (iv) to secrete siderophore into the growth medium (Srivastava et al. 2013). Put together, the enhancement of plant growth by PGPR does not only increase phytoextraction but also helps improve the health of crops during metal stress, increase yield and alter nutrient uptake.

### 17.3.2 Application of Non-PGPR Microbes

The use of the term 'non-PGPR' is comparatively new and is usually confused with control treatments with 'no-inoculum' added. The focus of the research so far has been to elucidate and describe microbial communities that can promote plant growth as well, in addition to alleviating metal toxicity. This has ignored/and belittled the importance of bacteria that can solely antagonize the effects of heavy metal stress on crop plants. Non-PGPR can produce microbial metabolites that could carry out biosorption and biocomplexation or affect the mobility of metal ions in the soil, thereby preventing metal uptake by plants. It is known that the production of extracellular polymeric substances (EPSs) by microbes is greatly enhanced under stress imposed by heavy metals. Hence, inoculating EPS-producing microbes helps in the bioremediation of heavy metals by the process of biosorption through its negatively charged components (Pal and Paul 2008). This helps in the promotion of plant growth not directly, but via alleviation of metal toxicity.

Cyanide is formed by a variety of bacteria as a secondary metabolite, e.g. *Chromobacterium violaceum* or *Pseudomonas fluorescens*, and it has the ability to carry out microbial mobilization of metals by forming cyanide complexes (Brandl et al. 2008). However, the recent research by Rijavec and Lapanje (2016) has found that HCN released by microbes is indirectly increasing the availability of phosphate, in addition to sequestering metals. This could possibly shift the role of microbial HCN from just a chelator to a plant growth promoter.

In general, the significance of non-PGPR in agriculture arises especially during the presence of heavy metals in excess. In cases of agricultural soil contamination with HM in greater amounts, the immediate focus would be to sequester/reduce the mobility of these ions, rather than plant growth promotion directly. Thus inoculation of non-PGPR that can hyperaccumulate, chelate and immobilize HMs needs further investigation.

## 17.4 Algal and Cyanobacterial Involvement in Metal Accumulation

Multicellular algae, microalgae and cyanobacteria are major parts of the soil's nitrogen assimilation along with the indigenous ability to tolerate and bioaccumulate a greater content of toxic elements compared to bacterial efficiency. With continuous industrial development, soil pollution peaks to its highest resulting in contamination of surrounding agricultural fields. Under different soil nutritional conditions and moisture content, soil algal and cyanobacterial community get altered. The remediation of HM by algae–microalgae follows the process of bioaccumulation by living cells and biosorption by cell-secreted by-products (Mitra et al. 2012; ul Hassan et al. 2017). Several reports have been shown efficient removal of elements with a very high ratio of the element to algal biomass (mg/g). Among microalgae, *Desmodesmus pleiomorphus* has been reported by Monteiro et al. (2009, 2010) for removal of Zn and Cd with a rate of 360.2 and 61.2 mg/g, respectively, whereas Tüzün et al. (2005) showed the ability of *Chlamydomonas reinhardtii* to remove Cd at 77.62 mg/g from the system involved. A series of experimentation has been conducted by Romera et al. (2007) showing various macroalgal species removing Cd (*Spirogyra insignis*) and both Cd–Zn mixture (*Fucus spiralis*, *Asparagopsis armata* and *Chondrus crispus*). Cu (115 mg/g, *Sargassum wightii*) and Zn (125.5 mg/g, *Ulva reticulata*) removal reported by Vijayaraghavan and Prabu (2006) and Senthilkumar et al. (2006), respectively.

Macro- and microalgae contain polyphosphate vesicles that act as inclusion bodies and can also act as storage for the HM like Pb, Cd, Hg, Ni, Cu, Zn, Mg etc. and can be detoxified gradually from the cell (Dwivedi 2012). In a comparative older study using a model of *Chlorella salina*, researchers have shown that the HM removal pattern of algae is biphasic, indicating the first phase of rapid biosorption independent of its growth parameters followed by a second slower elemental accumulation, which is dependent on the algal growth parameters and abiotic factors (Garnham et al. 1992). Microalgae can follow a series of cellular processes to manage HM tolerance and subsequent sorption to the cell surface and intracellular accumulation. Steps are briefly described by Monteiro et al. (2012) as attachment of metal ions to the algal surface and subsequent precipitation of ions, secretion of algal extra-cellular metabolites to make HM complex, a continuous flow of efflux takes place to balance the low concentration of ions inside the cell, conversion of HM oxidative state to the less toxic form intra-cellularly, chelation of metal ions to the cytoplasmic proteins or polysaccharides and finally release of toxic metal ions by forming volatile compound. Microalgal cell wall components play a crucial role in the development of such mechanistic uptake (García-Ríos et al. 2007; Gupta and Rastogi 2008). Some reports showed the detailed structural analysis of various microalgal cell wall components, which confers the ion-exchange mechanisms for metal ions adsorption and uptake (Kaduková and Virčíková 2005; La Rocca et al. 2009). Variable functional groups like carboxyl, hydroxyl, amino, phosphate and sulfhydryl contribute to the outer cellular charge distribution and provide the binding sites to the metal ions for further uptake (Bohumil 2007; Deng et al. 2007; Kaplan 2013). Algae can also detoxify arsenic (As) by producing glutathione and phytochelatins (Upadhyay et al. 2018), and reports suggest that algal inoculation can efficiently be used in remediation and detoxification (Pawlik-Skowrońska et al. 2004; Upadhyay et al. 2016).

Cyanobacteria, also known as blue-green algae, can be found predominantly in any agricultural soil with some stagnant water content and nutritional enrichment. These are well tested and reported for their multiple potential for environmental clean-up including pesticides (Megharaj et al. 1994), hydrocarbons (Al-Hasan et al. 1998), oil spillage (Cohen 2002) and heavy metals (Lefebvre et al. 2007; Gupta et al. 2013). The higher biomass is effective in the accumulation of toxic elements in both aqueous and soil systems (Roeselers et al. 2008; Pandey 2017). Cyanobacteria can produce a mucilaginous sheath of exopolysaccharides (EPS) outside their cell wall and that EPS can chelate exogenous metal ions (Ozturk and Aslim 2008) while some reports showed the effectiveness in the biotransformation of the toxic form of elements to the lesser toxic or non-toxic form of elements by changing the redox state of the HM (Faisal et al. 2005; Lefebvre et al. 2007). Like microalgae, cyanobacteria can also produce polyphosphate granules and some inorganic linear form of phosphate inclusions and sequester HM by synthesizing metallothioneins that are cysteine-containing metal-binding protein molecules (Pandey 2017). Although these are potential biomanagement agents, several environmental factors like pH, temperature, light flux rate, the redox potential of the growth system, presence of ions and nutrition, and salinity can affect the ability of the cyanobacterial HM biosorption (Cain et al. 2008).

## 17.5 Fungal Inoculation and Metal Uptake

Fungi represent a huge kingdom comprising moulds, mushrooms, lichens, smuts, rusts and yeasts. These are eukaryotes living as saprophytes, parasites or mutualists with each taxon exhibiting a markedly diverse life cycle. Besides contributing to nearly 75% of the total soil microbial biomass, they have developed a symbiotic relationship with roughly 80% of vascular plant species. These fungi possess extraordinary characteristics that render them suitable for detoxification of toxic heavy metal contaminations. While some fungi can survive extremities of temperature and pH, some can subsist in very low water activities and oxygen availability. Filamentous fungi growing in the soil possess extending hyphae that can easily spread across a vast region, thereby encountering toxic substances present even in very low concentrations with much higher contact frequencies (Harms et al. 2011). The fundamental roles played by fungi in geological processes and their interactions with metals and minerals are extensively reviewed and published in an excellent article by Gadd (2007). The principle strategies employed by fungi in resisting toxic heavy metal contaminations can be categorized into two: metal mobilization and immobilization. Mobilization of metals is achieved by their binding with metabolites or siderophores released from fungi; by their vaporization from the addition of methyl/organic groups by fungi; or by their change of oxidation states caused by fungal redox mechanisms, all of which potentially contribute towards bioremediation (Gadd 2007). Siderophores are iron chelators secreted by certain microorganisms and plants, which are later taken up as iron–siderophore complexes to compensate for the deficiency of iron within the cell/living system. It has been observed that the survivability of siderophore secreting taxa is considerably higher compared to their non-secreting companions along a natural heavy metal gradient. Additionally, in mixed communities of siderophore secreting and non-secreting groups, the latter seemed to significantly benefit from the binding of siderophores to toxic heavy metals in the environment (Hesse et al. 2018). Although primarily being iron chelators, siderophores can bind to a wide array of metals and heavy metals, albeit with lesser affinity. Unlike iron–siderophore complexes that are recognized and taken up by microbes and plants, siderophores bound to other metals are not recognized and hence do not enter the living system (Ahmed and Holmström 2014; O'Brien et al. 2014). Additionally, it was also reported that plants growing in heavy metal-contaminated soils that are often deficient in iron significantly benefit from siderophore secreting rhizobial microbes that maintain a consistent supply of iron through complexation and ensure better plant growth (He and Yang 2007). Other ways of mobilizing metals are by their transformation—either by reducing metalloid oxyanions to their elemental form as in the reduction of $SeO^{-2}{}_3$ and $SeO^{-2}{}_4$ to $Se^0$ or by methylating metalloids like $AsO^{3-}{}_4$ and $AsO^{-}{}_2$ to volatile $(CH_3)_3$ As species. Both these biotransformation processes hold huge significance in biogeochemistry as they aid in reducing the toxicity or in the vaporization of a potent toxin (White et al. 1997). Additionally, metals can be immobilized in several ways including adsorption onto fungal cell walls, pigments or extracellular polysaccharides; intracellular accumulation or sequestration by fungi; or precipitation of metals onto or around hyphae. Factors that influence the binding of metal ions to fungal hyphae include pH, amount of melanin and chitin, and the presence of extramatrical mycelium (Gadd 2007). In the case of intracellular sequestration, cells cannot allow the presence of free cytosolic heavy metals. Hence, toxic metals are either stored inside vacuoles or trapped by cytosolic chelators that are low molecular weight polypeptides rich in cysteine residues called metallothioneins (González-Guerrero et al. 2009). Extracellular precipitation is achieved by the secretion of organic acids like citrates or oxalates that react with metal ions. In recent times, glomalin, a glycoprotein produced excessively on hyphae and fungal spores, has gained attention by playing a prominent role in extracellular sequestration of toxic heavy metals (Wang et al. 2019a). Considering the ability of fungi to effectively use all these strategies for survival in heavily contaminated environments, they can be considered as a potent tool in bioremediation. In an interesting study, fungi were also reported to be capable of serving as a vector in the dispersion of pollutant degrading bacteria (Kohlmeier et al. 2005), which makes them all the more desirable for application.

Based on their ability or inability to associate in a symbiotic relationship with plant roots, fungi are categorized as mycorrhizal and non-mycorrhizal, respectively. In the following subsections, we have tried to cite the most recent advancements and findings on the roles of these two kinds of fungi in toxicity mitigation.

### 17.5.1 Role of Mycorrhizal Fungi in Toxicity Mitigation

Mycorrhiza is a specific symbiotic association between the roots of higher plants and fungi that extend into the rhizosphere and the surrounding soil. Depending on the kind of association with roots, mycorrhizal fungi are further classified as ecto-, endo- and ectoendomycorrhiza. The major difference between ecto- and endomycorrhiza is that the former is an obligatory symbiont that forms a mantle of hyphae around the roots completely covering the root tips but never entering into the cells, while the latter is more invasive where the hyphae besides growing inside the root also penetrate the root cell walls and become enclosed in their cell membranes. The third kind possesses characteristics of both ecto- and endomycorrhiza. Among the three, endomycorrhiza has gained particular attention in recent times for its highly beneficial association with plants. This is further categorized into five groups—arbuscular, ericoid, arbutoid, monotropoid and orchid endomycorrhizae. Among these, arbuscular mycorrhizae are predominantly studied due to their ubiquity in nature (Ganugi et al. 2019). Several discoveries have been made about the ability of arbuscular mycorrhizal fungi (AMF) in alleviating heavy metal toxicity. In lentils, it was observed that plants inoculated with AFM showed better growth and significantly less accumulation of arsenic in the root, shoot and pods compared to plants not inoculated with AFM under high arsenic conditions (Alam et al. 2019b). The same group of scientists in another study showed that AMF-treated pea plants showed the highest reduction in As accumulation in grain as compared to other treatments of selenium, silica gel, biochar and sulphur (Alam et al. 2019a). On the same line, several years back, it was observed that AFM significantly decreased arsenic uptake in rice plants by triggering the down-regulation of LSi1 and LSi2 mRNA expressions where LSi1,2 are silicon transporters that also transport arsenic, which is an analogue (Chen et al. 2012). Recently, it was observed that *Aspergillus* sp. and *Penicillium* sp. efficiently contributed towards the vaporization of As from contaminated soil by transforming them into gaseous tri-, mono- and dimethyl arsenates (Guimarães et al. 2019). In lead-contaminated soils, it was observed that the application of a combination of AFM and lignin biochar significantly reduced the bioavailability of Pb to barley crops. It was proposed that, with this combination of treatment, barley crops can safely be grown in lead-contaminated soils with no subsequent harm on human health since they observed the highest viability in HEK293 cells during this particular treatment (Khan et al. 2020). Remediation of Pb was also achieved using *Aspergillus niger* as a biological catalyst that mediated the dephosphorylation of phosphates from stable and potentially unavailable soil organic phosphates (OPs) by secretion of phytases that lead to the rapid formation of insoluble chloropyromorphites, which proved to be an efficient way of immobilizing toxic Pb from contaminated soil and water (Zhang et al. 2019). Concomitantly, AMF has recently been reported to mitigate the toxicity of several other heavy metals like cadmium (Rady et al. 2019; Shahid et al. 2019; Sidhu et al. 2019), manganese (Garcia et al. 2020), lead (Wang et al. 2019b), aluminium (Maldaner et al. 2020) etc., in crops and plants.

### 17.5.2 Role of Non-mycorrhizal Fungi in Toxicity Mitigation

Non-mycorrhizal fungi do not establish a close symbiotic relationship with the roots of vascular plants. Some examples are endophytic fungi that inhabit organs and tissues of plants without causing any damage contrary to mycorrhizal fungi that only associate with plant root cells, sometimes in an invasive relationship and epiphytic fungi that grow upon or attached to a living plant. Four strains of endophytic fungi, among 32 that were tested for mercury resistance, were able to successfully remove 100% of mercury from their culture medium (Pietro-Souza et al. 2020). Another interesting study on endophytic fungi *Penicillium funiculosum* revealed its ability to protect soybean from combined heavy metal stress by inducing the up-regulation of phytoprotective hormones/antioxidants and the down-regulation of heavy metal ATPases in plants (Bilal et al. 2019). The same group of scientists in another study stated that endophytic fungi enhance the tolerance of soybean against combined abiotic stresses of heavy metals, high temperature and drought (Bilal et al. 2020). Besides naturally occurring remediating fungi, yeast strains have been successfully engineered to produce sulphide species (Sun et al. 2020) and express metal transporters (Sun et al. 2019) for combined multimetal remediation that can be used as a safe and efficient biotechnological tool. On the same line, multimetal remediation was also achieved using live microbiomes, germinated spores and dead cell wall polysaccharides from new aquatic strains of *Mucor heimalis* (Hoque and Fritscher 2019).

It is clear that fungi that closely associate with plants—both mycorrhizal and endophytic (non-mycorrhizal) have a direct influence on plant nutritional status and defence (Eid et al. 2019; Ganugi et al. 2019) especially in the case of arbuscular mycorrhizal fungi that are hypothesized to have co-evolved with roots of vascular plants during the early adaptations of plants to terrestrial environments (Chen et al. 2018). Hence these fungi can play a substantial role in the remediation of contaminated agricultural fields where we are primarily concerned about plant growth and yield. That being said, other types of fungi that do not establish any kind of symbiosis with plants can still be used to effectively remove toxic heavy metals from industrial slurries, sewage and other contaminated sources as in the examples of yeast and mucor that were previously mentioned.

## 17.6 Biotransformation of Metals and Microbial Intervention

Biotransformation of metals refers to its chemical alteration by an organism, either by changing its oxidation state or by the addition of organic groups. Since oxidation states and chemical compositions of metals determine their physico-chemical characteristics, a change in either of the two factors causes a significant change in the toxicity of the metal species. Hence, this phenomenon holds environmental significance and can be exploited to mitigate the toxicity of heavily contaminated habitats in a safe manner without generating secondary contaminants. Addressing recent developments in biotransformation studies, six marine diatom species were lately identified to efficiently transform arsenates ($As^V$) to arsenites ($As^{III}$) followed by their methylation to form considerably less toxic methylated arsenic species, provided they were subjected to optimum temperatures and salinity (Papry et al. 2019). Corresponding to this study, it was discovered that biotransformation of arsenates ($As^V$) to arsenates ($As^{III}$) and their subsequent methylation occurred predominantly during the logarithmic phase of the growth curve while the transformation of arsenic into complex organic arsenic species occurred predominantly during the stationary phase in a diatom/phytoplankton (Hasegawa et al. 2019). Furthermore, scientists discovered the abundance of arsenic biotransformation genes in microbes isolated from marine sediments (Guo et al. 2019). Other microbes have also been observed to be able to transform arsenic species like microalgae in marine ecosystems (Al Mamun et al. 2019a, b) and bacteria (Kumari et al. 2019). Some recent findings on biotransformation of other heavy metal species include the transformation of chromium by root nodule bacteria (Jobby et al. 2019); of manganese by bacteria (Youngwilai et al. 2020); and of multiple metals like Pb, Cd, Cu and Zn by fungi (Chen et al. 2019).

**Acknowledgements** Authors are thankful to the IISER Kolkata, IISER Bhopal and LMU library facility for collection of scientific information and articles as a base of this chapter. Authors are also stating of no conflict of interest.

## References

Ahmad I, Akhtar MJ, Asghar HN et al (2016) Differential effects of plant growth-promoting rhizobacteria on maize growth and cadmium uptake. J Plant Growth Regul 35(2):303–315

Ahmed E, Holmström SJ (2014) Siderophores in environmental research: roles and applications. Microbial Biotechnol 7(3):196–208

Al-Hasan RH, Al-Bader DA, Sorkhoh NA et al (1998) Evidence for n-alkane consumption and oxidation by filamentous cyanobacteria from oil-contaminated coasts of the Arabian Gulf. Mar Biol 130(3):521–527

Al Mamun MA, Omori Y, Papry RI et al (2019a) Bioaccumulation and biotransformation of arsenic by the brown macroalga Sargassum patens C. Agardh in seawater: effects of phosphate and iron ions. J Appl Phycol 31(4):2669–2685

Al Mamun MA, Rahman IM, Datta RR et al (2019b) Arsenic speciation and biotransformation by the marine macroalga Undaria pinnatifida in seawater: a culture medium study. Chemosphere 222:705–713

Alam MZ, Hoque MA, Ahammed GJ et al (2019a) Arbuscular mycorrhizal fungi, selenium, sulfur, silica-gel and biochar reduce arsenic uptake in plant biomass and improve nutritional quality in Pisum sativum. BioRxiv 663120

Alam MZ, Hoque MA, McGee R et al (2019b) Arbuscular Mycorrhizal Fungus (AMF) and reduction of arsenic uptake in lentil crops. BioRxiv 522714

Andrade SAL, Silveira APD, Mazzafera P (2010) Arbuscular mycorrhiza alters metal uptake and the physiological response of Coffea arabica seedlings to increasing Zn and Cu concentrations in soil. Sci Total Environ 408(22):5381–5391

Ashour EH, El-Mergawi RA, Radwan SMA (2006) Efficacy of pseudomonas to phytoremediate nickel by canola (Brassica napus L.). J Appl Sci Res 2(7):375–382

Bai J, Xiao R, Cui B et al (2011) Assessment of heavy metal pollution in wetland soils from the young and old reclaimed regions in the Pearl River Estuary, South China. Environ Pollut 159 (3):817–824

Belimov AA, Hontzeas N, Safronova VI et al (2005) Cadmium-tolerant plant growth-promoting bacteria associated with the roots of Indian mustard (Brassica juncea L. Czern.). Soil Biol Biochem 37(2):241–250

Bilal S, Shahzad R, Khan AL et al (2019) Phytohormones enabled endophytic Penicillium funiculosum LHL06 protects Glycine max L. from synergistic toxicity of heavy metals by hormonal and stress-responsive proteins modulation. J Hazard Mater 379:120824

Bilal S, Shahzad R, Imran M et al (2020) Synergistic association of endophytic fungi enhances Glycine max L. resilience to combined abiotic stresses: heavy metals, high temperature and drought stress. Ind Crop Prod 143:111931

Bohumil V (2007) Biosorption and me. Water Res 41:4017–4029

Brandl H, Lehmann S, Faramarzi MA et al (2008) Biomobilization of silver, gold, and platinum from solid waste materials by HCN-forming microorganisms. Hydrometallurgy 94(1–4):14–17

Braud A, Jézéquel K, Bazot S et al (2009) Enhanced phytoextraction of an agricultural Cr-and Pb-contaminated soil by bioaugmentation with siderophore-producing acteria. Chemosphere 74 (2):280–286

Bruno LB, Karthik C, Ma Y et al (2020) Amelioration of chromium and heat stresses in Sorghum bicolor by Cr6+ reducing-thermotolerant plant growth promoting bacteria. Chemosphere 244:125521

Burd GI, Dixon DG, Glick BR (2000) Plant growth-promoting bacteria that decrease heavy metal toxicity in plants. Can J Microbiol 46(3):237–245

Cain A, Vannela R, Woo LK (2008) Cyanobacteria as a biosorbent for mercuric ion. Bioresource Technol 99(14):6578–6586

Chatterjee S, Sau GB, Mukherjee SK (2009) Plant growth promotion by a hexavalent chromium reducing bacterial strain, Cellulosimicrobium cellulans KUCr3. World J Microbiol Biotechnol 25(10):1829–1836

Chaudhuri SR, Mishra M, De S et al (2017) Microbe-based strategy for plant nutrient management. Biological wastewater treatment and resource recovery, 37

Chen X, Li H, Chan WF et al (2012) Arsenite transporters expression in rice (Oryza sativa L.) associated with arbuscular mycorrhizal fungi (AMF) colonization under different levels of arsenite stress. Chemosphere 89(10):1248–1254

Chen M, Arato M, Borghi L et al (2018) Beneficial services of arbuscular mycorrhizal fungi–from ecology to application. Front Plant Sci 9:1270

Chen Y, Chen Y, Li Y et al (2019) Changes of heavy metal fractions during co-composting of agricultural waste and river sediment with inoculation of Phanerochaete chrysosporium. J Hazard Mater 378:120757

Clemens S, Ma JF (2016) Toxic heavy metal and metalloid accumulation in crop plants and foods. Annu Rev Plant Biol 67:489–512

Cohen Y (2002) Bioremediation of oil by marine microbial mats. Int Microbiol 5(4):189–193

Crowley D (2008) Impacts of metals and metalloids on soil microbial diversity and ecosystem function. Revista de la ciencia del suelo y nutrición vegetal 8:6–11

Dąbrowska G, Hrynkiewicz K, Trejgell A et al (2017) The effect of plant growth-promoting rhizobacteria on the phytoextraction of Cd and Zn by Brassica napus L. Int J Phytoremed 19 (7):597–604

Dardanelli MS, Carletti SM, Paulucci NS et al (2010) Benefits of plant growth-promoting rhizobacteria and rhizobia in agriculture. In: Plant growth and health promoting bacteria. Springer, Berlin, Heidelberg, pp 1–20

Das S, Jean JS, Chou ML et al (2016) Arsenite-oxidizing bacteria exhibiting plant growth promoting traits isolated from the rhizosphere of Oryza sativa L.: implications for mitigation of arsenic contamination in paddies. J Hazard Mater *302*:10–18

de Souza LA, de Andrade SAL, de Souza SCR et al (2012) Arbuscular mycorrhiza confers Pb tolerance in Calopogonium mucunoides. Acta Physiol Plant 34(2):523–531

Dell'Amico E, Cavalca L, Andreoni V (2008) Improvement of Brassica napus growth under cadmium stress by cadmium-resistant rhizobacteria. Soil Biol Biochem 40(1):74–84

Deng L, Zhu X, Wang X et al (2007) Biosorption of copper (II) from aqueous solutions by green alga Cladophora fascicularis. Biodegradation 18(4):393–402

Di Gregorio S, Barbafieri M, Lampis S et al (2006) Combined application of Triton X-100 and Sinorhizobium sp. Pb002 inoculum for the improvement of lead phytoextraction by Brassica juncea in EDTA amended soil. Chemosphere 63(2):293–299

Dimkpa CO, Merten D, Svatoš A et al (2009) Metal-induced oxidative stress impacting plant growth in contaminated soil is alleviated by microbial siderophores. Soil Biol Biochem 41 (1):154–162

Dwivedi S (2012) Bioremediation of heavy metal by algae: current and future perspective. J Adv Lab Res Biol 3(3):195–199

Eid AM, Salim SS, Hassan SED et al (2019) Role of endophytes in plant health and abiotic stress management. In: Microbiome in plant health and disease. Springer, Singapore, pp 119–144

Ellis RJ, Morgan P, Weightman AJ et al (2003) Cultivation-dependent and-independent approaches for determining bacterial diversity in heavy-metal-contaminated soil. Appl Environ Microbiol 69(6):3223–3230

Faisal M, Hameed A, Hasnain S (2005) Chromium-resistant bacteria and cyanobacteria: impact on Cr (VI) reduction potential and plant growth. J Ind Microbiol Biotechnol 32(11–12):615–621

Gadd GM (2007) Geomycology: biogeochemical transformations of rocks, minerals, metals and radionuclides by fungi, bioweathering and bioremediation. Mycol Res 111(1):3–49

Ganesan V (2008) Rhizoremediation of cadmium soil using a cadmium-resistant plant growth-promoting rhizopseudomonad. Curr Microbiol 56(4):403–407

Ganugi P, Masoni A, Pietramellara G et al (2019) A review of studies from the last twenty years on plant–arbuscular mycorrhizal fungi associations and their uses for wheat crops. Agronomy 9 (12):840

Garcia KGV, Mendes Filho PF, Pinheiro JI et al (2020) Attenuation of manganese-induced toxicity in Leucaena leucocephala colonized by arbuscular mycorrhizae. Water Air Soil Pollut 231(1):22

García-Ríos V, Freile-Pelegrín Y, Robledo D et al (2007) Cell wall composition affects Cd2+ accumulation and intracellular thiol peptides in marine red algae. Aquat Toxicol 81(1):65–72

Garg N, Aggarwal N (2011) Effects of interactions between cadmium and lead on growth, nitrogen fixation, phytochelatin, and glutathione production in mycorrhizal Cajanus cajan (L.) Millsp. J Plant Growth Regul 30(3):286–300

Garnham GW, Codd GA, Gadd GM (1992) Kinetics of uptake and intracellular location of cobalt, manganese and zinc in the estuarine green alga Chlorella salina. Appl Microbiol Biotechnol 37 (2):270–276

Glick BR (1995) The enhancement of plant growth by free-living bacteria. Can J Microbiol 41 (2):109–117

Godwin CM, Cotner JB (2018) What intrinsic and extrinsic factors explain the stoichiometric diversity of aquatic heterotrophic bacteria? ISME J 12(2):598–609

Gontia-Mishra I, Sapre S, Sharma A et al (2016) Alleviation of mercury toxicity in wheat by the interaction of mercury-tolerant plant growth-promoting rhizobacteria. J Plant Growth Regul 35 (4):1000–1012

González-Guerrero M, Benabdellah K, Ferrol N et al (2009) Mechanisms underlying heavy metal tolerance in arbuscular mycorrhizas. In: Mycorrhizas-functional processes and ecological impact. Springer, Berlin, Heidelberg, pp 107–122

Guimarães LHS, Segura FR, Tonani L et al (2019) Arsenic volatilization by Aspergillus sp. and Penicillium sp. isolated from rice rhizosphere as a promising eco-safe tool for arsenic mitigation. J Environ Manage 237:170–179

Guo J, Chi J (2014) Effect of Cd-tolerant plant growth-promoting rhizobium on plant growth and Cd uptake by Lolium multiflorum Lam. and Glycine max (L.) Merr. in Cd-contaminated soil. Plant Soil 375(1–2):205–214

Guo T, Li L, Zhai W et al (2019) Distribution of arsenic and its biotransformation genes in sediments from the East China Sea. Environ Pollut 253:949–958

Gupta VK, Rastogi A (2008) Equilibrium and kinetic modelling of cadmium (II) biosorption by nonliving algal biomass Oedogonium sp. from aqueous phase. J Hazard Mater 153 (1–2):759–766

Gupta V, Ratha SK, Sood A et al (2013) New insights into the biodiversity and applications of cyanobacteria (blue-green algae)—prospects and challenges. Algal Res 2(2):79–97

Hadi F, Bano A (2010) Effect of diazotrophs (Rhizobium and Azatebactor) on growth of maize (Zea mays L.) and accumulation of lead (Pb) in different plant parts. Pak J Bot 42(6):4363–4370

Han HS, Lee KD (2005) Phosphate and potassium solubilizing bacteria effect on mineral uptake, soil availability and growth of eggplant. Res J Agric Biol Sci 1(2):176–180

Hansda A, Kumar V (2017) Cu-resistant Kocuria sp. CRB15: a potential PGPR isolated from the dry tailing of Rakha copper mine. 3 Biotech 7(2):132

Harms H, Schlosser D, Wick LY (2011) Untapped potential: exploiting fungi in bioremediation of hazardous chemicals. Nat Rev Microbiol 9(3):177–192

Hasegawa H, Papry RI, Ikeda E et al (2019) Freshwater phytoplankton: biotransformation of inorganic arsenic to methylarsenic and organoarsenic. Sci Rep 9(1):1–13

Hassan TU, Bano A, Naz I (2017) Alleviation of heavy metals toxicity by the application of plant growth promoting rhizobacteria and effects on wheat grown in saline sodic field. Int J Phytoremed 19(6):522–529

He ZL, Yang XE (2007) Role of soil rhizobacteria in phytoremediation of heavy metal contaminated soils. J Zhejiang Univ Sci B 8(3):192–207

Hesse E, O'Brien S, Tromas N et al (2018) Ecological selection of siderophore-producing microbial taxa in response to heavy metal contamination. Ecol Lett 21(1):117–127

Hinsinger P, Plassard C, Tang C et al (2003) Origins of root-mediated pH changes in the rhizosphere and their responses to environmental constraints: a review. Plant Soil 248 (1–2):43–59

Hoque E, Fritscher J (2019) Multimetal bioremediation and biomining by a combination of new aquatic strains of Mucor hiemalis. Sci Rep 9(1):1–16

Jobby R, Jha P, Gupta A et al (2019) Biotransformation of chromium by root nodule bacteria Sinorhizobium sp. SAR1. PLoS One 14(7)

Joshi PM, Juwarkar AA (2009) In vivo studies to elucidate the role of extracellular polymeric substances from Azotobacter in immobilization of heavy metals. Environ Sci Technol 43 (15):5884–5889

Kaduková J, Virčíková E (2005) Comparison of differences between copper bioaccumulation and biosorption. Environ Int 31(2):227–232

Kamran MA, Syed JH, Eqani SAMAS et al (2015) Effect of plant growth-promoting rhizobacteria inoculation on cadmium (Cd) uptake by Eruca sativa. Environ Sci Pollut Res 22(12):9275–9283

Kaplan D (2013) Absorption and adsorption of heavy metals by microalgae. Handbook of microalgal culture. Appl Phycol Biotechnol 2:602–611

Khan MS, Zaidi A, Wani PA et al (2009) Role of plant growth promoting rhizobacteria in the remediation of metal contaminated soils: a review. In: Organic farming, pest control and remediation of soil pollutants. Springer, Dordrecht, pp 319–350
Khan N, Seshadri B, Bolan N et al (2016) Root iron plaque on wetland plants as a dynamic pool of nutrients and contaminants. In: Advances in agronomy, vol 138. Academic Press, pp 1–96
Khan MA, Ramzani PMA, Zubair M et al (2020) Associative effects of lignin-derived biochar and arbuscular mycorrhizal fungi applied to soil polluted from Pb-acid batteries effluents on barley grain safety. Sci Total Environ 710:136294
Khanna K, Jamwal VL, Kohli SK et al (2019) Plant growth promoting rhizobacteria induced Cd tolerance in Lycopersicon esculentum through altered antioxidative defense expression. Chemosphere. 217:463–474
Kohlmeier S, Smits TH, Ford RM et al (2005) Taking the fungal highway: mobilization of pollutant-degrading bacteria by fungi. Environ Sci Technol 39(12):4640–4646
Kokyo O, Tao L, Hongyan C et al (2013) Development of profitable phytoremediation of contaminated soils with biofuel crops. J Environ Protect
Kumar KV, Srivastava S, Singh N et al (2009) Role of metal resistant plant growth promoting bacteria in ameliorating fly ash to the growth of Brassica juncea. J Hazard Mater 170(1):51–57
Kumari N, Rana A, Jagadevan S (2019) Arsenite biotransformation by Rhodococcus sp.: characterization, optimization using response surface methodology and mechanistic studies. Sci Total Environ 687:577–589
La Rocca N, Andreoli C, Giacometti GM et al (2009) Responses of the Antarctic microalga Koliella antarctica (Trebouxiophyceae, Chlorophyta) to cadmium contamination. Photosynthetica 47 (3):471–479
Lefebvre DD, Kelly D, Budd K (2007) Biotransformation of Hg (II) by cyanobacteria. Appl Environ Microbiol 73(1):243–249
Luo SL, Chen L, Chen JL et al (2011) Analysis and characterization of cultivable heavy metal-resistant bacterial endophytes isolated from Cd-hyperaccumulator Solanum nigrum L. and their potential use for phytoremediation. Chemosphere 85(7):1130–1138
Ma Y, Rajkumar M, Freitas H (2009) Inoculation of plant growth promoting bacterium Achromobacter xylosoxidans strain Ax10 for the improvement of copper phytoextraction by Brassica juncea. J Environ Manage 90(2):831–837
Ma Y, Rajkumar M, Vicente JAF et al (2010) Inoculation of Ni-resistant plant growth promoting bacterium Psychrobacter sp. strain SRS8 for the improvement of nickel phytoextraction by energy crops. Int J phytoremed 13(2):126–139
Ma Y, Rajkumar M, Luo Y, Freitas H (2011) Inoculation of endophytic bacteria on host and non-host plants – effects on plant growth and Ni uptake. J Hazard Mater 196:230–237
Madhaiyan M, Poonguzhali S, Sa T (2007) Metal tolerating methylotrophic bacteria reduces nickel and cadmium toxicity and promotes plant growth of tomato (Lycopersicon esculentum L.). Chemosphere 69(2):220–228
Majumdar A, Barla A, Upadhyay MK et al (2018) Vermiremediation of metal (loid) s via Eichornia crassipes phytomass extraction: a sustainable technique for plant amelioration. J Environ Manage 220:118–125
Majumdar A, Upadhyay MK, Kumar JS et al (2019) Ultra-structure alteration via enhanced silicon uptake in arsenic stressed rice cultivars under intermittent irrigation practices in Bengal delta basin. Ecotoxicol Environ Saf 180:770–779
Majumdar A, Kumar JS, Sheena et al (2020) Agricultural water management practices and environmental influences on arsenic dynamics in rice field. In: Arsenic in drinking water and food. Springer, Singapore, pp 425–443
Maldaner J, Steffen GPK, Saldanha CW et al (2020) Combining tolerant species and microorganisms for phytoremediation in aluminium-contaminated areas. Int J Environ Studies 77(1):108–121
Mallick I, Bhattacharyya C, Mukherji S et al (2018) Effective rhizoinoculation and biofilm formation by arsenic immobilizing halophilic plant growth promoting bacteria (PGPB) isolated

from mangrove rhizosphere: a step towards arsenic rhizoremediation. Sci Total Environ 610:1239–1250
Manoj SR, Karthik C, Kadirvelu K et al (2020) Understanding the molecular mechanisms for the enhanced phytoremediation of heavy metals through plant growth promoting rhizobacteria: a review. J Environ Manage 254:109779
Mathew DC, Ho YN, Gicana RG et al (2015) A rhizosphere-associated symbiont, Photobacterium spp. strain MELD1, and its targeted synergistic activity for phytoprotection against mercury. PLoS One 10(3)
Megharaj M, Madhavi DR, Sreenivasulu C et al (1994) Biodegradation of methyl parathion by soil isolates of microalgae and cyanobacteria. Bull Environ Contam Toxicol 53(2):292–297
Mitra N, Rezvan Z, Ahmad MS et al (2012) Studies of water arsenic and boron pollutants and algae phytoremediation in three springs, Iran. Int J Ecosys 2(3):32–37
Monteiro CM, Marques AP, Castro PM et al (2009) Characterization of Desmodesmus pleiomorphus isolated from a heavy metal-contaminated site: biosorption of zinc. Biodegradation 20(5):629–641
Monteiro CM, Castro PM, Malcata FX (2010) Cadmium removal by two strains of Desmodesmus pleiomorphus cells. Water Air Soil Pollut 208(1–4):17–27
Monteiro CM, Castro PM, Malcata FX (2012) Metal uptake by microalgae: underlying mechanisms and practical applications. Biotechnol Progress 28(2):299–311
Moreira H, Marques AP, Franco AR et al (2014) Phytomanagement of Cd-contaminated soils using maize (Zea mays L.) assisted by plant growth-promoting rhizobacteria. Environ Sci Pollut Res 21(16):9742–9753
Nogueira MA, Nehls U, Hampp R et al (2007) Mycorrhiza and soil bacteria influence extractable iron and manganese in soil and uptake by soybean. Plant Soil 298(1–2):273–284
O'Brien S, Hodgson DJ, Buckling A (2014) Social evolution of toxic metal bioremediation in Pseudomonas aeruginosa. Proc R Soc B: Biol Sci 281(1787):20140858
Oves M, Khan MS, Zaidi A (2013) Chromium reducing and plant growth promoting novel strain Pseudomonas aeruginosa OSG41 enhance chickpea growth in chromium amended soils. Eur J Soil Biol 56:72–83
Ozturk S, Aslim B (2008) Relationship between chromium (VI) resistance and extracellular polymeric substances (EPS) concentration by some cyanobacterial isolates. Environ Sci Pollut Res 15(6):478–480
Pal A, Paul AK (2008) Microbial extracellular polymeric substances: central elements in heavy metal bioremediation. Indian J Microbiol 48(1):49–64
Pandey VD (2017) Cyanobacteria-mediated heavy metal remediation. In: Agro-environmental sustainability. Springer, Cham, pp 105–121
Pandey S, Ghosh PK, Ghosh S et al (2013) Role of heavy metal resistant Ochrobactrum sp. and Bacillus spp. strains in bioremediation of a rice cultivar and their PGPR like activities. J Microbiol 51(1):11–17
Panhwar QA, Jusop S, Naher UA et al (2013) Application of potential phosphate-solubilizing bacteria and organic acids on phosphate solubilization from phosphate rock in aerobic rice. Sci World J
Papry RI, Ishii K, Al Mamun MA et al (2019) Arsenic biotransformation potential of six marine diatom species: effect of temperature and salinity. Sci Rep 9(1):1–16
Pawlik-Skowrońska B, Pirszel J, Kalinowska R et al (2004) Arsenic availability, toxicity and direct role of GSH and phytochelatins in As detoxification in the green alga Stichococcus bacillaris. Aquat Toxicol 70(3):201–212
Pietro-Souza W, de Campos Pereira F, Mello IS et al (2020) Mercury resistance and bioremediation mediated by endophytic fungi. Chemosphere 240:124874
Pramanik K, Mitra S, Sarkar A et al (2017) Characterization of cadmium-resistant Klebsiella pneumoniae MCC 3091 promoted rice seedling growth by alleviating phytotoxicity of cadmium. Environ Sci Pollut Res 24(31):24419–24437

Rady MM, Ahmed SM, El-Yazal MAS et al (2019) Alleviation of cadmium stress in wheat by polyamines. In: Cadmium tolerance in plants. Academic Press, pp 463–496
Rajkumar M, Freitas H (2008a) Influence of metal resistant-plant growth-promoting bacteria on the growth of Ricinus communis in soil contaminated with heavy metals. Chemosphere 71 (5):834–842
Rajkumar M, Freitas H (2008b) Effects of inoculation of plant-growth promoting bacteria on Ni uptake by Indian mustard. Bioresource Technol 99(9):3491–3498
Rajkumar M, Nagendran R, Lee KJ et al (2006) Influence of plant growth promoting bacteria and Cr6+ on the growth of Indian mustard. Chemosphere 62(5):741–748
Rani A, Shouche YS, Goel R (2008) Declination of copper toxicity in pigeon pea and soil system by growth-promoting Proteus vulgaris KNP3 strain. Curr Microbiol 57(1):78
Rattan RK, Datta SP, Chhonkar PK et al (2005) Long-term impact of irrigation with sewage effluents on heavy metal content in soils, crops and groundwater—a case study. Agric Ecosyst Environ 109(3–4):310–322
Rijavec T, Lapanje A (2016) Hydrogen cyanide in the rhizosphere: not suppressing plant pathogens, but rather regulating availability of phosphate. Front Microbiol 7:1785
Roeselers G, Van Loosdrecht MCM, Muyzer G (2008) Phototrophic biofilms and their potential applications. J Appl Phycol 20(3):227–235
Roman-Ponce B, Reza-Vázquez DM, Gutierrez-Paredes S et al (2017) Plant growth-promoting traits in rhizobacteria of heavy metal-resistant plants and their effects on Brassica nigra seed germination. Pedosphere 27(3):511–526
Romera E, González F, Ballester A et al (2007) Comparative study of biosorption of heavy metals using different types of algae. Bioresource Technol 98(17):3344–3353
Safronova VI, Stepanok VV, Engqvist GL et al (2006) Root-associated bacteria containing 1-aminocyclopropane-1-carboxylate deaminase improve growth and nutrient uptake by pea genotypes cultivated in cadmium supplemented soil. Biol Fertil Soils 42(3):267–272
Saluja B, Sharma V (2014) Cadmium resistance mechanism in acidophilic and alkalophilic bacterial isolates and their application in bioremediation of metal-contaminated soil. Soil Sediment Contam: Int J 23(1):1–17
Sarkar SR, Majumdar A, Barla A et al (2017) A conjugative study of Typha latifolia for expunge of phyto-available heavy metals in fly ash ameliorated soil. Geoderma 305:354–362
Senthilkumar R, Vijayaraghavan K, Thilakavathi M et al (2006) Seaweeds for the remediation of wastewaters contaminated with zinc (II) ions. J Hazard Mater 136(3):791–799
Seshadri B, Bolan NS, Naidu R (2015) Rhizosphere-induced heavy metal (loid) transformation in relation to bioavailability and remediation. J Soil Sci Plant Nutr 15(2):524–548
Shahabivand S, Maivan HZ, Goltapeh EM et al (2012) The effects of root endophyte and arbuscular mycorrhizal fungi on growth and cadmium accumulation in wheat under cadmium toxicity. Plant Physiol Biochem 60:53–58
Shahid M, Javed MT, Mushtaq A et al (2019) Microbe-mediated mitigation of cadmium toxicity in plants. In: Cadmium toxicity and tolerance in plants. Academic Press, pp 427–449
Sheng XF, Xia JJ (2006) Improvement of rape (Brassica napus) plant growth and cadmium uptake by cadmium-resistant bacteria. Chemosphere 64(6):1036–1042
Sheng XF, Xia JJ, Jiang CY et al (2008a) Characterization of heavy metal-resistant endophytic bacteria from rape (Brassica napus) roots and their potential in promoting the growth and lead accumulation of rape. Environ Pollut 156(3):1164–1170
Sheng X, He L, Wang Q et al (2008b) Effects of inoculation of biosurfactant-producing Bacillus sp. J119 on plant growth and cadmium uptake in a cadmium-amended soil. J Hazard Mater 155 (1–2):17–22
Sheng XF, Jiang CY, He LY (2008c) Characterization of plant growth-promoting Bacillus edaphicus NBT and its effect on lead uptake by Indian mustard in a lead-amended soil. Can J Microbiol 54(5):417–422

Shilev S, Naydenov M, Tahsin N et al (2007) Effect of easily biodegradable amendments on heavy metal solubilization and accumulation in technical crops-a field trial. J Environ Eng Landscape Manage 15(4):237–242
Sidhu GPS, Bali AS, Bhardwaj R (2019) Use of fungi in mitigating cadmium toxicity in plants. In: Cadmium toxicity and tolerance in plants. Academic Press, pp 397–426
Singh R, Tripathi RD, Dwivedi S et al (2010) Lead bioaccumulation potential of an aquatic macrophyte Najas indica are related to antioxidant system. Bioresource Technol 101 (9):3025–3032
Soni S, Jain S (2014) A review on phytoremediation of heavy metals from soil by using plants to remove pollutants from the environment. Int J Adv Res 2(8):197–203
Srivastava S, Verma PC, Chaudhry V et al (2013) Influence of inoculation of arsenic-resistant Staphylococcus arlettae on growth and arsenic uptake in Brassica juncea (L.) Czern. Var. R-46. J Hazard Mater 262:1039–1047
Sullivan TS, Gadd GM (2019) Metal bioavailability and the soil microbiome. In: Advances in agronomy. Academic Press Inc.
Sun GL, Reynolds EE, Belcher AM (2019) Designing yeast as plant-like hyperaccumulators for heavy metals. Nat Commun 10(1):1–12
Sun GL, Reynolds EE, Belcher AM (2020) Using yeast to sustainably remediate and extract heavy metals from waste waters. Nat Sustain:1–9
Tank N, Saraf M (2009) Enhancement of plant growth and decontamination of nickel-spiked soil using PGPR. J Basic Microbiol 49(2):195–204
Tripathi M, Munot HP, Shouche Y et al (2005) Isolation and functional characterization of siderophore-producing lead-and cadmium-resistant Pseudomonas putida KNP9. Curr Microbiol 50(5):233–237
Tüzün I, Bayramoğlu G, Yalçın E et al (2005) Equilibrium and kinetic studies on biosorption of Hg (II), Cd (II) and Pb (II) ions onto microalgae Chlamydomonas reinhardtii. J Environ Manage 77 (2):85–92
ul Hassan Z, Ali S, Rizwan M et al (2017) Role of bioremediation agents (bacteria, fungi, and algae) in alleviating heavy metal toxicity. In: Probiotics in agroecosystem. Springer, Singapore, pp 517–537
ul Islam E, Yang XE, He ZL et al (eds) (2007) Assessing potential dietary toxicity of heavy metals in selected vegetables and food crops. J Zhejiang Univ Sci B 8(1):1–13
Upadhyay AK, Singh NK, Singh R et al (2016) Amelioration of arsenic toxicity in rice: comparative effect of inoculation of Chlorella vulgaris and Nannochloropsis sp. on growth, biochemical changes and arsenic uptake. Ecotoxicol Environ Saf 124:68–73
Upadhyay MK, Yadav P, Shukla A et al (2018) Utilizing the potential of microorganisms for managing arsenic contamination: a feasible and sustainable approach. Front Environ Sci 6:24
Usman ARA, Mohamed HM (2009) Effect of microbial inoculation and EDTA on the uptake and translocation of heavy metal by corn and sunflower. Chemosphere 76(7):893–899
Vijayaraghavan K, Prabu D (2006) Potential of Sargassum wightii biomass for copper (II) removal from aqueous solutions: application of different mathematical models to batch and continuous biosorption data. J Hazard Mater 137(1):558–564
Wang Q, Mei D, Chen J et al (2019a) Sequestration of heavy metal by glomalin-related soil protein: implication for water quality improvement in mangrove wetlands. Water Res 148:142–152
Wang Y, Yi B, Sun X et al (2019b) Removal and tolerance mechanism of Pb by a filamentous fungus: a case study. Chemosphere 225:200–208
Wani PA, Khan MS, Zaidi A (2007) Effect of metal tolerant plant growth promoting Bradyrhizobium sp. (vigna) on growth, symbiosis, seed yield and metal uptake by greengram plants. Chemosphere 70(1):36–45
Wani PA, Khan MS, Zaidi A (2008) Effect of metal-tolerant plant growth-promoting Rhizobium on the performance of pea grown in metal-amended soil. Arc Environ Contam Toxicol 55(1):33–42

White C, Sayer JA, Gadd GM (1997) Microbial solubilization and immobilization of toxic metals: key biogeochemical processes for treatment of contamination. FEMS Microbiol Rev 20 (3–4):503–516

Wu SC, Cheung KC, Luo YM et al (2006) Effects of inoculation of plant growth-promoting rhizobacteria on metal uptake by Brassica juncea. Environ Pollut 140(1):124–135

Youngwilai A, Kidkhunthod P, Jearanaikoon N et al (2020) Simultaneous manganese adsorption and biotransformation by Streptomyces violarus strain SBP1 cell-immobilized biochar. Sci Total Environ:136708

Zaidi S, Usmani S, Singh BR et al (2006) Significance of Bacillus subtilis strain SJ-101 as a bioinoculant for concurrent plant growth promotion and nickel accumulation in Brassica juncea. Chemosphere 64(6):991–997

Zhang YF, He LY, Chen ZJ et al (2011) Characterization of ACC deaminase-producing endophytic bacteria isolated from copper-tolerant plants and their potential in promoting the growth and copper accumulation of Brassica napus. Chemosphere 83(1):57–62

Zhang L, Song X, Shao X et al (2019) Lead immobilization assisted by fungal decomposition of organophosphate under various pH values. Sci Rep 9(1):1–9

# Genetic Engineering to Reduce Toxicity and Increase Accumulation of Toxic Metals in Plants

18

Amit Kumar, Mohammad Israil Ansari, Sudhakar Srivastava, Gauri Saxena, and Kiran Gupta

**Abstract**

The most common heavy metals (HMs) contaminants are arsenic (As), cadmium (Cd), chromium (Cr), mercury (Hg), and lead (Pb). Unlike organic contaminants, HMs do not biodegrade, and persist in the environment forever. Hence, the solution to the cleanup of HMs contaminated site is their removal from the environment. Various conventional and physicochemical methods utilized for removal of HMs are expensive, inefficient, and non-eco-friendly. Nonetheless, nature and science provide new opportunities in the form of bioremediation and phytoremediation to remediate HMs contaminated soil. Phytoremediation is a perspective technology for several HMs polluted sites owing to inclusion of several integrative approaches in it, such as phytoextraction, phytostabilization, rhizofiltration, and phytovolatilization. Phytoremediation has emerged as an economic, eco-friendly, and popular remediation technology. Endophytic bacteria also play an interesting role in phytoremediation. However, phytoremediation can be limited by the small habitat range or the small size or long life cycle of plants. The precision of biotechnological approaches, mainly genetic engineering, holds potential to make rapid and significant change in plant's growth and development. Recent researches depicted the application of molecular biology, bioinformatics, omics, and next-generation DNA sequencing technologies to gain deeper insights into the mechanisms of HM accumulation in plants. Genetic engineering has also contributed greatly in understanding the stress responses of plants to a number of HMs through functional analysis of genes. Engineered endophytes and rhizospheric bacteria have also been tested for enhancing

A. Kumar · M. I. Ansari · G. Saxena · K. Gupta (✉)
Department of Botany, Lucknow University, Lucknow, India

S. Srivastava
Plant Stress Biology Laboratory, Institute of Environment and Sustainable Development, Banaras Hindu University, Varanasi, India



K. Mishra et al. (eds.), *Sustainable Solutions for Elemental Deficiency and Excess in Crop Plants*, https://doi.org/10.1007/978-981-15-8636-1_18

remediation of metals for phytoremediation purposes. The risk of gene escape with engineered hyperaccumulator lines can be reduced if plant material is applied in isolated industry- or mining-impacts areas instead of agricultural areas. A novel solution can be the application of cisgenesis and intragenesis in genetic engineering. The present article discusses various aspects of genetic engineering within the scope of phytoremediation and also sheds light on future prospects for designing plants for future demands.

**Keywords**

Arsenic · Cadmium · Endophytes · Genetic engineering · Heavy metals · Phytoremediation

## 18.1 Introduction

Among various components of soil, metals are persistent natural ingredients (Tangahu et al. 2011) (Table 18.1). Some of the metals are micronutrients (required at low concentrations) which are very significant for plant growth like as zinc (Zn), cobalt (Co), manganese (Mn), and copper (Cu). Other metals have no known biological functions viz., cadmium (Cd), lead (Pb), and mercury (Hg) (Appenroth 2010). Scientifically, heavy metals (HMs) are defined as the elements having metallic properties and atomic mass >20 along with specific gravity >5 g cm$^3$ (Rascio and Navari-Izzo 2011). According to United States Environmental Protection Agency (USEPA) regulation, eight HMs, namely, Pb, chromium (Cr), arsenic (As), Zn, Cd, copper (Cu), Hg, and nickel (Ni) are categorized under the most widespread HMs in the environment (Wang and Chen 2006; Selvi et al. 2019). According to coordination chemistry of HMs, these HMs come under class B metals and have higher affinity for binding to S-containing ligands (Rzymski et al. 2015). These HMs (As, Cr, Cd, Hg, and Pb) at higher concentration exert toxic effects on biota (Ha et al. 2014; Kumar et al. 2016; Rahman and Singh 2019; Gupta et al. 2020a, b; Kumar et al. 2020). Heavy metals are released into the environment via natural ways (e.g., geological erosion and saline seeps) and numerous anthropogenic activities like industrial processes, mining, construction activities, wastewater

**Table 18.1** Heavy metal European standards (EU) and concentration range in Indian soil

| S.N. | Metals | EU standard in soil (mg kg$^{-1}$) | Concentration range in soil (mg kg$^{-1}$) |
|---|---|---|---|
| 1. | Arsenic (As) | 20 | 0.1–102 |
| 2. | Cadmium (Cd) | 3 | 0.1–345 |
| 3. | Chromium (Cr) | 150 | 0.005–500 |
| 4. | Lead (Pb) | 300 | 1–690 |
| 5. | Mercury (Hg) | 270 | 0.001–1800 |
| 6. | Nickel (Ni) | 75 | 5–150 |

Source: Salt et al. (1995) and Awashthi (2000)

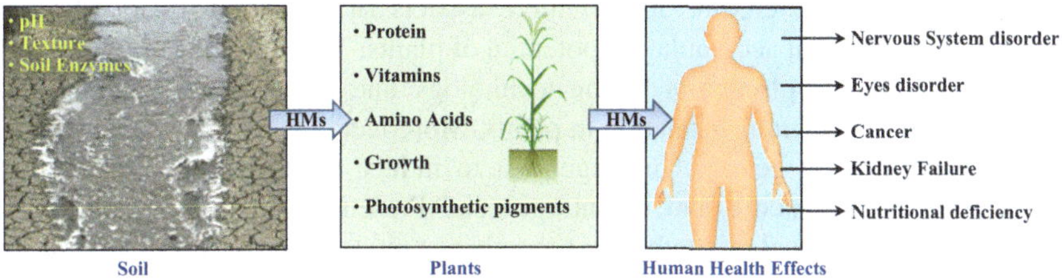


**Fig. 18.1** Heavy metal sources, plants uptake, and human health effects

treatment, and agriculture (Arthur et al. 2005; Pratap et al. 2006; Mishra et al. 2009). HMs contaminated sites show plant growth retardation due to deviation in their normal physiological and biochemical metabolisms (Oancea et al. 2005; Chibuike and Obiora 2014; Kumar et al. 2014a, b). Anatomical structures are also deteriorated by HMs (Gupta 2014, 2016). The HMs cause cyto-genotoxicity by decreasing mitotic index and enhancing various chromosomal abnormalities in the meristematic cells (Gupta et al. 2018, 2020a). Interphasic cells show toxic effects of HMs by having increased level of micronuclei in the cells (Gupta et al. 2012). Long term decline in plant growth results into yield reduction and this is ultimately linked to issues of food safety and food scarcity (Fig. 18.1). Unlike organic pollutants, HMs do not biodegrade and hence their removal from the environment is of great concern.

The conventional physicochemical methods like soil excavation, land filling and soil washing, thermal treatment, chemical extraction, and encapsulation were utilized for remediation of contaminated soil. Major drawbacks of these methods are that these are very costly, efficient only for small scale cleanup, and non-eco-friendly as they diminish biodiversity as well as alter landscape. In present scenario, apart from the conventional strategies including mechanical and/or chemical treatment for remediation of pollutants, science has provided new tools in form of bioremediation (utilizing microorganisms) and phytoremediation (using plants) to remove toxic pollutants from the environment (Kang 2014; Mani and Kumar 2014; Gavrilescu et al. 2015; Hare et al. 2020). Phytoremediation is a prospective strategy for remediation of HM pollutants from soil in which plants having high potential of stabilizing metals in soil itself or removing metals through accumulation are deployed in contaminated soil remediation. Phytoremediation techniques can be utilized in various modes for treating metal-contaminated sites like phytoextraction, phytostabilization, rhizofiltration, and phytovolatilization (Ali et al. 2013; Lee 2013). The phytoremediation strategy has emerged as an economic, eco-friendly, and very popular remediation technology for removal of toxic contaminants and/or to render them harmless in the environment as well as for $CO_2$ sequestration, stabilization of soil, betterment of biodiversity, watershed management, as energy sources and esthetics (Dickinson et al. 2009).

The lacuna of this technology is its slow pollutant removal rate, which throttles the phytoremediation process (Aken 2008). The incorporation of genetic engineering might overcome the negative aspects of phytoremediation through enhancement of metal accumulation potential of plants. The present review provides a detail account on phytoremediation technology and also sheds light on recent attainment of genetic transformation in plants. Moreover, a comprehensive information is presented on genomic management to induce phytoremedial potential of plants via gene silencing and cis/intragenesis. Present knowledge might act as stimulant to begin novel research arrays to enhance phytoremedial capacity in plants.

## 18.2 Phytoremediation for Toxic Metals

Phytoremediation is a prospective tool that utilizes plants for abatement of HMs from the contaminated sites. There are several approaches of phytoremediation for removal of HMs from contaminated sites.

### 18.2.1 Phytoextraction

Phytoextraction is the most popular mechanism of phytoremediation. It occurs through accumulation of metals in plants' shoot system followed by their harvesting. Plants used for phytoextraction usually possess the following characteristics: rapid growth rate, high biomass, extensive root system, ability to tolerate high amounts of HMs and high translocation of metals from roots to shoots. The ability to tolerate high concentration of HMs by phytoextractor plants leads to significant metal accumulation in shoots; this allows easy metal removal by harvesting the shoot (Marques et al. 2009). There are two approaches to phytoextraction depending on the characteristics of the plants involved in the process.

The first approach involves the use of natural hyperaccumulators, which are plants with very high metal-accumulation ability. The second approach involves the use of high biomass plants whose ability to accumulate metals is induced by the use of chelators, soil amendments with metal mobilizing capacity (Salt et al. 1998). The hyperaccumulators are very promising candidates of phytoremediation because they bioconcentrate metals 10–500 times more than nonhyperaccumulators plants in the harvestable shoot parts (Chaney et al. 1997). Plants can be defined as hyperaccumulator if they can concentrate metals in shoots beyond a certain threshold. The threshold for As, Cu, Ni, Cr, Al, Se, and Co is greater than 0.1% (Baker and Brooks 1989) while for Cd and Zn, metal concentration should be greater than 0.01 and 1.0%, respectively. Another feature is the root to shoot metal translocation should be higher than 1 (Chaney et al. 1997; Chibuike and Obiora 2014; Marques et al. 2009). Hyperaccumulator plants retain metals in their vacuolar compartments or cell walls so that metals do not affect vital activities like cell division, photosynthesis, and respiration (Chaney et al. 1997; Marques et al. 2009). Some hyperaccumulators plants are *Streptanthus polygaloides* and *Psycotria vanhermanni* for Ni, *Agrostis tenuis* and *Minuartia verna* for Pb, *Thlaspi caerulescens* for Cd, and *Pteris vittata* for As (Reeves et al. 2000). Some plants are hyperaccumulators of more than one metal. *Sedum alfredii* is a hyperaccumulator of Zn as well as Cd and accumulates both metals with almost same efficiency (Yang et al. 2004). Some hyperaccumulators like *Brassica rapa, Brassica napus,* and *Brassica juncea* show fast growth rate and high biomass and hence, these are more suitable for phytoextraction (Ebbs and Kochian 1997). Besides, the use of appropriate plants at the site of contamination, the soil conditions, and environmental features of the site are also the factors influencing phytoextraction of metals (Djingova and Kuleff 2000). However, slow growth rate, low biomass, and habitat restrictions of most of the hyperaccumulators do not allow their effective use for phytoextraction. To tackle this problem, the use of chelating agents has been studied and has been found to be effective in enhancing metal accumulation of both hyperaccumulator and non-hyperaccumulator plants like sorghum, alfalfa, and maize. Metals present in soil solution are taken up by plants if they are freely available to plants but some insoluble metals precipitate and become unavailable to plants. Such metals create hindrance in phytoremedial capacity of plants. The problems of unavailability of metals have been solved by the application of metal chelating substances. Chelators form metal chelate complexes, which are more soluble and thus available to plants for uptake (Blaylock et al. 1997; Salt et al. 1998; Marques et al. 2009). Both synthetic and natural chelates have been used (Salt et al. 1998). Some synthetic chelates include ethylene diamine tetra acetic acid (EDTA), ethylene glycol-bis (β-aminoethyl ether)-*N*,*N*,*N*′,*N*′-tetraacetic acid (EGTA), ethylenediamine disuccinic acid (EDDS), ethylenediamine-di-*o*-hydroxyphenylacetic acid (EDDHA), diethylenetriaminepentaacetic acid (DTPA), and *n*-hydroxy ethylenediaminetriacetic acid (HEDTA) (Chen et al. 2004; Lai and Chen 2005; Wu et al. 2006) . EDTA is inexpensive (Chaney et al. 2000) as well as very effective in enhancing metal bioavailability to plants. Several natural chemicals have also been used as chelators for increasing metal accumulation by plants. Chiu et al. (2006) used organic metals chelates like malic acid and citric acid for enhancing phytoremediation of toxic metals from contaminated sites. There are a few drawbacks associated with the use of chelators. The use of uncontrolled metal chelates may increase the metal bioavailability of toxic metals in soil solution and fasten their flow to the groundwater. Thus, metal chelates use has been found to be linked to groundwater pollution (Lombi et al. 2001). Further, chelates enhanced metal concentrations can be toxic to plants when their concentration exceeds plants' potential to tolerate metal stress (Chen et al. 2004).

### 18.2.2 Phytostabilization

Phytostabilization technique is useful for the highly contaminated sites where phytoextraction is not feasible and for metals which are sparsely bioavailable (McGrath and Zhao 2003). For some metals like Pb, it is easier to stabilize them in soil than to extract them even with plants and chelator combination. In phytostabilization, plants are utilized to reduce the mobilization of metals in soil and to also reduce their movement to groundwater via leaching (Marques et al.

2008). Several phenomena act in combination in the process of phytostabilization. These processes include precipitation, adsorption, change in speciation of HMs through oxidation and reduction, and plant chemical mediated complexation (Jadia and Fulekar 2009). In addition to external process, phytostabilization is also assisted through root system of plants as they restrict metal translocation to the shoots through effective complexation and storage within root tissues. Therefore, suitable plants for phytostabilization should have extensive root system (Laperche et al. 1997). Marques et al. (2008) reported that for phytostabilization also, some soil amendments can be used to assist plant mediated immobilization of toxic metals and to reduce their translocation in plants. Some of the soil amendments utilized in phytostabilization are organic compounds, litter, biosolids, phosphate, and lime (Adriano et al. 2004). The soil amendments should be non-toxic nature and also inexpensive for easy applicability in large contaminated areas also (Marques et al. 2008). The benefits of using organic amendments accrue from the fact they also improve the physical, chemical, and microbiological properties of soil and thus sustain good plant growth in due course of phytostabilization (Liu et al. 2017). Zn mobilization has been found to be reduced by 80% with the application of compost or manure in presence of *Solanum nigrum* cultivation (Marques et al. 2008).

### 18.2.3 Phytovolatilization

Phytovolatilization involves the transformation of metals to volatile species that are excluded from plants through vaporization into the atmosphere (Marques et al. 2009). This is observed in case of metals which have different organic forms including volatile species. Such metals include Hg, As, and Se (Arya et al. 2017). In this strategy, plants transform Hg to methylmercury and then to elemental Hg which is volatile (He et al. 2015). In case of As, it is converted to various methylated organic forms and ultimately to trimethylarsine (TMA), which is volatile (Ruppert et al. 2013). This strategy has been successfully utilized in field conditions with the use of plants. The drawback of this technique is that metals excluded to the atmosphere are redeposited back to soil or to aquatic bodies through precipitation (EPA 2000). Further, volatilization leads to distribution of a metal having localized contamination to large areas (Raskin et al. 1997). In addition, workers and farmers working in fields subjected to phytovolatilization get exposed to the toxic metal vapors (Khalid et al. 2017).

## 18.3 Advancements in Phytoremediation Techniques

The drawback of phytoremediation is its applicability for a narrow range of habitats as well as less size and biomass of most of the hyperaccumulator plant species (Arthur et al. 2005). Further, phytoremediation is a very time consuming process and hence, a field taken for the purpose of cleanup cannot be utilized for economic benefits by farmer or owner. One of the strategies to tackle this problem has been the utilization of crop plants having characteristics of high biomass. However, such crop plants must not have metal accumulation in edible or economically useable part. An example of this is the use of Indian mustard plants, which have short life cycle, gain high biomass, and metals are not found in oil extracted from seeds (Srivastava et al. 2009). Alternatively, plants having potential for biofuel production may be used for remediation of HMs to allow economic benefits during the process of phytoremediation (Pidlisnyuk et al. 2014; Mosa et al. 2016).

Apart from the use of chelators, other agricultural practices including fertilization and plant spacing, etc. have been utilized to support extensive plant growth and enhance phytoremediation (Chaney et al. 2000; Joner 2013). The utilization of plants in a unique more effective combination and succession has also been proposed. In case of As contaminated water, the use of appropriate combination of aquatic plants was studied and it was found that *Hydrilla verticillata* and *Ceratophyllum demersum* in combination could achieve greater As removal in the same time (45 d) than that achieved with the use of a single plant (Srivastava et al. 2014). Further, if the plants are to be used one at a time, an appropriate succession of plants was proposed for better results in a given time period (Poonam et al. 2017). There are some biological, physiological, and ecological limitations also. To overcome all these hurdles, the most popular strategy is the use of designer superhyperaccumulator plants generated through transgenic approaches, which can grow in highly contaminated site, generate high biomass in short time, and accumulate super high level of metals in their tissues (Shukla et al. 2013). Microorganisms assisted phytoremediation is also a useful approach, which is discussed in other chapters of this book. Here the focus is given to the use of endophytes, both natural and engineered, in assisted phytoremediation.

## 18.3.1 Endophyte-Assisted Phytoremediation

### 18.3.1.1 Natural Endophytes

Several researches reported the more advantageous role of endophytes as compared to that of rhizospheric bacteria in phytoremediation efficiency improvement (Newman and Reynolds 2005; Zhuang et al. 2007). The endophytic bacteria reside within plant tissues while rhizospheric bacteria grow around the roots of plants. Endophytes are well known to induce plant tolerance potential for the drought, pathogens as well as herbivores. In addition, endophytic microbes enhance the growth of plants (Selosse et al. 2004). Endophytes may provide plants with essential nutrients like fixed nitrogen due to diazotrophic nature (Reinhold-Hurek and Hurek 1998). Within the plant survival of endophytes also allows their better screening and monitoring. Further, such endophytes do not suffer competition from other microbes while the use of rhizospheric bacteria suffers from the competition from other rhizospheric microbes. Rhizospheric bacteria as well as endophytes present in the shoots of *Thlaspi goesingense* (Ni hyperaccumulator) showed varying efficiency toward phytoremediation of Ni (Idris et al. 2004). The endophytes have been found to possess greater resistance to high Ni concentrations in comparison to that of rhizospheric strains. There are some fungal species like arbuscular mycorrhizas (AMs) that also colonize and grow inside the plants. They also play significant role in phytoremediation. Mycorrhizae provide plants with nutrient elements like phosphorus and improve plant growth and in return receive sugars from plants (Upadhyay et al. 2018). Mycorrhizal colonization has been found to enhance metal tolerance potential as well as metal speciation inside the plants, such as in case of As (Upadhyay et al. 2018). They may themselves assist and increase metal uptake in plants through up-regulation of their own transporters (Trotta et al. 2006; Upadhyay et al. 2018). *Pteris vittata* having AM inoculation of *Glomus mosseae* showed increased translocation factor of 730 in comparison to plants devoid of AMs having translocation factor of 50 for As (Wang et al. 2002).

### 18.3.2 Engineered Endophytes

It is not necessary that bacteria with potentiality to degrade toxic pollutants flourishes in plants grown in polluted sites. Therefore, research aims to develop effective microbes via genetic engineering. These engineered endophytes could be more beneficial for phytoremediation of heavy metals. Previous research shows the utilization of innovative genetically engineered endophytes for phytoremediation of polluted soil as well as water resources from xenobiotics (Romantschuk et al. 2000; Wang and Dai 2011; Kang et al. 2012). For example, genes [*ncc* (nickel-cadmium-cobalt resistance)-*nre*] required for tolerance to Ni were transferred from *Ralstonia metallidurans* into two endophytic strains namely *Herbaspirillum seropedicae* and *Burkholderia cepacia*. *H. seropedicae* and *B. cepacia* having *ncc-nre* nickel tolerant system were inoculated onto seeds of *Lolium perenne* and *Lupinus luteus*. The engineered endophytes for Ni induced Ni resistance in the plants but not the translocation of Ni in inoculated plants (Lodewyckx et al. 2001). Luo et al. (2011) studied that endophytic bacterium LRE07 which is isolated from *Solanum nigrum* L. (Cd hyperaccumulator) has been used for resistance of HMs. LRE07 strain was capable to bound Cd (65%) and Zn (35%) from their respective solution after 72 h of inoculation. Moreover this strain also bound metals from multimetal solution (Cd+Zn). Likewise *Pseudomonas* sp. M6, *Pseudomonas jessenii* M15 in *Ricinus communis* could be efficient for abatement of Ni, Cu, and Cr (Rajkumar and Freitas 2008).

## 18.4 Phytoremediation Potential Enhancement of Plants Through Genetic Engineering Approaches

Genetic engineering and modern omics approaches including genomics, proteomics, metabolomics, as well as sophisticated computational and bioinformatics tools have allowed in depth analyses of plants responses to HMs, to identify candidate genes, and to transform plants through transfer of traits/genes for enhancing HM phytoremediation (Jagtap and Bapat 2015) (Fig. 18.2). Genetic engineering has

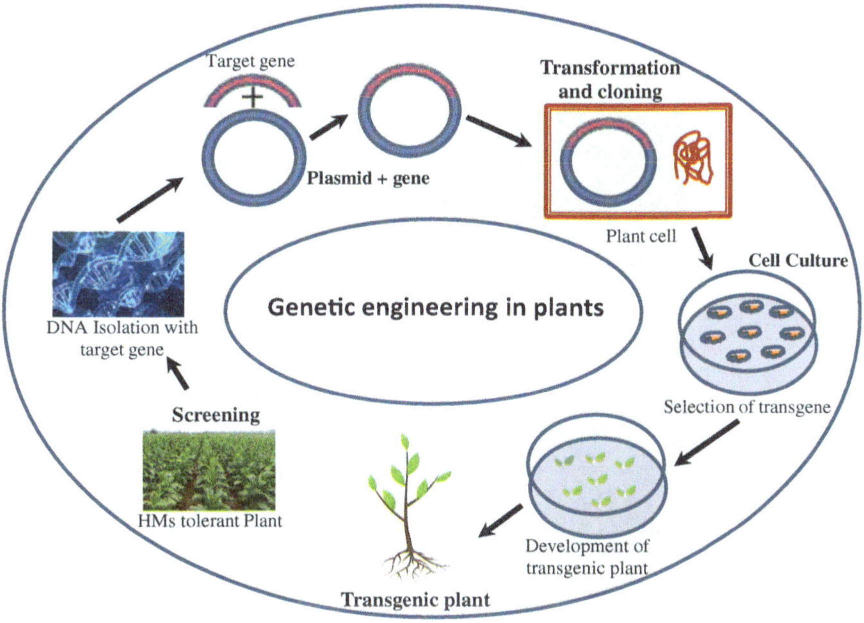


**Fig. 18.2** Genetic engineering in plants for heavy metal removal/tolerant

allowed successful manipulation of the genetic materials for the enhancement of metal accumulation capacity as well as tolerance potential in plants (da Conceição Gomes et al. 2016; Kumar et al. 2019) (Table 18.2). This technology enhanced the phytoextraction capacity of non-hyperaccumulator plants, which are much better than natural metal hyperaccumulators in other desirable traits like biomass, habitat, and life cycle (Leung 2013). Genetic engineering has been utilized for various phytoremediation strategies like phytoextraction and phytovolatilization. For instance, *Arabidopsis thaliana, Liriodendron tulipifera,* and *Nicotiana tabacum* have been developed into transgenic ones for phytovolatilization of Hg from the Hg contaminated sites (Rugh et al. 1998; Meagher et al. 2000). These transgenic plants possess bacterial genes *merA* for mercuric reductase and *merB* (organomercurial lyase) to detoxify methyl-Hg and to volatilize it to elemental Hg (USEPA 2000). However, regulatory authorities did not permit these engineered plants (USEPA 2000). On the other hand transgenic plants containing *merB* are approved by regulatory authority because this merB gene saves the food chain from entry of methyl-Hg (Meagher 1998). Soils contaminated with Se have also been cured by the application of phytovolatilization (Marques et al. 2009). In this technology, inorganic Se is transformed into organic selenoamino acids like selenocysteine and selenomethionine. Selenomethionine gets further methylated to dimethylselenide which is a volatile Se species (Terry et al. 2000). For the purpose of remediation of Se, promising candidates are *Brassica napus* and *Brassica juncea* that remove Se from the soil via phytovolatilization (Banuelos et al. 1997). In case of As also, gene

**Table 18.2** Genetic engineering in plants for improved heavy metal tolerance and accumulation in plants

| Plants | | Gene | Effects | References |
|---|---|---|---|---|
| *Physcomitrella patens* |  | *PpMT2* | Improved tolerance to high concentrations of $CuSO_4$ and $CdCl_2$ | Liu et al. (2020) |
| *Nicotiana tabacum* |  | *AtHMA4* | Restricted Cd transport from the roots to the shoots | Siemianowski et al. (2014) |
| *Manihot esculenta* |  | *AtZIP1, AtMTP1* | Zn accumulation in the roots, with restricted transport to the shoots | |
| *Brassica. juncea* |  | *AtATM3* | Increased accumulation of Cd and Pb in the shoots | Bhuiyan et al. (2011) |
| *Oryza sativa* |  | *OsMTP1* | Increased accumulation of Cd in the shoots | Das et al. (2016) |
| *Medicago sativa* |  | *ATP sulfurylase* | Increased tolerance to a mixture of vanadium (V) and lead (Pb) | Kumar et al. (2019) |
| *Sedum plumbizincicola* |  | *SpMTl* | Cadmium hyperaccumulation and hyper-tolerance | Peng et al. (2017) |
| *Nicotiana tabacum* |  | *BcMT1, BcMT2* | Enhanced tolerance to Cd and Cu | Lu et al. (2013) |
| *Oryza sativa* |  | *OsMT2c* | Improved tolerance to Cu stress and increased ROS scavenging ability | Liu et al. (2015) |

(continued)

**Table 18.2** (continued)

| Plants | | Gene | Effects | References |
|---|---|---|---|---|
| *Nicotiana tabacum* |  | *SaMT2* | Increased Cd tolerance and accumulation | Zhang et al. (2014) |
| *Linum usitatissimum* |  | *αMT1a* | Enhanced accumulation of cadmium | Vrbová et al. (2015) |
| *Tamarix Hispida* |  | *ThMT3* | Enhanced Cd, Zn, and Cu tolerance | Yang et al. (2011) |
| *Nicotiana tabacum* |  | *EhMT1* | Higher tolerance and accumulation copper in roots | Xia et al. (2012) |
| *Salix matsudana* |  | *ThMT3* | Increased tolerance to Cu stress and higher NO productions | Yang et al. (2015) |
| *Oryza sativa* |  | *OsHB4* | Cadmium accumulation and tolerance | Ding et al. (2018) |

responsible for As transformation to trimethylarsine (TMA), volatile As species, has been isolated from bacterial and fungal species and transformed into rice for As volatilization. Such transgenic rice plants were able to volatilize some portion of accumulated As into the atmosphere (Zhao et al. 2010; Srivastava et al. 2011).

The prominent approach for the phytoextraction augmentation in plants has been to overexpress the genes engaged in HMs uptake, sequestration, and translocation (Mani and Kumar 2014; Das et al. 2016). Since plants have complex metal metabolism, there are various mode to enhance metals sequestration from soil; uptake, translocation through apoplast/symplast, complex formation with ligands/chelators as well as metal detoxification via deposition in vacuoles (Nakamura et al. 2014; Das and Jayalekshmy 2015).

### 18.4.1 Genes Overexpression Encoding Metal Transporters

In plants genes encoding toxic metals transporters are usually designated by a large family. The manipulation of such genes might be suitable for augmentation of phytoremediation. An ABC (ATP-binding cassette) group of genes encodes the proteins which are accountable for metal detoxification mechanism in plants and ion-regulation process (Martinoia et al. 2002). Interestingly manipulation of ABC genes is much concerned with mobilization of metals; hence this gene family may be overexpressed to induce transportation of mobile ions viz., Cu and Cd or to bind the non-mobile ions viz., Pb. Therefore plants could become more suitable to reduce metal toxicity through all these ways. The expression of genes of cation diffusion facilitator (CDF) family encoding metal tolerance/transport protein (MTP) made cells more efficient to exclude excess ions from cytosol (Koźmińska et al. 2018). So through overexpression of these CDF genes, higher tolerance in plants and metal accumulation were reported. Das et al. (2016) reported that amid on exposure to Cd, overactivation of OsMTP1 gene of *Oryza sativa* L. cv. IR64 in *N. tabaccum* induced Cd hyperaccumulation as well as As accumulation. This finding represents wide substrate specificity of gene OsMTP1 and might be a potential strategy to induce co-resistance in plants toward diversified metals from polluted sites. A large gene family encoding ZRT/IRT-related proteins (ZIP) also form metal ion transporters that are involved in adjustment of cytoplasmic mobility of Zn and Fe. These ZIP genes have been applied for augmenting Zn and Fe level in plants (Tiong et al. 2015). Additionally the ZIP genes could be overexpressed to enhance phytoremediation of soils having high amount of toxic metals. Recent research concerns with the utilization of metal transporters of hyperaccumulator plants viz., *Noccaea caerulescens* that play a significant role in decontamination of polluted sites. Lin et al. (2016) reported that in *A. thaliana* hyperaccumulation of Zn and Cd could be enhanced through overexpression of transgene *NcZNT1* encoding Zn transporter.

### 18.4.2 Genes Overexpression Encoding Metal Chelators

Plants have mainly two classes of metal chelators: phytochelatins (PCs) and metallothioneins (MT). The metallothioneins are direct outcome of gene expression while phytochelatins are peptides which are enzymatically synthesized from glutathione (Grill et al. 2006). Apart from these two, low molecular organic acid and some amino acids are also involved in metal complexation (Koźmińska et al. 2018). Two main enzymes involved in phytochelatin synthesis are phytochelatin synthase (PCS) and *c*-glutamyl cysteine synthetase (*c*-GCS) (Hirata et al. 2005). There are numerous examples of enhanced tolerance of heavy metals in transformed plants overexpressing phytochelatin synthase gene (Srivastava et al. 2012). Therefore genes encoding the enzyme involved in PCs biosynthesis via manipulation could be helpful in reduction of metals toxicity. A number of transgenic studies have been done with reference to manipulation of genes code for enzymes involved in PCs synthesis. The transgenic tobacco plants (*Nicotiana glauca* and *Nicotiana tabacum*) showed higher accumulation of metals Pb and Cd and showed more resistance to these metals (Huang et al. 2012; Chen et al. 2015). A major complexity of this strategy is that different phytochelatins variously affect toxic metals movement within genetic engineered plant. PCS gene of *Populus tomentosa* inhibited Cd mobility to shoots in *Nicotiana* sp. and on the other hand PCS gene of *Ceratophyllum demersum* enhanced arsenic mobility in rice plant (Shukla et al. 2013). That is why productive approach might be co-transformation directed at diversified overexpression of gene engaged in metal homeostasis to get more potential hyperaccumulator plants. For instance *A. thaliana* has been made more efficient toward hyperaccumulation and represented co-tolerance of Cd and As through engineering of two genes namely PCS which was answerable for phytochelatin synthesis and other is YCF1 (an ABC metal transporter). Overexpression of ABC transporter gene decreased vulnerability toward accumulation of thiol-metal chelates that consequently improved plant resistance potential (Guo et al. 2012). The overexpression of glutamyl cysteine synthetase induced the response of PCS gene, so the co-transformation of PCS and glutamyl cysteine synthetase (GCS) gene was more productive over single transformation of PCS gene (Zhao et al. 2014). Lopez-Bucio et al. (2000) reported that enhanced synthesis of many organic acids and non-proteinogenic amino acids which are organic low molecular weight metal chelators made the plants more resistant against toxic metals. Moreover manipulation of gene encoding enzymes associated with these metal chelators organic acid as well as amino acid could also be useful in enhancing phytoremedial potential of plants. Kim et al. (2005) used this approach and incorporated HvNAS1 gene of *Hordeum vulgare* into *Arabidopsis.*

### 18.4.3 Strategies for Upgraded Heavy Metal Tolerance

Heavy metal tolerance in plants could be achieved through manipulation of metabolic pathways that are concerned with stress defense mechanisms. Resistance at cellular level prominently occurs through enhancement of reactions engaged in radical scavenging and DNA repair. If the plants' tolerance potential is enhanced through genetic moderation of these two characters, plants can tolerate metal stress in a better way and thus accumulate higher metal concentrations. The most prominent approach in this context is the adaptation of strategies which enhance antioxidant mechanism of the plants.

#### 18.4.3.1 Genes Overexpression Encoding Components of Antioxidant Machinery

Different types of abiotic stress cause oxidative damage due to accumulation of reactive oxygen species (ROS). The deteriorating effect of ROS on biomolecules like protein, lipid, and DNA could be declined by enhanced level of antioxidant defense system like superoxide dismutase (SOD), catalase (CAT), ascorbate peroxidase (APX), and glutathione *S*-transferase (GST) (Dixit et al. 2011; Hellou et al.

2012). Hence, phytoremedial potential of plants might be enhanced through genetic manipulation of genes encoding antioxidant enzymes so as to make plants more tolerant to metals. Transgenic plants have been developed for toxic metals stress with overexpression of genes concerning SOD, APX, and CAT. The plants had overproduction of antioxidative enzymes as well as inhibition of ROS in comparison to wild variety (Guan et al. 2009; Gao et al. 2016). However, genetic engineered plants frequently showed interruption in normal morphology and physiology of plants resulting growth retardation, root architecture anomaly and reduced photosynthetic performance (Iannone et al. 2015; Gao et al. 2016). Recent research revealed that plants could be improved for toxic metals resistance via overexpression of genes concerned with DNA repair and transcription. The strategy for toxic metals tolerance in plants are adjustments of genes encoding antioxidative enzymes like cytosolic and chloroplastic isoforms of SOD and CAT, metallothioneins and metal transporters, and enzyme encoding DNA repair system (Faè et al. 2014; Charfeddine et al. 2017).

### 18.4.4 Application of Gene Silencing in Plants to Overcome Heavy Metal Toxicity

HMs toxicity may also be overcome by the use of gene silencing. Biological basis of this restored defense mechanism is the utilization of very small molecule of RNA which subside gene expression through intervening the translation of targeted mRNA (Saurabh et al. 2014). This genetic tool has many advantages like stability, efficiency, flexibility, and precision (Saurabh et al. 2014). Gene silencing technique could promote heavy metals uptake consequently enhanced toxic metals hyperaccumulation in plants through RNA interference. Dhankher et al. (2006) studied that *Arabidopsis* plants could be made more capable for transferring metalloid As to the shoots in comparison to wild variety via silencing arsenate reductase gene. In contrary to this aspect this technique is also being utilized for avoiding toxic metal accumulation in plants. Cd accumulation was substantially inhibited by gene silencing of phytochelatin synthase gene (Li et al. 2007). Similarly, higher Cd translocation to shoot was observed in rice plants when *OsNRAMP5* was silenced (Takahashi et al. 2014). This gene silencing strategy ensures phytoextraction potential increase in plants. This gene silencing is also applicable for enhancing allergen exclusion, biofortification, and improved plant phenotype (Koźmińska et al. 2018).

## 18.5 Conclusions and Future Prospects

Phytoremediation technology as a whole possesses huge potential for the cleanup of contaminated sites. Some of its drawbacks may be overcome through the supportive use of endophytic microbes and genetic engineering approaches. Genetic engineering has the potential for the achievement of new plant genotype with required traits for phytoremediation purpose. However field testing in large scale is very problematic. There is also risk for their enormous growth as well as outcrossing with other undesirable genotypes. Although decline in required gene frequency of maneuvered hyperaccumulator can be overcome by their cultivation in isolated industrial field rather than agricultural area. However among various strategies of genetic engineering a recent tool like cis- and intragenesis may be considered as emerging valuable approaches to yield a tolerant genotype in response to various types of stresses including heavy metals (Espinoza et al. 2013; Koźmińska et al. 2018). This technology still demands more researches toward development of plant genotype for utilization in removal of HMs from polluted locations.

**Acknowledgements** The authors are thankful to Department of Botany, University of Lucknow, Lucknow for the facilities. Amit Kumar is thankful to SERB, DST, New Delhi for the award and financial assistance in form of SERB-NPDF. Kiran Gupta is thankful to University Grant Commission (UGC), India for the award of Post-doctoral fellowship for women candidate for the financial support.

## References

Adriano C, Wenzel WW, Vangronsveld J, Bolan NS (2004) Role of assisted natural remediation in environmental cleanup. Geoderma 122(2–4):121–142

Aken BV (2008) Transgenic plants for phytoremediation helping nature to clean up environmental pollution. Trends Biotechnol 26:225–227

Ali H, Khan E, Sajad MA (2013) Phytoremediation of heavy metals—concepts and applications. Chemosphere 91:869–881

Appenroth KJ (2010) Definition of heavy metals and their role in biological systems. In: Sherameti I, Varma A (eds) Soil heavy metals, vol 19, Soil biology. Springer, Berlin, pp 19–29

Arthur EL, Rice PJ, Rice PJ, Anderson TA, Baladi SM, Henderson KLD, Coats JR (2005) Phytoremediation - an overview. Crit Rev Plant Sci 24:109–122

Arya SS, Devi S, Angrish R, Singal I, Rani K (2017) Soil reclamation through phytoextraction and phytovolatilization. In: Volatiles and food security. Springer, Singapore, pp 25–43

Awashthi SK (2000) Prevention of Food Adulteration Act no. 37 of 1954. Central and State Rules as Amended for 1999. Ashoka Law House, New Delhi

Baker AJM, Brooks RR (1989) Terrestrial higher plants which hyperaccumulate metallic elements: a review of their distribution, ecology and phytochemistry. Biorecovery 1:81–126

Banuelos SH, Ajwa A, Mackey B et al (1997) Evaluation of different plant species used for phytoremediation of high soil selenium. J Environ Qual 26(3):639–646

Bhuiyan MSU, Min SR, Jeong WJ, Sultana S, Choi KS, Lee Y, Liu JR (2011) Overexpression of AtATM3 in Brassica juncea confers enhanced heavy metal tolerance and accumulation. Plant Cell Tiss Org Cult 107(1):69–77

Blaylock MJ, Salt DE, Dushenkov S et al (1997) Enhanced accumulation of Pb in Indian mustard by soil-applied chelating agents. Environ Sci Technol 31(3):860–865

Chaney RL, Malik M, Li YM et al (1997) Phytoremediation of soil metals. Curr Opin Biotechnol 8 (3):279–284

Chaney RL, Li YM, Brown SL, Homer FA, Malik M, Angle JS, Baker AJM, Reeves RD, Chin M (2000) Improving metal hyperaccumulator wild plants to develop commercial phytoextraction systems: approaches and progress. In: Terry N, Bañuelos G (eds) Phytoremediation of contaminated soil and water. Lewis, Boca Raton, pp 129–158

Charfeddine M, Charfeddine S, Bouaziz D, Messaoud RB, Bouzid RG (2017) The effect of cadmium on transgenic potato (*Solanum tuberosum*) plants overexpressing the StDREB transcription factors. Plant Cell Tiss Org Cult 128(3):521–541

Chen Y, Li X, Shen Z (2004) Leaching and uptake of heavy metals by ten different species of plants during an EDTA-assisted phytoextraction process. Chemosphere 57(3):187–196

Chen Y, Liu Y, Ding Y, Wang X, Xu J (2015) Overexpression of PtPCS enhances cadmium tolerance and cadmium accumulation in tobacco. Plant Cell Tiss Org Cult 121(2):389–396

Chibuike GU, Obiora SC (2014) Heavy metal polluted soils: effect on plants and bioremediation methods. Appl Environ Soil Sci:2014

Chiu KK, Ye ZH, Wong MH (2006) Growth of *Vetiveria zizanioides* and *Phragmites australis* on Pb/Zn and Cu mine tailings amended with manure compost and sewage sludge: a greenhouse study. Bioresour Technol 97(1):158–170

da Conceição Gomes MA, Hauser-Davis RA, de Souza AN, Vitória AP (2016) Metal phytoremediation: general strategies, genetically modified plants and applications in metal nanoparticle contamination. Ecotoxicol Environ Saf 134:133–147

Das R, Jayalekshmy VG (2015) Mechanism of heavy metal tolerance and improvement of tolerance in crop plants. J Glob Biosci 4(7):2678–2698

Das N, Bhattacharya S, Maiti MK (2016) Enhanced cadmium accumulation and tolerance in transgenic tobacco overexpressing rice metal tolerance protein gene OsMTP1 is promising for phytoremediation. Plant Physiol Biochem 105:297–309

Dhankher OP, Rosen BP, McKinney EC, Meagher RB (2006) Hyperaccumulation of arsenic in the shoots of *Arabidopsis* silenced for arsenate reductase (ACR2). Proc Natl Acad Sci U S A 103 (14):5413–5418

Dickinson NM, Baker AJM, Doronila A, Laidlaw S, Reeves RD (2009) Phytoremediation of inorganics: realism and synergies. Int J Phytoremediation 11:97–114

Ding Y, Gong S, Wang Y, Wang F, Bao H, Sun J, Cai C, Yi K, Chen Z, Zhu C (2018) MicroRNA166 modulates cadmium tolerance and accumulation in rice. Plant Physiol 177 (4):1691–1703

Dixit P, Mukherjee PK, Ramachandran V, Eapen S (2011) Glutathione transferase from *Trichoderma virens* enhances cadmium tolerance without enhancing its accumulation in transgenic *Nicotiana tabacum*. PLoS One 6(1):e16360

Djingova R, Kuleff I (2000) Instrumental techniques for trace analysis. In: Vernet JP (ed) Trace elements: their distribution and effects in the environment. Elsevier, London

Ebbs SD, Kochian LV (1997) Toxicity of zinc and copper to Brassica species: implications for phytoremediation. J Environ Qual 26(3):776–781

EPA (2000) A Citizen's guide to phytoremediation. EPA 542-F-98-011. United States environmental protection agency, p 6. Available at https://www.bugsatwork.com/XYCLONYX/EPA_GUIDES/PHYTO.PDF

Espinoza C, Schlechter R, Herrera D, Torres E, Serrano A, Medina C, Arce-Johnson P (2013) Cisgenesis and intragenesis: new tools for improving crops. Biol Res 46(4):323–331

Faè M, Balestrazzi A, Confalonieri M, Donà M, Macovei A, Valassi A, Carbonera D (2014) Coppermediated genotoxic stress is attenuated by the overexpression of the DNA repair gene MtTdp2α (tyrosyl- DNA phosphodiesterase 2) in *Medicago truncatula* plants. Plant Cell Rep 33 (7):1071–1080

Gao X, Ai WL, Gong H, Cui LJ, Chen BX, Luo HY, Qiu BS (2016) Transgenic NfFeSOD *Sedum alfredii* plants exhibited profound growth impairments and better relative tolerance to long-term abiotic stresses. Plant Biotechnol Rep 10(2):117–128

Gavrilescu M, Demnerova K, Aamand J, Agathos S, Fava F (2015) Emerging pollutants in the environment: present and future challenges in biomonitoring, ecological risks and bioremediation. New Biotechnol 32:147–156

Grill E, Mishra S, Srivastava S, Tripathi RD (2006) Role of phytochelatins in phytoremediation of heavy metals. In: Singh SN, Tripathi RD (eds) Environmental bioremediation technologies. Springer, Heidelberg, pp 101–146

Guan Z, Chai T, Zhang Y, Xu J, Wei W (2009) Enhancement of Cd tolerance in transgenic tobacco plants overexpressing a Cd-induced catalase cDNA. Chemosphere 76(5):623–630

Guo J, Xu W, Ma M (2012) The assembly of metals chelation by thiols and vacuolar compartmentalization conferred increased tolerance and accumulation of cadmium and arsenic in transgenic *Arabidopsis thaliana*. J Hazard Mater 199–200:309–313
Gupta K (2014) Bioaccumulation potential of *pistia stratiotes* and its response to tannery effluent exposure. Int J Phytopharm Res 5(1)
Gupta K (2016) Histological impact of tannery effluent on aquatic macrophyte. Res Environ Life Sci 9:1274–1277
Gupta K, Gaumat S, Mishra K (2012) Studies on phyto-genotoxic assessment of tannery effluent and chromium on Allium cepa. J Environ Biol 33(3):557
Gupta K, Mishra K, Srivastava S, Kumar A (2018) Cytotoxic assessment of chromium and arsenic using chromosomal behavior of root meristem in *Allium cepa* L. Bull Environ Contam Toxicol 100(6):803–808
Gupta K, Srivastava A, Kumar A (2020a) Arsenic: threat to water as well as soil. In: Contaminants and clean technologies, vol 10. Taylor and Francis Group, pp 166–187
Gupta K, Srivastava A, Srivastava S, Kumar A (2020b) Phyto-genotoxicity of arsenic contaminated soil from Lakhimpur Kheri, India on *Vicia faba* L. Chemosphere 241:125063
Ha H, Olson J, Bian L, Rogerson PA (2014) Analysis of heavy metal sources in soil using Kriging interpretation on principle components. Environ Sci Technol 48:4999
Hare V, Chowdhary P, Boudh S, Singh AK, Mani S, Kumar A (2020) Effect of arsenic on human health and its removal through physiochemical techniques. In: Contaminants and clean technologies, vol 6. Taylor and Francis Group, pp 93–104. https://doi.org/10.1201/9780429275852-6
He F, Gao J, Pierce E, Strong PJ, Wang H, Liang L (2015) In situ remediation technologies for mercury-contaminated soil. Environ Sci Pollut Res 22(11):8124–8147
Hellou J, Ross NW, Moon TW (2012) Glutathione, glutathione S-transferase and glutathione conjugates, complementary markers of oxidative stress in aquatic biota. Environ Sci Pollut Res 19:2007–2023
Hirata K, Tsuji N, Miyamoto K (2005) Biosynthetic regulation of phytochelatins, heavy metal-binding peptides. J Biosci Bioeng 100:593–599
Huang J, Zhang Y, Peng J-S, Zhong C, Yi H-Y, Ow DW, Gong Y-M (2012) Fission yeast HMT1 lowers seed cadmium through phytochelatin-dependent vacuolar sequestration in *Arabidopsis*. Plant Physiol 158:1779–1788
Iannone MF, Groppa MD, Benavides MP (2015) Cadmium induces different biochemical responses in wild type and catalase-deficient tobacco plants. Environ Exp Bot 109:201–211
Idris R, Trifonova R, Puschenreiter M, Wenzel WW, Sessitsch A (2004) Bacterial communities associated with flowering plants of the Ni hyperaccumulator *Thlaspi goesingense*. Appl Environ Microbiol 70:2667–2677
Jadia CD, Fulekar MH (2009) Phytoremediation of heavy metals: recent techniques. Afr J Biotechnol 8(6):921–928
Jagtap UB, Bapat VA (2015) Genetic engineering of plants for heavy metal removal from soil. In: Heavy metal contamination of soils. Springer, Cham, pp 433–470
Joner EJ (2013) Effects of biotic and abiotic amendments on phytoremediation efficiency applied to metal-polluted soils. In: Anjum NA, Pereira ME, Ahmad I, Duarte AC, Umar S, Khan NA (eds) Phytotechnologies: remediation of environmental contaminants. CRC Press Taylor and Francis Group, Boca Raton, pp 283–288
Kang JW (2014) Removing environmental organic pollutants with bioremediation and phytoremediation. Biotechnol Lett 36(6):1129–1139
Kang JW, Khan Z, Doty SL (2012) Biodegradation of trichloroethylene by an endophyte of hybrid poplar. Appl Environ Microbiol 78(9):3504–3507
Khalid S, Shahid M, Niazi NK, Murtaza B, Bibi I, Dumat C (2017) A comparison of technologies for remediation of heavy metal contaminated soils. J Geochem Explor 182:247–268

Kim S, Takahashi M, Higuchi K, Tsunoda K, Nakanishi H, Yoshimura E, Nishizawa NK (2005) Increased nicotianamine biosynthesis confers enhanced tolerance of high levels of metals, in particular nickel, to plants. Plant Cell Physiol 46(11):1809–1818

Koźmińska A, Wiszniewska A, Hanus-Fajerska E, Muszyńska E (2018) Recent strategies of increasing metal tolerance and phytoremediation potential using genetic transformation of plants. Plant Biotechnol Rep 12(1):1–4

Kumar A, Tripathi RD, Singh RP, Dwivedi S, Chakrabarty D, Mallick S, Trivedi PK, Adhikari B (2014a) Evaluation of amino acid profile in contrasting arsenic accumulating rice (*Oryza sativa* L.) genotypes under arsenic stress grown in hydroponic condition. Biol Plantarum 58 (4):733–742

Kumar A, Tripathi RD, Singh RP, Singh PK, Awasthi S, Trivedi PK, Chakrabarty D (2014b) Selenium ameliorates arsenic induced oxidative stress through modulation of antioxidant enzymes and thiols in rice (*Oryza sativa* L.). Ecotoxicology 23(7):1153–1163

Kumar A, Dixit G, Singh AP, Srivastava S, Mishra K, Tripathi RD (2016) Selenate mitigates arsenite toxicity in rice (*Oryza sativa* L.) by reducing arsenic uptake and ameliorates amino acid content and thiol metabolism. Ecotoxicol Environ Saf 133:350–359

Kumar A, Singh PK, Srivastava S, Dwivedi S, Tripathi RD, Awasthi G, Gupta K, Ansari MI (2020) A comparative study on effect of arsenic on thiolic ligands and phytochelatins in contrasting arsenic accumulating rice genotypes. Int J Plant Environ 6(2). (Accepted)

Kumar V, AlMomin S, Al-Shatti A, Al-Aqeel H, Al-Salameen F, Shajan AB, Nair SM (2019) Enhancement of heavy metal tolerance and accumulation efficiency by expressing Arabidopsis ATP sulfurylase gene in alfalfa. Int J Phytoremediation 21(11):1112–1121

Lai, Chen Z (2005) The EDTA effect on phytoextraction of single and combined metals-contaminated soils using rainbow pink (*Dianthus chinensis*). Chemosphere 60(8):1062–1071

Laperche VS, Traina J, Gaddam P, Logan TJ (1997) Effect of apatite amendments on plant uptake of lead from contaminated sail. Environ Sci Technol 30(10):1540–1552

Lee JH (2013) An overview of phytoremediation as a potentially promising technology for environmental pollution control. Biotechnol Bioprocess Eng 18(3):431–439

Leung DW (2013) From Arabidopsis thaliana to genetic engineering for enhanced phytoextraction of soil heavy metals. Recent Advances Towards Improved Phytoremediation of Heavy Metal Pollution 8:109

Li JC, Guo JB, Xu WZ, Ma M (2007) RNA interference mediated silencing of phytochelatin synthase gene reduce cadmium accumulation in rice seeds. J Integr Plant Biol 49(7):1032–1037

Lin Y-F, Hassan Z, Talukdar S, Schat H, Aarts MGM (2016) Expression of the ZNT1 zinc transporter from the metal hyperaccumulator Noccaea caerulescens confers enhanced zinc and cadmium tolerance and accumulation to *Arabidopsis thaliana*. PLoS One 11(3):e0149750

Liu J, Shi X, Qiana M, Zhenga Lm Lianb C, Yan Xiaa Y, Shen Z (2015) Copper-induced hydrogen peroxide upregulation of a metallothionein gene, OsMT2c, from Oryza sativa L. confers copper tolerance in Arabidopsis thaliana. J Hazard Mater 294:99–108

Liu YN, Guo ZH, Xiao XY, Wang S, Jiang ZC, Zeng P (2017) Phytostabilisation potential of giant reed for metals contaminated soil modified with complex organic fertiliser and fly ash: a field experiment. Sci Total Environ 576:292–302

Liu Y, Kang T, Cheng JS, Yi YJ, Han JJ, Cheng HL, Li Q, Tang N, Liang MX (2020) Heterologous expression of the metallothionein PpMT2 gene from Physcomitrella patens confers enhanced tolerance to heavy metal stress on transgenic Arabidopsis plants. Plant Growth Regul 90 (1):63–72

Lodewyckx C, Taghavi S, Mergeay M, Vangronsveld J, Clijsters H, van der Lelie D (2001) The effect of recombinant heavy metal resistant endophytic bacteria in heavy metal uptake by their host plant. Int J Phytoremediation 3:173–187

Lombi F, Zhao J, Dunham SJ, McGrath SP (2001) Phytoremediation of heavy metal-contaminated soils: natural hyperaccumulation versus chemically enhanced phytoextraction. J Environ Qual 30(6):1919–1926

Lopez-Bucio J, Nieto-Jacobo MF, Ramırez-Rodrıguez V, Herrera-Estrella L (2000) Organic acid metabolism in plants: from adaptive physiology to transgenic varieties for cultivation in extreme soils. Plant Sci 160(1):1–13

Lu Y, Rijzaani H, Karcher D, Ruf S, Bock R (2013) Efficient metabolic pathway engineering in transgenic tobacco and tomato plastids with synthetic multigene operons. Proc Natl Acad Sci U S A 110:E623–E632

Luo S, Wan Y, Xiao X, Guo H, Chen L, Xi Q, Zeng G, Liu C, Chen J (2011) Isolation and characterization of endophytic bacterium LRE07 from cadmium hyperaccumulator Solanum nigrum L. and its potential for remediation. Appl Microbiol Biotechnol 89(5):1637–1644

Mani D, Kumar C (2014) Biotechnological advances in bioremediation of heavy metals contaminated ecosystems: an overview with special reference to phytoremediation. Int J Environ Technol 11:843–872

Marques APGC, Oliveira RS, Rangel AOSS, Castro PML (2008) Application of manure and compost to contaminated soils and its effect on zinc accumulation by *Solanum nigrum* inoculated with arbuscular mycorrhizal fungi. Environ Pollut 151(3):608–620

Marques AP, Rangel AO, Castro PM (2009) Remediation of heavy metal contaminated soils: phytoremediation as a potentially promising clean-up technology. Crit Rev Environ Sci Technol 39(8):622–654

Martinoia E, Klein M, Geisler M, Bovet L, Forestier C, Kolukisaoglu Ü, Schulz B (2002) Multifunctionality of plant ABC transporters—more than just detoxifiers. Planta 214 (3):345–355

McGrath SP, Zhao F (2003) Phytoextraction of metals and metalloids from contaminated soils. Curr Opin Biotechnol 14(3):277–282

Meagher RB (1998) Phytoremediation: an affordable, friendly technology to restore marginal lands in the twenty-first century

Meagher RBC, Rugh L, Kandasamy MK, Gragson G, Wang NJ (2000) Engineered phytoremediation of mercury pollution in soil and water using bacterial genes. In: Terry N, Bañuelos G (eds) Phytoremediation of contaminated soil and water. Lewis Publishers, Boca Raton, pp 201–219

Mishra K, Gupta K, Rai UN (2009) Bioconcentration and phytotoxicity of chromium in Eichhornia crassipes. J Environ Biol 30(4)

Mosa KA, Saadoun I, Kumar K, Helmy M, Dhankher OP (2016) Potential biotechnological strategies for the cleanup of heavy metals and metalloids. Front Plant Sci 7:303

Nakamura M, Ochiai T, Noji M, Ogura Y, Suzuki K, Yoshimoto N, Yamazaki M, Saito K (2014) An improved tolerance to cadmium by overexpression of two genes for cysteine synthesis in tobacco. Plant Biotechnol 31:141–147

Newman LA, Reynolds CM (2005) Bacteria and phytoremediation: new uses for endophytic bacteria in plants. Trends Biotechnol 23:6–8

Oancea S, Foca N, Airinei A (2005) Effects of heavy metals on plant growth and photosynthetic activity. Analele Stiintifice ale Universitatii "AL.I. CUZA1 IASI, Tomul I, S. Biofizica, Fizica medicala S¸ Fizica mediului, pp 107–110

Peng JS, Ding G, Meng S, Yi HY, Gong JM (2017) Enhanced metal tolerance correlates with heterotypic variation in SpMTL, a metallothionein-like protein from the hyperaccumulator Sedum plumbizincicola. Plant Cell Environ. https://doi.org/10.1111/pce.12929

Pidlisnyuk V, Stefanovska T, Lewis EE, Erickson LE, Davis LC (2014) Miscanthus as a productive biofuel crop for phytoremediation. Crit Rev Plant Sci 33:1–19

Poonam SS, Pathare V, Suprasanna P (2017) Physiological and molecular insights into rice-arbuscular mycorrhizal interactions under arsenic stress. Plant Gene 11:232–237

Pratap S, Gupta K, Mishra KK (2006) Assessment of lead accumulation potential and its phytotoxic effects in *Eichhornia crassipes*. J Ind Pollut Control 22(2)

Rahman Z, Singh VP (2019) The relative impact of toxic heavy metals (THMs) (arsenic (As), cadmium (Cd), chromium (Cr)(VI), mercury (Hg), and lead (Pb)) on the total environment: an overview. Environ Monit Assess 191(7):419

Rajkumar M, Freitas H (2008) Influence of metal resistant-plant growth-promoting bacteria on the growth of *Ricinus communis* in soil contaminated with heavy metals. Chemosphere 71:834–842
Rascio N, Navari-Izzo F (2011) Heavy metal hyperaccumulating plants: how and why do they do it? and what makes them so interesting? Plant Sci 180:169–181
Raskin I, Smith RD, Salt DE (1997) Phytoremediation of metals: using plants to remove pollutants from the environment. Curr Opin Biotechnol 8(2):221–226
Reeves RD, Baker AJM, Raskin I, Ensley BD (2000) Phytoremediation of toxic metals. Using plants to clean up the environment. Wiley, New York, pp 193–229
Reinhold-Hurek B, Hurek T (1998) Life in grasses: diazotrophic endophytes. Trends Microbiol 6:139–144
Romantschuk M, Sarand I, Petanen T, Peltola R, Jonsson-Vihanne M, Koivula T, Yrjala K, Haahtela K (2000) Means to improve the effect of in situ bioremediation of contaminated soil: an overview of novel approaches. Environ Pollut 107:179–185
Rugh L, Senecoff JF, Meagher RB, Merkle SA (1998) Development of transgenic yellow poplar for mercury phytoremediation. Nat Biotechnol 16(10):925–928
Ruppert L, Lin ZQ, Dixon RP, Johnson KA (2013) Assessment of solid phase microfiber extraction fibers for the monitoring of volatile organoarsinicals emitted from a plant–soil system. J Hazard Mater 262:1230–1236
Rzymski P, Tomczyk K, Rzymski P, Poniedziałek B, Opala T, Wilczak M (2015) Impact of heavy metals on the female reproductive system. Ann Agric Environ Med 22:259–264
Salt DE, Blaylock M, Kumar PBAN, Dushenkov V, Ensley BD, Chet I, Raskin I (1995) Phytoremediation: a novel strategy for the removal of toxic metals from the environment using plants. Biotechnology 13:468–475
Salt DE, Smith RD, Raskin I (1998) Phytoremediation. Annu Rev Plant Biol 49:643–668
Saurabh S, Vidyarthi AS, Prasad D (2014) RNA interference: concept to reality in crop improvement. Planta 239(3):543–564
Selosse M-A, Baudoin E, Vandenkoornhuyse P (2004) Symbiotic microorganisms, a key for ecological success and protection of plants. C R Biol 327:639–648
Selvi A, Rajasekar A, Theerthagiri J, Ananthaselvam A, Sathishkumar K, Madhavan J, Rahman PK (2019) Integrated remediation processes toward heavy metal removal/recovery from various environments-a review. Front Environ Sci 7:66
Shukla D, Kesari R, Tiwari M, Dwivedi S, Tripathi RD, Nath P, Trivedi PK (2013) Expression of Ceratophyllum demersum phytochelatin synthase CdPCS1 in Escherichia coli and Arabidopsis enhances heavy metal(loid)s accumulation. Protoplasma 250:1263–1272
Siemianowski O, Barabasz A, Kendziorek M, Ruszczyńska A, Bulska E, Williams LE, Antosiewicz DM (2014) AtHMA4 expression in tobacco reduces Cd accumulation due to the induction of the apoplastic barrier. J Exp Bot 65(4):1125–1139
Srivastava S, Srivastava A, Suprasanna P, D'Souza SF (2009) Comparative biochemical and transcriptional profiling of two contrasting varieties of *Brassica juncea* L. in response to arsenic exposure reveals mechanisms of stress perception and tolerance. J Exp Bot 60:3419–3431
Srivastava PK, Vaish A, Dwivedi S, Chakrabarty D, Singh N, Tripathi RD (2011) Biological removal of arsenic pollution by soil fungi. Sci Total Environ 409(12):2430–2442
Srivastava S, Suprasanna P, D'souza SF (2012) Mechanisms of arsenic tolerance and detoxification in plants and their application in transgenic technology: a critical appraisal. Int J Phytoremediation 14(5):506–517
Srivastava S, Sounderajan S, Udas A, Suprasanna P (2014) Effect of combinations of aquatic plants (Hydrilla, Ceratophyllum, Eichhornia, Lemna and Wolffia) on arsenic removal in field conditions. Ecol Eng 73:297–301
Takahashi R, Ishimaru Y, Shimo H, Bashir K, Senoura T, Sugimoto K et al (2014) From laboratory to field: *OsNRAMP5*-knockdown rice is a promising candidate for Cd phytoremediation in paddy fields. PLoS One 9(6):e98816

Tangahu BV, Abdullah SRS, Basri H, Idris M, Anuar N, Mukhlisin M (2011) A review on heavy metals (As, Pb, and Hg) uptake by plants through phytoremediation. Int J Chem Eng 2011:1–31, Article ID: 939161

Terry NA, Zayed M, de Souza MP, Tarun AS (2000) Selenium in higher plants. Annu Rev Plant Biol 51:401–432

Tiong J, McDonald G, Genc Y, Shirley N, Langridge P, Huang CY (2015) Increased expression of six ZIP family genes by zinc (Zn) deficiency is associated with enhanced uptake and root-to-shoot translocation of Zn in barley (*Hordeum vulgare*). New Phytol 207(4):1097–1109

Trotta A, Falaschi P, Cornara L, Minganti V, Fusconi A, Drava G, Berta G (2006) Arbuscular mycorrhizae increase the arsenic translocation factor in the As hyperaccumulating fern *Pteris vittata* L. Chemosphere 65:74–81

United States Environmental Protection Agency (USEPA), "Introduction to phytoremediation," EPA 600/R-99/107, U.S. Environmental Protection Agency, Office of Research and Development, Cincinnati, Ohio, USA, 2000

Upadhyay MK, Yadav P, Shukla A, Srivastava S (2018) Utilizing the potential of microorganisms for managing arsenic contamination: a feasible and sustainable approach. Front Environ Sci 6:24

Vrbová M, Kotrba P, Horáček J, Smýkal P, Švábová L, Vétrovcová M, Smýkalová I, Griga M (2015) Enhanced accumulation of cadmium in Linum usitatissimum L. plants due to overproduction of metallothionein a-domain as a fusion to b-glucuronidase protein. Plant Cell Tiss Org Cult 112:321–330

Wang JL, Chen C (2006) Biosorption of heavy metals by Saccharomyces cerevisiae: a review. Biotechnol Adv 24:427–451

Wang Y, Dai CC (2011) Endophytes: a potential resource for biosynthesis, biotransformation, and biodegradation. Ann Microbiol 61(2):207–215

Wang J, Zhao FJ, Meharg AA, Raab A, Feldmann J, McGrath SP (2002) Mechanisms of arsenic hyperaccumulation in *Pteris vittata*. Uptake kinetics, interactions with phosphate, and arsenic speciation. Plant Physiol 130:1552–1561

Wu SC, Cheung KC, Luo YM, Wong MH (2006) Effects of inoculation of plant growth-promoting rhizobacteria on metal uptake by Brassica juncea. Environ Pollut 140(1):124–135

Xia Y, Qi Y, Yuan Y, Wang G, Cui J, Chen Y, Zhang H, Shen Z (2012) Overexpression of Elsholtzia haichowensis metallothionein 1 (EhMT1) in tobacco plants enhances copper tolerance and accumulation in root cytoplasm and decreases hydrogen peroxide production. J Hazard Mater 233–234:65–71

Yang XE, Long XX, Ye HB, He ZL, Calvert DV, Stoffella PJ (2004) Cadmium tolerance and hyperaccumulation in a new Zn-hyperaccumulating plant species (*Sedum alfredii* Hance). Plant Soil 259(1–2):181–189

Yang JL, Wang YC, Liu GF, Yang CP, Li CH (2011) Tamarix hispida metallothionein-like ThMT3, a reactive oxygen species scavenger, increases tolerance against $Cd^{2+}$, $Zn^{2+}$, $Cu^{2+}$, and NaCl in transgenic yeast. Mol Biol Rep 38:1567–1574

Yang JL, Chen Z, Wu S, Cui Y, Zhang L, Dong H, Li C (2015) Over-expression of the Tamarix hispidaThMT3 gene increases cop-per tolerance and adventitious root induction in Salix matsudana Koidz. Plant Cell Tiss Org Cult 121(2):469–479

Zhang J, Zhang M, Tian S, Lu L, Shohag MJI et al (2014) Metallothionein 2 (SaMT2) from Sedum alfredii Hance confers increased Cd tolerance and accumulation in yeast and tobacco. PLoS One 9(7):e102750

Zhao FJ, McGrath SP, Meharg AA (2010) Arsenic as a food chain contaminant: mechanisms of plant uptake and metabolism and mitigation strategies. Annu Rev Plant Biol 61:535–559

Zhao C, Xu J, Li Q, Li S, Wang P et al (2014) Cloning and characterization of a *Phragmites australis* phytochelatin synthase (PaPCS) and achieving Cd tolerance in tall fescue. PLoS One 9 (8):e103771

Zhuang X, Chen J, Shim H, Bai Z (2007) New advances in plant growth-promoting rhizobacteria for bioremediation. Environ Int 33:406–413

The manufacturer's authorised representative in the EU is Springer Nature Customer Service Centre GmbH, Europaplatz 3, 69115 Heidelberg, Germany. If you have any concerns regarding our products, please contact ProductSafety@springernature.com

Printed and bound by CPI Group (UK) Ltd, Croydon, CR0 4YY
30/09/2026
02225289-0016